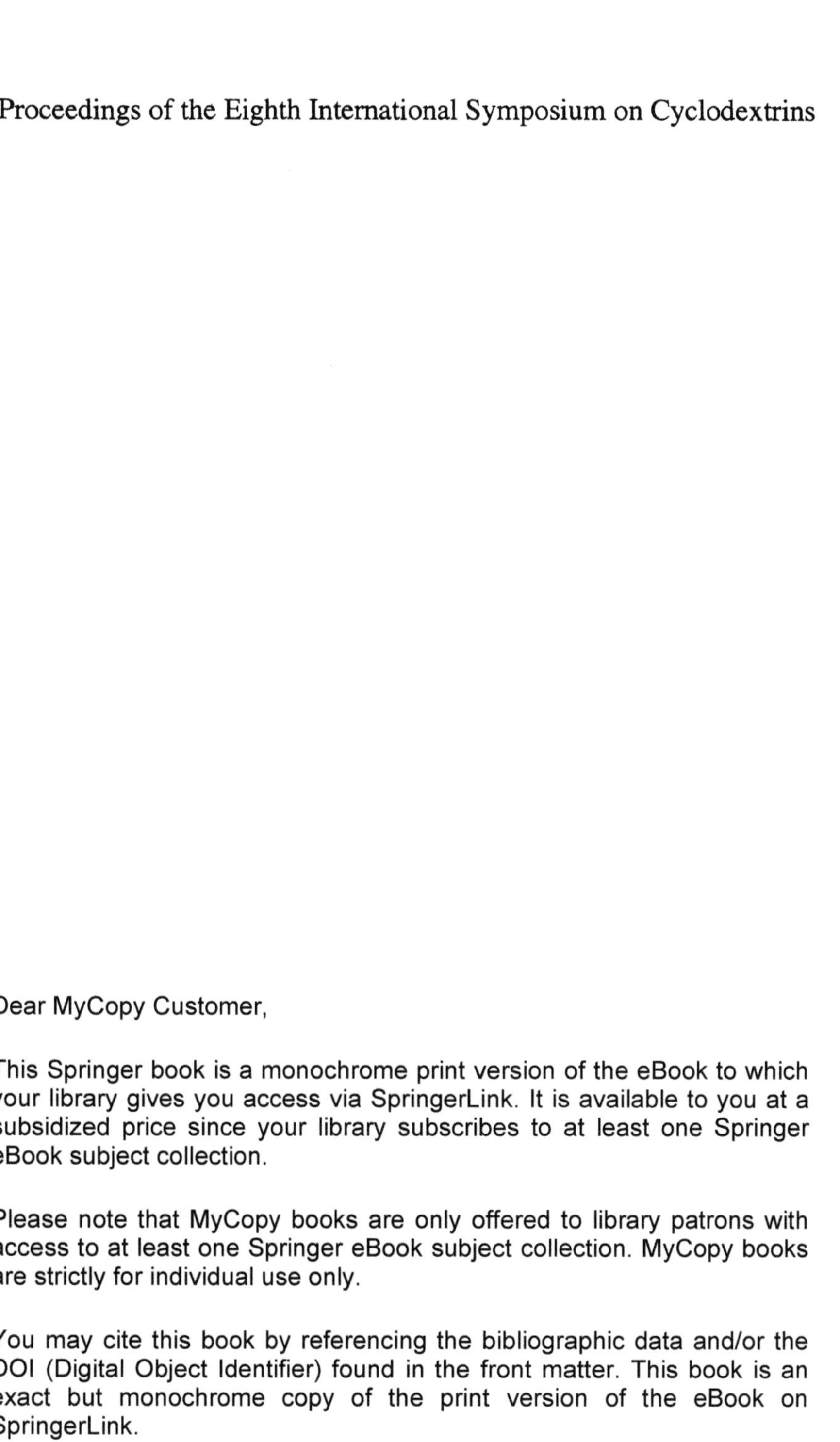

Proceedings of the Eighth International Symposium on Cyclodextrins

The *cover illustration* displays α-cyclodextrin with its MOLCAD program-generated contact surface, onto which the molecular lipophilicity patterns (MLPs) are projected in a color code ranging from blue (hydrophilic regions) to yellow (hydrophobic areas). The α-CD-donuts (top entries) are viewed through the larger opening of the conically shaped molecule (left), thereby exposing the intensely hydrophilic (blue) secondary hydroxyl face, or alternately, through the smaller cone opening (right) revealing the surface areas above the six 6–CH_2OH groups to be distinctly hydrophobic (yellow). The side view MLPs (lower entries), in closed form and bisected with a ball- and stick-model insert, are oriented with the 2-OH/3-OH side (larger opening of the torus) pointing upward; they clearly expose the distribution of hydrophillic (blue) and hydrophobic (yellow) surface regions outside and inside the cavity. For further comments see F. W. Lichtenthaler and S. Immel, *J. Incl. Phenom. Mol. Recogn.* **25**, 3–16 (1996), and pp. 3–16 of this volume.

Proceedings of the Eighth International Symposium on Cyclodextrins

Budapest, Hungary, March 31–April 2, 1996

Edited by

J. SZEJTLI and L. SZENTE
Cyclolab, Cyclodextrin Research and Development Laboratory Ltd.,
Budapest, Hungary

Partly reprinted from *Journal of Inclusion Phenomena*, Vol. 25, Nos. 1–3, 1996

SPRINGER-SCIENCE+BUSINESS MEDIA, B.V.

A C.I.P. Catalogue record for this book is available from the Library of Congress

DOI 10.1007/978-94-011-5448-2

Printed on acid-free paper

CONTENTS

* Papers marked with an asterisk have also appeared in the *Journal of Inclusion Phenomena*, volume 25/1-3.

WELCOME (again) TO BUDAPEST!

In 1980 the cyclodextrins had been known for a long time, and during the preceding 30 months 239 CD-related abstracts had been published by the Chemical Abstracts, i.e. the number of CD-related publications grew to 8/month. Several laboratories, companies, prepared CDs, but almost exclusively for research purposes. A group of scientists, scattered worldwide in various laboratories, however, suspected that the CDs were more, much more than only challenging curiosities, some of these people dreamed about or even prognosticated the production of thousands of tons of CDs, to be used in hundreds or thousands of products and technologies. At any rate, in 1980 the time seemed to be adequate for an international CD-Symposium.

I have sent letters to about 100 colleagues, working on cyclodextrins in the preceding years asking for their opinion: is it timely to organize a small CD-Symposium? The answers were unanimously YES. So we began to organize a CD-Symposium. I expected 25-30 participants outside Hungary, and in best case - including the presentations of Hungarian colleagues - about 35-40 lectures. The first surprise was that on 30th of September 1982, the first day of the Symposium from 17 countries more than 180 participants arrived and the submitted manuscripts filled a 544 page volume (Proceedings of the 1st International Cyclodextrin Symposium, Akadémiai Kiadó, Budapest and Reidel, Publ. Co., Dordrecht, 1982).

The second surprise on the last day of this Symposium was a question from a group of Japanese colleagues: "Do you intend to organize the second International Cyclodextrin Symposium, as well?" That was the very moment when the Budapest International Cyclodextrin Symposium was named by the serial number 1. The second CD Symposium was organized by Professor Nagai in 1984 in Tokyo, and since that time every second year the CD-Symposia have regularly been organized: Lancaster 1986 (Dr.Davies), Munich 1988 (Dr.Huber), Paris 1990 (Prof.Duchêne), Chicago 1992 (Dr.Hedges), Tokyo 1994 (Prof.Nagai) and again Budapest in 1996.

By now, the production of CDs worldwide has reached the several thousand ton per year level and is dynamically growing. Not only the α-, β- and γCDs are produced industrially, but also some derivatives, like random-methylated, hydroxypropylated βCD and branched βCD. CD-containing drugs, food and cosmetic products are approved, produced and marketed. The number of CD-related publications is more than 12700 (March 1996). This year (in average) every day four new papers are published somewhere in the world, patent applications, conference abstracts which are dedicated to some aspect of the cyclodextrins. It is a magnificent feeling to see, that the many efforts have not been wasted without rewards! The recent enormous development on CD-technology is reflected in the parameters of this symposium: *more than 230 participants from 32 countries, about 170 lectures and posters* represent a new record in the successful series of CD-Symposia. No doubt, that many new results which will be presented in this Symposium - will soon be realised: new, even parenterally administrable CD-derivatives, CD-complexed drugs with significantly improved biopharmaceutical properties, the use of CDs in environmental protection, in biotechnology, in food, cosmetics, chromatography, diagnostics, etc. will need continuously growing amounts of CDs and new CD-derivatives and complexes. I wish similar feeling of success to every mono-maniac scientists, who dedicate years or decades to some well-selected project. Please note, only the first 20 years are difficult, just those have to be survived!

Budapest, March 1996

Prof.J.Szejtli

The generous financial support of the following sponsors made possible to organize this Symposium

American Maize Product Co.	Hammond, Indiana/USA
Chiesi Farmaceutici S.p.A.	Parma, Italy
Chinoin-Sanofi Pharm. Chem. Co.	Budapest, Hungary
Ciba-Vision AG	Hettlingen, Switzerland
Cyclodextrin Tech. Dev. Inc.	Gaineswille, Florida/USA
Cyclolab Ltd.	Budapest, Hungary
CyDex, L.C.	Overland Park, Kansas/USA
Ensuiko Sugar Co.	Yokohama, Japan
Janssen Biotech N.V.	Beerse, Belgium
Kluwer Academic Publishers	Dordrecht, The Netherlands
Mane Fils Prod. Aromatiques	Le Bar-sur-Loup, France
Nagai Foundation	Tokyo, Japan
Nat. Committee for Tech. Dev.	Budapest, Hungary
Wacker-Chemie GmbH	Munich, Germany

Chapter 1.
CYCLODEXTRINS & DERIVATIVES

The Lipophilicity Patterns of Cyclodextrins and of Non-glucose Cyclooligosaccharides[1]

FRIEDER W. LICHTENTHALER and STEFAN IMMEL
Institute of Organic Chemistry, Technical University of Darmstadt, D-64287 Darmstadt, Germany

Abstract. Computer-aided generation of the 3D-geometries and the contact surfaces for the cyclodextrins and cyclooligosaccharides composed of mannose, altrose, galactose, and fructose provide a lucid picture of their molecular architecture, most notably their cavity dimensions. The MOLCAD program's computation of the molecular lipophilicity patterns (MLPs), projected in color-coded form onto the respective contact surfaces, for the first time allow a detailed localization of hydrophobic and hydrophilic domains, which to a substantial degree determine the capabilities of these cyclooligosaccharides for inclusion complex formation.

1. Introduction

The common, starch-derived cyclodextrins **4-6**, cyclic oligosaccharides composed of six, seven, or eight $\alpha(1\rightarrow4)$-linked D-glucopyranose units [2], have all been characterized by X-ray diffraction [3], thus unequivocally establishing their solid-state molecular geometries as depicted in Figure 1, together with their contact surfaces, shown as a dotted pattern [4, 5]. Of the 'small-ring cyclodextrins' **1-3**, which cannot be obtained from the action of *Bacillus macerans* amylase on starch, the cycloglucopentaoside **3** has yielded to chemical synthesis [6], and was shown by molecular modeling [7, 8] to exhibit a small cavity with only slight distortions of the glucose $^4C_1$ conformations. The cycloglucotetraoside **2** and its trimeric analog **1**, however, have no cavity (cf. Figure 1), and the pyranoid rings are forced into highly strained envelope conformations (E_1) [7]; it is therefore unlikely that they will become available by chemical synthesis.

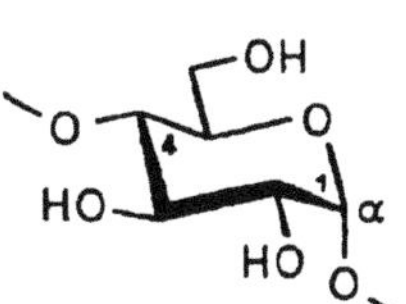

Small-ring cyclodextrins:

	1	*cyclo*[D-Glc*p* $\alpha(1\rightarrow4)]_3$
	2	*cyclo*[D-Glc*p* $\alpha(1\rightarrow4)]_4$
	3	*cyclo*[D-Glc*p* $\alpha(1\rightarrow4)]_5$

Native cyclodextrins:

(α-CD)	4	*cyclo*[D-Glc*p* $\alpha(1\rightarrow4)]_6$
(β-CD)	5	*cyclo*[D-Glc*p* $\alpha(1\rightarrow4)]_7$
(γ-CD)	6	*cyclo*[D-Glc*p* $\alpha(1\rightarrow4)]_8$
(δ-CD)	7	*cyclo*[D-Glc*p* $\alpha(1\rightarrow4)]_9$
(ε-CD)	8	*cyclo*[D-Glc*p* $\alpha(1\rightarrow4)]_{10}$
(η-CD)	9	*cyclo*[D-Glc*p* $\alpha(1\rightarrow4)]_{12}$

J. Szejtli and L. Szente (eds.), Proceedings of the Eighth International Symposium on Cyclodextrons, 3–16.

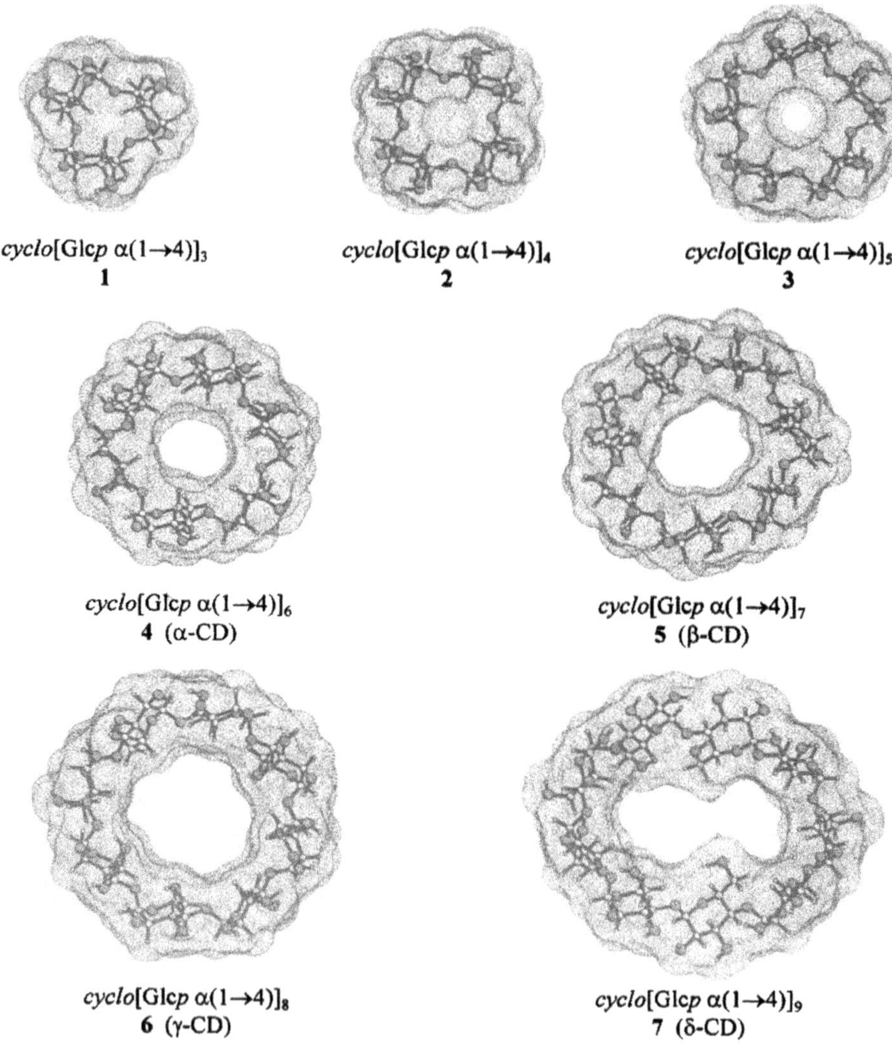

Figure 1. Ball-and-stick model representations of the minimum-energy structures for the small-ring cyclodextrins **1–3**, generated by the PIMM91 force-field program [7, 8], and the X-ray-derived [3, 10] solid-state structures of **4–7** [4, 5], together with their contact surfaces, shown as a dotted pattern. Structures are shown perpendicular to the mean ring plane of the macrocycles and are viewed through the large opening of the conically shaped molecules, i.e. the 2-OH/3-OH sides of the pyranoid rings point towards the viewer, and the primary 6-CH_2OH groups away from him towards the back; oxygen atoms are shaded.

Nomenclature: In accordance with the simplified designation recently proposed [7, 8], all cyclooligosaccharides composed of glucose units are given the generic name cyclodextrin, irrespective of their intersaccharidic linkage, and specific names in accordance with present carbohydrate nomenclature. Thus, α-CD is a cyclo-α(1→4)-glucohexaoside, abbreviated *cyclo*[D-Glc*p* α(1→4)]$_6$, rather than a 'cyclomaltohexaose' which entails confusion as to the number of maltose residues (six ?) and the presence of a free anomeric center by the ending -ose.

Indications of the existence of larger cyclodextrins, i.e. those composed of nine, ten and more glucose units, had already been obtained 30 years ago [9], yet only recently were they isolated and unequivocally characterized: δ-CD (**7**) [10], ε-CD (**8**) [11], and η-CD (**9**) [12] with 9, 10 and 12 α-D-glucopyranosyl residues, respectively. Their gross molecular shape, however, is no longer dictated by the strict necessity for the glucose moieties to adopt a common tilt within the macrocycle; the 'straitjacket' becomes too wide, allows higher flexibility, i.e. different canting of the pyranoid rings and, hence, substantial puckering of the torus. The nine-glucose unit δ-CD (**7**), in the solid state [10a], adopts a bowl-shaped rather than a planar structure: four of the nine glucose units are distinctly canted with their primary 6-OH groups towards the center cavity, while the others line up almost perpendicular to the macrocyclic ring or are even inclined towards the opposite direction (cf. Figure 1). The 2-OH/3-OH aperture is opened much wider than the opposite 6-CH_2OH side, and the range for the tilt angle (τ = 77–141°) is unusually large [5]. The same holds true for ε-CD (**8**) with 10 glucose moieties in the macrocycle: the X-ray structure [11] reveals a boat form or U shape with an elliptical cavity, caused by two glucotrioside units – of identical shape and situated on opposite sides – linked by two equally identical, yet differently canted glucodiosides. Undoubtedly, the possibilities for the construction of further unusually puckered macrocyclic conformations will increase with cyclodextrins composed of more than 10 glucose units.

Cyclooligosaccharides composed of sugars other than glucose have gained considerable attention recently [8, 13–18]. As the cyclodextrins derive their name from dextrose, an early synonym for glucose, it is an obvious extension to bestow upon cyclooligosaccharides with mannose, altrose, galactose, fructose, and rhamnose units the generic name cyclomannins, cycloaltrins, cyclogalactins, cyclofructins, and cyclorhamnins, respectively [8]. Figure 2 depicts the molecular geometries, contact surfaces and cross section contours for the respective hexameric species generated by molecular modelling, revealing subtle differences in their cavity dimensions, α-cyclofructin being devoid of a cavity.

The well-documented ability of cyclodextrins to incorporate a wide variety of hydrophobic molecules into their cavities infers those cavities to be hydrophobic a priori. However, just how hydrophobic they are, and to which extent hydrophobic interactions between guest and host play a role in the formation of inclusion complexes, is an open question, since it has never been possible to quantify the individual terms involved, nor to assess where exactly the hydrophobic and hydrophilic surface regions are, and how far they extend from or into the cavity, respectively. While such questions are as yet quite elusive to exact experimental characterization [19], they yield to computer modeling via generation of the **m**olecular **l**ipophilicity **p**atterns (MLPs) – an approach that provides a surprisingly lucid picture of the distribution of hydrophobic and hydrophilic 'patches' on the respective contact surfaces, particularly when transposed into a two-color-code representation: blue for hydrophilic, yellow for hydrophobic regions [4, 5, 7, 8]. The following account gives a review of the present state of our studies on this topic.

2. Materials and Methods

Calculation of the *molecular contact surfaces* [20] and the **m***olecular* **l***ipophilicity* **p***atterns* (MLPs) [21] was carried out by using the MOLCAD [22] molecular modeling program. Color-coded projection of the MLPs onto the corresponding contact surfaces was done by applying texture mapping strategies [23], using a two-color code graded into 32 shades, ranging from dark blue for the most hydrophilic areas to yellow for the most hydrophobic regions. Scaling of the hydrophobicity profiles was performed in arbitrary units and in relative terms for each molecule separately; no absolute values are displayed. Color graphics were photographed from the computer screen of a Silicon-Graphics workstation.

3. Results and Discussion

The lipophilicity patterns (MLPs) of α-, β-, and γ-CD (**4–6**) are depicted in Figure 4. They reveal the 2-OH/3-OH side of the macrocycles, i.e. the respective wider torus rim, to be distinctively hydrophilic as evidenced by the clearly defined blue areas (left entries). This is contrasted by hydrophobic (yellow) surface regions on the opposite narrower opening made up of the 6-CH_2OH groups (Figure 4, center models), obviously caused by the close spatial arrangement of the 6-CH_2- and O_5-C_5-H_5-fragments of the glucose residues [5]. Out of the calculated total surface area of $\approx$120 Å^2 per glucose unit (α-CD: 720, β-CD: 845, and γ-CD: 960 Å^2), only approximately 10–15% contributes to the inner surface of the central cavity (α-CD: 85, β-CD: 105, and γ-CD: 140 Å^2), yet the highest hydrophobicity is invariably to be found within this inner area of the molecular surface. Most notably, however – and this is vividly illustrated by the side view MLPs in bisected form (Figure 4, right entries) – *not the entire cavity surface areas are hydrophobic*, but only those regions in the narrower half of the cone structures, and these extend well out of the cavities towards the primary hydroxyl faces. Accordingly, the widespread view

Figure 2. (over page) Comparison of molecular geometries (ball-and-stick models), contact surfaces (shown in dotted form), and cross section contours of cyclooligosaccharides with six identical monosaccharide units each. To simplify their designation, all are given the prefix α – in analogy with α-cyclodextrin (α-CD, top left [7]), despite of the fact that the linkage geometries in α-cyclofructin (bottom left [17]) and in α-cyclogalactin (bottom right [8] – which can easily be infered from the more precise abbreviations given – are different from that realized in α-CD. The cavity dimensions of α-cyclomannin [8] and its all–6-deoxy-L-analog α-cyclorhamnin are very similar (top center and right); α-cyclofructin has a disk-shaped molecular shape with an indentation only on either side [17]; the gross molecular structure of α-cycloaltrin (bottom center [25] is derived from HTA simulations with an all-twist conformation of the pyranoid rings.

Cyclodextrins
cyclo[D-Glcp α(1→4)]6
Cyclomannins
cyclo[D-Manp α(1→4)]6
Cyclorhamnins
cyclo[L-Rhap α(1→4)]6
14.2 Å
5.4 Å
4.9 Å
14.9 Å
5.3 Å
4.6 Å
14.6 Å
5.6 Å
4.6 Å
Cyclofructrins
cyclo[D-Fruf β(1→2)]6
Cycloaltrins
cyclo[D-Altp α(1→4)]6
Cyclogalactins
cyclo[D-Galp β(1→4)]6
14.3 Å
5.7 Å
2.7 Å
14.8 Å
4.7 Å
5.2 Å
14.2 Å
5.5 Å
4.6 Å

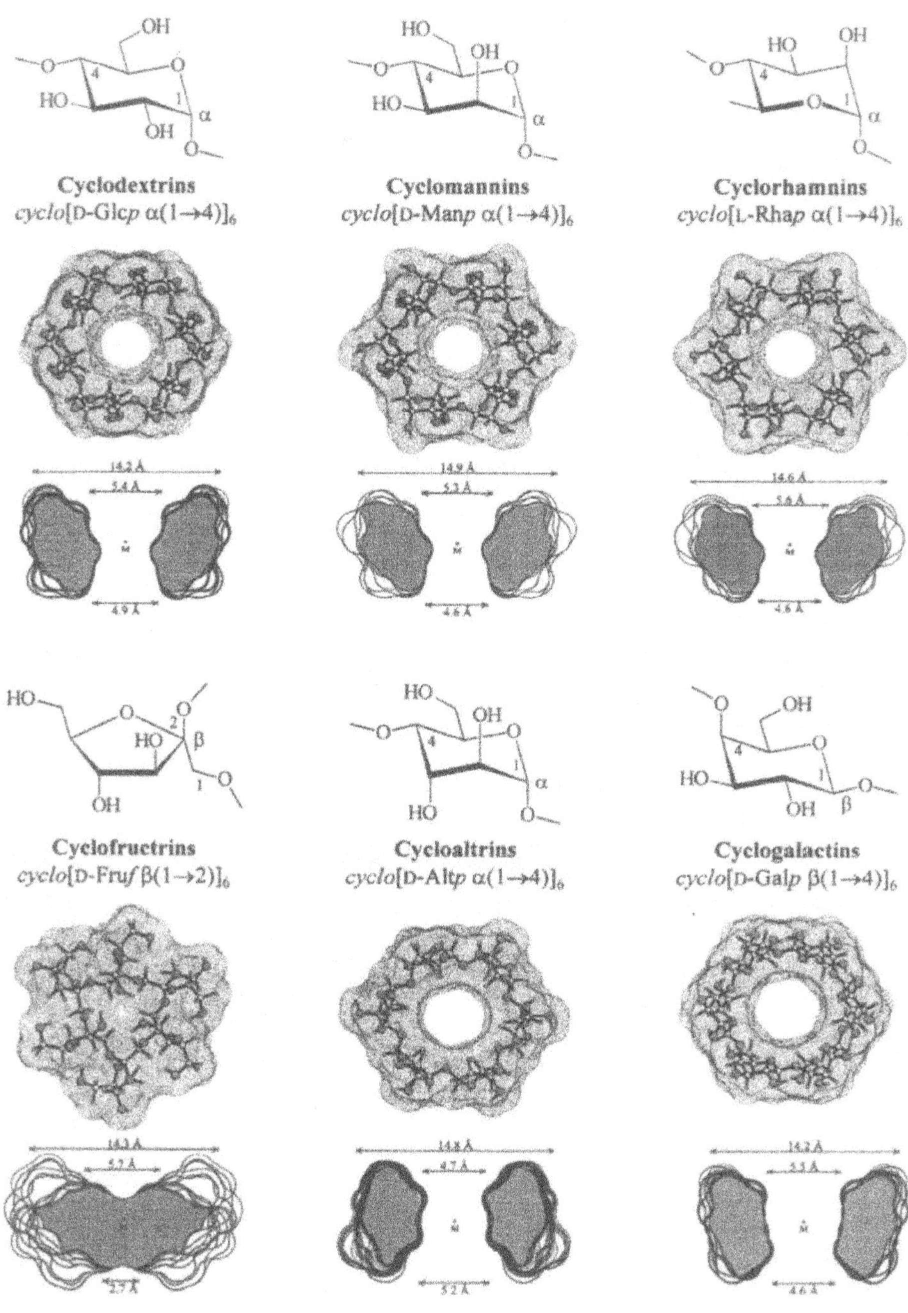

Jiph929/2

that only the cavity of the cyclodextrins is hydrophobic, or that "two hydrophilic faces surround a hydrophobic cavity [24]", is likely to be subject to revision.

In a similar fashion, the preferred conformations, the contact surfaces which reveal the cavity proportions, and the MLPs of non-glucose cyclooligosaccharides can be computed and evaluated in terms of their capabilities to form inclusion complexes.

The *α-cyclomannin*, which has been synthesized [13], but in such minute amounts as to exclude studies on its inclusion complex behavior, has a backbone structure similar to α-CD (cf. Figure 2), indicating that inversion of the glucosyl-2-OH in the pyranoid rings entails no principal changes, the somewhat smaller torus height being only a minor effect [8]. The cavity dimensions resemble those of α-CD, as do the MLP profiles (Figure 5, top entries), except for the finding that the outer surface areas of α-cyclomannin are more thoroughly hydrophilic than the ones in α-CD, obviously due to the more clearly cavity-localized hydrophobic regions (cf. side view in Figure 5). *In toto*, its behavior towards inclusion complex formation should be very similar to α-CD [8].

For the *α-cycloaltrin*, which has recently become accessible from α-CD by a straightforward three- step synthesis [15b], one would expect the altropyranoid rings to either adopt the all-$^4C_1$ (**A**) or the all-$^1C_4$ conformation (**B**), or – in aqueous solution more likely – a dynamic equilibrium between the two conformational extremes, i.e. **A** $\leftrightharpoons$ **B** (cf. Figure 3).

For vacuum boundary conditions, the all-*twist* or all- *'skew boat'* form **C** emerged from high temperature annealing (HTA) simulations [25] as the energetically most favorable one. In **C**, the steric constraints imposed on form **A** by the *syn*-1,3-diaxial interactions between the intersaccharidic oxygen (O^1) and O^3, and the congestion exerted by the six axial 3-OH projecting into the cavity, are released, yet it also causes the tilt of the altropyranose units to be inverted: the primary hydroxyl face becomes the larger opening, the torus rim carrying the secondary hydroxyls the smaller one (cf. cross-cuts in Figure 2). However, the alternative conical shape has only minor consequences for the lipophilicity distribution, as the secondary hydroxyl face is distinctly hydrophilic (blue, cf. Figure 5, center entries) versus a hydrophobic (yellow) cavity area reaching out over the wider torus rim lined by the six CH_2OH groups [25].

While this *all-twist* form **C** of α-cycloaltrin, if proved to be prevalent in solution, would be the first cyclooligosaccharide with *'inverse conicity'*, there are other forms conceivable, a most fascinating one being an 'internal hybrid form' between **A** and **B**, i.e. a continuous succession of $^4C_1$ and $^1C_4$ altropyranoid chairs around the macrocycle. It appears not unlikely that this hybrid form is adopted in the solid state.

In the case of *α-cyclogalactin*, requiring $\beta(1\rightarrow4)$-linkages to enable a CD-analogous cyclooligosaccharide, the pyranoid ring oxygens are directed towards the inside, entailing a cavity less congested (than in α-CD) by hydrogen atoms,

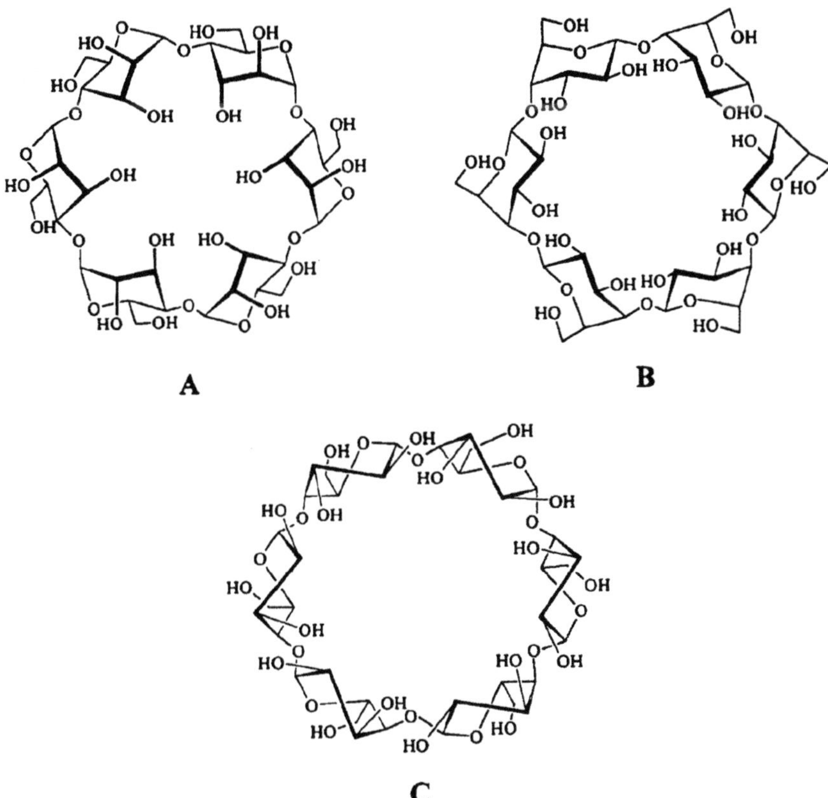

Figure 3. α-Cycloaltrin with different conformations of the pyranoid ring: the *all-*$^4C_1$*-form* **A**, which is encumbered with *syn*-1,3-diaxial interactions between the 3-OH and the intersaccharidic 1-oxygen, as well as with the steric congestion created by the six axial 3-OH sticking into the center cavity; the alternate *all-*$^1C_4$ *form* **B** appears to be less burdened by steric constraints, whilst the *all-twist form* **C** represents a conformational arrangement between the two extremes **A** and **B**. High-temperature annealing (HTA) simulations on the isolated molecules (i.e. for vacuum boundary conditions) led [26] to the *all-twist* (*skew boat*) geometry **C** as the energetically most favorable arrangement; in aqueous solution a dynamic equilibrium **A** $\leftrightharpoons$ **B** or a solvated intermediate form of type **C** appears to be likely.

hence wider by about 20%, and of more tube-like shape [8] (Figure 2). This '*inside-turned-out*' cyclic ribbon of pyranoid chairs is also evidenced in a lipophilicity distribution inverse to that of α-CD: substantially enlarged hydrophobic surface areas at the primary CH_2OH face, extending significantly out of the cavity over the rim towards the outside of the macrocycle (Figure 5, bottom entries) [8]. Accordingly, α-cyclogalactin – unlike α-CD – is surmised to be barely soluble in water, which will facilitate its isolation when the efforts towards its synthesis [26] materialize.

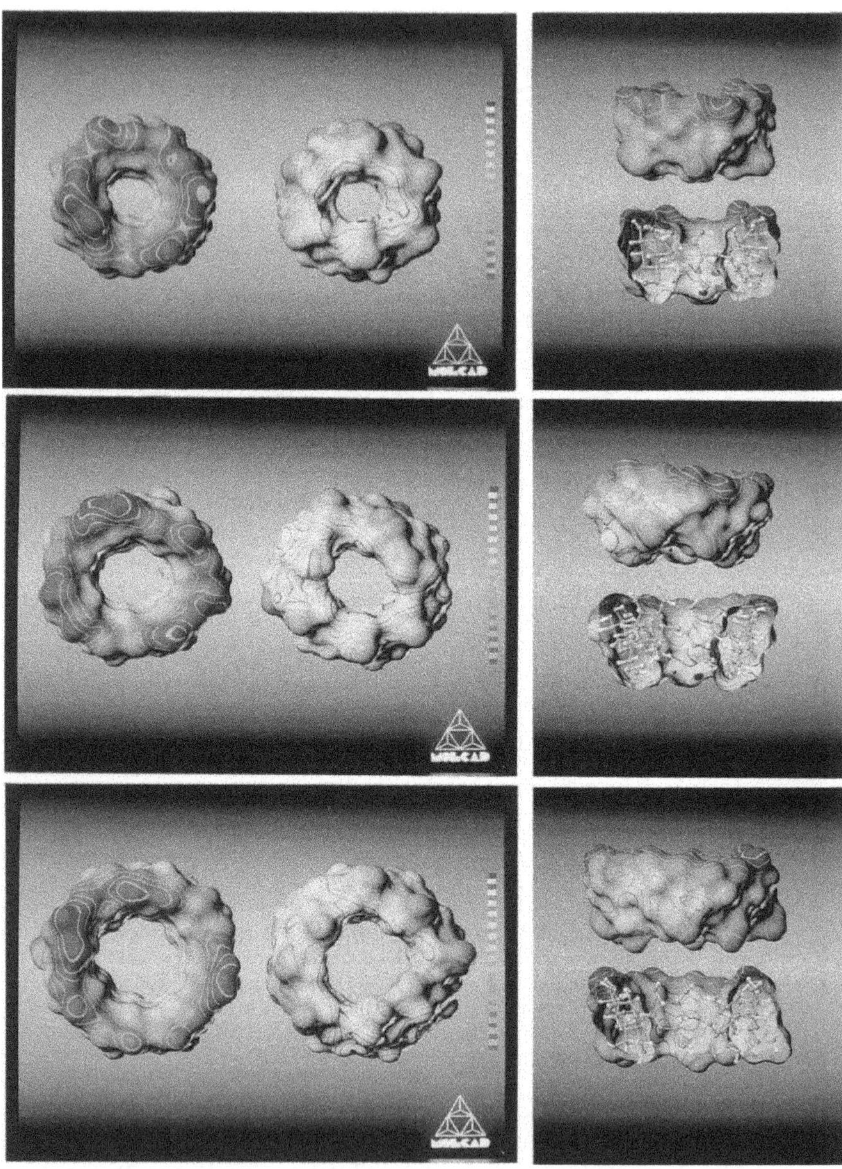

Figure 4. MOLCAD program-generated molecular lipophilicity patterns (MLPs), projected onto the respective contact surfaces (cf. Figure 1) of α-CD (**4**, top entry), β-CD (**5**, center), and γ-CD (**6**, bottom). The pictures on the left are viewed through the larger openings of the conically shaped molecules thus exposing the intensively hydrophilic (blue) 2-OH/3-OH side, whereas the representations in the middle depict the opposite side, i.e. the smaller opening with the 6-CH_2OH groups facing the viewer, thereby clearly exposing the hydrophobic (yellow) surface areas. The side view MLPs on the right, each in closed and bisected form, are oriented such that the 2-OH/3-OH side is aligned upward (larger opening of the torus) and the 6-CH_2OH groups are pointing down (smaller aperture). The similarities in the distribution of hydrophilic (blue) and hydrophobic (yellow) surface areas – most notably on the inside regions of the cavities of α-, β-, and γ-CD – are clearly apparent [5].

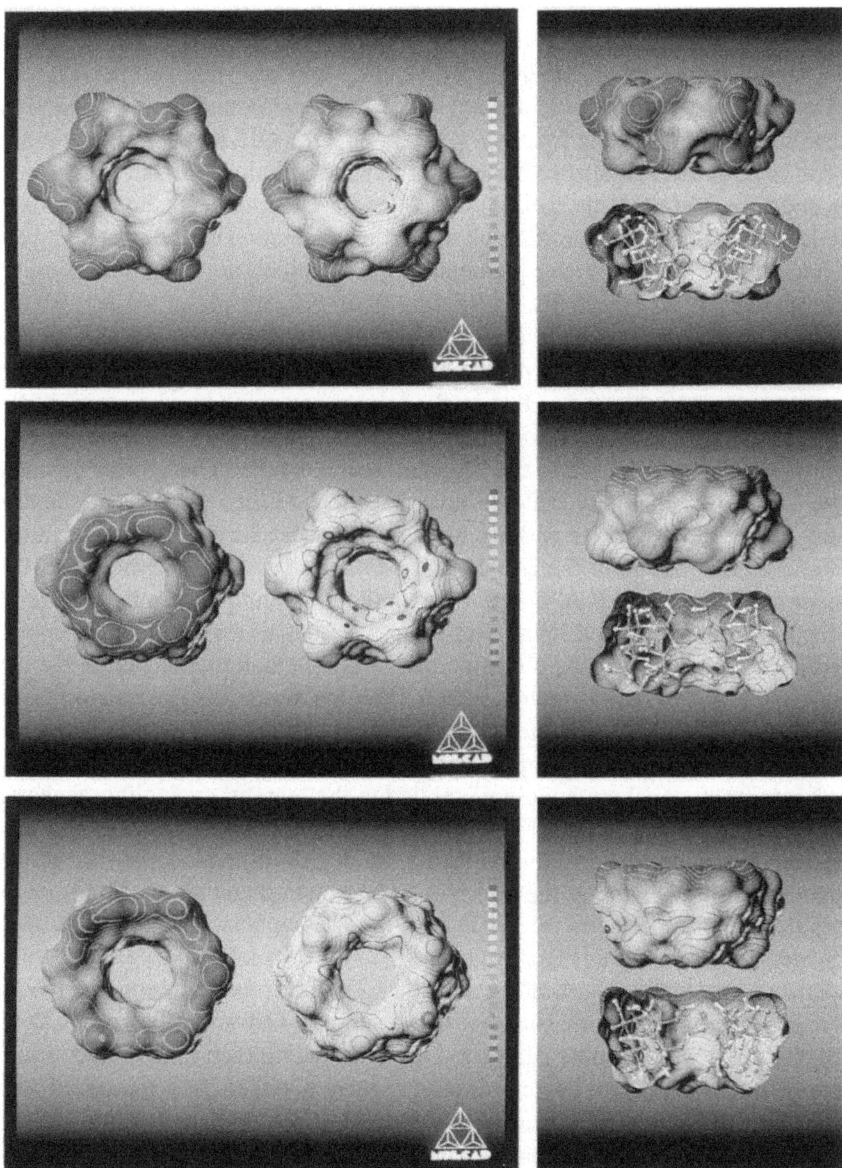

Figure 5. Molecular lipophilicity patterns (MLPs) for α-cyclomannin (top, cyclo-$\alpha(1\rightarrow4)$-mannohexaoside), α-cycloaltrin (cyclo-$\alpha(1\rightarrow4)$-altrohexaoside, center), and α-cyclogalactin (cyclo-$\beta(1\rightarrow4)$-galactohexaoside, bottom) [8], in order from left to right: hydrophilic side (blue, 2-OH/3-OH face oriented to the viewer) $\rightarrow$ hydrophobic primary hydroxyl face (yellow) $\rightarrow$ closed and bisected side views.

The inulin-derived *α-cyclofructin*, consisting of six $\beta(1\rightarrow2)$-linked fructofuranose units adding up to a crown ether backbone, reveals a contact surface topography devoid of an inner cavity [4, 17], and an MLP profile (Figure 6, top) with front-side/back-side separation of hydrophilic and hydrophobic areas. Due to the location of three fructosyl-oxygens (1-O, 3-OH, and 4-OH) on the same ('front') side of the disk-shaped molecule, this surface region is distinctly hydrophilic. The opposite side holds the 1-CH_2, 6-CH_2 and 5-CH fragments, and accordingly entails a distinctly hydrophobic 'back-side', the indentation in its center being open for potential binding with complementary guests, albeit not in an inclusion type fashion [17].

O-Alkylated or *O*-acylated derivatives of the cyclodextrins, as well as those where OH groups have been replaced by hydrogen, halogen, or amino functionalities, may also be subjected to this methodology, providing lucid pictures of their hydrophobicity/hydrophilicity distributions. Thus, the per-*O*-methylated α-CD and β-CD – not unexpected due to the 'lipophilization' of their OH-groups by methylation – show lipophilicity patterns inverse to that of their parent cyclodextrins: hydrophilic central cavities and the most hydrophobic surface regions located at the torus rims made up of the 2-OMe and 3-OMe at one side, and the 6-CH_2OMe moieties at the other [1]. A variety of experimental findings can be rationalized on the basis of the opposed lipophilicity profiles, such as for example the opposite orientation of benzaldehyde, *p*-nitrophenol, and 3-iodopropionic acid in the cavities of α-CD and of per-*O*-methyl-α-CD. Thus, the notion is substantiated that the operation of dispersive interactions between guest and CD-host cavities play a more dominant role in inclusion complex formation than has hitherto been appreciated.

The computational methodology applied to the 'empty' cyclooligosaccharides can, of course, be put to use for unraveling the lipophilicity patterns of their inclusion complexes, particularly for exploring the degree of conformity of hydrophobic and hydrophilic areas at the interface between guest and host [27]. Impressive examples are the complexes of β-CD with the adamantane-1-carboxylic acid and its 1-methanol derivative (Figure 6, center) [27]: in the solid state, either complex is characterized by head-to-head arranged β-CD dimers formed through an intense hydrogen bonding network between the secondary OH-groups of two CD tori, with the highly hydrophobic adamantyl moieties facing each other in the 'double' cavities. In the 1-carboxylic acid case, the guest is either fully immersed in the cavity with the hydrophilic head group located at the primary hydroxyl side, or is only included with its adamantyl portion (Figure 6, center left), both insertion modes providing an ideal disposition for the carboxyl group to engage in hydrogen bonding to CD-hydroxyl groups of the next dimer, or to water. In the adamantane-1-methanol case, the molecular assembly is basically similar with respect to the dimer formation, yet distinctly different in terms of the depths of immersion of the intensely hydrophobic adamantane residues. The 1-methanol group protrudes from either end, leaving empty space on the hydrophilic middle ribbon of the cavity,

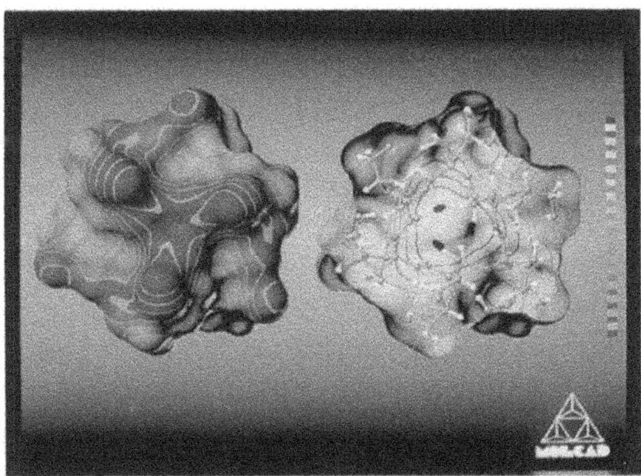

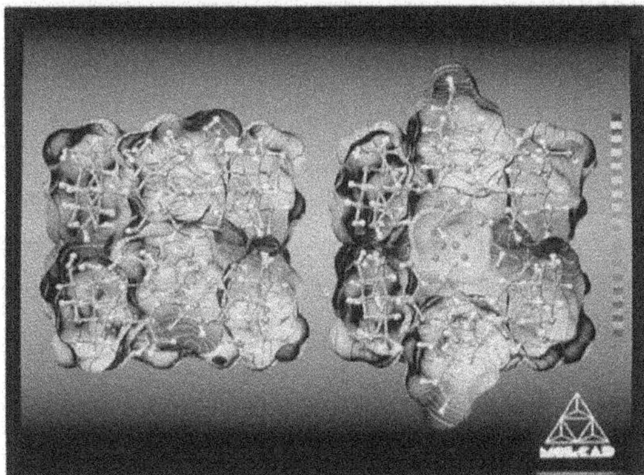

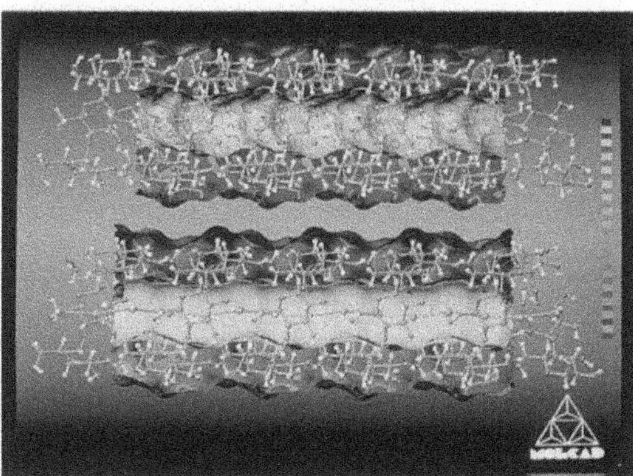

Figure 6. *Top:* Molecular lipophilicity pattern for α-cyclofructin (cyclo-$\beta(1\rightarrow2)$-fructohexaoside) [17]. *Center:* Inclusion complex of β-cyclodextrin with adamantane-1-carboxylic acid (left) and adamantane-methanol (right). While the carboxylic acid residue is fully immersed in the cavities, the hydroxymethyl group projects outwards, due to water molecules being locked in the center of the double cavity (two H_2O distributed over six crystallographic positions) [27]. *Bottom:* The single-stranded V_H-amylose with six glucose units per turn, in a section of $\approx$30 Å length; the half-opened model unveils the distinctly hydrophobic center channel, into which iodine is incorporated to form the dark-blue amylose-iodine complex [4, 30].

which is filled with two water molecules (Figure 6, center right), thereby achieving an optimum match of hydrophobic and hydrophilic surface regions [27].

The MOLCAD program methodology was also used to generate the contact surface and MLP profile for the V_H-amylose portion of starch, which on the basis of X-ray diffraction studies [28] forms a single-stranded left-handed helix with six glucose units per turn. As is clearly apparent from the color-coded representations in Figure 6 (bottom entries), the outside surface regions of V_H-amylose are intensely hydrophilic (blue) – corroborating its solubility in water – while its center channel is decisively hydrophobic (yellow) [4, 29]. This distinctive '*inside-outside*' separation of hydrophobic and hydrophilic domains preconditions the incorporation of iodide/iodine into the channel as a successive linear chain of I_2/I_3^- molecules [30], generating the dark-blue amylose–iodine complex [30], which thereby becomes conceptually understandable [4, 29].

Acknowledgment

We are most grateful to Prof. Dr. Jürgen Brickmann, Institute of Physical Chemistry of this University, for generously providing us access to the MOLCAD modelling software package [22] and his sophisticated computational facilities. Our thanks are also due to Dipl.-Ing. Guido E. Schmitt for expert help in the generation of some of the graphics.

References

1. 'Molecular Modelling of Saccharides', Part 14. – Part 13: S. Immel, and F.W. Lichtenthaler: 'Permethylated α- and β-CD: cyclodextrins with inverse hydrophobicity', *Starch/Stärke*, **48**, 225–232 (1996). This account is based on an invited lecture presented by F.W. L. at the 8th International Cyclodextrin Symposium, Budapest, March 31, 1996.
2. W. Saenger: 'Cyclodextrin Inclusion Compounds in Research and Industry', *Angew. Chem. Int. Ed. Engl. 19*, 344–362 (1980). J. Szejtli: *Cyclodextrins and their Inclusion Complexes*, Akademia Kiado, Budapest, 1982. *Cyclodextrin Technology*, Kluwer Acaddemic Publishers, Dordrecht, 1988. G. Wenz: 'Cyclodextrins as Building Blocks for Supramolecular Structures', *Angew. Chem. Int. Ed. Engl. 33*, 803–822 (1994), and literature cited therein.
3. K. Harata: 'Recent Advances in the X-ray Analysis of CD Complexes', in: J.L. Atwood and D.D. MacNicol (eds.), *Inclusion Compounds*, Oxford University Press, Oxford, **5**, 311–344 (1991).
4. F.W. Lichtenthaler, and S. Immel: 'Computer Simulation of Chemical and Biological Properties of Sucrose, the Cyclodextrins, and Amylose', *Internat. Sugar J.* **97**, 12–22 (1995).
5. F.W. Lichtenthaler, and S. Immel: 'On the Hydrophobic Characteristics of Cyclodextrins: Computer-aided Visualization of Molecular Lipophilicity Patterns', *Liebigs Ann. Chem.* 27–37 (1996).
6. T. Nakagawa, K. Ueno, M. Kashiwa, and J. Watanabe: 'Preparation of a Novel Cyclodextrin Homologue with d.p. Five', *Tetrahedron Lett.* **35**, 1921–1924 (1994).
7. S. Immel, J. Brickmann, and F.W. Lichtenthaler: 'Small-ring Cyclodextrins: Their Geometries and Hydrophobic Topographies', *Liebigs Ann. Chem.* 929–942 (1995).
8. F.W. Lichtenthaler, and S. Immel: 'Cyclodextrins, Cyclomannins, and Cyclogalactins with Five and Six (1→4)-linked Sugar Units: A Comparative Assessment of their Conformations and Hydrophobicity Potential Profiles', *Tetrahedron Asymmetry*, **5**, 2045–2060 (1994).

9. D. French, A.O. Pulley, J.A. Effenberger, M.A. Rougvie, and M. Abdullah: 'The Schardinger Dextrins: Molecular Size and Structure of the δ-, ε-, ζ-, and η-dextrins', *Arch. Biochem. Biophys.* **111**, 153–160 (1965).
10. T. Fujiwara, N. Tanaka, and S. Kobayashi: 'Structure of δ-cyclodextrin·13.75 H_2O', *Chem. Lett.* 739–742 (1990). I. Miyazawa, H. Ueda, H. Nagase, T. Endo, S. Kobayashi, and T. Nagai: 'Physicochemical Properties and Inclusion Complex Formation of δ-CD', *Europ. J. Pharm. Sci.* **3**, 153–162 (1995).
11. H. Ueda, T. Endo, H. Nagase, S. Kobayashi, and T. Nagai: 'Isolation, Purification, and Characterization of ε-CD', *8th Internat. Cyclodextrin Symp., Budapest*, **1996**, **p. 17-20.**
12. T. Endo, H. Ueda, S. Kobayashi, and T. Nagai: 'Isolation, Purification and Characterization of Cyclomalto-dodecaose (η-CD)', *Carbohydr. Res.* **269**, 369–373 (1995).
13. M. Mori, Y. Ito, and T. Ogawa: 'A Highly Stereoselective and Practical Synthesis of Cyclomannohexaose', *Carbohydr. Res.*, **192**, 131–146 (1989). M. Mori, Y. Ito, J. Izawa, and T. Ogawa: 'Stereoselectivity of Cycloglycosylation in *Manno*oligose Series Depends on Carbohydrate Chain Length: Synthesis of *Manno*-isomers of β- and γ-cyclodextrins', *Tetrahedron Lett.* **31**, 3191–3194 (1990).
14. M. Nishizawa, H. Imagawa, Y. Kan, and H. Yamada: 'Synthesis of Cyclo-L-rhamnohexaose by a Stereoselective Thermal Glycosylation', *Tetrahedron Lett. 32*, 5551–5554 (1991). M. Nishizawa, H. Imagawa, E. Morikuni, S. Hatekayama, H. Yamada: 'Synthesis of Cyclo-L-rhamnopentaose', *Chem. Pharm. Bull.*, **42**, 1356–1365 (1994).
15. (a) K. Fujita, H. Shimada, K. Ohta, Y. Nogami, K. Nasu, and T. Koga: 'β-Cycloaltrin: A Cyclooligosaccharide Consisting of Seven $\alpha(1\rightarrow4)$-linked Altropyranoses', *Angew. Chem. Int. Ed. Engl.* **34**, 1621–1622 (1995). (b) Y. Nogami, K. Fujita, K. Ohta, K. Nasu, H. Shimada, C. Shinohara, and T. Koga: 'CD-derived Cycloaltrins Made Up from $\alpha(1\rightarrow4)$-linked Altropyranoses', *8th Internat. Cyclodextrin Symp., Budapest*, **1996**, **p. 99-102.**
16. M. Sawada, T. Tanaka, Y. Takai, T. Hanafusa, T. Taniguchi, M. Kawamura, and T. Uchiyama: 'Crystal Structure of Cycloinulohexaose from Inulin', *Carbohydr. Res. 217*, 7–17 (1991).
17. S. Immel, and F.W. Lichtenthaler: 'The Electrostatic and Lipophilic Potential Profiles of α-cyclofructin: Computation, Visualization and Conclusions', *Liebigs Ann. Chem.*, 39–44 (1996).
18. P.R. Ashton, C.L. Brown, S. Menzer, S.A. Nepogodiev, J.F. Stoddart, D.J. Williams: 'Syntheses and Structural Porperties of Cyclo[(1→4)-α-L-rhamnosyl-(1→4)-α-D-mannosyl]trioside and -tetraoside', *Chem. Eur. J.* **2**, 580–591 (1996).
19. W. Blokzijl, and J. B. F. N. Engberts: 'Hydrophobic Effects, Opinions and Facts', *Angew. Chem. Int. Ed. Engl.* **32**, 2045–2060 (1993).
20. F.M. Richards: 'Areas, Volumes, Packing, and Protein Structure', *Ann. Rev. Biophys. Bioeng.* **6**, 151–176 (1977); *Carlsberg. Res. Commun.* **44**, 47–63 (1979). M.L. Connolly: 'Analytical Molecular Surface Calculation', *J. Appl. Cryst.* **16**, 548–558 (1983); *Science* **221**, 709–713 (1983).
21. W. Heiden, G. Moeckel, J. Brickmann: 'A New Approach to Analysis and Display of Local Lipophilicity/Hydrophilicity Mapped on Molecular Surfaces (MLP)', *J. Comput.-Aided Mol. Des.* **7**, 503–514 (1993).
22. J. Brickmann: *MOLCAD – MOLecular Computer Aided Design*, Technische Hochschule Darmstadt. The major part of the MOLCAD program is included in the SYBYL package of TRIPOS Associates, St. Louis, USA (1992). J. Brickmann: 'Molecular Graphics – How to See a Molecular Scenario with the Eyes of a Molecule', *J. Chim. Phys. 89*, 1709–1721 (1992). J. Brickmann, T. Goetze, W. Heiden, G. Moeckel, S. Reiling, H. Vollhardt, and C.D. Zachmann: 'Interactive Visualization of Molecular Scenarios with MOLCAD/SYBYL', in: J.E. Bowie (ed.), *Data Visualization in Molecular Science – Tools for Insight and Innovation* Addison-Wesley Publ., Reading, Mass. pp. 83–97 (1995).
23. M. Teschner, C. Henn, H. Vollhardt, S. Reiling, and J. Brickmann: 'Texture Mapping, a New Tool for Molecular Graphics', *J. Mol. Graphics* **12**, 98–105 (1994).
24. H. Parrot-Lopez, C.-C. Ling, P. Zhang, A. Baszkin, G. Albrecht, C. de Rango, and A.W. Coleman: 'Self-assembling Systems of Amphiphilic per-6-amino-β-cyclodextrin 2,3-dialkyl Ethers', *J. Am. Chem. Soc.* **114**, 5479–5480 (1992).
25. F.W. Lichtenthaler, S. Immel, and G.E. Schmitt: unpublished results.

26. M. Oberthür: *Aufbau β(1→4)-verknüpfter Galactooligosaccharide*, Diplomarbeit, Techn. Hochschule Darmstadt, (1995).
27. F.W. Lichtenthaler, and S. Immel: 'Towards Understanding Formation and Stability of Cyclodextrin Inclusion Complexes: Computation and Visualization of their Lipophilicity Patterns', *Starch/Stärke*, **48**, 145–154 (1996).
28. G. Rappenecker, and P. Zugenmaier: 'Detailed Refinement of the Crystal Structure of V_H-amylose', *Carbohydr. Res.* **89**, 11–19 (1981).
29. S. Immel, and F.W. Lichtenthaler: 'The Hydrophobic Topographies of Amylose and its Blue Iodine Complex', to be published.
30. T.L. Bluhm, and P. Zugenmaier: 'Detailed Structure of the V_H-Amylose-iodine Complex: A Linear Polyiodine Chain', *Carbohydr. Res.* **89**, 1–10 (1981).

ISOLATION, PURIFICATION, AND CHARACTERIZATION OF CYCLOMALTODECAOSE (ε-CD)

HARUHISA UEDA, TOMOHIRO ENDO, HIROMASA NAGASE,
*SHOICHI KOBAYASHI and TSUNEJI NAGAI
Hoshi University, 4-41, Ebara 2-chome, Shinagawa-ku, Tokyo 142, Japan,
**National Food and Research Institute, Ministry of Agriculture, Forestry and Fisheries, Tsukuba, Ibaraki 305, Japan*

ABSTRACT

ε-Cyclodextrin (ε-CD) is a cyclic oligosaccharide, composed of ten α-1,4-linked D-glucoses reported by French *et al.* in 1965[1], but has not been studied because of the difficulty in the preparation and purification of large-ring CDs composed of more than nine α-1,4-linked D-glucose units. This report describes the purification and characterization of ε-CD. Furthermore, the crystal and molecular structure of ε-CD hydrate (ε-CD $19H_2O$) was elucidated by X-ray analysis.

1. INTRODUCTION

We have already reported that one of the large-ring CDs, cyclomaltononaose(δ-CD), which is composed of nine α-1,4-linked D-glucose units, has a lower aqueous solubility than ither α-CD or γ-CD[2]. Large-ring CDs may have some unique characteristics in comparison with other conventional CDs. Furthermore, we have previously reported the isolation, purification, and characterization of η-CD (cyclomaltododecaose, composed of twelve α-1,4-linked D-glucose units)[3], ζ-CD (cyclomaltoundecaose, composed of eleven α-1,4-linked D-glucose units) and θ-CD (cyclomaltotridecaose, composed of thirteen α-1,4-

J. Szejtli and L. Szente (eds.), Proceedings of the Eighth International Symposium on Cyclodextrons, 17–20.

linked D-glucose units)[4]. Here, we isolated and crystallized ε-CD (cyclomaltodecaose, composed of ten α-1,4-linked D-glucose units) as its hydrate form for the first time and elucidated its structure by X-ray analysis.

2. MATERIALS AND METHODS

2.1. Materials

CD powder(Dexypearl K-50) was purchased from Ensuiko Sugar Refining Co., Ltd. All other chemicals were commercial sources and used without further purification.

2.2. Purification method of ε-CD

The large-ring CDs mixture was prepared in the same way as described previously.[2,3] Purification of ε-CD from the large-ring CD mixture powder(ca. 240g) was mainly carried out by HPLC, and Fig.1 shows its flow chart. The fractions containing pure ε-CD were collected and concentrated by vacuum evaporation. A prismatic precipitate separated out quickly, and then it was recrystallized with acetonitrile-water (65:35) solution. The final product(ε-CD) was obtained in a yield of ca.160mg as a fine crystal powder.

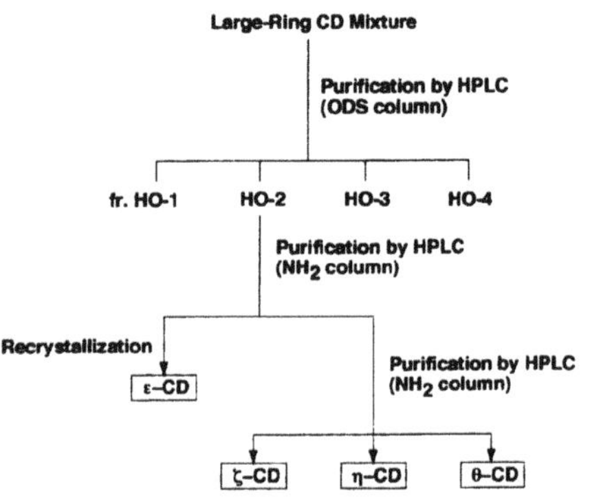

Figure 1. Purification Method of Large-Ring CDs.
ODS Column: Senshu Pak ODS-5251-SS,
Eluent: CH_3OH : H_2O=6:100, Flow Rate: 6.0mL/min.
NH_2 Column: Asahipak NH2P-50,
Eluent: CH_3CN : H_2O=58:42, Flow Rate: 2.0mL/min.

2.3. X-ray analysis of ε–CD hydrate

Crystallization of ε-CD $19H_2O$ was performed by slow evaporation of acetonitrile-water (50:50) solution containing ε-CD at room temperature. A transparent colorless, prismatic crystal of 0.60 x 0.50 x 0.40 mm in size was used for the X-ray analysis. The crystal data are: $C_{60}H_{100}O_{50}$ $19H_2O$, F.W. 1962.3, Monoclinic, C2, Z=2, a=29.338(3), b=9.982(2), c=19.340(2)Å, β=121.025(6)°, V=4853(1)Å^3, D_{calc} 1.356 g/cm^3, μ=11.07 cm^{-1}. X-ray intensity data were measured on a Rigaku automatic four circle diffractometer (AFC-7R,

Cu-K_α, λ=1.5418Å, ω-2θ scan with a 2θ<130.2°). In the structure determination by SIR88[5] and in the following refinements by full-matrix least-squares procedures 8746 independent reflections with $|F_0| > 3\sigma(|F_0|)$ were used. Hydrogen atoms were included but not refined. The final refinement was done by anisotropic temperature factors for oxygen and carbon atoms, and converged R to 0.108. All calculations were performed using a teXsan crystallographic software package of Molecular Structure Corporation.

3. RESULTS AND DISCUSSION

3.1. Identification of ε-CD

ε-CD was subjected to analytical HPLC columns. The purity of ε-CD was almost 100% by HPLC determinations. Fig.2 shows the HPLC chromatogram of large-ring CDs. The values of elution time increased with an increasing number of glucose units in the order of : δ-CD < CD < ζ-CD < η-CD < θ-CD. In the mass spectra of ε-CD, high parent ion peaks appeared at 1621.7, corresponding to the molecular weight plus proton$(M+H)^+$. The molecular weight of ε-CD was determined to be 1620, and this value corresponded to that of ε-CD $\{(C_6H_{10}O_5)_{10}\}$. ^{1}H-NMR, ^{13}C-NMR ^{13}C -^{1}H COSY two dimensional NMR were observed with a JEOL GX-400 spectrometer(400 MHz). The peaks of H1, H3, H2 and H4 of ε-CD were recognized, but the peaks of H5 and H6 were not completely assigned by the ^{1}H -NMR spectrum. The ^{13}C -NMR spectrum of ε−CD had six clear single peaks and as Table 1 shows its chemical shifts originated from the acyclic structure of CD itself in a similar manner as the other CDs. The measurement of ^{13}C -^{1}H COSY two dimensional NMR also supported this finding; the same peaks attributed to a cyclic structure of conventional CD could be obtained in ε-CD(data not shown in this proceedings).

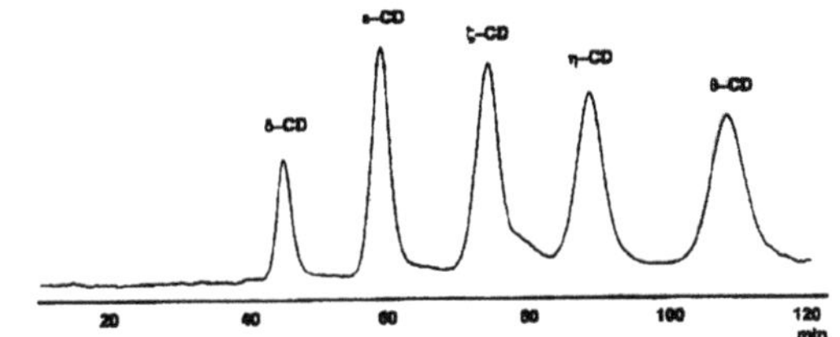

Figure 2. Elution Profile Large-Ring CDs on an Amino Column.
Column : Asahipak NH2P-50. Eluent : $CH_3CN:H_2O$=65:35
Flow rate : 0.7 mL/min

Table 1. ^{13}C-NMR Chemical Shifts of CDs.

Carbon	α–CD	β–CD	γ–CD	δ–CD	ε–CD
1	101.80	102.22	102.03	100.51	99.34
2	72.18	72.30	72.78	72.65	72.24
3	73.79	73.55	73.40	73.32	73.28
4	81.66	81.58	80.90	78.74	77.57
5	72.49	72.55	72.25	71.88	71.35
6	60.94	60.83	60.77	60.84	61.06

3.2. The crystal and molecular structure of ε-CD hydrate

Figs. 3 and 4 show the molecular structure of ε-CD. The overall shape of ε-CD is elliptic and its longer axis is parallel to a line from glucose unit G2 to G7 and shorter axis from G4 to G9 or from G3 to G8. The distance along the longer axis is about 12.58~14.02 Å(G1-O6•••O6-G6~G2-O2•••O2-G7), and that along the shorter axis is about 6.45~8.63 Å(G4-O6•••O6-G9~G3-O2•••O2-G8). Moreover, glucose units,G3,G4,G5,G8, G9 and G10 fall inside the ε-CD cavity. From the side view of ε-CD, the molecule takes a boat form or U character, that is, the glucosidic oxygen atoms(O4) connecting neighboring glucose unit by α-1,4-linkage,are not in a plane. All glucose units took a chair form. Bond lengths and angles of ε-CD are normal. In the ε-CD $19H_2O$, 19 water molecules are distributed over 20 positions, five in the cavity(6 water molecules), and fifteen in the interstices(13 water molecules).

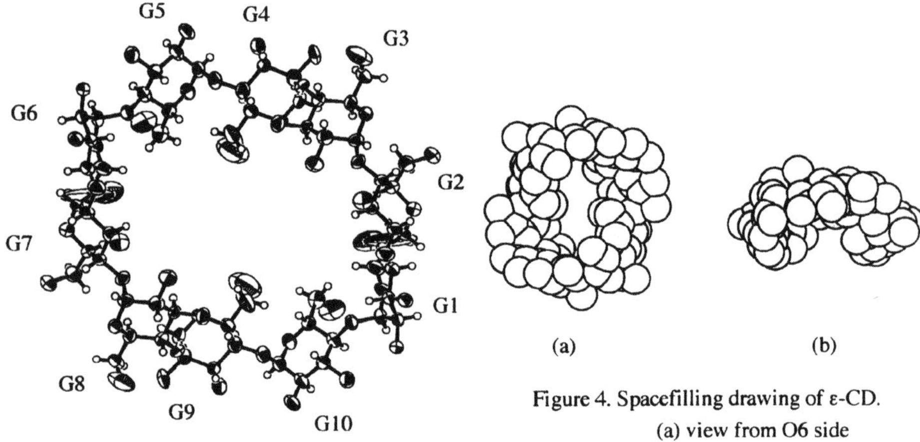

Figure 3. Molecular structure and numbering of ε-CD.

Figure 4. Spacefilling drawing of ε-CD.
(a) view from O6 side
(b) side view

REFERENCE

[1] French, D., Pulley, A.O., Effenberger, J.A., Rougvie, M.A., Abdullah, M., *Arch. Biochem. Biophys.*, **111**, 153-160(1965).

[2] Miyazawa, I., Ueda, H., Nagase, H., Endo, T., Kobayashi, S., Nagai, T., *Eur. J. Pharm. Sci.*, **3**, 153-162(1995).

[3] Endo, T., Ueda, H., Kobayashi, S., Nagai, T., *Carbohydr. Res.*, **269**, 369-373(1995).

[4] Ueda, H., 6th Int.Seminar on Inclusion Compounds, Istanbul, TURKEY, August 1995, p 45.

[5] SIR88: Burla,M.C., Camail,M., Cascarano,G., Giacovazzo,C., Polidori,G., Viterbo,D., *J. Appl. Cryst.*, **22**, 389-303(1989).

CALORIMETRIC INVESTIGATIONS OF BETA - CYCLODEXTRIN HYDRATION THERMODYNAMIC CONSEQUENCES OF WATER EXCHANGE IN INCLUSION REACTIONS.

C. DE BRAUER, P. CLAUDY, M. DIOT, P.GERMAIN, J.M LETOFFE.
Mineral Thermochemistry Laboratory, National Institute of Applied Sciences of Lyon 69621 Villeurbanne Cedex.
M. SERPELLONI
Roquette Frères, F. 62136 Lestrem France

ABSTRACT

The gravimetric study of Beta-Cyclodextrin hydration doesn't point up any definite hydrate. The energy of βCD dehydration is calculated from two distinct calorimetric results.The average value is ΔH = 10 kJ/mol.anh.βCD (to remove one mole of water). Adiabatic calorimetry measurements at very low temperatures applied to anhydrous βCD allow to discuss and to quantify intramolecular interactions between hydroxyl groups. From isothermal calorimetric results in pure ethanol, we conclude that four molecules of water are expelled from hydrated βCD upon the inclusion of ethanol.

1. INTRODUCTION

A significant number of scientific studies about the cyclodextrins have been published since their discovery, hundred years ago. In spite of that, it seems that the fondamental mechanism of inclusion by the cyclodextrins is not yet perfectly known. Many questions remain concerning the role of the cyclodextrin hydration water molecules during the transfer of a substrate from the solvent (usually water) to the cavity of the macrocyclic molecule.

The present paper concerns the Beta-Cyclodextrin molecule. We follow its hydration state according to the environment humidity rate. We analyse the effect of the number of hydration water molecules on the energetic behaviour of Beta-Cyclodextrin. Anhydrous βCD is prepared, stored and conditioned under strict conditions. Its use gives further important informations from a structural and fondamental point of view. The obtained values as a whole, allow to approach and to quantify the exchange of water molecules during the ethanol inclusion by Beta-Cyclodextrin.

J. Szejtli and L. Szente (eds.), Proceedings of the Eighth International Symposium on Cyclodextrons, 21–24.

2. MATERIALS AND METHODS

2.1. Materials

Beta-Cyclodextrin was supplied by Roquette under the reference βCD-16. The hydration rates are determined by thermogravimetric analysis. Anhydrous Beta-Cyclodextrin is prepared by slow dehydration, under vacuum, at 360 K during 48 hours. The samples are stored in a glove box under dry argon and strictly controled before use.

2.2. Methods

All the conclusions of this work are based on thermodynamic considerations, developed from direct measurements. The enthalpies of dissolution and of inclusion reactions are obtained with an isoperibolic calorimeter [1] or a flow micro-calorimeter. The thermal behaviour at very low temperature is followed and quantified with an adiabatic calorimeter built in the laboratory . Differential scanning calorimetry results are obtained using a Mettler TA 2000B apparatus. A MacBain-type thermobalance [1] is used to follow the hydration of βCD versus water vapour pressure .

3. RESULTS AND DISCUSSION

3.1. Hydration of βCD versus water vapour pressure

The experiments are carried out on a sample of βCD prealably dehydrated under vacuum and rehydrated (at 302 ± 1 K) under various determined water vapour pressure. From the gravimetric results, at this temperature, we assume that the percentage of water fixed at the equilibrium by βCD varies linearly versus the water vapour pressure of the environment [1]. In the range of water vapour pressure between 0.013 to 0.031 bar, the number of water molecules fixed on one βCD molecule varies between 10 to 12.

3.2. Energy of dehydration of βCD

From DSC measurements on eleven samples of βCD, we calculate the average value of the total thermal effect of dehydration. This thermal effect takes place between 323 to 483 K in our conditions of experiment. It includes the breaking of the βCD-H_2O bonds and the vaporization of the free water molecules. Knowing the initial hydration ratio of the samples, a simple calculation gives the average following value for the energy of the βCD-H_2O bond [1].ΔH = 9,6 kJ/mol.anh.βCD (to remove one mole of water).
From isoperibolic dissolution calorimetry (in water, at 298 K), we observe that the evolution of the dissolution energy for βCD,nH_2O is linear versus n in the range n = 0 - 12 water molecules [1].
In this range : $\Delta_{diss} H$ βCD,nH_2O = -91,2 + 10,5 n (in kJ/mol.anh.βCD).

To remove one mole of water from one mole of βCD : ΔH = 10,5 kJ/mol.anh.βCD.
Taking into account the uncertainties on DSC and on isoperibolic calorimetry measurements, the reliability of these two results is good. An energy closed to 10 kJ is necessary to break one mole of βCD-H_2O bonds and the bounded water molecules seems to be equivalent from an energetic point of view.

3.3. Peculiar behaviour of the dehydrated macrocyclic molecule

The specific heats (Cp) of anhydrous Beta-Cyclodextrin are determined by adiabatic calorimetry in the range of temperature 10 K - 300 K. The aim of this study is to specify the structural behaviour of the hydrophobic cavity free from hydration water molecules. The results are registrated during the increase of temperature, but they are largely conditioned by the "thermal story" of the sample. Three series of measurements (on a same sample) are represented on the figure 1 (Cp / T is plotted versus T)

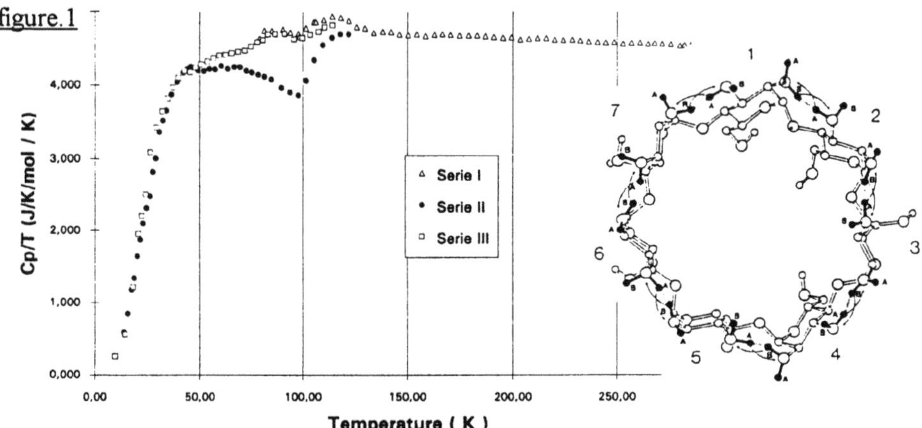

The series I and III correspond to a regular cooling respectively until 77 K and 10 K. We observe, a relaxation phenomena during the reheating stage between 75 K and 115 K. This behaviour id attributed to the random releasing of hydroxyl groups interactions [2] (scheme on fig.1). In the serie II, the sample is maintained at 80 K for several hours before cooling until 10 K. The reheating stage exhibit a different phenomena. At first, the increase of entropy is clearly weaker below 100 K, then, in a narrow range of temperature, we observe a "jump" and find again the upper curves around 115 K. In this case, we assume that the cooling treatment is responsible for the well-ordered freezing of the hydroxyl groups. The releasing takes place between 100 K and 115 K during the reheating. The corresponding entropy change is calculated form the area between upper and down curves. It comes : ΔSexp = 30.5 J/mol.anh.βCD. We assume that seven pairs of hydroxyl groups interact independantly in a molecule of βCD with only two positions (fig.1). The entropy change is calculated using the expression :

$$\Delta S_{calc} = 7R.\ln 2 = 40.3 \text{ J/mol.anh.}\beta\text{CD.}$$

This value must be considered as a limit value corresponding to the complete releasing of idealy well-ordered hydroxyl groups without any other interactions, both intra- and inter-molecular.

3.4. Energetic behaviour of βCD in ethanol and in ethanol/H_2O solutions.

The calorimetric study in the βCD/ethanol/H_2O systems doesn't clearly show the inclusion of ethanol by βCD.This is probably due to the strong interactions ethanol-H_2O for ethanol concentrations upper than 30 % (V/V). Taking into account this fact, we measure, by heat flow calorimetry, the heats of reaction in pure ethanol :

(1) βCD,n'H_2O + ethanol, Q_1 = + 32,6 kJ/mol.anh.βCD.

(2) βCD + ethanol, Q_2 = - 9,15 kJ/mol.anh.βCD.

We assume that n water molecules are expelled upon the reaction 1 and that Q_2 represents only the thermal effect of inclusion. The difference Q_1-Q_2 can be expressed as the sum of two energetic contributions from :

- The partial βCD dehydration, from n' to (n'-n)H_2O : $Q_{n\ deh} = 10{,}5n$ [see 3.2.]
- The dilution of ethanol by the n expelled water molecules $Q_{n\ dil} = -\ 0{,}2n$ [from mixing calorimetry]. It comes : $Q_1 - Q_2 = 10{,}5n - 0{,}2n$ (in kJ/mol.anh.βCD)

And finally $n = 4{,}05 \approx 4$

For the first time, we have a direct experimental approach of the water exchange during the inclusion process of a substrate. In this example the presence of water in the solvent is almost unfavourable for the inclusion of ethanol by βCD.

4. CONCLUSION

The structural and energetic behaviour of hydrated βCD is greatly influenced by the value of its hydration rate. The number of water molecules per mole of βCD directly depends on the features of the environment, especially the water vapour pressure. The breaking of some βCD-H_2O bonds is one major step in the inclusion process. The knowledge of the dehydration energy is therefore an important factor to provide or to calculate the thermodynamic parameters of an inclusion equilibrium. It seems that the hydration water molecules are energetically equivalent, but not structurally equivalent.This work will be continue in this way, to understand the role of each type of hydration water molecules in the inclusion mechanisms.

REFERENCES

[1] BILAL,M, DE BRAUER, C., CLAUDY, P., GERMAIN, P., LETOFFE,J.M., β-Cyclodextrin hydration : a calorimetric and gravimetric study, *Thermochimica Acta* **249**, 63-73 (1995)

[2] STEINER, T., SAENGER,W., Geometry of C-H...O Hydrogen bonds in Carbohydrate crystal structures. Analysis of neutron diffraction data. *J.Am.Chem.Soc* **114**, 10146-10154, (1992) and Ref.

HYDRATION AND DEHYDRATION PROCESSES OF β-CYCLODEXTRIN: A RAMAN SPECTROSCOPIC STUDY

A.M.G. MOREIRA DA SILVA
School of Agriculture, Polytechnical Institute of Coimbra, 3000 Coimbra Portugal

Th. STEINER, W. SAENGER
Institut of Crystallography, Free University of Berlin, 14195 Berlin, Germany

J.M.A. EMPIS
Laboratory of Biotechnology, Technical University of Lisboa, 1000 Lisboa Portugal

J.J.C. TEIXEIRA-DIAS
Molecular Physical Chemistry Research Unit, Faculty of Sciences and Technology, 3049 Coimbra Portugal

ABSTRACT

A linear relationship between the integrated intensity of the Raman OH stretching band (I_{OH}) and ambient humidity in equilibrium with a sample of a crystalline β-Cyclodextrin (βCD) hydrate, for humidities between 15% and 100%, was obtained. In addition, hydration and dehydration processes were monitored by measuring I_{OH} as a function of time. For the normally hydrated crystal, both hydration and dehydration processes present essentially continuous and similar variations with hysteresis for times shorter than 20 min.

1. INTRODUCTION

The water content of crystalline β-cyclodextrin (cyclomaltoheptaose, βCD) hydrate is known to establish fast equilibria with atmospheric humidities in the range 15%-100%. The overall crystal structure is conserved, though with a small percentual reduction of the cell volume from the 100% humidity level [1]. The number of water molecules per βCD varies continuously from about 9.4 to 12.3 in the humidity range 15-100% [1]. In fully hydrated βCD, 7 water molecules occupy the cyclodextrin cavity as a cluster, and 5.4 occupy intersticial spaces [1-3]. Below 15% humidity, the crystal structure collapses. At very low humidities, a distinct phase II is formed [4]. Although the crystal lattices of βCD hydrates do not have permanent channels, fast diffusion of water molecules occurs due to transient fluctuations in the lattices [1]. Exchange experiments carried out with water marked either with D or with ^{18}O showed that the H/D exchange is complete, hence extending also to sterically unaccessible O-H groups, and that the long-range transport of hydrogen takes place by diffusion of intact water molecules [5].

In this work, the relative integrated intensity under the Raman OH stretching band envelope (I_{OH}) is measured for βCD hydrates in equilibria with different relative humidities. From this, a relationship is established between I_{OH} and the relative ambient humidity, for

J. Szejtli and L. Szente (eds.), Proceedings of the Eighth International Symposium on Cyclodextrons, 25–28.

humidities between 15% and 100%. In addition, I_{OH} is measured as a function of time, during a hydration-dehydration sequence of experiments on crystalline βCD.

2. MATERIALS AND METHODS

2.1. Materials

βCD, kindly offered by Wacker Chemie, Munchen, Germany, was recrystallized by cooling concentrated aqueous solutions from *ca.* 80 °C in a Dewar flask. Crystals were taken from the solvent and exposed to atmosphere of moderate temperature and humidity (*ca.* 22°C and 70% relative humidity). After equilibration with the atmosphere, the crystals were grounded to obtain a micro-crystalline powder.

2.2. Methods

Two series of experiments were performed. The first aimed at correlating I_{OH} with atmospheric humidity, for βCD hydrates in equilibria with constant humidities. βCD powder was exposed to the atmosphere over suitable saturated salt solutions [1]. After equilibration, some powder was sealed in a Kimax capillary tube (inner diameter 0.8 mm).

In the second set-up, I_{OH} was followed as a function of time during the process of hydration and dehydration. In the hydration experiment, dehydrated βCD powder was introduced in a cell with the bottom filled with pure water. For dehydration, fully hydrated βCD was introduced in the same cell with water replaced by silica gel. The cell was positioned in the laser beam for Raman spectra recording.

The Raman spectra were recorded on a T64000 Jobin Yvon spectrometer, working in the subtractive configuration, with relevant slit widths set to 300 μm and intermediate slit between premonochromator and spectrograph wide open (14 mm). The detecting device was a CCD, and an integration time of 15 s was used. Spectral data for the 2800-3800 cm^{-1} overall region were collected in 10 subregions, hence corresponding to a total acquisition time of 150 s. An Ar ion laser (Innova 300-05 model with power track, from Coherent) provided *ca.* 100 mW at the sample position. Integrated intensities were determined using the machine software.

3. RESULTS AND DISCUSSION

3.1. Raman νOH intensity *vs.* ambient humidity

The Raman OH stretching band of crystalline hydrated βCD corresponds to a moderately intense and very broad band, centred around 3350 cm^{-1}. This band is assigned to the OH stretching vibration from βCD hydroxyl groups and from water molecules. The most intense band in the 2800-3800 cm^{-1} region is the peak intensity of the complex of νCH bands which occurs at a Raman shift of 2908 cm^{-1}. The peak intensity at this frequency was taken as internal reference for measurement of the relative νOH Raman band intensity (the corresponding frequency was found to be constant over the different conditions of the study). With this internal reference, I_{OH} was measured for βCD hydrates in equilibria with saturated salt solutions of different relative humidities, using the first set-up described in the Methods Section.

Fig.1 presents the Raman spectra of five samples of crystalline βCD hydrate in equilibrium with ambient humidities of 0, 58, 79, 88, 98%. The inset shows I_{OH} *vs.* ambient humidity. Except for the lowest humidity which should correspond to βCD in a different phase (phase II, mostly dehydrated form) [4], all points fall on a straight line (I_{OH} = 24*h+52, R=0.99). The constant term in this relationship corresponds to the hydroxyl groups in the βCD molecules and to water molecules possibly retained in the βCD structure at 0% humidity.

The data in Fig.1 show that in the humidity range 15%-100%, the water content of crystalline βCD hydrates exhibits a linear response to variations of ambient humidity. This is consistent with crystallographic data [1].

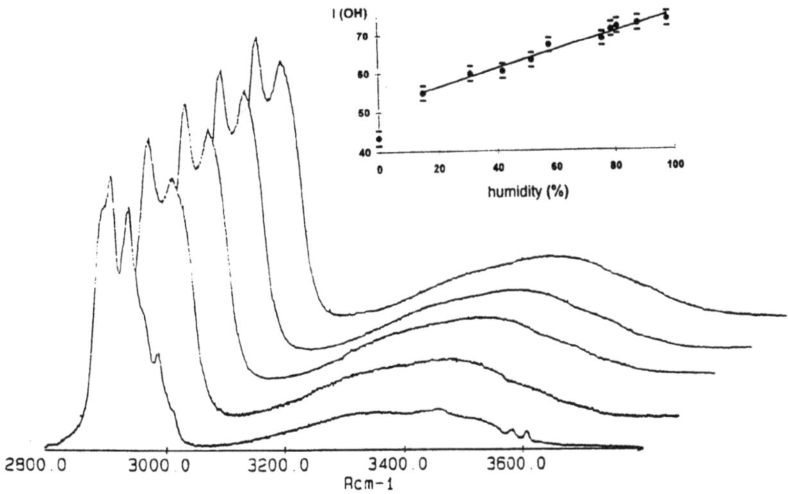

Figure 1. Raman spectra of crystalline βCD hydrates for different water activities. The inset shows I_{OH} *vs.* ambient humidity (h). I_{OH} is drawn on an arbitrary scale.

3.2. Hydration and dehydration processes

In the second experiment, I_{OH} was observed as a function of time during hydration and dehydration of βCD. During dehydration, the relevant kinetic variable is the *occupied fraction of removable water sites* (θ) in βCD, whereas for hydration the unoccupied fraction of removable water sites (1-θ) is the relevant quantity. These quantities (1-θ and θ) should be studied in their variations against time, for a comparison of hydration and dehydration on an equal footing.

On the reasonable assumption that all the OH oscillators have approximately the same intrinsic Raman intensity, the crystal water content per mole of βCD, *i.e.*, $n(H_2O)/\beta CD$, is linear in I_{OH}. However, I_{OH} contains a constant term $I_{OH}^{(min)}$ which includes the contributions from the 21 hydroxyl groups and of possibly present non-removable water molecules. If this constant contribution is subtracted from I_{OH}, the relative quantity $\mathit{I_{OH}}=[I_{OH}-I_{OH}^{(min)}]/[I_{OH}^{(max)}-I_{OH}^{(min)}]$ ranges from 0 to 1. Then, $\mathit{I_{OH}}$ represents the quantity $n=[n-n^{(min)}]/[n^{(max)}-n^{(min)}]$ that is the occupied fraction of the removable water molecules, θ. I_{OH} is therefore a suitable and observable variable in hydration/dehydration experiments. This redefinition of variables is all the more necessary since definite and precise values for $n^{(max)}$ and $n^{(min)}$ are not yet well defined, as they seem to depend on crystallization conditions, and $I_{OH}^{(max)}$ and $I_{OH}^{(min)}$ are expected to depend on current instrumental conditions.

Fig. 2 presents 1-θ *vs.* time (hydration set of experiments) and θ *vs* time (dehydration) for I_{OH} above the corresponding 15% ambient humidity value. This restriction excludes the initial value of the hydration experiment and long-time values of the dehydration experiment. As mentioned above, these points correspond to βCD in a different phase (phase II, mostly dehydrated form).
As it can be seen from Fig.2, both hydration (1-θ) and dehydration (θ) exhibit continuous and similar time variations. Hysteresis occurs for times between zero and 20 min., with the dehydration curve (occupancy curve) decaying faster than the hydration curve (unoccupancy curve), *i.e.*, the results show a different time behaviour for the occupied-to-unoccupied and the unoccupied-to-occupied changes, with the latter being slower for short times.

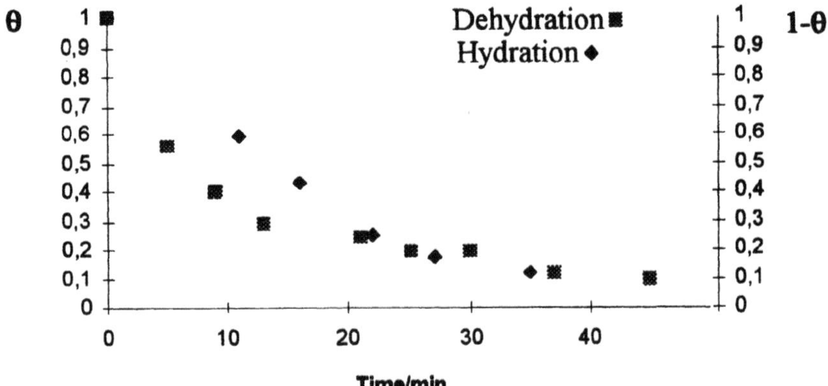

Figure 2. Quantities 1-θ (for the hydration experiment) and θ (for the dehydration experiment) shown as functions of time, for I_{OH} above the corresponding 15% ambient humidity value (see text).

Acknowledgement: This study was supported by J.N.I.C.T. (U.I.&D. #70/94), by the Deutsche Forschungsgemanschaft (Sa 196/26-1) and the DAAD (322-inida-dr).

REFERENCES

[1] Steiner, Th.; Koellner, G. *J. Am. Chem. Soc.*, ***116***, 5122 (**1994**).
[2] Lindner,K.; Saenger,W. *Carbohydr. Res.*, *99*, 103 (**1982**).
[3] Betzel, Ch.; Saenger, W.; Hingerty, B.E.; Brown, J.M. *J. Am. Chem. Soc.*, ***106***, 7545 (**1984**).
[4] Marini,A.;Berbenni,V.; Bruni,G.;Massarotti,V.;Mustarelli,P.; Villa, M.;*J.Chem. Phys.*,***103*** *(17)*, 7532 (1995).
[5] Steiner, Th.; Moreira da Silva, A.M.; Teixeira-Dias, J.J.C.; Müller, J.; Saenger, W. *Angew. Chem.*,***34*** *(13/14)*, 1452 (**1995**).

THERMAL PROPERTIES OF α-CYCLODEXTRIN CRYSTAL FORMS

G. P. BETTINETTI[1], C. NOVAK[2], M. RILLOSI[1], F. GIORDANO[3], P. MURA[4]

[1]*Dipartimento di Chimica Farmaceutica, Università di Pavia, Viale Taramelli 12, I-27100 Pavia*

[2]*Institute for General and Analytical Chemistry, Technical University of Budapest, Szt. Gellért tér 4, H-1521 Budapest*

[3]*Dipartimento Farmaceutico, Università di Parma, Viale delle Scienze, I-43100 Parma*

[4]*Dipartimento di Scienze Farmaceutiche, Università di Firenze, Via G. Capponi 9, I-50121 Firenze*

ABSTRACT

Differences in thermal properties of two α-cyclodextrin hexahydrate polymorphs appear in differential scanning calorimetry (DSC) (pattern of the dehydration endotherms, endothermal effect due to a phase transition of a dehydration product peculiar of a single crystal form) and in thermogravimetric (TGA) and thermomechanical (TMA) analyses.

1. INTRODUCTION

In a previous paper the physicochemical characterization of anhydrous and hydrated (6 and 7.57 H_2O) crystal forms of α-cyclodextrin was carried out on the basis of X-ray diffraction and thermal properties [1]. Since two α-cyclodextrin hexahydrate polymorphs, form I and form II, with different crystal packing schemes are described [2,3], a thermoanalytical study was undertaken to establish whether and how the structural differences were reflected by the thermal behaviour. The basic thermoanalytical techniques i.e. differential scanning calorimetry (DSC), thermogravimetric analysis (TGA) and thermomechanical analysis (TMA) were used. Simultaneous DSC-TGA runs were also done to complement the results of DSC and TGA experiments. Moreover, since the analysis of gaseous products during a thermal scan of a hydrate gives valuable information on its stability, the evolved gas analysis (EGA) curves of the α-cyclodextrin polymorphs were recorded.

2. MATERIALS AND METHODS

2.1. Materials

Commercial α-cyclodextrin[1] (purity grade >99%, αCd) was used. The hexahydrate polymorphs αCd6-I and αCd6-II were prepared by recrystallization from water [2] and from amorphous αCd under 71.5% relative humidity at 40 °C for 3 days [4], respectively.

J. Szejtli and L. Szente (eds.), Proceedings of the Eighth International Symposium on Cyclodextrons, 29–32.

2.2. Methods

Differential scanning calorimetry (DSC), thermogravimetric analysis (TGA) and evolved gas analysis (EGA) were carried out by using respectively a Du Pont 910 DSC cell, a Du Pont 951 Thermobalance, and a Du Pont 916 Thermal Evolution Analyser (TEA) equipped with a hydrogen-air flame ionization detector which is very sensitive for the organic components and gives no sign for inorganic compounds (water, carbon dioxide). 5 mg samples in open aluminium (DSC and EGA) or platinum (TGA) pans were heated (5 K min^{-1} in DSC and TGA, 8 K min^{-1} in EGA) under an argon (DSC and TGA, flow rate 10 L h^{-1}) or nitrogen (EGA, flow rate 1.8 L h^{-1}) atmosphere from room temperature to 300 °C (DSC), 500 °C (TGA) and 350 °C (EGA).
Simultaneous DSC-TGA was carried out with a Setaram Labsys^{TM} thermoanalyzer on 10-13 mg samples in a platinum crucible at a heating rate of 5 K min^{-1} from room temperature to 300°C under a nitrogen flow (flow rate 10 L h^{-1}). A blank was carried out in the same conditions and the results are given after correction.
Thermomechanical analysis (TMA) was done with a Mettler TA4000 apparatus equipped with a TMA 40 cell at 10 K min^{-1} in the 30-220 °C temperature range on samples (200 mg) compressed with an hydraulic press at 5,000 kg for 3 min, with the TMA instrument operating as a dilatometer to measure dimensional changes in the sample [1].

3. RESULTS AND DISCUSSION

α-cyclodextrin crystallizes from water as the hexahydrate in two crystal forms, $\alpha Cd.6H_2O$ form I (αCd6-I) which grows preferentially and $\alpha Cd.6H_2O$ form II (αCd6-II) which can be more easily obtained from amorphous αCd [4]. The polymorphs were characterized on the basis of theoretical and experimental X-ray powder diffraction patterns [1]. The main structural differences were in the packing scheme and in the inclusion stoichiometry, i.e. $\alpha Cd.(H_2O)_2.4H_2O$ for form I, with two water molecules included and four located in interstices between macrocycles, and $\alpha Cd.(H_2O.O(6)).5H_2O$ for form II, with one water molecule and one primary hydroxyl group of a neighbouring macrocycle included and five water molecules located outside [3]. The thermoanalytical curves of the polymorphs are displayed in Figs. 1, 2 and 3. As shown in Fig. 1, dehydration of αCd6-I can be related to a sharp DSC endotherm (t_{onset} = 34 °C, t_{peak} = 44 °C) and a similar thermal effect (t_{peak} = 72 °C) followed by a shallow endotherm peaked at 98 °C and stopping at about 110 °C, with a total dehydration enthalpy in the 29-110 °C temperature range of 265 J g^{-1}. The small endothermal effect (t_{onset} = 136 °C, t_{peak} = 142 °C, ΔH = 6.8 J g^{-1}) with no associated mass loss which appears after dehydration can be attributed to a phase transition of dehydrated αCd6-I. Dehydration of αCd6-II was associated only to the couple of sharp DSC endotherms peaked at 53 °C (t_{onset} = 46 °C) and 77 °C, with a total dehydration enthalpy in the 46-92 °C temperature range of 188 J g^{-1}. Thus distinct DSC profiles were recorded for the polymorphs, while the weight losses (each of about 9% as mass fraction of water) in the TGA temperature range of dehydration and the EGA profiles with no loss of organic component in the same temperature range were similar for both crystal forms. Simultaneous DSC-TGA (Fig. 2) showed a two-step dehydration for αCd6-I (t_{peak} = 59 and 94 °C on DTG, 58 and 98 °C on DSC) and a three-step dehydration for αCd6-II (t_{peak} = 35, 60 and 94 °C on DTG, 60 and 96 °C on DSC). The DSC endothermal effect peaked at 138 °C for αCd6-I confirmed the occurrence of a phase transition peculiar of its dehydration product. The TMA curve of αCd6-I displayed two expansion steps associated to water loss between 55-85 °C and 115-130 °C (respectively 3.9% and 2.8% increase of the initial thickness of 1.1264 mm) (Fig. 3). αCd6-II expanded only a little between 30-95 °C (1.9% increase of the initial thickness of 1.0385 mm) during transformation to the respective anhydrate under the same experimental conditions.

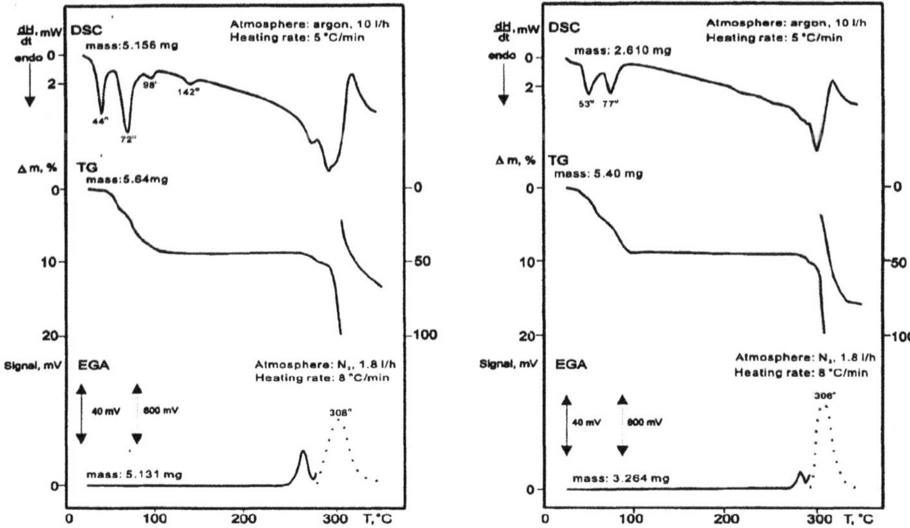

Figure 1. DSC, TGA and EGA curves of αCd6-I (left) and αCd6-II (right).

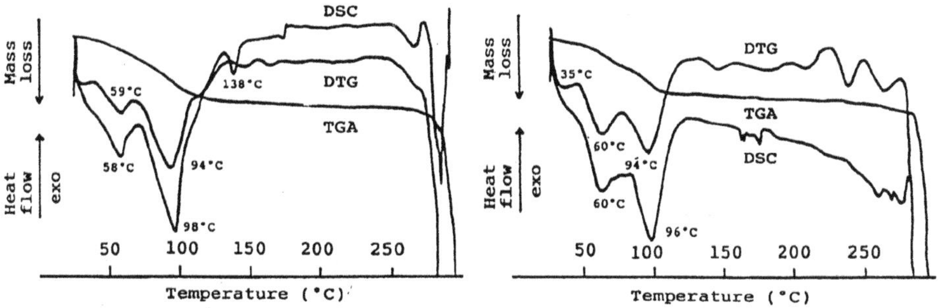

Figure 2. Simultaneous TGA-DSC with DTG (1st derivative of TGA) curves of αCd6-I (left) and αCd6-II (right).

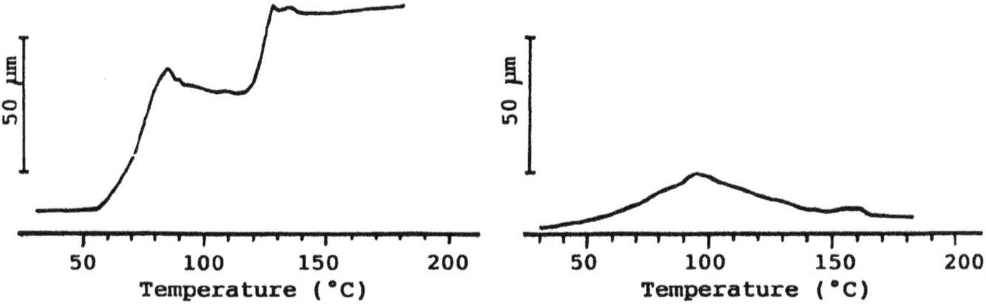

Figure 3. TMA curves of αCd6-I (left) and αCd6-II (right).

4. CONCLUSION

Differences in the DSC and TGA dehydration patterns between the αCd hexahydrate polymorphs are not so sharp as those observed between the hexahydrate and the hydrate containing 7.57 water molecules per αCd molecule, which is characterized by two overlapping DSC endotherms, peaked respectively at about 60 °C and 82 °C, and a broad, shallow endothermal effect stopping at 170 °C [1]. The macrocyclic ring is equally distorted in both the hexahydrate polymorphs [2,3] while it assumes an unstrained, symmetrical shape in the crystal form with 7.57 hydration water mol/asymmetric unit. Five water molecules are located outside the αCd cavity and the remaining 2.57 water molecules are distributed within the cavity and disordered [5]. Hence the pattern of water elimination from the crystal lattice of αCd hydrates is influenced more deeply by conformational factors than by the herring-bone type crystal packing characteristic of these crystal forms of αCd. Thermoanalytical differentiation between the αCd hexahydrate polymorphs can be based on the profiles of DSC dehydration endotherms, the absence or the presence of the DSC endothermal effect at about 140 °C, the number of dehydration steps, i.e. of temperature maxima on DTG curves, and TMA behaviour.

ACKNOWLEDGEMENTS

Financial supports from MURST-60% founds and CNR (Italy) and from the Hungarian National Research Foundation (OTKA, Grant No. T 014 550) are gratefully acknowledged.

NOTES

[1]Donated by Wacker-Chemie GmbH, Hanns-Seidel-Platz 4, D-81737 München.

REFERENCES

[1] Bettinetti G. P., Conte U., Maggi L., Rillosi M., Setti M., *Solid state physicochemical characterisation of α-cyclodextrin, a possible excipient in tablet technology*, in Proc. 14th Pharmaceutical Technology Conference, Barcelona, 4-6 April 1995, Vol. 2a.Pp. 390-398.

[2] Klar, B., Hingerty, B., Saenger, W., Topography of cyclodextrin inclusion complexes. XII. Hydrogen bonding in the crystal structure of α-cyclodextrin hexahydrate: the use of a multicounter detector in neutron diffraction. *Acta Cryst.*, **B36,** 1154-1165 (1980).

[3] Linder, K., Saenger, W., Topography of cyclodextrin inclusion complexes. XVI. Cyclic system of hydrogen bonds: structure of α-cyclodextrin hexahydrate, form (II): comparison with form (I). *Acta Cryst.*, **B38,** 203-210 (1982).

[4] Bettinetti, G. P., Rillosi, M., Setti, M., Mura, P., *The amorphous state of α-cyclodextrin,* in Proc. European Symposium on Formulation of Poorly-available Drugs for Oral Administration, Paris, 5-6 February 1996, (in press).

[5] Chacko, K. K., Saenger, W., Topography of cyclodextrin inclusion complexes. 15. Crystal and molecular structure of the cyclohexaamylose-7.57 water complex, form III. Four- and six-membered circular hydrogen bonds. *J. Am. Chem. Soc.*, **103,** 1708-1715 (1981).

NMR STUDY ABOUT THE STRUCTURE AND BEHAVIOR OF N-PERALKYLAMINO-CYCLODEXTRINS IN AQUEOUS AND NON-AQUEOUS SOLVENTS

Hiroshi Nakanishi*†, Kenji Kanazawa*, Toru Yamagaki*, Yasuko Ishizuka*, and Waichiro Tagaki**
*Biomolecules Department, National Institute of Bioscience and Human-Technology, 1-1 Higashi, Tsukuba 305 Japan
**Department of Applied Chemistry, Faculty of Engineering, Osaka City University, Sugimoto-cho 3, Sumiyoshi-ku, Osaka 558 Japan

ABSTRACT

^{1}H and ^{13}C NMR spectra of the N-alkylamino-cyclodextrin(CD)s were measured in organic and aqueous solvents. It was found by the spectral analyses of the very broad signals of the glucose rings and N-alkyl groups that these CD derivatives strongly aggregate and the aggregation form is a reversed micelle type in organic solvents and is a normal micelle type in aqueous solvents.

1. INTRODUCTION

It is well known that CD is insoluble in organic solvents. Therefore, we introduce the long N-alkylamino chain to the 6 position of the glucose ring in the CD in order to be soluble in both organic and aqueous solvents. The N-peralkylamino-CDs show an amphiphilic property.

Recently, we report the syntheses of N-peralkylamino-β-CDs and their host-guest phenomena [1, 2]. On the way of the NMR analyses of these substituted CDs, we have found the unusual broad signals. Thus, we made various kinds of NMR measurements in order to get detailed information about the aggregation mode in these amphiphilic molecules.

J. Szejtli and L. Szente (eds.), Proceedings of the Eighth International Symposium on Cyclodextrons, 33–36.

In this paper, we report the interesting results of the NMR spectral analyses.

2. MATERIALS AND METHODS

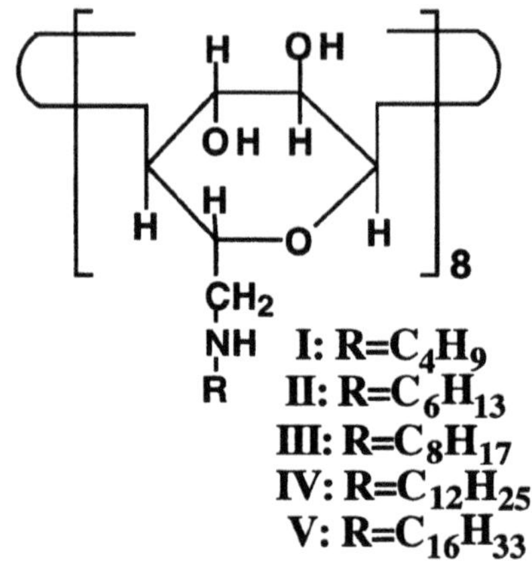

Substituted γ-CDs (I-V) were chemically synthesized from bromo-γ-CD and N-alkylamines. The purification was done by the column chromatographic method. The purity was checked by ^{1}H and ^{13}C NMR spectroscopy. The smaller compounds were analyzed by TOF-MS spectra. The purity is ca. 99 %. NMR spectra were measured by a JEOL α-500 spectrometer (for ^{1}H: 499.65 MHz and for ^{13}C: 122.5 MHz).

3. RESULTS AND DISCUSSION

3.1. NMR spectra of N-alkylamino CDs in organic solvents

γ-CD derivatives with eight alkylamino groups in the 6 position of the glucose skeleton of the CD were chemically synthesized. The alkylamino group (RNH) is; R: C_4H_9 (I), C_6H_{13} (II), C_8H_{17} (III), $C_{12}H_{25}$ (IV) and $C_{16}H_{33}$ (V). ^{1}H NMR spectra of these compounds (I - V) were measured in a chloroform ($CDCl_3$) solution at room temperature. As is shown in Fig.1(a), ^{1}H NMR signals are very broad in all regions. Interestingly, the degree of the broadness in the signals of the glucose groups is very larger than that of the alkyl groups, whereas the latter is much larger than that of the signals in the unsubstituted γ-CD. These line-broadenings can be explained by considering the existence of strong aggregation of the N-alkylamino- γ-CD and the presence of isomers in different aggregation modes. The careful observation of the degree of the line broadenings in I clearly indicates that the molecular motion of γ-CD ring is very different from the motions of the eight N-butyl groups. The same tendency was also found in the spectra of II and

III (Fig. 1). These suggest the existence of the intermolecular hydrogen bondings between the glucose groups of the different γ-CD rings.

When the length of the chain in the N-alkyl group in I - V is larger, the ^{1}H NMR spectral pattern changes dramatically as shown in Fig. 1, which indicates the large change of the mode and the degrees of the aggregations of the molecules. The spectral changes are especially outstanding, when the alkyl group changes from n-hexyl group to n-octyl group (II→III). The ^{1}H NMR signals of V with the longest alkyl groups are very broad, especially in the region of the glucose protons (6~3 ppm from TMS). It is reasonable to consider a very high aggregation mode such as the high polymers. These interesting results strongly suggest that the degree of the aggregations in these N-alkylaminated CD (I ~V) in chloroform solution becomes larger, as the length of the alkyl group becomes larger and that the change occurs very strongly, when the alkyl group changes from C_6H_{13} to C_8H_{17} .

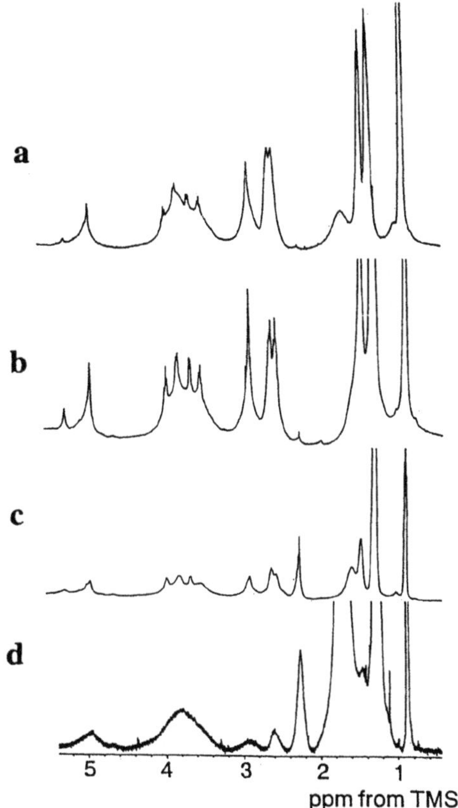

Fig. 1. ^{1}H NMR spectra at 25.0 °C in $CDCl_3$ solution
a) I: R=C_4H_9, b) II: R= C_6H_{13}, c) IV: R= C_8H_{17}, d) V: R=$C_{16}H_{33}$

^{1}H NMR spectra were also recorded at various kinds of conditions changing the concentrations, temperatures and organic solvents. ^{13}C NMR spectra were also recorded and analyzed in $CDCl_3$ solution. Following important results were obtained:

1) NMR analyses of I ~V indicate that these compounds form the reversed micelle type of aggregation mode in organic solutions. 2) The ratio of the monomer to the reversed micelle in the CD's increases as an increase in temperature. 3) An increase in the concentration of the CD's results in the larger amount of their aggregations in chloroform solution.

3.2. NMR spectra changes in I-V in an aqueous solutions.

In order to get the informations about the molecular structures and the behaviors in an aqueous solution, the ^{1}H and ^{13}C NMR spectra were also measured in various kinds of aqueous solutions. It was made clear that, in the aqueous solution, these N-alkylamino-γ-CDs form the normal micelle type of aggregation. This is a very contrasting result to the case in organic solvents. The detailed analyses in these aggregations are in progress in our laboratory.

REFERENCES

[1] Tanaka M., Ishizuka Y., Matsumoto M., Nakamura T., Yabe A., Nakanishi H., Kawabata Y., Takahashi H., Tamura S., Tagaki W., Host Guest Complexes of Amphiphilic β-Cyclodextrin and Azobenzene Derivatives in Laugmir-Blodgett Films, Chem. Lett., 1307-1310 (1987)

[2] Nakanishi H., Kanazawa K., Ishizuka Y., Tagaki W., Interaction between Fullerene (C60) and Long Chain N-alkylamino Substituted Cyclodextrins, Materials Science & Engineering C2, 83-86 (1994)

AGGREGATION BEHAVIOR OF HYDROPHOBICALLY MODIFIED β-CYCLODEXTRINS IN AQUEOUS SOLUTION

F. Witte, H. Hoffmann
Department of Physical Chemistry I, University Bayreuth
Universitätsstraße 30, 95440 Bayreuth, Germany

ABSTRACT

Three amphiphilic β-cyclodextrins in aqueous solution were investigated upon their aggregation behavior. Surface methods, conductivity and scattering methods were used to determine the cmc and radius of gyration of the aggregates. Depending on the cotenside it was possible to shift the cmc to lower concentrations. Above the cmc small spherelike aggregates existed.

1. INTRODUCTION

Modified CDs are widely used in industry e.g. as drug carriers. The substitution leads to increased water solubility and can enhance inclusion complex formation. Substitution of the primary hydroxyl face by alkylamino- or thio-groups leads to amphiphilic cyclodextrins, which are able to form stable langmuir layers and lyotropic liquid crystals in organic solvents. On the other hand by hydrophbic substitution of the secondary hydroxyl face it's possible to get mixed vesicles with phospholipids. Watersoluble surface active cyclodextrins might furthermore aggregate in aqueous solution.

2. MATERIALS AND METHODS

2.1. Materials

Three partially substituted β-CD derivatives form WACKER Company were investigated:

• W7 CM EHGE (Carboxymethyl-ethylhexylglycidyl-β-CD, **D1015**
  substitution degree : 0.6 / 0.4
  molar weight : ≈ 1900 g/mol

J. Szejtli and L. Szente (eds.), Proceedings of the Eighth International Symposium on Cyclodextrons, 37–40.

- **W7 HP PGE (Hydroxypropyl-phenylglycidyl-β-CD, D1066**
  substitution degree : 0.5 / 0.4
  molar weight : ≈ 1700 g/mol
- **W7 HP HH (Hydroxyhexyl-hydroxypropyl-β-CD, D1069**
  substitution degree : 0.4 / 0.4
  molar weight : ≈ 1600 g/mol

2.2. Methods

To investigate the interface activity of the β-CD derivatives, surface tension and interfacial tension were measured with a ring tensiometer and a drop volume tensionmeter. The aggregate dimensions were determined by light scattering and neutron scattering. Furthermore conductivity of one derivative was investigated. Viscosity was measured with a capillary viscosimeter.

3. RESULTS AND DISCUSSION

3.1. Surface and interfacial tension

All three CDs show a cmc in surface tension. While the cmc of D1069 is not affected by the decane molecules in interfacial tension, the aggregation concentration of D1015 and D1066 is lowered significantly. At the same time the head group area rises.

TABLE 1. cmc data from surface and interfacial tension

Sample	cmc[wt-%] (surface tension)	head group area [Å] (surface tension)	cmc [wt-%] (interfacial tension)	head group area [Å] (interfacial tension)
D1015	0.25	224	0.07	300
D1066	20	149	1.3	180
D1069	0.6	125	0.6	124

The origin of this behavior must be the interaction between CD-substituents and the decane molecules. To investigate this interaction, several surface and interfacial tension experiments with D1066 were carried out. The interfacial tension depends on the media next to the aqeous phase. An alkane lowers the cmc more than an alcohol. Furthermore it's possible to lower the cmc in surface tension by addition of alcohol to a D1066 solution.

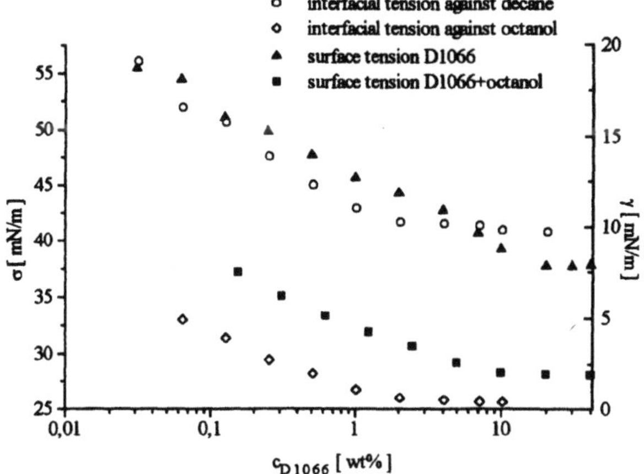

Fig. 1:Surface and interfacial tensions of D1066

3.2. Light scattering, neutron scattering and conductivity

The aggregate molar weight, determined by static light scattering, indicates very small aggregates consisting of only few monomers. This is confirmed by neutron scattering, which shows aggregates with radii of about 15 Å. The cmc of D1015 in light scattering (0.8 wt-%) corresponds to that measured by conductivity, but it differs significantly from that given by surface tension measurement (0.25 wt-%).

3.3. Viscosity

The viscosities of all three derivatives show a similar course with a divergence at high concentrations. To determine the aggregate shape from viscosity data the theoretical viscosity for hard spheres is calculated. The case of hard spheres with a viscosity divergence at the maximum packing volume is described by the Dougherty-Krieger formular:

$$\eta/\eta_0 = \left(1 + x - \frac{\Phi}{\Phi_m}\right)^{-2.5\cdot\Phi_m} \qquad (1)$$

with Φ_m = maximum packing volume fraction and x = thickness of hydrate shell in % of total aggregate radius.

The displayed fit, calculated for Φ_m=0.5 and x=0.35 corresponds well to the experimental data.

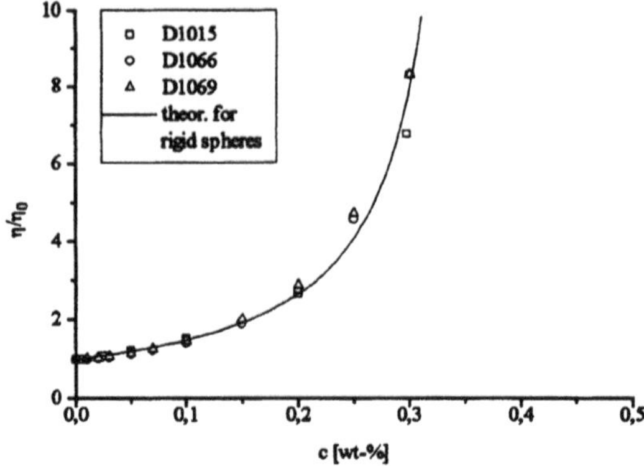

Fig. 2 Experimental viscosity data and calculated viscosity (Dougherty-Krieger)

4. CONCLUSIONS

All investigated β-CD derivatives have a critical aggregation concentration. While the cmc of derivative D1069 isn't affected by the nature of the molecules at the interface, in case of D1015 and D1066 they have a rather strong influence on the aggregation. Because the lowering of the cmc is accompanied by a significant increase of head group area, the following model is proposed: The hydrophobic molecules at the interface insert into the space between the hydrophobic chains of the β-CD derivative and enhance the amphiphilic character of the CD. Hydrophobisation by inclusion complex forming wouldn't explain the strong increase in head group area. The aggregates above the cmc are small and spherelike. The driving force for aggregation should be hydrophobic interaction of the substituents like in classical micelles, not inclusion of hydrophobic chains from neighbour CDs. Not already explainable are the different cmc values from interface methods (e.g. surface tension) and volume methods (e.g. conductivity).

Fig. 3 Interaction between β-CD derivative and hydrophobic molecules

COMPARISON OF SEMIEMPIRICAL AND MOLECULAR MECHANICS OPTIMIZED GEOMETRIES OF CYCLODEXTRINS

Imre BAKÓ[1], László JICSINSZKY[2]
[1] CRIC- HAS, Budapest, H-1525, P. O. Box: 17 Hungary
[2] CYCLOLAB Ltd. Budapest, H-1525 P. O. Box: 435, Hungary

ABSTRACT
The structures of cyclic oligomers of glucopyranose, small size, theoretical cyclic oligomers up to the γ-cyclodextrin have been optimized by AM1 and PM3 semiempirical (SE) methods, and MM2, AMBER, and CHARMM Molecular Mechanics (MM) force fields. *Ab initio* optimization for glucose was also done at STO-3G level. In order to study the differences in the heat of formations of the various optimized geometries more detailed calculations were performed on the various CH_2OH rotamers of glucose.

INTRODUCTION

The most popular program packages offer various quantum chemical and molecular mechanics methods. Unfortunately, presently there is no absolute method for all types of molecules. The parametrization of various calculation methods is difficult and usually the published datasets are not up to date. The transferability of the calculations from the small molecules to larger ones is not unambiguous. The previous quantum chemical calculations showed [1] e. g. the partial atomic charges can be transferred from glucose to the cyclic, maltose type oligosaccharides. This presentation compares the different calculation methods.

CALCULATION METHODS:

The AM1 semiempirical geometry optimizations of cyclodextrins were performed on an IBM RS6000/m350 computer using MOPAC 6.0 [1-2]. MM geometry optimization of calculations and geometry characterizations were performed on an i586-100MHz computer using HyperChem® (Release 4.5) [3]. MM calculations were performed by enhanced MM2 method implemented in HyperChem® using both the bond dipole and the point charge option. AMBER and CHARMM type force fields were used the most recent public parametrizations [4, 3]. In order to decrease the occurrences of false minima, the energy criteria for the optimizations were chosen enough low for the optimization algorithm (usually less than 0.001 Kcal/mol/Å) to get sensible minima among the lowest energy states.

Calculations on the various rotamers of the primary hydroxyls of α-D-Glc*p* were performed by restricted optimization of the molecule. The O5-C5-C6-O6 dihedral angle was varied in 2.5° steps and kept constant during the geometry optimization.

RESULTS AND DISCUSSION

The MM geometry optimizations were carried out on structures where the atoms have partial charges and the results were compared with the chargeless calculations. The charges of crystalline cyclodextrins (with and without the crystal water) were calculated

J. Szejtli and L. Szente (eds.), Proceedings of the Eighth International Symposium on Cyclodextrons, 41–44.

by both mentioned SE methods. The MM geometry optimization using the SE calculated partial atomic charges gave satisfactory results in comparison of the SE methods, and *ab initio* results, as well. The obtained geometries were in good agreement with the AM1 SE results though the calculated heat of formations showed relatively large deviations. Optimized geometries by PM3 SE method gave worse geometries but the partial atomic charges were closer to the *ab initio* values of α-D-glucopyranose. Comparison of various methods are demonstrated in Table I.

The calculations showed that though only small differences can be observed in geometry the corresponding heat of formation values may differ up to 10 Kcal/mol. In general the differences for bond length are less than 0.01 Å, for bond angles less than 5° and for dihedrals less than 5° comparing the different force fields to the SE and ab initio optimized geometries.

CONCLUSION

The heat of formations of the molecules suggest that:

i. The water molecules stabilize the cyclodextrins on a higher energy level. The crystalline structures without the water molecules give the energetically worst geometries.

ii. The MM and semiempirical optimizations gave slightly different structures resulting in that MM optimized structures have a slightly higher heat of formation. The MM2 method gives the most comparable heat of formations in comparison with the AM1 optimized structures. The PM3 optimized geometries are worse than either the MM optimized or AM1 optimized ones;

iii. In case of glucose the *ab initio* optimized structures differs non-significantly from the semiempirical ones;

iv. The combination of MM and semiempirical methods may accelerate the optimizations without decreasing the accuracy (both in geometry and heat of formation).

v. From the detailed investigation of the energetic relationship of the glucose primary hydroxyl rotamers one can conclude that the charge estimation for MM calculations have many pitfalls, independently from the used force fields. The charge estimation is a crucial point of the modeling of carbohydrates.

Comparing the data [5] with the *ab initio* calculations one can conclude that some simplification of calculations may lead to better approximation of the natural behavior of molecules. SE calculations on more artificial small-size cyclic oligosaccharides gave similar results to the MM calculations [6].

ACKNOWLEDGMENT. The support of this work by the OTKA Foundation of the Hungarian Academy of Sciences (F4309) is appreciated.

REFERENCES

1. I. Bakó, L. Jicsinszky, Semiempirical Calculations ... *J. Incl. Phenom.* **18**, 275-289 (1993)
2. QCPE:455
3. HyperChem® is a product of Hypercube Inc., Waterloo, Ontario, Canada
4. S. W. Homans, *Biochemistry*, **29** 9110 (1990)
5. J. W. Brown, B. D. Wladkowski, *Ab Initio* ..., *J. Am. Chem. Soc.* **118** 1190-1193 (1996)
6. S. Immel, J. Brickmann, F. W. Lichtenthaler, Small-Ring ..., *Liebigs Ann.* **1995** 929-942

Table I.: Comparison of Semiempirical and MM Energies of the MM Optimized Cyclodextrins (with AM1 Calculated Charges on the Starting Geometries) [Heat of Formation and Total energies in Kcal/mol, Gradients in Kcal/mol/Å, in parentheses]

Molecule	***Heat of Formation***	***Dipole***	***AMBER Total Energy***	***CHARMM***	***Total Energy***	***MM2 Total Energy***		***MM2 Total Energy***
			with charge	**without charge**	**with charge**	**without charge**	**with charge**	**bond dipole**
αDGlcp*H_2O (Cryst.)	-179 (110)	3.8	134 (75)	189 (67)	200 (68)	118 (70)	129 (71)	115 (70)
αDGlcp (Cryst.)	-265(15)	4.1	44 (27)	92 (30)	106 (29)	35 (14)	50 (13)	35 (14)
αDGlcp (AM1)	-303.9 (0.04)	2.6	15 (12)	12 (8)	75 (14)	14 (8)	19 (7)	12 (8)
αDGlcp (PM3)	-269.4 (0.04)	0.8	18 (10)	66 (11)	78 (11)	14 (7)	26 (7)	11 (7)
αDGlcp gg (*ab initio* opt.)	-294.0 (17)	3.2	17.1 (17)	73.3 (19)	85.6 (19)	22.7 (15.6)	34.9 (15.1)	20.0 (16)
αDGlcp gt (*ab initio* opt.)	-292.5 (17)	1.6	17.4 (16)	73.4 (19)	86.6 (19)	22.1 (15.3)	35.8 (14.8)	20.1 (15)
αDGlcp tg (*ab initio* opt.)	-292.4(17)	2.6	17.9 (17)	76.0 (19)	87.9 (19)	24.3 (15.4)	36.2 (14.8)	20.6 (15)
cyclo[D-Glcp α(1→4)]$_2$ (AM1)	-444.0 (0.08)	3.8	113 (14)	160 (16)	185 (17)	73 (8)	98 (8)	72 (8)
cyclo[D-Glcp α(1→4)]$_2$ (AMBER.)	-415 (18)	3.1	86.7 ($9*10^{-3}$)					
cyclo[D-Glcp α(1→4)]$_2$ (Charmm)	-409 (22)	3.2		131.0 (1)	157.5 ($1*10^{-3}$)			
cyclo[D-Glcp α(1→4)]$_3$ (AM1)	-686.5 (0.07)	5.2	131 (14)	199 (17)	229 (15)	90 (8)	119 (7)	84 (8)
cyclo[D-Glcp α(1→4)]$_3$ (AMBER)	-645 (16)	4.5	94.8 ($8*10^{-3}$)					
cyclo[D-Glcp α(1→4)]$_3$ (Charmm)	-633 (20)	3.4		154.0 (1.5)	184.4 ($4*10^{-3}$)			
cyclo[D-Glcp α(1→4)]$_4$ (AM1)	-922.2 (0.08)	1.3	147 (13)	238 (16)	228 (16)	238 (16)	133 (7)	91 (7)
cyclo[D-Glcp α(1→4)]$_4$ (AMBER)	-874 (15)	0.4	100.1 ($9*10^{-3}$)					
cyclo[D-Glcp α(1→4)]$_4$ (Charmm)	-854 (19)	0.5		152.1 (1.3)	191.9 ($4*10^{-3}$)			
cyclo[D-Glcp α(1→4)]$_5$ (AM1)	-1172.7 (0.01)	8.0	127 (12)	229 (15)	274 (15)	92 (8)	137 (7)	85 (8)
cyclo[D-Glcp α(1→4)]$_5$ (AMBER)	-1121 (14)	7.2	77.1 ($8*10^{-3}$)					
cyclo[D-Glcp α(1→4)]$_5$ (Charmm)	-1092 (19)	8.2		139.8 (1.3)	189.0 ($5*10^{-3}$)			
αCD*7.5H_2O (Cryst.)	-840 (104)	3.3	745 (71)	955 (63)	1054 (63)	649 (69.0)	748 (69.2)	644 (69)
αCD*6H_2O (Cryst. form I.)	-1563 (36)	6.3	265 (34)	530 (32)	590 (32)	227 (26.8)	287 (26.7)	213 (27)
αCD*6H_2O (Cryst. form II.)	-966 (77)	4.8	1656 (121)	1657 (100)	1715 (100)	1195 (85.1)	1051 (85.1)	1180 (85)

Continuation of ***Table I.***

αCD (Cryst. 7.5H_2O)	2623 (293)	6.7	$8.7*10^9$	$4.9*10^{10}$	$4.9*10^{10}$	6309 (1014)	5295 (1038)	6317 (1014)
αCD (Cryst. Form I.)	1067 (172)	6.8	$1.1*10^{14}$	$4.1*10^{15}$	$4.1*10^{15}$	5619 (2531)	6056 (2829)	5619 (2531)
αCD (Cryst. Form II.)	2426 (291)	3.2	$8.7*10^9$	$4.9*10^{10}$	$4.9*10^{10}$	6110 (1014)	6116 (1038)	6115 (1014)
αCD (AM1)	-1411.5 (0.08)	7.1	150 (12)	403 (16)	458 (15)	110 (8)	165 (7)	100 (8)
αCD (PM3)	-1241.0 (0.08)	6.1	201 (18)	481 (30)	571 (30)	126 (7)	215 (7)	118.4 (7)
βCD*12H_2O (Cryst.)	-2172 (31)	29.0	198.7 (22)	500.3 (22)	594.1 (21)	212.3 (22)	306.0 (22.3)	199.8 (22.4)
βCD*12H_2O (AMBER)	-2252 (13)	20.2	77.1 (0.003)	376.4 (9)	466.2 (9)	121.5 (9)	211.4 (9)	114.3 (9)
βCD*12H_2O (Charmm)	-2281 (16)	8.1	113.8 (5)	308.6 (2)	381.1 (0.02)	156.7 (12)	229.3 (12)	134.0 (12)
βCD*12H_2O (MM2, AM1 Charge)	-2366 (12)	10.4	132.3 (10)	447.9 (12)	514.6 (12)	85.6 (2)	152.2 (0.03)	57.7 (1)
βCD (Cryst.)	+1780 (270)	3.1	$1.3*10^9$	$1.1*10^9$	$1.1*10^9$	4221.1	4082.6	4206.2
βCD (Cryst.) (AMBER)	-1461 (15)	2.3	94.7 (0.001)	233 (10)	499.3 (9)	128.2 (10)	192.8 (9)	121.7 (10)
βCD (Cryst.) (Charmm)	-1456 (19)	4.9	205.7 (8)	404.0 (1)	510.8 (0.006)	193.8(114)	300.6 (14)	191.4 (14)
βCD (Cryst.) (MM2, AM1 Charge)	-1510 (13)	6.4	230.2 (11)	564.5 (15)	650.3 (15)	132.9 (1)	218.7 (0.01)	127.3 (1)
βCD (AM1)	-1654.4 (0.07)	2.0	191.2 (12)	483.2 (16)	532.8 (15)	138.1 (8)	187.7 (7)	126.0 (8)
βCD (PM3)	-1454.1 (0.07)	7.5	218 (13)	656 (42)	748 (42)	150 (7)	242 (6)	141 (7)
βCD (Charmm, AM1 Charge)	-1536 (18)	1.9	143.1 (6)	330.8 (1.2)	401.8 (0.007)	163.7 (13)	234.7 (13)	158.4 (13)
βCD (MM2 , AM1 Charge)	-1615 (11)	4.0	145.6 (10)	458.0 (13)	510.6 (13)	104.7 (1)	157.3 (0.02)	94.4 (1)
γCD*14H_2O (Cryst.)	-1963 (53)	17.2	1019.3 (131)	1348.2 (88)	1451.7 (89)	646.5 (42)	750.0 (43)	639.6 (42)
γCD*14H_2O (AMBER)	-2632 (13)	6.1	77.1 (0.004)	471.4 (11)	543.6 (11)	122.5 (9)	194.7 (9)	116.4 (9)
γCD*14H_2O (Charmm)	-2666 (16)	4.6	115.0 (6)	341.7 (2)	397.6 (0.03)	159.6 (12)	215.4 (12)	134.0 (12)
γCD*14H_2O (MM2, AM1 Charge)	-2768 (11)	8.5	142.1 (10)	510.6 (12)	553.9 (12)	89.1 (2)	132.4 (0.01)	58.7 (1)
γCD (Cryst.)	-2198 (90)	7.2	$1.0*10^6$	$5*10^6$	$5*10^6$	1952.8 (93)	1941.9 (92)	1946.4 (93)
γCD (Cryst.) (AMBER)	-1666 (15)	7.4	108.5 (0.001)	489.6 (8)	556.9 (8)	152.0 (10)	219.3 (9)	140.6 (10)
γCD (Cryst.) (Charmm)	-1644 (19)	7.5	240.0 (7)	527.4 (1)	655.9 (0.02)	237.7 (14)	369.3 (14)	235.3 (14)
γCD (AM1)	-1896.7 (0.09)	3.0	199.4 (12)	552.7 (16)	610.6 (15)	152.9 (7)	210.7 (7)	137.6 (7)
γCD (PM3)	-1675.6 (0.07)	5.0	226 (14)	628 (28)	730 (28)	165 (7)	266 (7)	153 (7)
γCD (Charmm, AM1 Charge)	-1774 (17)	4.4	150.9 (5)	384.1 (1)	459.7 (0.005)	181.6 (13)	257.3 (13)	173.8 (13)
γCD (MM2, AM1 Charge)	-1859 (11)	7.2	162.8 (10)	532.4 (13)	584.5 (13)	118.0 (1)	170.1 (0.03)	105.8 (1)

TRANSFER AND DISPROPORTIONATION REACTION OF THE CYCLO-DEXTRIN GLUCANOTRANSFERASE FROM *KLEBSIELLA PNEUMONIAE*

E. FLASCHEL,[#] E. R. ANDRIAMBOAVONJY, A. RENKEN
Institute of Chemical Engineering, Swiss Federal Institute of Technology, EPFL-Ecublens, CH-1015 Lausanne, Switzerland
[#] *University of Bielefeld, Department of Technical Sciences, P. O. Box 10 01 31, D-33501 Bielefeld, Germany*

ABSTRACT

Cyclodextrin glucanotransferases not only transform starch into cyclodextrins, but also catalyse the transfer of cyclodextrins to appropriate acceptors and the disproportionation of linear dextrins. These three reactions are manifestations of a single catalytic activity. The transfer of α-cyclodextrin to maltitol as well as the disproportionation of functionalised dextrins carrying maltitol at the reducing end are studied in the presence of the cyclodextrin glucanotransferase of *Klebsiella pneumoniae*. The transfer reaction is a fast process followed by the disproportionation of the primary product. The primary functionalised dextrin is degraded in the presence of excess acceptor mostly by the exchange of maltose units. Studies on the disproportionation of purified functionalised linear dextrins reveals that the acceptor is the most prominent product. Apparently, disproportionation leads to the synthesis of starch.

1. INTRODUCTION

The cyclodextrin glucanotransferase (CGT) of *Klebsiella pneumoniae* is the most selective enzyme for α-cyclodextrin production [1]. In addition, it catalyses the reverse reaction: the transfer of α-cyclodextrin to appropriate acceptors yielding linear (functionalised) dextrins - again with outstanding selectivity with respect to the cyclodextrin and the acceptors [2]. A prerequisite of the transfer reaction seems to be the presence of α-cyclodextrin and a glucosidic acceptor with the C1-position blocked [2]. Both reactions, cyclisation and transfer are accompanied by the disproportionation of linear (functionalised) dextrins. Linear functionalised dextrins are the ideal substrates for studying the disproportionation reaction. However, these substrates have to be produced by the transfer reaction and have to be separated from a complex mixture of similar substances. Nevertheless, this may be achieved by preparative HPLC on reversed phase columns using water as the eluent [3].
Results are shown for the transfer of α-cyclodextrin to maltitol as well as for the disproportionation of linear maltitol dextrins. Unfortunately, this can only be done here by means of some examples.

J. Szejtli and L. Szente (eds.), Proceedings of the Eighth International Symposium on Cyclodextrons, 45–50.

2. MATERIALS AND METHODS

The cyclodextrin glucanotransferase of *Klebsiella pneumoniae* M5 al was obtained as a lyophilised powder and was a gift of the Consortium für Elektrochemische Industrie, Munich, Germany. The enzyme was taken from a stock solution containing 10 mM TRIS-HCl, 5 mM $CaCl_2$ of pH 7. The enzymic activity was standardized against a cyclisation assay reported by Landert and Flaschel [4].

Maltitol (63415) and α-cyclodextrin (28705) were purchased from Fluka. All other reagents were at least of analytical grade.

α-Cyclodextrin was analyzed by HPLC on a Spherisorb 5 NH2 column. Aqueous acetonitrile of 65 vol.% was used as eluent. Quantitative analysis was achieved by differential refractive index detection.

Linear functionalised dextrins were analyzed by HPLC on a Spherisorb 5 ODS2 column. Distilled water served as the eluent. Quantitative analysis was achieved by measuring the absorbance at 190 nm [4]. The response factors of the different maltitol dextrins were obtained by means of a multiple-regression analysis [5].

Reactions were carried out in batch reactors consisting of small jacketed stirred vessels. Reaction media were kept at 40 °C by thermostatting. A pH of 7 was established due to the presence of buffer in the enzyme stock solution. Samples for HPLC analysis were immediately deluted in 0.01 M hydrochloric acid in order to inactivate the enzyme.

3. RESULTS AND DISCUSSION

The disproportionation reaction of the cyclodextrin glucanotransferase from *Klebsiella pneumoniae* has been investigated by means of purified linear functionalised dextrins. Such functionalised dextrins have the advantage over maltodextrins that the reducing end is blocked. This makes it easier to follow the reaction of disproportionation by analysing the development of products. Functionalised dextrins (AG_n) consisting of an acceptor (A) and a number of glucose units (G_n) are easily obtained by means of the transfer reaction of cyclodextrin glucanotransferases. Unfortunately, the transfer of α-cyclodextrin to appropriate acceptors leads to a complex mixture of products, which have to be separated in order to obtain pure substances.

Maltitol was chosen as an acceptor for α-cyclodextrin, because maltitol is a potent acceptor [2] and the products of the reaction may conveniently be separated by preparative HPLC [3]. A typical transfer reaction of α-cyclodextrin and maltitol is shown in Figures 1 to 3. The reaction in the presence of an excess of acceptor leads to a fast drop of the concentration of cyclodextrin - as shown in Figure 1. The primary product of the transfer reaction is the functionalised dextrin consisting of the acceptor in addition to maltohexaose (AG_6). Commonly, its concentration rises steeply at the beginning of the reaction, but falls at longer reaction periods due to the parallel reaction of disproportionation - as shown in Figure 2. Disproportionation leads to the appearance of secondary products, of which AG_4 and AG_2 are the most prominent members - as can be seen in Figures 2 and 3. Secondary products, therefore, seem mainly to evolve from the transfer of maltose units to the acceptor. Products with odd numbers of glucose units do appear as well, but to a much lesser extent - as shown in Figures 2 and 3.

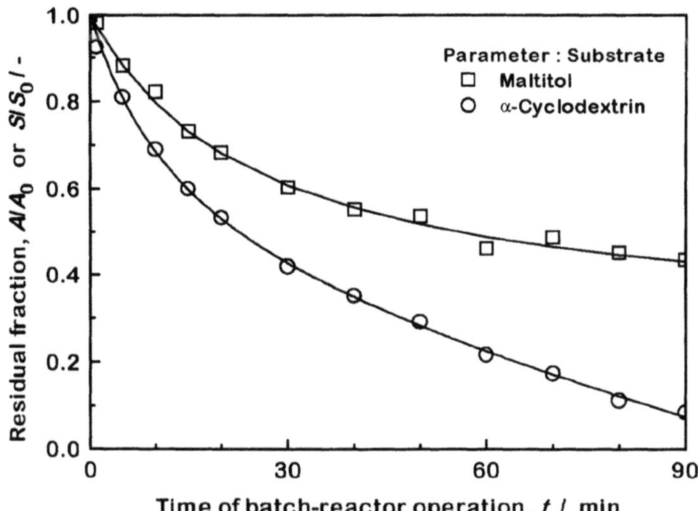

Figure 1. Transfer of α-cyclodextrin to maltitol. The disappearance of substrates. (E_0 = 2.2 kU L^{-1}, S_0 = 10.08 mM, A_0 = 22.15 mM, 40 °C, pH 7)

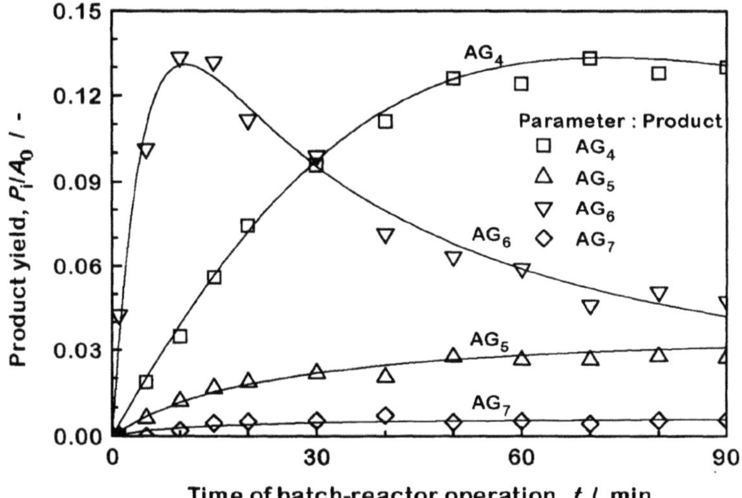

Figure 2. Transfer of α-cyclodextrin to maltitol. The appearance of transfer products. (E_0 = 2.2 kU L^{-1}, S_0 = 10.08 mM, A_0 = 22.15 mM, 40 °C, pH 7)

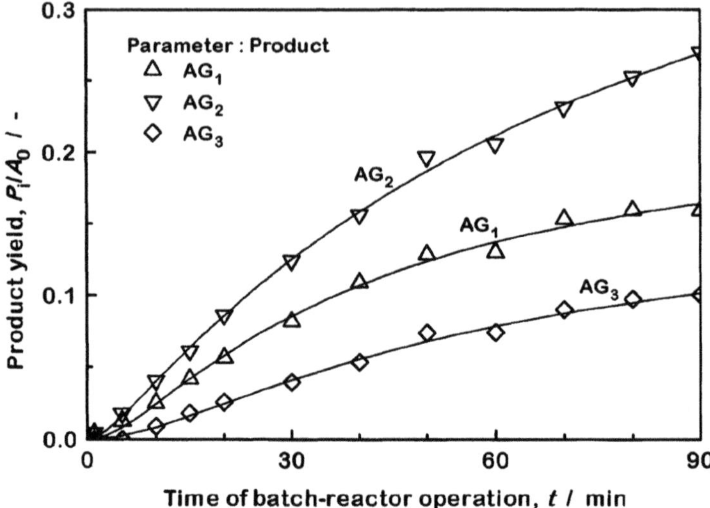

Figure 3. Transfer of α-cyclodextrin to maltitol. The appearance of secondary transfer products. (E_0 = 2.2 kU L^{-1}, S_0 = 10.08 mM, A_0 = 22.15 mM, 40 °C, pH 7)

Such reaction mixtures where subjected to reversed phase HPLC in order to separate the constituents of the mixture. The chromatographic separation had to be repeated twice with each fraction for achieving a satisfactory purity of the functionalised dextrins [3,5]. An example of the disproportionation reaction of maltitol dextrins is shown in Figures 4 and 5.

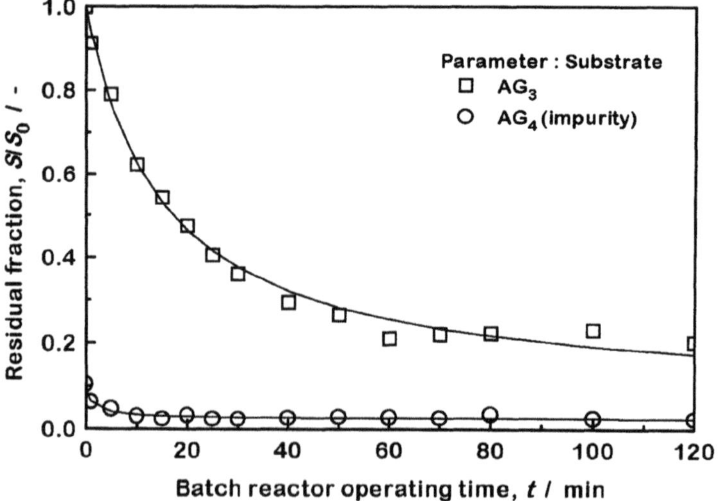

Figure 4. Disproportionation of the maltitol dextrin AG_3. The disappearance of substrate. (E_0 = 0.93 kU L^{-1}, S_0 = 3.95 mM, 40 °C, pH 7)

Figure 4 shows the rapid disappearance of the substrate AG_3 as well as of the small impurity AG_4. Figure 5 summarizes the appearance of the reaction products.

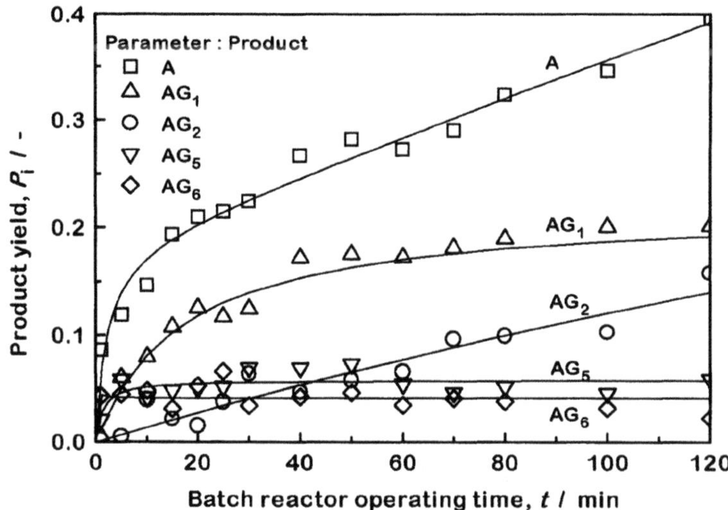

Figure 5. Disproportionation of the maltitol dextrin AG_3. The appearance of reaction products. (E_0 = 0.93 kU L^{-1}, S_0 = 3.95 mM, 40 °C, pH 7)

The most striking fact is the appearance of the acceptor as the most prominent product of the disproportionation reaction. The yield of acceptor can be rather high. This feature shows that maltotriose units are primarily exchanged. The appearance of AG_1 as the second major product may evolve from the exchange of maltose units. Apparently, the substrate itself functions as the main acceptor of the maltosaccharides transferred by disproportionation.

If the molecular species are counted, the balance is nicely fulfilled with respect to the acceptor - as shown in Figure 6. However, if the glucose units of the substrate as well as of the reaction products from AG_1 until AG_6 are counted, a rather significant disappearance of glucose units is observed - as can be deduced from Figure 6 as well.

Since cyclodextrins have not been detected as products of the disproportionation, the only explanation can be that high molar mass (functionalised) dextrins are built up by the enzyme. Therefore, the disproportionation of functionalised dextrins may lead to *de novo* synthesis of considerable amounts of starch at the equilibrium of reaction.

4. CONCLUSION

The transfer of α-cyclodextrin to acceptors like maltitol leads to functionalised dextrins. The transfer reaction is always accompanied by the disproportionation of the primary transfer product. This disproportionation is characterized mainly by the exchange of maltose units. The disproportionation of single functionalised dextrins shows an unexpected feature: the appearance of the acceptor as the main product. An analysis of the balance with respect to the acceptor as well as the glucose units present in the reaction system leads to the conclusion that high molar mass dextrins are produced during disproportionation.

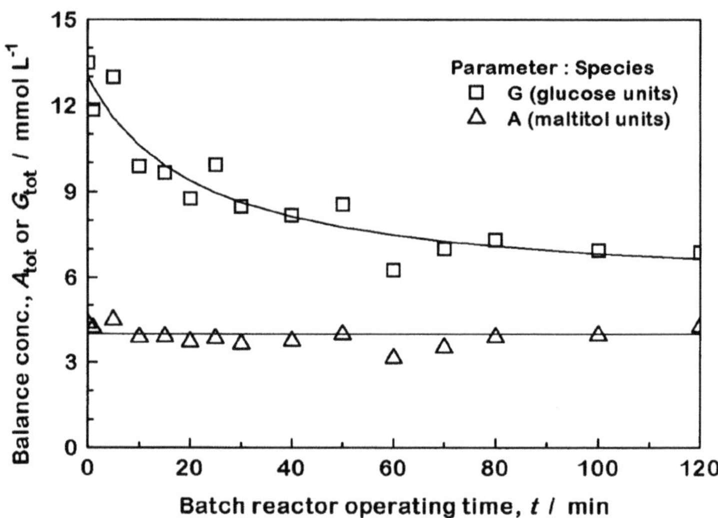

Figure 6. Disproportionation of the maltitol dextrin AG_3. The balance of maltitol and glucose units. (E_0 = 0.93 kU L^{-1}, S_0 = 3.95 mM, 40 °C, pH 7)

REFERENCES

[1] E. Flaschel, J.-P. Landert, A. Renken : Process Development for the Production of α-Cyclodextrin, in : Advances in Inclusion Science, Proceedings of the First International Symposium on Cyclodextrins, J. Szejtli (Ed.), 41-49, Budapest : Akademiai Kiado, 1982

[2] E. Flaschel, D. Spiesser, A. Renken : Production of Functionalized Dextrins by Enzymatic Transfer of α-Cyclodextrin, in : *DECHEMA Biotech. Conf. 1* (1988) 161-66

[3] E. Andriamboavonjy, E. Flaschel, A. Renken : Separation of functionalized dextrins by reversed-phase high-performance liquid chromatography, *J. Chromatogr. 587* (1991) 288-291

[4] J.-P. Landert, E. Flaschel, A. Renken : A Photometric Test for the Cyclisation Activity of Cyclodextrin-Glycosyltransferases, in : Advances in Inclusion Science, Proceedings of the First International Symposium on Cyclodextrins, J. Szejtli (Ed.), 89-94, Budapest : Akademiai Kiado, 1982

[5] E. Andriamboavonjy, Cinétique des réactions de disproportion des dextrines fonctionnalisées catalysées par la cycloglycosyltransférase issue de la souche *Klebsiella pneumoniae* M5 al, PhD Thesis No. 1342, Swiss Federal Institute of Technology, Lausanne, 1995

ENHANCED CYCLODEXTRIN PRODUCTION BY FERMENTATION

H. O. S. LIMA, G. M. ZANIN and F. F. DE MORAES
Department of Chemical Engineering,
State University of Maringá; Av. Colombo, 5790, Bloco E46-09
87020-900 - MARINGÁ - PR, BRAZIL

ABSTRACT

The aim of this work was to investigate the possibility of enhanced cyclodextrin production by simultaneous action of the CGTase enzyme and fermentation with *Saccharomyces cerevisiae* of a liquefied starch solution. A 10% w/v cassava starch suspension was liquefied with α-amylase up to DE 22.3. Then the solution's pH was corrected to 6.0, which is a compromise for the optimum for the yeast and the CGTase enzyme. The temperature was set to 38°C. Results have clearly shown the advantage of the simultaneous cyclization and fermentation (SCF) process. While the test with CGTase without ethanol produced 0.6 mM of β-CD in 22 hours, the SCF produced more than the double (1.5 mM of β-CD).

1. INTRODUCTION

Literature on cyclodextrin (CD) production from liquefied starch recommends the use of lower dextrose equivalent (DE) since CD yield decreases with an increase in DE [1]. On the other hand, it is also known that the addition of certain solvents and surfactants leads to an increase in the yield of β-cyclodextrin (β-CD). For example acetone, ethanol, glycerol, propanol, and the surfactant Triton-X have shown this property [2,3]. Mattsson et al. [4] observed an yield increase in β-CD of 2.4 times when they produced CD in the presence of 15% ethanol with a solution of 30% starch. These authors concluded that ethanol and other solvents exclude water molecules from the active site of the CGTase (cyclodextrin glycosyltransferase) enzyme therefore preventing hydrolytic reactions.

The objective of our work was to associate these two ideas: lower DE and the presence of ethanol, to increase β-CD yield. We called the process SCF: Simultaneous Cyclization and Fermentation, since the idea was to use simultaneously the CGTase enzyme to produce CDs and yeast (*Saccharomyces cerevisiae*) to remove glucose and maltose from the liquefied starch solution. The yeast ferments these small saccharides to ethanol and expected benefits are doubled since the removal of glucose and maltose lowers the DE

J. Szejtli and L. Szente (eds.), Proceedings of the Eighth International Symposium on Cyclodextrons, 51–54.

and consequently increases CD yield, while the production of ethanol in the reaction medium contributes to higher β-CD production. Another advantage of the SCF process is that we may then use higher DE in the starting liquefied starch so that we may have higher ethanol levels, and working with higher DE liquefied starch is easier since it is more soluble and less prone to retrogradation.

2. MATERIALS AND METHODS

2.1. Materials

The substrate was a liquefied cassava starch solution 10% w/v hydrolysed with α-amylase (NOVO Termamyl) at pH 6.5, 95°C, during 45 min so that the final DE was about 20. The hydrolysis reaction was stopped with 0.27% v/v of 1N HCl and boiled for 10 min. Then the pH was corrected to 6.0 using 1N NaOH. The CGTase enzyme used was provided by WACKER-Gmb (Germany) and it is obtained from *E. coli* engineered with CGTase from an alkalophilic *Bacillus* sp.1. Specific enzyme activity was 70.5 U/mg of protein, determined with an 1% w/v soluble starch solution, pH 6.0, 35°C, and where each unity (U) corresponds to the production of a micromol of β-CD per min. Enzyme concentration in the stock solution was 0.997 mg of protein/mL. Dried yeast FLEISCHMANN normally employed for breadmaking, containing *Saccharomyces cerevisiae* cells, was used. As a source of nitrogen and phosphorus for fermentation, the commercial products ammonitm sulfate and triple superphosfate were used each in the concentration of 1 g/L of substrate solution.

2.2. Methods

Protein concentration was determined by the Coomassie blue method of Bradford [5]. The DE was determined through the method of Somogyi [6]. The quantity of β-CD produced in the reaction medium was assayed by the dye extinction method of Vikmon [7] as modified by Mäkelä et al. [8] and Hamon and Moraes [9]. The concentration of ethanol produced by fermentation was determined by gas chromatography after ethanol separation by distillation.

2.3. Experimental Details

Substrate solution was divided into five aliquots of 150 mL, and each one was added into a 250 mL erlenmeyer. The SCF process was conducted at pH 6.0 with the erlenmeyers continuously agitated in a thermostatic bath at 38° C for a period of 22 hours. Samples were taken at regular intervals, of 15 min in the first 2 h, of 30 min within 2 and 4 h, and intervals of 1 h thereafter. Reaction in the samples was stopped by boiling them in bain-marie for 10 min, then the samples were cooled and stocked at 4°C until analysis. Each of the five substrate aliquots received different reagent additions which are shown in Table 1. Samples from tests 2 and 4 were centrifuged for cells separation before being assayed for β-CD.

TABLE 1. Reagents added to the reaction medium

Test / Erlenmeyer	Reagents added to 150 mL of substrate
1 / control	0.12 mL of stock CGTase
2 / SCF	as 1 + 18 g of yeast + nutrients
3 / no fermentation	as 1 + 3% of ethanol
4 / no yeast	as 1 + nutrients
5 / only fermentation	18 g of yeast + nutrients

3. RESULTS AND DISCUSSION

The substrate solution had a DE equal to 22.3. The ethanol produced in the tests 2 and 5 was equal to 5.6% v/v. Figure 1 presents the results for tests 1 to 4, whereas for test number 5 no β-CD was produced as expected, since only fermentation by yeast occurred in it. The comparison of tests 1 and 3 confirms that the addition of ethanol enhances β-CD production, it increased from 0.6 mM β-CD without ethanol (test 1) to 1.2 mM β-CD with 3% v/v of ethanol (test 3).

Test 2, which is representative of SCF process, shows that a maximum yield of β-CD was produced (1.5 mM of β-CD) which is 2.5 times greater than in the control test 1 and 20% higher than in test 3. The higher yield of test 2 in comparison with test 3 is probably due to the higher ethanol concentration of the former, respectively 5.6% v/v as opposed to 3% v/v of test 3. It can be observed in Figure 1 that the addition of nutrients in the reaction medium does not cause CGTase inhibition since the yield of β-CD with nutrients (test 1) or without them (test 4) was practically equal.

These results emphasise that the premises behind the SCF process are correct, that is, the removal of glucose and maltose by fermentation with concomitant production of ethanol does help to increase β-CD yield during CD production from liquefied starch. The β-CD concentration produced might seem meagre at the moment but reaction conditions have not been optimised so far.

4. CONCLUSION

Experimental results demonstrated clearly the advantages of the SCF process which even with unoptimised conditions produced 2.5 times more β-CD than in the case of cyclodextrin production without yeast fermentation.

ACKNOWLEDGEMENTS

The authors thank companies COPAGRO, NOVO NORDISK DO BRASIL and WACKER-Gmb for providing respectively the starch, α-amylase and CGTase. Also, financial support is acknowledged from CNPq, CAPES and FUEM.

REFERENCES

[1] Szejtli, J., Cyclodextrin Technology, Kluwer Academic Publishers, Dordrecht, p. 36, 1988.

[2] Lee, Y. D., Kim, H. S., Enhancement of enzymatic production of cyclodextrins by organic solvents, *Enz. Microb. Tecnhol.*, **13**, 499-503 (1991)

[3] Tomita, K., Mitsutoshi, K., Kawamura, K., Nakanishi, K., Purification and properties of a cyclodextrin glucanotransferase from *Bacillus autolyticus* 11149 and selective formation of β- cyclodextrin, *J. Ferment.Bioeng.*, **75**, 89-92 (1993)

[4] Mattsson, P., Korpela, T., Paavilainen, S., Mäkelä, M., Enhanced conversion of starch to cyclodextrins in ethanolic solutions by *Bacillus circulans* var. *alkalophilus* cyclomaltodextrin glucanotransferase, *Appl. Biochem. Biotechnol.*, **30**, 17-28 (1991)

[5] Bradford, M. M., A rapid and sensitive method for the quantitation of microgram quantities of protein utilizing the principle of protein-dye binding, *Anal. Biochem.*, **72**, 248-254 (1976)

[6] Somogyi, M., A new reagent for the determination of sugar, *J. Biol. Chem.*, **160**, 61-68 (1945)

[7] Vikmon, M., Rapid and simple spectrophotometric method for determination of micro-amounts of cyclodextrins, in Proc. 1st Int. Symp. on Cyclodextrins, (Ed. Szejtli, J.), D. Riedel Publishing Company, Dordrecht, 1982, Pp. 69-74

[8] Mäkelä, M., Korpela, T., Laasko, S., Colorimetric determination of β-cyclodextrin: two assay modifications based on molecular complexation of phenolphthalein, *J. Biochem. Biophys. Meth.*, **14**, 85-92 (1987)

[9] Hamon, V., Moraes, F.F. de, Etude preliminaire a l'immobilisation de l'enzyme CGTase WACKER, report, Laboratoire de Tecnologie Enzymatique, Université de Tecnologie de Compiègne, 1990

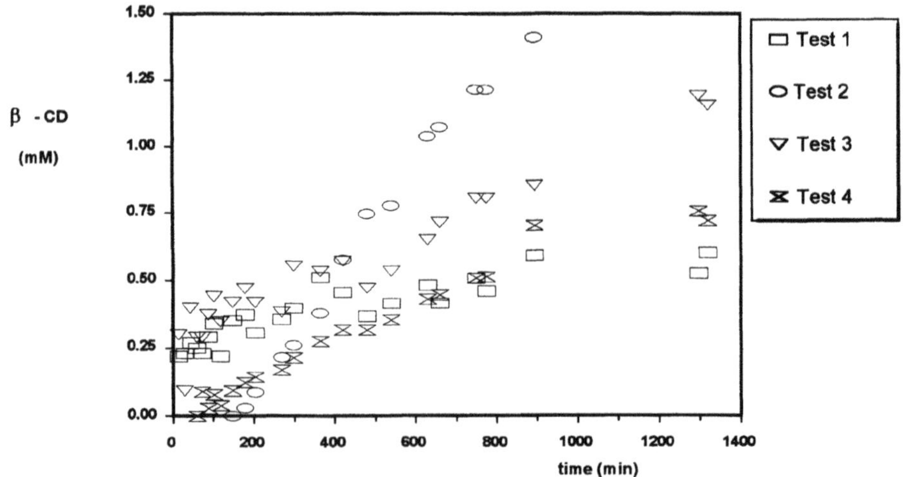

Fig. 1 Production of cyclodextrin as a function of time

SCREENING FOR GAMMA-CGTase

G. MATIOLI[1], G.M. ZANIN[2], M.F. GUIMARÃES[3] and F.F.DE MORAES[2]
[1]Department of Pharmacy and Pharmacology,
[2]Department of Chemical Engineering,
State University of Maringá; Av. Colombo, 5790, Bloco E46-09
87020-900 - MARINGÁ - PR, BRAZIL
[3]Department of Biochemistry,
Federal University of Paraná, P. O. Box 19046,
81531-990 - CURITIBA - PR, BRAZIL

ABSTRACT

CGTase positive alkalophilic microorganisms were selected by parallel application of two screening techniques based on colorimetric distinction of the colonies at the cultivation plate level. One of the techniques uses Congo red and xylene cyanole dyes which form a clear halo only in the presence of γ-cyclodextrin (γ-CD), whereas the other uses phenolphthalein that is specific for β-cyclodextrin (β-CD). The CGTase produced by strains selected up to now yields, however, mainly β-CD. The production of β-CD by the CGTase of strain 37 is, nevertheless, very interesting since β-cyclodextrin concentration reached 16 mM at 50°C, pH 8.0, 10% w/v initial substrate concentration.

1. INTRODUCTION

The potential application by the pharmaceutical industry of large drug molecules encapsulated by γ-cyclodextrin (γ-CD) has driven research interest into: (i) better ways of separating γ-CD from a mixture of cyclodextrins normally produced by action of CGTase enzymes upon starch [1]; (ii) increase the production of γ-CD by complexation during these reactions [2]; or (iii) search for a microorganism whose CGTase kinetically favours γ-CD production at the beginning of the reaction [3]. In this work the third option was adopted.

J. Szejtli and L. Szente (eds.), Proceedings of the Eighth International Symposium on Cyclodextrons, 55–58.

2. MATERIALS AND METHODS

Selection of alkalophilic microorganisms that produce CGTase enzyme was accomplished using the Horikoshi II plate screening medium [4]: soluble starch (1%), polypetone (0.5%), yeast extract (0.5%), K_2HPO_4 (0.1%), $MgSO_4$ (0.02%), Na_2CO_3 (1%), agar (1.5%), final pH 10.3, modified by inclusion of the following dyes: Congo red (0.01%) and Xylene cyanole FF (0.001%). These dyes were used by Hamaker and Tao [5] with Luria-Bertani neutral media (pH 7.0) containing 1% starch, to detect γ-CD production by recombinant *E. coli*. The presence of γ-CD is revealed by a clear halo around the colonies. A drop of solution containing either of each cyclodextrin (α, β or γ) has shown that also for modified Horikoshi medium, only γ-CD produces a clear halo. Samples of soils used for cultivation of wheat, corn, potato and cassava, were suspended in distilled water; screening plates were inoculated with these suspensions and incubated for 24 h at 37°C.

To determine the CGTase activity of positive microorganisms, these were cultivated in a liquid medium composed of the same substances aforementioned with exception of the dyes and agar. Agitated cultivation proceeded for up to five days at 37°C. Then cells were separated in a centrifuge, and CGTase activity for production of β and γ-CD was determined in the supernatant by colorimetric methods [6], and by the method of successive dilution s of the enzyme and characterisation of CD produced by precipitation with TCE-trichloroethylene [7].

The CGTase enzyme was obtained from the supernatant by precipitation with ammonium sulfate, and further purified by bioespecific affinity chromatography according to László et al. [8] using either β-CD or γ-CD as affinants [3].

Purified CGTase was concentrated by ultrafiltration, washed with Tris-HCl buffer five times to remove molecules with MW smaller than 30,000, and used thereafter for CD production. These tests, run at 50°C, used as substrate a 10% solution of Dextrin 10 Fluka, Tris-HCl buffer pH 8.0, 0.01 M and 5mM $CaCl_2$. Enzyme concentration in the reaction medium was set around 1 mg/L. the concentration of β and γ-CD produced was followed by a period of 24 h, being measured by the colorimetric methods of Hamon and Moraes [6].

3. RESULTS AND DISCUSSION

Fifty five CGTase positive strains were selected, being strain n° 37 which gave the strongest precipitation with TCE at 2^7 crude enzyme dilution. The purified enzyme has not shown, however, the expected characteristic of favouring the production of γ-CD at the beginning of reaction. On the contrary, β-CD concentration was always higher and final concentrations reached 16 mM β-CD and 2.5 mM γ-CD, a molar ratio β to γ-CD equal to 6.4 (Figure 1).

It was decided then to use the above screening technique in parallel with the method of Park and Kim [9] which uses phenolphthalein plus methyl orange and Horikoshi II plate medium for β-CGTase activity detection. The idea was that a microorganism which showed as initial large halo in the Congo red medium while displaying a small halo in the phenolphthalein plate would be a good candidate for producing a CGTase which would kinetically favour γ-CD production for short reaction times. Two new microorganisms were then selected according to this new scheme, but again expectations were frustrated by a cyclodextrin production similar to the first strain, although at lower ratio of β to γ-CD, 4.9 and 5.36 respectively for strains 41 and 47.

The CGTase enzyme of strains 37, 41 and 47 was successfully purified by affinity chromatography only in the column which had β-CD as affinant showing characteristics in accord to that of β-CGTase.

Sabioni and Park [10,11] have also used the phenolphthalein screening technique of Park and Kim [9] with Brazilian soil from Campinas - SP and arrived at an alkalophilic β-CGTase strain which formed CDs in the proportion of 1:67:1.6 respectively for α, β and γ-CD.

4. CONCLUSION

In conclusion, the strain selecting scheme so far developed is good for detecting microorganisms that produces CGTase enzyme, but it fails to be selective enough to separate only the strains that produce γ-CGTases, that is, those that kinetically favour the production of γ-CD for short reaction times.

The production of β-CD by the CGTase of strain 37 is, however, very interesting since β-CD concentration reached 16 mM.

ACKNOWLEDGEMENTS

The authors thank the financial support obtained from CNPq, CAPES and FUEM.

REFERENCES

[1] Mattson, P., Mäkelä, M., Korpela, T., Isolation and purification of gamma-cyclodextrin by affinity chromatography, in Proc. of the Fourth Intern. Symp. Cyclodextrins,(Ed. Huber, O.,Szejtli, J.), Kluwer Academic Publishers, London, 1988, Pp. 65-70

[2] Schmid, G., Huber, O.S., Eberle, H. J., Selective complexing agents for the Production of γ-cyclodextrins, in Proc. of the Fourth Intern. Symp.Cyclodextrins,(Ed. Huber, O. ,Szejtli, J.), Kluwer Academic Publishers, London, 1988, Pp. 87-92

[3] Englbrecht, A., Harrer,G., Lebert, M., Schmid,G., Biochemical and genetic characterisation of a CGTase from an alkalophilic bacterium forming primarily γ-cyclodextrin, in Minutes of the Fifth Intern. Symp.on Cyclodextrins,(Ed. Duchene, D.) Paris 28-30 / March / 1990, Editions de Santé, Paris, 1990, Pp. 25-31

[4] Nakamura, N., Horikoshi, K., Characterization and some cultural condition of a cyclodextrin glycosyltransferase-producing alkalophilic *Bacillus* sp., *Agr. Biol. Chem.*, **40**, (4), 753-757, 1976

[5] Hamacker, K., Tao, B.Y, Screening of gamma - cyclodextrin - producing recombinant *E. coli* using Congo red dye on solid complex media, *Starch / Stärke*, **45**, 181-182, 1993.

[6] Hamon, V., Moraes, F.F. de, Etude preliminaire a l'immobilisation de l'enzyme CGTase WACKER, report, Laboratoire de Tecnologie Enzymatique, Université de Tecnologie de Compiègne, 1990

[7] Nomoto, M., Shew, D.C., Chen, S.J., Yen, C.W.L., Yang, C. P., Cyclodextrin glucanotransferase from alkalophilic bacteria of Twain, *Agr. Biol. Chem.*, **48**, 1337-1338, 1984

[8] László,E., Bánky,B., Seres,G., Szejtli, J., Purification of cyclodextrin glycosyl transferase enzyme by affinity chromatography, *Starch / Stärke*, **33**, 281-283, 1981.

[9] Park, C. S., Park, K. H., Kim, S. H., a rapid screening method for alkaline β-cyclodextrin glucanotransferase using phenolphthalein-methyl orange-containing-solid medium, *Agr. Biol. Chem.*,**53**, 1167-1169, 1989

[10] Sabioni, J. G., Park, Y. K., cyclodextrin glycosyltransferase production by alkalophilic *Bacillus lentus, Rev. Microbiol.*,**23**, (2), 128-132, 1992

[11] Sabioni, J. G., Park, Y. K., Production and characterization of cyclodextrin glycosyltransferase from *Bacillus lentus*, *Starch/Stärke*, **44**, 225-229, 1992

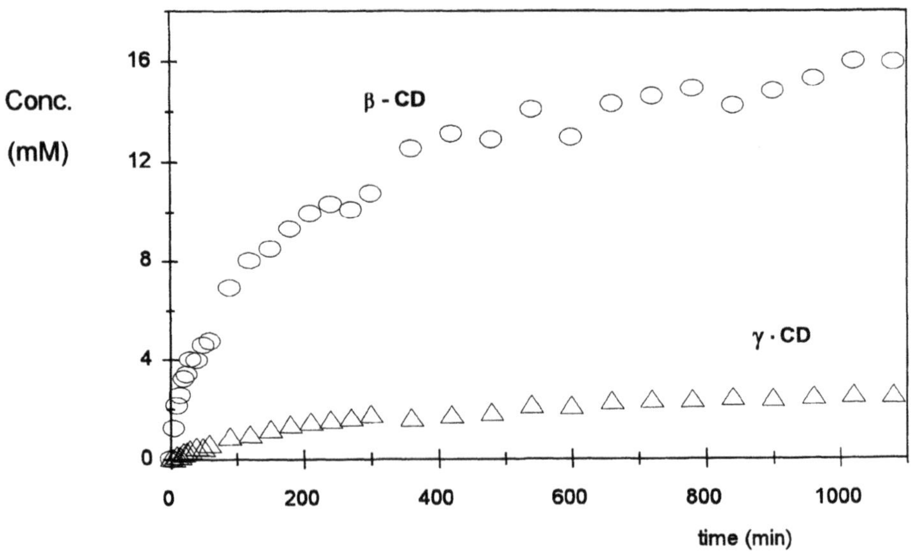

Figure 1. Cyclodextrin production by CGtase obtained from strain 37

EFFECTIVE METHOD FOR PRODUCING BRANCHED-α-CYCLODEXTRINS BY PULLULANASE AND GLUCOAMYLASE

Noriyasu WATANABE,[1] Kazutaka YAMAMOTO,[2] Wakako TSUZUKI,[2] Takaichi OYA,[1] and Shoichi KOBAYASHI[2]

[1]*Amano Pharmaceutical Co.,Ltd, Nishishiroyashiki 51, Ohazakunotsubo, Nishiharu-cho, Nishikasugai-gun, Aichi 481, Japan*

[2]*National Food Research Institute, Ministry of Agriculture, Forestry, and Fisheries, 2-1-2 Kannondai, Tsukuba, Ibaraki 305, Japan*

ABSTRACT

An effective method was developed to produce branched-α-cyclodextrins (Br-α-CDs) such as glucosyl-α-CD (G_1-α-CD) and di-glucosyl-α-CD ((G_1)$_2$-α-CD). The method is an one-step process comprising of both (1) condensation of maltose (G_2) and CDs by reverse-synthesis with a debranching enzyme (pullulanase, PUL) and (2) hydrolysis of intermediate maltosyl-α-CDs (G_2-α-CDs) with a hydrolyzing enzyme (glucoamylase, GA). Effects of reaction conditions on the branching were investigated : total substrate content, initial molar ratio of G_2 to CD, enzyme unit ratio of PUL to GA. Effective condensation by our method indicates that accumulation of the intermediate G_2-α-CDs inhibiting condensation could be reduced by GA-catalyzed hydrolysis of the intermediates.

1. INTRODUCTION

Branched cyclodextrins, glucosyl-CD (G_1-CD), di-glucosyl-CDs ((G_1)$_2$-CDs), and maltosyl-CD (G_2-CD), are generally more soluble in water than parent CDs.[1] Maltosyl-CDs (G_2-CDs) are conventionally produced by a condensation method using PUL (Fig. 1(A))[2,3]. However, in this method, it is difficult to obtain a high yield of G_2-CDs, since G_2-CDs accumulate in the system and inhibit the synthesizing reaction (product inhibition).

Hence, an effective branching method has been devised to achieve a higher yield of G_1-α-CDs (G_1-α-CD and (G_1)$_2$-α-CDs) as shown in Fig. 1(B). The method is a one-step process which includes condensation of G_2 and α-CD with pullulanase (PUL) and hydrolysis of intermediate G_2-α-CD with glucoamylase (GA). Branching of α-CD would proceed effectively, provided the intermediate compounds, viz., G_2-α-CDs, are readily hydrolysed. In this method condensation and hydrolysis take place simultaneously by PUL and GA, hence it can be called "pullulanase-glucoamylase method". An attempt has been made to investigate operational conditions.

J. Szejtli and L. Szente (eds.), Proceedings of the Eighth International Symposium on Cyclodextrons, 59–62.

(A) Conventional branching method

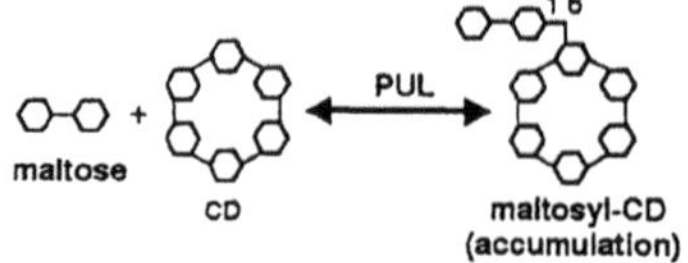

(B) Novel branching method

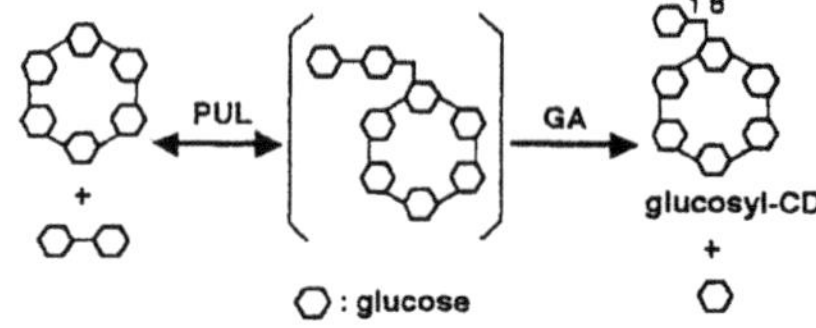

Fig. 1 Conventional (A) and Novel (B) Branching Methods.

2. MATERIALS AND METHODS

2.1. Materials

α-CD was supplied from Ensuiko Sugar Refining Co., Ltd. (Japan). Pullulanase "pullulanase Amano" from *Klebsiella pneumoniae* and glucoamylase "Gluczyme NL" from *Aspergillus niger* were manufactured by Amano Pharmaceutical Co., Ltd. (Japan). Pullulan and authentic samples of G_1-α-CD and G_2-α-CD were obtained from Wako Pure Chemical Industries Co., Ltd. Crystalline GA from *Rhizopus niveus* was purchased from Seikagaku Kogyo Co., Ltd. GA from *Endomyces* sp. was supplied from Godoshusei Co., Ltd. All other chemicals were of reagent grade.

2.2. Methods

One unit of GA (1Ug) activity was defined as the amount of enzyme releasing 1 mmol of G_1 per min. One unit of PUL activity (1Up)was defined as the amount of enzyme which hydrolyzes pullulan, liberating reducing sugar with a reducing power equivalent to 1 mmol G_1 per min.

Desired amount of G_2 and 0.500 g (0.514mmol) of α-CD were mixed in 50 mM acetate buffer (pH5.5) and dissolved by heating in boiling water for 15 min. Total substrate content (TSC, w/w%) was expressed by 100 x [substrates] / [mixture]. In a conventional method in this study, 90 Up of PUL was added to the mixture, and the reaction mixture was incubated at 50°C. In our method, a desired amount of PUL and 20 Ug of GA were added to the mixture, and the reaction mixture was incubated at 55°C. Products in the mixture were analyzed by HPLC.

Yield of products was evaluated by branched ratio (mol/mol%) defined as 100 x [Br-α-CDs] / [total CDs]. Products by the conventional method were G_2-α-CD, di-maltosyl-α-CD ($(G_2)_2$-α-CD). In addition, G_1-α-CD, di-glucosyl-α-CD ($(G_1)_2$-α-CD), and glucosyl-maltosyl-α-CD (G_1-G_2-α-CD) were produced by our method. These Br-α-

CDs was identified by HPLC with authentic samples.

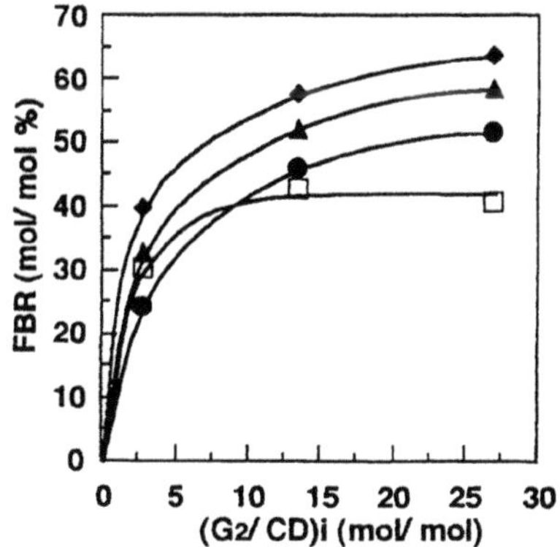

Fig. 2 Effects of Total Substrate Content (TSC) and (G2/CD)i on branched ratio by the novel method.
The reaction conditions were as follows : PUL = 180 Up, GA = 20 Ug, reaction temperature = 55°C. The other conditons were the same as described in materials and methods.
TSC (w/w%) : (●) 40, (▲) 50, (◆) 60, (□) 70.

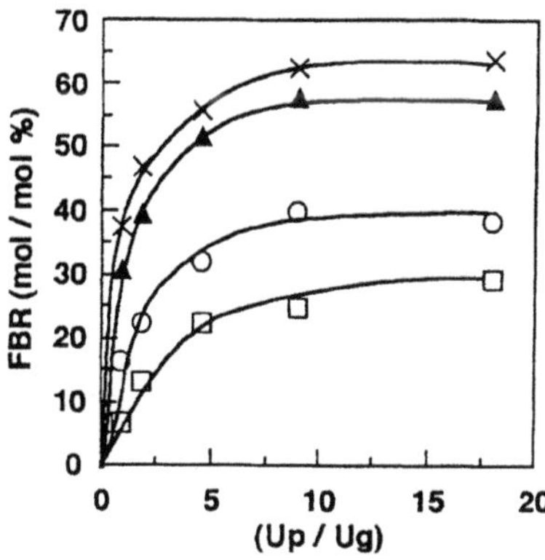

Fig. 3 Effects of [PUL]/[GA] (Up/Ug) and (G2/CD)i on branched ratio by the novel method.
The reaction conditions were as follows : Total Substrate Content (TSC) = 60w/w%, reaction temperature = 55°C. The other conditions were the same as described in materials and methods.
(G2/CD)i (mol/mol) : (□) 1.03, (O) 2.7, (▲) 13.5, (X) 27.

RESULTS AND DISCUSSION

In the conventional method using only PUL (Fig.1(A)), G_2-α-CDs were produced, and the maximum of final branched ratio (FBR, branched-CD / total CD mol/mol%) was 53%.

In our method using PUL combined with GA (Fig. 1(B)), GA hydrolyses not only G_2-α-CDs but also G_2. Therefore, GA hydrolysing G_2-α-CDs faster than G_2 was expected to achieve an effective branching. Accordingly, hyrolysing rates of G_2-α-CD relative to those of G_2 were measured with commercially available GAs. GA from *Aspergillus oryzae* showed the highest hydrolysing rate among the GAs. As the optimum temperatures of PUL and GA from *Aspergillus oryzae* are 50-55°C and around 65°C, respectively, an optimum temperature for the mixed enzyme system in the effective method was examined and selected at 55°C. Effects of reaction conditions such as G_2 or CD concentrations, PUL or GA amounts, total substrate content (TSC) on final (equilibrated) branched ratio (FBR) were studied. A maximum FBR, 64mol/mol%, was obtained at TSC=60w/w%, $(G_2/CD)i$, initial molar ratio of G_2 to CD, = 27mol/mol (FIg. 2). Furthermore, effects of (Up/Ug), enzyme unit ratio of PUL to GA, and $(G_2/CD)i$ were examined. The maximum FBR in this study, 64%, was obtained over (Up/Ug) = 9 (FIg. 3). The higher FBR might be caused by decreased accumulation of the pullulanase-catalysed product, G_2-α-CDs, which leads to a product inhibition in the conventional method. Finally, the commercially available GAs were applied to our method. As can be expected from the hydrolysing rates of G_2-α-CD relative to those of G_2, the higher the hydrolysing rate was, the more branched CDs were obtained. It was suggested that a hydrolysing enzyme which hydrolyses G_2-α-CD much faster than G_2 would much catalyse the branching in the effective method. Further study on screening for such enzymes is under way.

CONCLUSION

1. Branched-α-cyclodextrins (CD) were efficiently produced by our method in which pullulanase and glucoamylase are used in combination.
2. The maximum branched ratio by pullulanase-glucoamylase method was 64%, whereas that by a conventional method with the action of only pullulanase was 53%.
3. It was suggested that accumulation of the pullulanase-catalysed product, G_2-α-CDs, might be reduced by the glucoamylase-catalysed hydrolysis of G_2-α-CDs and the branching was efficiently promoted.
4. The higher the hydrolysing rate of GA for G_2-α-CD relative to that for G_2 was, the more branched-α-CD were produced by the effective method.
5. An enzyme which hydrolyses G_2-α-CD much faster than G_2 is expected to be much effective in our method.

REFERENCES

[1] Okada, Y., Kubota, Y., Koizumi, K., Hizukuri, S., Ohfuji, T., Ogata, K., Some properties and inclusion behavior of branched cyclodextrins, *Chem. Pharm. Bull.*, **36**, 2176-2185 (1988)
[2] Sakano, Y., Japanese Patent, 61-70996 (1986)
[3] Kobayashi, S., Kainuma, K., Japanese Patent, 61-92592(1986)

ISOLATION AND CHARACTERIZATION OF A *B. CIRCULANS* STRAIN AND ITS CGTase

D. GLICKSTEIN and R. BAR*,
Dept. of Applied Microbiology, The Hebrew University, Jerusalem , Israel

ABSTRACT

A CGTase-secreting bacterial strain was isolated from soil and bacteriologically identified to be *Bacillus circulans* . Its enzyme was found to be a β-CD type CGTase. The production of this enzyme was undertaken in a 2L-fermentor following medium optimization in terms of carbon and nitrogen sources. The enzyme was characterized in terms of its activities at various temperatures and pH values.

1. INTRODUCTION

A study was conducted in view of isolating and identifying a spore-forming CGTase-producing microorganism and optimizing the production of the enzyme and characterizing its activities. Bacterial species particularly from the *Bacillus* genus are known to secrete CGTase and they include *B. macerans, B. circulans, B. megaterium, B. subtilis* as well as alkalophilic *Bacillus* species [1].

2. MATERIALS AND METHODS

2.1. Materials

Soluble starch from BDH (England) was used as a substrate for enzymatic activities. Corn steep liquor which contains 25% solids was obtained from Galam Ltd (Israel). α–CD (98.7%) and γ–CD (100%) were purchased from Chinoin (Hungary) while β–CD (100%) - from Roquette Freres (France).

2.2. Methods

2.2.1. Microbe Isolation

Soil samples (1 g) were suspended in 1 ml peptone (0.1%) water and after heating at 80 °C for 15 min to kill vegetative cells, were spread on agar-starch plates for a subsequent incubation at 37 °C. The solid medium consists of yeast extract (0.5%), peptone (0.5%),

* Present address: TEVA Pharmaceutical Indusries Ltd., R@D Lab, P.O.Box 353, Kfar-Sava, Israel.
Fax: 972-9-479525.

J. Szejtli and L. Szente (eds.), Proceedings of the Eighth International Symposium on Cyclodextrons, 63–66.

soluble starch (1%), KH_2PO_4 (0.1%), $MgSO_4$ (0.02%) and agar (1.5%) at pH=7. The amylolytic colonies were identified by flooding the plates with an iodine solution (0.2% w/w KI and 0.02% w/w I_2). Colonies were then cultivated in liquid cultures in the same medium (no agar) at 37 °C and the preliminary screening for CGTase-secreting colonies was performed by detection of precipitating solid CD-complexes upon addition of toluene or trichloroethylene. The positive fermentation broths were then tested for dextrinogenic and CD-forming activities.

2.2.2. Microbe maintenance

As the bacterial strain was a spore former , it was cultivated on a sporulaion [2] agar medium for 48 h to induce a rich spore formation. The spores were scratched from the surface and suspended in skim milk preparation (5%). The concentrated spore suspension was then absorbed into sterilized silica grains which were subsequently allowed to dry in the incubator . One silica grain was sufficient to successfully inoculate a 250 ml culture flask.

2.2.3. Cultivation in a fermentor

A New Brunswick Bioflo 2L-fermentor containing 1L of fermentation medium was used for cultivation of the microbe and enzyme production. The fermentor was seeded with 10 ml of a 24h-grown culture in a medium identical wit h that in the fermentor. The fermentor was operated (300 rpm) at 37 °C with an aeration rate of 0.3 vvm and pH controlled at 7.0. The fermentation broth was subjected to measurements of enzymatic activities and served for CGTase isolation.

2.2.4. Determination of enzymatic activities

The dextrigonenic activity was measured by the ability of the enzyme to degrade soluble starch to dextrins as reflected by the reduced blue color intensity of the starch-iodine complex. One ml of a soluble starch (0.2% w/v) in Tris buffer (pH=6) was placed in a test tube containing a known volume of the centrifuged broth and after the overall volume was made up with buffer to 1.5 ml, the test tube were incubated without agitation at 45 °C for 5 min. The reaction was halted by adding 3.5 ml HCl solution (0.2M) and after adding 0.5 ml of iodine solution, the O.D. of the mixture was measured at 700 nm. The control was a test tube into which the acid solution was added with the broth. One unit of dextrinogenic activity is defined as the percentage of O.D. decrease with respect of the control.

To a test tube containing 1 ml of a soluble starch (BDH, England) in Tris buffer (pH=6, I=0.01) , a known volume of the centrifuged broth was added and the mixture was incubated without agitation at 60 °C for 1 h. CDs were quantitated by HPLC . One unit of CD-forming activity is defined as the amount of all CDs (mmoles) produced by 1 ml of the enzyme broth after 1 hour of incubation.

3. RESULTS AND DISCUSSION

Seven months of search and screening of various sources led ultimately to a successful isolation of a CGTase-secreting bacterial colony . This isolate was subjected to a series of bacteriological and biochemical tests according to Bergey's manual [2] and was identified as a strain of ***Bacillus circulans*.** The enzyme of this strain, surnamed GB strain, was then tested for its product distribution. Figure 1 shows that the enzyme converts starch (5% w/v) to a mixture of the three natural CDs, β-CD being clearly the predominant one. A representative product distribution is that of a molar α: β:γ ratio of 1-2:10:1 respectively, thus classifying the enzyme as a β–CD CGTase type.

Medium optimization was subsequently undertaken in order to establish the ability of the isolate to grow in an industrial type of medium. To this purpose, various carbon sources such as soluble corn starch from several suppliers, potato starch and maltodextrins, as well as mono- and disaccharides were tested. Similarly, nitrogen sources such as urea, ammonium sulphate, and corn steep liquor were investigated for their effect on CGTase production. Ultimately, an industrial type of medium consisting of soluble starch (1%), KH_2PO_4 (1%), $MgSO_4.7H_2O$ (0.02%) and Corn steep liquor (5%) at pH=7, was adopted for microbe cultivation in a fermentor. Figure 2 shows enzyme production in such a medium given in a stirred fermentor as determined by the dextrinogenic and CD-forming activities of the fermentation broth.

The enzyme was subsequently isolated and purified from the fermentation broths by either ultrafiltration, or precipitation by ammonium sulfate or ethanol. The enzyme was found to be most active in the range of 50 - 60 ^{0}C and between pH=6.5 and pH=8.5. The enzyme thermostability was investigated and the enzyme was found to be unstable above 60 ^{0}C. Electrophoresis of the enzyme allowed an estimation of a molecular weight of 73 kD.

REFERENCES

[1]. Szejtli, J. Cyclodextrin Technology, Kluwer Academic Publishers, Dordrecht, 1988.

[2]. Sneath, P.H.A.,Endospore-forming Gram positive rods and cocci, in: Bergey's Manual, (Ed. Transil, B.), Williams and Wilkins II, pp. 1105-1139.

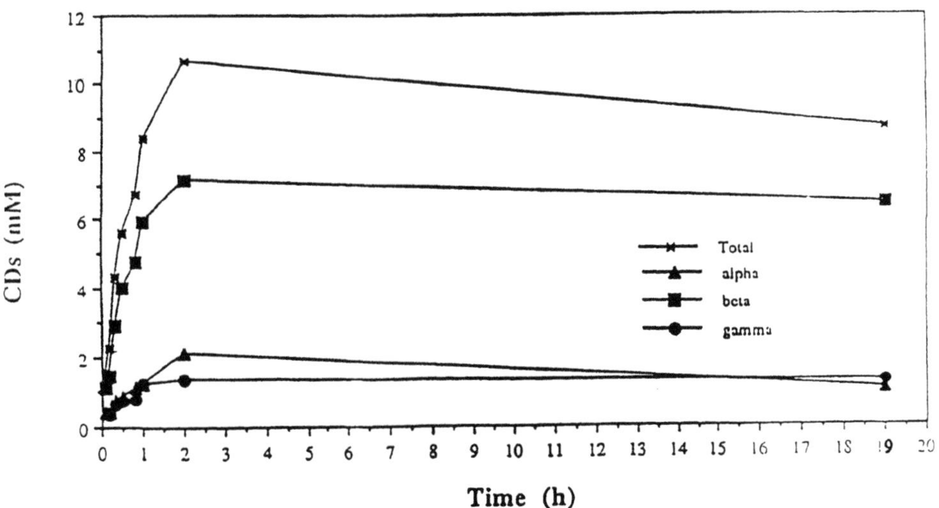

Fig. 1 Conversion of corn soluble starch (5%) to CDs by the *B. circulans* GB at pH=6 and 37 ^{0}C.

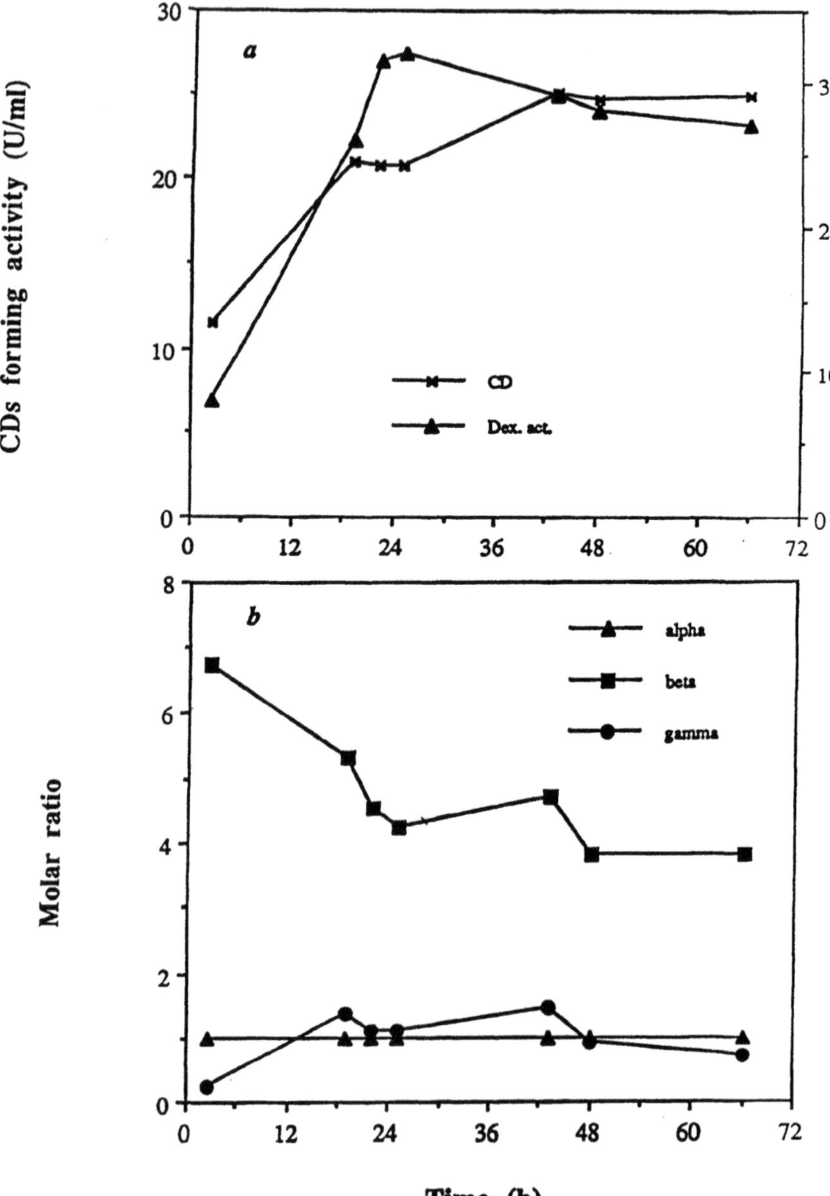

Fig. 2 Enzyme production in a 2L-fermentor (300 rpm, 0.3 vvm, pH=7.0 and 37 ^{0}C) by *B. circulans* GB (a) and product distribution of CDs produced from starch by aliquots of the fermentation broth withdrawn at various time intervals (b).

THE PREPARATIONS OF A NOVEL CYCLODEXTRIN HOMOLOGUE WITH D.P. FIVE, [5]CD

Toshio NAKAGAWA, Koji UENO, Mikio FUJII, Yasuyuki KOGA, and Yasuhiro KAJIHARA

Department of Chemistry, Yokohama City University, 22-2 Seto, Kanazawa-ku, Yokohama 236, Japan

ABSTRACT

Three kinds of acyclic precursors **3**~**5** were prepared from commercially available maltopentaose and their cyclization conditions to the title compound **1** ([5]CD) were examined. Secondary, a chemo-enzymic new route to [5]CD in a preparative scale starting from *inexpensive* β-CD is described. A HPLC profile of [5]CD is also presented comparing with that of α-CD ~ γ-CD.

1. INTRODUCTION

Recently we reported the first synthesis of the title compound **1** ([5]CD) via two routes using, as the starting materials, D-glucose and maltose [1] and commercially available maltopentaose [1, 2], respectively. Here are described (i) further elaborated results of the latter route using **3**~**5** as the cyclization precursors, (ii) a chemo-enzymic new route to [5]CD starting from *inexpensive* β-CD, and (iii) characterization of [5]CD by HPLC.

2. METHODS AND RESULTS

2.1. Preparation of precursors 3~5 and their cyclization to [5]CD

Three kinds of acyclic precursors **3**~**5** were prepared from commercially available maltopentaose as shown in Scheme 1. Their cyclization conditions and results (yields and ratios of [5]CD to its β-linked isomer) are given in Table 1.

Abbreviation: [n]CD, cyclodextrin with d.p.n; *for example*, [6]CD means α-CD [Lit.2].

J. Szejtli and L. Szente (eds.), Proceedings of the Eighth International Symposium on Cyclodextrons, 67–70.

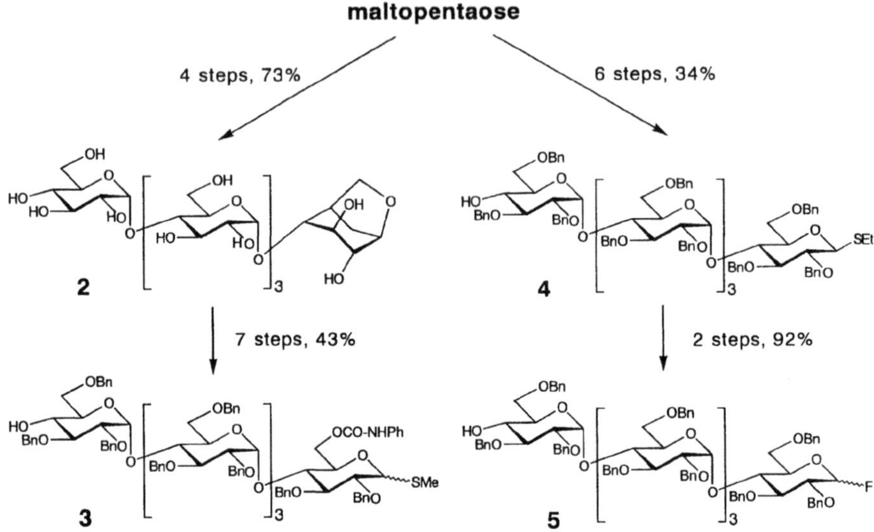

Scheme 1

2.2 A chemo-enzymic route to [5]CD starting from β-CD

The two routs to [5]CD mentiond above are not suitable for the preparative method, because the former takes many steps and the latter needs very expensive maltopentaose. Now we have tried the chemo-enzymic third route to [5]CD in a preparative scale starting from inexpensive β-CD, as shown in Scheme 2. Acetolysis of β-CD yielded per-*O*-acetylmaltoheptaose[3] which was transformed into 1,6-anhydromaltoheptaose(**6**) in the usual manner[4]. Then, a maltosyl residue was selectively splitted off by *beta*-amylase

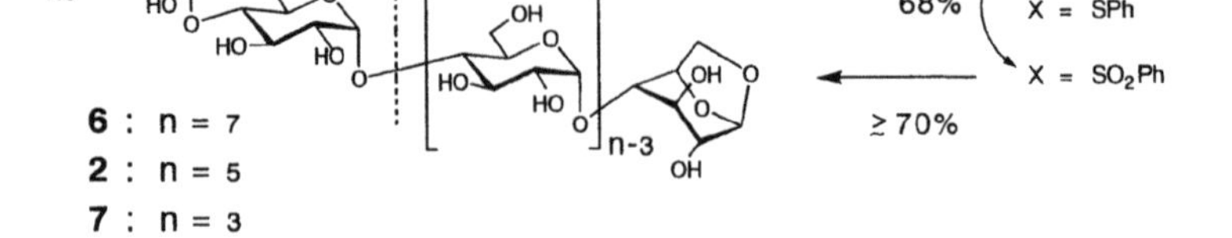

β-CD → Acetolysis [Lit.3] 41%

beta-amylolysis

X = OAc

X = SPh

X = SO_2Ph

68%

≥ 70%

6 : n = 7

2 : n = 5

7 : n = 3

Scheme 2

giving the key-intermediary 1,6-anhydromaltopentaose (**2**). In this case a small amount of 1,6-anhydromaltotriose (**7**) was inevitably formed by splitting of one more maltosyl residue. Establishing the best condition of *beta*-amylolysis as well as better substrates are now under way in our laboratory.

2.3. Table

TABLE 1. Cyclization of compounds **3** ~**5**

Entry	Compoud	Activator (equiv.)	Solvent	Yield(%)	α/β
1	**3**	MeOTf (5)	$(CH_2Cl)_2$	28	5.2
2	**4**	MeOTf (5)	$(CH_2Cl)_2$	17	1.1
3	**4**	MeOTf (5)	Et_2O	8.6	2.1
4	**4**	NIS(5),TfOH(0.5)	$(CH_2Cl)_2$	14	2.0
5	**4**	NIS(5),TfOH(0.5)	Et_2O	5.8	3.4
6	**4**	IDCP(5)	$(CH_2Cl)_2$	17	2.3
7	**4**	IDCP(5)	$Et_2O/(CH_2Cl)_2(5/1)$	24	6.1
8	**5**	Cp_2HfCl_2(5), AgOTf (10)	$(CH_2Cl)_2$	21	3.5
9	**5**	Cp_2HfCl_2(5), AgOTf (10)	Et_2O	8.2	14
10	**5**	Cp_2HfCl_2(5), AgOTf (5)	$(CH_2Cl)_2$	20	7.5
11	**5**	Cp_2HfCl_2(5), AgOTf (5)	Et_2O	11	β:trace

All reactions were carried out in the presence of M.S.4A for 12h under Ar atmosphere.
Thioglycosides **3** and **4** were reacted at r.t.; glycosyl fluoride **5** was done at -35°C.
NIS: *N*-iodomosuccinimide; IDCP: iodium di-*s*-collidine perchlorate;
Cp_2HfCl_2: dicyclopentadienylhafnium dichloride.

2.4. Figure

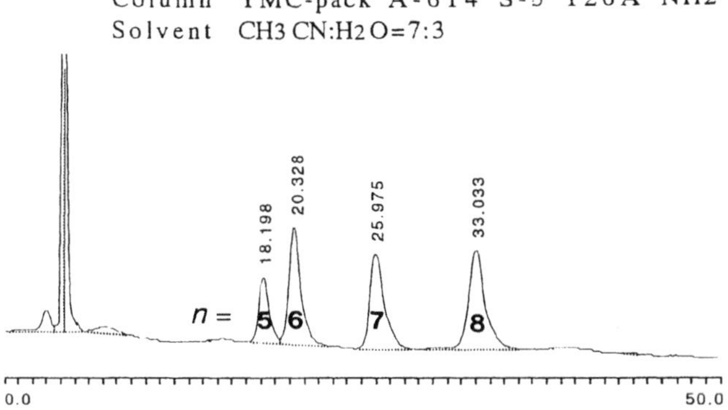

Fig.1 HPLC profile of cyclodextrins [*n*]CD (*n* = 5 ~ 8)

REFERENCES

[1] Nakagawa,T., Ueno,K., Kashiwa,M., Watanabe,J., The stereoselective synthesis of cyclomalto-pentaose. A novel cyclodextrin homologue with d.p. five, *Tetrahedron Lett.*, **35**,1921-1924 (1994)

[2] Nakagawa,T., Ueno,K., Kashiwa,M.,Watanabe,J., Stereoselective synthesis of cyclomaltopentaose. A novel cyclodextrin homologue with d.p. five, [5]CD, *Proceedings of the 7th International Cyclodextrins Symposium, Tokyo,* 114-117 (1994)

[3] Sakairi, N., Wang, L., Kuzuhara, H., Insertion of a D-glucosamine residue into the α-cyclodextrin skeleton; A model synthesis of chimera cyclodextrins, *J.Chem.Soc., Chem.Commun.*, 289-290 (1991)

[4] Funabashi, M., Nagashima, H., Synthesis of some 1,6-anhydro disaccharides via aryl glycosyl sulfones, *Chem. Lett.*, 2065-2068 (1987)

CYCLODEXTRIN DIMERS USED TO PREVENT SIDE EFFECTS OF PHOTOCHEMOTHERAPY AND GENERAL TUMOR CHEMOTHERAPY

J.G.MOSER, A.RUEBNER, A.VERVOORTS, B.WAGNER
Institute of Laser Medicine, Heinrich-Heine-University,
Universitaetsstr. 1, D-40225 Duesseldorf (Germany)

ABSTRACT

Recent developments of polyphasic tumor detection and tumor therapy including applications of dimeric cyclodextrin inclusion complexes of high affinity for Photodynamic Tumor Therapy (PTT) porphyrinoids are discussed in the light of future applications of similar methodology to general tumor chemotherapy. These approaches hopefully will end the martyrium of patients undergoing tumor chemotherapy today.

1. INTRODUCTION

Up to now, there are only two approaches seriously to be discussed for a tumor selective targeting of substances either to detect or to destroy tumors.

One is devoted to radiological detection of tumors using biotinylated monoclonal antibodies and a biotinylated chelating agent to transport γ-radiating isotopes. The connecting member is the egg protein avidin or the bacterial streptavidin which bind with high affinity (10^{15} l/mole) to biotin. After injection of the monoclonal antibody into the blood stream and its fixation to the tumor considerable amounts of antibody molecules remain in the circulation. These are chased off by an injection of avidin or streptavidin which label by the same time the tumor bound antibody. Finally the radioisotope carrying chelate is injected which binds exclusively to the tumor. By this method *Fazio a. Paganelli*[1] were able to localize just liver metastases which are normally hidden by a heavily scintillating background in this organ.

The second method uses an ingenious biotechnological construct consisting of the active center of a mouse monoclonal antibody directed against the cell bound tumor marker CEA (*C*arcino-*E*mbryonic *A*ntigen), F_c-parts of human immunoglobulin, and a normally only intracellularly displayed enzyme, β-glucuronidase, to label adenocarcinoma and its

J. Szejtli and L. Szente (eds.), Proceedings of the Eighth International Symposium on Cyclodextrons, 71–76.

metastases. After injection, the construct labels the tumor and is slowly washed out from the circulation. Then the natural detoxification product of the antibiotic adriamycin, adriamycin glucuronide, is injected which is split and therefore retoxified exclusively at the tumor site. By this method *Bosslet and his coworkers*[2] were able to kill completely adenocarcinomas without the dangerous side effects of adriamycin to the heart which normally prevents a longer and high-dose application of this drug. This approach is now in phase II clinical investigations.

When we intend to prevent the side effects of tumor chemotherapy which lead to a martyrium of the patients we have to learn from these both successful approaches. The first we can learn is that an antibody recognizing a tumor site must be humanized or labelled only with a minimum structure like biotin to slip through the rigorous control of the reticuloendothelial system preventing foreign molecules to reside for a longer time in the human body. The second lesson is that only drugs with an excretion signal or, broader speaking, a high hydrophilicity preventing uptake into any living cell combined with an attachment signal structure will allow for a successful specific labelling of tumors.

This in mind we tried to form inclusion complexes of porphyrinoid drugs for use as photosensitizers in Photodynamic Tumor Therapy (PTT) with a shell of dimeric or trimeric cyclodextrins in order to prevent any uptake into living cells[3]. We attached biotin with a long spacer molecule to the shell or to the porphyrinoid itself as a signal structure. And we begun first experiments to localize such drugs at tumor sites.

2. MATERIALS AND METHODS

Cyclodextrins and methyl β-cyclodextrin (1.8 saturation) were obtained from *Wacker-Chemie (Burghausen, Germany)*. β-Cyclodextrin dimers connected with alkyl spacers of different lengths .were synthesized according to *Ruebner et al.*[4] (see the accompagnying poster paper). Complex stability constants with tert.butyl phenoxy substituted porphyrinoids were determined by fluorimetry using guest exchange between p-toluidino naphthalene sulfonic acid and the respective porphyrinoids[5]. Tert.butyl phenoxy substituted porphyrinoids were kindly gifted by *Dr. M. Müller (Laboratory of Prof. Dr. E. Vogel, Organic Chemistry, University of Cologne, Germany),and by Andrea Weitemeyer (Laboratory of Prof. Dr. D. Wöhrle, Organic and Macromolecular Chemistry, University of Bremen, Germany)*. Attachment of spacered biotin followed the route of reaction of the free acid groups with carbonyl diimidazole and coupling with biotin cadaverine in dimethyl formamide. Similar products can be obtained by coupling via dicyclohexyl carbodiimide or - on NH_2 groups - with biotin-X-NHS.

Absorption spectra were obtained with a Shimadzu UV 200 instrument at 2 nm resolution, fluorescence spectra with a Shimadzu 1501 instrument at 10 nm resolution. Mass spectra were of MALDI type obtained on a LAMMA 1000 spectrometer with 2,5-dihydroxy benzoic acid as the matrix.

3. RESULTS AND DISCUSSION

γ-Cyclodextrin was sometimes used to solubilize photosensitizing porphyrinoids, like tin etiopurpurins, for injection purposes[6]. We found monomeric β-cyclodextrin and its methylated derivative much more suitable to solubilize e.g. pheophorbide a, phthalocyanines[4], or porphycenes[3]. Complex binding occurs in a two step reaction where the first step, i.e. desaggregation of dye polymers, is velocity-limiting , and so masks the real complexation process[3]. As a result we obtained very low binding constants[7] which could be corrected only by the guest exchange procedure using fluorimetry of p-toluidino naphthalene sulfonic acid[4,5]. Injection of porphyrinoids bound to monomeric cyclodextrins with binding constants in the order of $3x10^3$ [l / mol] leads to exchange with the lipoprotein fraction in blood plasma[7, 8]. In cell culture we see an enhanced uptake into tumor cells possibly due to a directing of the dye molecule perpendicularly to the cell membrane with its lipophilic end leading to a substantial incorporation of the drug into lipophilic parts of the membrane and, consequently, to a considerably better uptake into the cell. An example shows fig. 1 for the uptake of glutaryl amido tetra propyl porphycene (GlamTTPn).

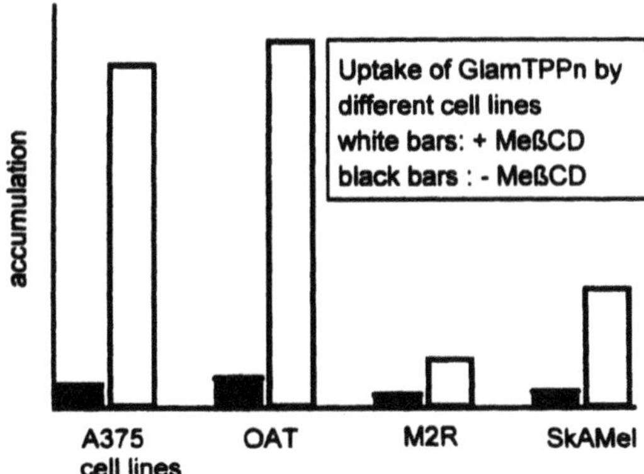

Fig. 1 Uptake of GlamTTPn into different cell lines with and without prior complexation to methyl β-cyclodextrin.

For high affinity inclusion only dimeric taylored β-cyclodextrins are suitable. The prerequisite are modified peripheries of the porphyrinoids since the normally less voluminous peripheral side-groups of native porphyrins cannot dive deeply enough into the lipophilic cavity of the cyclodextrin torus. *Breslow a. Chung*[9] have synthesized model compounds containing voluminous tert.butyl benzoic acid groups on both ends allowing for binding constants of about 10^{10}[l/mol] with sterically hindered cyclodextrin dimers.

The substitution of phthalocyanines, *meso*-tetraphenyl porphyrins, and pheophorbides with tert.butyl phenoxy or tert.butyl benzoic acid substituent led to affinity constants of 10^7 l/mol upon inclusion into β-cyclodextrin dimers with spacers of 11-13 C-atoms lengths[10,11]. Stability of these inclusion complexes against the lipoprotein system of blood plasma has been proved by gel permeation chromatography and absorption spectroscopy[12]. Stability against transfer of the guests into cell membranes are under investigation. The drugs synthesized up to now are model compounds which have to be replaced further by effective photosensitizing drugs alike the one shown in fig. 2.

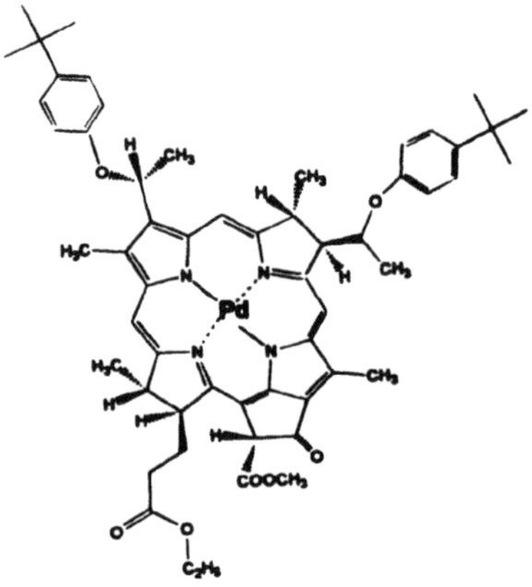

Fig. 2 Ideally substituted bacteriopheophorbide sensitizer as prepared now from bacteriochlorophyll g of a recently discovered *Heliobacterium*[13]

Such photosensitizers are easily biotransformed into inert splitting products which overcome the longtermed photosensitation in patients treated by PDT[14].

To bind such sensitizers exclusively onto tumor cells a polyphasic system based on biotin-avidin interaction has been formulated which is now under investigation in model animal experiments[15]. Such systems were successfully introduced into radiological detection of tumors[1] and seem to work as well with biotinylated photosensitizers included into cyclodextrin dimers so long as the spacer structures involved are not too extended[16]. An example of such a construct is shown in fig.3

To set free the included drugs at the tumor site preformed breaking points inside the spacer structures were provided. For $^1O_2$-producing porphyrinoids such a structure can be a simple tartric acid inset sensitive to activated oxygen species (fig.4). Upon opening the bridge the affinity constant drops by 3-4 orders of magnitude and the drug can entry into the next-neighboured cell. Several other types of laser sensitive bridges were conceptually possible and are just synthetized in part[17] for other purposes.

(3)

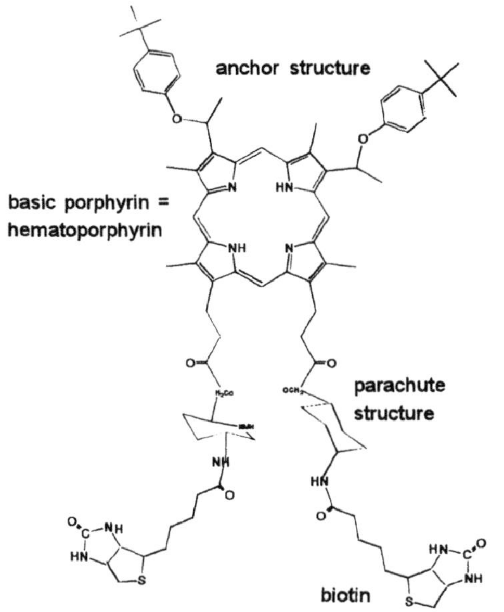

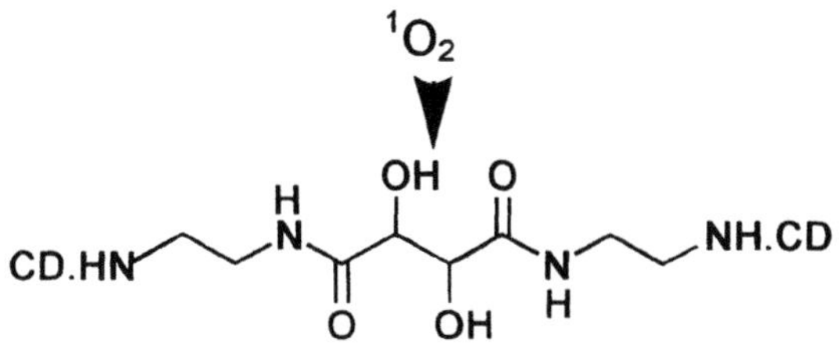

(4)

Fig. 3 Biotinylated di-tert.butyl phenoxy hematoporphyrin as included into a taylored β-cyclodextrin dimer

Fig.4 A $^1O_2$-sensitive breaking point structure allows for setting free the guest molecule.

We should realize that photodynamic therapy is only a very small cutting out of the manifold of methods in general chemotherapy with its far-reaching side effects. The long-term neurotoxic and genotoxic events have to be avoided in future. We have learned a lot from photodynamic therapy drugs to transfer this knowledge now to general chemotherapeutic drugs which work up to now according to the rule:"We poison the whole patient and hope that the tumor will die a little bit earlier as its carrier". There are several highly poisonous drugs which cannot be used for tumor treatment because of its high general toxicity. Overviewing the chemical structures of such drugs there are a manifold which presumably can be included into inert carriers alike those developed for PDT drugs, i.e. cyclodextrin dimers. One of the most interesting of these is the 25 years old microtubule hyperstabilizing drug, paclitaxel or taxol. This drug can be given to patients resistant to all other chemotherapeutic regimes due to its uncommon mechanism of action. This drug with its 3 phenolic groups is a model case of possible inclusion into a taylored cyclodextrin trimer which is now under investigation in our laboratory. First experiments show a consistent binding into cyclodextrin monomers.

4. CONCLUSION

Dimeric and trimeric cyclodextrins will be useful tools to detoxify and to direct anti-tumor drugs to its target. These coming developments should be of interest not only for medical doctors but for the pharmaceutical industry. Experimental designs are not easily projected into industrial realization. So far, these should be accompagnied in an early stage by industrial partnership.

REFERENCES

[1] **Fazio F**, Paganelli G (1994): Eur.J.Nucl.Med **20**, 1138-1140
[2] **Bosslet K**, et al .(1992): Br.J.Cancer **65**, 234-238
[3] **Moser J.G.**, et al (1992): SPIE Biomedical Optics **2523**, 92-99
[4] **Ruebner** et al.(1996):Tetrahedron (in press)
[5] **Johnson MD**, Reinsborough VC (1992): Austr. J. Chem. **45**, 1961-1966
[6] **Kessel D**, et al.(1991): Photochem.Photobiol **54**, 193-196
[7] **Moser, J.G.**, et al. (1991): Abstr. 4th Congr. Europ. Soc. Photobiol. **A12**
[8] **Jori, G.**,Reddi E.(1993): Int. J. Biochem. **25**, 1369-1375
[9] **Breslow R**, Chung S(1990): J. Am. Chem. Soc.**112**, 9659-9660
[10] **Roehrs S.**, et al (1995): SPIE Biomedical Optics **2625**, 333-338
[11] **Kliesch**, et al: Liebig's Ann. **1995**, 1269-1273
[12] **Moser J.G.**,et al.(1995): SPIE **2625**, 138-145
[13] **Van de Meent,E.J.**: The antenna-reaction center complex of *Heliobacteria* and of green sulfur bacteria. PhD Thesis,1992
[14] **Henderson,BW**,et al (1991): J.Photochem.Photobiol **10**, 303-313
[15] **Vervoorts A**, etal.(1996): SPIE Biomedical Optics **2678**, -51- (in press)
[16] **Moser, J.G.**, et al. (1996): SPIE Biomedical Optics (delivered)
[17] **Venema** F.(1996): Monofunctionalised cyclodextrins as building blocks for supramolecular systems. PhD-thesis. p.40

ACKNOWLEDGEMENTS

The authors thankfully acknowledge support by grants (# 13N6291) of the Ministry of Education and Research (BMBF) of the Federal Republic of Germany and by the German Research Council (DFG) (# Mo 115/13-2 and # 436RUS17/5/95).

SYNTHESIS OF β-CYCLODEXTRIN DIMERS AS CARRIER SYSTEMS FOR PHOTODYNAMIC THERAPY OF CANCER

A. RUEBNER, J. G. MOSER, D. KIRSCH, B. SPENGLER,
S. ANDREES, S. ROEHRS
Institute of Laser Medicine, University of Duesseldorf, Universitaetsstr. 1, D-40225 Duesseldorf (Germany)

ABSTRACT

The aim of our investigation was the development of carrier systems for an application of inert drugs in polyphasic photodynamic tumor therapy. As carrier systems, β-cyclodextrin dimers linked at their primary and secondary faces by spacers of varying lengths were synthesized. Cyclodextrins are known to form stable inclusion complexes with porphyrinoïd photosensitizers. The influence of spacer length on the β-cyclodextrin dimer inclusion complexes with porphyrinoïd photosensitizers was studied.

1. INTRODUCTION

Photodynamic therapy (PDT) is a cancer treatment that uses a combination of light-activated drugs (photosensitizers, e.g. porphyrins) and laser light to create highly reactive forms of oxygen that destroy tumor cells. Photosensitizers, however, do not stain tumor tissue exclusively. This is the main drawback in PDT. Antibody-directed targeting of photosensitizers have produced promising results only in *in-vitro* systems [1,2], but have failed to work *in-vivo*. Only a recombination of photosensitizers with specific antibodies into a new class of drugs might fulfill the desired specificity of photodynamic therapy [3]. The first prerequisite is a stable inclusion complex of the cyclodextrin dimer and porphyrinoïd photosensitizers, which prevents photosensitizer transport on the lipoprotein pathway. This would avoid any unwanted targeting of organs except the tumor, provided that a bond between drug and tumor specific antibodies can be established. Since β-cyclodextrins are known to form stable inclusion complexes with porphyrinoïd photosensitizers, we have used them as carrier systems.

2. MATERIALS AND METHODS

2.1. Synthesis

The synthesis of mono-2-(ω-aminopropylamino)-2-deoxy-ß-cyclodextrin **5a**, mono-2-(ω-aminobutylamino)-2-deoxy-ß-cyclodextrin **5b**, mono-6-(2-aminoethylthio)-6-deoxy-β-cyclodextrin **6**, 2,2′N-β-cyclodextrin dimers **7a-c** and 6,6′S-β-cyclodextrin dimer **8a** were described elsewhere [4].

J. Szejtli and L. Szente (eds.), Proceedings of the Eighth International Symposium on Cyclodextrons, 77–80.

2.2. Determination of binding constants

Binding constants of porphyrinoïds with CD-dimers were studied by means of competitive spectrofluorometry and determined as described elsewhere [4].

3. RESULTS AND DISCUSSION

3.1. Synthesis

Various primary and secondary face linked cyclodextrin dimers were successfully prepared by a multistep procedure, starting from monosubstituted β-cyclodextrin (Scheme 1). Alkylation of linear α,ω-diaminoalkanes with mono-2-tolylsulfonyl-2-deoxy-β-cyclodextrin **3** (route 1) resulted in ω-aminoalkylamino substituted β-cyclodextrins **5a, b**. These were reacted with N-hydroxysuccinimide esters **9a-c** to produce 2,2′N-β-cyclodextrin dimers **7a-c**. Ion exchange chromatography was used to purify crude dimers. Side products which eluted last, having only one β-cyclodextrin bound to the spacer were isolated, activated with N-hydroxysuccinimide and reacted with **5a** or **5b** respectively to produce the desired dimers. The three fractions of ion exchange chromatography were characterized by MALDI-MS and are shown in Fig. 1. Reaction of mono-6-iodo-6-deoxy-β-cyclodextrin **4** (route 2) with 2-amino-ethanthiol resulted in mono-6-(ω-aminoethylthio)-6-deoxy-β-cyclodextrin **6**. This was reacted with N-hydroxysuccinimide ester **9b** to produce 6,6′S-β-cyclodextrin dimer **8a**.

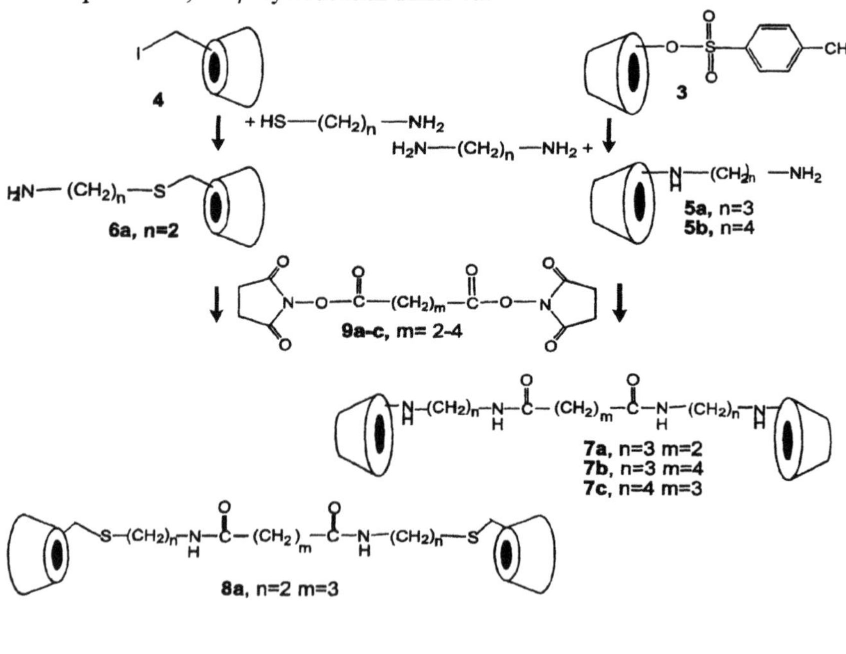

Scheme 1. Synthesis of β-cyclodextrin dimers.

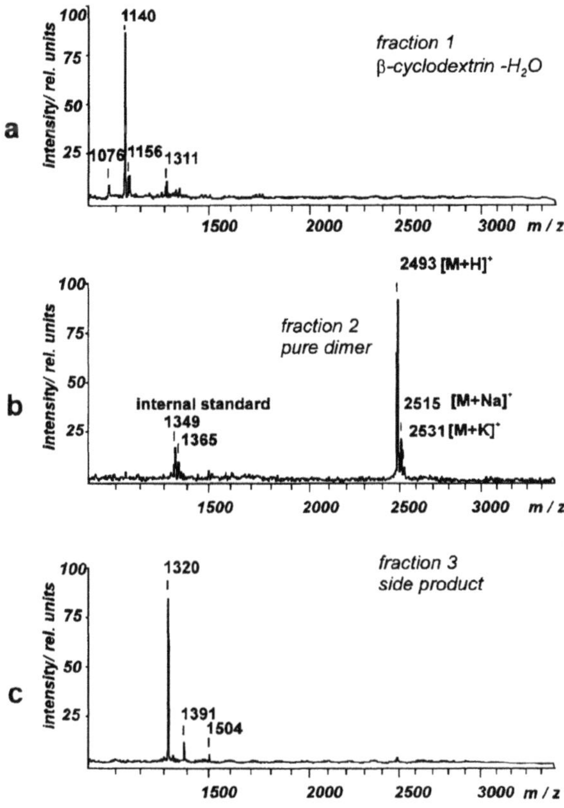

Fig. 1. Fractions of ion exchange chromatography. a: β-cyclodextrin-H_2O [M+ Na], b: dimer **7b** [M+H], c: side products with only one β-cyclodextrin bound to the spacer [M+H].

3.2. Binding constants

The binding constants of the inclusion complexes of β-cyclodextrin dimers with taylor-made porphyrinoïd derivatives are summarized in Table 1. We examined the porphyrinoïd derivatives **1** and **2** as guest molecules. They are similar in that they are substituted with one or more t.-butylphenyl group. Breslow and co-workers [5,6] showed that guest molecules with t.-butylphenyl groups bind tightly into the cavity of β-cyclodextrins. Our investigations yielded binding constants of β-cyclodextrin dimer inclusion complexes with porphyrinoïd derivatives ranging from 10^5 to 10^7 l/mol. It was shown that all secondary face linked dimers exhibit larger binding constants with porphyrinoïd derivatives as do primary face linked dimers. By comparing the two porphyrinoïd derivatives **1** and **2** (Fig. 2) it can be seen that the insertion of more than one t-butylphenyl residues enhanced the binding constant. The largest binding constant found was that of Zn-tri-tert.butylphenoxy-mono-sulfophenoxy-phthalocyanine **2** in combination with dimer **7c**.

TABLE 1. Binding constants [l/mol] of β-cyclodextrin dimers with porphyrinoïd derivatives. 3^1-tert.-butylphenoxy-ethyl-pyropheophorbide-ethylester **1**, Zn-tri-tert.-butylphenoxy-mono-sulfophenoxy-phthalocyanine **2**.

hosts	**guests**	**porphyrinoïd derivatives**	
cyclodextrins	TNS	**1**	**2**
7a	$3.6 \pm 0.2 \times 10^3$		$3.3 \pm 0.5 \times 10^5$
7b	$3.8 \pm 0.3 \times 10^3$	$2.9 \pm 0.4 \times 10^5$	$1.3 \pm 0.3 \times 10^5$
7c	$5.5 \pm 0.3 \times 10^3$	$5.7 \pm 0.5 \times 10^6$	$1.5 \pm 0.3 \times 10^7$
8a	$8.3 \pm 0.4 \times 10^3$		$4.0 \pm 0.7 \times 10^6$

1

2

Fig. 2. **1** = 3^1-tert-butylphenoxy-ethyl-pyropheophorbide-ethylester, **2**= Zn-tri-tert-butylphenoxy-mono-sulfophenoxy-phthalocyanine

4. CONCLUSION

As a result of synthesizing β-cyclodextrin dimers of varying spacer lengths, we were able to establish the optimal spacer length for a strong binding to taylor-made porphyrinoïd derivatives. Using these results, it was possible to outline a β-cyclodextrin dimer suitable for use as a carrier for photosensitizers in PDT.

ACKNOWLEDGEMENTS

Financial support by the BMBF (Bundesministerium fuer Bildung, Wissenschaft Forschung und Technologie, grant #13N6291) is gratefully acknowledged.

REFERENCES

[1] Mew, D., Wat, C.-K., Towers, G. H. N., Levy, J. G. *J. Immunol.*, **130**, 1473-1477 (1983)
[2] Oseroff, A. R., Ara, G., Ohuoha, D., Aprille, J., Bommer, J. C., Yarmush, M. L. *Photochem. Photobiol.*, **46**, 83-96 (1987)
[3] Moser, J. G., Ruebner, A., Vervoorts, A., Wagner, B., this issue.
[4] Ruebner, A., Kirsch, D., Andrees, S., Decker, W., Roeder, B., Spengler, B., Kaufmann, R., Moser, J. G., Dimeric cyclodextrin carriers with high binding affinity to porphyrinoïd photosensitizers. submitted to Tetrahedron.
[5] Breslow, R., Halfon, S., Zhang, B. *Tetrahedron*, **51**, 377-388 (1995)
[6] Breslow, R., Greenspoon, N., Guo, T., Zarzycki, R. *J. Am. Chem. Soc.*, **111**, 8296-8297 (1989)

THREE-DIMENSIONAL CYCLODEXTRIN: A NEW CLASS OF HOSTS BY TREHALOSE CAPPING OF β-CYCLODEXTRIN

VINCENZO CUCINOTTA[a], GIULIA GRASSO[b], SONIA PEDOTTI[b], ENRICO RIZZARELLI[b,c], GRAZIELLA VECCHIO[b]

a Dipartimento di Scienza dell'Alimentazione, Universita' di Napoli "Federico II", via Universita' 100, 80055 Portici, Napoli, Italy.

b Istituto per lo Studio delle Sostanze Naturali di Interesse Alimentare e Chimico-Farmaceutico, CNR, V.le A.Doria 8, 95125 Catania, Italy.

c Dipartimento di Scienze Chimiche, Universita' di Catania, V.le A.Doria 8, 95125 Catania, Italy.

ABSTRACT

In order to obtain better abiotic receptors for analytical and kinetic applications, the new capped derivative of cyclomaltoheptaose (CDTH) was synthesized by reaction of 6,6'-dideoxy-6,6'-di(S-cysteamine)-α,α'-trehalose with 6A,6D-dideoxy-6A,6D-diiodo-cyclo-maltoheptaose. The CDTH-ACS (anthraquinone-2-sulfonic acid sodium salt) system was investigated. by ^{1}H NMR spectroscopy and by i.c.d. (induced circular dicroism), and a deep inclusion of ACS inside the CDTH cavity, with an association constant about six times larger with respect to ACS - β-CD system was found.

1. INTRODUCTION

The functionalization of cyclodextrin can modify and improve some features of this class of molecules, such solubility, stability and selectivity, when forming inclusion complexes. By replacing one or more -OH groups at a desired position and with appositely designed substitution group, multisite recognition systems have been obtained [1]. An improvement can be obtained by introducing groups able to coordinate metal ions: the metal can be one more recognition site when coordinating guest are recognized by the cavity. In this class of complexes, some of them have shown to behave as chiral discriminating agents, and their use in LEC chromatography has been described [2, 3].

Capped cyclodextrins have been synthesized as well and their properties have been investigated. Among the capping groups, porphyrins [4], cyclopeptides [5], the azobenzene [6] and other aromatic moieties have been used [7]. The expansion of the hydrophobic region in comparison with the CD parents have been underlined. This increase in the hydrophobicity has resulted in larger association constants with different guests, in comparison to the CD parents [6, 7]. However, in all the capped cyclodextrins reported until now, the carbohydrate cavity hasbeen coupled with capping moieties with completely different features in comparison to CDs, as can also be seen by the short list above reported.

J. Szejtli and L. Szente (eds.), Proceedings of the Eighth International Symposium on Cyclodextrons, 81–84.

Here, we report the synthesis of the new capped derivative of cyclomaltoheptaose (CDTH). The synthesized CDTH may be considered the first example of a three-dimensional cyclodextrin for the presence of a disaccharide, which extends the carbohydrate system to a plane perpendicular to the main cavity, with a presumably increased inclusion ability with respect to the CD parent. Furthermore the presence of heteroatoms (N and S) in the two bridges between CD and TH could increase the selectivity of this class of compounds.

2. MATERIALS AND METHODS

2.1 Synthesis

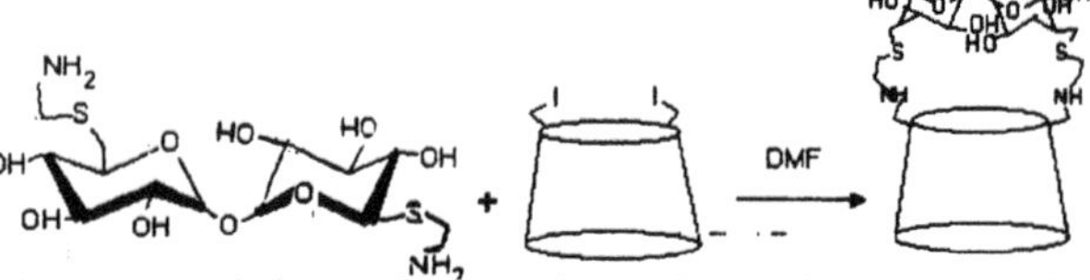

The synthesis of 6A,6D-dideoxy-6A,6D-[6,6'-dideoxy-6,6'-di(S-cysteamine)-α,α'-trehalose]-β-cyclodextrin (CDTH) was carried out by reaction of 6,6'-dideoxy-6,6'-di(S-cysteamine)-α,α'-trehalose (TH) with 6A,6D-dideoxy-6A,6D-diiodo-cyclomaltoheptaose ($ADCDI_2$) [8]. according to the following scheme. The synthesis details are reported elsewhere [9]

2.2 Spectroscopic measurements

^{1}H NMR spectra (400MHz) were recorded on a Varian Unity 400 spectrometer and ^{13}C NMR spectra (50.33 MHZ) on a Bruker AC-200 spectrometer on D_2O solutions without a reference compound.
Circular Dichroism spectra were recorded on a JASCO J-600 spectropolarimeter at 25°C on freshly prepared aqueous solutions in phosphate buffer (0.015M) pH = 6. Quartz cuvettes of 0.1 cm pathlength were used. The ACS concentration was kept constant ($4.0\ 10^{-4}$ mole dm^3), whereas the CDTH concentration was varied such as to obtain, respectively, 1:1.5, 1:2.5, 1:4.5, 1:8, 1:15, 1:25 ACS/CDTH molar ratios. Results are reported in terms of $\Delta\varepsilon$ (molar CD coefficient) in $dm^3\ mol^{-1}\ cm^{-1}$.

2.4 Calculations

For the determination of the association constants, a graphical method was used, by the equation, not neglecting the analytical concentration of the observed component (ACS in our case) with respect to that of the other component [10].

3. RESULTS AND DISCUSSION

The ^{1}H NMR spectrum of CDTH is reported in fig. 1a, together with its assignment, obtained by COSY spectrum. The trehalose substitution has an overall slight effect on the cavity protons, with the obvious exeception of the methilene protons of the substituted rings (6A, 6D, 6'A and 6'D). Analogously, large shifts are observed on the 6-H of trehalose, as well as on all the protons of the two cysteamine bridges. Among these eight protons, one for each bridge is at a sharply lower field with respect to the others, presumably for a cavity effect. In the 1-protons region only a peak is showed for all the seven glucopyranosinic rings of β-CD, while the two

1-protons of trehalose appear at lowerfield as two separated doublets, thus giving evidence of a loss of equivalence in this symmetric molecule, following the reaction with the cavity.

The CDTH has two secondary amino groups and can form a mono- and a di-protonated species. The CDTH was titrated by DCl and the variation was followed by ^{1}H NMR spectroscopy. These experiments shows that the two protonation constants are very similar and quite low for an amino group (log K is about eight), as tipically observed for this kind of derivatives. The ^{1}H NMR spectrum of the diprotonated species appears more complex than the unprotonated species, showing an increased asymmetry of the cavity, probably due to the formation of hydrogen bonds between the protonated nitrogens of cysteamine moities and upper rim hydroxyl groups.

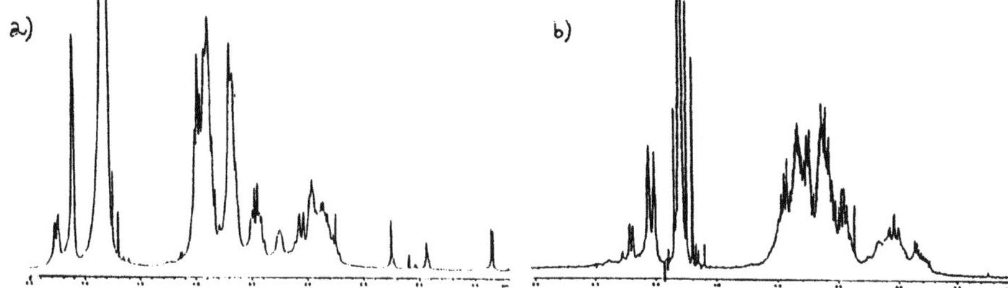

Fig. 1: a) ^{1}H NMR spectrum of CDTH; b) ^{1}H NMR spectrum of CDTH-ACS system.

^{1}H NMR spectrum of CDTH - ACS system is reported in fig. 1b. The comparison with the spectrum in fig. 1a, shows the dramatic effect that the ACS inclusion causes on the host proton chemical shifts. As expected, the 3- and 5- protons of β-CD are very affected. Furthermore, interestingly, also the trehalose protons chemical shifts are influenced, suggesting a deep inclusion of ACS inside the CDTH cavity. ROESY spectrum shows correlations between the inner protons (3,5) of the cavity with the protons of the ACS unsubstituted benzene ring, suggesting that this ring is deeply included in the CD cavity. The sulphonic group seems to be near the CD secondary OH rim, outside the cavity.

CD spectrum of CDTH shows a small positive Cotton effect ($\Delta\varepsilon = 0.54$) at 222 nm, due to the chirality of the cysteamine moieties, induced by the CD cavity. The ACS does not show a c.d. spectrum; however, if CDTH is added to an ACS solution, an i.c.d. spectrum is obtained.

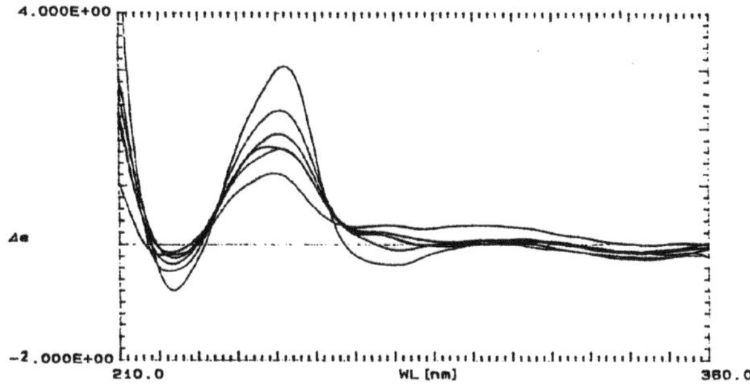

Fig. 2 - I.c.d. spectrum of CDTH - ACS system. From curve *a* to curve *f*, increasing concentrations of CDTH (see text).

In order to determine the CDTH - ACS association constant, a titration of the guest with the host was carried out, and the c.d. spectra obtained are reported in fig. 2. The positive band at 252 nm, (where the Cotton effect is stronger) corrected for the slight CDTH contribution, were used, as above described, to obtain the equilibrium constant value (Kass = 3589). This constant value is much higher in comparison to the value calculated for the ACS-β-CD complex (Kass = 598).

4. CONCLUSION

The preliminary results here reported on the new host CDTH, show its strong inclusion properties. This host can be considered the prototype of a three-dimensional cyclodextrin. By a suitable choice of the "bridge" groups, is possible to modulate the strength and the selectivity of these receptors towards specific substrates, as well as their coordinating ability towards metal ions. Work is in progress in our laboratories in order to more explore the features of CDTH, as well as to synthesize other compounds of this class.

Acknowledgements. We thank Dr. Carla Isernia (University of Napoli) for the spectra carried out on Varian Unity 400.

REFERENCES

[1] Tabushi, I., Kuroda, Y., Mizutani, T., Artificial Receptors for Amino Acids in Water. Local Environmental Effect on Polar Recognition by 6A-Amino-6B-carboxy and 6B-Amino-6A-carboxy-β-cyclodextrins, *J.Am.Chem.Soc.*, **108**, 4514-4518 (1986).
[2] Corradini, R., Dossena, A., Impellizzeri, G., Maccarrone, G., Marchelli, R., Rizzarelli, E., Sartor, G., Vecchiio, G., Chiral Recognition and Separation of amino Acids by means of a Copper(II) Complex of Hiatamine Monofunctionalized β-Cyclodextrin, *J.Am.Chem.Soc.*, **116**, 10267-10274 (1994) and some references therein.
[3] Cucinotta, V., D'Alessandro, F., Impellizzeri, G., Vecchio, G., The Copper(II) Complex with the Imidazol-bound Histamine Derivatived of β-Cyclodextrin as a Powerful Chiral Discriminating Agent, *J.Chem.Soc.Chem.Commun.*, 1743-1744 (1992).
[4] Kato, T., Nakamura, Y., Convenient Synthesis and Properties of Water Soluble Cyclodextrin capped Mesoporphyrinatoiron, *Heretocycles* **27**, 973-979 (1988).
[5] Bonomo. R.P., Impellizzeri, G., Pappalardo, G., Rizzarelli, E., Vecchio, G., cyclo-L-Histidyl-L-Hystidyl capped β-cyclodextrin. A New Potentially Enzyme-mimicking Compound or Artificial Receptor with two Recognition Sites, *Gazzetta Chimica Italiana* 123, 593-595 (1993).
[6] Ueno, A., Takahashi, Osa, T., Photocontrol of Catalytic Activity of Capped Cyclodextrin, *J.Chem.Soc., Chem.Commun.*, 94-96, (1981).
[7] Tabushi, I., Shimokawa, K., Shimuzu, N., Shirakata, H., Fujita, K., Capped Cyclodextrin, *J.Am.Chem.Soc.*, **98**, 855-856 (1976).
[8] Cucinotta, V., D'Alessandro, F., Impellizzeri, G., Vecchio, G., Synthesis and Conformation of Dihistamine derivatives of Cyclomaltoheptaose, *Carbohdr.Res.*, **224**, 95-102 (1992).
[9] Cucinotta, V., Grasso, G., Rizzarelli, E., Vecchio, G., submitted.
[10] Perkampus, H.-H., UV-Vis Spectroscopy and Its Applications. Springer Verlag 1992, pg 131.

6-HYDROXYALKYLAMINO-6-DEOXY-CYCLODEXTRINS: TOWARDS DENDRIMERIC HOST-MOLECULES

C.AHERN, R. DARCY, F. O'KEEFFE AND P. SCHWINTÉ

National University of Ireland

Laboratory for Carbohydrate and Molecular Recognition Chemistry, Department of Chemistry, University College Dublin, Dublin 4, Ireland.

ABSTRACT

6-Perhydroxyalkylamino-6-perdeoxy-β–cyclodextrins have been synthesised by treating 6-perbromo-6-perdeoxy-cyclodextrins with hydroxyalkylamines. The products (**1**, **2**) are precursors of dendrimeric cyclodextrins in which the cavity provides access for the guest to interact with the branches. A fluorescence study has demonstrated the effects of the branches on binding of anilinonaphthalene sulfonate probes. The hosts show selectivity towards guests, and pH-dependence of binding, consistent with polar interaction between guest sulfonate anions and the protonated amino groups of the dendrimeric structure.

1. INTRODUCTION

Dendrimers [1] are highly-branched macromolecules which have potential as hosts because of the spaces between the branch stems. Such structures, if grafted onto cyclodextrins, might be expected to modulate the cyclodextrin's host properties, and in turn, the cyclodextrin might act as a channel to the dendrimer core. As a development of our studies on branched cyclodextrins [2], we have synthesised the heptakis(6-hydroxyalkylamino-6-deoxy)-β-cyclodextrins **1** and **2**, and evaluated the effects of the branched structures on their ability to bind naphthalene sulfonate probes.

J. Szejtli and L. Szente (eds.), Proceedings of the Eighth International Symposium on Cyclodextrons, 85–88.

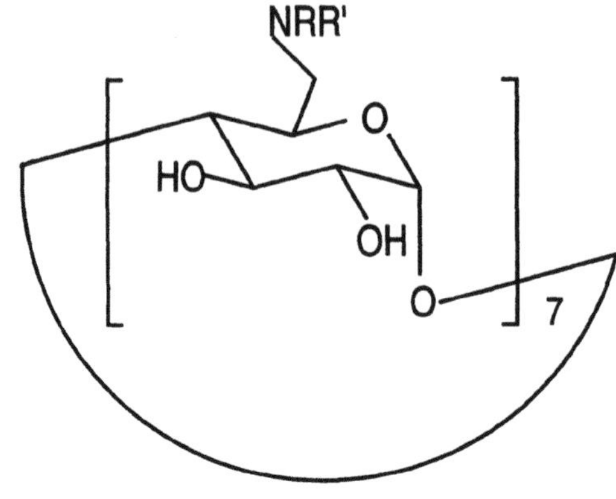

1 R = H, R' = CH_2CH_2OH

2 R = R' = CH_2CH_2OH

2. MATERIALS AND METHODS

2.1 Syntheses

The heptakis(6-hydroxyethylamino-6-deoxy)-β-cyclodextrins **1** and **2** were prepared from heptakis(6-bromo-6-deoxy)-β-cyclodextrin [3] by dissolving the latter at 65° in ethanolamine (15 molar equivalents)(for **1**), or in diethanolamine (for **2**), and heating at the same temperature for 48 hours. The reagent/solvent was removed under vacuum, and the residue was dissolved in hot methanol and precipitated by addition of this solution to stirred acetone. The precipitate was collected by gravity filtration, then dissolved in water, and the solution was treated with basic ion-exchange resin. Lyophilisation yielded **1**, $[\alpha]_D$+110°(*c* 0.1, water), or **2**, $[\alpha]_D$+109°(*c* 0.1, water)(60-65%).

Persubstitution was confirmed by elemental analysis and by the simple NMR spectra, although the ^{1}H-spectrum of the more highly branched **2** showed fluxional broadening: (**1**) δ^1H (500 MHz, D_2O) 5.10 (d, 1H, $J_{1,2}$ 4Hz, H-1), 3.98-3.93 (m, 2H, H-3, H-5), 3.73-3.63 (m, 3H, CH_2O, H-2), 3.51 (t, 1H, $J_{3,4}$=$J_{4,5}$=9Hz, H-4), 3.02 (dd, 1H, $J_{5,6a}$ = 2Hz, $J_{6a,6b}$ 13Hz, H-6a), 2.88 (dd, 1H, $J_{5,6b}$ = 7.5 Hz, H-6b), 2.7 (m, 2H, NCH_2); $\delta^{13}C$ (68 MHz, D_2O) 101.5 (C-1), 82.7 (C-4), 72.6 (C-3), 71.8 (C-2), 70.1 (C-5), 59.7 (CH_2O), 50.0 (NCH_2), 48.8 (C-6); (**2**) δ^1H (270 MHz, D_2O) 5.25 (d, 1H, $J_{1,2}$ 2Hz, H-1), 3.98-3.94 (m, 2H, H-3, H-5), 3.6-3.52 (m, 4H, CH_2O, H-2, H-4), 2.97-2.73 (m, H-6a, H-6b, NCH_2); $\delta^{13}C$ (68 MHz, $CHCl_3$) 99.2 (C-1), 79.7 (C-4), 72.9 (C-3), 71.8 (C-2), 70.4 (C-5), 59.0 (CH_2O), 56.4 (NCH_2), 55.7 (C-6).

2.2 Fluorescence measurements

Fluorescent probes were obtained from Molecular Probes Europe BV. Solutions were buffered with HCl-carbonate. Binding constants were calculated from double reciprocal plots of variations in fluorescence intensity with host concentrations [4].

3. RESULTS AND DISCUSSION

3.1 Fluorescence measurements

Binding of these probes is complicated by the possibility of inclusion of either the naphthalene or anilino systems [5], however it is probable that both are at least partially included, and there may not be a clear kinetic distinction between complexes of different configuration. From a double reciprocal plot there is evidence for a 2:1 (CD: probe) complex at high host concentrations in the case of the methylanilino probe but only at pH 10. This indicates non-inclusion of the (protonated) methylanilino group by a second CD molecule at low pH. It also favours the view that those probes which show the stronger binding at lower pH, do so because of favourable interaction between the protonated cyclodextrin amino branches and probe sulfonate group. This effect is evident also for the simple naphthalene sulfonate probe. It is consistent with results for nucleoside phosphate binding [6] where the polar interaction is between protonated methylamino groups and phosphate ion.

TABLE 1. Association constants of hydroxyalkylamino-CD's with naphthalene sulfonate probes

CD	probe	K_a (M^{-1})		
		pH 2	6	10
1	2,6-anilino (ANS)	2300	700	1200
1	2,6-tolylamino (TNS)	200	170	300
1	2,6-methylanilino (MANS)	1000		5300
1	1,8-anilino	200		
1	naphthalene sulfonate	9000		3000
2	2,6-methylanilino (MANS)			2500

Strong binding (ANS and MANS) (Table) is accompanied by blue shifts in fluorescence emission wavelengths of 30-45nm, compared with 20nm for β-cyclodextrin, which confirm inclusion in the hydrophobic cavity [7]; while weak binding is associated with red shifts of 20-40nm.

The electrostatic binding by the branches also prevents the guests from threading through the cavity, and this results in shallower and weaker binding of MANS at lower pH. Polar binding by the amino branches does not always therefore result in higher K_a values, but there is very selective modulation of these values compared with unmodified cyclodextrin [5]. The similar ANS and TNS structures are undifferentiated by β-cyclodextrin, but with **1** they show binding constants differing by a factor of ten at pH2 (Table). Also, large variations in K_a with pH are observed here for some guests, which are not observed with β-cyclodextrin until it itself is deprotonated above pH10 [5]. An obvious design improvement however would be to have binding groups further removed from the cavity.

4. CONCLUSION

The concept of extending the hydrophobic cavity of cyclodextrins by means of dendritic structures has been realised with increased solubility in spite of the extended structures. Access to the core groups of the branches is provided, and these modulate binding, although the ultimate advantages of dendrimeric cyclodextrins have yet to be realised.

ACKNOWLEDGEMENTS

We thank Forbairt, the Irish Science and Technology Agency, for the award of a grant, SC/95/237.

REFERENCES

[1] Tomalia, D.A., Naylor, A.M., Goddard, W.A., Starburst dendrimers: molecular level control of size, shape, surface chemistry, topology and flexibility from atoms to macroscopic matter, *Angew. Chem. Int. Ed. Engl.*, **29**, 138-175 (1990)

[2] Ling, C.-C., Darcy, R., 6-*S*-Hydroxyethylated 6-thiocyclodextrins: expandable host molecules, *J. Chem. Soc. Chem. Commun.*, 203-205 (1993)

[3] Gadelle, A., Defaye, J., Selective halogenation at primary positions of cyclomaltooligosaccharides and a synthesis of per-3,6-anhydro cyclomaltooligosaccharides, *Angew Chem. Int. Ed. Engl.*, **30**, 78-80, (1991)

[4] Connors, K.A., *Binding Constants*, Wiley, New York, 1987

[5] Catena, G.C., Bright, F.V., Thermodynamic study on the effects of β-cyclodextrin inclusion with anilinonaphthalenesulfonates, *Anal. Chem.* *61*, 905-909 (1989)

[6] Eliseev, A.V., Schneider, H.-J., Molecular recognition of nucleotides, nucleosides and sugars by aminocyclodextrins, *J. Am. Chem. Soc.*, **116**, 6081-6088 (1994)

[7] Kosower, E.M., Kanety, H., Intramolecular donor-acceptor systems. 10. Multiple fluorescences from 8-(phenylamino)-1-naphthalene sulfonates, *J. Am. Chem. Soc.*, **105**, 6236-6243 (1983)

SYNTHESIS AND PROPERTIES OF A NEW FAMILY OF CYCLODEXTRIN ANALOGUES

S.A. NEPOGODIEV, G. GATTUSO, AND J. F. STODDART
School of Chemistry, University of Birmingham
Edgbaston, Birmingham B15 2TT, UK

ABSTRACT

The chemical synthesis of a series of cyclic oligosaccharides built up from (1→4)-linked alternating D- and L-pyranosidic units is described for the first time. Key intermediates employed were disaccharides representing minimal repeating units. These disaccharides ('monomers') have been prepared in specifically modified forms so that they bear both 'glycosyl donor' (cyanoethylidene group) and 'glycosyl acceptor' (trityloxy group) functions. Polycondensation-cyclisation of these disaccharide monomers, catalysed by $TrClO_4$ under normal conditions of dilution, has led to series of homologous cyclic oligosaccharides with an even number of sugar residues (6, 8, 10, 12, *etc.*) in each case. Cyclic hexa- and octa-saccharides, based on L-rhamnose and D-mannose as the alternating monosaccharides units, have been deprotected to produce analogues of α- and γ-cyclodextrins (CDs) and the X-ray crystal structure of the cyclic octasaccharide has been determined.

1. INTRODUCTION

A large number of chemical modifications have been carried out on the native CDs.[1, 2] With few exceptions, these modifications do not alter the constitution or the configuration of the repeating α-D-glucopyranose residues in the CDs leaving their gross molecular shape essentially the same. Another entry into cyclic oligosaccharides is by total chemical synthesis. Obviously, cyclodextrin analogues, which differ more substantially from the parent CDs, can be obtained by this route. [3] However, taking into account the usual difficulties associated with oligosaccharide synthesis and the problem of efficient macrocyclisation which arises in addition, the chemical synthesis of cyclic oligosaccharides remains something of a challenge. Here, we report a novel cyclo-oligomerisation process which has been applied to the construction of α-(1→4)-linked cyclic oligosaccharides containing either mannose or rhamnose residues or a combination of them so that adjacent units have the opposite D- and L-configurations.

2. RESULTS AND DISCUSSION

2.1. Synthetic Strategy

The chemical synthesis of cyclic oligosaccharides implies the intramolecular glycosylation of linear oligosaccarides incorporating both glycosyl donor and acceptor functions. Provided that the target cyclic molecule is symmetrical, a laborious construction of a long-chain linear precursor can be overcome by using oligomerisation, prior to cyclisation, in a one-pot synthesis (Fig. 1).

J. Szejtli and L. Szente (eds.), Proceedings of the Eighth International Symposium on Cyclodextrons, 89–94.

Figure 1. A schematic representation of the cyclo-oligomerisation approach to the chemical synthesis of cyclic oligosaccharides

There are several types of glycosyl donor which can be introduced into potential monomers in this kind of process. In fact, only 1,2-*O*-cyanoethylidene derivatives of sugars have been extensively studied in polycondesation reactions, affording polysaccharides.[4] Relying upon the efficiency of this glycosylation methodology, as well as on the appropriate preorganisation of the growing oligosaccharide chains, we have planned the syntheses of cyclic oligosaccharides incorporating saccharides with the α-*manno* configuration, starting from the 'monomers' **1–3** (Fig. 2).

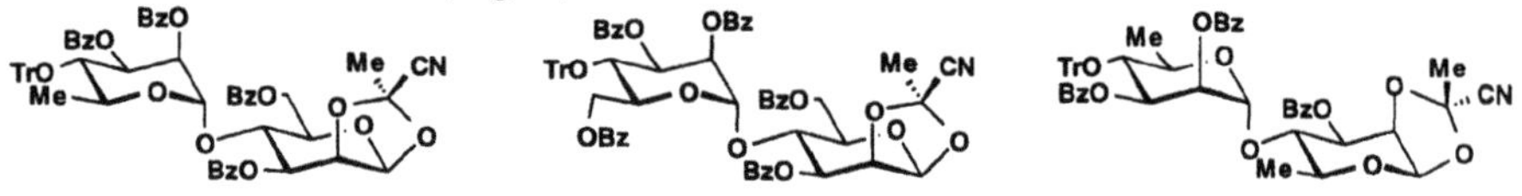

Figure 2. Structures of the disaccharide precursors of the cyclic oligosaccharides

2.2. Construction of the disaccharide precursors and their cyclo-oligomerisation

On account of the structural similarities of the target cyclic oligosaccharides, the syntheses of the disaccharide precursors **1–3** have been performed using a single methodology. The glycosyl donor parts (**6** and **9**) of the monomers **1–3** have been prepared from known compounds **4**[5] and **7**[5] by methanolysis, followed by selective benzoylation (Fig. 3).

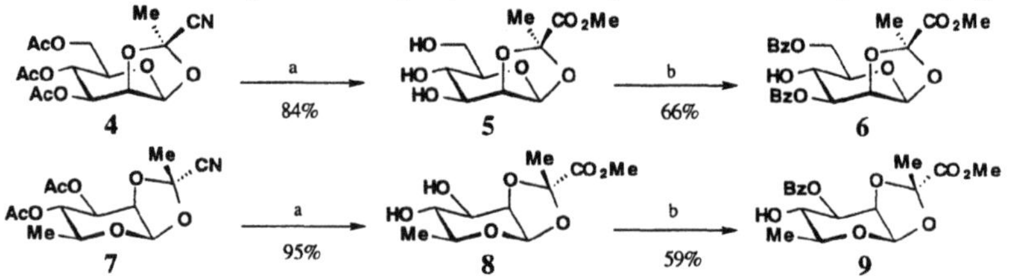

Figure 3. Reagents and conditions: a) MeONa/MeOH, 20°C, 6 h; b)BzCN/Py, 20°C, 17 h.

Selective benzoylation was also used in the first step of the preparation of the glycosyl bromides **12** and **16** (Fig. 4) from commercial L-rhamnose **10** or L-mannose **14**. It gave intermediate 4-hydroxy derivatives which were chloroacetylated and converted into compounds **12** and **16** by a standard bromination procedure. The D-rhamnose[6] derivative **18** has been converted into the bromide **20** using these same two reactions, as well as some additional manipulations with protecting groups at OH-2 and OH-3 as shown in Fig. 4. Coupling of bromides **12**, **16**, and **20**, with alcohols **6** or **9** was accomplished successfully (Fig. 4) using an $AgOSO_2CF_3$–promoted condensation to give the fully-protected disaccharides **13**, **17**, and **21**. These compounds were dechloroacetylated, and successively tritylated by the action of Ph_3CClO_4/collidine to afford the derivatives **22**, **24**, and **26**. The final conversion of methoxycarbonyl groups in **22**, **24**, and **26** to cyano-groups was performed in a two-step procedure involving ammonolysis and dehydration by the action of BzCl/Py (Fig. 5) to afford the desired 'monomers' **1–3**. The structures of these disaccharide derivatives have been confirmed by the presence of characteristic signals for both the cyanoethylidene and trityl groups in their ^{13}C NMR spectra.

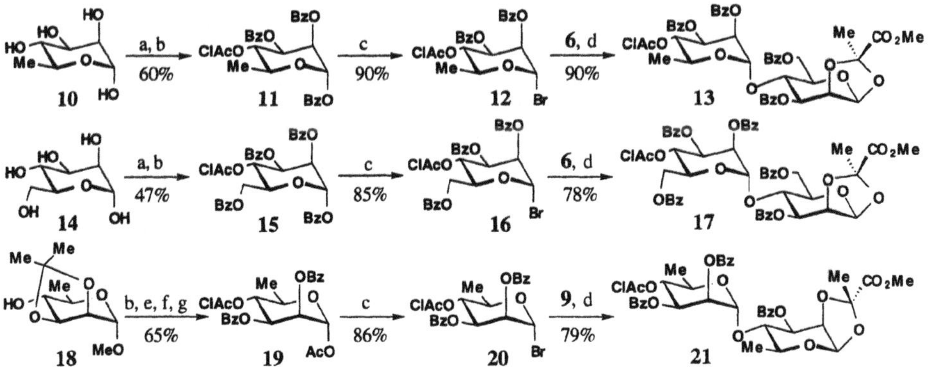

Figure 4. Reagents and conditions: a) BzCl/Py, -30°C, 3 h; b) $ClCH_2COCl/Py/CH_2Cl_2$, 0°C, 1 h; c) $HBr/AcOH/CH_2Cl_2$, 20°C, 8 h; d) $AgOTf/collidine/CH_2Cl_2$, -30–0 °C, 1-3 h; e) $CF_3COOH/CH_2Cl_2/H_2O$, 20 °C, 3 h; f) BzCl/Py, 20°C, 17 h; g) Ac_2O/H_2SO_4, 20°C, 3 h.

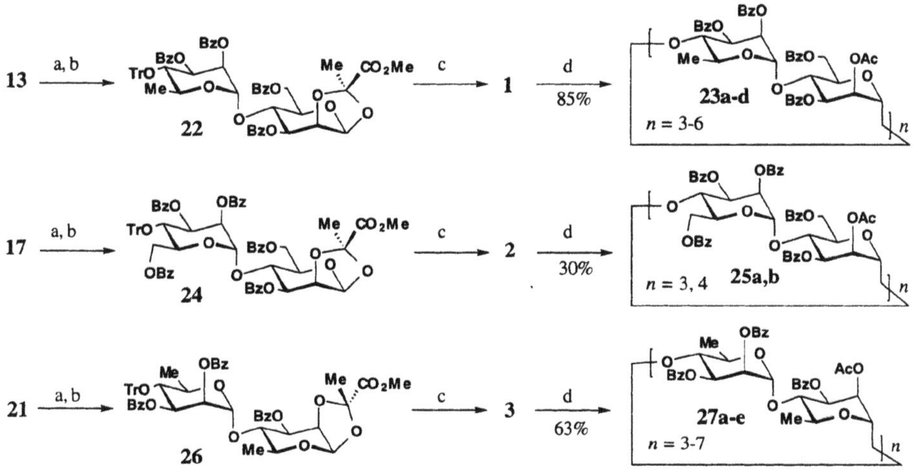

Figure 5. Reagents and conditions: $(NH_2)_2CS/MeCN/H_2O$, 20°C, 20 h; b) $TrClO_4$/collidine/ CH_2Cl_2, 20°C, 5 h, c) $NH_3/MeOH$, 20°C, 17 h, then BzCl/Py, 20°C, 18 h; d) $TrClO_4/CH_2Cl_2$, 20°C, 40 h, concentration of the disaccharide substrate and $TrClO_4$ = 0.01M.

The crucial cyclo-oligomerisation reactions of the disaccharide monomers **1-3** have been carried out under identical conditions (Fig. 5, d) involving activation of the glycosyl donor (1,2-*O*-cyanoethylidene group) by $TrClO_4$ using moderately dilution conditions. As the carbocation intermediates generated in these reactions are extremely sensitive to moisture, it was important to use ultra-dry conditions to avoid chain breaking and the formation of linear oligomers. The overall yields of cyclic products varied from 85 down to 30%, depending on the structure of the disaccharide monomers. The ratio between homologous cyclic compounds was also different. Thus, the major products in the case of the cyclo-oligomerisation of **1**, were the cyclohexa- and cycloocta-saccharides **23a** and **23b**, separated in 34 and 31% yields, respectively, whereas **3** produced the analogous compounds **27a** and **27b** in only 14 and 17% yields, respectively, and starting from **2**, less than 10% of each of **25a** and **25b**.

The symmetrical structures of the newly formed cyclic oligosaccharides are obvious from both their ^{1}H and ^{13}C NMR spectra which, according to the C_n symmetry of the compounds, showed only *one* set of signals, corresponding to the disaccharide repeating units. The reactions proceeded stereospecifically and, as a consequence, in the anomeric region of the ^{13}C NMR spectra, there are only *two* signals, which can be assigned to the C-1 of α-rhamnosyl and the C-1 of α-mannosyl residues. The precise determination of the ring size of the synthetic cyclic oligosaccharides was accomplished by LSIMS and MALDI-TOFMS (Table 1). The combination of mass-spectrometry and NMR spectroscopy seems to provide comprehensive evidence for the cyclic structure of products obtained by the cyclo-oligomerisation of **1-3**.

TABLE 1. Mass spectrometric data (*m/z* values) for acylated synthetic cyclic oligosaccharides [a]

Compound	a, *n = 3*	b, *n = 4*	c, *n = 5*	d, *n = 6*	e, *n = 7*
23	2321	3087	3852	4619	
25	2683	3569			
27	1965	2613	3260	3904	4563

[a] All data correspond to [M+Na]$^+$ ions.

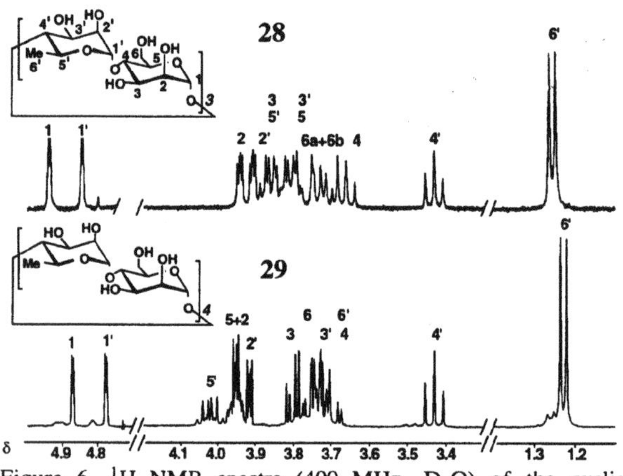

Figure 6. ^{1}H NMR spectra (400 MHz, D_2O) of the cyclic oligosaccharides **28** and **29**.

Deprotection of the cyclo-oligomerisation products can be achieved by saponification with NaOMe in $MeOH/CH_2Cl_2$, followed by treatment with NaOH in aqueous MeOH. The structures of the 'free' cyclic oligosaccharides **28** and **29** have been analysed by ^{1}H NMR (Fig. 6) and ^{13}C NMR spectroscopies and confirmed by MALDI-TOFMS data which reveals peaks with *m/z* values of 947 and 1255 corresponding to [M+Na]$^+$ ions arising from the cyclohexa- and cycloocta-saccharides. The latter compound crystallised from aqueous solution as needles that were suitable for X-ray crystallography.

2.3 Crystal structure of cyclo{→4-[α-L-Rha*p*-(1→4)-α-D-Man*p*]$_4$-(1→}

The X-ray crystal structure analysis of the **29** reveals (Fig. 7a) the presence of two crystallographically independent C_4 symmetric molecules in the asymmetric unit. The conformational difference between these molecules is very small and both of them contain the alternating 1,4-linked α-L-rhamnopyranosyl and α-D-mannopyranosyl residues, adopting normal $^1C_4$ and $^4C_1$ chair conformations, respectively. These individual pyranose rings are disposed almost orthogonally towards the plane of the cyclic oligosaccharide. The intra-annular diameter of **29**, measured as a distance between the diametrically opposite glycosidic oxygen atoms is about 11.2 Å. The highly symmetical conformation observed for **29** is in sharp contrast with the much more distorted geometries observed for the closely-related hydrated γ-cyclodextrin.[7] Due to the different natures of the monosaccharide components, both rims of **29** differ significantly in their structures from the cyclodextrin rims. Whereas the cyclodextrins are characterised by the presence of 'primary' and 'secondary' faces with an intrinsic intramolecular hydrogen bonding network, compound **29** is almost identical on both faces (exluding C-6), and no intramolecular hydrogen bonding has been detected. In contrast, intermolecular hydrogen bonds play an important role in determining the crystal lattice. Both crystallographically independent molecules form discrete stacks (Fig. 7b) that extend *via* a lattice translation (7.92 Å repeat) in the *c* direction. The molecules within each stack are perfectly in register with each other and form large open channels. Stacks of molecules exist in a cubic close-packed array with adjacent stacks which are cross-linked *via* pair of [O–H...O] hydrogen bonds. The space within the intra- and inter-stack cavities, associated with the close-packed arrangement of stacked molecules, is filled by many H_2O molecules.

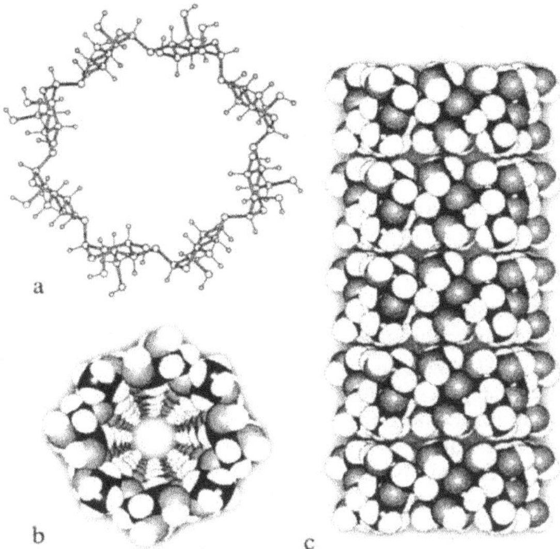

Figure 7. (a) Ball-and-stick representations of the solid state structure of **29** in a plan view; discrete stacks of **29** in a space-filling representation – (b) a view looking down one of the stacks, and (c) a side-on view of the stack.

2.4. Molecular modelling

Since we had solid state structural information available for **29**, we decided to carry out some molecular simulations, based on the semi-empirical method AM1, with the objective of comparing theoretical and experimental data for this cyclic oligosaccharide. The data for the calculated structure of **29** and solid state structure discussed above demonstrate the relatively close correspondence between most of their geometrical and physical characteristics. This high degree of correlation allows us to use computational methods to predict the structures of other cyclic hexa- and octa-saccharides, as well as one decasaccharide, listed in Table 2. As the saccharides in these three series differ only by the substituents at C-6, it is not surprising that cyclic oligosaccharides of the same size showed very similar conformational features. Thus, intra-annular diameters of cyclic hexa- and octa-saccharides, which have cylindrical shapes, were 8.3 Å and 11.4 Å, respectively. The space-filling molecular models of cyclic DL rhamno-oligosaccharides are shown in Fig. 8.

TABLE 2. Synthetic cyclic oligosaccharides used for semi-empirical simulations.

cyclo{→4-[α-L-Rha*p*-(1→4)-α-D-Man*p*]$_3$-(1→}	(**28**)
cyclo{→4-[α-L-Rha*p*-(1→4)-α-D-Man*p*]$_4$-(1→}	(**29**)
cyclo{→4-[α-L-Man*p*-(1→4)-α-D-Man*p*]$_3$-(1→}	(**30**)
cyclo{→4-[α-L-Man*p*-(1→4)-α-D-Man*p*]$_4$-(1→}	(**31**)
cyclo{→4-[α-D-Rha*p*-(1→4)-α-L-Rha*p*]$_3$-(1→}	(**32**)
cyclo{→4-[α-D-Rha*p*-(1→4)-α-L-Rha*p*]$_4$-(1→}	(**33**)
cyclo{→4-[α-D-Rha*p*-(1→4)-α-L-Rha*p*]$_5$-(1→}	(**34**)

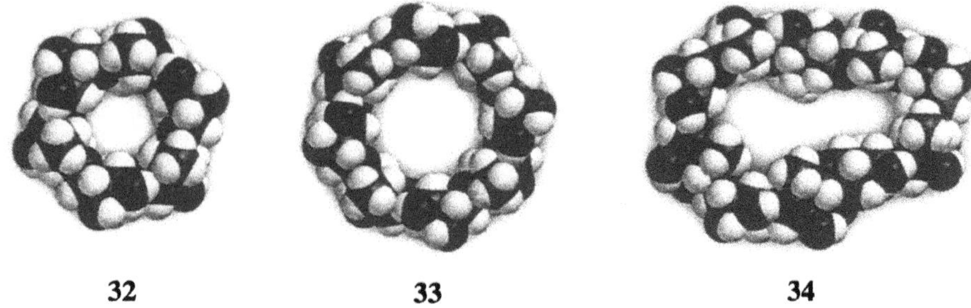

Figure 8. Space-filling representations of **32** (n = 3), **33** (n = 4), and **34** (n = 5).

3. CONCLUSION

A number of homologous cyclic oligosacharides with even numbers of sugar units in their rings have been synthesised by using disaccharide precursors playing the roles of 'monomers' in polycondensation–cycloglycosylation processes. The success of these syntheses results from *i)* the efficiency of the glycosylation procedure, *ii)* the similarities of the constitutions of the synthetic cyclic oligosaccharides to those of the cyclodextrins giving the advantage of the desired helical preorganisation of the linear chain prior to cyclisation, and *iii)* the α-D-*manno*-configurations of the newly-formed intersaccharide bonds which are relatively easy to construct. The yields of the cyclic oligosaccharides are not the same in the reactions of the different 'monomers', but cyclic products predominate very much in the cases of **1** and **3**. The low yield of cyclic oligosaccharides from **2** can be explained in terms of the low reactivity of the 4-*O*-trityl group in mannose (not 6-deoxymannose) structures. In the solid state, the cyclic oligosaccharide **29** forms nanotubes with a diameter of *ca.* 1 nm.

ACKNOWLEDGEMENTS

We thank Dr. D. J. Williams and Dr. S. Menzer for determining of the X-ray crystal structure, Dr. C. L. Brown for carrying out the molecular modelling study, and Mr. P. R. Ashton for recording the mass spectra. This research was supported in the UK by the Royal Society, the Biotechnology and Biological Sciences Research Council, and the Engineering and Physical Sciences Research Council.

REFERENCES

[1] Wenz, G., Cyclodextrins as building blocks for supramolecular structures and functional units, *Angew. Chem. Int. Ed. Engl.*, **33**, 803-822 (1994)

[2] Croft, A. P., Bartch, R. A., Synthesis of chemically modified cyclodextrins, *Tetrahedron*, **39**, 1417-1474 (1983)

[3] Kuyama, H., Nukada, T., Ito, Y., Nakahara, Y., Ogawa, T., Cyclo-glycosylation of a (1→4)-linked glycooctaose and glycodecaose: Synthesis of cyclo-*lacto*octaose and cyclo-*lacto*decaose *Carbohydr. Res.*, **268**, C1-C6 (1995)

[4] Kochetkov, N. K., Synthesis of polysaccharides with a regular structure, *Tetrahedron*, **43**, 2389-2436 (1987)

[5] Betaneli, V. I., Ovchinnikov, M. V., Backinowsky, L. V., Kochetkov, N. K., Synthesis of 1,2-*O*-cyanoalkylidene derivatives of carbohydrates, *Izv. Akad. Nauk USSR., Ser. Khim.*, 2751-2758 (1978)

[6] Tsvetkov, Yu. E., Backinowsky, L. V., Kochetkov, N. K., Synthesis of a common antigen of *Pseudomonas aerugenosa* as the 6-aminohexyl glycoside, *Carbohydr. Res.*, **193**, 75-90 (1989)

[7] Harata, K., The structure of the cyclodextrin complex. XX. Crystal structure of uncomplexed hydrated γ-cyclodextrin, *Bull. Soc. Chem. Jpn.*, **60**, 2763-2767 (1987)

NEW TYPE OF BRIDGED MONOAMINO-β-CYCLODEXTRINS

J. KOVÁCS[1], Fl. SALLAS[2], I. PINTÉR[1], A. MARSURA[2] and L. JICSINSZKY[3]
[1]*Central Research Institute for Chemistry, H.A.S., H-1525 Budapest, P.O.B. 17, Hungary*
[2]*G.E.V.S.M. EA N° 1123, Université Henri Poincaré, Nancy-1, 5 rue A. Lebrun, B.P. 403, F-54000 Nancy, France*
[3]*Cyclolab, Cyclodextrin Res. and Dev. Laboratory Ltd., H-1525 Budapest, P.O.B. 435, Hungary*

ABSTRACT

The synthesis of new urea-bridged ß-cyclodextrin dimers (CDs) has been successfully achieved by a one-pot transformation of 6-monoazido-6-monodeoxy-ß-CDs (**1** and **2**) *via* the phosphinimines. Pseudo-first-order rate constants for hydrolysis of bis-(*p*-nitrophenyl)-phosphate (BNPP) by metal-complexes of 7 have been measured.

1. INTRODUCTION

Although a large number of synthetic compounds have been designed as biomimetic host molecules [1], only few of them reproduce characteristics of enzymatic action.

In the design of artificial enzymes cyclodextrin hosts are highly available compounds and have interesting properties. Hydrophobic binding of non-polar substrates in water can be achieved with hydrophobic cavities of natural or modified cyclodextrins (CDs). Among numerous published cyclodextrin conjugates [2] cyclodextrin dimers have a special status and notably can bind appropriate substrates very strongly [3].

Another important feature of CD dimers is that the doubly-bound substrate is normally stretched along the linker. In the case of linkers containing a catalytic group, this leads to striking rate acceleration. A representative example was recently reported by Breslow [4]. The dimer including bipyridine moiety is able to coordinate La^{3+} ion and hydrogen peroxide molecule to result in oxidative hydrolysis of phosphate anion or neutral phosphate triester with high rate acceleration.

Considering the increasing interest of this field and the necessity to develop new systems to improve the properties of the previously described enzyme mimics (*e.g.*, selectivity, rate acceleration, chelation) we decided to bring a new contribution with the synthesis of a full family of novel ß-cyclodextrin dimers.

J. Szejtli and L. Szente (eds.), Proceedings of the Eighth International Symposium on Cyclodextrons, 95–98.

2. MATERIALS AND METHODS

6-Monoazido-6-deoxy-ß-cyclodextrin (**1**) was prepared according to the literature [5, 6] but with only a slight excess of sodium azide (1.2 mol). Acetylation of **1** with pyridine/ acetic anhydride at 80°C afforded the peracetylated **2** as white powder (m.p. 155-157°C) in 91% yield. Phenanthroline-2,9-bis-(2-aminoethyl)-carboxamide (**5**) was obtained as yellow powder from phenantroline-2,9-dicarboxylic acid dimethyl ester with ethylene diamine in 90% yield. Hydrocortison, Ketoconazole and Itraconazole were isolated from commercially available drug formulations by extraction with organic solvents.
Solubility isotherms were measured at 25 ± 1°C, with 48 h equilibration in a Millipore reverse osmotic purified water. Stability constants were determined for 1 : 1 complex composition by linear curve fitting to the solubility isotherms (Table 1). Kinetic measurements were carried out in HEPES buffer (pH 7.06) at room temperature, using dimer **7** (0.2mM), $CuCl_2$, $ZnCl_2$ or $EuCl_3$ (0.2mM), H_2O_2 (48mM) and bis-(*p*-nitrophenyl)-phosphate (BNPP) (0.06mM). Concentration of *p*-nitrophenol, resulted by the hydrolysis, was determined by UV/vis spectrometer at λ_{max} = 400 nm and the rate constants were calculated (Table 2).

3. RESULTS AND DISCUSSION

Hitherto, CD dimers bridged with carbon-sulfur single bonds were obtained by direct condensation of dithiols with 6-monoiodo-6-mono-deoxy-ß-cyclodextrin [7].
We report here the synthesis and characterization of four original symmetrical CD dimers with urea linkage or with a spacer which contains diamino and urea moieties.
For the synthesis we used the sugar phosphinimine reaction, *i. e.*, conversion of azido sugars with triphenylphosphine and CO_2. From protected azidosugars the reaction leads to the formation of carbodiimides [8] in which two sugar units are linked with carbodiimide bridge. In contrast, from unprotected sugar azides cyclic carbamates [9] were obtained. Both syntheses were carried out either in two steps or under one-pot conditions. Recently, we have found that similar transformation of azidosugars, in the presence of amines, furnishes urea derivatives of the aminosugars corresponding to the starting azides [10].
Application of the reaction to 6-monoazido-6-monodeoxy-ß-cyclodextrin (**1**) gave readily the dimer **3** in which the two CD units are bound with urea linkage (Scheme 1). Similarly, the peracetyl compound **2** afforded the acetylated urea-linked dimer (**4**) and not the corresponding carbodiimide. Formation of **4** can be attributed, very probably, to the presence of water molecules complexed by ß-CD. Dimer **4** was also obtained by acetylation of **3** in pyridine/acetic anhydride at 80°C providing evidence for the structure of both molecules (Scheme 1).
Under similar conditions, but in the presence of phenantroline-2,9-bis-(2-aminoethyl)-carboxamide (**5**) **2** gave a new dimer (**6**) in which the two ß-CD units are linked symmetrically to the phenantroline-bis-carboxamide moiety with urea units. Deacetylation of **6** by the Zemplén method gave the hydroxy derivative **7** (Scheme 2).
The structures of the new dimers were corroborated by NMR showing the characteristic signals of the CD, phenantrolyl and urea moieties. Data of FABMS analyses matched

also the proposed structures.

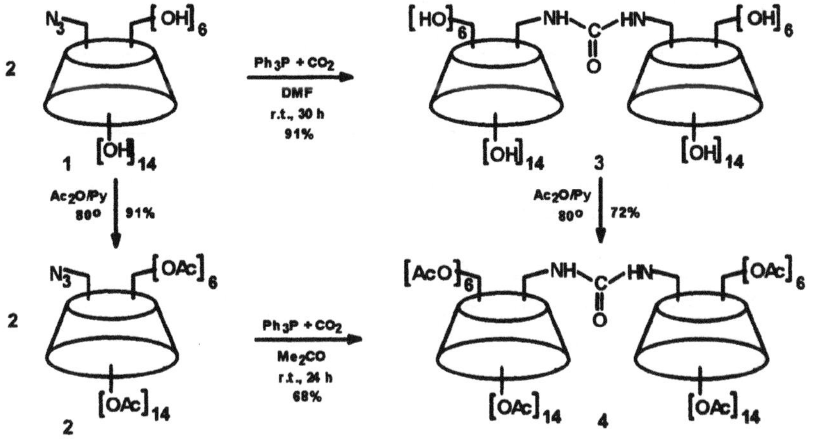

Scheme 1

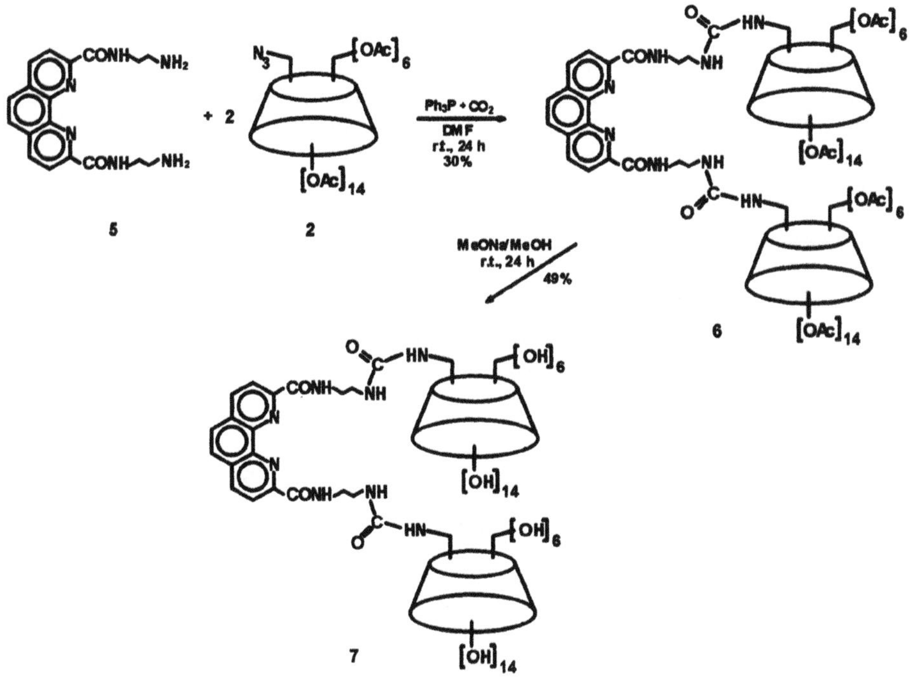

Scheme 2

Stability constants of dimers **3** and **7** were determined and compared with those of ß-CD (Table 1).

Table 1. Stability constants of urea-bridged ß-CDs and ß-CD with various drugs

CD-compound	K [$dm^3.mole^{-1}$]		
	Ketoconazole	Itraconazole	Hydrocortisone
3	131	101	2481
7	55	28	
ß-CD	939	250	6111

Complexes of dimer **7** with cations were formed and the rate constants of BNPP hydrolysis were measured under pseudo-first-order conditions (Table 2).

Table 2. Rate constants of BNPP hydrolysis by dimer 7

Compounds	Time(s)	$k_1(s^{-1})$	Time(s)	$k_2(s^{-1})$
7-Cu^{2+}	0<t<180	$9.55x10^{-4}$	$180<t<7.2.10^3$	$1.60x10^{-5}$
Cu^{2+}	$0<t<10.8x10^3$	$8.62x10^{-6}$	$64.8x10^3<t<97.2x10^3$	$5.69x10^{-8}$
7-Eu^{3+}	0<t<240	$3.10x10^{-4}$	$300<t<90x10^3$	$2.10x10^{-6}$
Eu^{3+}	$0<t<70.2x10^3$	$6.45x10^{-7}$	$75.6x10^3<t<163.8x10^3$	$2.39x10^{-7}$
7-Zn^{2+}	$0<t<7.2x10^{-3}$	$8.03x10^{-6}$	$9.0x10^3<t< 14.4x10^3$	$1.80x10^{-7}$
Zn^{2+}	no hydrolysis after 72 h			

4. CONCLUSION

Structures of the new CD dimers allow to conclude positively for the scope of this methodology in the synthesis of such derivatives. The calculated complex constants for 1:1 complex compositions are in a similar range to those of ß-CD. Complexation of Hydrocortison by the dimer **7,** however, is more complicated. The low solubility of the complexes are the limits of their application to steroids and conazole-like drugs.

5. ACKNOWLEDGEMENTS

The National Fund for Scientific Research (Hungary, OTKA T014458 and T014939) and the ARC association are acknowledged for financial support and Roquette S.A. (Lestrem France) for providing ß-cyclodextrin.

6. REFERENCES

[1] Bender, M.L., Komiyama, M., in *Cyclodextrin Chemistry,* Springer Verlag, Berlin, 1978; Gutsche, C.D., *Calixarenes,* in *Monographs in Supramolecular Chemistry,* (Ed. Stoddart, J.F.) The Royal Society of Chemistry, Cambridge, 1989; Gokel, G.W., in *Crown Ethers and Cryptands,* The Royal Society of Chemistry, Cambridge, 1991
[2] Breslow, R., *Acc. Chem. Res.*, **28**, 146-153 (1995)
[3] Breslow, R., *Rec. Trav. Chim. Pays-Bas*, **13**, 493-498 (1994)
[4] Breslow, R., Zhang, B., *J. Am. Chem. Soc.*, **116**, 7893-7894 (1994)
[5] Tsujihara, K., Kurita, H., Kawazu, M., *Bull. Chem. Soc. Jpn.*, **50**, 1567-1571 (1977)
[6] Jicsinszky, L., *J. Inclusion Phenom. Mol. Recogn. Chem.*, **18**, 247-254 (1994)
[7] Breslow, R., Halfon, S., Zhang, B., *Tetrahedron,* **51**, 377-388 (1995)
[8] Messmer, A., Pintér, I., Szegő, F., *Angew. Chem.*, **76**, 227-228 (1964); Kovács, J., Pintér, I., Messmer, A., Tóth, G., Duddeck, H., *Carbohydr. Res.*, **166**, 101-111 (1987)
[9] Kovács, J., Pintér, I., Messmer, A., Tóth, G., *Carbohydr. Res.*, **141**, 57-65 (1985)
[10] Pintér, I., Kovács, J., Tóth, G., *Carbohydr. Res.*, **273**, 99-108 (1995)

A NEW SYNTHETIC STRATEGY OF CYCLOOLIGOSACCHARIDES.

CYCLODEXTRIN-DERIVED CYCLOALTRINS MADE UP FROM α (1→4)-LINKED ALTROPYRANOSES

Y. NOGAMI,[1] K. FUJITA,[2] K. OHTA,[2] K. NASU,[1] H. SHIMADA,[2]
C. SHINOHARA,[1] and T. KOGA[1]

[1] *Daiichi College of Pharmaceutical Sciences, Fukuoka 815, Japan*
[2] *Faculty of Pharmaceutical Sciences, Nagasaki University, Nagasaki 852, Japan*

ABSTRACT

An effective synthetic strategy for preparing a new type of cyclooligosaccharide is proposed and along this plan, α-, β-, and γ-cycloaltrins, made up from six to eight α (1→4)-linked D-altropyranoses, have been prepared in 36, 52, and 37% overall yields from the corresponding cyclodextrins.

1. INTRODUCTION

For the syntheses of the novel cyclooligosaccharides uniformly composed of sugars other than glucose, there are two typical strategies. One is enzymatic cyclization reaction of naturally occurring open chain oligosaccharides and another is chemical cyclization of oligosaccharides which were elongated stepwise synthetically.
According to the former strategy, cyclofructans *etc.* have been prepared.[1] But, there are intrinsic disadvantages of the usual enzymatic reaction. Namely, specificity of the enzyme lowers the general applicability of this method and furthermore, enzymatic cyclization often lacks control of the reaction concerning the ring size of the products making various homologues which are difficult to isolate.
According to the latter strategy, α- , β- and γ- cyclomannins *etc.* are synthesized.[2] But, in these cases, there are difficulties of multi-step preparation. Furthermore, isolation of the pure oligomer in each steps requires a considerable endeavor.
To overcome these difficulties in the syntheses, we evolved a new synthetic strategy of cyclooligosaccharides using cyclodextrins as starting materials. This route might have a high possibility of the shortest approach to the aimed host molecules, because CDs are easily available in large amount and are free from the fear of contamination of different sized analogues.
Along this project, an approach to the cycloaltrins, which were made up from α-D-altropyranoses, was intended to go *via* per-2β,3-epoxy-cyclodextrins (**4a-c**)[3] as shown in Scheme 1.

J. Szejtli and L. Szente (eds.), Proceedings of the Eighth International Symposium on Cyclodextrons, 99–102.

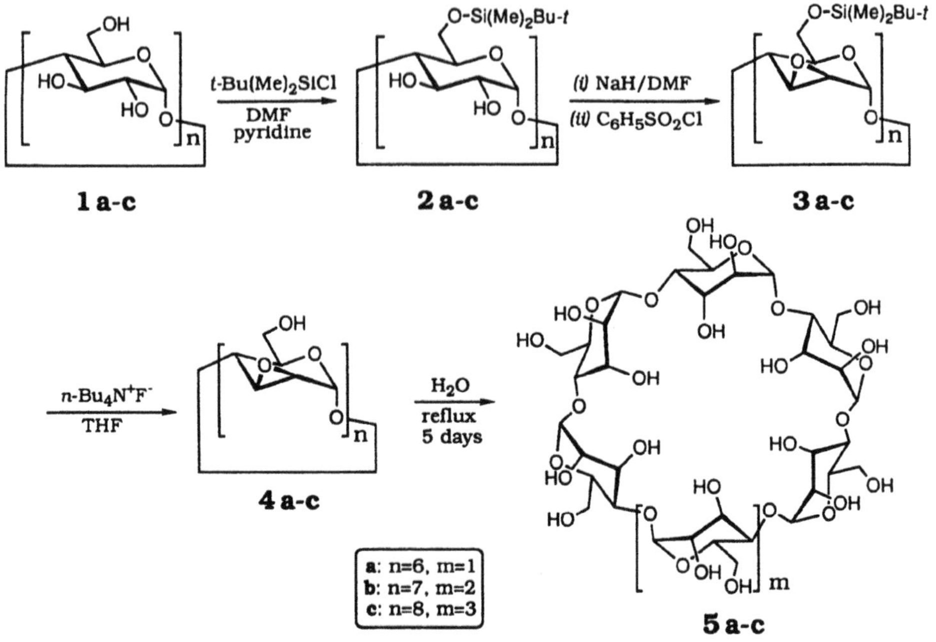

Scheme 1

2. MATERIAL AND METHODS

Compounds **2a-c** were obtained from corresponding cyclodextrins according to the reported methods[4] in 90%, 82%, and 85% yields, respectively.
Compound **2** was treated with NaH in dry dimethylformamide and the evolution of hydrogen gas ceased in 2 h (for **2a**) , 4 h (for **2b**) and 5.5 h (for **2c**) at 60°C. To the resulting oxyanion solution, benzenesulfonyl chloride was injected. The reaction was completed almost in 30 min at room temperature. After filtration, water was added to precipitate **3**. Compounds **3a-c** were characterized by elemental analyses and ^{1}H and ^{13}C NMR spectra.
Compound **3** was desilylated by treatment with tetrabutylammonium fluoride in tetrahydrofuran solution at 40°C for 4 h. Flash column chromatography of the crude product gave a pure sample of **4**. The yield of this step was almost quantitative. These compounds were characterized by FAB mass spectra and ^{1}H, ^{13}C NMR, C-H COSY, and H-H COSY spectra. Thus, per-2,3-anhydro-(2*S*)-α-, β-, or γ-cyclodextrin (**4a-c**) was obtained from **2a-c** in two steps in 59, 87, or 63% yields, respectively.
A solution of compound **4a-c** in distilled water was refluxed and the progress of the reaction was monitored by thin-layer chromatography. After five days, the starting material disappeared in favor of only one major product, which had distinctly different mobility from that of **4**. Reversed-phase column chromatography gave the novel cyclooligosaccharide, cycloaltrins **5a-c** in 68, 73 and 70 % yields, respectively.

3. RESULTS AND DISCUSSION

There are several methods[5] in the literature for converting glucoside units into altroside units. But, the yields of the conversion reported in these papers are too poor to employ, since, for our present purpose, six to eight glucoside units coexisting in one molecule must be converted by a single trial. We have already reported a general and convenient method for preparing altropyranoside from 2,3-manno-epoxide.[6] The transformation of this reaction is almost quantitative. By this method, an effective and useful synthesis of α-, β- and γ-cycloaltrins was achieved for the first time. (Overall yields of **5a-c** were 36, 52, and 37% from the corresponding CDs **1a-c**).

The FAB mass spectra showed **5** to have the same molecular weight as **1**; however, the NMR data are clearly different (Figure 1). The assignment of the signals of **5** relied on ^{1}H, ^{13}C, ^{1}H-^{13}C, and ^{1}H-^{1}H COSY (Figure 2) NMR spectra and indicates that **5** consists of altropyranoses. The major product of acid hydrolysis of **5** was a mixture of altrose and altrosan (1,6-anhydroaltropyranose), which is produced by acid treatment of altrose.

The coupling constants (e.g. $J_{H-1,H-2} = 4.5$Hz) indicate that **5b** is a mixture of at least two rapid interconverting conformations, $^{1}C_{4}$ and $^{4}C_{1}$ chair conformations, since the vicinal coupling constant $J_{H-1,H-2}$ is intermediate between that of α-methylaltropyranoside (2.0 Hz; $^{4}C_{1}$ chair conformation) and that of the α-altropyranose unit in $2^{A}(S),3^{A}(R)$-β-cyclodextrin (6.6 Hz;[6] mainly $^{1}C_{4}$ chair conformation). Alternatively, **5** assumes a single fixed conformation, for example, in which the altropyranose unit have a twist conformation.

Further studies on the conformation, the inclusion properties, and the chemical modification of **5a-c** are currently in progress.

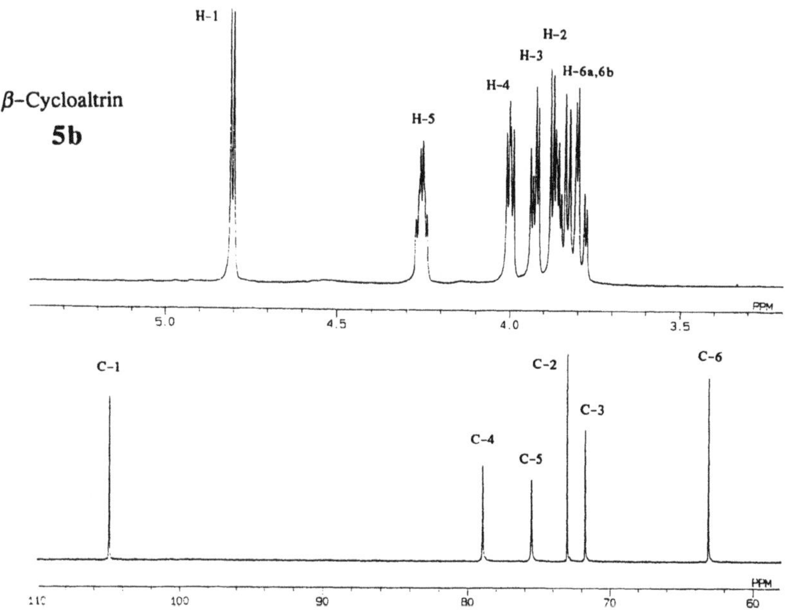

Figure 1. ^{1}H (500 MHz) and ^{13}H (125 MHz) NMR spectra of β-cycloaltrin **5b** in D_2O at 40°C

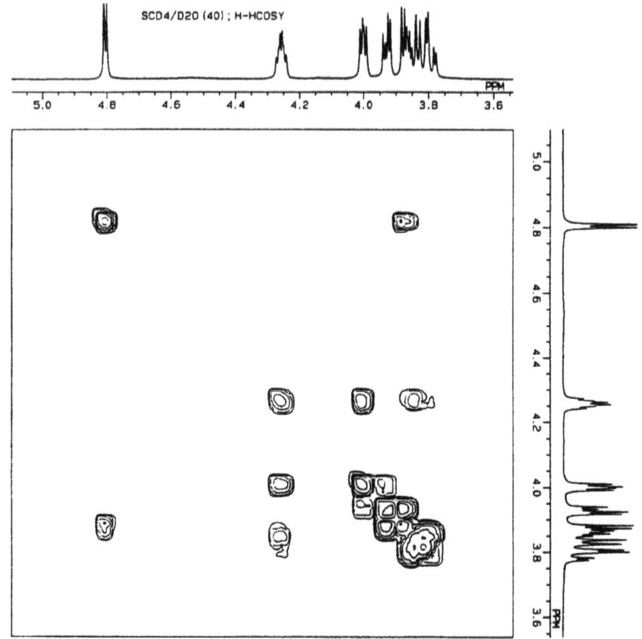

Figure 2. H-H COSY NMR spectrum of β-cycloaltrin **5b** (500 MHz, in D_2O at 40°C)

REFERENCES

[1] Sawada, M., Tanaka, T., Takai, Y., Hanafusa, T., Hirotsu, K., Higuchi, T., Kawamura, M., Uchiyama, T., Crystal Structure of Cycloinulinohexaose, *Chem. Lett.*, 2011-2014 (1990)

[2] (a) Mori, M., Ito, Y., Uzawa, J., Ogawa, T., A Highly Efficient and Stereoselective Cycloglycosylation, *Tetrahedron Lett.*, **30**, 1273-1281 (1989). (b) Nishizawa, M., Imagawa, H., Morikuni, E., Hatakeyama, S., Yamada, H., Synthesis of Cyclorhamunopentaose, *Chem. Pharm. Bull.*, **42**, 1365-1366 (1994). (c) Houdier, S., Vettero, P.J.A., Synthesis of Benzylated Cycloisomaltohexaoside, *Angew. Chem. Int. Ed. Engl.*, **33**, 354-356 (1994). (d) Collins, P.M., Ali, M.H., A New Cycloglucohexaose Derivative, *Tetrahedron Lett.*, **31**, 4517-4720 (1990)

[3] (a) Khan, A.R., Barton, L., D'Souza, V.T., Heptakis-2,3-epoxy-β-cyclodextrin, A Key Intermediate in the Synthesis of Custom Designed Cyclodextrins, *J. Chem. Soc., Chem. Commun.*, 1112-1114 (1992). (b) Coleman, A.W., Zhang, P., Ling, C-C., Mahuteau, J., Parrot-Lopez, H., Miocque, M., Synthesis and Properties of Cyclo-α-1,4-manno-2,3-epoxides, *Supramolecular Chem.*, **1**, 11-14 (1992)

[4] (a) Takeo, K., Uemura, K., Mitoh, H., Cyclodextrin-based Enzyme Models, *J. Carbohydr. Chem.*, **7**, 293-308 (1988). (b) Fugedi, P., Synthesis of Heptakis(6-*O*-*tert*-butyldimethylsilyl)cyclomaltoheptaose and Octakis(6-*O*-*tert*-butyldimethylsilyl)cyclomaltooctaose, *Carbohydr. Res.*, **192**, 366-369 (1989)

[5] (a) Robertson, G.J., Griffith, C.F., Conversion of Derivatives of Glucose into Derivatives of Altrose by Simple Optical Inversion, *J. Chem. Soc.*, 1193-1201 (1935). (b) Pedersen, C., Reaction of Sugar Esters with Hydrogen Fluoride, *Acta Chem. Scand.*, **16**, 1831-1836 (1962). (c) Evans, M., Facile Syntheses of Methyl α-D-altropyranoside and Its Derivatives, *Carbohydr. Res.*, **30**, 215-217 (1973). (d) Espinosa, F.G., Maruri, J. M., Disaccharide Acetoxonium Salts, *An. Quim.*, **70**, 1088-1095 (1974)

[6] Fujita, K., Ohta, K., Ikegami, Y., Shimada, H., Tahara, T., Nogami, Y., Koga, T., Saito, K., Nakajima, T., General Method for Preparing Altrosides from 2,3-Manno-epoxides and Its Application to Syntheses of Alternative β-Cyclodextrin with an Altroside as the Constituent of Macrocycle Structure, *Tetrahedron Lett.*, 9577-9580 (1974)

SELECTIVE ONE-POINT CLEAVAGE OF CAPPED β–CYCLODEXTRIN BY TAKA-AMYLASE A. TWO-STEP SYNTHESIS OF GLYCOPHANE FROM β-CYCLODEXTRIN

K. FUJITA, M. MIZUOCHI, K. KOGA, and K. OHTA

Faculty of Pharmaceutical Sciences, Nagasaki University,

Bunkyo-machi, Nagasaki 852, Japan

ABSTRACT

A cyclic compound, glycophane, which is composed of glucoses and biphenyl was prepared in 21% yield through regiospecific cleavage of the glucose-wall of $6^A,6^D$-di-*O*-(*p,p'*-biphenyldisulfonyl)-β-cyclodextrin by Taka-amylase A.

1. INTRODUCTION

The direct and practical preparation of linear maltooligosaccharides from cyclodextrins (CDs) has been difficult. The hydrolysis of a parent CD catalysed by acid or by Taka-amylase A (TAA) gives glucose or glucose and maltose as the only isolable products. Simple modification of the CD hydroxyls can provide the enzymatic hydrolysis with some selectivity. Mono- and disubstituted CDs can be selectively transformed to the corresponding mono- or disubstituted linear maltooligosaccharides.[1,2,3] Nevertheless, this reaction still suffers from the fast decompose of some glucosidic bonds in the linear maltooligosaccharides formed by direct ring-opening of the macrocycle of CDs. As the result, the isolated products are no longer those of the one-point cleavage. We report here selective, enzymatic one-point cleavage of the macrocycle of capped CD (Scheme 1).

2. MATERIALS AND METHODS

$6^A,6^D$-Di-*O*-(*p,p'*-biphenyldisulfonyl)-β-CD **1** was prepared by the reported

J. Szejtli and L. Szente (eds.), Proceedings of the Eighth International Symposium on Cyclodextrons, 103–106.

method.[4] The hydrolysis with TAA was carried as follows: acetate buffer solution (0.2 M, pH 5.6) containing 0.02 M $CaCl_2$ was added to a solution of the capped CD **1** in a minimum amount of DMSO followed by addition of TAA at 40 °C.

3. RESULTS AND DISCUSSION

Capped β-CD **1** (200 mg) was hydrolyzed enzymatically by TAA at 40°C for 4 h to give **2** (42 mg, 21%) and a mixture of **3** and **4** (21 mg, **3**/**4** = 3.3 mol/mol) together with the recovery of **1** (50%). The FAB mass (FABMS) spectrum of **2** showed the expected ions $[M+H]^+$ and $[M+Na]^+$, indicating a structure of $6^Y,6^Z$-di-*O*-(*p,p'*-biphenyldisulfonyl)-maltoheptaose. The ^{13}C-NMR spectrum (Fig. 1) showed that **2** existed as a mixture of α- and β-anomers, demonstrating the cleavage of macrocycle.

The positions Y and Z were determined by the conversion of **2** to the acetate **6** and analysis of FABMS spectral fragmentation (Scheme 2) which suggested that the two phenylthio-groups were attached to the third and sixth glucose units respectively from the reducing end.[5] Therefore, **2** is 6",6'""-di-*O*-(*p,p'*-biphenyldisulfonyl)-maltoheptaose.

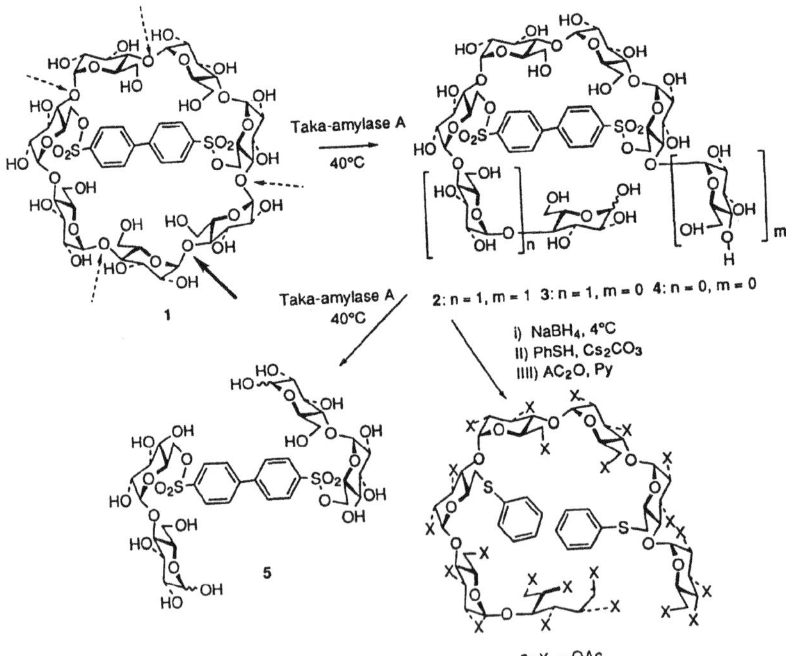

Scheme 1. Selective One-point Cleavage of $6^A,6^D$-Di-*O*-(*p,p'*-biphenyldisulfonyl)-β-CD by TAA

The assignment of 6"-phenylthio-group was confirmed by the conversion of **2** to di-*O*-(*p,p'*-biphenyldisulfonyl)-maltohexaose (**3**) by enzymatic hydrolysis with *Rhyzopus niveus* glucoamylase (an exo-splitting enzyme that consecutively removes the glucose units from the nonreducing end of starch and glycogen) in acetate buffer solution (pH 4.7) at 40 °C.

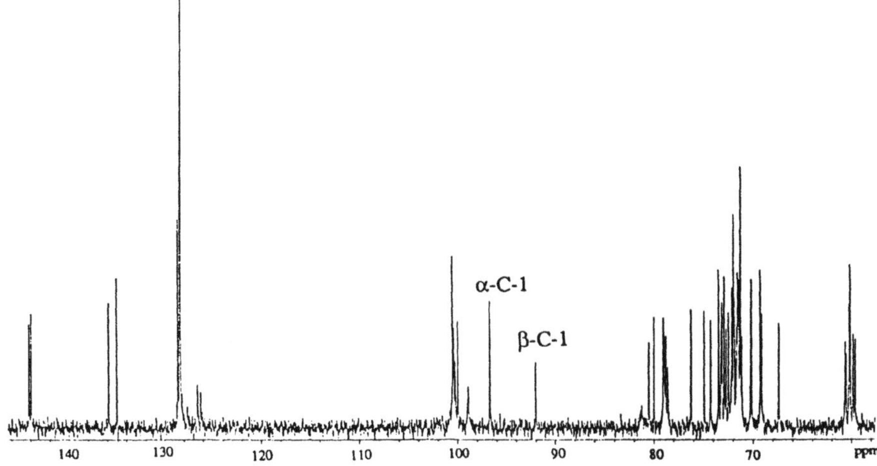

Figure 1. ^{13}C NMR Spectrum of Glycophane **2** in DMSO-d_6

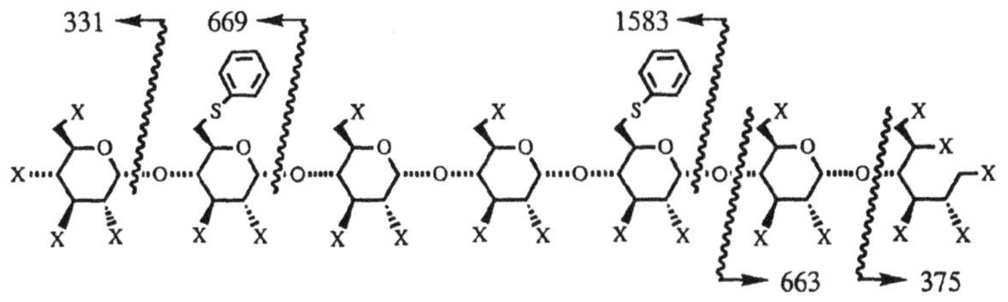

Scheme 2. FABMS Spectral Fragmentation Pattern of the Peracetate **6** (X = OAc) (The parent ion peaks, m/z 2263 $[M+H]^+$ and 2285 $[M+Na]^+$, were observed, too.)

Monitoring progress of the hydrolysis with TAA demonstrated that **2** produced **3** and **4** which were shown to be di-*O*-(*p,p'*-biphenyldisulfonyl)-maltohexaose and -pentaose by the FABMS spectra. Their structures were also assigned as shown in Scheme 1.

The prolonged reaction time (2 d) of the hydrolysis of **1** with TAA led to **5** as the final major product which can be reasoned from the previous observations in the cases of 6-*O*-tosyl-CDs and the related compounds. The enzymatic hydrolyses of 6-*O*-tosyl-α- or β-CDs, $6^A,6^D$-di-*O*-ditosyl-β-CD, and the related compounds with TAA gave the 6'-substituted maltoses as the final products,[2,3] whereas longer linear oligosaccharides were too small in amount to be isolated,[2] because their cleavages were much faster than those of the macrocyclic rings.[3] The present success in isolation of the intermediate **2** in moderate yield is resulted from the cyclic structure of **2** which retards the succeeding cleavages.

4. CONCLUSION

In compound **1**, only one given point (shown by bold arrow in Scheme 1) was cleaved by TAA mainly although other four points (dotted arrows) were potentially possible. This demonstrates that there is an obvious order of acceptability for enzymatic hydrolysis in the glucose-wall of capped cyclodextrin. Therefore, variation of the capped position, the capping group and the kind of CD can give various analogues of glycophane **2**. The specifically activated oligosaccharide **2** can be easily converted to the corresponding specifically substituted linear oligosaccharides which are difficult to be prepared by stepwise chemical method.

ACKNOWLEDGMENT*:* We thank Japan Maize Products Co. Ltd. (Nihon Shokuhin Kako) for a generous gift of β-cyclodextrin.

REFERENCES AND NOTES

[1] Fujita, K., Tahara, T., Ohta, K., Nogami, Y., Koga, T., Yamaguchi, M., A complete set of $2^A,6^X$-di- *O*-diactivated α-cyclodextrin, *J. Org. Chem.*, **60**, 3643-3647 (1995)

[2] (a) Melton, L.D., Slessor, K.N., Preparation of 6'-substituted maltoses, *Can. J. Chem*, **51**, 327-332 (1973). (b) Fujita, K., Matsunaga, A., Imoto, T., 6^A6^B-, 6^A6^C-, and 6^A6^D-disulfonates of α-cyclodextrin, *J. Am. Chem. Soc.*, 106, 5740- 5741 (1984). (c) Fujita, K., Matsunaga, A., Imoto, T., 6^A6^B, 6^A6^C, and 6^A6^D-ditosylates of β-cyclodextrin, *Tetrahedron Lett.*, **25**, 5533-5536 (1984). (d) Tabushi, I., Nabeshima, T., Fujita, K., Matsunaga, A., Imoto, T., Regiospecific A,B capping onto β-cyclodextrin. Characteristic remote substituent effect on ^{13}CNMR chemical shift and specific Taka-amylase hydrolysis, *J. Org. Chem.*, **50**, 2638-2643 (1985)

[3] Fujita, K., Tahara, T., Koga, T., Imoto, T., Enzymatic synthesis of specifically modified linear oligosaccharides from γ-cyclodextrin derivatives. Study on importance of active sites of Taka amylase A, *Bull. Chem. Soc. Jpn.*, **62**, 3150-3154 (1989)

[4] Tabushi, I., Kuroda, Y., Yokota, K., Yuan, L.C., Regiospecific A,C- and A,D-disulfonate capping of β–cyclodextrin, *J. Am. Chem. Soc.*, **103**, 711-712 (1981)

[5] This procedure for regiochemical determination is already established. For example, see ref. 2.

NEW ASSOCIATING POLYMER SYSTEMS INVOLVING WATER SOLUBLE β-CYCLODEXTRIN POLYMERS.

C. AMIEL, B. SEBILLE

Laboratoire de physico-Chimie des Biopolymères, UMR 27 CNRS

2-8 rue H. Dunant, 94320 Thiais France.

ABSTRACT

Original associating systems have been obtained by mixing hydrophobically end-capped polyethylene oxide and water soluble β-cyclodextrin polymers in aqueous solutions. The hydrophobic ends of the PEO polymers, naphtyl and adamantyl groups, have been chosen in order to match the β-cyclodextrin cavities. Inclusion complex formation between the PEO terminal groups and β-cyclodextrin are at the origin of polymolecular associations. Complexation constants have been determined by fluorescence methods, using a fluorescent probe 1-8 ANS as a competitor for complexation against the adamantyl groups or directly checking the fluorescence of the naphytl groups by fluorescence anisotropy measurements. The onsets of the polymolecular associations have been monitored by viscosimetry.

1. INTRODUCTION

Associating polymers have recently become the subject of extensive research with the use of hydrophobically modified polymers [1]. These amphipathic macromolecules are generally obtained from water soluble polymers modified with relatively low amounts of hydrophobic comonomer (1-5 mol%). They exhibit unusual aqueous solution behavior with the appearance of thickening properties above certain polymer concentrations, due to polymolecular associations. The hydrophobic interactions that occur in order to minimize water hydrophobe contacts are at the origin of the polymolecular associations. In this work, a new class of associating polymer systems is proposed. The principle is to add a water soluble β-cyclodextrin polymer to an aqueous solution of amphiphilic polymer.

The β–cyclodextrin polymers are promoting the associations between the amphiphilic polymers by forming inclusion complexes between the hydrophobic moieties and the β-cyclodextrin cavities. In this study, the water-soluble β-cyclodextrin polymers are formed by polycondensation with epichlorhydrin under strongly alkaline conditions. The amphiphilic polymers are hydrophobically end-capped polyethylene oxide where the hydrophobic ends have been chosen in order to match the β-cyclodextrin cavities. Two kinds of terminal groups have been used: naphtyl groups substituted in position 1 or 2, whose interactions can be easily studied by fluorescence methods, and adamantyl groups which possess higher affinities for the β-cyclodextrin cavities.
The complexation constants have been determined by fluorescence methods: fluorescence anisotropy measurements have been developped for naphtalene end-capped PEO while a fluorescent probe, 1-8 ANS, has been used for adamantane end-capped PEO. In the last case, the measurements involved a competitive complexation between the ANS probes and the adamantyl groups.
The onsets of the polymolecular associations have been monitored by viscosimetry, by comparing the viscosity of the system at a certain concentration to the one of a solution containing the same amount of precursor PEO chain and β–CD polymer. The influences of the nature of the end- capping group as well as the PEO chain length have been

J. Szejtli and L. Szente (eds.), Proceedings of the Eighth International Symposium on Cyclodextrons, 107–113.

studied.

2. MATERIALS AND METHODS

2.1. Materials

2.1.1. Hydrophobically end-capped PEO

Polymers of polyethylene oxide, molecular weight 6000, 20000 and 35000 have been purchased from Merck, France. PEO-N1 and PEO-AD were obtained by reaction of the OH terminal functions with isocyanate groups, 1-naphtyl isocyanate and 1-adamantyl isocyanate respectively (Aldrich, France). PEO-N2 was obtained by reaction of the OH groups with the acid chloride group of 2-chlorocarbonylnaphtyl (Aldrich, France). N-Naphtyl ethyl carbamate (model-N1) was obtained by reaction of 1-naphtyl isocyanate with ethanol. The structural formula of the modified polymers are schematically represented on figure 1.

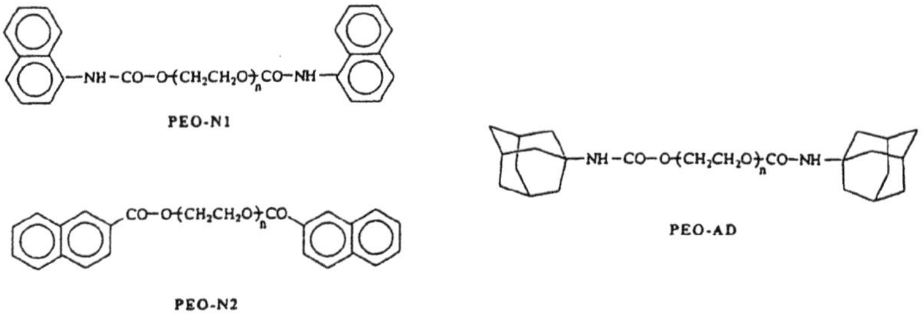

Fig.1 Structural formula of the end-capped PEO.

The extent of the end grafting reactions were determined with ^1H NMR spectroscopy and the results are reported in table 1.

TABLE 1. Characterization of the end-capped PEO

Polymer	Molecular weight	% grafting
PEO-N1	6000	98
PEO-N1	20000	96
PEO-N2	6000	76
PEO-AD	6000	97
PEO-AD	20000	84
PEO-AD	35000	75

2.1.2. β-cyclodextrin polymers (β-CD/EP)

β-cyclodextrin was a gift from Orsan company, France. The β-cyclodextrin polymers (β-CD/EP) were prepared by crosslinking β-cyclodextrin under strongly alkaline conditions with epichlorhydrin (Prolabo, France), the molar ratio being 1/10. The reaction was conducted until the immediate vicinity of the gelation point was reached, in order to obtain relatively high molecular weight of water soluble β-CD polymers. The synthesis and the characterization of these polymers have been described in details elsewhere [2].The molecular weight distribution, monitored by size exclusion chromatography, showed a characteristic pre-gelation pattern with two populations having approximately the same weight ratio: a population of oligomers, mean molecular weight around 5000 and a population of polymers , molecular weights higher than 200,000. The β-cyclodextrin content , titrated by ^1H NMR spectroscopy, was 65 weight %.

2.2. **Methods**

2.2.1. Fluorescence anisotropy

The fluorescence studies were performed on a SLM AMINCO 8000 spectrofluorimeter eqipped with a xenon lamp and with a polarization device. The fluorescence anisotropy measurements were carried on at an excitation wavelength of 285 nm and emission wavelength of 360 nm. The inclusion of the naphtyl probes inside the cavities of the β-CD/EP polymers does not induce a marked change in the spectral character of the probes: the ratio of the quantum yields of the complexed form over the free one is R=1.86, determined from fluorescence lifetime measurements [3]. However, the complexation of the probes greatly reduces their rotationnal diffusion time and consequently largely increases their fluorescence anisotropy [4]. From measuring the anisotropy of the free probe in the absence of β-CD/EP , the complexed probe with an excess of β-CD/EP and the anisotropy at an intermediate β-CD level, one can easily determine the complexation constant K_C [5]. The concentration of PEO-N was held constant at a value of 2 g/l and the concentration of β-CD/EP was varied in the range 10 - 80 g/l.

2.2.2. Competitive complexation using a fluorescent probe 1,8 ANS

Since the adamantyl groups are not fluorescent, a fluorescent probe 1-anilino 8-naphtalene sulfonic acid 1-8 ANS (purchased from Sigma, France) has been used as a competitive inhibitor [6][7]. The association constant of 1-8 ANS with β-CD/EP polymers were determined from dependencies of fluorescence intensity of ANS on the concentration of β-CD. For this, the concentration of ANS was fixed at 10^{-5} M . In order to have a medium of constant polarity, unmodified PEO has been added at a fixed concentration of 17.5 g/l .The concentration range of β-CD was 0.2 - 8 mM. The association constants of the PEO adamantyl groups with β-CD were determined by following the fluorescence intensity of the ANS - β-CD/EP solutions at fixed adamantyl concentration (1mM) as a function of the β-CD concentration . In these solutions, the total concentration in PEO was held constant at 17.5 g/l by adding the appropriate quantity of unmodified PEO.

2.2.3 Viscosimetry

The viscosity mesurements were carried out with an Ubbelohde type viscometer. The temperature was fixed at 25.0 ± 0.05 °C. Solvent and solutions were filtered prior to any mesurements.

3. RESULTS AND DISCUSSION

3.1 **Complexation constant measurements**

The complexation constants have been measured by the two fluorescence methods and the values are reported on table 2.

TABLE 2: Complexation constants between PEO end capped polymers and β-CD/EP

	PEO molecular weight	K_c (M^{-1})
Model-N1	44	740
PEO-N1	6000	26
PEO-N1	20000	12
PEO-N2	6000	400
PEO-ADAM	6000	3200
PEO-ADAM	20000	2300
PEO-ADAM	35000	2000

The results obtained for PEO-N1 and PEO-N2 of the same PEO molecular weight (6000) show the same trend as reported in the literature for naphtyl groups substituted in position 1 or position 2 [8]: the complexation constant is much higher for PEO-N2 than for PEO-N1.The inclusion of the naphtyl groups being axial , this effect has been attributed to a deeper penetration of the 2-naphtyl group into the cyclodextrin cavities as it does not experience any hindrance from the substituant unlike the 1-naphtyl group.
The model-N1 molecule has been synthesized and its complexation constant measured in order to study the influence of the PEO chain on the complexation behavior. The results show that the complexation constants are sharply decreasing as a function of the PEO chain length: K_C has decreased by a factor of the order of 60 between model-N1 and PEO-N1 20000. This behavior can be attributed to an excluded volume effect reducing the entropy of the chain when one or both ends are linked to cyclodextrins. The PEO chain length influence has also been studied for the PEO-AD polymers with 3 PEO molecular weights: 6000, 20000 and 35000. Again, the K_C values are decreasing as a function of the PEO chain length but the rate is lower. At this point, we should recall that the K_C measurements are done at different experimental conditions for the polymer bearing naphtyl or adamantyl end groups: the PEO concentration (2g/l) was lower than the chain overlapping concentrations for the first ones, while for the other ones the PEO concentration was fixed at a higher value (17.5 g/l) which is in the semi-dilute range for PEO 20000 and 35000 .

3.2. Viscosity measurements

The hydrophobic ends of the modified PEO polymers can give them associating properties, as it is usually observed with the HEUR polymers which are PEO end-capped with alkyl chains of different lengths (C10 to C20) [9]. Specific viscosities of the modified PEO have been measured as a function of the concentration in the range 5 to 40 g/l and compared to the ones of the precursor PEO chains. No thickening effects have been observed . Moreover the viscosities of the modified polymers are equal to the ones of the unmodified polymers within the range of uncertainty of the measurements. The viscosity measurements being done at relatively high shear rates imposed by the capillary viscometer, these results do not necessarily imply that there is no autoassociations of the end groups. Nevertheless, these eventual autoassociations give negligible viscosity effects compared to the associations with the β-CD/EP polymers, as will be seen below.

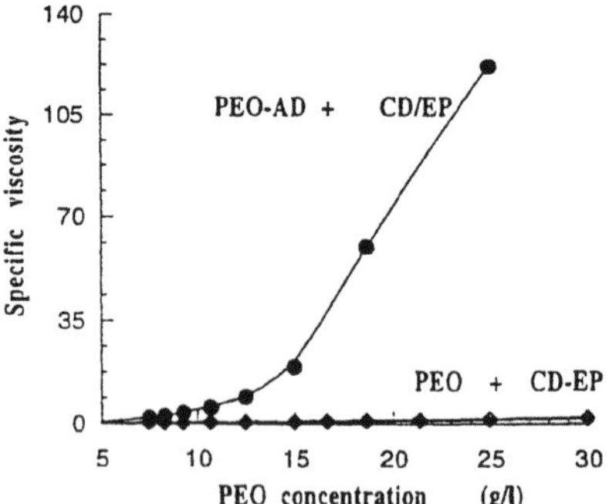

Fig.2 Specific viscosities of mixtures PEO-AD 20000 and β-CD/EP having a weight ratio 1.5. Comparison with the specific viscosities of mixtures containing the precursor PEO 20000.

Thickening effect are obtained when the two polymers are put together, as is shown on figure 2 for PEO-AD 20000 mixed with β-CD/EP in a weight ratio 3/2. Specific viscosities are plotted as a function of the PEO concentration and compared to the specific viscosities of the equivalent mixtures containing unmodified PEO 20000. As it has been observed for associating polymers, the viscosity increases sharply above a certain critical concentration (around 10 g/l for PEO-AD 20000) which corresponds to the onset of extended polymolecular associations. The probability to form extended structures by associations of the two polymers is controlled both by the percentage of links established by inclusion complex formation and by the connectivity of the system. The percentage of links can be easily calculated from the complexation constant K_C and the concentrations of the two polymers. For instance, one expects that a majority of the terminal PEO groups form links when both concentrations of terminal groups and β–CD are higher than $1/K_C$, that is for PEO concentrations higher than 4.5 g/l in the case of PEO-AD 20000. On the other hand, the probability to link the two terminal groups of a PEO chain to two different β-CD/EP polymers is increased when the chains are close to contact, that is for concentrations higher that critical overlapping concentrations of the heteropolymer system.

The influence of the Kc values has been studied on PEO 6000 grafted with three different groups: 1-naphtyl, 2-naphtyl and adamantyl giving complexation constants 26, 400 and 3200 M^{-1} respectively. The specific viscosities of the mixtures of modified PEO and β-CD/EP, weight ratio 3/2, have been reported on figure 3 as a function of the PEO concentration. Very low thickening effects are observed for PEO-N1 which forms only 30% links at 30g/l while higher viscosities have been obtained for PEO-AD which possess a high enough K_C value to form more than 80% links in the concentration range explored. From these results, it is clear that better associating properties are obtained with PEO bearing adamantyl end groups.

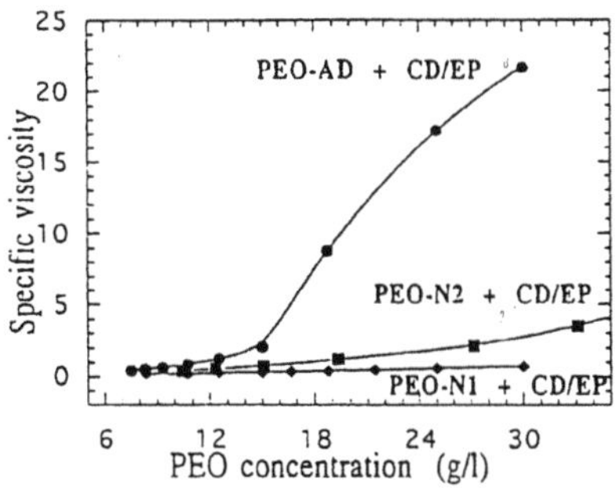

Fig. 3 Specific viscosities of mixtures β-CD/EP and PEO 6000 modified with three different end groups 1-naphtyl (N1), 2-naphtyl (N2) and adamantyl (AD).

The influence of the PEO chain length has been studied using three different molecular weights 6000, 20000 and 35000 of PEO-AD. The specific viscosities of their mixtures with β-CD/EP (weight ratio 3/2) have been measured as a function of the PEO concentration, as well as the specific viscosities of the equivalent mixtures with unmodified PEO. Relative viscosities which are the viscosity ratios of the mixtures with PEO-AD over mixtures with PEO are reported on figure 4. The higher viscosity enhancement is obtained for PEO 20000 and the enhancement is much lower for PEO 35000 than for PEO 6000. Obviously, there is an optimum of the thickening properties for PEO molecular weights around 20000, resulting from the occurence of different opposing effects. The connectivity of the system is increased as the PEO chain length is increased giving higher probability of network formation, but conversely the concentration in end groups is decreased giving lower elastic modulous to the structures formed. On the other hand, the percentage of links calculated from the complexation constants (table 2) are qualitatively comparable (around 80%) for the 3 PEO molecular weights in the concentration range 10 - 30 g/l, but the lower grafting ratio (75%) of adamantyl groups on PEO 35000 (table 1) could partially explain the unexpected behavior of PEO-AD 35000.

One can notice on figure 4 that the relative viscosities seem to reach plateau values at high PEO concentrations for PEO 6000 and 20000. Viscosity measurements have been reproduced in this concentration range using an other experimental set-up allowing low shear rates (0.01 - 10 s^{-1}) and the same viscosities have been obtained. Thus non-newtonian behavior of the polymolecular agregates do not seem to be at the origin of this phenomenom. The saturation of the viscosities could rather be interpreted as a levelling at high concentration of the sizes of the agregates. Unmodified PEO and β-CD/EP are known to be incompatible polymers which form two-phase systems at high concentration [10] . Thus the formation of structures of growing size might not be favored by the vicinity of the binary transition and rather limiting sizes might be obtained similarly to what is observed in micellar solutions.

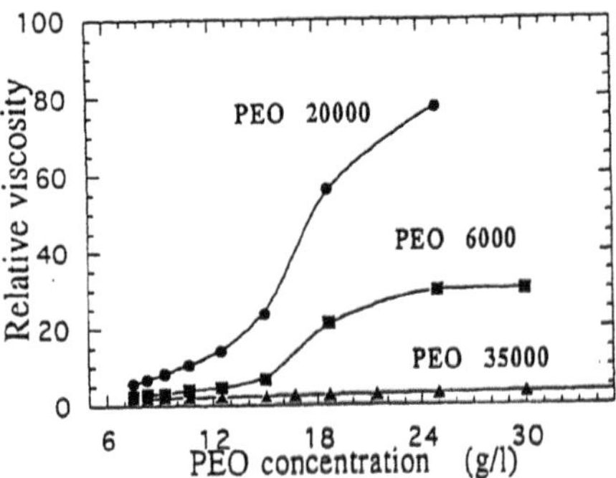

Fig.4 Influence of the PEO molecular weights.Relative viscosities are viscosity ratio of mixtures (3/2 w/w) β-CD/EP and PEO-AD over the equivalent mixtures with unmodified PEO.

4 . CONCLUSION

An original associating polymer system has been described in this study. Inclusion complex formation are at the origin of the associations between hydrophobically end-capped PEO and water soluble β-cyclodextrin polymer. Associations have been demonstrated through both microscopic and macroscopic analysis of the properties of the medium: fluorescence spectroscopies for the determination of the complexation constants and viscosimetry for the onset of the polymolecular associations. The results have shown that the complexation constants are not only dependant on the nature of the end groups (higher values are obtained with adamantyl than with naphtyl groups) but also on the length of the PEO chain: the complexation constants are a decreasing function of the chain length. Thickening properties have been observed by viscosimetry above certain polymer concentrations. The polymolecular associations, which are at the origin of this unusual behavior are both controlled by the connectivity of the system (increasing function of the PEO chain length) and the percentage of complexed groups (decreasing function of the PEO chain length). An optimum of the thickening properties have been observed for a PEO chain length of ca 20000. No network formation has been observed and the limiting growth of the polymolecular agregates has been attributed to the repulsive interactions of PEO and β-CD/EP polymers.

ACKNOWLEDGEMENTS

The authors are indebted to J.-C. Poccheschi, A. Sandier, E. Renard, P. Valat and V. Wintgens for their experimental help.

REFERENCES

[1] Glass, J. E. , Polymers in Aqueous Media: performance through associations, Advances in Chemistry Series 223, (American Chemical Society, Washington DC, 1989).
[2] Renard, E., Deratani, A., Sebille, B., Preparation and characterization of water soluble high molecular weight β-cyclodextrin - epichloridrin polymers, *European Polymer Journal*, in press.
[3] Amiel, C., Sandier, A., Sebille, B., Valat, P., Wintgens, V., Associations between hydrophobically end-capped polyethylene oxide and water soluble β-cyclodextrin polymers, *Int. J. Polymer Analysis & Characterization*, **1**, 289-300 (1995).
[4] Lakowitcz, J. R., Principles of fluorescence spectroscopy, Plenum Press, New York (1983).
[5] Nakashima, K., Anzai, T., Fujimoto, Y., *Langmuir*, **10**, 658 (1994).
[6] Park, J. W., Song, H. J., *J. Phys. Chem.*, **93**, 6454-6458 (1989).
[7] Catena, G. C., Bright, F. V., *Analytical Chemistry*, **61**, 905-909 (1989).
[8] Pitchumani, K., Vellayappan, M., Complex formation of nitrobenzoic acids and some naphtalene derivatives with β-cyclodextrin, *Journal of Inclusion Phenomena and Molecular Recognition in Chemistry*, **14**, 157-162 (1992).
[9] Annable, T., Buscall, R., Ettelaie, R., Whittlestone, D., The rheology of solutions of associating polymers: comparison of experimental behavior with transient network theory, *J. Rheol.*, **37**, 695-726 (1993).
[10] Ekberg, B., Sellergren, B., Olsson, L., Mosbach, K., *Carbohydrate Polymers*, **10**, 183 (1989).

CHARACTERIZATION AND STRUCTURE OF CYCLODEXTRIN - EPICHLOROHYDRIN POLYMERS - EFFECTS OF SYNTHESIS PARAMETERS

E. RENARD,[1] G. BARNATHAN,[2] A. DERATANI,[3] and B. SEBILLE[1]
[1] *Laboratoire de Physico-Chimie des Biopolymères, UMR 27, CNRS - Université Paris Val de Marne, 2 rue H. Dunant, 94320 Thiais France*
[2] *Institut des Substances et Organismes de la Mer, Faculté de Pharmacie, Université de Nantes, BP 1024, 44035 Nantes cedex 1 France*
[3] *Laboratoire des Matériaux et Procédés Membranaires, UMR 9987, CNRS - ENSCM - Université Montpellier II, 2 place E. Bataillon, 34095 Montpellier cedex 5 France*

ABSTRACT

A series of water-soluble polymers were prepared by reacting α- and β-cyclodextrin with epichlorohydrin in NaOH solutions. Depending on the experimental conditions, polymers having very high molecular weight averages ($M_w > 10^6$ g.mol^{-1}) can be obtained. However, low M_W fractions ($M_w < 10000$) were always present. The substitution patterns were determined by ^{13}C NMR analysis and standard methylation analysis of polysaccharides. The formation of side chains and bridges consisting of sequences of poly(2-hydroxypropyl)ether units were demonstrated by electrospray ionization mass spectrometry and gas liquid chromatography - mass spectrometry. The predominant position of substitution depends on the NaOH concentration.

1. INTRODUCTION

The reaction of cyclodextrins (CDs) with epichlorohydrin (EP) in NaOH solutions gives polymers with hyperbranched structures consisting of CD moieties, linking bridges and side-chain substituents (tails) [1,2]. Insoluble crosslinked compounds are generally obtained. Interestingly, the CD cavities keep their complex-forming ability so that beads of these materials have found applications in extraction [3] and separation processes [4]. Water-soluble polymers can be also produced under controlled conditions in order to avoid the gelation of the reaction mixture [5-7]. The compounds obtained are much more soluble than the corresponding CDs and can be used for solubilizing drugs and stabilizing active species. Recently, we have developped new HPLC chiral stationary phases (CSP) based on the adsorption of soluble β-CD - EP polymers onto porous silica beads [8,9]. The CSP obtained exhibit markedly different enantiomeric discriminations than those of supports where the CD cavity is linked to an organic or an inorganic polymer chain through a spacer arm.

J. Szejtli and L. Szente (eds.), Proceedings of the Eighth International Symposium on Cyclodextrons, 115–120.

This prompted us to start a study on the structure of β-CD - EP polymers. A series of samples were synthesized by changing the molar ratios of reactants β-CD:EP:NaOH. The M_w distribution of the compounds prepared was determined by aqueous size exclusion chromatography. Their structure was studied by ^{13}C NMR and by mass spectrometry (MS) of the alditol acetates produced by degradative standard methylation analysis.

2. MATERIALS AND METHODS

2.1. Synthesis of soluble β-CD - EP polymers

5 g of anhydrous β-CD (4.4 mmol) were dissolved in 8 ml NaOH solution under mechanical stirring overnight. The desired amount of EP was rapidly added to the solution heated to 30°C. The polymerization was stopped by addition of acetone. After removal of the organic layer, the aqueous phase was heated to 50°C overnight, then neutralized with 6N HCl and ultrafiltered using membranes of M_w cut-off 1000. The polymer samples were finally isolated by freeze drying.

2.2 Standard methylation analysis [10,11]

Dried polymer samples (0.65 g) were first permethylated at room temperature by methyl iodide (3 ml) in dry DMSO (12 ml) using powdered NaOH (1.2 g) as base. After 4 days of reaction under nitrogen, $CHCl_3$ (20 ml) was added and the resulting mixture washed three times with water. Evaporation of solvents yielded the methylated polymer. The procedure had to be repeated to achieve the complete permethylation of samples (disappearance of the OH absorbance at 3400 cm^{-1}).
Permethylated compounds (24 mg) were hydrolyzed in 2M trifluoroacetic acid (6 ml) at 115°C for 24 h in a sealed tube. After removal of the acid, the samples were reduced with $NaBH_4$ (or $NaBD_4$) (90 mg) overnight at room temperature. Acetylation was carried out with acetic anhydride in pyridine (100°C, 2 h). The alditol acetates were extracted with $CHCl_3$ and subjected to subsequent analyses.

2.3 Methods

Absorbances were measured on a Perkin-Elmer 1760 FTIR spectrometer. NMR spectra were recorded with a Brüker A2000 instrument. Size exclusion chromatography was performed on PL aquagel-OH 40 (Polymer Laboratories) and SW 3000 G (Beckmann) columns using the dual detection refractive index (RI-71 Shodex) and multi-angle laser light scattering (mini Dawn Wyatt Technologies). 0.1 M $LiNO_3$ with a flow rate of 0.5 $ml.mn^{-1}$ was the eluent. Electrospray ionization MS was carried out in positive mode using a Fisons V-6-Trio 2000 instrument. Extracting cone voltages were usually about 31 V. Scanning was performed in the multichannel analyzer mode from m/z 0 to 2000. Solutions in H_2O:MeOH:AcOH (10 μl) were introduced in the spectrometer at a constant flow rate of 10 $\mu l.mn^{-1}$. GLC - MS was performed on a Hewlett - Packard 5890 system with helium as the carrier gas using a DB1 capillary column. Methane, isobutane and ammonia were successively applied in chemical ionization mode.

3. RESULTS AND DISCUSSION

3.1 Synthesis

First, we investigated the influence of the molar ratio EP:CD on the polymer synthesis. Results are summarized in Tab. 1 for β-CD. Ratios EP:CD < 10 gave water soluble samples with good chemical yields. The molar ratios EP:CD in polymer samples were higher than that of starting monomers. For higher EP:CD values, the reaction must be stopped before completion to obtain soluble polymers because of the gelation of reaction mixtures (t_{gel} ca 4 h). Under these conditions, the chemical yields were only fair. Besides, the polymer composition remained nearly constant (β-CD content about 55%).

TABLE 1. Influence of the molar ratio EP:β-CD on the polymer characteristics (molar ratio NaOH:CD=15, T=30°C)

Starting EP:β-CD	1	3	5	7	10	12	15
Chemical yield (%)	36	76	86	75	50	52	54
EP:β-CD in polymer	1	4	7	11	14	13	14
β-CD content (%)	91	84	74	57	52	55	53

We found that α-CD is less reactive than β-CD: no significant amounts of polymer were formed for EP:CD < 10 and for higher ratios it took more than 20 h for gelation.
The M_w distribution of samples prepared from both CDs showed that only low M_w polymers (5000-12000 g.mol^{-1}) were obtained when no gelation ocurred (Fig. 1).

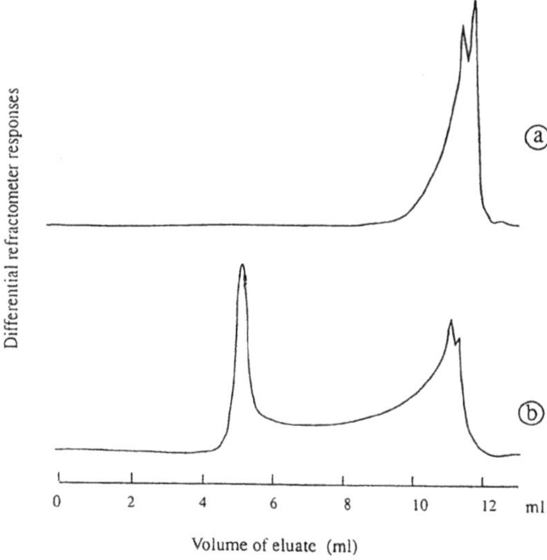

Fig. 1. Size exclusion chromatograms of β-CD - EP polymers. (a) EP:β-CD = 7; (b) EP:β-CD = 10

Conversely, the higher EP:CD ratios gave rise to samples having broadly polydispersed M_w and weight-average M_w values higher than 10^6 g.mol^{-1}. We examined the change of the weight-average M_w during the course of the reaction (Tab. 2): it initially increased very slowly to ca 10^4 g.mol^{-1} and then increased rapidly to values > 10^6 g.mol^{-1} just before the gelation time.

TABLE 2. Dependence of the weight-average M_w on the reaction time (molar ratio NaOH:CD=15, T=30°C)

Reaction time (h)	1	2.5	3.5	3.8	4
Weight-average M_w (g.mol^{-1})	5.10^3	10^4	3.10^4	2.10^6	gel

The molar ratio NaOH:CD was varied from 4.5 to 23. The gelation time was taken as an indication of the reaction kinetics. As can be seen in Table 3, the faster rates were observed for intermediate values of NaOH concentration. Surprisingly, no gelation happened for NaOH:CD ratios > 15. In a corresponding way, polymers with chromatographic profiles similar to that of (b) in Fig.1 were only obtained for NaOH:CD ratios from 7 to 15.

TABLE 3. Influence of the molar ratio NaOH:β-CD (molar ratio EP:β-CD=10, T=30°C)

NaOH:β-CD	4.5	7	10	15	23
Gelation time (h)	-	2.7	2.7	4	-

3.2 Structural analysis

The determination of the structure of CD - EP polymers is rather complex due to the polyfunctionality of CDs with three possible positions *O*-2, *O*-3 and *O*-6 on which the substitution can take place, and by the fact that EP can react with grafted (2-hydroxypropyl)ether residues leading to polytails and polybridges (Fig.2).

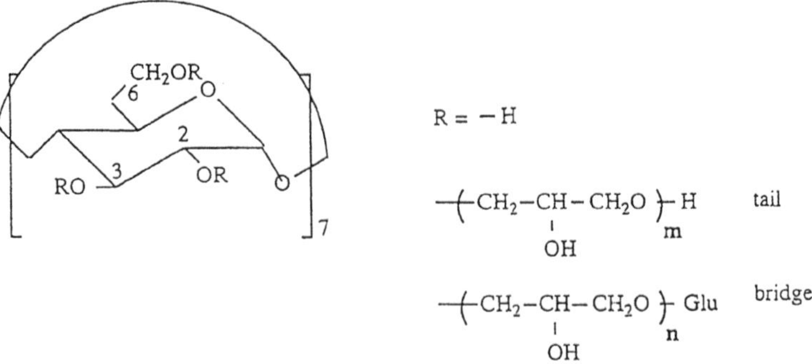

Fig. 2. Schematic structure of CD units in β-CD - EP polymers

^{13}C NMR is a useful tool to obtain information on the location of substituent residues. Thus, substitution on *C*-2 and *C*-3 produces a downfield chemical shift of the carbon bearing the substituent of about 5 ppm and on *C*-6 of about 9 ppm [12]. The results obtained confirmed that the patterns of substitution are governed by the alkali concentration as reported for non-polymeric CD derivatives [13]. At high NaOH concentrations (NaOH:β-CD > 10), the reaction takes mainly place at the more accessible primary hydroxyls whereas all the hydroxyls are affected under intermediate conditions. There was a marked difference in the case of α-CD since only substitution at the secondary hydroxyls was evidenced for NaOH:α-CD < 9.
The occurrence of mono- and poly- tails and bridges was examined by electrospray ionization MS of alditol acetates produced from the hydrolysates of permethylated polymer samples (Tab. 4). Under the experimental conditions used, the single charged sodium adducts $[M+Na+H]^+$ were the predominant species formed.

TABLE 4. Selected adducts from the electrospray ionization MS of alditols acetates obtained from standard methylation analysis of β-CD - EP polymer (molar ratio EP:β-CD=10, OH:β-CD=15)

$[M+Na+H]^+$	374	462	550	638	726	766	854	942	1030
M	350	438	526	614	702	742	830	918	1006
Relative intensity (%)	21	100	35	15	9	16	11	8	5
Tail	-	mono	di	tri	tetra	-			
Bridge	-	-	-	-	-	mono			

The peak at m/z 374 can be unambigously assigned to the alditol corresponding to unsubstituted glucoses. The series of "tail" alditols can easily be identified on the basis of the increase of 88 mass units for each 2-methoxypropyl ether residue. Four adducts were listed in Table 4 but the spectra showed more highly substituted compounds with relative intensity < 5 %. Taking into account a maximum of three substituents by glucose, these data demonstrated the formation of poly-tails. The peak at m/z 766 was assigned to the simplest "bridge" alditol. A series of bridge adducts can be similarly identified by the 88 mass units increment arising either from one tail or one bridge segment. This implies that we cannot be confident that poly-bridges are present. However, we assume that they are probably formed in the same way as poly-tails.
Combined GLC-MS analyses were performed to precise the substitution patterns of alditol acetates. Unfortunately this method is limited to derivatives having M_w less than 800 $g.mol^{-1}$. Up to this value, the structure of the alditols reported in Table 4 was clearly identified by their electron-impact fragmentation (Tab. 5).

TABLE 5. Substitution patterns from GLC -MS data of alditols acetates obtained from standard methylation analysis of β-CD - EP polymer (molar ratio EP:β-CD=10, OH:β-CD=15)

M	350	438	438	438	526	526	614	702	742
Relative intensity (%)	100	5	8	25	2	5	3	<1	2
Tail	-	2	3	6	33	66	666	6666	-
Bridge	-	-	-	-	-	-	-	-	66'

The distribution of substituents in the sample studied shows an almost exclusive reaction at the *O*-6 position. This result agrees well with the high NaOH concentration used for the synthesis (NaOH:CD = 15). As can be seen, we found only derivatives having substituents at one point of attachment thus confirming the existence of poly-tails. Moreover, it appears that the number of substitution per β-CD (NS) infered from the relative intensity of peaks would be about 3. In fact, it is probably higher because only compounds with $M_w < 800$ g.mol^{-1} were taken into account. This NS value should mean that substituents consist of sequences with a length-average of about four (2-hydroxypropyl)ether segments. This value is likely to be overestimated as mentioned above.

From the preceding results, we can conclude that the probability for EP to react with either CD or a hydroxyl of substituent residues is nearly identical. We propose the following mechanism of reaction. At the first stage of polymerization, substitution on CD occurs. At the same time, the polycondensation starts giving low M_w compounds. Then EP preferentially adds to substituents giving poly-tails. Finally, low M_w compounds condense together leading to high M_w polymers and gels. Poly-bridges are likely to form during this step. This can account for the minimum molar ratio EP:CD necessary to the formation of the high M_w polymers.

For intermediate NaOH concentrations, the reactivity of secondary hydroxyls increases, involving significant substitutions on both sides of the β-CD cavity. The polycondensation is easier and the gelation time decreases.

REFERENCES

[1] Wiedenhof, N., Lammers, J.N.J.J., Van Panthaleon Van Eck, C.L., Properties of cyclodextrins. Pt. III. Cyclodextrin-Epichlorohydrin resins: preparation and analysis, *Staerke*, **21**, 119-123 (1969)

[2] Szejtli, J., Cyclodextrin Technology, Kluwer Academic Publishers, Dordrecht, 1988

[3] Shaw, P.E., Wilson, C.W., Reduction of bitterness in grapefruit with β-cyclodextrin polymer in a continous -flow process, *J. Food Sci.*, **50**, 1205-1207 (1985)

[4] Zsadon, B.,Szilasi, M.,Otta, K.H., Tudos, F., Fenyvesi, E., Szejtli, J., Characterization and chromatographic behaviour of cyclodextrin polymers, *Acta Chim. Acad. Hung.*, **100**, 265-273 (1979)

[5] Cserhati,T., Oros, G., Fenyvesi, E.,Szejtli, J., Inclusion complexing by water-soluble β-cyclodextrin polymers, *J. Incl. Phenom.*, **1**, 395-402 (1984)

[6] Szeman, I., Fenyvesi, E., Szejtli, J., Ueda, H.,Machida, Y., Nagai, T., Water-soluble cyclodextrin polymers: their interaction with drugs, *J. Incl. Phenom.*, **5**, 427-431 (1987)

[7] Ekberg, B., Sellergren, B., Olsson, L.,Mossbach, K., The synthesis of water-soluble polymers of β-cyclodextrin and their use in aqueous two-phase systems, *Carbohydr. Polym.*, **10**, 183-188 (1989)

[8] Thuaud, N., Sébille, B., Deratani, A., Lelièvre, G., Retention behavior and chiral recognition of β-cyclodextrin derivative polymer adsorbed on silica for warfarin, structurally related compounds and Dns-amino acids, *J. Chromatogr.*, **555**. 53-64 (1991)

[9] Thuaud, N., Sébille, B., Deratani, A., Pöpping, B., Pellet, C., Enantiomeric separations with chromatographic supports based on β-cyclodextrin polymers immobilized on porous silica. Role of the polymer structure on the separation ability, *Chromatographia*, **36**, 373-380 (1993)

[10] Ciucanu, I.,Kerek, F., A simple and rapid method for the permethylation of carbohydrates, *Carbohydr. Res.*, **131**, 209-217

[11] Mischnick, P.; Determination of the patterns of substitution of hydroxyethyl and hydroxypropyl cyclomaltoheptaoses, *Carbohydr. Res.*, **192**, 233-241

[12] Pöpping, B., Deratani, A., Synthesis of cyclodextrins with pendant chlorinated groups. Reaction of β-cyclodextrin with epichlorohydrin in acidic medium, *Makromol. Chem. Rapid Commun..*, **13**, 327-330 (1992)

[13] See for example: Reuben, J.,Rao, C.T., Pitha, J., Distribution of substituents in carboxymethyl ethers of cyclomaltoheptaose, *Carbohdr. Res.*, **258**, 281-285 (1994)

THE SYNTHESIS OF OLIGOSACCHARIDE-BRANCHED CYCLODEXTRINS AND THEIR INTERACTION WITH CONCANAVALIN A

Kenjiro HATTORI, Hideo IMATA, Kohji KUBOTA, Keisuke MATSUDA, Masaaki AOYAGI[1)], Keiko YAMAMOTO[1)], Chiaki JINDOH[1)], Takashi YAMANOI[2)], and Toshiyuki INAZU[2)]

Faculty of Engineering, Tokyo Institute of Polytechnics
1583 Iiyama, Atsugi, Kanagawa 243-02, Japan
Scientific Instrument Sales Laboratory, Nissei Sangyo Co., Ltd.[1)]
20-1 Morinosato Aoyama, Atsugi, Kanagawa 243-01, Japan
The Noguchi Institute[2)]
1-8-1 Kaga, Itabashi-ku, Tokyo 173, Japan

ABSTRACT

New oligosaccharide-branched β-cyclodextrins having a various oligosaccharides in a primary hydroxyl group of β-cyclodextrin(CD) were synthesized and examined for the interaction with the immobilized concanavalin A(ConA) compared with 6-*O*-glucosyl and maltosyl CD. Oligosaccharides were converted to lactones at the reducing end, which was connected with 6-*mono*amino-β-CD forming an amide bond. For the analysis of the interaction between ConA immobilized on an aminosilane-hydrogel surface and various oligosaccharide CDs, a biosensor of the FISONS IAsys apparatus based on a resonant mirror detector (RMD) was used. As a result, it interacted with immobilized ConA and both maltosyl-β-CD(**3**) and maltosyl-γ-CD(**6**), and glucosylglucono-amide-β-CD(**7**) with the association constants, K_a of 134, 833 and 8730 M^{-1}, respectively.

1. INTRODUCTION

Biological recognition and adhesion processes often involve the formation of saccharide-protein complexes. Elucidation of the carbohydrate-binding specificity of lectins has become important in connection with the structure analysis of carbohydrates and the biological roles of the sugar chains and lectins. ConA is widely used as a biological tool to investigate the membrane properties of both normal and transformed cells[1)]. In this study, the molecular interaction between immobilized ConA and oligosaccharide-branched CDs were measured using a resonant mirror detector through the change in reflected light(arc sec) *vs*. time.

2. MATERIALS AND METHODS

2.1 Materials

Glucosylgluconolacone, galactosylgluconolactone, and gluconolactone were prepared according to the literature[2)]. These lactones were connected with 6-amino-β-CD.

J. Szejtli and L. Szente (eds.), Proceedings of the Eighth International Symposium on Cyclodextrons, 121–124.

The crude products were purified through an ion exchange column and preparative HPLC. The obtained glucosylglucono-amide-β-CD(**7**), galactosylglucono-amide-β-CD(**8**) and gluconoamide-β-CD(**9**) were identified from NMR and MS spectra[3]. Glucosyl-β, γ-CD(**3**,**4**), maltosyl-β, γ-CD(**5**,**6**), and β, γ-CD(**1**,**2**) were available from Ensuiko Co., Ltd. and ConA (Wako Co., Ltd.) was dissolved in *p*H 6.5 phosphate buffer solution (PBS, SIGMA), giving a 10 mM solution of *p*H 7.4 in ultrahigh purity water (filtered by Mill-Q, Millipore) with 2.7 mM KCl and 0.14 M NaCl. An IAsys cuvette coated with aminosilane was supplied by FAST. Bissulfosuccinimidyl suberate(BS3) was purchased from Pierce & Warriner. One mg/ml BS3 in PBS was made with 0.05 % v/v Tween 20(Pierce & Warriner). Acetate buffer (*p*H 5.4, 10 mM) was made by titrating sodium acetate with 2 M acetic acid.

2.2 Immobilization of ConA on the optical sensor

To immobilize ConA, aminosilane biosensor surfaces were activated with 10 mM PBS (*p*H 6.5). BS3(1 mM, 0.2 ml) of the crosslinker solution was injected into the aminosilane cuvette. After it was washed with 10 mM PBS (*p*H 6.5) and 10 mM acetate buffer (*p*H 5.4), ConA was immobilized by a reaction with the amino group. Moreover, its cuvette was washed with NaOH (*p*H 8.90) and 10 mM acetate buffer, blocking ConA with 1 M ethanolamine.

2.3 Determination of the association constant K_a

Using the IAsys apparatus (FISONS) with a resonant mirror detector.[4,5,7], the association rate constant (k_a) and the dissociation rate constant (k_d) can be calculated in the elapsed time from the intercept of a plot of dR/dt *vs*. R through the response in reflected light, R (arc sec), at the concentration of saccharide-branched CD as C if the maximum binding response R_{max}, was known and the interactions were carried out under pseudo first order conditions ($k_a >> k_d$) for which the relation is:

$$dR/dt=k_aR_{max}C-(k_aC+k_d)R$$

However, if R_{max} is not known, the value of k_a and k_d can be obtained by plotting the slope of the plot dR/dt *vs*. R changing C.

3. RESULTS AND DISCUSSION

3.1 Immobilization of ConA on the optical sensor

The conditions of the immobilization for ConA were investigated based on the *p*H, the presence of epitope and metal ion, recycle condition, and the reproducibility. The typical immobilization conditions of ConA on an IAsys cuvette were described in the of Materials and Methods section.

3.2 The interaction analysis of the CDs with ConA.

The biosensor was applicable for investigating for the interaction analysis, such as the association constant between the various oligosaccharide-branched CDs and ConA. Fig. 1 shows the response of the optical sensor for the oligosaccharide-branched CD binding to the immobilized ConA. Fig. 2 presents the interaction curve of the newly synthesized oligosaccharide-branched CDs with ConA using the optical biosensor IAsys. From the result of Fig. 2, only glucosylglucono-amide-β-CD(**7**) shows an interaction but not galactosylglucono-amide-β-CD(**8**), gluconoamide-β-CD or β-CD(**1**). The results of the association constants are summarized in Table 1. As a result, it interacted with immobilized ConA and both the maltosyl-β-CD(**3**) and maltosyl-γ-CD(**6**).

However, β-CD(**1**), γ-CD(**4**), glucosyl-β-CD(**2**) and glucosyl-γ-CD(**5**) did not interact with ConA. The association constants, K_a, of maltosyl-β, γ-CD(**3**,**6**) and glucosylglucono-amide-β-CD(**7**) were measured to be 134, 833 and 8730 M^{-1}, respectively. The reported K_a values between ConA and glucose or maltose were approximately the same as the present results in the range of 10^3-10^4 M^{-1}. However oligosaccharides of triantena type having sialyl unit[5] or mannose unit[6] showed more than 10^7 M^{-1} of K_a values. It was suggested that only CD derivatives that have a glucose unit at the non-reducing end interacted with ConA.

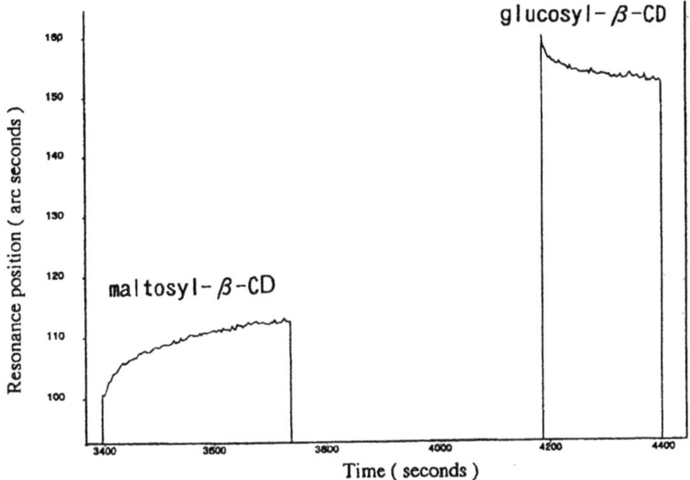

Fig. 1 Profiles of ConA binding to glucosyl-β-CD and maltosyl-β-CD by Iasys
25°C, *p*H 5.3, in acetate buffer, [maltosyl-β-CD]= [glucosyl-β-CD]= 2.0 x 10^{-2} M

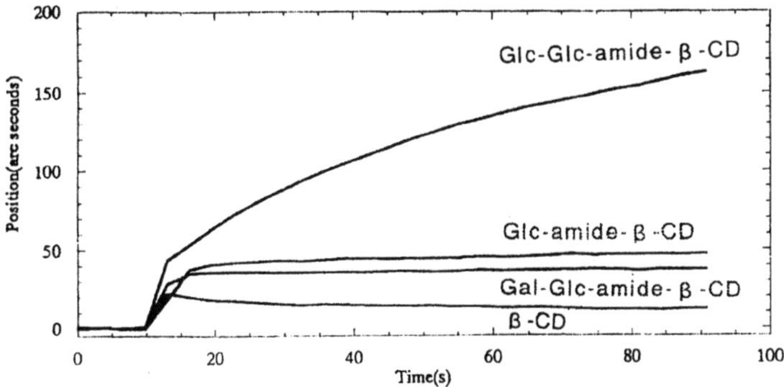

Fig. 2 Interaction between ConA and oligosaccharide-branched CDs
25°C, pH 7.02 in Tris-HCl buffer with 1 mM $CaCl_2$,,100 mM NaCl, [CDs]= 1.7 x 10^{-3} M

The length of the spacer between the non-reducing end and CD cavity seems vitally important for the interaction with ConA. The non-reducing end of these newly synthesized oligosaccharide-branching CDs(**7**) would have freer movement than that of the commercially available 6-*O*-glucosyl and maltosyl CDs(**3**,**6**). The behavior of the interaction of various oligosaccharide-branched α-CD derivatives was also examined and found to be different in complex formation from the derivatives of β- and γ-CD. It is suggested that the ring size of the CD cavity was related to the recognition for ConA. This investigation has demonstrated the applications of IAsys using a resonant mirror detector to determine the association constants for the interaction of the oligosaccharides-branched CD with immobilized ConA on the surface of the aminosilane cuvette.

Table 1 Association constant for the oligosaccharide-branched CDs with optical biosensor IAsys

Scheme A,	n= 6,	m= 0	(1) β-CD	no complex
		1	(2) Glucosyl-β-CD	no complex
		2	(3) Maltosyl-β-CD	134
	n= 7,	m= 0	(4) γ-CD	no complex
		1	(5) Glucosyl-γ-CD	no complex
		2	(6) Maltosyl-γ-CD	833
Scheme B,	X= -OH,	Y= -H	(7) Glucosyl-glucono-amide-β-CD	8730
	-H,	-OH	(8) Galctosyl-glucono-amide-β-CD	no complex

Scheme A:

Scheme B:

CONCLUSION

The behavior of the interaction of various oligosaccharide CD derivatives was summarized as follows: β- and γ-CD derivatives that have a glucosyl unit at the non-reducing end on the oligosaccharide branch, which is longer than the maltosyl group, show association with immobilized ConA using the optical biosensor. The larger the spacer between CD and the glucose unit, the larger association constant for the oligosaccharide-branched CDs. By using this method, it will be possible to obtain the association constant for the precise investigation of the interaction of ConA and various oligosaccharide derivatives of CD.

REFERENCES

[1] Goldstein, I.J., Poretz, R. D., The Lectin, 1986, p51

[2] Kobayashi, K., Sumitomo, H., Ina, Y., *Polymer Journal*, **17**, 567(1985)

[3] Hattori,K., Takahashi, K., Koshikawa, T., *Synthesis of oligosaccharide-branched cyclodextrins*, Proceedings of 7th International Cyclodextrins Symposium, (Ed. Osa, T.) Business Center for Academic Societies Japan, Tokyo, 1994, pp. 90-93.

[4] Z. Salamon, Wang, Y., Brown, M.F., Macleod, H.A., Tollin G., Conformational changes in rhodopsin probed by surface plasmon resonance spectroscopy, *Biochemistry*, **1994**, *33*,.

[5] Yamamoto, K., Ishida, C., Shinohara Y., Hasegawa, Y., Konami, Y., Osawa, T., Irimura, T., Interaction of immobilized recombinant mouse C-type macrophage lectin with glycopeptides and oligosaccharides, *Biochemistry*, **1994**, *33*, 8159-8166.

[6] Mega, T. and Hase, S., Characterization of carbohydrate-binding specificity of concanavalin A by competitive binding of pyridylamino sugar chains, *J. Biochem.*, **1992**, *111*, 396-400

[7] Macleod, H. A., Tutorials in Optics (Ed., Moore, D. T.) **1992**, p121 Optical Society, Washington, DC.

CONJUGATES BASED ON CYCLODEXTRINS AND POLY(ETHYLENE OXIDE) AS COMPLEXATION AND TRANSPORT AGENTS

I. N. TOPCHIEVA, S. V. ELEZKAYA, V. A. POLYAKOV, K. I. KAREZIN

Department of Chemistry, Moscow State University

119899 MGU, Lenin Hills, Moscow, Russia

ABSTRACT

A new series of branched derivatives of CDs is described. They result from the polymerization of ethylene oxide initiated with secondary hydroxyl groups of CDs. These conjugates are water soluble, amorphous liquids. Complexation properties of these compounds with 4-nitrophenol and calcium acetylhomotaurinate (CAHT) are studied. It was shown that CAHT as a drug combined with conjugates shows a prominent anticonvulsive effect in the experiments in vivo. This effect is due to the complex crosses the brain endotelium barrier with the following release of a drug.

1. INTRODUCTION

Cyclodextrins (CD) have been extensively studied due to the unique properties of these cyclic starches to form dynamic molecular inclusion complexes with many compounds, including pharmaceuticals. Inclusion complexes of drugs with natural CD have been used for pharmaceutical purposes, but the solubilization procedures are not straightforward and physico-chemical studies must be performed prior to use.
Among the variety of derivatives our attention was drawn to the group of bouquet-shaped molecules based on CD and oligomers of ethylene oxides. These molecules are incorporated in vesicle membranes and display transmembrane transport of alkaline metals. At the same time these compounds are not soluble in water and can not be used as solubilizers. One can assume that a novel type of CD-derivatives in which a bundle of polymer chains are grafted onto one side of core of CD may serve as a new object for drug incapsulation. With this aim we developed a simple method of synthesis of such structures and tested their complexation properties.

2. MATERIALS AND METHODS

α– and β–Cyclodextrins (CD) were purchased from "Cyclolab" (Hungary); ethylene oxide (EO), 4-nitrophenol (NP) and CAHT $(CH_3CONH(CH_2)_3SO_3^-)_2Ca^{2+}$ were analitically grade reagents. Differential absorption spectra were measured by a Hitachi

J. Szejtli and L. Szente (eds.), Proceedings of the Eighth International Symposium on Cyclodextrons, 125–128.

557 UV-spectrophotometer; ^{1}H and ^{13}C-NMR spectra were recorded on CXR-200 Bruker spectrometer. Chemical shifts in ppm rev.to DSS as internal reference. Glass-transition temperatures were recorded with a Mettler TA 400 thremoanalyzer using a DSC-30 unit.

3. RESULTS AND DISCUSSION

3.1 Synthesis of semi-bouquet structures

Conjugates in which several chains of PEO originate from a central core of α and β-CD were synthesized in one-pot manner by initiating the polymerization of ethylene oxide with secondary hydroxyl groups of multifunctional core of CD.

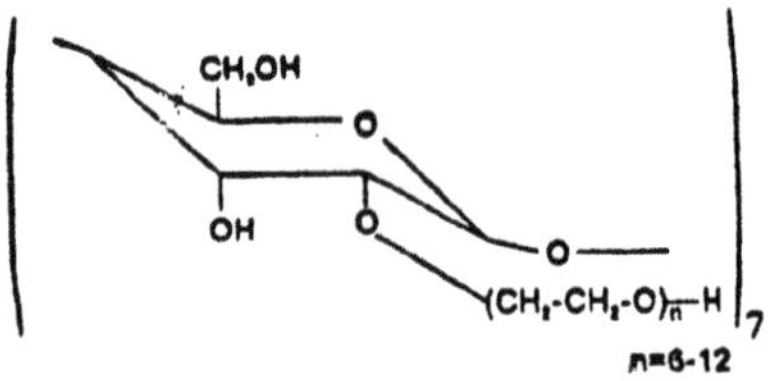

CH_2OH
OH
$(CH_2-CH_2-O)_n-H$ 7
n=6-12

This reaction was conducted in water containing the additive of NaCl at 60-80°C under the pressure 0.15-0.20 MPa. The structure and composition of conjugates were established using NMR-spectroscopy. It was shown that values of the degree of polymerization of ethylene oxice in a series of conjugates differs from 6 to 12.
These substances are amorphous and water soluble compounds. The evidence of the amorphous state was obtained from the study of their glass transition behaviour.

TABLE 1. Characteristics of conjugates based on cyclodextris and poly(ethylene oxide)

N°	CD	number of monomer unit per one glucoside residue	Glass transition temperature T_g °C
1	α	8	-62,6
2	β	6	-74.7
3	β	10	-73.3
4	β	12	-63.6

3.2 Complexation properties of conjugates

Fig.1 shows binding curves for α-CD and a series of conjugates with NP. One may conclude that :1) the binding curve for PEO-α-CD has a S-shaped form due to the formation of autocomplex in which one oligooxyethylene chain coupled to the CD core is included in the cavity of the same molecule; 2) for complexes based on conjugates in parallel with trivial steric effects of the reduction of the cavity size of CD, polymer chains favor the adaptation due to additional fixation of the ligands in polymer surrounding.

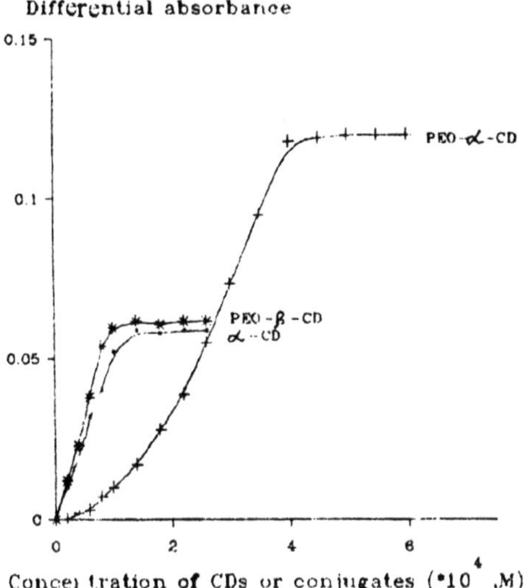

Fig.1 Binding of NP by CDs or conjugates

Complexation properties of conjugates with CAHT were studied using ^{1}H-NMR spectroscopy. The values of the constant of complex formation were determined from the binding curves (change of chemical shifts of methyl protons of CAHT versus of concentration of CD or its polymer derivatives) by a curve fitting analysis.

TABLE 2. Complexation properties of system based on CDs and their PEO derivatives and CAHT

N°	CDs or the derivatives	Temperature, K	K_a l/mole
1	α–CD	313	40.8
2	PEO-α–CD	313	1.9
3	β–CD	303	7728
4	PEO-β–CD	303	373 (16.5)

Value in bracket corresponds to complex with 1:2 stoichiometry.

It is seen that K_a for PEO-derivatives of α-CD as compared to parent CD changes in the same way as for the system with NP. As for β-CD K_a was increased forc more than two orders compared to the corresponding system based on α-CD.

The modelling experiments shows that this fact may be explained by the fact that the cavity of β-CD is sufficiently large and may include two anions of CAHT (Fig.2a). At the same time the binding of CAHT with conjugate PEO-β-CD was described by two constant corresponding to 1:1 and 1:2 stoichiometry (Fig.2b).

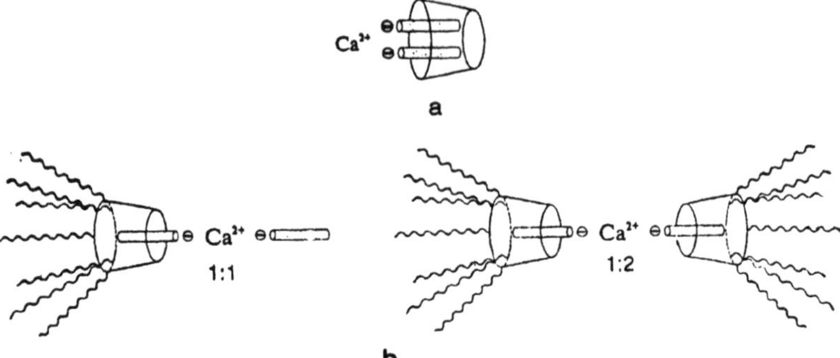

Fig.2 Structure of complexes of CAHT with β-CD and PEO-β-CD

3.3 Conjugates as drug delivery system

Based on the chemical structure of conjugates one may suppose that their complexes with drugs will be able to transport across the cellular membranes.

CAHT was choosen as a drug, as this compound is used in clinic to prevent relapses in weaned alcoholics. However due to their low permeability across the blood-brain barrier their anticovulsive effect is sometimes underestimated. The experiments with CDs and CAHT were carried out on mail mice. Convulsion were induced by i.p. injection of corasol. CAHT and its complexes with CDs and conjugates were injected 15 to 60 min before corasol injection. The injection of complex CAHT-PEO-β-CD resulted in increase of the latent period from 28 to 300 second and decreases the mortality from 95% to 40%. The long latency is likely to be due to the complex crosses the brain endotelium barrier with the following release of CAHT.

4. Conclusion

The structure of the conjugates obtained using the polymerization of EO in the presence of NaCl presents a semi-bouquet in which a bundle of polymer chains are grafted onto secondary hydroxyl groups of cyclic core of CD. This substances are amorphous and water soluble. The investigation of binding properties of PEO-modified CDs reveals some characteristic features of these hosts. In parallel with trivial steric effects of the reduction of the cavity of CD it was shown that PEO-chains grafted to the CD favour the adaptation due to additional fixation of the ligand in polymer surrounding. Using CAHT, molecule of which consists from two anion moieties, allows to demonstrate new types of topologic structures of inclusion complexes. Conjugates based on cyclodextrins and poly(ethylene oxide) have a high permeability across the blood-brain barrier and may be effective as transport materials for drug delivery system.

STRUCTURAL ASPECTS OF COMPLEXATION BETWEEN CDs AND POLY(ALKYLENE OXIDES)

I.N.Topchieva, I.G.Panova, V.I.Gerasimov
Department of Chemistry Moscow State University
119899 Moscow, Vorobievy Hills, MGU, Russia

ABSTRACTS

The mixing of di- and threeblock copolymers of ethylene and propylene oxides (proxanols, pluronics, polaxomers) with CDs results in the formation of crystalline complexes - molecular necklaces. The regularities of complexation between CDs and these copolymers are studied. The stoichiometry of complexes based on diblock copolymers corresponds to one CD molecule per two monomer units of poly(alkylene oxide) as for corresponding homopolymers. At the same time the composition of complexes based on threeblock copolymers depends on the concentration of the solution and varies from completely filled to half filled molecular necklaces. Thus the complexation reaction result in the production of new types of block copolymes of diverse architecture.

1. INTRODUCTION.

The complexation between linear synthetic polymers and cyclodextrins (CD) leading to the formation of molecular necklaces is well documented. From the work of A.Harada and his colleges is known that α-CD forms complexes with poly(ethylene glycol), β-CD with poly(propylene glycol), and γ-CD forms double-standed complexes with chains of poly(ethylene glycol). Using block copolymers instead of homopolymers allows to produce a host of new molecular shapes of inclusion complexes. This investigation covers the complexation between block copolymers of ethylene and propylene oxides (proxanols, pluronics, polaxomers) with α-, β- and γ-CD which are able selectively threaded onto poly(ethylene oxide) (PEO) or poly(propylene oxide) (POP) blocks of copolymer.

2. MATERIALS AND METHODS

2.1. MATERIALS

Block copolymers were synthesized at the pilot plant of the Moscow Research and Production Assotiation NIOPIK. α-, β-, γ- cyclodextrins were produced by "CYCLOLAB" firm.

J. Szejtli and L. Szente (eds.), Proceedings of the Eighth International Symposium on Cyclodextrons, 129–132.

2.2. METHODS

The structure of crystal water insoluble copolymers was studied using optical microscopy, X-ray analysis and thermoanalytical methods (differential scanning calorimetry and thermogravimetry). The composition of complexes was determined by the method IR-spectroscopy and polarimetry.

3.1. CHARACTERISTICS OF BLOCK COPOLYMERS

The characteristics of used copolymer is presented in table 1.

Table 1. The characteristics of block-copolymers (PEO-block of poly(ethylene oxide), POP-block of poly(prolene oxide)).

The type of copolymers	M	The content of propylene oxide, %	The degree of polymerization	
			n	m
POP-PEO	3000	50	30	28
"-"	2000	60	18	20
PEO-POP-PEO	2700	45	18 (of one block)	19
"-"	6000	23	52 (of one block)	24
"-"	2000	90	3 (of one block)	30
POP-PEO-POP	3000	40	40	10 (of one block)
"-"	4500	18	69	14 (of one block)

When aqueous solutions of CDs and poly(alkylene oxide) are mixed the precipitation takes place. Depending on the conditions of the experiment (concentration, temperature) this precipitate presents either a crystalline gel-phase or a crystalline sediment. Both forms of participation are thermoreversible: they dissolve at the heating up to 70-80^0 C and precipitate at the cooling. The formation of gel-phase is probably caused by the interaction between rod-like crystallites of molecular necklaces.

3.2. COMPLEXES BASED ON DIBLOCK COPOLYMERS

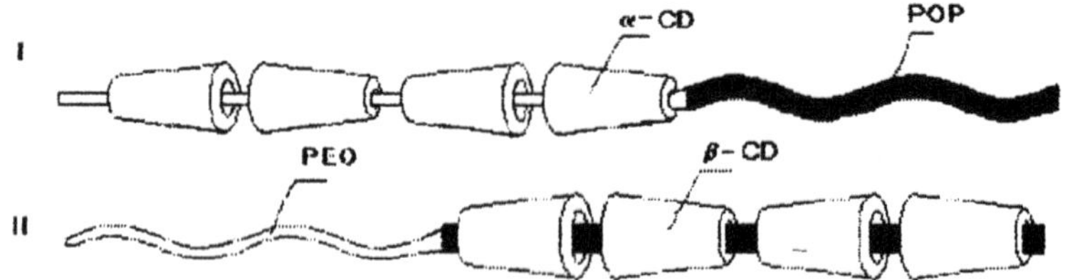

Both complexes are characterized by the general composition, corresponding to one molecule of CD per two monomer units of poly(alkylene oxide). Complexes are water insoluble crystal compounds with the type of crystalline lattice differs from each other and that of the initial components. It should be noted that PEO block is a crystal and POP block is amorphous parts of copolymer. The complexation of CD with block copolymer results in the formation of two new block copolymers comprising the rigid molecular necklace and flexible blocks of free poly(alkylene oxide). X-ray pattern of complexes did not contain reflexes characteristic for PEO block and amorphous halo characteristic for POP blocks. The same picture was observed from DSC data (Fig. 1): the absence of melting peak of PEO and glass transition of POP block.

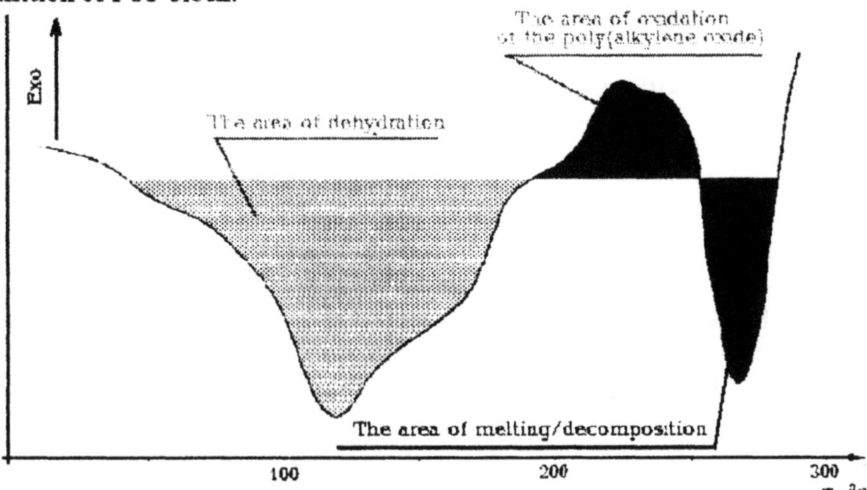

Fig.1. DSC curver of complex of α- CD with diblock copolymer

Thus the complexation of diblock copolymer with α- and β-CD results in the change of morphological properties of both blocks.

3.3 COMPLEXES BASED ON THREEBLOCK COPOLYMERS

The characteristic properties of the complexation between these copolymers and CD is the dependence of the stoichiometry on the location of the interacting block. If CDs interact with internal block of copolymer the stoichiometry of complex is the same as for homo- and diblock copolymer (Structure III).
If CDs interact with end blocks of copolymer the degree of complexation depends on the conditions. In saturated solutions half filled complexes are formed (Structure IV). Using diluted solutions of CDs results in the performing of completely filled molecular necklaces (Structure V).
Thus the composition of complexes depends on the ratio between the rate of complexation and crystallization. In saturated solutions the values are close whereas in diluted solutions the rate of crystallization is essentially lowered as compared to the rate of complexation. In the last case both end blocks have managed to fill with CDs before the precipitation will start.

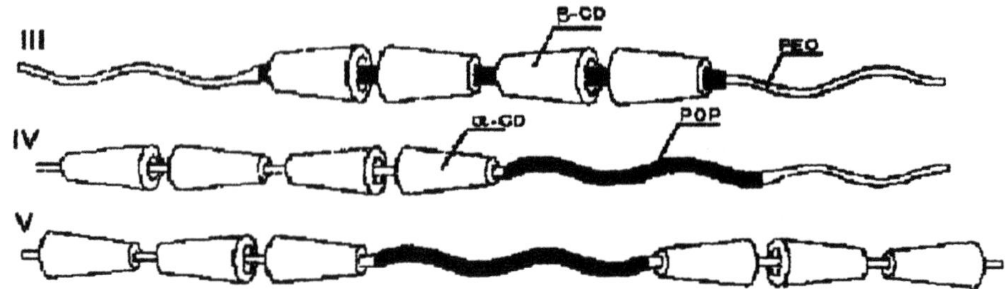

Block copolymers were used as the basis for the synthesis of new types of double-standed structures with γ-CD.

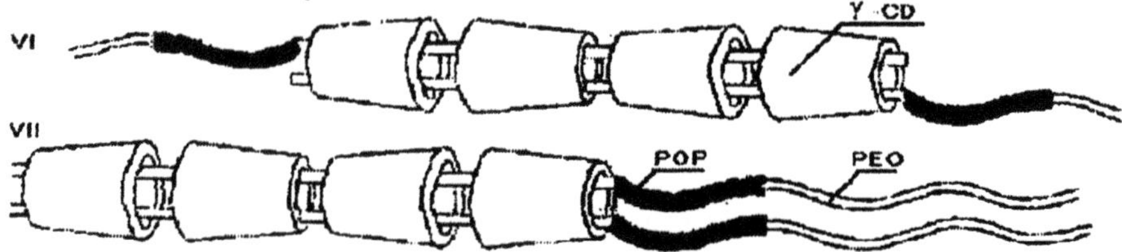

Two probable complexes based on threeblock copolymers PEO-POP-PEO and γ-CD. To discriminate between these structures the interaction between the crystal complex (structures VI or VII) and α-CD was investigated. Methods of HPLC and X-ray analysis show that the interaction between PEO-POP-PEO and γ-CD results in the structure VII characterized by two free blocks arranged at one side of inclusion complex. The complexation between γ-CD and PEO-POP-PEO in diluted solution results in the following structure:

4. CONCLUSIONS

1. The interaction of CDs with complementary blocks of diblock copolymers results in the formation of new block copolymers comprising molecular necklace and free poly(alkylene oxide).
2. Both complexes are characterized by the general composition: one molecule of CD per two monomer units of poly(alkylene oxide).
3. Depending on the experimental conditions the complexation between threeblock copolymers and CDs results in either symmetrical copolymer with molecular necklaces as end blocks, or unsymmetrical copolymer with three different blocks. These systems are excellent examples of the influence of crystallization process on the composition and structure of the complexes.
4. Using the systems based on threeblock copolymer PEO-POP-PEO and γ-CD allows to perform double-stranded inclusion complexes of diverse architecture.

CHARACTERIZATION OF SULFOBUTYL ETHER β-CYCLODEXTRIN MIXTURES

E.A. LUNA[1], E.R. BORNANCINI[1], D.O. THOMPSON[2], R.A. RAJEWSKI[1] and V.J. STELLA[1].
[1]Department of Pharmaceutical Chemistry and the Center for Drug Delivery Research, The University of Kansas, Lawrence, KS - 66047, USA; [2]CyDex, L.C., 8675 W. 96th St., Overland Park, KS - 66212, USA.

KEY WORDS: anion exchange chromatography, sulfobutyl ether cyclodextrin, fractionation

1. INTRODUCTION

The recently patented SBE-β-CD is currently being developed as a parenterally safe solubilizing agent. These derivatives have shown to have potential applications as solubility and stability enhancers, as well as chiral selective analytical reagents [1-9].

The SBE-β-CDs mixtures are the product of the reaction of 1,4-butane sultone with β-CD in alkaline solution [8,9]. Since the introduction of the SBE moiety may occur on any of the 21 available hydroxyl sites of the β-CD the number of compounds possibly present in the reaction product is very high. Thus, the product mixture includes regio and positional isomers distributed over a range of substitution levels. These mixtures are generally characterized by an average degree of substitution (DS), calculated on the basis of elemental analysis (DS_{EA}) and/or nuclear magnetic resonance (DS_{NMR}) data, which expresses the average number of substituted hydroxyl groups per cyclodextrin molecule. The insert in Figure 1 provides a generalized structure of the $(SBE)_{nm}$-β-CD derivatives (n = average DS, m = mixture of DS).

Capillary electrophoresis (CE) has proved to be a useful analytical tool to demonstrate differences in the composition of the SBE-β-CD mixtures [10,11]. Figure 1 illustrates a typical electropherogram of a $(SBE)_{4m}$-β-CD preparation. From the CE profile, it can be observed that the mixture resolved into eleven peaks with the fastest eluting compound corresponding to unreacted β-CD. The SBE derivatives with DS of 1 and 2 (CE peaks 1b and 2b) have been previously identified [10,11]. This information together with FAB-MS of the mixtures allowed the assignment of the peaks with increasing DS. Each of these peaks should correspond to a mixture of isomers bearing the same number of charges per CD molecule. However, the assignment of the identity of the peaks with DS greater than 2 needs to be confirmed.

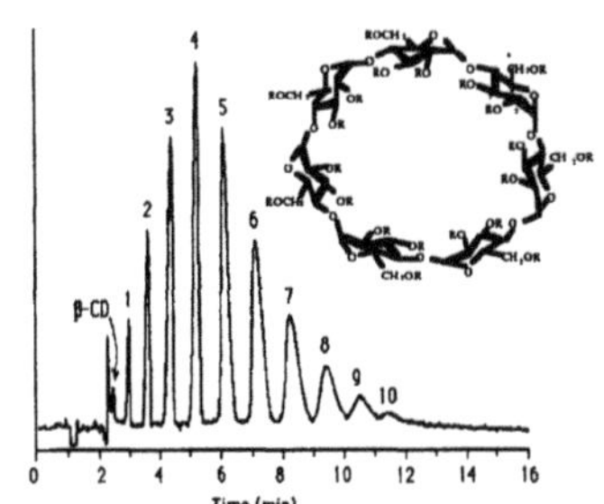

Figure 1. Electropherogram obtained for a mixture of $(SBE)_{4m}$-β-CD, with the general structure of $(SBE)_{nm}$-β-CD displayed in the inset.

Although the FAB-MS spectra and CE analysis produce a distribution profile of the SBE-β-CD mixtures, and EA and NMR data provide an average DS, none of these methods give information regarding the individual contribution of the different substitution levels to the mixture. Therefore, a SBE-β-CD mixture was fractionated for characterization and evaluation of the mass contribution of each band to the mixture.

J. Szejtli and L. Szente (eds.), Proceedings of the Eighth International Symposium on Cyclodextrons, 133–136.

2. MATERIALS and METHODS

The SBE-β-CD mixtures were synthesized according to the general procedure of Stella and Rajewski [8,9].

The SBE-β-CD mixture was fractionated by anion exchange chromatography (AEC) using Sephadex DEAE A-25 (Pharmacia LKB, S-75182, Uppsala, Sweden) and the experiment was conducted under the following conditions: column dimensions: 28 cm x 4.1 cm I.D.; fraction size: 125 ml; sample loading: SBE-β-CD mixture (25 gm dissolved in H_2O, 75 ml); mobile phase: step gradient from H_2O to 0.5 M Na_2SO_4 at pH 3.5-4.0 and 5.5. The elution was by gravity flow at a constant solvent height of 30 cm. The flow rate was approximately 2 ml/min. The fractions were analyzed by CE without treatment. After desalting and lyophilization, the isolated bands were analyzed by CE, NMR and FAB-MS.

The Capillary Electrophoresis analyses (CE) were conducted on an automated capillary electrophoresis system (P/ACE 2210, Beckman) using a 365 mm O.D., 50 mm I.D. uncoated fused silica capillary with an on-column detection window. Samples were introduced by pressure injections, 5 sec at 0.5 psi. The separations were conducted at 25°C and the detection was at 230 nm. The applied voltage was 30 kV.

The proton NMR spectra were obtained at ambient temperature (QE-300 NMR spectrometer, General Electric) at 300.7 MHz. Standard pulse sequences were employed. The sample concentrations were 40-60 mg/ml of D_2O.

Mass spectra were obtained on a AUTOSPEC-Q tandem hybrid mass spectrometer equipped with an OPUS data system. FAB-MS experiments were performed using a cesium gun operated at 20 keV energy and 2 μamp emission. The samples were dissolved in distilled water to a concentration of 10-12 mg/ml and added to glycerol or triethanolamine as the matrix. Exact mass FAB experiments were carried out at 1:10,000 resolution using linear voltage scans under data system control and collecting continuum data mode. Matrix and matrix/potassium adduct (or polyethylene glycol) ions served as bracketing calibration ions.

3. RESULTS and DISCUSSION

3.1. Fractionation and Characterization of SBE-β-CDs

Figure 2 displays the AEC elution profile of the SBE-β-CD containing from β-CD up to deca substituted material and the Na_2SO_4 concentrations used to elute the compounds. The mixture was well resolved at each substitution level as analyzed by CE. Since the group bearing the charge in the SBE moiety is removed from the β-CD by the butyl chain, it seems very likely that the ion exchange process alone is responsible for the separation obtained.

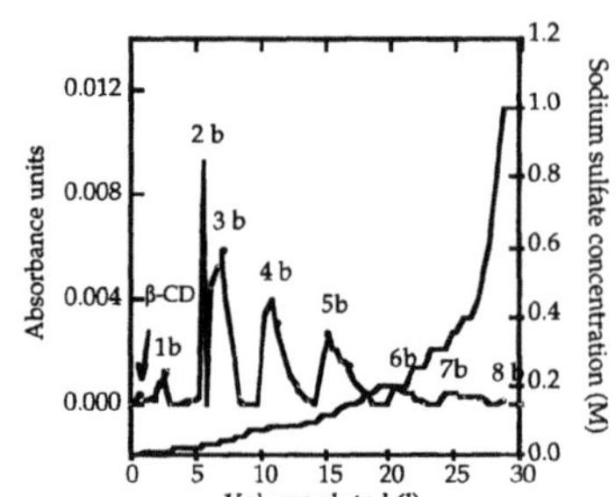

Figure 2. Fractionation of the SBE-β-CD mixture

Figure 3 depicts the CE profile obtained with each sample and the purity of the compounds as compared to the initial mixture (Figure 3-a). The results demonstrate that eight of a total of ten initial components of the $(SBE)_{4m}$-β-CD have been isolated.

Figure 4 displays the ^{1}H-NMR spectra of bands with DS equal to 1, 4 and 7 and the spectrum of the original $(SBE)_{4m}$-β-CD. Although not all the ^{1}H-NMR spectra are included in the figure, the spectrum of each band contained the same set of signals as the original mixture.

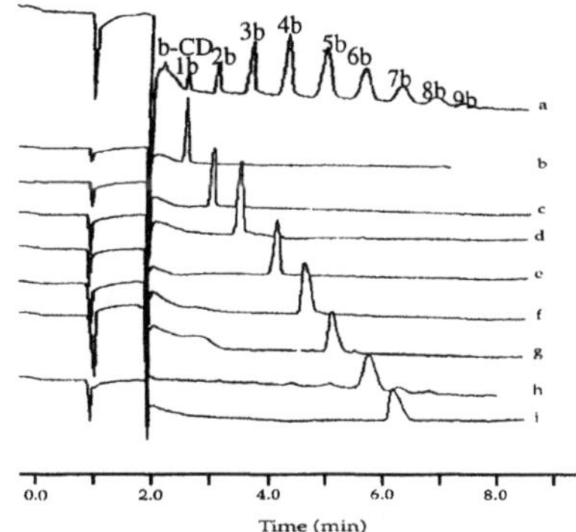

Figure 3. Capillary electrophoresis analysis of the SBE-β-CD bands isolated by anion exchange chromatography.

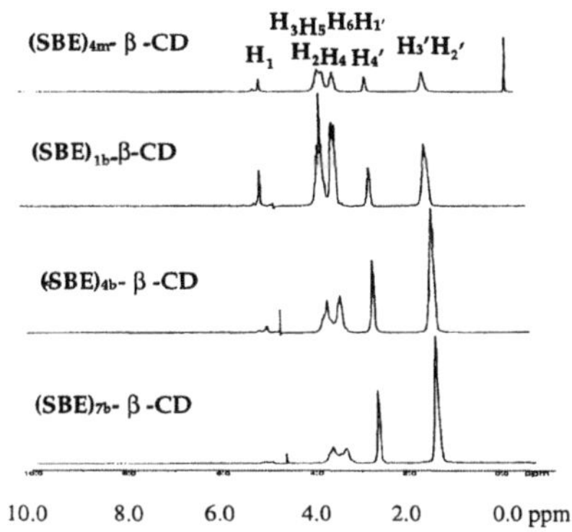

Figure 4. ^{1}H-NMR of SBE-β-CD mixture and isolated bands.

Figures 5-a and 5-h depict the FAB-MS spectrum of $(SBE)_{4m}$-β-CD obtained in two different matrices glycerol (GLY) and triethanolamine (TEA), respectively. FAB-MS in the negative ion mode using TEA or GLY, as a liquid matrix, gave well-defined mass spectra and the molecular ions of the SBE-β-CDs were clearly separated. The FAB-MS of each band is presented in the figure, where bands with DS from 1 to 6 were obtained in GLY and bands with DS of 7 and 8 in the TEA matrix. In each case the major peak corresponds to mono anions formed by the loss of one sodium atom from the neutral compound.

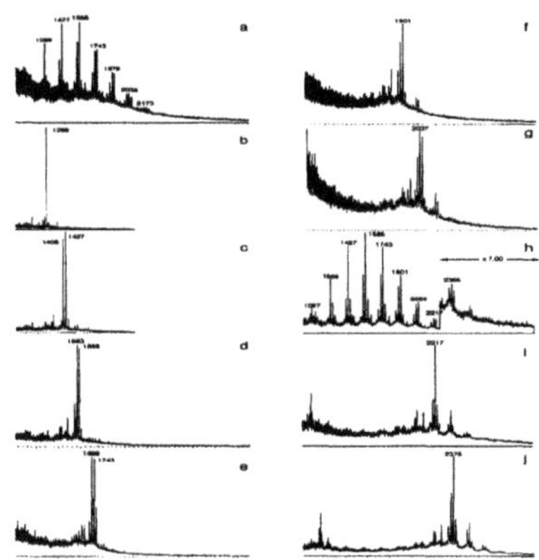

Figure 5. FAB-MS of the SBE-β-CDs

3.2. Composition of the SBE-*β*-CD Mixture

The molar response factors were determined by CE analysis under the experimental conditions described. The results are presented in Table I. The linear correlation was analyzed by the least-squares method. The CE molar response (slope) obtained with each band clearly increases with the increase of the DS of the SBE-β-CD.

DS	MW	n	slope	intercept	R2
0	1134	4	7.71E+07	-1.96E+03	0.999
1	1292	5	2.04E+08	1.08E+04	0.997
2	1450	5	4.04E+08	-7.85E+04	0.989
3	1608	4	5.34E+08	-1.63E+05	0.985
4	1768	4	6.87E+08	-7.49E+04	0.994
5	1924	5	9.71E+08	-2.03E+04	0.997
6	2082	4	1.13E+09	-5.41E+04	0.998
8	2398	4	1.75E+09	-1.71E+05	0.997

Table I. CE Molar Response Factors.

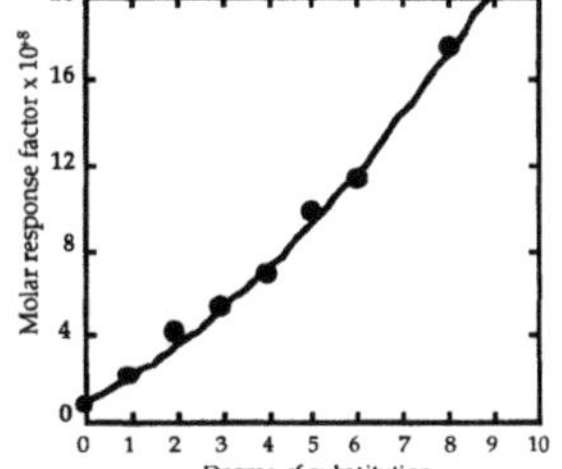

Figure 6. Dependence of the molar response factors on the degree of substitution of the $(SBE)_{nb}$-β-CD.

Figure 6 displays the dependence of the molar

factors with the degree of substitution. This dependence was found to follow a second degree polynomial. The molar response factor for $(SBE)_{7b}$-β-CD was obtained from this expression.

In order to estimate the actual mass contribution of the individual bands, a $(SBE)_{4m}$-β-CD sample was analyzed under the CE conditions used to obtain the molar response factors and the molar composition was then calculated.

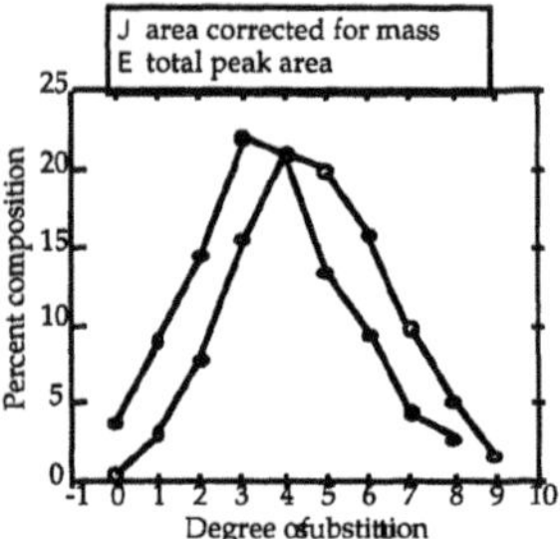

Figure 7. $(SBE)_{4m}$-β-CD Distribution Pattern:

Figure 7 presents the comparison of the pattern of distribution of the $(SBE)_{4m}$-β-CD calculated based on the percentage total peak area and the pattern of distribution as calculated from the percentage of the molar contribution of the individual bands corrected for the molar response factors. The percent total peak area, which represents the percent contribution of each peak to the total peak area of a given sample [11], assumes the same response factor for all the components of the mixture. As can be observed from the figure, the actual mass percent distribution of the mixture is not symmetrical but skewed toward the lower degrees of SBE substitution.

4. CONCLUSIONS

- The preparative fractionation of mixtures of SBE-β–CDs to produce the substitution bands was successfully accomplished by AEC with Na_2SO_4 step gradient elution and pH control.
- The components of the mixture were well resolved based on the differences in charge.
- The degree of substitution of each band was clearly assigned by FAB-MS and the purity of the samples as compared to the original mixture was established by CE analysis.
- A better characterization of the $(SBE)_{4m}$-β-CD was accomplished through a mass contribution definition of the composition of the product mixture.
- The values for the average degree of substitution derived form this method have a better agreement with the ones calculated from the elemental analysis (DS_{EA}).

ACKNOWLEDGMENTS

We wish to thank the Kansas Technology Enterprise Corporation through the Centers of Excellence Program for the financial support.

REFERENCES

[1] V. J. Stella, H. Y. Lee and D. O. Thompson. *Int. J. Pharm.* **120**, 189-195 (1995).

[2] V. J. Stella, H. Y. Lee and D. O. Thompson. *Int. J. Pharm.* **120**, 197-204 (1995).

[3] T. Jarvinen, K. Jarvinen, D. O. Thompson and V. J. Stella. *Cur. Eye Res.*, **13**, 897-905 (1994).

[4] T. Jarvinen, K. Jarvinen, A. K. Artti, D. O. Thompson and V. J. Stella. *J. Ocul. Pharmacol.*, **11**, 95-106(1994).

[5] B. A. Gorechka, D. S. Bindra, Y. D. Sanzgiri and V. J. Stella. *Int. J. Pharm.* 125, 55-61 (1995).

[6] T. Jarvinen, K. Jarvinen, N. Schwarting, and V. J. Stella. *J. Pharm. Sci.* **84**; 295-299 (1995).

[7] R. A. Rajewski, G. Traiger, J. Bresnahan, P. Jaberaboansari, V. J. Stella and D. O. Thompson. *J. Pharm. Sci.*, **85** (8), 927-932 (1995).

[8] R. A. Rajewski, "Development and Evaluation of the Usefulness and Parenteral Safety of Modified Cyclodextrins", Ph. D. Dissertation, University of Kansas, Lawrence, KS, USA, 1990.

[9] V. J. Stella and R.A. Rajewski, "Derivatives of Cyclodextrins Exhibiting Enhanced Aqueous Solubility and Pharmaceutical Uses Thereof", US Patent 5, 134,127, 1992.

[10] R. J. Tait, D. J. Skanchy, D. O. Thompson, N. C. Chetwyn, D. A. Dunshee, R. A. Rajewski, V. J. Stella and J. F. Stobaugh, *J. Pharm. Biomed. Analysis*, **10** (9), 615-622 (1992).

[11] E. A. Luna, E. R. N. Bornancini, R. J. Tait, D. O. Thompson, J. F. Stobaugh, R. A. Rajewski and V. J. Stella, *J. Pharm. Biomed. Analysis*, submitted.

DEVELOPMENT OF A SENSITIVE AND ACCURATE METHOD FOR THE QUALITATIVE ANALYSIS OF ENCAPSIN ^{TM} HPB.

S. S. DE KOCK and L. J. PENKLER
Druggists Group Research, Division of SA Druggists Ltd,
P.O. Box 6014, Korsten 6014, Port Elizabeth, South Africa.

ABSTRACT

A newly developed method is described by which the quality of Encapsin ^{TM} HPB (commercial 2-hydroxypropyl beta cyclodextrin) can be sensitively and accurately determined. The method involves the use of electrospray mass spectrometry in the negative ion detection mode for the determination of the degree of substitution and distribution of the 2-hydroxypropyl substituents. This method is an improvement on other mass spectrometric methods with regard to sensitivity and the lack of the formation of trace impurity ion adducts.

1. INTRODUCTION

Currently considerable attention is focused on the utilization of cyclodextrins and cyclodextrin derivatives. In the pharmaceutical industry cyclodextrins are used in drug formulations as they act as solubility enhancers for many small hydrophobic drugs by forming cyclodextrin- drug inclusion complexes.

The beta cyclodextrin molecule (BCD) is most idealy used, as the cavity size is highly suited for strong complexation with many drug entities. It has, however been shown that BCD is toxic, limiting its pharmaceutical use.

Because of this hydroxypropylated and other alkylated derivatives of beta cyclodextrin have become available. Of these the 2-hydroxypropylated beta cyclodextrin (Encapsin^{TM} HPBCD) derivative shows no toxicity, and can be used, even in parenteral formulations.

J. Szejtli and L. Szente (eds.), Proceedings of the Eighth International Symposium on Cyclodextrons, 137–140.

However, as the 2-hydroxypropylation reaction is difficult to control, the quality of this derivative may vary considerably between batches with regard to the average degree of substitution (DS) as well as the distribution of the degree of substitution of the hydroxypropyl substituents.

Incomplete derivatisation may lead to residual amounts of the toxic parent beta cyclodextrin being present, and the degree of substitution further affects solubility and stability of the cyclodextrin- drug complex. Knowledge of the DS is furthermore required as it is directly related to the molecular weight, a parameter which is of importance as the molar ratio of cyclodextrin to host effects the stability of the complex and resultant drug formulation.

While nuclear magnetic resonance spectrometry (NMR) can give an accurate indication of the DS of these molecules, determination of the distribution of the degree of substitution is difficult.

Mass spectroscopy is most often applied for the qualitative analysis of hydroxypropylated derivatives, most popularly plasma desorption (^{252}Cf PD-MS), fast atom bombardment (FAB-MS) and electrospray mass spectrometry (ES-MS).
It has however been shown that results obtained by these MS techniques can differ extensively, which may indicate that not all of these techniques are accurate under the given analysis conditions.

With this presentation a method is described by which the DS, mean molecular weight and distribution of the degree of substitution of a hydroxypropyl cyclodextrin samples can easily, sensitively and accurately be determined by electrospray mass spectrometry.

2. RESULTS AND DISCUSSION

The ES-MS analysis of cyclodextrins is almost always undertaken in the positive ion mode. In the spectra the many different HPB derivatives of different mass give rise to signals representing these mass differences and the quality of the HPB sample.
While the cyclodextrin molecule carries no positive charge, analysis conditions are normally selected which results in the formation of ion adducts which gives the cyclodextrin a positive charge. Most often protonation is effected to form $(M+H)^+$ adducts by acidifying the solvent or by adding ammonium ions to form $(M+NH_4)^+$ adducts.
The addition of sodium ions to form $(M+Na)^+$ adducts is not effective in ES-MS as sensitivity is greatly reduced due to the presence of a high sodium ion concentration.

The presence of trace amounts of cations such as Na^+ and K^+ in commercial samples of cyclodextrins derivatives however also results in the formation of $(M+Na)^+$ and $(M+K)^+$ adducts. In the positive ion detection mode the presence of trace amounts of these non volatile ions also reduces sensitivity markedly. Signals due to these Na^+ and K^+ adducts, as well as from adducts due to solvent, complicate the spectra and it is often difficult to obtain a true representation of the quality of cyclodextrin derivatives.

Results obtained from the analysis of the same batch of 2-HPB by two different MS techniques.

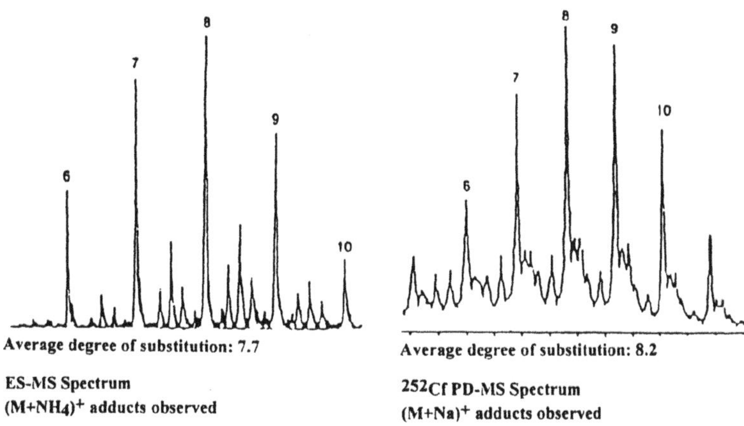

Average degree of substitution: 7.7

ES-MS Spectrum
$(M+NH_4)^+$ adducts observed

Average degree of substitution: 8.2

^{252}Cf PD-MS Spectrum
$(M+Na)^+$ adducts observed

When analysing HPB with ES-MS in the presence of either a volatile acid such as formic acid or ammonium formate in a range of different solvents we observed strong responces for the $(M+Na)^+$ and $(M+K)^+$ adducts, which greatly complicated the spectra and which could not be displaced even at high acid or ammonium ion concentrations.

None of the above methods gave satisfactory results in our study of commercial samples of 2-hydroxypropyl beta cyclodextrin and it was clear that an improvement on the currently used ES-MS methods was required.

It was reasoned that as the HPB molecule possess hydroxyl groups on the glucose residues as well as the hydroxypropyl substituents, high pH values should effect deprotonation to yield negatively charged species. When the HPB samples were analysed in negative ion mode using a solvent and mobile phase of isopropanol containing 1% ammonia, signals corresponding to $(M-H)^-$ species were observed.

Acetonitrile and water were then added to the solvent and mobile phase which increased sensitivity and signal. The best result was obtained when the solvent as well as mobile phase consisted of 50% isopropanol, 25% acetonitrile, 24% water and 1% ammonia.

The sensitivity was increased in comparison with previous analyses in the positive ion mode, while adduct ions were almost completely absent. From the spectra the DS and the distribution of the degree of substitution could easily be determined.

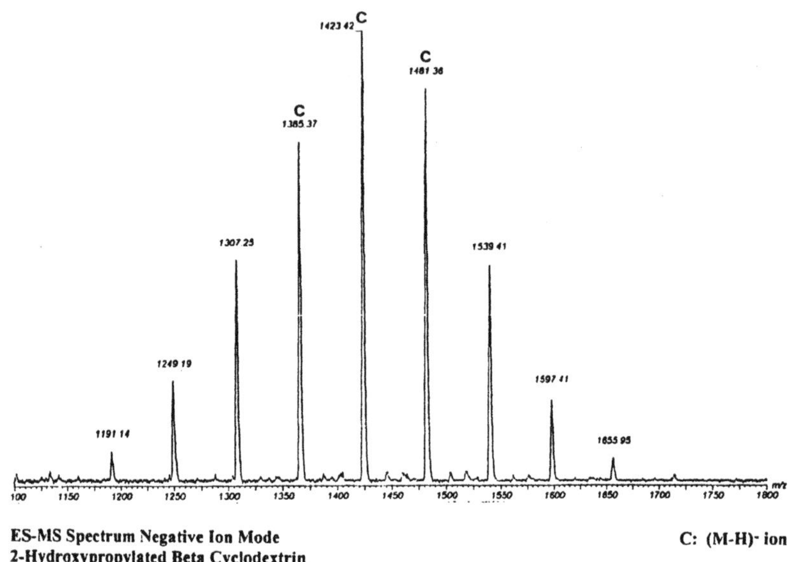

ES-MS Spectrum Negative Ion Mode
2-Hydroxypropylated Beta Cyclodextrin

C: $(M-H)^-$ ion

3. CONCLUSION

The accuracy of the method was investigated by comparing the DS results with NMR results for a range of HPB samples of different DS (Table 1)

TABLE 1. A comparison degree of substitution (DS) results as obtained by ES-MS with that obtained by NMR for different HPB batches.

Batch	DS (ES-MS)	DS (NMR)
Janssen 03J 281/1	4.79	4.82
Amaizo E8026	5.00	4.84
Amaizo 92/1	4.19	4.38
Amaizo D8037	5.78	5.87
Amaizo D8038	5.16	4.95

The results showed a strong correlation between the two sets of data, presenting some proof that this ES-MS method can be accepted as an accurate method for determination of DS.

SYNTHESIS, WATER SOLUBILITY, AND INCLUSION PROPERTIES OF THIOUREIDO β-CYCLODEXTRINS

C. ORTIZ MELLET, J.M. GARCIA FERNANDEZ, S. MACIEJEWSKI AND J. DEFAYE
CNRS and CEA, Département de Recherche Fondamentale sur la Matière Condensée, Centre d'Etudes de Grenoble, F-38054 Grenoble, France

ABSTRACT

The synthesis of 6^{I}-deoxy-6^{I}-thioureidocyclomaltoheptaose derivatives is described. A marked increase in the water solubility of these adducts as compared to the native β-CD is observed, even for derivatives incorporating hydrophobic substituents. Saccharide as well as glycopeptide antennae have been also efficiently appended to the β-CD core through thiourea spacers. By using these thioureido—β-CD conjugates, the water solubility of the anticancer agent Taxotère® was raised up to 4.7 g L^{-1}.

1. INTRODUCTION

A large array of covalently modified cyclodextrin derivatives has been prepared over the last few years with the aim to improve the properties of these host molecules for the solubilisation, stabilisation and, in some instances, targeting of drugs. When considering the more accessible β-cyclodextrin (β-CD) entity, a main feature to be improved is water solubility. A general strategy to face this problem consist in the binding of hydrophilic substituents onto the β-CD core through appropriate spacers. In addition, the use of such derivatives for the vectorized delivery of medicinally useful compounds would require incorporation of biological markers. With this problem in mind, we have now devised a more comprehensive approach based on the use of the thiourea functionality as highly hydrophilic spacer arm. The methodology relies on the high selectivity and efficiency of the condensation reaction of amines and isothiocyanates. Alkylthioureido, as well as more elaborated derivatives incorporating saccharide or glycopeptide substituents, have been thus prepared following a relatively simple synthetic methodology and, remarkably, with no need for hydroxyl protection.

J. Szejtli and L. Szente (eds.), Proceedings of the Eighth International Symposium on Cyclodextrons, 141–144.

2. MATERIALS AND METHODS

2.1. Materials

^{13}C NMR spectra (D_2O, 50.3 MHz) were recorded at 323 K using a Bruker AC200 instrument. LC was carried out using a Perkin-Elmer 250 pump fitted to a LC-30 refractive index detector. A Nucleosil C-18, 5 μ column was used with 12:88 MeOH-H_2O as eluent.

2.2. Methods

2.2.1. 6^I-Amino-6^I-deoxycyclomaltoheptaose (**1**): To a solution of 6^I-azido-6^I-deoxycyclomaltoheptaose [1] (1.49 g, 3.69 mmol) in DMF (1 mL) was added 1,3-propanedithiol (62 mL, 6.15 mmol) and ethyldiisopropylamine (0.65 mL, 6.15 mmol), and the reaction mixture was stirred for 16 h at room temperature. Acetone (100 mL) was then added, the solid precipitate was collected, dissolved in water (100 mL), filtered through Celite and freeze-dried. Yield 1.34 g (92%). Compound **1** showed the spectroscopical properties already reported [2].

2.2.2. 6^I-Deoxy-6^I-isothiocyanatocyclomaltoheptaose (**2**): To a solution of **1** (1.13 g, 1 mmol) in water-acetone (3:2, 75 mL) was added $CaCO_3$ (0.9 g, 3.0 mmol) and $CSCl_2$ (0.15 mL, 1.5 mmol). The mixture was stirred overnight at room temperature, then concentrated to half volume, diluted with water (30 mL), demineralized with Amberlite MB-6113 mixed resine (H^+,OH^-; 15 mL, 15 min), filtered and freeze-dried. Yield 0.76 g (65%), $[\alpha]_D$ +112.1 (*c* 0.6, pyridine).

2.2.3. General procedure for the preparation of 6^I-deoxy-6^I-thioureidocyclomaltoheptaose derivatives (**3-7**): To a solution of **1** or **2** in pyridine was added the corresponding isothiocyanate or aminonucleophile, respectively. The solution was stirred for 48 h at room temperature, then concentrated, the residue was dissolved in water, washed with $CHCl_3$, and freeze-dried. The yield of the condensation reactions was >90%, as seen from LC chromatograms. Pure samples were obtained after LC purification. The structures of **3-7** were confirmed by ^{13}C NMR, FABMS and analytical data; **3**, $[\alpha]_D$ +114.3 (*c* 0.8, H_2O); **4**, $[\alpha]_D$ +108.7 (*c* 1.0, H_2O); **5**, $[\alpha]_D$ +96.9 (*c* 0.7, H_2O); **6**, $[\alpha]_D$ +116.1 (*c* 0.6, H_2O); **7**, $[\alpha]_D$ +103.4 (*c* 0.9, H_2O).

2.2.4. 6^{I}-Deoxy-6^{I}-[3-(*N*-glycyl-β-D-glucopyranosylamine-6-yl)thioureido]cyclomaltoheptaose (8): A solution of **7** (77 mg, 51 μmol) in TFA-H_2O (9:1, 1 mL) was stirred at room temperature for 1 h, then concentrated. Traces of acids were eliminated by coevaporation with water and further treatment with Amberlite IR-904 (OH^-, 2 mL) anion exchange resin. Yield 66 mg (91%), $[\alpha]_D$ +82.2 (*c* 1.8, H_2O).

3. RESULTS AND DISCUSSION

Isothiocyanation of the monoamine derivative **1** was effected with $CSCl_2$ following the procedure previously reported for per-6-deoxy-6-isothiocyanato-CDs [3]. The title compounds were synthetized by one of the two different routes given in Scheme 1. The condensation was efficiently achieved in pyridine at r. t., and the crude products showed a high degree of purity (>90%). Although route (a) is preferable for commercially available isothiocyanates (e.g. MeNCS), the use of the β-CD—NCS reagent **2** (route b) was advantageous for the reaction with unprotected amino sugars, the reverse procedure (a) leading to formation of cyclic thiocarbamates in this case [4].

All the prepared thioureido—β-CDs, excepting **3**, exhibited a water solubility several times higher as compared to the native β-CD (Table 1). The observed solubility values also compare favourably with those reported for monosubstituted glycosyl—β-CDs or their *S*-linked counterparts [5]. Actually, they are similar to those for branched β-CDs persubstituted at the primary face, with the a priori advantage of the higher accessibility of the cyclodextrin cavity in monosubstituted derivatives. In agreement with that, the water solubility of the anticancer drug Taxotère® (~0.004 g L^{-1}) was raised up to 4.7 g L^{-1} in a 64.2 mM solution of the methylthioureido derivative **4**.

From this set of results, it can be concluded that the increase in water solubility in **4-8** lies in the intrinsic hydrophilicity of the thiourea functionality, since similar values were measured for thioureido—β-CDs incorporating hydrophilic (e.g. **5** and **8**) or hydrophobic substituents (e.g. **4** and **6**). The simplicity of the synthetic methodology, the selectivity and high yield of the coupling reaction between amines and isothiocyanates, compatible with the presence of the free hydroxyl groups, and the stability of the thiourea functionality allows incorporation of relatively complex haptens suitable for biological recognition in a controlled and very convergent fashion. This possibility is illustrated by the preparation of the glycosylamino acid—β-CD conjugate

8. Reaction of **2** with the selectively functionalized *N*-glycyl-β-D-glucopyranosylamine derivative precursor afforded the Boc-protected compound **7** (70% yield) which after deprotection with 90% aqueous TFA afforded **8** in almost quantitative yield.

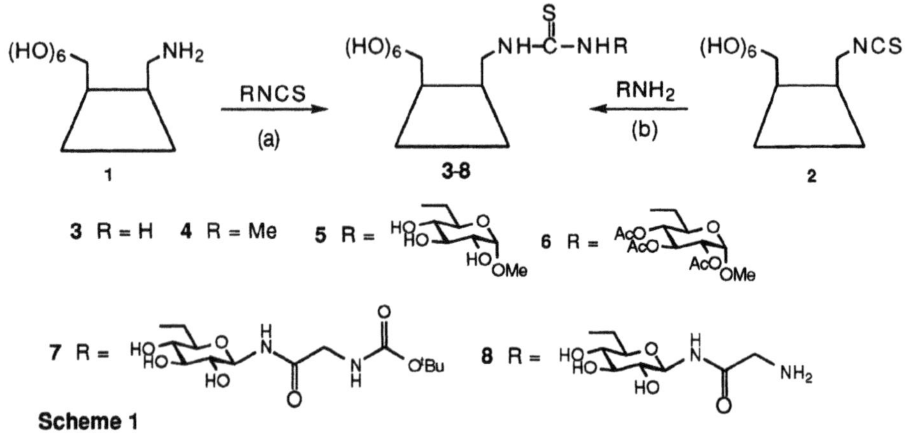

Scheme 1

TABLE 1. Water solubility at 25 °C (mmol L^{-1}) of thioureido—β-CD derivatives.

β-CD	**4**	**5**	**6**	**8**
15	640 (x 42.6)	518 (x 34.5)	373 (x 24.8)	>567 (x 38.0)

ACKNOWLEDGEMENTS

This work was supported by the European Commission DG12 under the FAIR programme contract no. FAIR-0300.

REFERENCES

[1] Petter, R.C., Salek, C.T., Sikorski, G., Kumaravel, G., Lin, F.-T., Cooperative binding by aggregated mono-6-(alkylamino)-β-cyclodextrins, *J. Am. Chem.* Soc., **112,** 3860-3868 (1990).

[2] Brown, S.E., Coates, J.H., Coghlan, D.R., Easton, C.J., van Eyk, S.J., Janowski, A.L., Lincoln, S.F., Luo, Y., May, B.L., Schiesser, D.S., Wang, P.; Williams, M.L., Synthesis and properties of 6^A-amino-6^A-deoxy-α- and β-cyclodextrin, *Aust. J. Chem.*, **46,** 953-958 (1993).

[3] García Fernández, J.M., Ortiz Mellet, C., Jiménez Blanco, J.L., Fuentes Mota, J., Gadelle, A., Coste-Sarguet, A., Defaye, J., Isothiocyanates and cyclic thiocarbamates of α,α'-trehalose, sucrose, and cyclomaltooligosaccharides, *Carbohydr. Res.*, **268,** 57-71 (1995).

[4] García Fernández, J.M., Ortiz Mellet, C., Fuentes, J., Chiral 2-thioxotetrahydro-1,3-O,N-heterocycles from carbohydrates. 2. Stereocontrolled synthesis of oxazolidine pseudo-*C*-nucleosides and bicyclic oxazine-2-thiones, *J. Org. Chem.*, **58,** 5192-5197 (1993).

[5] Lainé, V., Coste-Sarguet, A., Gadelle, A., Defaye, J., Perly, B., Djedaïni-Pilard, F., Inclusion and solubilization properties of 6-*S*-glycosyl-6-thio derivatives of β-cyclodextrin, *J. Chem. Soc., Perkin Trans. 2*, 1479-1487 (1995).

ONE STEP SYNTHESIS OF BRANCHED CYCLODEXTRINS

J. DEFAYE AND J.M. GARCIA FERNANDEZ
CNRS and CEA, Département de Recherche Fondamentale sur la Matière Condensée, Centre d'Etudes de Grenoble, F-38054 Grenoble, France

ABSTRACT

Branched cyclodextrins with enhanced solubility in water and of interest as therapeutic carrier agents have been prepared in a single step by hydrogen fluoride-mediated glycosylation of a cyclomaltooligosaccharide substrate with monosaccharides or reducing oligosacharides. The reported examples include α-, β-, and γ-cyclodextrin derivatives incorporating D-glucose, D-mannose, 2-acetamido-2-deoxy-D-glucose, and maltose substituents. Glycosylation occurs preferentially at the primary hydroxyl rim and, interestingly, with almost exclusively α-anomeric configuration.

1. INTRODUCTION

6-O-Glycosyl cyclomaltooligosaccharides (branched cyclodextrins) represent an attractive improvement with regard to the parent cyclodextrins (CDs) for the solubilisation in water of hydrophobic molecules [1]. These derivatives are, however, produced as complex mixtures with fairly low yields by biotechnological processes, and their purification, especially from linear oligosaccharide by-products, is tedious. Their chemical synthesis involving Köenigs-Knorr type glycosylation methodologies requires a number of steps and can therefore not be scaled-up realistically for industrial applications. Incorporation of an oligosaccharide cell recognition signal in the CD core, as would be needed for drug targeting using these molecular vessels, is still then problematic on a reasonable scale. Alternative procedures involving pyrolysis of CDs alone or in admixture with another sugar substrate yield many side decomposition products which make the isolation and purification difficult [2, 3]. We have previously found [4] that glycosylation of reducing sugars in anhydrous hydrogen fluoride (HF) at high sugar concentration occurs without solvolysis of the existing glycosidic linkages,

J. Szejtli and L. Szente (eds.), Proceedings of the Eighth International Symposium on Cyclodextrons, 145–148.

non-reducing oligosaccharides remaining stable under these conditions. This approach has now been extended to the preparation of glycosides of CDs in a single step, at r. t., and with high glycosylation yields by using HF as both solvent and catalyst [5].

2. MATERIALS AND METHODS

2.1. Materials

Anhydrous HF was a commercial product obtained in steel cilinders. Prior to use, it was distilled and kept in polyethylene bottles at -25 °C. ^{13}C NMR spectra (D_2O, 50.3 MHz) were recorded using a Bruker AC200 instrument. FAB-mass spectra (Cs, acceleration potential 8 kV) were measured in the positive mode with a VG ZAB-SEQ instrument. Glycerol-thioglycerol (unacetylated products) and *m*-nitrobenzyl alcohol (peracetylated derivatives) were used as the liquid matrix. NaI or CsI was added as cationizing agent.

2.2. Methods

2.2.1. General procedure for the preparation of poly(glycosyl) cyclodextrins: All reactions were carried out in capped polyethylene bottles. A mixture of the corresponding cyclomaltooligosaccharide (α-, β-, or γ-CD; 2.8 mmol) and a monosaccharide or reducing oligosaccharide, in different relative proportions, was carefully triturated in a mortar and cooled down in a dry ice-acetone bath. HF (total sugar-HF ratio 2:1 w/v) was added and the mixture was vigorously stirred with a steel spatula and then allowed to reach room temperature. A homogeneous stiff solution was obtained within 5 min. The bottle was then closed and the mixture was stored at room temperature for 1 h. The reaction was quenched (liquid nitrogen), ether (3 x 50 mL) was then added and decanted from the solid white precipitate which was collected and dried.

2.2.2. General procedure for the preparation of poly(2-acetamido-2-deoxy-D-glucopyranosyl) cyclodextrins: The above protocole was followed starting from a CD and 2-acetamido-2-deoxy-D-glucose. After Addition of HF , the reaction mixture was stored in the open air overnight to allow evaporation of HF. The residue was dissolved in water, neutralized by addition of solid $CaCO_3$, filtered, and freeze-dried.

3. RESULTS AND DISCUSSION

Preliminary experiments showed that all three CDs considered in this study (α-, β-, and γ-CD) were transformed into D-glucose and oligosaccharides thereof in HF solutions using CD-HF ratios higher than 1:1, similarly to previous reported results in the HF-solvolysis of starch [6]. In contrast, CDs remained unchanged after treatment with HF at r. t. for 16 h when the CD-HF ratio was kept lower than 2:1. Addition of a reducing mono- or oligo-saccharide to these solutions resulted in its complete incorporation onto the CD core, as inferred from the absence of signals arising from reducing anomeric carbon atoms in the ^{13}C NMR spectra of the crude reaction mixtures. A 2:1 (w/v) total sugar-HF ratio was found to be optimal in terms of homogeinity. After 1 h, the branched product was recovered by precipitation with ether. The overall structure of the reaction mixtures was assessed by ^{13}C NMR and FABMS techniques. Thus, ^{13}C signals for α-glycopyranosides were almost exclusively observed (C-1 at 95.6 ppm for D-Glc derivatives). No signals from furanoid derivatives were detected, whereas a low intensity signal at 103 ppm in the spectra of D-Glc—CD products probably account for β-glucopyranosyl residues (C-1 β, less than 5% of C-1 α, see above). Glycosylation occurs preferentiallly at the primary hydroxyl rim of the CD core (Scheme 1) and further glycosylation may follow at the primary alcool positions of the already grafted residues. The degree of branching was dependent on the initial CD-glycosylating sugar molar ratio, as seen from FABMS spectra of the crude or peracetylated mixtures, ranging from 0-4 glycosyl substituents for 1:1 ratios, to 0-15 for 1:3 ratios.

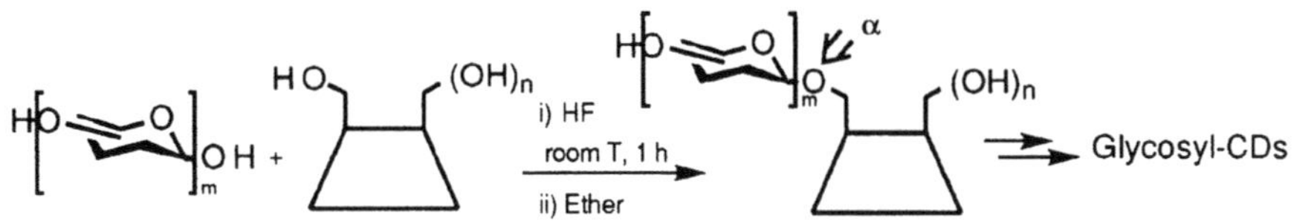

HO (OH)n OH + i) HF room T, 1 h ii) Ether α (OH)n Glycosyl-CDs

Scheme 1

An important aspect of these resulting products relates to their water solubility (Table 1), which was found to be closely related to the number of glycosyl substituents grafted onto the CD core as well as to their mono- or di-sacharide nature. Thus, the solubility in water at 25 °C of the branched product arising from HF treatment of a β-CD—D-Glc mixture change from 91 g L^{-1} to 1320 g L^{-1} for 1:1 to 1:3 molar ratio of reactants, respectively, whereas a value of 1200 g L^{-1} was measured for a 1:1 β-CD—maltose mixture. It must be noticed that the α-(1→4) pre-existing glycosidic linkage of maltose is not affected under the stated reaction conditions, activation occurring exclusively at the anomeric position of the reducing end. The above protocol has been extended to 2-

acetamido-2-deoxy-D-glucose as glycosyl donor. Slow evaporation of HF was found advantageous in this case, with glycosylation yields of ~60%.

The potential application of these poly(glycosyl)CDs as hydrophobic drug carriers in biological media is ilustrated by the increase in solubility observed for the anticancer drug Taxotère® which rises from 0.004 g L^{-1} in neat water to 1.5 g L^{-1} in a 65 g L^{-1} solution of the branched β-CD product obtained from HF treatment of a 1:1 β-CD—D-Glc mixture.

TABLE 1. Degree of branching and water solubility at 25 °C (g L^{-1}) of poly(glycosyl)CDs.

CD-reducing sugar (molar ratio)	Degree of branching	water solubility (25 °C, g L^{-1})
β-CD—D-Glc (1:1)	0-4	91
β-CD—D-Glc (1:3)	0-15	1320
β-CD—D-Man (1:3)	0-17	1400
β-CD—D-GlcNAc (1:2)	0-3	1100
β-CD—maltose (1:1)	0-6	1200
α-CD—maltose (1:1)	0-8	1450
γ-CD—maltose (1:1)	0-8	1580

ACKNOWLEDGEMENTS

This work was supported by the European Commission DG12 under the FAIR programme contract no. FAIR-0300.

REFERENCES

[1] Hashimoto, H., Preparation, structure, properties and applications of branched cyclodextrins, in New Trends in Cyclodextrin Derivatives (Ed. Duchêne, D.), Éditions de Santé, Paris, 1991, pp. 99-156.

[2] Ammeraal, R.N., Deboer, E.D., Method for making branched cyclodextrins and product produced thereby, U.S. patent 4904307 (1990).

[3] Kobayashi, S., Kadoma, M., Toshiba, O., Okemoto, T., Hara, K., Ishigami, H., Mikuni, K., Osawa, T., Hashimoto, H., Heterosaccharide-modified cyclodextrin manufacture, Japanese patent 03192101 (1991).

[4] Defaye, J., García Fernández, J.M., Synthesis of dispirodioxanyl pseudo-oligosaccharides by selective protonic activation of isomeric glycosylfructoses in anhydrous hydrogen fluoride, *Carbohydr. Res.*, **251**, 1-15 (1994).

[5] Defaye, J., García Fernández, J.M., Method for the preparation of branched cyclomaltooligosaccharides, WO patent 95 21870 (1995).

[6] Defaye, J., Gadelle, A., Pedersen, C., The behavior of cellulose, amylose, and β-D-xylan towards anhydrous hydrogen fluoride, *Carbohydr. Res.*, **110**, 217-227 (1982).

TOXICOLOGICAL COMPARISON OF CYCLODEXTRINS

G. ANTLSPERGER AND G. SCHMID
Wacker - Chemie GmbH, Hanns-Seidel -Platz 4
D-81737 München, Germany

ABSTRACT

Analysis of the toxicological properties of cylodextrins reveals that these very similar molecules possess surprisingly different efficiency. CDs differ in absorption rates as important parameter for systemic activity. α-and β-CD can only be degraded by the intestinal flora and therefore intact molecules are absorbed in the upper intestines at rates of 2% or 4-6% respectively.
However, the absorption of γ-CD is negligible, because it is easily digestible by pancreatic enzymes.
β-CD shows already at relatively low oral doses reactions at the primary target organ, the kidneys. Osmotic nephrosis and release of proteins of the globulin fraction as indicator for glomeruli damage, are induced.
In comparison, α- and γ-CD did not damage neither tubules nor glomeruli, even at the highest oral doses applicable of 20% in the daily diet of rats and dogs.
γ-CD only causes osmotic nephrosis within a 1 month study at the very high *intravenous* dosage of 2000 mg/kgBW and day, 600 mg/kg for 90 days result in marginal effects. Orally, γ-CD does not affect a target organ. Only i.v. administration in extreme doses additionally provokes alveolar histiocytosis, erosion of epithelia of the urinary tract and higher reticulocyte counts.
Comparison of the NOAELs makes obvious that γ-CD is by far the most favourable CD.

1. INTRODUCTION

In recent years toxicological data on CDs have accumulated significantly, performed for supporting pharmaceutical and food approvals. Now, there are subchronic and chronic studies for β-CD [1][2][4] and for HP-β CD [3] available for the public. Wacker-Chemie has subchronic studies on α-, γ- and Methyl-β-CD DS 1.8 [5] and respective ADME data [6]. With this review we intend to provide a better understanding of the different characteristics of the toxicity of CDs.

J. Szejtli and L. Szente (eds.), Proceedings of the Eighth International Symposium on Cyclodextrons, 149–155.

The article is concentrating on comparison of systemic effects. However factors influencing absorption are very important and are significant parameters for differentiation.

2. DEGRADABILITY AND DIGESTIBILITY

Degradability was investigated by us [7] by means of the Zahn-Wellens test (OECD 302B), and some examples we summarized in Fig.1.
Parent CDs are degraded after a lag phase of 24 h with identical rate (slope) to reach 90% mineralisation within 4 days.
Ester-bond CD derivatives like Acetyl-CDs show a prolonged lag-phase of 2 or 3 days followed by a similar degradation rate. Obviously, at first the esters has to be hydrolysed to reach a low degree of substitution, then ring cleavage occurs.
Ether-bond CD derivatives are far more stable. For instance, low substituted Methyl-β-CD 0.6 is degraded only up to 40% and Methyl-β-CD with an average substitution degree of 1.8 is completely stable. In case of the Me-ß-CD 0.6 it appears that only the lower substituted fraction of the product mixture can be metabolized.

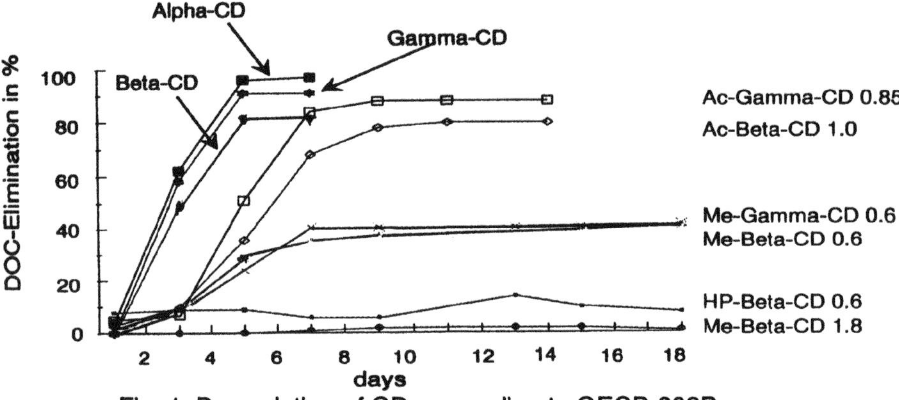

Fig. 1: Degradation of CDs according to OECD 302B

Interestingly, low substituted HP-ß-CD can be degraded only to 10-15 %. The more voluminous and/or numerous the substituents are, the more resistant are ether-bond derivatives against biological breakdown. A possible explanation is reduced accessibility of enzymes by steric hindrance. This could be confirmed using the more potent soil system [7], being superior to fecal bacteria [8].

Digestibility is shown in Fig. 2 as 14-CO_2 exhalation graphs for labeled α and γ-CD [6]. The peak of CO2 exhalation for α-CD is reached after 9-10 hours of oral administration indicating that α-CD is only degraded in caecum and colon by the intestinal flora. When germ-free rats are fed with α-CD, no radioactive CO_2 is liberated [6]. We postulate the same results for β-CD [9].

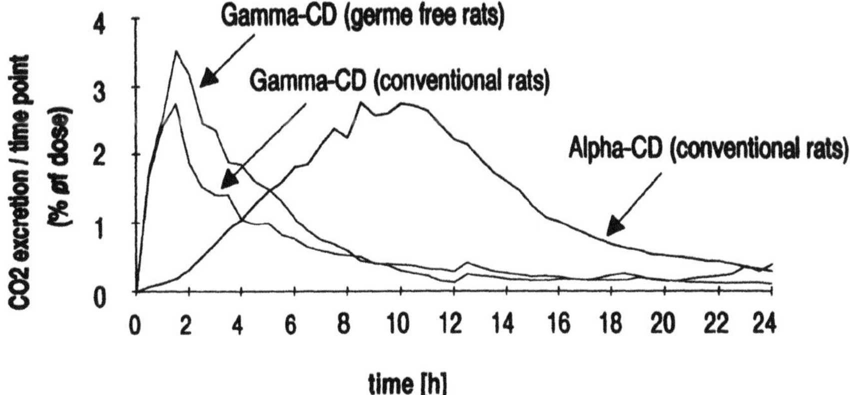

Fig.2: CO2-Excretion after an oral dose of 1000 mg/kgBW

In contrast, if germ-free rats are fed with γ-CD, the CO_2 release is identical to that of conventional rats [6] and to that of starch [9].
This gives evidence γ-CD to be digested in the upper intestinal tract by pancreatic enzymes and intact γ-CD is hardly to appear in the large intestines and to cause caecal enlargement, acidification ect.. Furthermore, digestion of starch will not be disturbed by competitive inhibition of amylases [10]. This favourable behaviour is confirmed in feeding studies in rats and dogs [11].
In summary, adaptive phenomena of the low digestible carbohydrates type [12] [13] are avoided and direct systemic effects of γ-CD by oral uptake can be excluded as proven by ADME studies [6].

3. ABSORPTION

Absorption rates are important parameters for systemic activity of CDs after oral dosage. In the studies available [1][2][6][14][15][16], absorption is monitored by recovery in urine. Because CDs and derivatives do not undergo enterohepatic circulation and as there is no metabolism during the systemic passage, the assumption above may be allowed [6]. There is a minimal excretion by the salivary glands which is estimated to be less than 10 % of the recovery in urine [6].

		% of Dose	Species	Reference
1	α-CD	≤ 2 %	Rat	Wacker-Chemie
2	β-CD	1,3 - 6,2 %	Dog	WHO, 44. JECFA
3	γ-CD	≤ 0,1 %	Rat	Wacker-Chemie
4	Me-β-CD 1.8	0,5 - 11,5 %	Rat	Wacker-Chemie
5	DM-β-CD	6,3 - 9,6 %	Rat	Szatmari u. Vargay
6	HP5-β-CD	3 - 6 %	Rat	Gerlozy et al
7	HP?-β-CD	3 %	Dog	Monbaliu et al

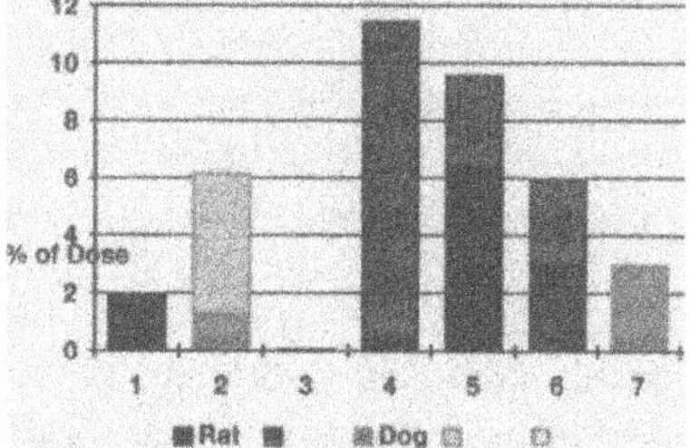

Fig.3: Absorption rates of CDs and derivatives in rats and dogs

4. KIDNEY EFFECTS

CDs induce in kidneys, preferably in the proximal tubules, a reaction, well known as osmotic nephrosis. This phenomenon is triggered by i.v. dosed dextrans and glucose as well [17].
We have estimated the quantities of intact CDs in the blood per kg BW and day causing reabsorptive vacuolization and put them in correlation to those (column 10), which correspond to the no-effect levels of toxicological studies. The figures differ clearly. But the situation becomes really obvious when realistic time frames are chosen: We use a model, where uptake and excretion of all *intact* CDs occurs within 2.4 h after a single meal. This is close to the results we obtained regarding the kinetics of α- and γ-CD. For gavage studies, in general, this is real, for feeding studies in dogs the simplification may be allowed. The estimation is further simplified by an average excretion per hour, disregarding the development of concentrations (C_{max}).Results are listed in Table 1. This model may be justified, because (i)the determination of the bioavailability of *intact* parent CDs causes a high financial input, (ii) practicability is questionable and (iii) it is likely that pharmacokinetics in this range are linear and the decisive data are of the same magnitude.

						Quantitiy absorbed and excreted daily					
CD	Duration	% in diet	Species	Uptake g/kg BW xd	Resorptionrate	total mg/kg BW xd	total mg/animal	mg/kgBW xh for 2.4 h	NEL mg/kgBW xh for2,4 h	NELoral mg/kg/ BW/d	% in diet
α	90 d	20	rat	14	2 %	280	56	110	110	14 000	20
α	90 d	20	dog	8	(2 %)	160	1600	67	67	8 000	20
β	90 d	5	rat	2,5	5 %	125	25	50	11,4	570	1,25
β	52 w	5	rat	1,28	5 %	64	26	26	13,5	650	1,25
β	52 w	5	dog	2.0	6 %	120	1200	50	11,3	470	1,25
γ	90 d	20	rat	14	0,1 %	14	3	6	250	14 000	20
γ	90 d	20	dog	8	0,1 %	8	80	3,3	?	8 000	20
γ	90 d	i.v.	rat	0,6	100%	600	150	250	250	equiv.600 g/kg BW xd oral	–
Me-β 1.8	26 w	gavage	rat	0.5	10%	50	25	20	4.2	100	–

Table 1: Estimation of systemic charge in relation to kidney damage

Histopathological consequences do not appear when β-CD structures do not exceed quantities of 10-14 mg/kgBW per hour to be excreted. α-CD may reach the 10-fold and γ-CD the 25- to 50-fold of that charge. In reality, the kidneys excreted in the 90-day i.v. γ-CD study within 1 hour (3x t/2) 525 mg/kg BW daily without being damaged.
Regarding the habit of rats in feeding studies to divide their daily dose in multiple portions, the figures in column 9 and 10 would reduce according to the elongated time frame which mimics an unknown steady state level.
Whole body autoradiographic data of RAMEB show (Fig. 4) that Me-β-CD DS1.8 is persisting in kidneys for at least 147 h after i.v. application in nearly unchanged concentrations mainly in the cortex [6]. We speculate that there is a binding site for CDs, with far highest affinity to β-CD and derivatives, triggering

continuously reabsorptive vacuolization at not measurable concentrations of β-CD in blood.

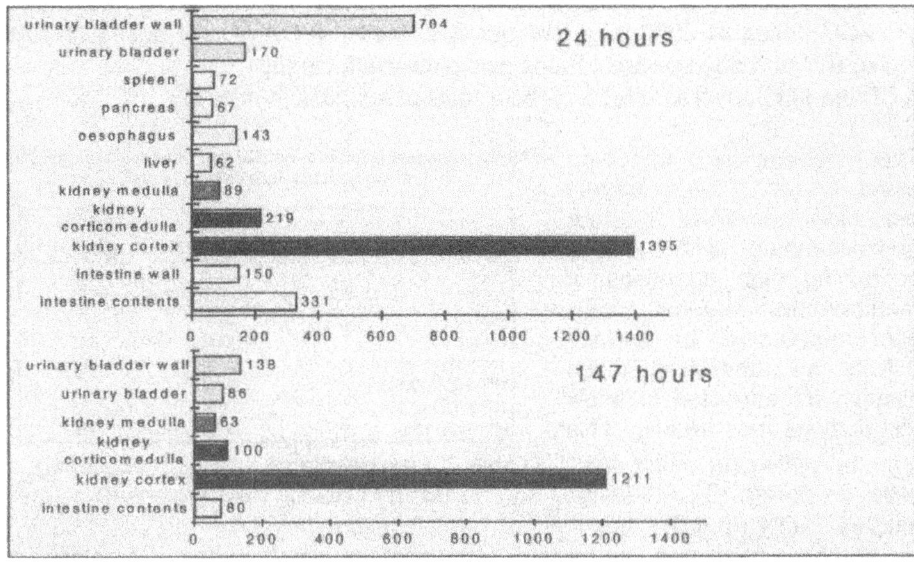

Fig.: Concentration of radioactivity after a single i.v. dose of 10 mg/kgBW (3MBq/kg)

A further target of β-CD in kidneys are the glomeruli. It is reported for β-CD that a daily oral application of 1300 mg/kg BW and day led to loss of proteins of the globulin fraction found in urine. i.m. injections of β-CD of 50 mg/kg BW to rabbits cause considerable damage of the glomeruli within 12 days [18]. Surprisingly, Me-β-CD DS1.8 even at daily oral doses of 1000 mg/kg BW (equiv. to 50-100 mg/kgBW x d parenteral) does not effect the glomeruli.

Intravenous γ-CD-doses of 2000 mg/kg BW for 30 days do not induce effects in glomeruli, neither suspicious quantities of protein were released nor histopathological alterations were detected.

5. MEMBRANE EFFECTS

Hemolysis of CDs is often reported [ ][ ], however, the toxicological relevance has to be considered as not important. The i.v. toxicity of α- and β-CD to the kidneys does not allow to apply quantities to provoke hemolysis.

A daily 2000 mg/kg BW bolus of γ-CD for 30 days causes hemolysis to an only negligible extent, but the erythrocyte membrane gets artificially aged. Consequently, the reticulocyte concentration increases slightly as result of a higher turn over of erythrocytes [5].

The effect on membranoid materials causes reactions of complete epithelia: in the case of the *alveolar histiocytosis*, it does not result in fibrosis and therefore complete reversibility could be observed. But the *urinary tract epithelia* erode at i.v. γ-CD doses of 2000 g/kg BW per day. Methyl-β-CD at oral doses of 1000 mg/kg BW and day upward causes also similar effects [5].
β-CD leads to adverse effects of the intestinal epithelia in mice [2]

The repeatedly observed effect of *macrophage activation* correlates for our under-standing with tissues containing high quantities of membranoid material. CDs alter membranes by surface activity and therefore those tissues are expected to show this findings most readily. This immune activating effect was also re-ported in feeding studies with β-CD, most prominent as activated appearance of mesenteric lymph nodes and patches of Peyer [1][2][5]. β-CD is likely to change membrane surfaces most effectively. These findings correlate with theory and results of the bacterial luminescence test [19] as shown in Table 2.

Inhibitory concentrations for natural and modified cyclodextrins in the bacterial luminescence test

Cyclodextrin	IC20 % w/w	x fold [β-CD]
α-CD	0.16	5
β-CD	0.031	1
γ-CD	2.0	65
Dimeb	0.0063	0.2
Trimeb	0.16	5
HP α-CD (0.6)	0.08	2.5
HP β-CD (0.6)	0.063	2
HP γ-CD (0.6)	0.63	20

Table 2: Luminescence inhibition according to Bar and Ulitzur 1994 (selected)

6. SUMMARY

- Digestibility and different absorption rates result in decisive different toxicological characteristics of CDs.
- Systemically, β-CD induces effects at much lower concentrations than α- and γ-CD.
- The systemic effects are likely to be connected in some cases with tissues containing high quantities of membranoid material. The capability of CDs to alter membranes is different, γ-CD needs excessive concentrations, but causes at least some of the same reactions.
- In the case of triggering osmotic nephrosis we postulate a binding site, as already very low concentrations of β-CD and some of its derivatives are sufficient; on the other hand, the extreme concentrations reached during i.v. application of γ-CD, are provoking the same but do not result in precipitation of any complex.
-Sum and combination of these effects find their expression in the possibilities of use and in proposed and theoretical ADIs as indicated on table 3.

Simplified over all comparison

CD oral	sensitve marker	NEL mg/kg BW/d	ADI	g/d man
α	carbohydrate effects	1900	19*	1.4
β	carbohydrate effects, protein in urine histopathology: liver, kidney, epithelia	500	5°	0.35
γ	marginal carb.hydr.eff.	14000	140*	10
Me-β	histopathology: kidney, liver	100	1*	0.07

* **theoretical ADI, based on subchronic studies**

° **proposed ADI by FAO/WHO**

Table 3: Primary target organs and estimated dose range for human use, parent CDs and Me-β-CD

7. REFERENCES

[1] *WHO Food Additives Series: 32* Toxicological evaluation of certain food additives and contaminants 41. Meeting of the joint FAO/WHO Expert Committee on Food Additive, Geneva 1993 IPCS techn. information

[2] *WHO Food Additives Series:.35* Toxicological evaluation of certain food additives and contaminants 44. Meeting of the joint FAO/WHO Expert Committee on Food Additive, Rome 1995

[3] Coussement W, Van Cauteren H, Vandenberghe J, Vanparys P, Teuns G, Lampo A, and Marsboom R, Toxicological profile of hydroxypropyl β-cyclodextrin (HP β-CD) in laboratory animals, p 522-4, in Duchene D, Editor, *Mins 5th Int Symp Cyclodextrins:* Mar 28-30, 1990, Paris, Fr., Editions de Sante; Paris, Fr., 1990.

[4] Wacker-Chemie GmbH, Chronic (52 weeks) feeding study in dogs with β-CD, unpublished
Multigeneration study with β-CD in rats, unpublished

[5] Wacker-Chemie GmbH, 1987 - 1995:Toxicity Studies on α-CD, γ-CD and Methyl-β-CD, unpublished

[6] Wacker-Chemie GmbH: ADME- Studies with α-, γ-CD and Methyl-β-CD DS1.8 (RAMEB) in rats, unpublished

[7] Wacker-Chemie GmbH, Degradation tests on CDs in sewage and soils, unpublished

[8] Personal communication on incubation of feces of adapted rats

[9] Andersen, G.H.; Robbins, F.M.; Domingues, F.J., Moores & Long, C.L.(1963)
The utilization of Schardinger dextrins by the rat. *Toxicol. Appl. Pharmacol.*, ***5***, 257 - 266

[10] Fukuda K.et al. Specific inhibition by Cyclodextrins of raw starch digestion by fungal glucoamylase *Biosci.Biotech.Biochem.,**56**(4),556-559.1992*

[11] Antlsperger G. New Aspects in Cyclodextrin Toxicology, *Min 6th CD-Symposium, Chicago, April 1992,*Ed.Santé

[12] de Groot, A.P., Biological effects of low digestibility carbohydrates. *Proc.TNO-CIVO Workshop* 1986, Zeist, NL

[13] Lina B. Suzuki T.90-day oral toxicity study with AO-128 in rats. *Jap. Pharmacology &Therapeutics,* **19,**10,1991

[14] Szatmári I, and Vargay Z, Pharmacokinetics of dimethyl-beta-cyclodextrin in rats, p 407-413, *Proc 4th Int Symp Cyclodextrins:* April 20-22, 1988, Munich, W.Ger., Kluwer Academic Publishers; Dordrecht, Neth., 1988.

[15] Gerlóczy A, Antal S, Szatmari I, Müller-Horvath R, and Szejtli J, Absorption, distribution and excretion of 14-C-labeled HP-β-CD in rats following oral administration, p 507-13, *Mins 5th Int Symp Cyclodextrins*: Mar 28-30, 1990, Paris, France, Editions de Sante; Paris, Fr. 1990.

[16] Monbaliu J, Van Beijsterveldt L, Meuldermans W, Szathmary S, and Heykants J, Dispositon of HP-β-CD in experimental animals, p 514-17, *Mins 5th Int Symp Cyclodextrins*: Mar 28-30, 1990, Paris, Edition Sante;1990.

[17] Maunsbach A B, Madden S C, and Latta H, Light and electron microscopic changes in proximal tubules of rats after administration of glucose, mannitol, sucrose or dextran, *Lab Invest,* **11**, 421-432, 1962.

[18] Serfozo J, Szabo P, Ferenczy T, and Toth-Jakab A, Renal effects of parenterally administered methylated CDs on rabbits, p 123-32, in Szejtli J, Editor, *Proc 1st Int Symp Cylcodextrins:* Sep 30-Oct 2, 1981, Budapest, Hungary, Kluwer Academic Publishing Group: D. Reidel Publishing Co; Dordrecht, Neth., 1981.

[19] Bar R,Ulitzur S, Bacterial toxicity of CDs:luminous E. coli as a model, *Appl.Microbiol.Biotechnol.*, **41**, 574-7,1994.

DIETARY FIBER-LIKE EFFECTS OF ORALLY ADMINISTERED CYCLODEXTRINS IN THE RAT

FUJIKO SHIZUKA [1], KOZO HARA [2], and HITOSHI HASHIMOTO [2]

[1] *Nagano Prefectural College, 8-49-7 Miwa, Nagano 380, Japan*
[2] *Ensuiko Sugar Refining Co., Ltd., 13-46 Daikoku-cho, Turumi-ku, Yokohama 230, Japan*

ABSTRACT

Molecular difference of physiological effects of cyclodextrins (CD) was compared with those of dietary fiber in the rat. Animals were fed diets containing either 5 or 10% of CD supplemented with 1% cholesterol. α CD as well as pectin prevented diet-induced hypercholesterolemia, whereas β CD increased the weight of small intestine. Modified CD, maltosyl-α, maltosyl-β and hydroxypropyl-β, showed similar effects observed in their parent CD. Dietary fiber-like effects of α and β CD might be related to their characteristics of digestibility, being highly indigestible and highly fermentable, respectively.

1. INTRODUCTION

Since considerable amounts of CD are used in food and pharmaceutical industries, physiological effects of orally consumed CD have been received considerable attention. Most of CD are indigestible oligosaccharide, which effects are supposed to be similar to those of dietary fiber. Some studies with β CD revealed hypolipidemic effect[1-2] but such information about the other CD has not yet been available. In addition to three types of ordinary α, β, and γ CD, various types of modified CD have been developed. Compared with ordinary CD, most of modified CD possess beneficial characteristics for industrial application, being improved solubility in the solvent. Such advantage make it possible to apply more modified CD for industrial usage. The purpose of present study was to clarify the health benefit of CD containing products by examining the structural difference of physiological effects of popularly used CD.

2. MATERIALS AND METHODS

2.1. Animals and diets

Six-week-old male SD rats, six rats per dietary group, were given diets *ad libitum* for two weeks. Experimental diets contained either 5 or 10% of CD. Six types of CD used are ordinary α, β, γ CD, and modified CD, maltosyl (G2) - α, maltosyl (G2) - β, hydroxypropyl (HP) - β CD. Control diets contained dietary fiber instead of CD. Two

J. Szejtli and L. Szente (eds.), Proceedings of the Eighth International Symposium on Cyclodextrons, 157–160.

different types of dietary fibers, water insoluble cellulose and soluble pectin, are used. All the diets are supplemented with 1% cholesterol. Compositions of the diets are shown in Table 1.

TABLE 1. Composition of experimental diets

Dietary levels of dietary fiber (DF) or cyclodextrin (CD)	5%	10%
	g/100 g	
α-Starch	44.5	39.5
Sucrose	20.0	20.0
Casein	20.0	20.0
Corn oil	5.0	5.0
Mineral mix.	3.5	3.5
Vitamin mix.	1.0	1.0
Cholesterol	1.0	1.0
DF or CD	5.0	10.0
1. Cellulose 2. Pectin		
3. α-CD 4. β-CD 5. γ-CD		
6. Maltosyl-α-CD (G2-α-CD)		
7. Maltosyl-α-CD (G2-β-CD)		
8. Hydroxypropyl-α-CD (HP-β-CD)		
Total	100.0	100.0

2.2. Analyses

On the following morning of the completion of feeding period, blood was withdrawn from inferior vena cava under pentobarbital anesthesia. Plasma lipids were determined enzymatically by using commercial kits. The entire small intestine was removed, cut into equal length of three segments, weighed after removing the luminal contents by flushing with ice cold saline. Data are expressed as mean±SEM. Statistical analysis was carried out by ANOVA followed by Fisher's protected LSD. Values of $p<0.05$ were considered significant.

3. RESULTAND DISCUSSION

Fig. 1 shows plasma lipid concentration in rats given dietary fiber or CD diets. Animals fed diets containing 5% of cellulose and pectin developed hyperlipidemia due to dietary cholesterol. When dietary pectin was increased to 10%, plasma lipid concentrations decreased. Among the CD tested, α and G2 α CD similarly affected as pectin. Pectin-like effect was not seen in other CD. Those of β, γ, G2 β, HP β CD rather increased plasma total cholesterol concentration. Pectin is gel-forming soluble non-starch polysaccharide. The most feasible hypolipidemic mechanism of such compounds as pectin is acceleration of bile acids and sterols elimination into feces[3]. Since α CD is highly indigestible[4], the characteristic nature of the CD to form inclusion complex with lipids might affect as similarly as pectin. On the other hand, hypocholesterolemic effect of β CD in the rat[1-2], which is inconsistent with our result, has been reported. In those experiments, contribution of fecal elimination of bile acids has been suggested for the hypocholesterolemic mechanism of β CD. Since indigestibility and fermentability are the characteristic feature of α and β CD, respectively, it is possible that indigestible α CD exert more potent hypolipidemic effect than β CD as shown in the present study. Further studies are needed to evaluate the hypolipidemic effect of α and β CD.

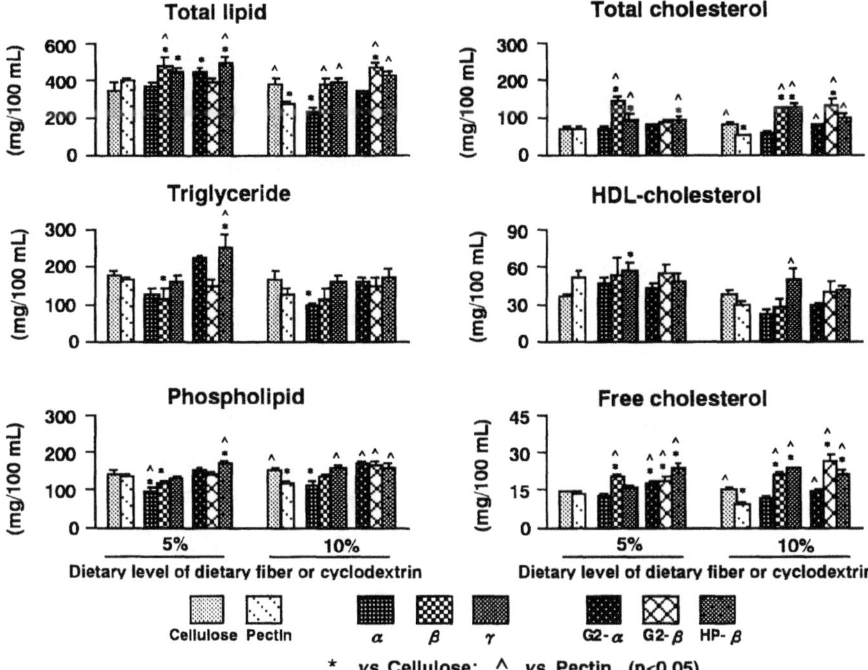

Fig. 1 Plasma lipid concentration

Segmental small intestine weight is shown in Fig 2. Pectin increased small intestine weight than cellulose. All of the CD tested, except highly digestible γ CD, affected as similarly as pectin. Small intestine weight of β CD and modified G2 α, G2 β, HP β CD fed rats was significantly heavier than that of pectin. The effect of β and modified CD are marked in the upper and lower third segments of small intestine. Two possible mechanisms to explain the trophic effect by dietary fiber were suggested. One is the physical stimulation by indigestible fibers such as cellulose. The other is the chemical stimulation by fermentable fibers such as pectin. Present result showed fermentable fiber exert more potent trophic effect than indigestible fiber. Trophic effect observed in α CD might be related to physical stimulation by this indigestible CD, whereas the effect observed in fermentable β CD might related to chemical stimulation by fermentation products. Contribution of both physical and chemical stimulation to the trophic effect was confirmed by the fact that the most potent effect was observed in fermentable but most indigestible HP β CD. The principle products of luminal fermentation by microorganisms are short chain fatty acids, which are utilized by epithelial cell as a fuel, thereby stimulating the epithelial cell proliferation[5]. Therefore, trophic effect by fermentable β CD might be the result of stimulation of epithelial cell proliferation. In the present experiment, trophic effect observed in small intestine was not confirmed in the colon. It is possible, therefore, the trophic effect is the secondary effect to increase intestinal blood flow, in which mechanism trophic hormones such as enteroglucagon might be involved. Apart from the question of mechanism, application of CD in food and pharmaceutical products may augment the beneficial effect to maintain gut function.

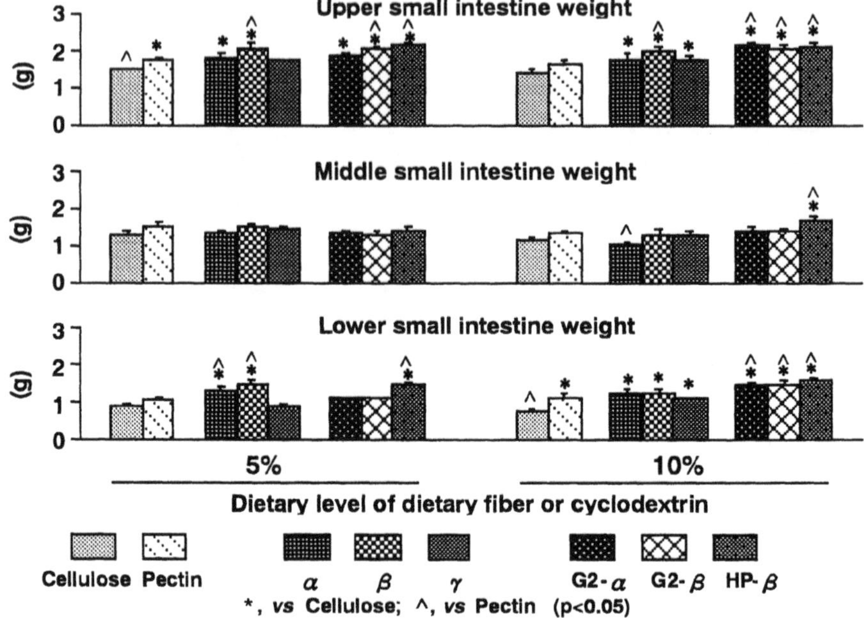

Fig. 2 Segmental weight of small intestine

4. CONCLUSION

Present study showed that orally consumed CD could provide dietary fiber-like health benefits and the effects were structure depended. Structural difference of CD's effect seemed to be related to the digestibility of the CD. Indigestible α CD might exert physical effect in the gut serving to moderate lipid absorption, resulting in hypolipidemic effect. Fermentable β CD might promote gut function by supplying the fermentation products to affect epithelial cell metabolism.

REFERENCES

[1] Riottot, M., Olivier, P., Huet, A., Coboche, J.-J., Parquet, M., Khallou, J. and Lutton, C., Hypolipidemic effects of β-cyclodextrin in the hamster and in the genetically hypercholesterolemic Rico rat, *Lipids*, **28**, 181-188 (1993)

[2] Levrat, M.-A., Favier, M.-L., Moundras, C., Remesy, C., Demigne, C. and Morand, C., Role of dietary propionic acid and bile acid excretion in the hypocholesterolemic effects of oligosaccharides in rats, *J. Nutr.*, **124**, 531-538 (1994)

[3] Miettinen, T. A., and Tarpila, S., Effect of pectin on serum cholesterol, fecal bile acids and biliary lipids in normolipidemic and hyperlipidemic individuals, *Clin. Chim. Acta,*, **79**, 471-477 (1977)

[4] Suzuki, M. and Sato, A., Nutritional significance of cyclodextrin: indigestibility and hypolipemic effect of α-cyclodextrin, *J. Nutr. Sci. Vitaminol.*, **31**, 209-223 (1985)

[5] Sakata, T., Effects of indigestible dietary bulk and short-chain fatty acids on the tissue weight and epithelial proliferation rate of the digestive tract in rats, *J. Nutr. Sci. Vitaminol.*, **32**, 355-367 (1986)

Chapter 2.

CYCLODEXTRIN INCLUSION COMPLEXES

METALLO - CYCLODEXTRINS.

N.R. RUSSELL and M.McNAMARA.
Dublin Institute of Technology, Kevin Street,
Dublin 8.

ABSTRACT

Metallo-CDs have potential as metallo enzyme models. CD has relatively poor coordinating ability which can, however be enhanced if (i) the pH is well above 7 or (ii) the CD is functionalised with an appended moiety containing active coordinating sites. Exploitation of these concepts leads to the formation of first sphere metallo-CDs having a range of applications in catalysis and molecular recognition. On the other hand the host-guest activity of CDs leads to the formation of second-sphere adducts with metal complexes containing hydrophobic ligands. These adducts can form the basis for drug delivery processes and for stabilisation of complexes otherwise subject to oxidation or reduction.

1. INTRODUCTION.

The presence of a chiral hydrophobic cavity and an adjacent metal ion centre within the same molecule are undoubtedly desireable features of any species designed for molecular recognition phenomena. Metallo-CDs, despite having these features, have not received the level of attention afforded to other such systems. The principal aim of this paper is to put into context the different facets of metallo-CD chemistry as developed over the last two decades. The ability of CDs to include hydrophobic species together with the presence of a multitude of active sites on the cavity rims with potential for hydrogen bonding and, under appropriate conditions, coordination, renders CDs ideal candidates as both first sphere and second sphere ligands for metal ions.

2. CYCLODEXTRINS AS SECOND SPHERE LIGANDS.

2.1 Nature of second sphere Interaction.

Second sphere interaction was recognised by Alfred Werner[1] when he described solvation affects on the physical properties of metal complexes. Host compounds can be regarded as a series of solvation sites joined by covalent bonds. First sphere ligands are attached covalently to the metal ion whereas second sphere ligands are attached non-covalently to the first sphere ligands, resulting in the formation of a second coordination sphere.

J. Szejtli and L. Szente (eds.), Proceedings of the Eighth International Symposium on Cyclodextrons, 163–169.

The nature of this phenomenon can be described as being driven by intermolecular forces, such as, hydrogen bonding, charge transfer dispersion, Van der Waal's forces, hydrophobic and electrostatic interactions.

Over the past twenty years molecular hosts have been recognised as potential second sphere ligands [2,3,4]. CD's in particular because of their increased availability, their ability to include a wide range of guest molecules, their hydrophilic exterior and, of course, their relatively non-toxic character, should have an important role to play in this regard.

2.2 Organometallic complexes.

Because of the nature of the CD cavity it is not surprising that the first reports of second sphere coordination adducts of CDs involved the hydrocarbon type ligands of ferrocene and its derivatives. Breslow[5] reported a 1:1 complex of ferrocene with β-CD. Further studies of hydrolysis of p-nitrophenyl esters of ferrocene carboxylic acids showed dramatic rate accelerations of ~$5.9x10^6$ particularly in the case of ferrocene acrylic acid ester [6-10]. Harada and Takahashi [11-13] prepared crystalline samples of ferrocene adducts with ratio 2:1 for α-CD and 1:1 for both β- and γ-CD. On the basis of ICD spectra the various different orientations of ferrocene were proposed and these were later supported by solid state nmr[14]. The inclusion of ferrocene in CDs was the subject of many studies including cyclic voltammetry [15,16,17] and ICD [18,19]. Finally it was determined[20] that the inclusion of ferrocene in β- and γ-CD is axial and equatorial respectively and that ferrocene is axial to a bimolecular cavity in the case of the α-CD adduct, the α-CD molecules being joined via hydrogen-bonding.

For a series of complexes of the type, $[(\eta^6arene)Cr(CO)_3]$, it was shown [21] that adducts with CDs were formed selectively on the basis of the steric bulk of guests and of the cavity size.

A detailed crystallographic study[22] carried out on the mixed sandwich complex adduct, $[(\eta^5Cp)Fe(\eta^6C_6H_6)]$ $(\alpha CD)_2$ (PF_6) revealed that the cation is encapsulated in a cavity formed by an α-CD dimer, the PF_6^- ion residing at the primary rim of the CD moiety containing the benzene ring.

A study of the percentage conversion in insertion reactions of $[\eta^5Cp)Fe(CO)_2R]^+$ with CO and SO_2 in the presence of CD[23], showed that insertion activity is controlled by the size of the CD cavity, γ>β>>α-CD.

2.3 Metallo-organic complexes.

The concept of using cyclodextrins as drug delivery agents is, of course, an attractive one, not least for the fact that they are relatively non-toxic.

A good example of this is the 1:1 adduct of carboplatin (diammine-1,1-cyclobutane dicarboxyatoplatinum(II)) with α-CD. X-Ray[24] and ^{1}H nmr[25] results show clearly that the hydrophobic cyclobutane moiety resides in the α-CD cavity, extra stabilisation being achieved via hydrogen-bonding between the two ammino ligands and the secondary hydroxy sites on the CD rim. The effectiveness of the drug was not significantly altered by adduct formation, however, the considerable side effects were reduced. A further development of this system is the so called 'oyster' complex[26], in which the carboplatin is encapsulated by α-CD functionalised with a crown ether moiety. The cyclobutane ring resides in the α-CD cavity, while the ammino ligands are hydrogen-bonded to the crown ether cavity.

Some of the highest values for free energy of adduct formation (~20kJmol^{-1}) were recorded for the adduct of trans-$[Pt(PMe_3)Cl_2(NH_3)]$ with β-CD and DMβ-CD[27], in which PMe_3 moiety enters the cavity at the primary face. Metal-rotaxane adducts have been reported for long chain hydrocarbons (η=8-12) with coordinating groups such as, ammino or bipyridyl at either end, extrusion being inhibited by metallo-complex formation[28].

3. CYCLODEXTRINS AS FIRST-SPHERE LIGANDS.

3.1 Metallo-complexes with unmodified CDs.

Matsui et al[29] first reported metallo-complexes which were prepared at pH~10 and which had metal ions directly coordinated to the CD hydroxy sites. These authors reported the preparation of Cu(II) complexes with α- and β-CD which analysed for a Cu:CD ratio of 2:1. Electronic spectra showed a shift in λmax from 800nm for aqueous copper(II) to 667nm for the metallo-CD complex. On the basis of this evidence and further results obtained from potentiometric experiments, a binuclear hydroxy bridged structure was proposed with Cu(II) ions coordinated by O(2) and O(3) positions on adjacent pyranose rings. A manganese(III)-CD complex also having a metal:CD ratio of 2:1 was reported by Nair and Dismukes[30] and a similar structure proposed. A series of transition metal complexes were prepared[31] as pale coloured powders (Cr(III), Mn(III), Co(II), Ni(II) and Cu(II)) all having an M:CD ratio of 2:1. An Fe(III)-CD complex analysed for an M:CD ratio of 4:1. Electronic spectroscopy, magnetic susceptibility, FTIR and Raman studies supported the proposed structures. Other authors[32] obtained similar FTIR results, however, their interpretation did not involve a bridged structure. More recently, Klufers et al[33] isolated blue crystals, formulated as, $Li_4[Li_7Cu_4(\beta\text{-CDH}_{11.5})_2]\ \eta H_2O$. The 2:1 ratio for Cu:CD is obtained, but the X-Ray data is interpreted in terms of a β-CD double torus with bridges formed between four copper(II) ions. Seven lithium ions are also coordinated within the double torus cavity.

Yamanari et al reported[34] a cobalt(lll) complex formulated as $[Co(en)_2CD]ClO_4$ for α- and β-CD. ^{13}C nmr and ICD evidence suggested that the Co(lll) ion is coordinated at O(2) and O(3) of the same glucose unit. Chromatographic columns prepared[35] from SP-sephadex C-25 resin and individual enantiomers of the β-CD complex were successful in resolving cis and trans isomers of a Co(lll) anionic complex containing an aminoacidato ligand. Evidence suggests that chiral recognition is mainly responsible for this process.

3.2 Metallo-complexes with functionalised CDs.

The very limited coordination ability of CDs particularly at pH≤7 can be overcome using functionalised CDs. The concept of a vacant chiral cavity and an adjacent metal ion site in one molecule prompted Breslow et al to prepare[36] a nickel(ll) chelate of the α-CD ester of pyridine-2,5-dicarboxylic acid as a model metalloenzyme. A development of this was the preparation of a nickel(ll) complex with CD-picolinate and 2-pyridinealdoxime as ligands, known as 'Breslow's Enzyme'. The presence of this complex accelerated the rate of hydrolysis of p-nitrophenyl acetate by a factor of $\sim 10^3$.

A complex of copper(ll) having a Cu:CDen ratio of 1:2 (CDen=mono(C(6)-NH-$(CH_2)_2$-NH_2)-6 deoxy-βCD) was reported[37] as an effective catalyst for the oxidation of furoin at pH~10.5. This complex consists of two hydrophobic cavities joined by a catalytically active site, a geometry well suited to the furoin molecule which also contains two hydrophobic moieties joined by an active site. The catalyst is poisoned in the presence of cyclohexanol indicating that inclusion plays an important role in the catalytic mechanism. Electronic spectral evidence indicates that the Cu(ll) is coordinated by N's of CDen. A series of polyammine functionalised β-CDs have been used[38] as ligands for metal ions such as, Cu(ll), Zn(ll) and Mg(ll). In all cases evidence suggests that the metal ion is located outside the cavity and is coordinated by the appended moiety. A zinc(ll) complex of a diethylenetriamine functionalised β-CD[39] is an effective catalyst for the hydrolysis of adenosine cyclic phosphate.

Models for immobilised metallo-enzyme systems have been prepared[40] by soaking copper(ll) exchanged montmorillonite in an aqueous solution of CDen. Analyses indicate a 1:1 Cu:CDen ratio in the solid. ESR studies indicate that the first sphere of the copper ion is a planar $[Cu(en)(H_2O)_2]^{2+}$ species, where the en is provided by the CDen molecule.

Mixed inclusion ligand exchange chromatography by a histamine functionalised β-CD copper(ll) complex was shown[41] to be successful - separating D- and L-aminoacids. The deoxy-6-N-histamino-βCD formed a 1:1 complex with copper(ll) thus giving rise to an ideal receptor for enantiomeric recognition of aromatic amino acids.

A supramolecular assembly formulated as $[\beta\text{-CD(aza)Eu}]^{3+}$ where aza=1,4,10,13-tetraoxa-7, 16-diazacyclo-octadecane, containing the luminescent europium(lll) ion was characterised[42] by ^{1}H and ^{13}C nmr, FTIR, FABMS and UV-vis spectroscopy. Enchanced luminescence of the Eu^{3+} active site in the presence of benzene over that of $[Eu(aza)]^{3+}$ is attributed to the unimolecular absorption-energy transfer - emission process, involving benzene included in the βCD cavity as donor and the Eu^{3+} ion residing in the aza cavity as the energy acceptor.

4. CONCLUSION.

Although this paper is not meant to be comprehensive it does provide a snapshot of the activity in the field of metallo-CD chemistry over the past two decades. This promises to be a most fertile area of research forming part of the ever widening field of supramolecular chemistry. It requires close cooperation between organic and inorganic chemists, biochemists and instrumental analysts to cope with the new challenges in synthesis and structural investigation necessary to further probe the nature of supramolecular phenomena.

REFERENCES.

[1] Werner, A, *Z. Inorg. Chemic.*, **3**, 267 (1893).

[2] Colquhoun, H.M., Stoddart, J.F., and Williams, D.J., *Angew. Chem., Int. Ed. Engl.,* **25**, 487 (1986).

[3] Pedersen, C.J. *J. Amer. Chem. Soc.,* **89**, 2495, 7017 (1967); *Aldrichimica Acta*, **4**, 1 (1971).

[4] Popov, A.I. and Lehn, J-M, in '*Coordination Chemistry of Macrocycle Compounds'*, ed. G.A. Melson, Plenum Press, New York (1979), pp. 537; Szejtli (ed.), *'Cyclodextrin Technology'*, Kluwer Academic Publishers, Dordrecht (1988). Bender, M.L. and Komiyama, M. (ed.), *'Cyclodextrin Chemistry'*, Springer-Verlag, New York (1978). Atwood, J.L., Davies, J.E.D., and Osa, T. (eds.), *'Clathrate Compounds, Molecular Inclusion Phenomena and Cyclodextrins'*, D. Reidel, Dordrecht (1984).

[5] Siegel, B. and Breslow, R., *J. Am. Chem. Soc.*, **97**, 6869 (1975).

[6] Czarniecki, M.F. and Breslow, R., *J. Am. Chem. Soc.*, **100**, 7771 (1978).

[7] Breslow, R., Czarniecki, M.F., Emert, J. and Hamaguchi, H. *J. Am. Chem. Soc.,* **102,** 762 (1980).

[8] Trainor, G.L. and Breslow, R., *J. Am. Chem. Soc.,* **103**, 154 (1981).

[9] Breslow, R., Trainor, G.L. and Ueno, A., *J. Am. Chem. Soc.*, **105**, 2739 (1983).

[10] Le Noble, W.J., Srivastava, S., Breslow, R. and Trainor, G.L., *J. Am. Chem. Soc.*, **105**. 2745 (1983).

[11] Harada, A. and Takahashi, S., *J. Chem. Soc., Chem. Commun.*, 645 (1984).

[12] Harada, A. and Takahashi, S., *J. Incl. Phenom.*, **2**, 791 (1984).
[13] Harada, A. and Takahashi, S., *Chem. Lett.*, 2089 (1984).
[14] Clayden, N.J., Dobson, C.M., Heyes, S.J., and Wiseman, P.J., *J. Incl. Phenom.*, **5**, 65 (1987).
[15] Matsue, T., Akiba, U., Suzufuji, K. and Osa, T., *Denki Kagaku.*, **53**, 508 (1985).
[16] Matsue, T., Evans, D.H., Osa, T. and Kobayashi, N., *J. Am. Chem. Soc.*, **107**, 3411 (1985).
[17] Ryabor, A.D., Tyapochkin, E.M., Varfolomeer, E.D., and Karyakin, A.A., *J. Electroanal. Chem., Bioelectrochem. Bioenerg.*, **24**, 257 (1990). McCormack, S., Russell, N.R. and Cassidy, J.F., *Electrochimica Acta.*, **37(11),** 1939 (1992).
[18] Ueno, A., Moriwaki, F., Osa, T., Hamada, F. and Murai, K. *Tetrahedron Lett.*, **26**, 899 (1985).
[19] Kobayashi, N. and Osa, T., *Chem. Lett.*, 421 (1986).
[20] Kobayashi, N. and Opallo, M., *J. Chem. Soc., Chem. Commun.*, 470 (1990).
[21] Harada, A., Saeki, K and Takahashi, S., *Chem. Lett.*, 1157 (1985).
[22] Klingert and Rihs, G., *Organometallics,* **9(4),** 1136 (1990).
[23] Shimada, M., Harada, A., Takahashi, S., *J. Chem. Soc., Chem. Commun.*, 263 (1991).
[24] Alston, D.R., Slawin, A.M.Z., Stoddart, J.F., and Williams, D.J., *J. Chem. Soc., Chem. Commun.*, 1602 (1985).
[25] Alston, D.R., Lilley, T.H., and Stoddart, J.F., *J. Chem. Soc., Chem. Commun.*, 1600 (1985).
[26] Stoddart, J.F., and Zarzycki, R., *Rec. Trav. Chem. Pays-Bas*, **107**, 515 (1988).
[27] Alston, D.R., and Slawin, A.M.Z., Stoddart, J.F., Williams, D.J., and Zarzycki, R., *Angew. Chem., Int. Ed. Engl.*, **27**, 1184 (1988).
[28] Ogino, H., *J. Am. Chem. Soc.*, **103**, 1303 (1981) ; Ogino, H. and Ohata, K., *Inorg. Chem.*, **23,** 3312 (1984).
[29] Matsui, Y., Kurita, T., and Date, Y., *Bull. Chem. Soc. Jpn.*, **45,** 3229 (1972) ; Matsui, Y, Kurita, T., Yagi, M., Okayama, T., Mochida, K., and Date, Y., *Bull. Chem. Soc. Jpn.*, **48**, 2187 (1975) ; Matsui, Y. and Suemitsu, D., *Bull. Chem. Soc. Jpn.*, **58**, 1658 (1985).
[30] Nair, B.U. and Dismukes, G.C., *J. Am. Chem. Soc.*, **105**, 124 (1983).
[31] McNamara, M., and Russell, N.R., *J. Incl. Phen. and Mol. Recog. Chem.*, *10*, 485 (1991).
[32] (a) Egyed, O., and Weiszfeiler, V., *Vibrational Spectroscopy* (submitted 1992).
(b) Fuchs, R., Habermann, N. and Klüfers, *Angew. Chem. Int. Ed. Engl.*, **32(b),** 852 (1993).
[34] Yamanari, K., Nakamichi, M. and Shimura, Y., *Inorg. Chem.*, **28,** 248 (1989).
[35] Yamanari, K. and Nakamichi, M., *J. Chem. Soc., Chem. Commun.*, 1723 (1989).
[36] Breslow, R. and Overman, L.E., *J. Am. Chem. Soc.*, **92 (4),** 1075 (1970), Szejtli and Pagington, J., *Cyclodextrin News*, **5(10),** 143 (1991).
[37] Matsui, Y., Yokoi, T., and Mochida, K., *Chem. Lett.* 1037 (1976).

[38] Tabushi, I., Shimiza, N., Sugimoto, T., Shiozuka, M., and Yamamura, K., *J. Am. Chem. Soc.*, **99(2),** 7100 (1977).

[39] Yoichi, M. and Makota, K., *J. Mol. Catal.*, **61(1),** 129 (1990).

[40] Matsui, Y., Kijima, T., Tanaka, J. and Gota, M., *Nature*, **310**, 45 (1984).

[41] Corradini, R., Impellizeri, G, Maccarrone, R., Rizzarelli, E. and Lecchio, G., *Dep. Mol. Organ.*, **8**, 209 (1991).

[42] Pikramenou, Z. and Nocera, D.G., *Inorg. Chem.*, **31**, 532 (1992).

Enhanced Mass Spectrum of the 2:1 γ-CD-C-60 complex

The detection of a sandwich-type complex (Pb-γ-CD) and the observation of cyclodextrin oligomers in the gas phase by FAB-MS is also presented

Thomas Andersson, Gunnar Westman, Gunnar Stenhagen
and Olof Wennerström
Department of Organic Chemistry, Chalmers University of Technology, S-412 96 Göteborg, Sweden

ABSTRACT

The FAB/LSIMS mass spectrum of the 2:1 γ-CD/C-60 complex is greatly enhanced with addition of $PbNO_3$ to the sample. Fast atom bombardment (FAB/LSIMS) has also been used for the detection of a new type of complex in cyclodextrin chemistry. These complex are the multinuclear sandwich-type complexes of deprotonated cyclodextrins and metal ions. Characterisation of hydrogen bounded cyclodextrin oligomers by FAB/LSIMS and linked scan collision experiments are also reported.

1. INTRODUCTION

Cyclodextrin derivatives as well as their host-guest complexes, have been characterised by X-ray, NMR, UV and CD spectroscopic methods.

Another method which is much less explored, but indeed shows a great potential, is mass spectrometry (MS). Of the ionisation-techniques for cyclodextrin used today, FAB/LSIMS is the most utilised. Since 1990, a few host guest complexes has been studied by MS.[1,2] Our interest in cyclodextrin complexes started with the observation of a 2:1 complex between γ-CD and C_{60} which was studied by negative FAB/LSIMS.[2] Our mass spectra also clearly shows the presence of the negative ion corresponding to the empty dimer of γ-CD minus one proton. Although, clusters or aggregates of various hydrogen-bonded matrices and solutes are commonly observed by FAB/LSIMS the specific structure of the cyclodextrines, an empty cone with two rims with many hydroxyl groups, made us pursue this observation further. As a part of this study we examined briefly the stability of the cyclodextrin dimer.

J. Szejtli and L. Szente (eds.), Proceedings of the Eighth International Symposium on Cyclodextrons, 171–174.

Recently, a new type of complex, multinuclear sandwich-type complexes of deprotonated cyclodextrins and metal ions, has been prepared. The first of these stable cage compounds was reported to consist of a double torus of two β-CD units joined together by copper and lithium ions.[3] The second example, a complex between two γ-CD units and lead ions has a similar cage structure.[4] Both these complexes were reported by Klüfers, who obtained well resolved X-ray structures. We now which to report a mass spectrum of the complex consisting of two γ-CD held together by lead ions. This is the largest cyclodextrin containing molecule which has been studied by MS (m/e 5877) and it is a nice example of the potential of the MS-technique. During our study of cyclodextrin-lead complexe we found a method to enhance the intensity of the MS of the γ-CD/C_{60} complex considerable. This might be a general method to obtain good FAB-MS spectra when their is low abundance of the complex ion.

2. MATERIALS AND METHODS

Mass spectra were recorded on a VG ZabSpec (Fison Instruments) in both positive and negative FAB/LSIMS mode. A Caesium gun with an accelerating voltage of 20 kV was used to generate the high energy ions. The chemicals used including the matrices were all commercial. The γ-CD lead complex was prepared as previously described.[4]
For the enhancement of the 2:1 γ-CD/C_{60} peak, the complex, prepared as described previously,[2] was dissolved in water by gentle heating. When $Pb(NO_3)_2$ was added to the solution a brown solid appeared. This solid was then applied to the probe target on the MS. The various cyclodextrins were used and applied as solids to the matrix already on the probe target. The spectra were recorded after calibration with CsI up to m/e 8000 at a resolution of 2000 (10% valley). The spectra were collected as the sum of several scans. The FAB/LSIMS linked scan experiments were performed with the same parameters as in the FAB/LSIMS mode. The collision gas (nitrogen) was added in first field-free region and was added so that the selected peak was reduced by 50%.

3. RESULTS AND DISCUSSION

3.1. Enhancement of the γ-CD/C_{60} peak

Additives in FAB/LSIMS can help to improve the intensity of the molecular ion. We accidentally enhanced the molecular ion peak of the 2:1 γ-CD/C-60 complex by adding a salt, $Pb(NO_3)_2$, to the complex. The intensity m/e (3314) increased dramatically, about 60-70 times and became the largest peak in the spectrum above m/e 1000 (see Fig 1.). The peak from the 1:1 γ-CD/C_{60} complex is also clearly seen in this spectrum. The intensity of this peak is about 20% of the molecular ion peak. The intensities of the peaks

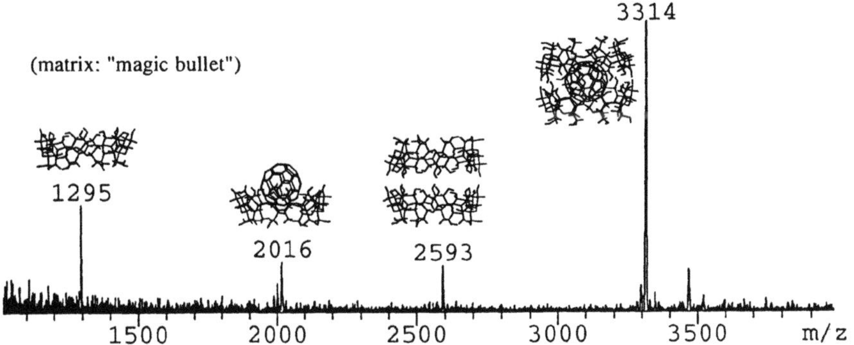

Figure 1. Negative ion FAB/LSIMS mass spectrum of γ-CD/C_{60} complex with addition of $Pb(NO_3)_2$

from γ-CD (m/e 1296) and the dimer of γ-CD (m/e 2592), are not affected by the addition of $PbNO_3$.

3.2. Observation of adducts

The observation of adducts, is a common phenomena in MS. However, the distinct peaks from various CD:s indicate certain stability of oligomeric species. In the gas phase and in absence of guest molecule the CD can stack to form multiple hydrogen bonds between the hydroxyl groups along the rims of the CD:s. By using negative FAB/LSIMS, we observed up to seven α-CD (m/e 6808) held together. With γ-CD oligomers as many as six CD:s were observed (m/e 6484).

We also recorded mass spectra of the mixtures of cyclodextrins in the negative mode. All possible dimers and trimers were observed in the three mixtures: α-CD and β-CD, α-CD and γ-CD, β-CD and γ-CD, respectively. Starting with the relative size of the signals for the molecular ions of the monomers and assuming a purely statistical formation of dimers and trimers, it is possible to calculate the relative peak sizes of the latter. The experimental results are in good agreement with this approach. Somewhat to our surprise there seems to be no clear selectivity for the formation of homostructural dimers and trimers of the cyclodextrins. The mixed dimers and trimers are abundant even in the mixtures of α-CD and γ-CD. The mass spectrum of a mixture of all three cyclodextrins show all possible dimers and trimers which have different masses. Matrix adducts are usually small and doubly charged ions were not commonly observed in these experiments.

To learn about the stability of the adducts, we performed linked scan experiments (positive FAB/LSIMS) on the molecular ions of α-CD (m/e 973) and the dimer (m/e 1945) after colliding the ions with nitrogen. The monomer showed the typical fragments from loss of different numbers of glucose units (m/e 162) and confirms earlier experiments.[5] The dimer also lost one, two etc. up to eight glucose units besides fragmentation into two α-CD:s (Fig. 2.). This is somewhat surprising, because it means that the hydrogen bonds between the two cyclodextrins are strong enough to keep the macrocycles together even also after the loss of several glucose units. The intensity was approximately the same for all the fragments and about one third of the intensity of the monomer (m/e 972).

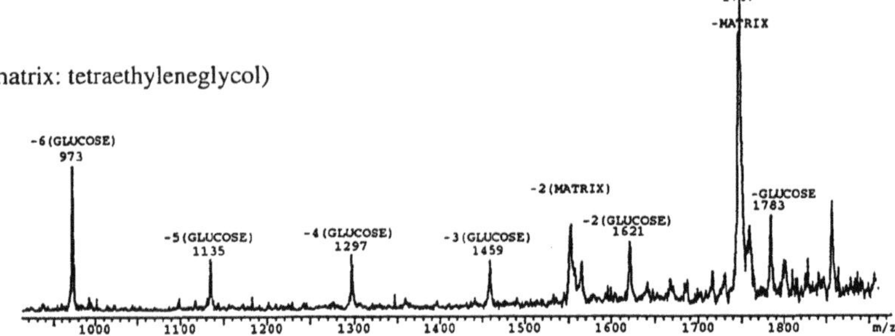

Figure 2. Collision experiment of the dimer of α-CD obtained by positive-FAB/LSIMS

3.3. The γ-CD-Pb-γ-CD molecule

Positive-FAB/LSIMS and glycerol as matrix afforded the mass spectra of the sandwich type complex of cyclodextrin and lead as shown in Fig. 1. The spectrum shows distinct peaks for the molecule ion (m/e 5877) and for the doubly charged ion (m/e 2938). Weak peaks of the molecule ion minus one lead (m/e 5670) and of the molecule ion plus one matrix molecule is also seen (m/e 5969).

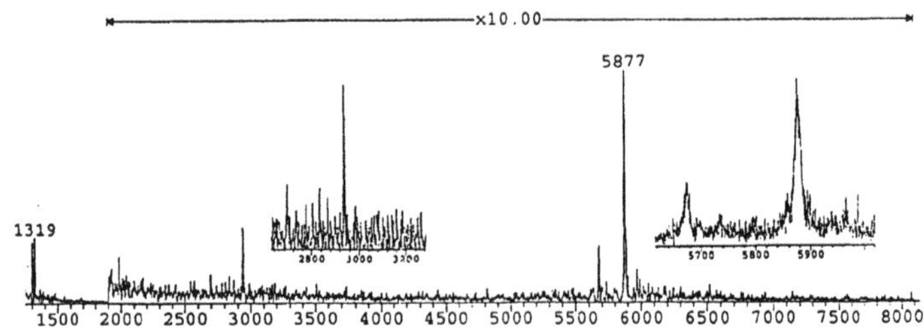

Figure 3. Positive-ion FAB/LSIMS mass spectrum of the γ-CD-Pb complex (matrix: glycerol)

(1) a) Z. H. Qi, V. Mak, L. Diaz, D. M. Grant and C. Chang, *J. Org. Chem.*, **56**, 1537-1542 (1991); b) S. Kurono, T. Hiranao, K. Tsujimoto, M. Ohashi, M. Yoneda and Y. Ohkawa, *Org. Mass Spectrom.*, **27**, 1157-1160, (1992); c) A. Selva, A. Redenti, M. Zanol, P. Ventura and B. Casetta, *Org. Mass Spectrom.*, **28**, M. Kolodziejczyk, R. Kupfer, M. Rosenberg, M. D. Poliks, M. Orlando and M. L. Gross, *Angew. Chem. Int. Ed. Engl.*, **32(9)**, 1344- 1345 (1993)

(2) T. Andersson, G. Westman, G. Stenhagen, M. Sundahl and O. Wennerström, *Tetrahedron Letters*, **36(4)**, 597-600 (1994)

(3) R. Fuchs, N. Habermann and P. Klüfers, *Angew. Chem. Int. Ed. Engl.*, **32**, 852-854 (1993)

(4) P. Klüfers and J. Schuhmacher, *Angew. Chem. Int. Ed. Engl.*, **33**, 1863-1865 (1994)

(5) E. C. Huang and J. D. Henion, *Rap. Commun Mass Spectrom.*, **4(11)**, 467-471 (1990)

STRUCTURE AND DYNAMIC STABILITY OF CYCLODEXTRIN INCLUSION COMPLEXES WITH 1,4-DISUBSTITUTED BICYCLO-[2.2.2]OCTANES

ULF BERG,* MARCIA BERGLUND, NINA BLADH, ANETTE SVENSSON AND MAGNUS STÖDEMAN[$]

Division of Organic Chemistry 1 and Division of Thermochemistry[$], Chemical Center, Lund University, P.O.Box 124, S-221 00, Lund, Sweden

ABSTRACT

The properties of inclusion complexes of 1,4-di-R-bicyclo[2.2.2]octanes (R = H (**1**), Me (**2**), Cl (**3**), Br (**4**), and OH (**5**)) with cyclodextrins have been studied by NMR, microcalorimetry, and force-field computations. The compounds **2** and **3** (but not the other compounds) give dynamically stable 1:2 guest-host complexes with α-cyclodextrin. Microcalorimetry of **5** in water indicates a moderately strong 1:1 complex with β- but weak complexes with α- or γ-cyclodextrin. The behaviour depends on the subtle interplay of size, polarity, hydrophobicity and type of solvent.

1. INTRODUCTION

Cyclodextrins (CD) act as hosts for a variety of small molecules in water solution and have found use in many fields, such as chromatography, pharmaceutical industry and as potential enzyme mimics,[1] and the phenomenon has proven to be an excellent model system for studying the nature of noncovalent bonding in aqueous solution. In order to study the structure and dynamic stability of such inclusion complexes 1,4-disubstituted bicyclo[2.2.2]octane turned out to possess interesting properties as guest molecules. Thus, the 1,4-dimethyl derivative was found to give a 1:2 complex with α-cyclodextrin in $D_2O/CD_3OD/DMF\text{-}d_7$, which exchanges with "free" species with remarkably high barrier.[2] We here present an extended study of a series of analogues in order to gain information of the origin of the unusual dynamic stability of the complexes. The complexes of the analogues, shown in the Scheme, with α-, β- and γ-cyclodextrin have been examined by both experimental and computational methods.

J. Szejtli and L. Szente (eds.), Proceedings of the Eighth International Symposium on Cyclodextrons, 175–178.

R = H (1), CH_3 (2),
Cl (3), Br (4),
OH (5),

2. MATERIALS AND METHODS

The compounds **1-5** have been described earlier.[3,4] The molecular mechanics calculations were performed using the MM2(91) force field implemented in the MacMimic program package.[5] The microcalorimetric titration technique has been described earlier.[6]

3. RESULTS AND DISCUSSION

3.1 NMR Spectroscopy

The ^1H NMR study of **2** and **3** and α-CD in $D_2O/CD_3OD/DMF\text{-}d_7$ exhibited broad singlets at room temperature for the methyl and methylene protons, respectively. On lowering the temperature the signals broadened further and at ca. 10 °C decoalesced to three sets of signals as described earlier.[2] With **3** (and even more pronounced with **4**) crystallization occurred at low temperatures, but otherwise it behaved as **2.** The rate constants for the observed exchanges were evaluated by bandshape simulations,[7] and the corresponding free energies of activation were calculated as 13.4 ± 0.1 kcal/mol for **2** and 13.2 ± 0.2 kcal/mol for **3.** None of the other bicyclooctanes showed similar behaviour, nor did any of them with β- or γ-CD. We interpret the phenomenon as the slow exchange between "free" species and 1:2 guest-host complex (Scheme). Evidence in terms of peak intensities, ROESY spectra, behaviour of α-CD 2-monotosylate and computations have been presented.[2] The ROESY spectrum of **2** allowed a precise determination of the structure of the complex in which half of the guest molecule has penetrated into each of the CD cavities.[2] Thermodynamic parameters of **2** for the process in the Scheme derived from NMR peak intensities were: $\Delta H^o_{app} = -16.3 \pm 1.0$ kcal/mol and $\Delta S^o_{app} = -41 \pm 5$ cal/K·mol, values far from the pattern for the classical hydrophobic effect.

3.2 Microcalorimetry

Isothermal microcalorimetric titration of **5** in water at 25 °C gave the results shown in Table 1. The other compounds in the series could not be studied in water due to solubility problems. The calorimetric data was fitted to a 1:1 model using nonlinear regression methods. A complex of moderate stability was found with β-CD and complexation with α- and γ-CD was very weak. There was no indication of 1:2-complexation. Experiments to determine the temperature dependence of ΔH^o (ΔC^o_p) are in progress.

TABLE 1. Results from the microcalorimetric titration of **5** at 25.01 °C using a 1:1 model.

CD	K_c (M^{-1})	ΔG° (kcal/mol)	ΔH° (kcal/mol)	ΔS° (cal/K·mol)
α	< 100	–	–	–
β	1460 ± 72	– 4.32 ± 0.03	– 4.27±0.04	0.17±0.13
γ	< 50	–	–	–

3.3 Force-field modelling

The gas-phase 1:1 complexes of the compounds **1-5** with α-CD were calculated by molecular mechanics using the MM2(1991) force field. A snug fit of the bicyclo[2.2.2]octane molecule accomponied with large negative enthalpy was obtained. However, the 1,4-substituents prevent penetration of the guest such that half of the guest is exposed to the exterior for **2, 3** and **4**, enabling for complexation with another host molecule. On the other hand, **1** and **5** penetrate too deep into the cavity to allow for 1:2-complexation. The results are shown in Table 2 and in Figure 1.

TABLE 2. Molecular mechanics calculations for gas-phase docking with α-CD. The distance is between the centre of the CD cavity and the rear bridgehead carbon.

Guest	Distance (Å)	ΔE (kcal/mol)
1	4.14	−10.5
2	5.20	−11.3
3	5.14	−11.0
4	5.45	−12.2
5	4.47	−14.5

4. CONCLUSION

The computations indicate a favourable interaction between the guests and α-CD, in the cases of **1** and **2** solely as a result of dispersion forces. The structure of **2**-α-CD is in excellent agreement with the ROESY data and with the stability of the 1:2 complexes for **2** and **3** observed in the NMR experiments. However, computations indicate strong 1:1 complex of **5** in the gas phase. Experiments in water show a different picture: weak interactions with α-CD and γ-CD, but a relatively strong 1:1 binding to β-CD.

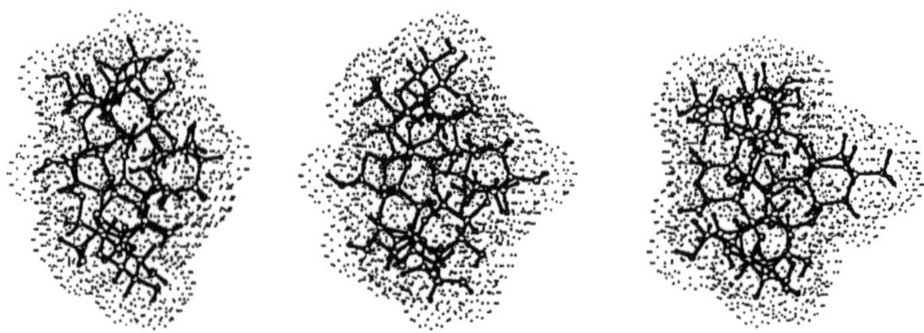

Figure 1. MM2 minimized structures of α-CD complexes with from left **1**, **5** and **2**, with added van der Waals radii, showing the variation of the non-covered parts of the guests.

We have observed 1:2 complexes only with **2** and **3** with α-CD, which is reasonable considering the degree of penetration of the guests in the cavity (Fig. 1). Less obvious is the origin of the high dynamic stability of these complexes; small uncharged molecules notoriously exchange with much higher rate. Considering the available data, we propose that the main cause to the high barrier to complex formation is the dissymmetrical development of enthalpy and entropy along the reaction coordinate. The entropy contribution ($-T\cdot\Delta S$) increases earlier than the enthalpy term decreases. Probably, the solvent mixture also plays a crucial role, the finer details of which we have not so far been able to settle, although further addition of methanol or DMF increases the exchange rate.

ACKNOWLEDGEMENTS

We are grateful for support from the Swedish Natural Science Research Council.

REFERENCES

[1] Breslow, R. *Acc. Chem. Res.* **13**, 170 (1980); *Science* **218**, 532 (1982).

[2] Berg, U., Gustavsson, M., and Åström, N. *J. Am. Chem. Soc.* **117**, 2114 (1995).

[3] Kopecky, J. and Smejkal, J. *Collect. Czech. Chem. Commun.* **45**, 2965 (1980).

[4] Kopecky, J., Smejkal, J. and Hanus, V. *Collect. Czech. Chem. Commun.* **46**, 1370 (1981).

[5] Burkert, U. and Allinger, N.L. *Molecular Mechanics*, American Chemical Society, Washington D.C., 1982; MacMimic is available from Instar Software AB, Ideon Research Park, S-223 70 Lund, Sweden.

[6] a) Bäckman, P., Bastos, M., Hallén, D. and Wadsö, I. *J. Biochem. Biophys. Meth.* **28**, 85 (1994); b) Gómez-Orellana, I., Hallén, D. and Stödeman, M. *J. Chem. Soc. Faraday Trans.* **90**, 3397 (1994).

[7] Sandström, J. *Dynamic NMR Spectroscopy*, Academic Press, London & New York, 1982.

THE BINDING OF PYRENE AND OTHER PROBES TO CD POLYMERS

T. C. WERNER, J. L. IANNACONE AND M. N. AMOO
Department of Chemistry
Union College
Schenectady, NY 12308
U. S. A.

ABSTRACT

The binding interactions of commercial epichlorohydrin (EP)-linked CD polymers (CDPs) with pyrene and p-substituted phenols are compared with those of the respective CD monomers for these probes. Shifts in pK_a for the phenols are consistent with a more open site on the CDPs than the CDs. The relative affinity for pyrene exhibited by β-CDP is estimated by a competition experiment with pyrene / β-CD binding. Spectral studies on synthesized EP-linked sucrose polymers are employed to gain insight into the role of glyceryl-linker units in guest binding.

1. INTRODUCTION

Commercially-available water-soluble cyclodextrin polymers (CDPs), formed by linking CDs with epichlorohydrin (EP), have the general formula $[CD\text{-}(CH_2\text{-}CHOH\text{-}O)_nX]_p$, where CD is α-, β- or γ–CD; X is H or a CD; p is >1 and the average n value is 12-15. The CDPs contain about 55% of the CD monomer, and GFC data indicate a broad range of molecular weights (MW) is observed for each CDP, with prominent peaks at MW ~ 2000 (one CD/polymer chain) and MW = 9-10,000 (4-5 CDs/polymer chain).

We have reported systematic studies of the relative binding interactions and environments of three naphthalene-based fluorescence probes and pyrene with all three commercial CDP / CD combinations [1,2]. Our results are consistent with a significant role for the glyceryl linker units in the binding interaction with the CDPs. We also find evidence against the existence of cooperative binding of pyrene by two CDs on the polymers[2]. Such an interaction, called a "clam shell" arrangement, has been reported for pyrene bound to β-CD [3,4].

We report herein a continuation of our studies on the binding of spectral probes to CDs and CDPs, including the measurement of pKa's for p-nitrophenol (p-NP) and methylparaben (MPB, p-hydroxybenzoic acid, methyl ester)) bound to α-CD and α–CDP, a comparison of the relative strengths of pyrene binding to β-CD and β-CDP, and an evaluation of pyrene interaction with polymers containing the glyceryl linkers but no CD units.

J. Szejtli and L. Szente (eds.), Proceedings of the Eighth International Symposium on Cyclodextrons, 179–182.

2. MATERIALS AND METHODS

2.1 Materials

The commercial CDPs were obtained from Cyclolab R&D Laboratory Ltd., while the CDs were gifts from the American-Maize Products Company. All other chemicals were the highest grade available from Aldrich Chemical Company. Pyrene was recrystallized twice from ethanol, and β-CD was recrystallized from water.

2.2 Methods

Absorption spectra were obtained using a Hewlett-Packard 8452A Diode Array Spectrophotometer, while fluorescence spectra were obtained with a Perkin-Elmer Lambda 5B Spectrofluorometer. GFC data were acquired using a Perkin-Elmer HPLC equipped with a TSK G2000SW column, which is employed with an aqueous mobile phase for separation of molecules in the 500-100,000 MW range.

The pK_a's of p-NP and MPB in the presence of α-CD and α-CDP were determined from the intercepts of plots of eq.(1), where A is the absorbance at the wavelength maximum for the anion form at a given pH and A_t is the absorbance at this same wavelength at a pH where only the anion form exists.

$$pH = pK_a + \text{Log}\,[A / (A_t - A)] \tag{1}$$

The pH was adjusted by adding small amounts of 3M HCl to solutions containing 0.010 M Na_3PO_4. The concentration of α-CD or α-CDP was 0.020 M. A clear isosbestic point was observed in all cases.

3. RESULTS AND DISCUSSION

3.1 pK_a's of p-NP and MPB bound to α-CD and α-CDP

The pK_a's for free p-NP and free MPB and when both are bound to α-CD and α–CDP are given in Table 1.

Table 1. pK_a's of p-NP and MPB when free and when bound to α-CD and α–CDP

Probe	Free	0.020 M α-CD	0.020 M α-CDP
p-NP	7.21 +/- 0.01	6.14 +/-0.04	6.34 +/- 0.02
MPB	8.50 +/- 0.01	8.13 +/- 0.02	8.44 +/- 0.01

Values for free p-NP and for p-NP bound to α-CD are in agreement with literature values [5]. The pK_a values in the presence of the polymer are intermediate between those for the free probes and those when the probes are bound to the monomer. This suggests a different, more open binding site exists for these probes in the CP polymer than in the CD monomer. A similar conclusion was ascertained for pyrene based on fluorescence lifetime data in the presence of quenchers [2]. Thus, benzene derivatives, like naphthalenes and pyrene, appear to have distinctly different binding environments in CDs and CDPs.

3.2 Competitive Binding of Pyrene to β-CD and β-CDP

Since ambiguity exists about the nature of the pyrene binding sites on β-CDP, a binding constant can not be determined [2]. Instead, we have chosen to compare the relative strengths of pyrene / CDP and pyrene / CD binding by observing the effects of a fixed [β-CDP] on the pyrene I/III emission ratio (I = 373 nm, III = 384 nm) in the presence of increasing [β-CD]. In the absence of β-CDP, the I/III ratio decreases from 1.75 to 0.82 as [β-CD] goes from 0 to 0.010 M. The limiting value for this ratio is 0.78 at [β-CD] ~0.015 M. As a result, we can calculate the % bound pyrene from eq.(2), where I/III is the observed ratio at a given [β-CD]:

$$\% \text{ bound pyrene} = \{[1.75 - I/III] / [1.75 - 0.78]\} \times 100 \qquad (2)$$

If β-CDP is present at 0.0025 M, this ratio changes from 1.55 to 1.12 over the same [β-CD] concentration range. Thus, we can use eq.(2) to determine % bound pyrene in the presence of 0.0025 M β–CDP by replacing 1.75 in eq.(2) with 1.55. Using these relations, we find that % bound pyrene decreases from 96% in the presence of 0.010 M β-CD but no β-CDP to 56% in the presence of 0.010 M β-CD and 0.0025 M β-CDP. Since the K value for the 2:1 β-CD : pyrene complex is quite large (6.76 x 10^4, [6]), we can infer that binding must also be also quite strong between pyrene and β-CDP. For the above example, a 40% decrease in pyrene bound to the monomer is observed in the presence of the polymer, even though the ratio of the β-CD binding site concentration from the monomer to that from the polymer is four. Unusually favorable 2:1 β-CD : pyrene complex formation with the polymer, with two CDs coming from the same polymer chain [4], could account for this but is unlikely for two reasons. First, the limiting I / III ratio for pyrene bound to β-CDP (1.55) is much higher than that for pyrene bound to β-CD (0.78). Second, the I /III ratio for pyrene bound to β-CDP is unaffected by the addition of 2,2,3,3,3-pentafluoro-1-propanol (PFP, [2]). When PFP is added to pyrene in the presence of β-CD, binding is enhanced and the I / III ratio drops to 0.38 [6]. These results suggest a quite different binding environment exists for pyrene with β-CDP than with β-CD and supports the notion of glyceryl linker participation in the former case.

3.3 Synthesis of an EP-linked sucrose polymer

An EP-linked polymer were synthesized, using the procedure of Xu et al. [4], starting with 25% sucrose and a 10:1 initial mole ratio of EP to sucrose [25%(10:1)SP]. GFC data for this polymer are shown in Fig.1, along with GFC data for commercial β-CDP. The sucrose polymer, which has a MW distribution intermediate between the two major components for commercial β-CDP (see Fig.1), was prepared to see if the CD cavity is necessary for pyrene binding. The pyrene I/III ratio in water is 1.75, while the limiting values for this ratio in the presence of 25%(10:1)SP and commercial β-CDP are 1.69 and 1.55, respectively. Clearly, the CD units must also be somewhat involved, but the latter I/III ratio is much greater than the limiting ratio for pyrene bound to β-CD (0.78), thereby indicating a more open, hydrophilic site for β-CDP than for β-CD. It seems likely that bound pyrene has contact with both the cavity and glyceryl linker units in the case of the commercial CD polymer.

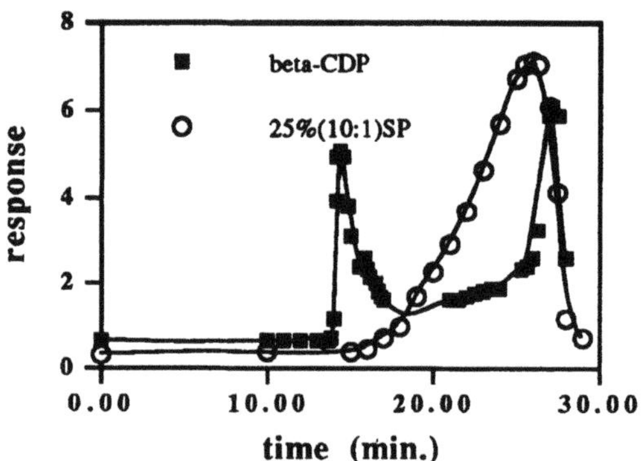

Figure 1 GFC Data

CONCLUSION

We have shown that spectral probes often show binding behavior to CDs that depend on whether the CD is monomeric or part of an EP-linked polymer. While these probes can bind very strongly to the CDPs, their environment is more open and hydrophilic than what is observed with the CD monomers, suggesting a role for the EP linker units.

ACKNOWLEDGMENTS

TCW acknowledges the support of the Petroleum Research Fund for support of this work and the American Maize-Products Company for donation of the CDs.

REFERENCES

[1] Werner, T., Warner, I., The use of naphthalene fluorescence probes to study the binding sites on cyclodextrin polymers, *J. Incl. Phen.* **18**, 385-396 (1994)

[2] Werner, T., Colwell, K., Agbaria, R., Warner, I., The binding of pyrene to cyclodextrin polymers, *J. Incl. Phen.*, in press

[3] Munoz de la Pena, A., Ndou, T., Zung, J., Warner, T., Stoichiometry and formation constants of pyrene inclusion complexes with β- and γ-cyclodextrin, *J. Phys. Chem.* **95**, 3330-3334 (1991)

[4] Xu, W., Demas, J., DeGraff, B., Whaley, M., Interactions of pyrene with cyclodextrins and polymeric cyclodextrins, *J. Chem. Phys.* **97**, 6546-6554 (1993).

[5] Connors, K., Lipari, J., Effect of cycloamyloses on apparent dissociation constants of carboxylic acids and phenols, J. *Pharm. Sci.* **65**, 379-383 (1976).

[6] Elliott, N., Ndou, T., Warner, I., Influence of fluorinated alcohols on cyclodextrin : pyrene complexation, *J. Incl. Phen.* **16**, 99-112 (1993)

DETERMINATION OF BINDING ENERGIES BETWEEN CYCLODEXTRINS AND AROMATIC GUEST MOLECULES BY MICROCALORIMETRY

Th. HÖFLER, G. WENZ*
Polymer-Institut der Universität Karlsruhe
Hertzstr. 16, D-76187 Karlsruhe, Germany

ABSTRACT

The inclusion of substituted benzoic acids in β-CD or selectively methylated β-CDs was investigated by titration microcalorimetry. All thermodynamic functions of the inclusion process ΔG°, ΔH° and ΔS° could be obtained very accurately within one experiment. A very strong influence of the substitution pattern at both the host and the guest on the stability of the inclusion compounds was found.

1. INTRODUCTION

The search for a highly selective and strong inclusion of guests by CD hosts became a very important issue, as many applications, e.g. separation processes, stabilization, solubilization and release of active substances make use of molecular recognition. While slim linear alkane derivatives are included by α-CD, benzene and cyclohexane derivatives fit in β-CD. γ-CD is already large enough to accomodate two anthracene moieties [1]. Binding constants can be determined by spectroscopic methods, e.g. UV, fluorescence, NMR, if a suitable chromophore is present in the host or the guest. Also indirect methods, e.g. solubility, competitive binding, chromatography were used in some cases for this purpose. A very versatile and accurate method for the determination of all thermodynamic binding data is titration microcalorimetry [2,3]. We want to present here the microcalorimetric investigation of the inclusion of substituted benzoic acids by β-CD and its methyl derivatives.

2. MATERIALS AND METHODS

Substituted benzoic acids were obtained from *Aldrich*, β-CD **1** from *Wacker AG*, München, Germany, heptakis-(2-*O*-methyl)-β-CD **2** and heptakis-(2,3-di-*O*-methyl)-β-CD **3** from K. Petzold and D. Klemm, Universität Jena, Germany, heptakis-(2,6-di-*O*-methyl)-β-CD **4** from *Cyclolab*, Budapest, Hungary and heptakis-(6-*O*-methyl)-β-CD **5** and heptakis-(2,3,6-tri-*O*-methyl)-β-CD **6** from G. Weseloh and W. A. König, Universität Hamburg.

A titration microcalorimeter OMEGA from *Microcal. Inc.*, Northampton, MA, USA was used. The reference cell was filled completely with buffer. The sample cell

J. Szejtli and L. Szente (eds.), Proceedings of the Eighth International Symposium on Cyclodextrons, 183–186.

(volume 1.3 mL) was completely filled with a 2-5 mM solution of one of the hosts **1-6** in aqueous 0.1 M phosphate buffer, pH 7.2. The 20-50 mM solution of one of the guests in the same buffer was injected as 20 portions of 5 μL into the sample cell at 25°C. The heat flow q caused (fig. 1a) was integrated for every peak to give the heat Q which was corrected by the corresponding heat of dilution of the guest Q_0. $\Delta H° = Q - Q_0$ was calculated per mol of added guest and plotted versus the ratio of total concentrations of guest and CD, [G]/[CD]. The data points were fitted by non linear regression, assuming a 1:1 stoichiometry as reported elsewhere [2] (Fig. 1b).

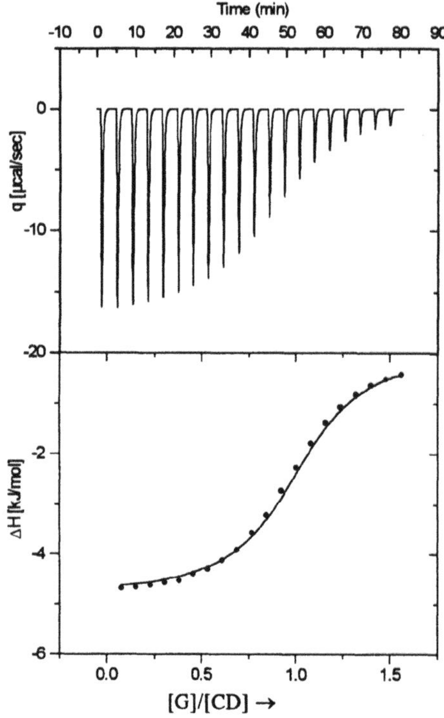

Fig. 1a

Heat flow q detected by an OMEGA microcalorimeter upon the addition of 5 µL portions of a 20 mM solution of 4-t-butylbenzoate to 1.3 mL of a 1 mM solution of heptakis-(2,6-*O*-dimethyl)-β-CD at 25°C.

Fig. 1b

Inclusion enthalpies, calculated by the integration of the heat flows q for the consecutive peaks,

$$\Delta H = Q - Q_0 = \int q\,dt - \int q_0\,dt$$

curve calculated for $\Delta H° = -4.8$ kcal/mol and $K_s = 28'000\ M^{-1}$.

3. RESULTS AND DISCUSSION

An excellent fit of the experimental data by the binding curve, calculated for an 1:1 complex, was found. Therefore we concluded that in any case one guest was included in one CD ring. The stabilities of the inclusion compounds of benzoic acids in β-CD strongly depend on the position and size of substituents at the benzene ring. Free energies of inclusion range from $\Delta G° = -7$ kJ/mol ($K_s = 20$ L/mol) for benzoic acid to $\Delta G° = -24$ kJ/mol ($K_s = 18'400$ L/mol) for 4-t-butyl-benzoic acid (fig. 2). Consequently, a benzene ring alone is too small to fill the β-CD cavity completely.

Hydrophobic substituents at the para position, especially t-butyl, contribute to fill the cavity. A methyl substituent in the meta position or more hydrophilic substituents like i-propoxy, are less suited to increase the binding constant.

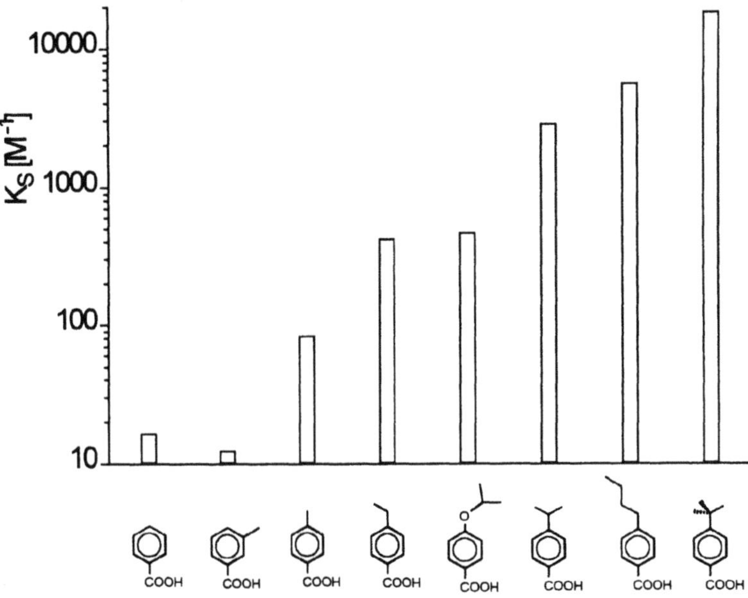

Fig. 2 Stability constants of the inclusion compounds of substituted benzoic acids and β-CD at pH 7.2

We also investigated the influence of methyl substituents at β-CD on its binding capability. 4-t-butylbenzoic acid was chosen as the guest. We found that the binding constant K_s highly depends on the position of the methyl groups. Those CD derivatives methylated at the secondary face of β-CD showed reduced binding constants (fig. 3). Especially methylation at O-3 causes a severe reduction of K_s. This reduction might be caused by a loss of rigidity of the host due to a loss of intramolecular hydrogen bonds. On the other hand, methylation at O-6 increases the binding constant to a remarkable value of K_s = 42'000 L/mol. This may be due to a prolongation of the hydrophobic cavity which allows an additional hydrophobic interaction between the t-butyl group of the guest and the methyl groups of the host. This topography would be in accordance with the crystal structure found for the inclusion compound of β-CD and 4-t-butyl-benzoic acid [4].

4. CONCLUSION

Very high stability constants can be reached for CD inclusion compounds by appropriate attachment of hydrophobic substituents at both host and guest.

5. ACKNOWLEDGEMENTS

We thank the *Bundesministerium für Bildung, Wissenschaft, Forschung und Technologie* (BEO 22 / 0310059A), Bonn and *Wacker AG*, München for support. We thank K. Petzold and D. Klemm, Universität Jena and G. Weseloh and W. A. König, Universität Hamburg for the donation of the methylated CD derivatives.

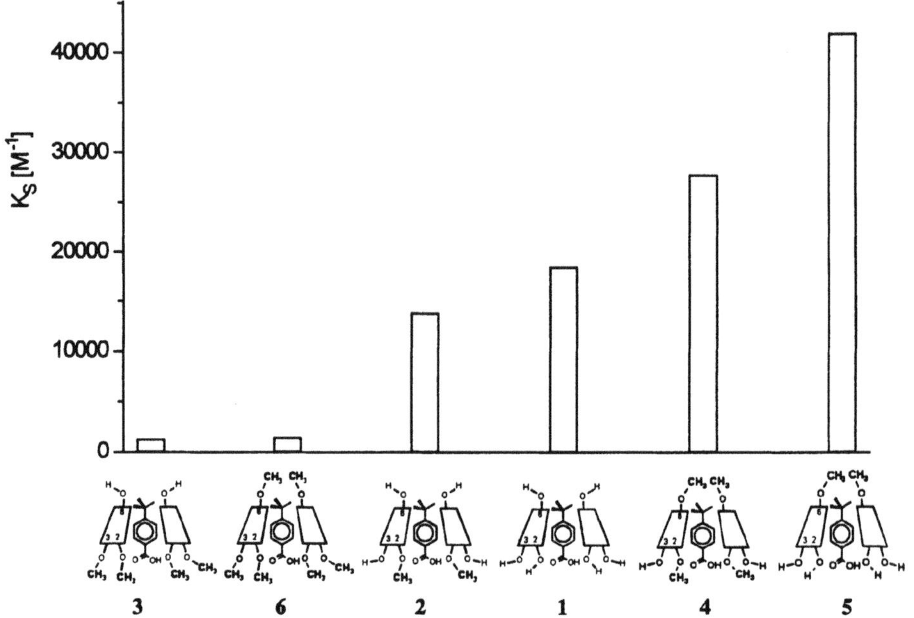

Fig. 3 Infuence of the pattern of methylation of β-CD on the binding constant K_S of t-butyl benzoate at pH 7.2

6. REFERENCES

[1] Wenz, G., Cyclodextrins as building blocks for supramolecular structures and functional units, *Angew. Chem. Int. Ed. Engl.*, **33**, 803-822 (1994)

[2] Wiseman, T., Williston, S., Brandts, J. F, Lin, L.-N., Rapid measurement of binding constants and heats of binding using a new titration calorimeter, *Anal. Biochem.*, **179**, 131-137 (1989)

[3] Inoue, Y., Hakushi, T., Liu, Y., Tong, L., Shen, B., Jin, D., Thermodynamics of molecular recognition by cyclodextrins 1. Calorimetric titration of inclusion complexation of naphthalenesulfonates with α-, β-, and γ-cyclodextrins: enthalpy-entropy compensation, *J. Am. Chem. Soc.*, **115**, 475-481 (1993)

[4] Rontoyianni, A., Mavridis, I. M., Hadjoudis, E., Duisenberg, A. J. M., The crystal structure of the inclusion complex of cyclomaltoheptaose (ß-cyclodextrin) with 4-tert-butylbenzoic acid, *Carbohydr. Res.*, **252**, 19-32 (1994)

COMPLEX FORMATION OF HYDROXYPROPYL-β-CYCLODEXTRINS WITH P-NITROPHENOL

(Preliminary communication)

ÁGNES BUVÁRI-BARCZA, ESZTER RÁK, ÁGNES MÉSZÁROS AND LAJOS BARCZA
Inst. Inorg. Anal. Chem., L. Eötvös Univ.
Budapest 112, POB 32, 1518 Hungary

ABSTRACT

The p-nitrophenol/phenolate conjugate has been investigated as the model of small guest molecules with respect to inclusion complex formation with different hydroxypropyl-β-cyclodextrin hosts. The stability constants measured seem to pass a maximum in function of increasing average degree of substitution.

1. INTRODUCTION

Hydroxypropyl-β-cyclodextrin (HP-CD) products have different average degree of substitution (DS), moreover the pattern of substitution depends also on the conditions of preparation (first of all on the alkalinity of the batch). It was proved, that samples of identical DS could give inclusion complexes of rather different formation constants [1]. This effect has been measured using phenolphthalein as a model for relatively large guest molecules. The investigation of a smaller molecule which can fit easier into the cavity of the host seemed to be reasonable, and p-nitrophenol has been chosen for this purpose. The interaction of p-nitrophenol (or p-nitrophenolate, depending on the pH of the solution) with β-CD proved to be a very interesting case, because the stability of inclusion complex is increased by hydrogen bonding in the case of protonated species, while the resonance charge delocalization in the p-nitrophenolate anion promotes an unusual strong interaction by London dispersion forces [2].

J. Szejtli and L. Szente (eds.), Proceedings of the Eighth International Symposium on Cyclodextrons, 187–188.

2. EXPERIMENTAL

The data have been measured spectrophotometrically in a series of different concentrations and at different pH values at least at three wavelengths, as published elsewhere [2]. HP-CD products of different DS (DS=3-14) as well as dimethyl-β-CD have been investigated.

3. RESULTS AND DISCUSSION

The computed stability constants prove that both p-nitrophenol and p-nitrophenolate form stable complexes and those of dimethyl-β-CD are of highest stability among the CDs investigated. It is well worth mentioning that the formation constants of both species are increasing (nearly parallel to each other) with increasing DS of HP-CD at the beginning, then passing a maximum (at DS ≈ 8), they decrease at higher DS values. The individual constants computed show this general trend very roughly since the values measured with products of identical DS can differ from each other significantly.

The results confirm that, also with smaller guest molecules, the stability constants are influenced not only by DS but by the pattern of substitution, too, i.e. the fit into the cavity of HP-CD can be disturbed by steric factors. The relatively high and increasing values of formation constants point to the better steric conformity of p-nitrophenol/phenolate species to the substitution, while H-bonded interaction is also possible with HP substituent(s).

ACKNOWLEDGEMENT

We thank Cyclolab Ltd. (Hungary) for supplying HP-CD samples.

REFERENCES

[1] Buvári-Barcza, Á., Bodnár-Gyarmathy, D., Barcza, L.: Hydroxypropyl-β-cyclodextrins: correlation between the stability of their inclusion complexes with phenolphthalein and the degree of substitution, *J. Incl. Phenom.*, **18,** 301 (1994)

[2] Buvári, Á., Barcza, L.: Complex formation of phenol, aniline and their nitro derivatives with β- cyclodextrin, *J. Chem. Soc., Perkin Trans. 2*, 543 (1988)

ENERGETICS OF PROTEIN-CYCLODEXTRIN INTERACTIONS

ALAN COOPER, MICHELLE LOVATT & MARGARET A. NUTLEY

Chemistry Department,
Glasgow University,
Glasgow G12 8QQ,
Scotland, U.K.

ABSTRACT

The energetics of interaction of a range of cyclodextrins with folded and unfolded proteins has been examined by sensitive microcalorimetry techniques. Weak interaction with exposed amino acid residues promotes unfolding and dissociation of proteins. The possibility that such interactions may facilitate the use of cyclodextrins as "chaperone-mimics" in the refolding of denatured protein has been explored with the enzyme phosphoglycerate kinase. Up to 40% regain of activity can be achieved in some cases.

1. INTRODUCTION

The interaction of cyclodextrins with amino acid groups on proteins can have several consequences. Firstly, binding of cyclodextrins to exposed side-chains on unfolded polypeptides will de-stabilise the native folded form of the protein and lead to denaturation at lower temperatures [1,2]. Alternatively, interactions with groups on oligomeric folded proteins can lead to dissociation of these protein aggregates, especially if the complexation occurs at sites in the protein-protein interface [3]. Thirdly, combining these effects, cyclodextrin interaction with unfolded proteins may enhance the solubility of denatured protein by masking the exposed hydrophobic residues, thereby possibly assisting the refolding of the polypeptide. In this way cyclodextrins might act as small chaperone-mimics in the protein folding process in cases where re-folding is inhibited by poorly-reversible aggregation or entanglement. The energetics of all these processes have been examined by sensitive differential scanning (DSC) and isothermal titration microcalorimetry (ITC) yielding information on the thermodynamics and stoichiometry of protein-cyclodextrin interactions in various systems. The effects of cyclodextrins on regain of enzyme activity following thermal denaturation are also described here.

J. Szejtli and L. Szente (eds.), Proceedings of the Eighth International Symposium on Cyclodextrons, 189–192.

2. MATERIALS AND METHODS

Calorimetric measurements of protein stability and dissociation energetics were done using Microcal MC2-D and Omega titration calorimeters respectively, following standard procedures [1,2]. Proteins, enzymes and cyclodextrins were obtained from Sigma or Aldrich chemical companies, as appropriate.

3. RESULTS AND DISCUSSION

3.1 Thermal Stability

The thermal stability of a range of globular proteins in solution is reduced in the presence of cyclodextrins. DSC experiments (Fig.1) show that the energetics of this process are consistent with binding of cyclodextrins to the unfolded form of the protein.

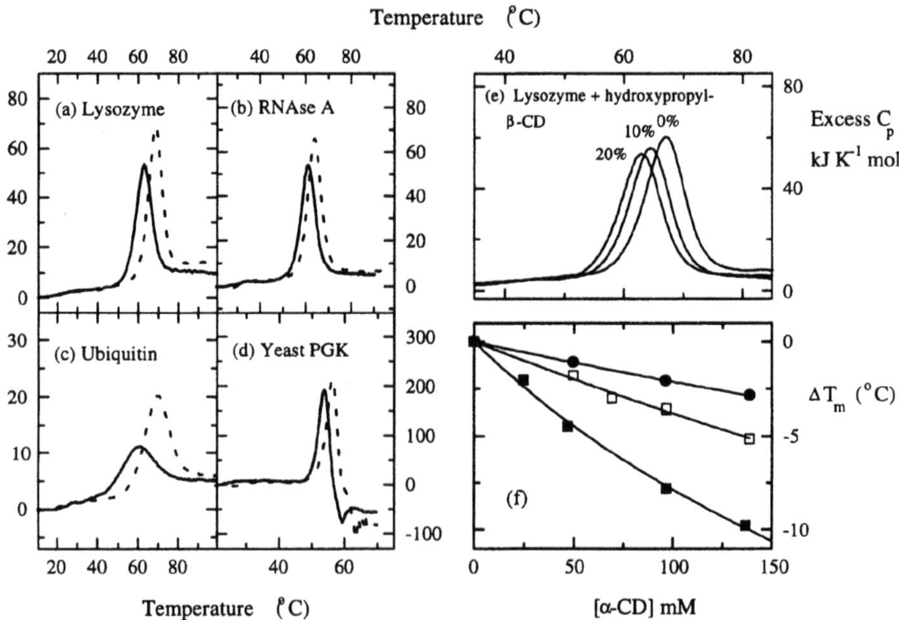

Fig.1 Effects of cyclodextrins on thermal unfolding of proteins. (a)-(d) DSC data showing excess heat capacity (ΔC_p) of various proteins in buffer alone (dotted) or in the presence of ca.100mM α-cyclodextrin. (e) Effect of increasing concentrations of hydroxypropyl-β-CD on lysozyme (pH 3). (f) Variation of thermal unfolding temperature (ΔT_m) with α-CD concentration for RNAse (filled circles), lysozyme (open squares), and ubiquitin (filled squares). Curves show theoretical fits to the model described in the text.

The data can be analysed to give both the enthalpy of unfolding of the protein and the binding affinity of cyclodextrins for groups on the unfolded polypeptide. Assuming a simple unfolding equilibrium:

$$N \rightleftharpoons U \qquad \Delta H_{unf,0}$$

with cyclodextrin binding to n identical sites on the unfolded protein, the decrease in thermal unfolding temperature (T_m) can be written [2]:

$$\Delta T_m / T_m = -(nRT_{m0} / \Delta H_{unf,0}).\ln(1 + [C]/K_C)$$

where K_C = [C][U]/[UC] is the dissociation constant for cyclodextrin (C) binding to unfolded protein (U). Values obtained are compatible with known numbers of aromatic residues in these proteins (n = 4-12) with $\Delta H_{unf,0}$ = 250-500 kJ mol^{-1}. K_C values indicate weak free energies of CD binding, with $-\Delta G^o \approx$ 2-6 kJ mol^{-1} at T_m.

3.2 Protein Subunit Dissociation

Interaction of cyclodextrins with amino acid residues on the surface of folded proteins will affect specific aggregation and other properties, as has been shown in the case of the insulin dimer (see accompanying paper [3] for details). This can be modelled simply as follows. Assuming simple monomer-dimer equilibrium for insulin (I), or other protein:

$$I_2 \rightleftharpoons 2I \quad ; \quad K_{diss,0} = [I]^2/[I_2] \quad ; \quad \Delta H_{diss,0}$$

with sequential binding of cyclodextrins (C) only to monomer:

$$I + C \rightleftharpoons IC \quad ; \quad K_1 = [IC]/[I][C] \quad ; \quad \Delta H_1$$

$$I + IC \rightleftharpoons IC_2 \quad ; \quad K_2 = [IC_2]/[I][IC] \quad ; \quad \Delta H_2$$

..... and so on for subsequent binding steps.

the apparent dissociation constant may be written:

$$K_{diss} = [I_{tot}]^2/[I_2] = K_{diss,0}(1 + K_1[C] + K_1K_2[C]^2 + ...)^2$$

Despite uncertainties regarding the number of potential binding sites and their binding affinities, this polynomial expression does form a reasonable basis for empirical modelling of the observed effects of cyclodextrins on insulin dissociation thermodynamics. The energetics of dimer dissociation have been determined by titration microcalorimetry [3] and data fit to the above model.

3.3 Protein Refolding - Chaperone Mimicry

The regain of enzyme activity after thermal denaturation of phosphoglycerate kinase (PGK) is significantly enhanced in the presence of cyclodextrins. PGK solutions (pH 7.5, 1mM DTT) were heated to 60°C for 10 min. then cooled to room temperature for 3.5 hours prior to enzyme assay. Results (Table 1) show that increasing concentrations of various cyclodextrins can promote refolding and regain of biological function in this normally irreversible process. Similar results are obtained regardless of whether cyclodextrins are added before or after the denaturation step. Unfolded protein is sticky stuff, and PGK normally forms entangled aggregates or precipitates when thermally unfolded, as indicated by the noisy exothermic post-transition DSC baseline in Fig.1(d). Complexation of cyclodextrins with exposed groups on the unfolded protein will enhance solubility and encourage refolding rather than non-specific aggregation. In this way, cyclodextrins may mimic naturally occurring chaperone molecules that perform similar functions *in vivo*. It is important that the CD-protein interactions are relatively weak and reversible, otherwise removal of the cyclodextrins during refolding might be too difficult.

TABLE1: Regain of PGK activity after thermal denaturation.

% CD (w/v):	0	2	5	12
		%	recovery	
Without CD	ca.10	-	-	-
α-CD	-	21	26	38
methyl-β-CD	-	11	15	39

4. CONCLUSION

Weak interaction between cyclodextrins and protein groups can be measured by sensitive calorimetric techniques. This provides a useful probe of protein folding and subunit interactions, with potentially important biotechnology applications.

ACKNOWLEDGEMENTS

The UK Biological Calorimetry facility in Glasgow is supported by funds from BBSRC and EPSRC. ML is supported by SKB and a studentship from EPSRC.

REFERENCES

[1] Cooper , A., Effect of cyclodextrins on the thermal stability of globular proteins. *J.Amer.Chem.Soc.* **114**, 9208-9209 (1992).

[2] Cooper. A. & McAuley-Hecht, K.E., Microcalorimetry and the Molecular Recognition of Peptides and Proteins. *Phil.Trans.R.Soc. Lond.* **A 345**, 23-35 (1993).

[3] Lovatt, M., Cooper, A. & Camilleri, P. Energetics of Cyclodextrin-Induced Dissociation of Insulin. - accompanying paper, this symposium (1996).

INTERACTION OF SUPERCRITICAL FLUIDS WITH DRUG/CYCLODEXTRIN INCLUSION COMPOUNDS AND PHYSICAL MIXTURES

F. GIORDANO[1], M. RILLOSI[2], G.P. BETTINETTI[2], A. GAZZANIGA[3], W. MAJEWSKI[4], M. PERRUT[4]
[1]*Pharmaceutical Department Viale delle Scienze 78, 43100 Parma, Italy*
[2]*Pharmaceutical Chemistry Department, Viale Taramelli 12, 27100 Pavia, Italy*
[3]*Pharmaceutical Chemistry Institute, Viale Abruzzi 42, 00131 Milano, Italy*
[4]*Separex Chimie Fine, B.P. 9, F 54250 Champigneulles, France*

ABSTRACT

The interaction of supercritical carbon dioxide with a model drug/betacyclodextrin system, tested as solid mixture or as inclusion compound, was investigated. The influence of temperature, particle size and water content on the interaction behavior was also assessed. The inclusion compound proved to be fairly stable in the experimental conditions adopted.

1. INTRODUCTION

Pharmaceutical applications of supercritical fluids technology are presently concerned with extractions from natural matrices, residual solvent elimination, and purification of active principles. Recently, supercritical fluids have demonstrated their potentiality also in the preparation of solid phases by recrystallization and impregnation processes [1,2]. It seemed worth to investigate the utilization of supercritical fluids in the drug/cyclodextrin systems, in particular for the preparation of the inclusion compounds from solid mixtures, their purification from residual solvents, recovery of components, etc.. The aim of this work was to investigate the interaction of supercritical CO_2 with the model system Acetaminophen/Betacyclodextrin, tested as physical mixture or inclusion compound at different temperatures and on powders of different particle size and water contents.

2. MATERIALS AND METHODS

2.1. Materials

Acetaminophen (AMF, Carlo Erba, Italy); two granulometric fractions (> 300 μm and < 44 μm) of commercial betacyclodextrin (BCD, Roquette Frères, Lestrem, France), water content 14%, were used. Dry BCD was prepared by heating BCD at 120 °C for 2 hours under

J. Szejtli and L. Szente (eds.), Proceedings of the Eighth International Symposium on Cyclodextrons, 193–196.

vacuum. The inclusion compound AMF:BCD (1:1 mol:mol) was obtained by kneading [3]. Dry CO_2 (purity > 99%) and methanol of analytical grade were used throughout this work.

2.2. Methods

2.2.1. Supercritical fluid extraction (SFE)

The experiments were carried out in a pilot plant S FE 200 (Separex, France). The capacity of the stainless steel extractor was 200 cm^3, equipped with three separators. A pump (DOSAPRO-MILTON Roy, Pont-St Pierre, France) maintained CO_2 at the selected pressure (300 bar); a series of valves regulated the pressure in the extractor and separators, while the temperature was controlled (40-75 °C) by means of thermostated iackets.

The amount of cosolvent was changed throughout the experiments within the range 10-25 %. 20-30 g of BCD-drug complex were inserted in the internal compartment of the extractor. Two glass fiber filters separated the complex from the stainless steel filling material positionned at the bottom and at the top of the extractor in order to assure an homogeneous flux of the supercritical fluid through the sample. In another set of experiments with the same apparatus we investigated the possible inclusion compound formation induced by SFE. In the extractor two distinct layers of the drug and BCD (overall amount 20-30g) were deposited (Figure 1.). At the end of the experiment the extractor content was carefully divided in three portions (upper, middle, lower) prior to subsequent analyses. Each run was performed for 1 hour.

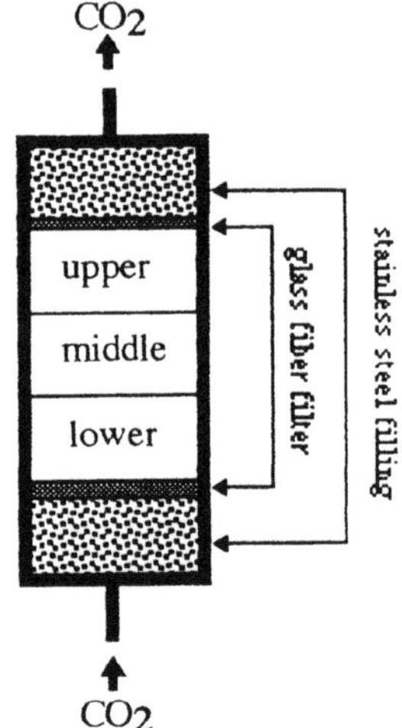

Figure 1. Schematic representation of the extractor.

2.2.2. Instrumental analyses

The solid phases collected from the extractor and those obtained after elimination of methanol from the liquid phases recovered in the separators were analyzed by means of differential scanning calorimetry (DSC, Mettler TA 4000 equipped with a DSC 25 and TGA 50 cells). Samples (3-5 mg) were weighed in aluminium pans and scanned between 30 and 250 °C at the heating rate of 10 K/min under static air atmosphere. Samples obtained from the extractor and separators after SFE were also spectrophotometrically assayed for AMF contents with a Perkin Elmer u.v. spectrophotometer (λ_{max} 242 nm).

3. RESULTS AND DISCUSSION

In Figure 2 the DSC traces of the AMF/BCD inclusion compound before and after SFE are reported.

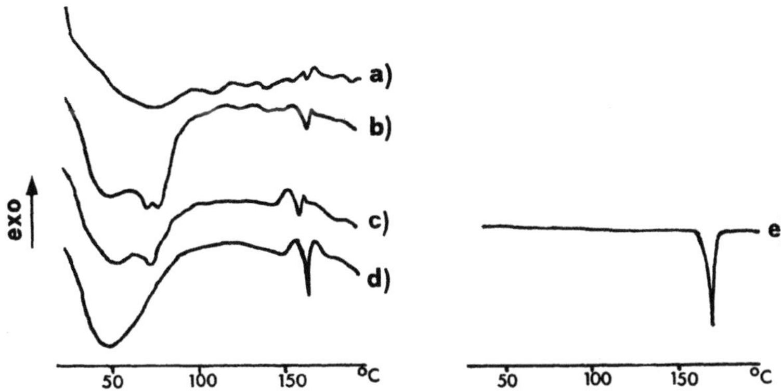

Figure 2. DSC curves of AMF/BCD inclusion compound: a) before and after SFE process: b) upper, c) middle, d) lower portion; e) solid phase recovered in the separators.

The thermal analysis of the solid phases recovered in the extractor show an endothermal effect between 35-120 °C related to the dehydration of BCD, and the peak of fusion of AMF at 168-170 °C, which is more evident in samples from the lower portion of the extractor. In the separators only pure AMF was recovered (Fig. 2 e). The UV analysis proved that the SFE causes a substantial decomposition of the complex (see the following Table).

TABLE. UV analysis of the solid phases recovered in the extractor after SFE of the AMF/BCD complex. The results are presented in terms of AMF % of the initial title (AMF ≈ 10% by weight).

Layer	Experimental conditions	
	$T_{extr.}$ 40°C 10% of methanol	$T_{extr.}$ 75°C 20% of methanol
upper	35,7	80,76
middle	> 100	> 100
lower	> 100	> 100

Surprisingly, the active ingredient content assessed on samples taken from the upper portion was invariably lower than in the other two portions, as also confirmed by DSC profiles showing the presence of higher amounts of uncomplexed (crystalline) AMF in the lower portions of the extractor. This common trend can possibly be ascribed to a segregation caused by a difference in density of particles and the consequent floating of lower density particles (BCD) generated by the extraction of AMF from the complex, when the resulting mixture (BCD + complex) is suspended in supercritical CO_2.

In Figure 3 the DSC curves of the AMF/BCD physical mixture before and after SFE are shown.

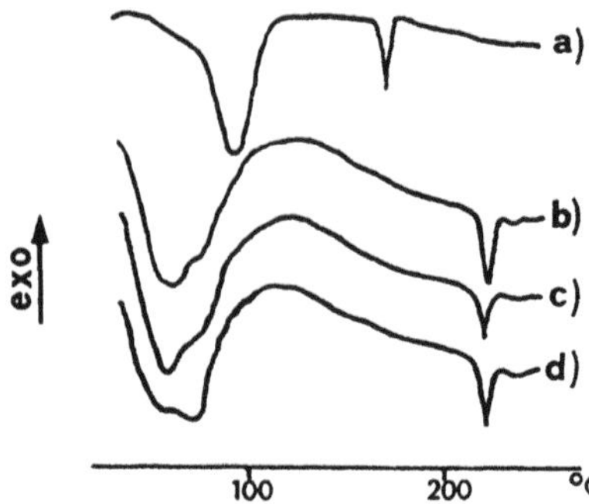

Figure 3. DSC curves of AMF/BCD (< 44 μm) physical mixture: a) before and after SFE process: b) upper, c) middle, d) lower portion.

In the solid phases recovered in the extractor after SFE the fusion peak of AMF was absent because the drug was totally solubilized in the CO_2-methanol mixture, as confirmed by UV analysis on these samples. The endothermal effect over 200 °C is related to the solid-solid transition of BCD. Similar trends were found when dry BCD (or BCD > 300 μm) was used.

4. CONCLUSIONS

A series of experiments has been designed to determine the ability of SFE method to extract the drug from its complex with BCD. Results show that the cyclodextrinic matrix is pratically insoluble in the experimental conditions adopted. When portions of methanol are added as cosolvent to carbon dioxide, the AMF-BCD inclusion compound is not stable. Moreover, the formation of inclusion compounds could not be achieved because AMF, once dissolved, showed higher affinity toward the supercritical phase. Work is in progress also on other drug/cyclodextrin systems to elucidate mechanisms and properties of this ecologically appreciable alternative to classical extraction procedures.

REFERENCES

[1] Schmitt W.J., US Patent 88-253849 (1988). World Patent WO 9003782 (1990).

[2] Bertucco A., Guarise G.B., Pallado P., Corain B., Proc. 2nd Symposium on Supercritical Fluids, p. 189, (Eds. Reverchon E. & Schiraldi A.), Ravello, Italy (1993).

[3] Gazzaniga A., Sangalli M.E., Benelli P., Conte U., Bettinetti G.P., Giordano F., *S.T.P. Pharma Sciences*, 4, 421-424 (1994).

CYCLODEXTRINS IN THE ROLE OF HOST AND SIMULTANEUSLY GUEST COMPONENT

P. Mondík, A. Sopková[1], H. Viernstein[2], B. Légendre[3]

Department of Chemistry, Technical University, SK-042 00 Košice, Slovakia
[1]Department of Inorg. Chem., Fac. Sciences, P. J. Šafarik's University, private: Vihorlatská 80, 040 01 Košice
[2]Institut f. Pharmazeutiche Technologie, Univesität Wien, A-1090 Wien, Austria
[3]Laboratoire de Chimie Physique, Minerale et Bioinorganique, Faculté de Pharmacie, Université Paris - Sud, F-92296 Chatenay Malabary, France

ABSTRACT

The preparation and characteristics of new inclusion compounds, prepared by combination of natural zeolite of clinoptilolite (CT) type with cyclodextrin (CD) and also with enclosed second guest component (salicylic acid, resp. spironolactone) is reported.

1. INTRODUCTION

Inclusion phenomena have excited great interest among chemists and biochemists [1-5] but little attention has been paid to combinations or complexes containing two or more inclusion compounds or their hosts or guest components [6]. Although some models have been proposed, only recently has their true nature been recognised [7,8]. Natural [9] or synthetic [10] minerals are often used for preparation of complex combinations. Various molecules being intercalated in layered crystals or enclosed in some manner in various combinations of host lattices [11-14] or other inclusion compounds or very large molecules, thus forming supramolecules [3]. They are used as solid supports in GC or LC, as microcapsulating agents [4,5]. The thermal stability of the enclosed species and the electrical conductivity values [10,14] are always increased [12]. Newly formed complexes can become more accessible [6] to other guest components - as has been shown by Mondík P. for the clinoptilote [15].

2. MATERIALS AND METHODS

2.1. Materials

Natural zeolitic material of Clinoptilolite type (CT) from East Slovakian deposit in its Cu(II) form [16,17] was used.
Cyclodextrin derivates used: -dimethyl (DMBCD, Roquette), hydroxypropyl- (HPBCD, Roquette), -ß-cyclodextrin (BCD, Roquette).

J. Szejtli and L. Szente (eds.), Proceedings of the Eighth International Symposium on Cyclodextrons, 197–200.

Salicylic acid (Lachema Brno) and spironolactone (SP, Roquette) used in pure form.

2.2. Methods

In the zeolite being the host component the molecules of CDs were sorbed. The latter molecules contained either their original guest molecules of water or the salicylic acid, iodine, spirolactone or iodine-instead ofthe original water. Cyclodextrin plays a specific role (guest and host) Therefore the prepared compounds can be written by both formulas as follows: $H_1.G_1.G_2$ or $H_1.H_2.G_2/H_1$: zeolite, G_1 and H_2: cyclodextrin, G_2: - e.g.c.
Two methods of the complex preparation, were used:
1. sorption of e.g.c. into the previously prepared zeolite - cyclodextrin complex
2. sorption of (cyclodextrin - e.g.c.) complex into the zeolitic host.
Following groups of inclusion compounds were prerared:
CTCu- DMBCD (1A) (CTCu- DMBCD)- sal (2A)
CTCu- HPBCD (1B) (CTCu- HPBCD)- sal (2B)
(DMBCD - sal.acid) complex in sollution sorbed into the CTCu in four different ways: (3A),(3B),(3C),(3D).
CTCu-(BCD-SP) (4A),(4B) (CTCu- DMBCD)- SP (6)

3. RESULTS AND DISCUSSION

3.1. IR spectra

All characteristic bands of zeolite in the products were conserved as strongest because of the role of zeolite as the host component. CDs show their characteristics [17] too.

3.2. UV spectra, concentration and desorption tests

The quantity of sorbed guest components were determined by the UV spectra. In the case of compounds containg spirolactone as the e.g.c., they were prepared from the solution of (BCD-SP) complex in molar weights of 3:2, resp. 1:2. Simlarly as in to case of ther products formed by the natural CT zeolite and unmodified synthetic NaY zeolite [15] the product with lower molar proportion showed better sorptive abilities.
The product prepared by direct sorption of SP into the zeolite included higher quantities of the e.g.c. than in the case of the complex with CD. From desorption tests it can be seen that the e.g.c. is in some way stabilised in the CT-CD host (better than in the zeolite alone). In the case of the salicylic acid product (3C) and in the case of spironolactone the product (4B) was exhibited as the best. From the (3C) product 55% of salicylic acid, from (4B) 72% of SP was dissolved after 4 hours into the solution [16]. The e.g.c. in the host is stabilised after its enclosure in CD.
The tests also showed that although by the former method of preparation (e.g.c. into CT-CD complex) a higher quantity of the e.g.c. can be sorbed, it

TABLE 1 The absorbed amounts of guest components in the products according to UV spectra

Compound	Content of CD [%]	Content of e.g.c. [%]
CTCu-DMBCD	24.05	-
CTCu-HPBCD	14.92	-
CTCu-sal	-	26.15
CTCu-SP	-	29.32
[CTCu-DMBCD]-sal (2A)	6.73	13.08
[CTCu-HPBCD]-sal (2B)	5.28	10.34
CTCu-[DMBCD-sal] (3A)	16.13	2.92
CTCu-[DMBCD-sal] (3B)	23.76	4.18
CTCu-[DMBCD-sal] (3C)	10.80	3.84
CTCu-[DMBCD-sal] (3D)	13.24	2.76
CTCu-[BCD-SP] (4A)	5.58	0.14
CTCu-[BCD-SP] (4B)	26.80	4.36
[CTCu-DMBCD]-SP	5.07	11.67

is better stabilised and slowier released from the product prepared by the second method (e.g.c.-CD complex into the zeolite).

3.3. Thermanal analysis

The shape of the TG curves was depending on the number of the guest components in the product and on the preparation. DSC measurements showed also differentiation of products.

3.4. X-Ray powder diffraction

The system zeolite-guest component in the products can be looked upon as a dispersed system with the distribution of the guest troughout the dispersed zeolite phase. According to our measurements, the quantity of absorbed CD in relation to zeolite is on the average 1:6. Therefore only weak interactions among the CD molecules in the products and we can interprete it as a dispersion of CD in the mineral phase.

3.5. Molecular modelling

Molecule of DMBCD was taken [18] as model, the structural data of CT from japanese deposit were used. From the results [15,16] it can be seen that the space dimensions of zeolite cavities and BCD molecule do not allow a full inclusion of CD. We may assume the cyclodextrins anchored by a chemisorption in the pores of the zeolites by their functional groups.

4. CONCLUSION

The second method of preparation (CD complex into the zeolite) was advantageous, because of higher quantity od sorbed cyclodextrin in the product, but on the other hand, a lower quantity of the e.g.c. was sorbed. In these cases the quantity of the e.g.c. has to be decisive. CDs are only ammeliorating the utility of the e.g.c.

REFERENCES

[1] Cram, D., J., Cram, J., M., Container molecules and their guests, Royal Soc. of Chem., Cambrigde, 1994

[2] Weber, M.,Josel I., *J. Incl. Phenom.*, **1**, 79 (1983)

[3] Comprehensive Supramolecular Chemistry (Eds. Lehn, J., M., Atwood, J., Davies, I., E., D., MacNicol, D., D., Vögtle, E.,) Vol. 1-11, Elsevier N.Y. (1995-1996)

[4] Szejtli, J., Cyclodextrin Technology, Kluwer Academic Publishers, Amsterdam, 1988

[5] Cyclodextrins and their Industrial Uses (Eds. Duchěne, D.,) Editions de Santé, Paris, (1991)

[6] Sopková, A., Mondík, P., Reháková, M., Thermanal Behaviour of more Complex Compounds prepared from Inclusion Compounds, ICTAC News, 1995 (rewiew)).

[7] Kijima, T., Kobayashi, M., Matshui, Y., *J. Incl. Phenom.*, **3-4**, 807 (1984)

[8] *J.Incl. Phenom.*, **1-25**, (1983-1996)

[9] Occurence, Properties and Utilization of natural Zeolites (Eds. Kalló, D., Sherry, H., S.,), Akadémiai Kiadó, Budapest (1988)

[10] Reháková, M., Casciola, M., Krogh Andersen, I., G., Bastl, Z., Preparation and characterisation of a composite of silver iodide and synthetic zeolite ZSM-5, *J. Incl. Phenom.*, in press (1996)

[11] Intercalation Chemistry (Eds. Wittinghham, M., S., Jacobson, A., J.,) Academic Press, N. Y., 1982

[12] Kijima., T., Nakazawah., Kobayashi., M., Bull., *Chem. Soc. Japan*, **61**, 4277 (1988)

[13] Beneš., L., Zima., V., *J. Incl. Phenom.*, **20**, 38 (1995)

[14] Casciola., M., Palombari., R., *Solid State Ionocs*, **47**, 155 (1991)

[15] Mondík., P., unpublished data

[16] Mondík., P., Sopková., A., Vierstein., H., Légendre., B., Supramolecular Chemistry, in press (1996)

[17] Mondík., P., Sopková., A., Suchár., G., Wadsten., T., *J. Incl. Phenom.*, **13**, 109 (1992)

[18] Szejtli., J., Cyclodextrins and their Inclusion Complexes, Akadémiai Kiadó, Budapest 1982

COMPLEX FORMATION OF UNSATURATED CYCLODEXTRIN SOLUTIONS WITH VARIOUS POLYMERS

N. KILDEMARK, K.L. LARSEN, AND W. ZIMMERMANN
Biotechnology Laboratory, Department of Civil Engineering,
Aalborg University, Sohngaardsholmsvej 57, DK-9000 Aalborg, Denmark

ABSTRACT

By using methyl orange as a competitive complexant, it was shown that solutions of poly(ethylene glycol), poly(propylene glycol) and poly(N-vinylpyrrolidone) equivalent to a concentration of 1.67 mM monomer are able to form soluble inclusion complexes with 0.20 mM solutions of α-, β- and γ-cyclodextrins. The complex formation is independent of polymer chain length. Poly(ethylene glycol) solutions complexed, on average with 40%, 45% and 75% of α-, β-, and γ-cyclodextrin, respectively. A poly(propylene glycol) solution complexed 55% α-cyclodextrin, 80% β-cyclodextrin and 90% γ-cyclodextrin. Poly(N-vinylpyrrolidone) solutions complexed on average with 90% of α-, β-, and γ-cyclodextrin.

1. INTRODUCTION

Insoluble inclusion complexes can be formed with saturated solutions of cyclodextrins (CD) and various synthetic polymers[1,2]. The insoluble inclusion complexes can be dissolved in water or by the addition of a stronger complexant than the polymer[1]. The polymers are selective in the formation of insoluble inclusion complexes with respect to the type of CD (α-CD, β-CD or γ-CD)[1]. Poly(ethylene glycol) (PEG) precipitates as an inclusion complex with α-CD and to a lesser extent with γ-CD[1,3]. Poly(propylene glycol) (PPG) forms insoluble inclusion complexes with β-CD and γ-CD but not with α-CD, while no insoluble inclusion complexes are formed between poly(N-vinylpyrrolidone) (PVP) and CDs[1]. By formation of insoluble inclusion complexes with PPG it is possible to purify β-CD and γ-CD[4]. Increase in the production yield of CDs by addition of PEG in enzyme synthesis reactions has been reported[5,6]. For the purpose of isolating individual types of CDs from dilute solutions, the interaction between the above mentioned polymers and unsaturated CD solutions was investigated.

J. Szejtli and L. Szente (eds.), Proceedings of the Eighth International Symposium on Cyclodextrons, 201–204.

2. MATERIALS AND METHODS

2.1. Materials

PEGs with average molecular weights of 2000, 4000, 5000, 6000, 10000, 15000 and 20000 were obtained from Merck Co. PPG with an average molecular weight of 425 and PVPs with an average molecular weight of 10000, 24000 and 40000 were obtained from Fluka Co. The polymers were of technical grade. Methyl orange was obtained from Aldrich Co. CDs were purchased from Merck Co. Methyl orange and the CDs were of analytical grade.

2.2. Methods

2.000 g of an aqueous polymer solution equivalent to a 20 mM monomer concentration was weighed into cuvettes and 0.40 ml of a 0.3 mM methyl orange solution in 50 mM phosphate buffer (pH 6.80) containing 1.25 mM CD was added. After mixing, the absorbance was measured at 515 nm, 505 nm or 480 nm for α-CD, β-CD, γ-CD, respectively, with a lambda2 spectrometer (Perkin-Elmer Co.). Samples containing 2.000 g polymer solution and 0.40 ml of a 0.3 mM methyl orange in 50 mM phosphate buffer (pH 6.80) without CD was measured at the same wavelengths. The experiments were carried out at ambient temperature and repeated four times.

3. RESULTS AND DISCUSSION

3.1. Competitive Complexation of α-CD, β-CD, and γ-CD with Methyl Orange and Polymers

Methyl orange and CDs form inclusion complexes with 1:1 stochiometry[7]. A difference in the absorbance between the free and the complexed form of methyl orange can be found. This difference depends on the type of CD forming the inclusion complex with methyl orange. At pH 6.80, the largest difference was found at 515 nm, 505 nm and 480 nm for α-CD, β-CD, and γ-CD, respectively. Addition of a complex-forming polymer to the solution containing methyl orange and CD will change the equilibrium between methyl orange and CD. The inclusion complexes between methyl orange and CD will dissociate resulting in the release of methyl orange and a change in the absorbance of the solution. The equilibrium between methyl orange and CD will depend on the equilibrium constant and not on the amount of CD complexed by the polymer. The difference in absorbance between a solution containing methyl orange and polymer and a solution containing methyl orange, polymer and CD indicates the amount of released methyl orange due to the complexation of CD with the polymer. A standard curve was prepared by measuring the absorbance of methyl orange and CD as a function of the CD concentration. Comparing the absorbance difference of the samples with the standard curve gave the amount of CD not complexed with the polymer.

3.2. Soluble Inclusion Complexes

From the results in Table 1 it can be seen that all the investigated polymers formed soluble inclusion complexes with CDs. A necessary condition for complex formation to occur is the inclusion of the polymer into the CD cavity. In the case of insoluble inclusion complexes it has been reported that α-CD and γ-CD are threaded on PEG[3,8] and γ-CD can

accommodate two PEG chains[3]. Soluble inclusion complexes could be formed in a similar way. The cavity of β-CD has been regarded as too large to fit PEG since no insoluble inclusion complexes could be formed[1]. PEG can be accommodated by the CD cavity since soluble inclusion complexes could be formed between PEG and β-CD.
Insoluble inclusion complexes of α-CD with PPG have not been observed[1]. Through molecular modelling it has been shown that α-CD cannot be stringed on PPG due to steric hindrance of the methyl substituents on the polymer chain[1]. Soluble inclusion complexes can however be formed with PPG and α-CD. Binding of α-CD to the methyl substituents of the polymer could lead to a complexation without threading the polymer. PPG can be precipitated in saturated β-CD or γ-CD solutions[1]. The binding of CD has not been shown to occur by threading. Therefore it is possible that CDs could bind in more than one way to polymers with substituents which is supported by the observation that α-CD can complex with PPG.
Molecular modelling has indicated that poly(methylvinyl ether) (PMVE) is not capable of stringing α-CD or β-CD[1]. In spite of having larger substituents than PMVE, PVP was shown to form complexes with α-CD and β-CD. This could be explained by binding of the CDs to the polymers substituents as suggested above. It has not been shown that γ-CD can be stringed by PVP.
Comparing the preference of the polymers to α-CD, β-CD, and γ-CD, the highest percentage of complexation was found with γ-CD, followed by β-CD and α-CD, except for PVP that formed complexes with the CDs in equal amounts. As suggested above, inclusion complex formation is a flexible process where a large CD cavity provides a better possibility for complexing with a polymer by either threading one or more polymer chains or complexing with the substituents of the polymer.
The various polymers showed differences in their ability to form inclusion complexes with CDs. It is generally believed that the hydrophobicity of a molecule enhances the complex formation with CDs because of favourable interaction with the hydrophobic interior of the CD molecule[9]. In accordance with this, PEG, which has no hydrophobic substituents, showed the lowest ability of the investigated polymers to form complexes with CDs.

TABLE 1. Polymer complexation in 0.20 mM CD solutions

Polymer	Complexed CD (mM)		
	α-CD	β-CD	γ-CD
PEG 2000	0.08 ± 0.02	0.08 ± 0.03	0.13 ± 0.02
PEG 4000	0.07 ± 0.01	0.09 ± 0.01	0.15 ± 0.01
PEG 5000	0.06 ± 0.02	0.08 ± 0.02	0.15 ± 0.01
PEG 6000	0.07 ± 0.01	0.09 ± 0.01	0.15 ± 0.01
PEG 10000	0.08 ± 0.01	0.09 ± 0.01	0.15 ± 0.01
PEG 15000	0.08 ± 0.01	0.09 ± 0.01	0.17 ± 0.02
PEG 20000	0.08 ± 0.01	0.09 ± 0.01	0.16 ± 0.01
PPG 425	0.11 ± 0.01	0.16 ± 0.01	0.18 ± 0.01
PVP 10000	0.19 ± 0.01	0.17 ± 0.01	0.18 ± 0.01
PVP 24000	0.18 ± 0.01	0.17 ± 0.01	0.18 ± 0.01
PVP 40000	0.18 ± 0.01	0.16 ± 0.01	0.18 ± 0.01

Inclusion complex formation of polymers with CDs appears to be a frequently occurring event which in some cases can lead to the formation of an insoluble inclusion complex. One of the factors influencing insolubility of the complex could be the ratio between stringed CDs and the chain length of the polymer. Every stringed CD replaces a part of the shell of ordered water molecules surrounding the polymer. At one point, most of the water shell will be displaced and the solubility will be lost. In our experiments, the polymer was present in excess so complete threading of polymer chains with CDs did not occur. CDs could therefore have interacted with the ends of the polymer chains only replacing a small fraction of their water shell resulting in the formation of soluble inclusion complexes. This could explain the observation that the various chain lengths of PEG and PVP had no effect on the complex formation.

4. CONCLUSION

Soluble inclusion complexes can be formed between CDs and polymers. The selectivity of the polymers for the different types of CDs in forming insoluble inclusion complexes cannot be observed in the case of soluble inclusion complexes. It is therefore not possible to separate individual types of CDs by formation of soluble inclusion complexes. The polymers have preferences with respect to the type of CD with which they are complexed. It appears that the formation of soluble inclusion complexes is independent of the molecular weight of the polymer. Polymers containing substituents (PPG and PVP) are better at forming inclusion complexes than an unsubstituted polymer (PEG).

REFERENCES

[1] Harada, A., Macromolecular recognition: Inclusion complexes of polymers with cyclodextrins and preparation of polyrotaxanes, *Polym. News*, **18**, 358-363 (1993)

[2] Wenz, G., Keller, B., Threading cyclodextrins on polymer chains, *Angew. Chem. Int. Ed. Engl.*, **31**, 197-199 (1992)

[3] Harada, A., Li, J., Kamachi, M., Double-stranded inclusion complexes of cyclodextrin threaded on poly(ethylene glycol), *Nature*, **370**, 126-128 (1994)

[4] Japan Organo Co., Ltd., Separation of β- and/or γ-cyclodextrin from mixture via their inclusion compound formation with polypropylene glycols, Japanese Patent 4013701 (1992)

[5] Delbourg, M.F., Drouet, Ph., De Moraes, F., Thomas, D., Barbotin, J.N., Effect of PEG and other additives on cyclodextrin production by *Bacillus macerans* cyclomaltodextrin-glycosyl-transferase, *Biotechnol. Lett.*, **15**, 157-162 (1993)

[6] Hayashida, K., Kawakami, K., Enhancement of enzymatic production of cyclodextrins by adding polyethylene glycol or polypropylene, *J. Ferment. Bioeng.*, **73**, 239-240 (1992)

[7] Hirsch, W., Choy, C.K., Ng, K.W., Fried, V., Determination of cyclodextrin-guest association constant by competition with indicator dyes, *Anal. Lett.*, **22**, 2861-2869 (1989)

[8] Harada, A., Li, J., Kamachi, M., The molecular necklace: A rotaxane containing many threaded α-cyclodextrins, *Nature*, **356**, 325-327 (1992)

[9] Szejtli, J., Cyclodextrin technology, Kluwer Academic Publishers, Dordrecht, 1988

SELF ORGANIZATION OF FLUORESCENT MOLECULAR NECKLACES IN AQUEOUS SOLUTION

I. Kräuter, W. Herrmann, G. Wenz*
Polymer-Institut der Universität Karlsruhe,
Hertzstr. 16, D-76187 Karlsruhe, Germany

ABSTRACT

N-Fluoresceinyl-N'-(mono-6-desoxy-6-β-cyclodextrinyl)-thiourea **2** was synthesized from mono-6-amino-6-desoxy-β-CD **1** and fluorescein isothiocyanate. The fluorescent CD derivative **2** was threaded on a water soluble polymer, poly(N,N-dimethylammoniumhexamethylene-N',N'-dimethylammoniumdecamethylene dibromide) **3**. The existence of the molecular necklace was visualized by gel electrophoresis.

1. INTRODUCTION

Fluorescent CD derivatives are well known by the work of Ueno et al. for the detection of various guest molecules [1-4]. So, chromophores like naphthalene, pyrene, anthracene were attached to CDs. These fluorescent CDs were used to detect a great variety of guest molecules, e. g. steroids. Up to now, the detection of polymers by fluorescent CDs was never reported. Threading of unsubstituted CDs on polymer chains is well known since the year 1990 by the work of Harada and us. For example α-CD was threaded on poly(ethylene glycol), β-CD on poly(propylene glycol) and γ-CD on poly(methyl vinyl ether) or poly(isobutylene) to form polymer inclusion coumpounds which are insoluble in water [5-8]. We found that CDs form watersoluble polymer inclusion compounds with poly(N,N-dimethylammoniumoligomethylene)s **3** and poly(iminooligomethylene)s **4** if the alkyl segments are longer than eight methylene groups [9-13]. Here, we report of threading of a fluorescent β-CD derivative on a polymer chain to form watersoluble "shining" molecular necklaces.

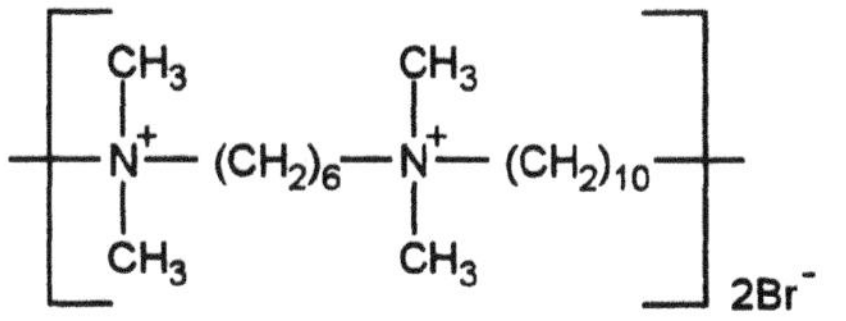

$$\left[-N^{+}(CH_3)_2-(CH_2)_6-N^{+}(CH_3)_2-(CH_2)_{10}- \right] 2Br^- \qquad \left[-N^{+}H_2-(CH_2)_k-N^{+}H_2-(CH_2)_l- \right] 2Ac^-$$

3 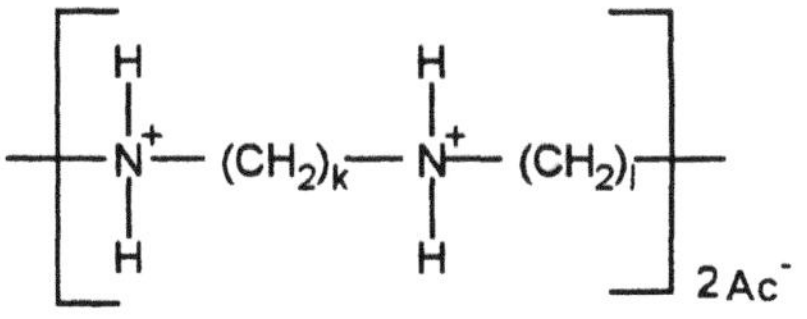 **4**

J. Szejtli and L. Szente (eds.), Proceedings of the Eighth International Symposium on Cyclodextrons, 205–208.

2. MATERIALS AND METHODS

Mono-6-amino-6-desoxy-β-CD **1** was synthesized starting from tosyl-β-CD via the mono-6-desoxy-6-azido-β-CD according to the literature [14]. The reduction of the azido derivative to the amine was performed by hydrogenation using the catalyst Pd/C in nearly quantitative yield. Fluorescein isothiocyanate (isomer I) was from *Aldrich* and β-CD from *Wacker AG*, München. Poly(N,N-dimethylammoniumhexamethylene-N',N'-dimethylammoniumdecamethylene) **3** was prepared by the *Menschutkin* analogous reaction of N,N,N',N'-tetramethyl-1,6-diaminohexane and 1,10-dibromodecane (both from *Aldrich*) in N-methylformamid / methanol 1:1 at 50°C for 14 days and purified by dialysis.

For the electrophoresis gel plates of a size of 10x10 cm^2 were casted from a 1.2 % solution of agarose (15510-019, *Gibco* BRL) in 50 mL 1 x TAE buffer at pH = 8.3. The electrophoresis was run at 90 V and 70 mA using a standard equipment.

3. RESULTS AND DISCUSSION

The fluorescent β-CD derivative **2** was synthesized by addition of mono-6-amino-6-desoxy-β-CD **1** to fluorescein isothiocyanate in 85 % yield. The addition reaction was followed by IR spectroscopy by means of the rising thiourea band at 1614 cm^{-1} and the decaying isothiocyanate band at 2041 cm^{-1}. The structure of **2** was proved by NMR spectroscopy. The product shows narrow bands in both the absorption spectrum at λ = 493 nm and the emission spectrum at λ = 519 nm.

O O OH COO S=C NH NH NH_2 ß FIT Py/25°C ß

1 **2**

In contrast to previous fluorescent CD derivatives, **2** is quite well soluble in water at pH > 9. The inclusion of a guest causes only very small changes of the UV/VIS spectra (fig. 1). Consequently, the hydrophilic fluorescein substituent seems not to interact strongly with the β-CD cavity.

Fluorescent β-CD **2** can be threaded on the polymer **3** in aqueous solution. To prevent the labeled rings **2** from sliding off the polymer chain, α-CD was threaded afterwards. From earlier studies we know that α-CD propagates very slowly along the polymer chain. Therefore it can be used as an efficient blocking entity. The resulting structure is an example of a *pseudopolyrotaxane* (fig.2).

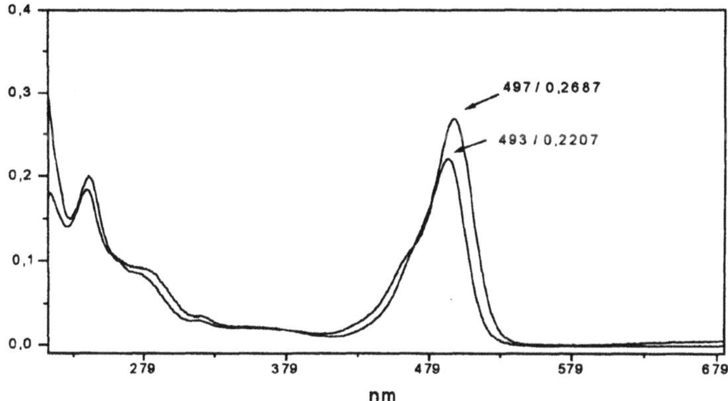

Fig. 1 UV/VIS spectra of a 3 * 10^{-6} M solution of the fluorescent ß-CD 2, free (493 nm) and threaded on the ionene 3 (497 nm).

The existence of this supramolecular structure was proved by horizontal gel electrophoresis in a straightforeward way. This proof was based on the fact that free fluorescent β-CD **2** has a negative charge, while threaded fluorescent β-CD **2·3** has a net positive charge due to the excess positive charge of the polymer **3**. Therefore free compound **2** should migrate to the anode and threaded **2·3** to the cathode. This was really the case. To exclude a loose association of fluorescent β-CD **2** at the polymer **3**, we also performed a control experiment: α-CD was threaded first on the polymer chain **3** to block the chain ends. Indeed nearly no fluorescent β-CD **2** could be threaded afterwards. Consequently, the formation of fluorescent pseudopolyrotaxanes (fig. 2) was proved unambiguously.

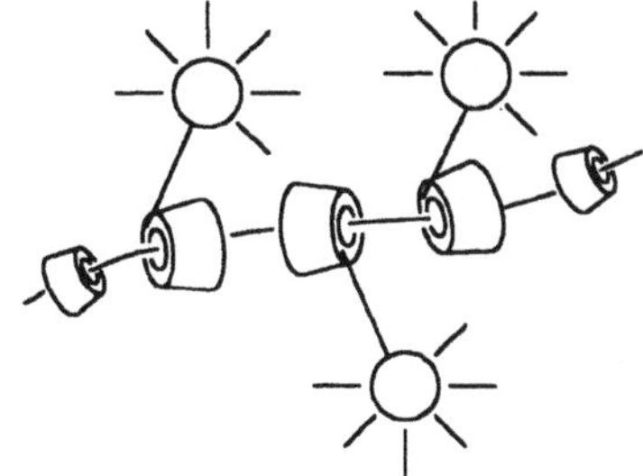

Fig. 2 Schematic drawing of the fluorescent pseudopolyrotaxane

4. CONCLUSION

A new watersoluble fluorescent CD **2** derivative was synthesized. Polymer chains can be made visible by inclusion in fluorescent CD **2**. Gel electrophoresis is well suited in general to detect the inclusion of polymer chains.

5. ACKNOWLEDGEMENTS

We thank the *Bundesministerium für Bildung, Wissenschaft, Forschung und Technologie* (BEO 22 / 0310059A), Bonn, and *Wacker AG*, München, for support.

6. REFERENCES

[1] Ueno, A., Moriwaki F., Azuma A., Osa T., Exciton coupling, intramolecular self-complexation and host-guest complex formation of 6A,6X-bis(anthracene-9-carbonyl) derivatives of cyclomaltooctaose, *Carbohydr. Res.* **192**, 173-180 (1989)

[2] Ueno, A., Minato, S., Suzuki, I., Fukushima, M., Ohkubo M., Osa T., Hamada F. Murai K., Host-guest sensory system of dansyl-modifies β-cyclodextrin for detecting steroidal compounds by dansyl fluorescence, *Chem. Lett.* **4**, 605-608 (1990)

[3] Ueno, A., Suzuki, I. Osa, T., Host-guest sensory systems for detecting compounds by pyrene excimer fluorescence, *Anal. Chem.* **62**, 2461-2466 (1990)

[4] Ueno, A., Minato, S. Osa, T., Host-guest sensors of 6A,6B-, 6A,6C-, 6A,6D-, and 6A,6E-bis(2-naphthalinylsulfenyl)-γ-cyclodextrins for detecting organic compounds by fluorescence enhancements, *Anal. Chem.* **64**, 1154-1157 (1992)

[5] Harada, A. Kamachi, M., Complex formation between poly(ethylene glycol) and α-cyclodextrin, *Macromolecules* **23**, 2821-2823 (1990)

[6] Harada, A. Kamachi, M., Complex formation between cyclodextrin and poly(propylene glycol), *J. Chem. Soc. , Chem. Commun.* **19**, 1322-1323 (1990)

[7] Harada, A., Li, J. Kamachi, M., Complex formation between poly(methyl vinyl ether) and γ-cyclodextrin, *Chem. Lett.* , 237-240 (1993)

[8] Harada, A., Li, J. Kamachi, M., Formation of inclusion complexes of monodisperse oligo(ethylene glycol)s with α-cyclodextrin, *Macromolecules* **27**, 4538-4543 (1994)

[9] Wenz, G., Keller, B., Threading cyclodextrin rings on polymer chains, *Angew. Chem. Int. Ed. Engl.*, **31**, 197-199(1992)

[10] Wenz, G., Cyclodextrins as building blocks for supramolecular structures and functional units, *Angew. Chem. Int. Ed. Engl.*, **33**, 803-822 (1994)

[11] Keller, B., Wenz, G., *Synthesis of polyrotaxanes from cyclodextrins* in: Minutes of the 6th International Symposium on Cyclodextrins (Ed. A. R. Hedges), Editions de Santé Paris, 1992, 62

[12] Wenz, G., Keller, B., Synthesis of polyrotaxanes or how to thread many cyclodextrin rings on a polymer chain, *Polym. Prep. (Am. Chem. Soc., Div. Polym. Chem.)* **34**, 62-64 (1993)

[13] Wenz, G., Keller, B., Speed control for cyclodextrin rings on polymer chains, *Macromol. Symp.* **87**, 11-16 (1994)

[14] Parrot-Lopez, H., Djedaini, F., Perly, B., Coleman, A. W., Galons, H., Micoque, M., An approach to vectorisation of pharmacologically active molecules: the covalent binding of Leu-enkephalin to a modified β-cyclodextrin, *Tetrahedron Lett.* **31**, 1999-2002 (1990)

THEORETICAL AM1 STUDIES OF INCLUSION COMPLEXES OF α- AND β-CYCLODEXTRINS WITH METHYLATED BENZOIC ACIDS AND PHENOL, AND γ-CYCLODEXTRIN WITH BUCKMINSTERFULLERENE

NICHOLAS BODOR, MING-JU HUANG, AND JOHN D. WATTS
Center for Drug Discovery
POB 100497, Health Science Center
Gainesville, FL 32610 USA

ABSTRACT

Semiempirical AM1 calculations have been performed on the inclusion complexes of α- and β-cyclodextrin with benzoic acid and phenol and β-cyclodextrin with methylated benzoic acids in the "head first" and "tail first" positions. The results show that α-cyclodextrin complexes with phenol and benzoic acid guests in the "head first" position are more stable than in the "tail first" position, while β-cyclodextrin complexes with the same guests prefer the "tail first" position. The preferred orientation for β-cyclodextrin with methylated benzoic acids is determined by the position of the methyl substituent(s). In general, para-methyl benzoic acid derivatives prefer the "tail first" position. γ-cyclodextrin forms a slightly unstable 1:1 complex with C_{60} (3.4 kcal/mol), but two γ-cyclodextrins provide enough stabilization by about 10 kcal/mol to "cage-in" the C_{60}.

1. INTRODUCTION

The hydrophobic character of the central cavity of cyclodextrins (CDs) enables them to form inclusion complexes with many different molecules. As a result, many poorly water soluble drugs can be administered in solution in the complex form by taking advantage of the well-established low toxicity of most CDs. Therefore, investigations of the driving forces of complexation and the structures of inclusion complexes are very important. Modern computers with graphical capabilities are valuable tools for studying the structures of inclusion complexes.

Various investigators have applied theoretical methods such as molecular mechanics, molecular dynamics, and CNDO with fixed-geometry to CD inclusion complexes. There are very few quantum mechanical calculations on CDs using the most advanced semi-empirical method, AM1 [1] with fully unrestricted geometry optimization [2].

2. MATERIALS AND METHODS

Phenol, benzoic acid, and methylated benzoic acids were studied using the MOPAC programs on a Tektronix CAChe workstation. The AM1 fully optimized geometries

J. Szejtli and L. Szente (eds.), Proceedings of the Eighth International Symposium on Cyclodextrons, 209–214.

with no symmetry constraints for α-CD and β-CD were taken from our previous calculations [2]. AM1 calculations on all the inclusion complexes of α- and β-CD with phenol and benzoic acid, β-CD with methylated benzoic acids, γ-CD, buckminsterfullerene, γ-CD dimer, and γ-CD with buckminsterfullerene were performed using a modified version of the AMPAC program from QCPE [1]. In all cases, considerable computer power is required in order to obtain reliable fully optimized structures.

3. RESULTS AND DISCUSSION

AM1 calculations for the inclusion complexes of α- and β-CD with phenol and benzoic acid and β-CD with methylated benzoic acid have been performed for the "head first" and "tail first" orientations (Fig. 1).

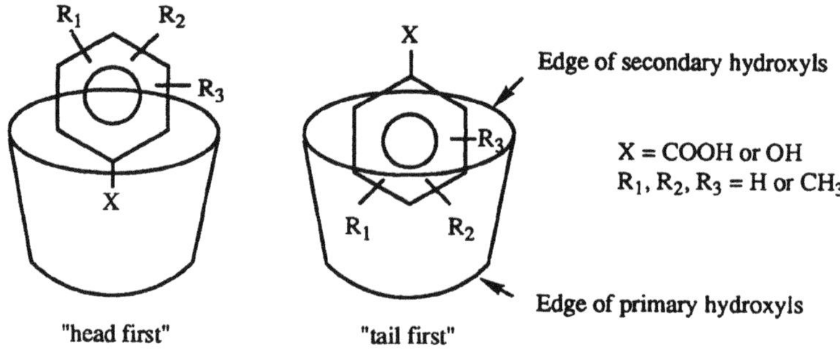

Figure 1. The two possible penetration pathways for benzoic acid, phenol and methylated benzoic acids.

The calculated partition coefficients, volume, surface area, and ovality are based on our previously developed BLOGP program [3]. These properties, the heats of formation, dipole moment, HOMO energy, and the lengths of the guest molecules are listed in Table 1. As expected, increased molecular complexity introduced by homologue extension increases the lipophilicity of the benzoic acid and decreases the water solubility. From our previous studies [2], α-CD and β-CD have calculated heights of 6.2-6.5 Å, heats of formation -1414.0 and -1647.5 kcal/mol, dipole moment 10.4 and 7.5 Debye, and HOMO energies of -10.21 and -10.35 eV respectively.

The AM1 heats of formation, dipole moment, HOMO energy, and stabilization energy for different inclusion complexes in the "tail first" and "head first" positions are shown in Table 2. From Table 2, we can draw a few conclusions: (1) phenol and benzoic acid form more stable complexes with α-CD in the "head first" position and form more stable complexes with β-CD in the "tail first" position; (2) among o-, m-, and p-methyl benzoic acid, o- and m-methyl benzoic acid form more stable complexes with β-CD in the "head first" position and p-methyl benzoic acid forms a more stable complex with β-CD in the "tail first" position; (3) among all the dimethyl benzoic acids, 2,3-dimethyl,

2,5-dimethyl, 2,6-dimethyl, and 3,4-dimethyl benzoic acid form more stable complexes with β-CD in the "head first" position, and 2,4-dimethyl and 3,5-dimethyl benzoic acid form more stable complexes with β-CD in the "tail first" position; (4) of the trimethyl benzoic acids, only 2,3,5-trimethyl, 2,4,5-trimethyl and 3,4,5-trimethyl benzoic acids form stable complexes with β-CD. The first one prefers the "head first" position and the others prefer the "tail first" position.

TABLE 1. Heats of formation (ΔH_f, kcal/mol), dipole moment (D, Debye), HOMO energy (eV), the calculated partition coefficients, volume (V, Å^3), surface area (S, Å^2), ovality (O), and length (Å) of phenol, benzoic acid, all the possible one, two, and three methylated benzoic acids (BA).

Compound	ΔH_f	D	HOMO	LogP	V	S	O	Length
phenol	-22.2[a]	1.2	-9.11	1.30[b]	91.84	120.50	1.22	5.662
benzoic acid (BA)	-68.0[a]	2.4	-10.08	1.61[b]	110.74	143.02	1.28	7.012
2-methyl BA	-73.6	2.4	-9.73	2.04	126.83	160.07	1.31	7.077
3-methyl BA	-75.6	2.7	-9.75	2.02[b]	127.39	164.42	1.34	7.011
4-methyl BA	-75.8	2.8	-9.82	2.02[b]	127.49	164.57	1.34	7.922
2,3-dimethyl BA	-79.5	2.6	-9.55	2.39	143.45	180.84	1.36	7.039
2,4-dimethyl BA	-81.5	2.7	-9.63	2.43	143.57	181.82	1.37	7.981
2,5-dimethyl BA	-81.3	2.4	-9.45	2.45	143.60	181.77	1.37	7.075
2,6-dimethyl BA	-79.3	2.0	-9.53	2.42	143.42	179.57	1.36	7.024
3,4-dimethyl BA	-81.0	3.0	-9.61	2.41	143.62	182.78	1.38	7.910
3,5-dimethyl BA	-83.1	2.7	-9.58	2.47	144.18	185.98	1.40	7.009
2,3,4-trimethyl BA	-86.0	2.9	-9.50	2.74	159.76	199.18	1.40	7.989
2,3,5-trimethyl BA	-87.2	2.6	-9.33	2.81	160.01	201.57	1.41	7.052
2,3,6-trimethyl BA	-85.8	2.1	-9.32	2.78	160.15	201.16	1.41	7.013
2,4,5-trimethyl BA	-88.5	2.7	-9.37	2.82	160.14	201.63	1.41	7.987
2,4,6-trimethyl BA	-87.1	2.3	-9.49	2.81	160.03	200.50	1.41	7.994
3,4,5-trimethyl BA	-89.1	3.1	-9.56	2.81	160.35	202.98	1.42	7.954

[a] Experimental ΔH_f for phenol and benzoic acid are -23.0 kcal/mol and -70.3 kcal/mol (Pedley, J. B.; Rylance, G. "Sussex-N.P.L. Computer Analysed Thermochemical Data: Organic and Organometallic Compounds", Sussex University, 1977). [b] Experimental LogP values for phenol, benzoic acid, 3-methyl BA, and 4-methyl BA are 1.49, 1.95, 2.37, and 2.27 respectively (Hansch, C. and Leo, A. "Substituent Constants for Correlation Analysis in Chemistry and Biology"; Wiley: New York, 1979).

The hydroxyl groups of phenol and carboxylate group of benzoic acid in the "head first" position in α-CD inclusion complexes could possibly form hydrogen-bond with the primary OH-s. Using cut-off criteria from Steiner and Saenger [4-7], namely a hydrogen bond is an O-H...O interaction in which the H...O distance is less than or equal to 3.00 Å and the angle at H is larger tha 90°, we found that there are two possibilities for hydrogen bonds between α-CD and phenol in "head first" position and no hydrogen bond with phenol in "tail first" position. There are three possible positions for hydrogen bonds between β-CD and phenol in the "tail first" position, and no hydrogen bonds with phenol in the "head first" position. There is one possible hydrogen bond between β-CD and benzoic acid in the "tail first" position and no hydrogen bond with benzoic acid in the "head first" position. For the inclusion complex of α-CD with benzoic acid, we found there is one hydrogen bond in both "head first" and "tail first". Beyond the hydrogen bond factor, there is a strong dipole-dipole interaction between the host and guest in the "head first" orientation of the α-CD (10.4 Debye) with benzoic acid (2.4 Debye), which is in agreement with the previous CNDO studies [8-11] which indicated that dipole moments of guest molecules are antiparallel to the dipole moment of host α-CD in the crystalline state.

TABLE 2. Heats of formation (ΔH_f, kcal/mol), dipole moment (D, in Debye), HOMO energy (eV) for AM1 optimized geometries of which are complexes of α- and β-CD with phenol and benzoic acid (BA), and β-CD with methylated benzoic acids in "tail first" and "head first" positions. The stabilization energies ($\Delta\Delta H_f$) are in kcal/mol.

Compound	$\Delta H_f^t(\Delta\Delta H_f)$	D^t	$HOMO^t$	$\Delta H_f^h(\Delta\Delta H_f)$	D^h	$HOMO^h$
α-CD + phenol	-1434.2(2.0)	8.9	-9.32	-1439.4(-3.2)	6.6	-9.28
β-CD + phenol	-1673.2(-3.5)	8.6	-9.46	-1672.6(-2.9)	8.1	-9.42
α-CD + BA	-1482.0(0.0)	5.1	-10.26	-1489.1(-7.1)	11.4	-10.08
β-CD + BA	-1716.9(-1.4)	8.3	-10.30	-1716.1(-0.6)	6.8	-10.34
β-CD + 2-Me BA	-1721.8(-0.7)	6.6	-10.03	-1728.1(-7.0)	7.2	-10.00
β-CD + 3-Me BA	-1723.4(-0.3)	8.6	-10.15	-1729.8(-6.7)	8.2	-9.99
β-CD + 4-Me BA	-1725.2(-1.9)	8.3	-10.21	-1723.5(-0.2)	8.2	-10.12
β-CD + 2,3-diMe BA	-1726.2(0.8)	6.1	-9.78	-1729.6(-2.6)	6.5	-9.74
β-CD + 2,4-diMe BA	-1735.9(-6.9)	6.8	-9.97	-1729.6(-0.6)	8.1	-9.95
β-CD + 2,5-diMe BA	-1726.8(2.0)	5.9	-9.76	-1734.0(-5.2)	6.2	-9.66
β-CD + 2,6-diMe BA	-1728.5(-1.7)	6.5	-9.85	-1730.2(-3.4)	6.4	-9.82
β-CD + 3,4-diMe BA	-1735.1(-6.6)	5.1	-9.94	-1735.9(-7.4)	7.0	-9.89
β-CD + 3,5-diMe BA	-1740.4(-9.8)	5.9	-9.98	-1738.4(-7.8)	8.1	-9.80
β-CD + 2,3,4-triMe BA	-1731.7(1.8)	7.3	-9.76	-1731.4(2.1)	5.5	-9.68
β-CD + 2,3,5-triMe BA	-1731.6(3.1)	6.5	-9.62	-1737.3(-2.6)	5.8	-9.50
β-CD + 2,3,6-triMe BA	-1729.4(3.9)	7.8	-9.62	-1727.8(5.5)	5.6	-9.35
β-CD + 2,4,5-triMe BA	-1743.9(-7.9)	3.8	-9.72	-1734.3(1.7)	6.9	-9.57
β-CD + 2,4,6-triMe BA	-1733.0(1.6)	6.8	-9.79	-1726.3(8.3)	6.5	-9.68
β-CD + 3,4,5-triMe BA	-1740.0(-3.4)	5.6	-9.87	-1737.7(-1.1)	7.8	-9.59

The guest HOMO energies are somewhat affected by complexation. For example isolated 3,4,5-trimethyl benzoic acid has a HOMO energy of -9.56 eV, while in "tail first" and "head first" complexes with β-CD the HOMO energies are -9.87 and -9.59 eV, respectively.

The reasons for the orientational preferences of the methylated guests are a combination of hydrogen bonding, steric hindrance, van der Waals interactions, and hydrophobic interactions. All the mono-, di-, and trimethyl benzoic acids can form weak hydrogen bonds between the carboxylate group and the primary or secondary hydroxylic rim of the β-CD. The H-bonding affects the relative positions. Methyl substituent in the ortho and meta position cause more steric hindrance than in the para position. Since the energy of interaction between the guest and host is a van der Waals-London dispersion forces interaction, maximization of contact within the attraction distance between the benzoic acid derivatives and the β-CD cavity will enhance complexation. For increased van der Waals interaction, the 2- and 3-methyl benzoic acids would prefer the "head first" position, since when the methyl group of the guest molecule is near the primary hydroxylic rim, the distance between the cavity wall and the substrate is smaller than when it is near the secondary hydroxylic rim and consequently there is a larger van der Waals interaction. In the case of the di- and trimethyl substituted benzoic acids, a balance between the various forces will determine the optimum position.

Since the first report [12] on a water soluble γ-CD and C_{60} fullerene complex, we were intrigued by the potential structure of it. The complex was reportedly prepared by boiling aqueous C_{60} with with a 0.08 mol dm^{-3} aqueous solution of γ-CD for at least 48 h. The X-ray structure of the complex is not available. Indirect evidence for complexation is from UV-VIS and NMR spectroscopy [12-13] and photophysical [14-15] studies. The most likely complex is suggested to be a 2 : 1 adduct of γ-CD and C_{60}, in which both γ-CD's interact with C_{60} at their secondary hydroxyl sides [16-18].

The heats of formation from AM1 fully unrestricted geometry optimization for γ-CD, C_{60}, and their 1 : 1 complex are -1890.4, 973.4, and -913.6 kcal/mol respectively, which shows that the 1 : 1 complex, although a minimum on the potential surface, is slightly unstable by 3.4 kcal/mol. Our previous studies [2] suggested that intermolecular interactions between β-cyclodextrin molecules are at an about 34 kcal/mol level, explaining the stability of the crystals. Relatively strong interactions between CD molecules could also exist in solution, leading to some special spacial arrangements around the water insoluable C_{60}. Two γ-CD-s could surround C_{60} and interact with each other to form a stable 2 : 1 complex. In order to support this concept, we have calculated the stabilization energy between two γ-CD-s arranged with their secondary OH-s facing each other. The dimer is by 13.8 kcal/mol (γ-CD dimer ΔH_f = -3668.0 kcal/mol and monomer ΔH_f = -1827.1 kcal/mol) more stable than two γ-CD-s and would allow to accomodate a C_{60} molecule "cage-in" between them.

4. CONCLUSION

In conclusion, we have studied the inclusion complexes of α-CD and β-CD with benzoic acid and phenol, and β-CD with methylated benzoic acids in the "head first" and "tail first" positions. The preferred position for the guest compound inside the cavity depends on a combination of steric, hydrogen bonding, van der Waals, and hydrophobic effects. The stabilization energy from the dimerization of γ-CD through the secondary hydroxylic rims may render the 2 : 1 adduct of γ-CD and C_{60} more stable than the 1 :1 adduct or isolated molecules.

REFERENCES

[1] Dewar, M. J. S., Zoebisch, E. G., Healy, E. F., Stewart, J. J. P., AM1: A New General Purpose Quantum Mechanical Molecular Model[1], *J. Am. Chem. Soc.*, **107**, 3902-3909 (1985)

[2] Bodor, N. S., Huang, M.-J., Watts, J. D., Theoretical Studies on the Structures of Natural and Alkylated Cyclodextrins *J. Pharm. Sci.*, **84**, 330-336 (1995)

[3] Bodor, N. S., Huang, M.-J., An Extended Version of a Novel Method for the Estimation of Partition Coefficients, *J. Pharm. Sci.*, **81**, 272-281 (1992)

[4] Jeffrey, G. A., Saenger, W., *Hydrogen Bonding in Biological Structures*, Springer Verlag, Berlin, 1991

[5] Steiner, Th., Saenger, W., Geometric Analysis of Non-Ionic O-H...O Hydrogen Bonds and Non-Bonding Arrangements in Neutron Diffraction Studies of Carbohydrates, *Acta Crystallogr., Sect. B*, **48**, 819-827 (1992)

[6] Steiner, Th., Saenger, W., Role of C-H...O Hydrogen Bonds in the Coordination of Water

Molecules. Analysis of Neutron Diffraction Data, *J. Am. Chem. Soc.*, **115**, 4540-4547 (1993)
[7] Steiner, Th., Saenger, W., Reliability of assigning O-H...O hydrogen bonds to short intermolecular O...O separations in cyclodextrin and oligosaccharide crystal structures, *Carbohydr. Res.*, **259**, 1-12 (1994)
[8] Kitagawa, M., Hoshi, H., Sakurai, M., Inoue, Y., Chujo, R., The large dipole moment of cyclomaltohexaose and its role in determining the guest orientation in inclusion complexes, *Carbohydr. Res.*, **163**, C1-C3 (1987)
[9] Sakurai, M., Kitagawa, M., Hoshi, H., CNDO-Electrostatic Potential Maps for α-Cyclodextrin, *Chem. Lett.*, 895-898 (1988)
[10] Kitagawa, M., Hoshi, H., Sakurai, M., Inoue, Y., Chujo, R., A Molecular Orbital Study of Cyclodextrin Inclusion Complexes. I. The Calculation of the Dipole Moments of α-Cyclodextrin-Aromatic Guest Complexes, *Bull. Chem. Soc. Jpn.*, **61**, 4225-4229 (1988)
[11] Sakurai, M., Kitagawa, M., Hoshi, H., Inoue, Y., Chujo, R., A Molecular Orbital Study of Cyclodextrin Inclusion Complexes. II. The Structure Analysis of α-Cyclodextrin Inclusion Complex with m-Nitrophenol in Aqueous Solution Based on the Quantum-Chemical Solvation Theory, *Bull. Chem. Soc. Jpn.*, **62**, 2067-2069 (1989)
[12] Andersson, T., Nilsson, K., Sundahl, M., Westman, G., Wennerström, O., C_{60} Embedded in γ-Cyclodextrin: a Water-soluble Fullerene, *J. Chem. Soc. Chem. Commun.*, 604-606 (1992)
[13] Andersson, T., Westman, G., Wennerström, O., Sundahl, M., NMR and UV-VIS Investigation of Water-soluble Fullerene-60-γ-Cyclodextrin Complex, *J. Chem. Soc. Perkin Trans.*, **2**, 1097-1101 (1994)
[14] Priyadarsini, K. I., Mohan, H., Mittal, J. P.Guldi, D. M., Asmus, K.-D., Pulse Radiolysis Studies on the Redox Reactions of Aqueous Solutions of γ-Cyclodextrin/C_{60} Complexes, *J. Phys. Chem.*, **98**, 9565-9569 (1994)
[15] Boulas, P. Kutner, W., Jones, M. T., Kadish, K. M., Bucky(basket)ball: Stabilization of Electrogenerated $C_{60}^{\cdot-}$ Radical Monoanion in Water by Means of Cyclodextrin Inclusion Chemistry, *J. Phys. Chem.*, **98**, 1282-1287 (1994)
[16] Yoshida, Z.-i., Takekuma, H., Takekuma, S.-i., Matsubara Y., *Angew. Chem. Int. Ed. Engl.*, **33**, 1597-1599 (1994)
[17] Priyadarsini, K. I., Mohan, H. Tyagi, A. K., Mittal, J. P., Inclusion Complex od γ-Cyclodextrin-C_{60}: Formation, Characterization, and Photophysical Properties in Aqueous Solutions, *J. Phys. Chem.*, **98**, 4756-4759 (1994)
[18] Kuroda, Y., Nozawa, H., Ogoshi, H., Kinetic Behaviors of Solubilization of C_{60} into Water by Complexation with γ-Cyclodextrin, *Chemistry Letters*, 47-48 (1995)

STRUCTURE AND SPECTROSCOPIC PROPERTIES OF CYCLODEXTRIN INCLUSION COMPLEXES.

G. KÖHLER[1)], G. GRABNER[1)], C. TH. KLEIN[1)], G. MARCONI[2)], B. MAYER[1)], S. MONTI[2)], K. RECHTHALER[1)], K. ROTKIEWICZ[3)], H. VIERNSTEIN[4)] AND P. WOLSCHANN[1)]

[1)]Institute for Theoretical Chemistry and Radiation Chemistry, University of Vienna, Althanstraße 14, 1090 Vienna, Austria; [2)]Institute of Photochemistry and High Energy Radiation, Italian Council for Research (CNR), Via P. Gobetti 101, 40129 Bologna, Italy; [3)]Institute for Physical Chemistry, Polish Academy of Sciences, Kasprzaka 44-52, PL-01-224 Warsaw, Poland; [4)]Institute for Pharmaceutical Technology, University of Vienna, Althanstraße 14, 1090 Vienna, Austria;

ABSTRACT

We compare spectroscopic properties of higher order complexes of organic guests (e.g. naphthalene, phenols, indole, C_{60}-fullerene) with cyclodextrins (CDx) to results of molecular modeling investigations. Naphthalene 1:2 complexes with α-CDx show high spectral resolution and peculiar triplet properties. Molecular simulations and calculation of the experimentally measured induced circular dichroism (ICD) provide detailed structural information.

1. INTRODUCTION

Self-assembly of macromolecular systems is of great importance in chemistry and molecular biology. A valuable model system to study molecular self-assembly is inclusion complex formation of cyclodextrins as hosts with appropriate organic guest molecules. CDx's are most suitable for spectroscopic studies extending into the far u.v., as they show no absorption in this region. Besides electronic absorption and fluorescence, circular dichroism induced in the u.v. absorption band of an aromatic guest by the chirality of the glucose subunits of the macromolecular host is used to study complexation.

Additionally to spectroscopic studies molecular docking between CDx and various guests is performed by a Dynamic Monte Carlo (DMC) approach. Solvation effects are included by a continuum model. Sign and strength of the resulting ICD of complexes are calculated by the Kirkwood-Tinoco exciton chirality method for the obtained geometries.

J. Szejtli and L. Szente (eds.), Proceedings of the Eighth International Symposium on Cyclodextrons, 215–220.

2. MATERIALS AND METHODS

2.1. Experimental Section

Experimental details are given in [1]. Triplet-Triplet transient absorption spectra and triplet quantum yields were measured by laser flash photolysis, using a frequency quadrupled Nd-YAG laser (QuantaRay DCR-1) at 266 nm under right angle conditions.

2.2. Theoretical Methods

Molecular Mechanics calculations are performed using Allinger's MM3-92 force field within the program package MolDoc[2]. The reference geometries for α- and β-CDx are obtained by minimizing crystallographic geometries. γ-CDx is modeled as a symmetric cone and optimized within [2]. Low energy complex geometries are located by calculating complexation pathways (defined stepwise transfer of guest molecules through the CDx cavity) and by Dynamic Monte Carlo simulations. Solvation effects are considered by a continuum approximation. Energetic contributions based on both, potential energies and solvation energies, are represented within a modified Metropolis criterion. For a detailed description of the methods see reference [3-6].

Induced circular dichroism spectra of complexes are calculated by the Kirkwood-Tinoco exciton chirality method. ICD is the most direct method to determine association constants and is used to study structural properties of host-guest systems. It results from dipole-dipole interactions between the transition moment of the guest and induced transient dipoles in the bonds of the chiral groups of the host, i.e. of the glucose subunits of CDx. This technique provides, therefore, direct information on the relative position of the achiral guest in the CDx cavity. Details of the formalism of ICD calculations are reported in reference [3].

3. RESULTS AND DISCUSSION

3.1. Experimental Results

Absorption and fluorescence spectra of aqueous naphthalene become highly structured upon inclusion in α-CDx, similar to those obtained in hydrocarbon solutions (see fig. 1). Two lifetimes are recovered, one in agreement with that in aqueous solution and a second long one, identical to non-aqueous solvents. The α-CDx concentration dependence of their relative weight suggests the formation of 1:2 guest:host complexes.

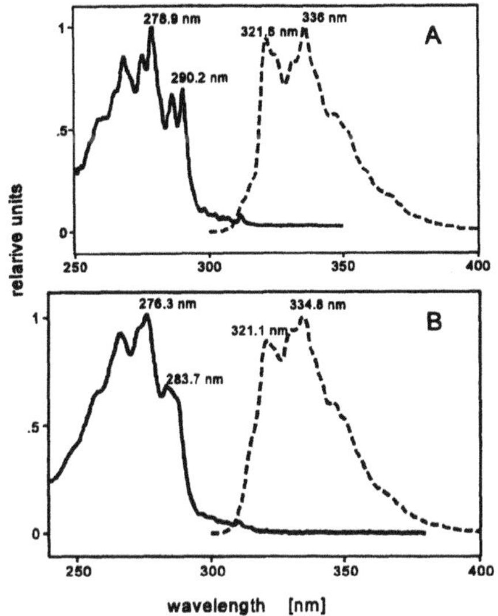

Fig. 1: Absorption (full line) and fluorescence (broken line) spectra of naphthalene 1:2 complexes with α-CDx (A) and of naphthalene in saturated aqueous solution (B).

Transient absorption spectroscopy has been used to study the effect of complexation with cyclodextrins (α-CDx and ß-CDx) on spectroscopic and kinetic properties of naphthalene triplets. The main results are:

- CDx complexation induces a marked red shift and line narrowing of T-T absorption. Narrowing is very marked in α-CDx complexes, the band width being strongly temperature dependent (see fig. 2). A value of 5.7 nm is found at 12 ^{0}C, which is significantly below that obtained in hexane solution (8.1 nm).
- Triplet decay is slowed down in complexes, in particular with α-CDx. This fact can be used to study the complexation equilibrium. Predominance of a 1:2 naphthalene-α-CDx complex is obtained over a wide CDx concentration range.
- Both, association constant and triplet decay kinetics are strongly temperature-dependent. An activation energy of 66 kJ mol^{-1} is obtained for the quenching of α-CDx-complexed triplets by molecular oxygen.

1:2 complex formation with α-CDx is found generally for molecules with appropriate shape and cause comparable effects on spectroscopic and kinetic properties. Further examples are 2-naphthol, p-substituted phenols and indole [1,7-8].

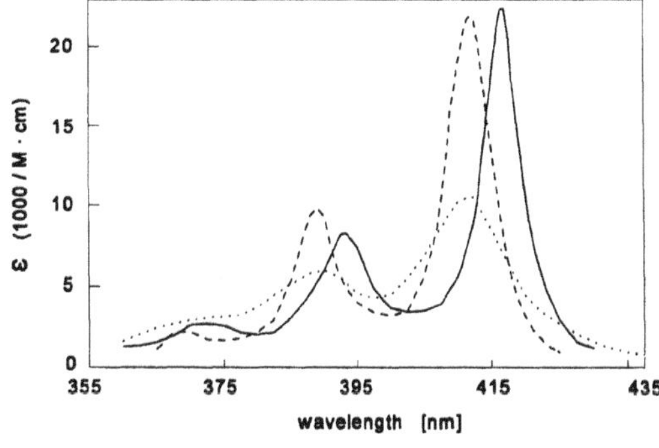

Fig. 2: Triplet-triplet absorption spectra of naphthalene in hexane (broken line), water (dotted line), and aqueous 0.06 M α-CDx (full line), measured at T=22°C.

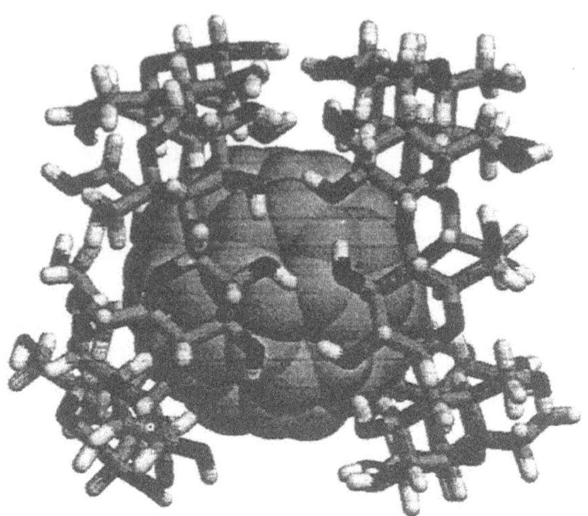

Fig. 3: Low energy 1:2 C_{60}-fullerene - γ-CDx complex

3.2 Molecular Modeling and Calculation of Spectral Properties:

Complexation between a host and a guest molecule can formally be characterized by a two step process: surface properties of host and guest entail molecular recognition and the dynamic properties of the complex defines the final inclusion geometry and its structural distribution. In the case of cyclodextrins, recognition is generally unspecific, despite of possible intermolecular hydrogen bond formation as shown for the phenol - β-CDx system [3] and results mainly from electrostatic and van der Waals interactions. However, CDx molecules as hosts are highly dynamic species, and especially β-CDx shows a strong internal asymmetry [5]. As the complexation process imposes strong constraints on the dynamics of the host, the conformational space of the complex differs completely from the respective space of the host molecule alone, and this was shown for the 2-naphthol - β-CDx complex in some detail [5].

Dynamic Monte Carlo (DMC) simulations proved a good tool for the determination of correct complex structures when solvation effects, especially hydrophobic interactions, are taken into account [3,6-9]. Figure 3 shows exemplarily a low energy geometry of a 1:2 C_{60}-fullerene complex with γ-CDx. This structure is the result of a 1000 step DMC simulation performed at 300 K. Potential energy is calculated by the force field, and solvation energy of hydrophobic surfaces is considered. This model follows a classical view of the hydrophobic effects, i.e. the conception of 'minimizing' the hydrophobic contact surface with water molecules, and reaches a structure prediction capacity for CDx inclusion complexes of over 90 %. This is verified by the calculation of the respective rotatory strength of the final, low energy structures, which were found in agreement with value and sign of the experimentally determined ICD signal [3,4].

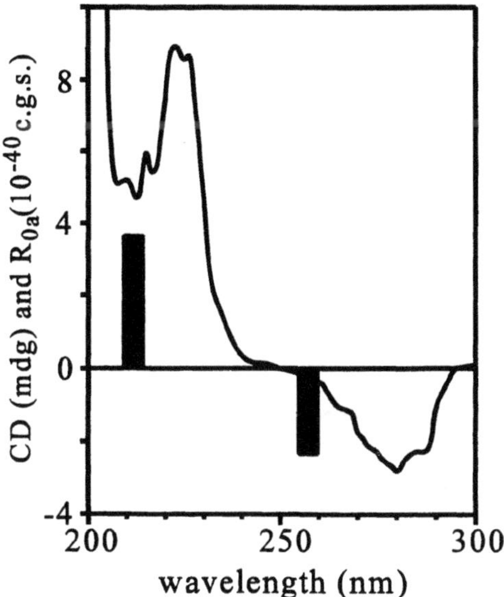

Fig. 4: Induced circular dichroism (CD) and calculated rotatory strength (R_{0a}) of p-cresol in ß-CDx complexes (for more details see text).

4. CONCLUSION

The main results are summarized as follows:

- Optical absorption, induced circular dichroism and fluorescence spectra of inclusion complexes of simple aromatic compounds with cyclodextrins show generally more vibronic features than in aqueous solutions or in other polar solvents. A single fluorescence lifetime is found for these complexes, and its value varies in accordance with environmental effects observed when water is replaced by alcohols or ethers as solvent. Spectroscopic and photophysical properties of higher order complexes, like the 1:2 naphthalene - α-CDx complex, have more similarities to gas phase spectra than to those observed in hydrocarbon solution.
- The sign of induced circular dichoism spectra of such inclusion complexes can be predicted. Solvation reduces the highly complex conformational space of inclusion complexes and selects these structures which yield the correct ICD spectra. Another important dynamic feature is the stability of the intramolecularly hydrogen bonded, secondary hydroxylic rim of CDx.
- These complexes have well defined structures within narrow geometrical limits.
- The main structural constraint arises from hydrophobic interactions in aqueous solution, which can be described by minimization of the hydrophobic molecular surface area.

Higher order complexation, i.e. formation of complexes composed of at least one guest and two host molecules, appears as an intrinsic feature of inclusion complex formation resulting from the unspecific recognition process with appropriately shaped guests. These complexes gain essential stabilization energy from hydrogen bonding between the rim of the CDx subunits and formation of even larger structures, as e.g. self-assembled tubes, might be proposed from the experimental and theoretical results. Figure 5 shows such a structure of a molecular, self-assembled tube, consisting of four α-CDx and five indoles as obtained from computer simulation.

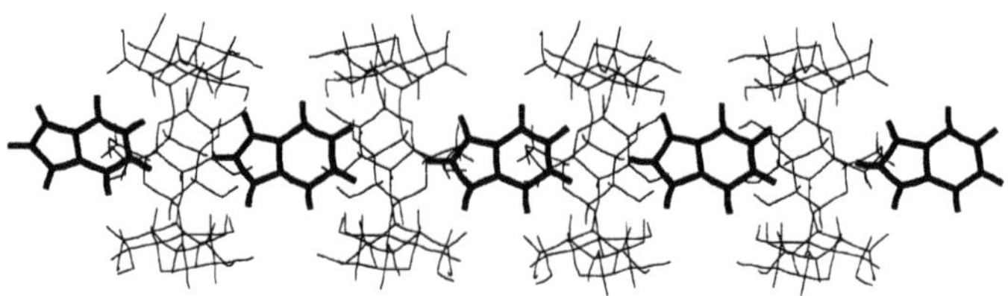

Fig. 5: Molecular tube based on indole - α-CDx complexes

ACKNOWLEDGEMENTS

G. K., B. M., and K. R. thank the *Fonds zur Förderung der wissenschaftlichen Forschung* (P09750-CHE) in Austria for generous financial support. C. Th. K. thanks the *Von Schadsche Stiftung Frankfurt* for a scholarship.

REFERENCES

[1] Park, H.-R., Mayer B., Wolschann, P., Köhler, G., Excited-State Proton Transfer of 2-Naphthol Inclusion Complexes with Cyclodextrins, *J. Phys. Chem.*, **98**, 6158-6166 (1994).

[2] Mayer, B., Program Package MolDoc-V1.2 (1995)

[3] Marconi, G., Monti, S., Mayer, B., Köhler, G., Circular Dichroism of Methylated Phenols Included in β-Cyclodextrin. An Experimental and Theoretical Study, *J. Phys. Chem.*, **99**, 3943-3950 (1995)

[4] Mayer, B., Marconi, G., Klein, C. Th., Köhler, G., Wolschann, P., Structural Analysis of Host-Guest Systems. Methyl-substituted Phenols in β-Cyclodextrin, *J. Phys. Chem.*, submitted.

[5] Mayer, B., Köhler, G., Structural Properties of α- and β-Cyclodextrin and of a β-Cyclodextrin - 2-naphthol Complex. A Case Study., *J. Mol. Struct.*,THEOCHEM, in the press.

[6] Klein, C. Th., Köhler, G., Mayer, B., Mraz, K., Reiter, S., Viernstein, H., Wolschann, P., Solubility and Molecular Modeling of Triflumizole - β-Cyclodextrin Inclusion Complexes, *J. Incl. Phen. Mol. Rec. Chem.*, **22**, 15-32 (1995)

[7] Monti, S., Marconi, G., Grabner, G., Mayer, B., Klein, C. Th., Köhler, G., A Spectroscopic and Photochemical Study of Dimethoxybenzene - Cyclodextrin Inclusion Complexes., preprint.

[8] Grabner, G., Köhler, G., Mayer, B., Rechthaler K., Rotkiewicz K., Spectroscopic Properties and Triplet Behaviour of Naphthalene Inclusion Complexes with Cyclodextrins. To be published.

[9] Marconi, G., Mayer, B., Klein, C. Th., Köhler, G., Higher Order C_{60}-Fullerene - γ-Cyclodextrin Inclusion Complexes. To be published.

INFLUENCE OF DEGREE AND PATTERNS OF SUBSTITUTION ON THE INCLUSION PROPERTIES OF ETHYLATED β-CYCLODEXTRINS

D. WOUESSIDJEWE[1], V. LEMESLE-LAMACHE[1], D. DUCHENE[1], B. PERLY[2]

[1] *Laboratoire de Physico-Chimie, Pharmacotechnie et Biopharmacie, URA CNRS 1218, Faculté de Pharmacie, Université de Paris-Sud, 92290 Châtenay Malabry, France*

[2] *CEA, DRECAM, Service de Chimie Moléculaire, Centre d'Etudes Atomiques, 91191 Gif sur Yvette, France*

ABSTRACT

Different batches of ethylated derivatives of β-cyclodextrins (Et-βCD) obtained by various synthetic routes were used as host molecules to prepare inclusion compounds of salbutamol. Previous results showed that these complexes were suitable to achieve variable sustained-release behaviour of salbutamol. The mechanism of sustained-release was related to the inclusion capacity and physicochemical properties of Et-βCD. Among the various analytical techniques carried out to characterize them, high resolution NMR afforded relevant information in terms of degree and patterns of subtitution, which could explain the differences observed between CD derivatives.

1. INTRODUCTION

The chemical modifications of natural cyclodextrin by a regioselective substitution of the hydroxyl groups located on the C2, C3 and C6 positions is very complicated to achieve, due to the differences in reactivity between the hydroxyl groups and steric factors of the cyclodextrin torus [1]. Two different strategies of synthesis, varying by the choice of reagent and operating condition, were used to obtain Et-βCD .These CD derivatives were successfully used to achieve inclusion compounds of a drug model, salbutamol. The complexes were submitted to dissolution assays *in vitro,* and the results showed variable sustained release of salbutamol [2]. It was obvious that these differences in behaviours of the complexes could be related to the properties of the Et-βCD. A combination of many characterization methods, such as solubility study, X-ray diffraction patterns and electrospray ionization mass spectrometry was carried out on the CD derivatives [3].

This paper deals with high resolution ^{1}H NMR and ^{13}C NMR of ethylated β-cyclodextrins, which appeared as an ultimate tool allowing the determination of the homogeneity of CD derivatives from batch to batch.

2. MATERIALS AND METHODS

2.1. Materials

Four batches of ethylated β-cyclodextrin were used in this study. Batch N1 was purchased from CYCLOLAB (Budapest, Hungary). Briefly, this product was synthetized by reacting dried βCD with ethyl iodide in alkaline-DMSO media at room

J. Szejtli and L. Szente (eds.), Proceedings of the Eighth International Symposium on Cyclodextrons, 221–224.

temperature. The final Et-βCD was separated from the reactive medium by a series of crystallizations in different media.

Batches N9A, N10 and N10A were generous gifts from ORSAN (Les Ulis, France). The synthetic pathway consisted in ethylation of βCD using an excess of ethyl sulphate in strong alkaline organic medium. Ethylated β-cyclodextrins were isolated from the reactive medium using suitable procedures. These three different batches were prepared with a view to obtaining various patterns of substituted βCD derivative.

2.2 Methods

All NMR experiments shown here were performed using a dual ^{13}C-^{1}H probe on a Bruker AMX 500 spectrometer operating at 500.13 and 125.77 MHz respectively for ^{1}H and ^{13}C. The length of 90° pulse was ca 11 and 6 μs for ^{1}H and ^{13}C. All spectra were collected in deuterium chloroform at 298 K.

3. RESULTS AND DISCUSSION

3.1. Determination of degree of substitution

^{1}H NMR spectra of ethylated β-cyclodextrins, batches N1 and N9A, are displayed in Figure 1. Comparison of the integration of the signals in the 1.0 to 1.5 ppm spectral region (ethyl group only) and of 3.0 to 4.5 ppm region (all signals from glucose units except H-1 and methylene groups) directly provides the DS, that is the number of ethyl chains by glucose unit. Although the spectra appear rather different, they give a very similar DS of 2, confirming diethyl βCD stuctures. However, only the sample N1 is consistent with a single symmetrical structure and correspond to the expected 2,6 diethyl βCD. The sample N9A (as N10 and N10A) reveal a large molecular heterogeneity induced by the variations on the substitution positions.

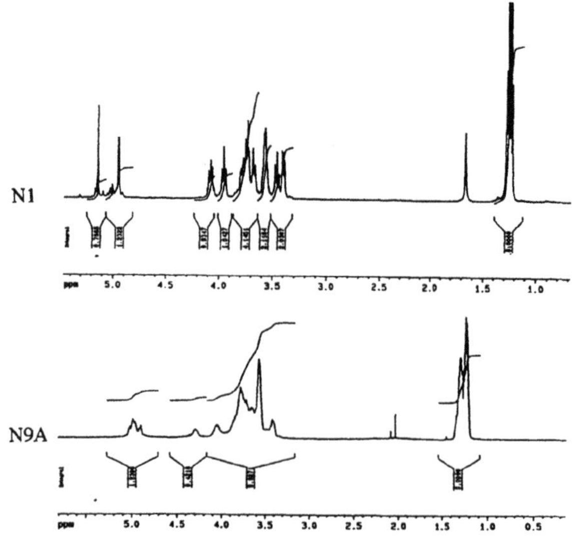

Fig. 1 ^{1}H NMR spectra of Et-βCDs (samples N1 and N9A) in $CDCl_3$ at 298 K

3.2. Complete assignment of ^{1}H and ^{13}C spectra

Only the sample N1 was subjected to this complete procedure. Figure 2 shows the ^{1}H NMR spectrum achieved by a double-quantum correlation experiment. All the signals were assigned subsequently starting from the ^{1}H signals of anomeric protons at 4.9 ppm.

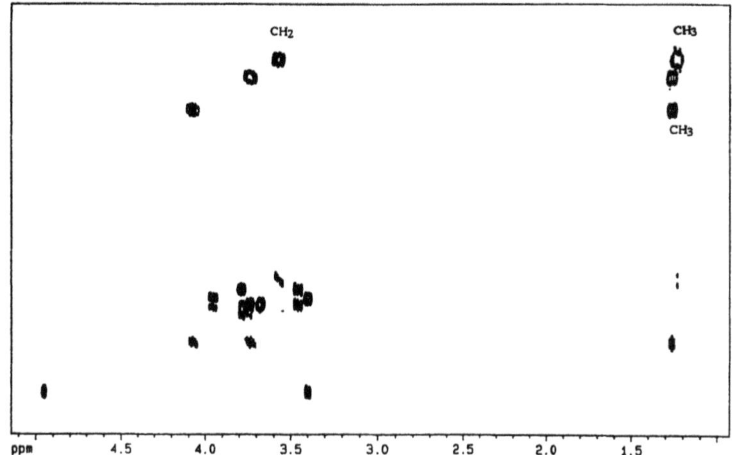

Fig. 2 Double-quantum correlation experiment for sample N1 performed in $CDCl_3$ at 298 K

Figure 3 displays the ^{13}C NMR spectrum of sample N1. The signals were assigned from the heteronuclear correlation procedure starting from information gathered for protons (see Figure 4).

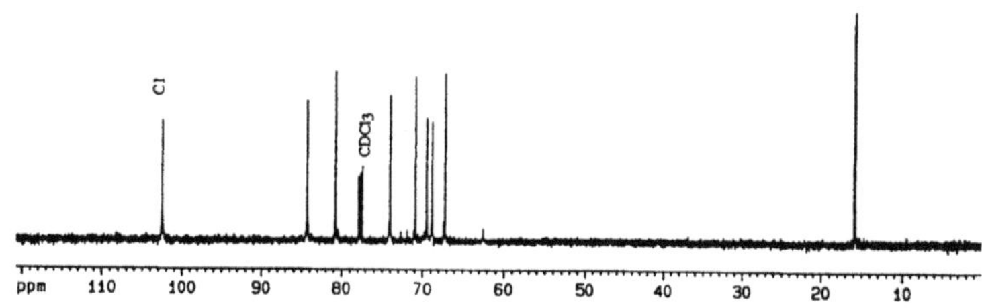

Fig. 3 ^{13}C NMR spectrum of sample N1 in $CDCl_3$ at 298

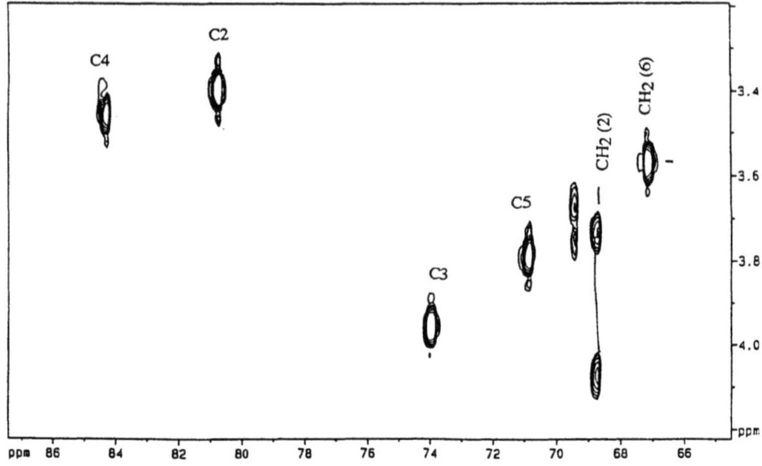

Fig. 4 ^{13}C-^{1}H 2D correlation experiment for sample N1

For the other samples N9A, N10 and N10A, the heterogeneity of signals from C-6, and the presence of a signal at 4.35 ppm suggest that ethylation may occur on the 2, 3 and 6 positions. However, from the standpoint of pharmaceutical use, these heterogeneous derivatives, diethyl β-cyclodextrins, revealed interesting properties on the sustained release of salbutamol.

CONCLUSION

Different batches of ethyl β-cyclodextrins obtained from various synthetic routes were used to achieve inclusion compouuds of salbutamol. The variable sustained-release behaviour of the complexes were related to the physicochemical properties of cyclodextrin derivatives. High resolution ^{1}H NMR and ^{13}C NMR were used to characterize these diethlyl β-cyclodextrins in terms of average substitution ratio and location of substituents on the CDs glucose units. This approach is more realistic and very useful to understand the CD properties from batch to batch.

REFERENCES

[1] Wenz, G., Cyclodextrins as building blocks for supramolecular structures and functional units, Angew. Chem. Int., Ed. Engl., 33, 803-822, (1994).

[2] Lemesle-Lamache, V., Wouessidjewe, D., Chéron, M., Duchêne, D., Study of β-cyclodextrin and ethylated β-cyclodextrin salbutamol complexes, in vitro evaluation of sustained-release behaviour of salbutamol, Int. J. Pharm., submitted.

[3] Lemesle-Lamache, V., Wouessidjewe, D., Taverna, M., Ferrier, D., Perly, B., Duchêne, D., Physicochemical characterization of different batches of ethylated β-cyclodextrins, J. Pharm. Sci., submitted.

^{1}H NMR AND UV SPECTROSCOPIC STUDY OF INCLUSION COMPLEX FORMATION BETWEEN PYRIDOXINE AND β- AND γ-CYCLODEXTRINS

M. COTTA RAMUSINO, M. BARTOLOMEI, L. RUFINI
Laboratorio di Chimica del Farmaco, Istituto Superiore di Sanità
Viale Regina Elena 299, 00161 Roma (Italy)

ABSTRACT

Inclusion complex formation between pyridoxine and β- and γ-cyclodextrins in aqueous solution has been investigated by ^{1}H NMR and UV spectroscopy. Both complexes exhibited a 1:1 stoichiometry and the inclusion process has been shown to perturb the equilibrium between the lactim and the lactam tautomer of pyridoxine, with a preferential inclusion of the former, less polar tautomer.

1. INTRODUCTION

Pyridoxine (2-methyl-3-hydroxy-4,5-bis(hydroxymethyl)pyridine) and structurally related compounds play a relevant role in a large number of biological systems. Many reactions catalyzed by the coenzymes of the pyridoxal group are known to take place in a rather hydrophobic environment inside the enzyme active site [1]. In the study of these systems it is therefore of fundamental importance to devise a biomimetic model which can simulate such hydrophobic interactions. α-, β- and γ-Cyclodextrins (CD_S), which exhibit a toroidal shape with a hydrophobic cavity, may afford suitable biomimetic models in the study of hydrophobic interactions of biological relevance.

2. MATERIALS AND METHODS

2.1. Materials

β-Cyclodextrin (β-CD) and γ-cyclodextrin (γ-CD), both from Sigma, were used without further purification; pyridoxine·HCl (> 98.5% purity) was purchased from Carlo Erba. D_2O (> 99.5% isotopic purity) was obtained from Fluka. All solutions for UV absorption measurements were made in phosphate buffer, pH 6.8±0.1. The pD of the solutions for ^{1}H NMR experiments was set at 7.4 ± 0.1 (pH-meter reading plus 0.4 to correct for isotopic effects).

2.2. Methods

UV absorption spectra were recorded at 298 ± 0.5 K on a Perkin Elmer Lambda 16 spectrometer. Due to the known instability of pyridoxine in neutral aqueous medium [2], all solutions were freshly prepared immediately before recording the spectra and protected from light. ^{1}H NMR spectroscopy was performed on a Bruker AMX spectrometer operating at 400.13 MHz. The probe temperature was set at 298 ± 1 K using a Haake

J. Szejtli and L. Szente (eds.), Proceedings of the Eighth International Symposium on Cyclodextrons, 225–228.

control system. Chemical shifts were measured relative to external sodium 4,4-dimethyl-4-silapentane-1-sulphonate (DSS) at 0 ppm.

3. RESULTS AND DISCUSSION

3.1. UV spectroscopic study

The UV absorption spectrum of pyridoxine has been shown to be markedly solvent dependent. Such dependence is also related to the tautomeric equilibrium between the lactam and the lactim forms of pyridoxine. The apparent tautomerization constant K_T in aqueous medium (buffered solution, pH 7) has been reported to be 3.92 [3], with a preferential stabilization of the more polar lactam form.

A spectral change is therefore to be expected following inclusion of pyridoxine in the hydrophobic cavity of the cyclodextrins. This is indeed the case as depicted in Fig. 1 which shows the UV spectra of pyridoxine both in the absence and in the presence of β-CD.

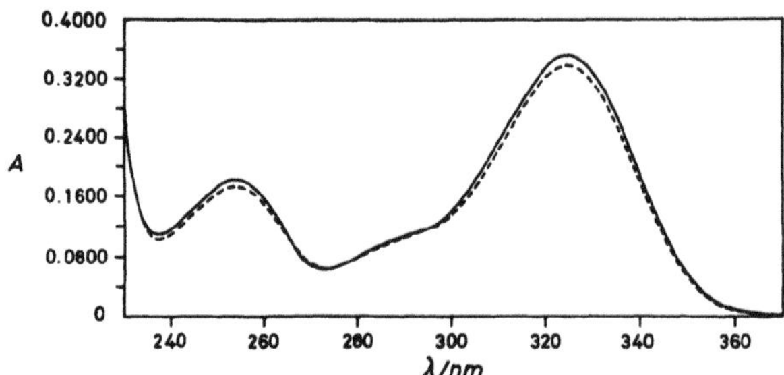

Fig. 1. UV absorption spectra of pyridoxine ($5.0 \cdot 10^{-5}$ M) in phosphate buffer solution (pH = 6.8) at 298 K in the absence (——) and in the presence of $1.0 \cdot 10^{-2}$ M β–CD (----).

It can be seen that the absorption centered at 324 nm, which is attributed to the long wavelength $\pi \rightarrow \pi^*$ transition of the lactam tautomer, decreases . Similar effects were observed with γ-CD while α-CD did not perturb the pyridoxine spectrum to any noticeable extent.

A 1:1 inclusion mechanism has been assumed throughout the present study because pyridoxine presents only one interacting moiety (the aromatic ring) and its molecular dimensions seem to preclude inclusion in more than one CD molecule. Moreover the guest is not known to give rise to autoassociation processes from which a 2:1 (guest:host) inclusion complex may arise.

On the assumption that the molar absorptivity and λ_{max} of the complexed pyridoxine do not differ from those of the free guest, only the relative concentration of the different tautomers being affected by the inclusion process [4] the following equation was derived ($[CD] \gg [B_6]$):

$$\frac{A \cdot 1/K_T}{[B_6] \cdot \varepsilon - A \cdot (1 + 1/K_T)} = \frac{1}{[CD]} \cdot \frac{1}{K_1 - K_2} + \frac{K_1}{K_1 - K_2} \qquad (1)$$

where A and ε are the absorbance value and the molar absorptivity, respectively, at $\lambda = 324$ nm; $[B_6]$ is the total concentration of the guest which was kept constant at $5 \cdot 10^{-5}$ M while the CD concentrations were set at 0.002, 0.004, 0.006, 0.008 and 0.01 M (β-CD) and at 0.01, 0.02, 0.03, 0.04 and 0.05 M (γ-CD); K_1 and K_2 are the apparent equilibrium constants for inclusion complex formation between the cyclodextrins and the lactim and lactam tautomer of pyridoxine, respectively. The apparent association constants are reported in Table 1.

Table 1. UV determined apparent association constants for the inclusion complex formation between pyridoxine and β- and γ-cyclodextrins at 298 K in aqueous solution.

Host	K_1 /(M^{-1})	K_2 /(M^{-1})	K_{Tc}
β-CD	31.5 ± 10.5	7.6 ± 3.5	0.94 ± 0.4
γ-CD	12.9 ± 4.2	2.6 ± 1.3	0.80 ± 0.4

The K values found confirm a preferential inclusion of the less polar lactim tautomer of pyridoxine. The tautomerization constant (K_{Tc}) values of pyridoxine in β-CD (0.94) and γ-CD (0.8) are similar to those found in water-dioxane mixtures where the volume fraction of the less polar solvent is ~ 0.3.

3.2. ^{1}H NMR study

Fig. 2 shows the ^{1}H NMR spectra of β-CD in the absence and in the presence of pyridoxine.
More pronounced upfield shifts of the H-3 and H-5 signals are observed while smaller effects are experienced by the other hydrogens of the host compound, thus supporting the hypotesis of true inclusion complex formation. Similar conclusions could be drawn for inclusion complex formation between pyridoxine and γ-CD. The 1:1 stoichiometry of the last complex was obtained by the continuous variation (Job) method. For the β-CD·pyridoxine complex we were unable to obtain a Job plot due to the very small chemical shift differences observed when both the guest and the host concentrations were of the order of 0.01 M which approaches the solubility limit of β-CD. The stoichiometry of the complex was thus inferred from the UV measurements.

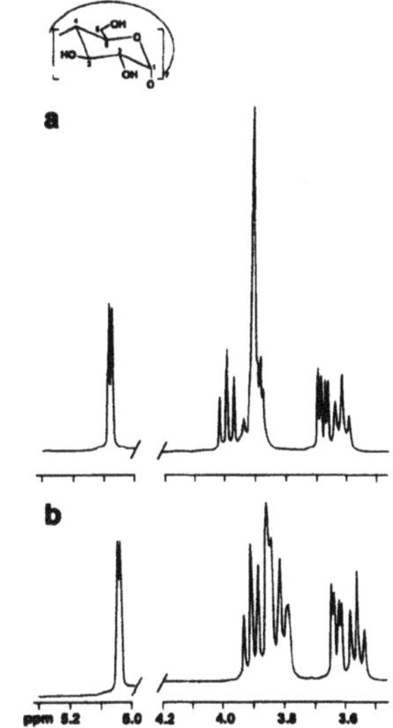

Fig.2. 400-MHz ^{1}H NMR spectra in D_2O: (a) $5 \cdot 10^{-3}$ M β-CD; (b) $5 \cdot 10^{-3}$ M β-CD + $2.0 \cdot 10^{-1}$ M pyridoxine.

For the determination of the apparent association constants, K_1 and K_2, β-CD and γ-CD concentrations were kept at $5 \cdot 10^{-3}$ M while the guest concentration was set at 0.05, 0.1, 0.15, 0.20 and 0.25 M (with β-CD) and at 0.1, 0.2, 0.3, 0.4 and 0.5 M (with γ-CD). In devising an equation for estimating the apparent association constant, the presence of comparable amounts of the two pyridoxine tautomers had to be taken into account. Therefore, two equations were constructed and solved simultaneously:

$$K_1 = \frac{[C_1] \cdot (1 + K_T)}{([B_6]_t - [C_1] - [C_2]) \cdot ([CD]_t - [C_1] - [C_2])} \tag{2}$$

$$K_2 = \frac{[C_2] \cdot (1 + 1/K_T)}{([B_6]_t - [C_1] - [C_2]) \cdot ([CD]_t - [C_1] - [C_2])} \tag{3}$$

where $[C_1]$ and $[C_2]$ are the concentrations of the complexes with the lactim and the lactam tautomer, respectively . Moreover $[C_2] = [C_1] \cdot K_2 \cdot K_T/K_1$. $[C_1]$ and $[C_2]$ can be related to the observed chemical shift differences of CD_s H-3, $\Delta\delta_{obs}$, through the relationship $\Delta\delta_{obs} \cdot [CD]_t = [C_1] \cdot \Delta\delta_1 + [C_2] \cdot \Delta\delta_2 = [C_1] \cdot (\Delta\delta_1 + \Delta\delta_2 \cdot K_2 \cdot K_T/K_1)$, where $\Delta\delta_1$ and $\Delta\delta_2$ are the chemical shift differences of the pure complexes with the lactim and the lactam tautomer, respectively. Substituting for $[C_1]$ into eq. 2 and assuming that $[B_6]_t - [C_1] - [C_2] \cong [B_6]_t$ and defining $\Delta\delta_{ap}$ as $(K_1 \cdot \Delta\delta_1 + \Delta\delta_2 \cdot K_2 \cdot K_T)/(K_1 + K_2 \cdot K_T)$, we obtain :

$$\frac{1}{\Delta\delta_{obs}} = \frac{1}{\Delta\delta_{ap}} + \frac{1 + K_T}{\Delta\delta_{ap} \cdot (K_1 + K_2 \cdot K_T)} \cdot \frac{1}{[B_6]_t} \tag{4}$$

The apparent association constants are reported in Table 2; they support the hypotesis of a preferential inclusion of the less polar lactim tautomer of pyridoxine both in β-CD and in γ-CD.

Table 2. NMR determined apparent association constants and $\Delta\delta_{ap}$ for inclusion complex formation between pyridoxine in β- and γ-cyclodextrins at 298 K in aqueous solution.

Host	K_1 / (M^{-1})	K_2 / (M^{-1})	$\Delta\delta_{ap}$ / p.p.m.
β-CD	20.2 ± 7.5	4.9 ± 2.0	0.100 ± 0.003
γ-CD	5.5 ± 2.0	1.2 ± 0.5	0.166 ± 0.003

4. CONCLUSION

The formation of 1:1 inclusion complexes between β- and γ-CD and pyridoxine has been demonstrated both by ^1H NMR and UV spectroscopic techniques. The apparent equilibrium constants for these processes are characteristic of rather weak intermolecular interactions,which are nevertheless able to perturb the tautomeric equilibrium of pyridoxine with a preferential stabilization of the lactim form .

REFERENCES

[1] Cortijo , M., Shaltiel, S. , On the microenvironment of the pyridoxamine 5-phosphate residue in $NaBH_4$-reduced glycogen phosphorylase b, *Eur. J. Biochem.*, **29** 134-142 (1972)

[2] Hochberg, M., Melnick, D., Siegel , L., Oser, B. L., Destruction of vitamin B_6 (pyrdoxine) by light, *J. Biol. Chem.*, **148** 253-254 (1943)

[3] Sanchez-Rui, J. M., Llor , J., Cortijo, M., Thermodinamic constants for tautomerism and ionization of pyridoxine and 3-hydroxypyridine in water-dioxane, *J. Chem. Soc. Perkin Trans. II* , 2047-2051 (1984)

[4] Cotta Ramusino, M., Rufini , L., Mustazza, C., UV spectroscopic study of the interaction between α-, β-, and γ- cyclodextrins and pyridine derivatives, *J. Incl. Phen. and Mol. Rec. in Chem.* , **15** 359-368 (1993)

ELECTROSPRAY IONIZATION MASS SPECTROMETRY OF CYCLODEXTRIN COMPLEXES OF AMINO ACIDS AND PEPTIDES

Laszlo Prokai,[1] R. Ramanathan,[1,3] J. Nawrocki,[2] and J. Eyler[2]

[1]University of Florida, Center for Drug Discovery, Gainesville, FL 32610-0497,

[2]University of Florida,Department of Chemistry, Gainesville, FL 32611, U.S.A.

([3]Present Address: Washington University, St. Louis, MO 63130, U.S.A.)

ABSTRACT

Appropriate complex-forming side-chains of peptides and proteins can bind cyclodextrins. We have characterized interactions of amino acids with cyclodextrins by electrospray ionization mass spectrometry. The stoichiometry of various α-cyclodextrin/protonated tryptophan complexes sampled from solution to the gas phase by ESI was revealed by taking advantage of the extended mass range and high mass resolution of a Fourier-transform ion cyclotron resonance instrument. Binding of β-cyclodextrin to recombinant human insulin has been demonstrated.

1. INTRODUCTION

Large molecules such as peptides and proteins cannot be fully entrapped into inclusion complexes of cyclodextrins (CDs). Appropriate complex-forming side-chains can, however, bind cyclodextrins, resulting in modified solubility, stability and/or membrane transport properties [1]. The lack of adequate techniques has hindered the proper characterization of peptide - cyclodextrin interactions, because of the complexity of their association involving multiple binding sites for cyclodextrins on these large biomolecules.

Electrospray ionization (ESI) can be used effectively to study noncovalent complexes involving CDs. Our recent studies[2] on amino acids and CDs have revealed that complexation equilibria in solution are reflected by ESI mass spectra. Amino acid - CD complexes were stable for characterization by ESI mass spectrometry; thus, their relative abundances and the stoichiometry could be determined. The preferential formation of CD- aromatic over CD- aliphatic amino acid complexes have been confirmed, and the

J. Szejtli and L. Szente (eds.), Proceedings of the Eighth International Symposium on Cyclodextrons, 229–232.

relative gas-phase stabilities were also determined by collisionally-induced dissociation (CID). Our preliminary results[3] for peptides, together with recent investigations by others on CD complexes of small [4] and large peptides[5], have shown the potential of the technique to characterize their formation and stability. However, the high molecular weight and complexity of the systems may benefit from the use of high-performance mass spectrometry, such as Fourier-transform ion cyclotron resonance (FTICR) [6], that provides an extended mass range and increased resolution, compared to the commonly used quadrupole mass analyzers [2-4]. We present results of a detailed study on complexes between amino acids and CDs detected and identified by ESI-FTICR, and report preliminary data on the binding of CDs to a therapeutic peptide, insulin.

2. MATERIALS AND METHODS

ESI - quadrupole mass spectrometry was done by using a Vestec 200 ES instrument (PerSeptive Biosystems/Vestec, Houston, TX). Ionization was achieved by applying 2.5-2.7 kV to a flat tipped (120 μm ID) stainless steel needle. The sample solution was delivered via a fused silica capillary by a syringe pump at 2-5 μL/min flow rate. A stainless steel plate with a 0.4-mm orifice functioned as the counterelectrode. The source was heated to 250 °C to reach 55-60 °C in the spray chamber. The collimator and repeller potentials were 10 V and 15 V, respectively, to avoid CID. Sample concentrations were 1×10^{-4} M for insulin, and 5×10^{-5} M in 45/45/10 (v/v/v) methanol/water/acetic acid for β-CD.

FTICR experiments were performed on a 4.7-tesla instrument (Bruker Instruments, Inc., Billerica, MA). Ions were produced via an ESI source equipped with a hexapole ion guide/trap and heated capillary (Analytica of Branford, Branford, CT). ESI was achieved at a flow rate of 2 ul/min with a needle voltage of 3 kV. The capillary was held at ground potential while the temperature was maintained at 100 °C. Sample concentrations in 45/45/10 (v/v/v) methanol/water/acetic acid were 1×10^{-6} M for the CDs and 5×10^{-5} M for the amino acids and peptides.

3. RESULTS AND DISCUSSION

Electrospray ionization (ESI) mass spectrometry offers some unique advantages for studying inclusion complexes. With this “soft” ionization technique, ions existing in solution can be transferred into the gas phase without breaking noncovalent interactions, allowing their subsequent investigation by various mass spectrometric techniques. The stoichiometry of an inclusion complex may, among others, be revealed unequivocally.

Fig. 1 shows the ESI mass spectrum of a solution containing α-CD and tryptophan obtained by the FTICR technique. Besides the already reported 1:1 α-CD/protonated

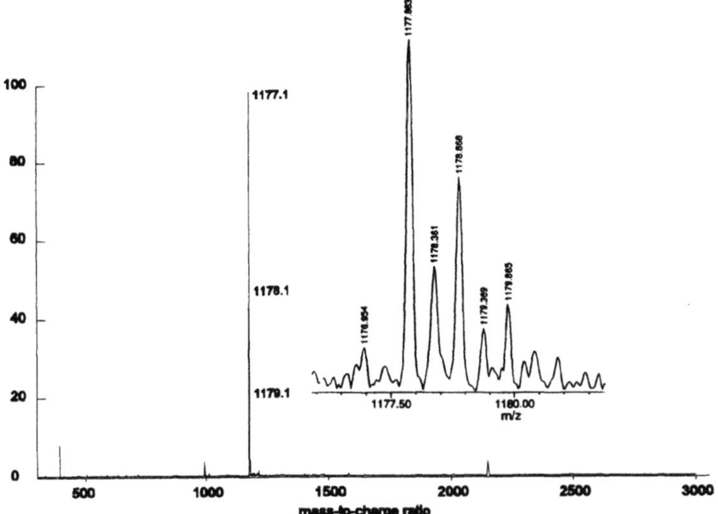

Figure 1. ESI-FTICR mass spectrometry of α-CD and tryptophan complexes.
(The inset shows the expanded region between 1174 to 1183.)

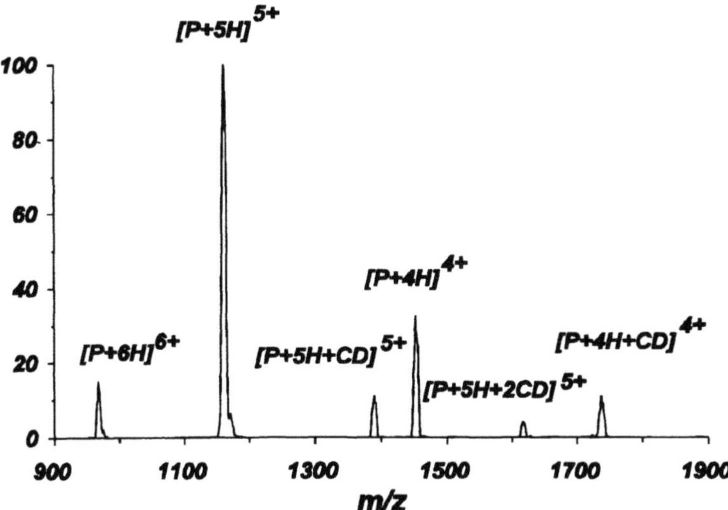

Figure 2. ESI-MS of human recombinant insulin solution containing excess β-CD.

tryptophan complex [2], the fully resolved isotope profiles made the identification of a 2:2 complex (a proton-bound dimer) possible. Note that mass spectrometry determines mass-to-charge (*m/z*) values; therefore, the doubly-charged ions show isotope ions with about 0.5 Da apart, as opposed to the about 1.0 Da difference for singly-charged ions. The 1:1 and 2:2 adducts that contain only ^{12}C, ^{1}H, ^{16}O, and ^{14}N isotopes have identical *m/z* values. The mass resolution ($M/\Delta M$) required to resolve this profile is >2,400, which is not met by most commercial quadrupole analyzers, but easily obtained on FTICR instruments. (The actual mass resolution exceeded 10,000.) In addition, the extended mass-range acquisition (up to m/z 3,000) has allowed us to detect a less abundant 2:1 α-CD/protonated tryptophan complex (*m/z* 2,149.3).

For the studies on binding of CDs to a large peptide such as recombinant human insulin (Mr 5807.6), a large molar excess of CD was necessary to detect the formation of intense adduct ions. We used the routine ESI - quadrupole mass spectrometer system for a preliminary study. The β-CD adducts, together with the multiply-charged insulin molecular ions (due to acid-base equilibria involving the basic amino acid residues of the peptide), were indicated in Fig. 2. (P represents the peptide, CD abbreviates β-CD, and H indicates the number of protons). Under this solution conditions, the formation of isobaric ions is unlikely, but ESI-FTICR is necessary to confirm this hypothesis.

In conclusion, ESI mass spectrometry has been a powerful technique for the analysis of binding of CDs with amino acids and peptides. The technique is ideal for the characterization of binding CDs to amino acids and peptides, because they are present mostly as ions in aqueous solutions, which are the preferred solvents for ESI.

REFERENCES

1. Szejtli, J. , Medicinal applications of cyclodextrins, *Med. Res. Reviews*, **14**, 353-386 (1994).
2. Ramanathan, R., Prokai, L., Electrospray ionization mass spectrometric study of encapsulation of amino acids by cyclodextrins, *J. Am. Soc. Mass Spectrom.* **6**, 866-871 (1995).
3. Ramanathan, R., Prokai, L., Electrospray ionization mass spectrometric study of host-guest complexes, *Proceedings of the 43rd ASMS Conference on Mass Spectrometry and Allied Topics*, Atlanta, GA, May 21-26, 1995, p. 253.
4. Cunniff, J.B., Vouros, P., False positives and the detection of cyclodextrin inclusion complexes by electrospray mass spectrometry, *J. Am. Soc. Mass Spectrom.*, **6**, 437-447 (1995).
5. Camilleri, P., Haskins, N.J., Howlett, D.R., β-Cyclodextrin interacts with the Alzheimer amyloid β-A4 peptide, *FEBS Lett.* **341**, 256- (1994).
6. Buchanan, M.V. (Ed.), Fourier Transform Mass Spectrometry: Evolution, Innovation, and Applications, ACS Symposium Series, ACS Books, Washington, DC, 1987.

CHEMICALLY SWITCHED DNA INTERCALATORS USING MODIFIED CYCLODEXTRIN COMPLEXES

T. IKEDA, T. TAKAHASHI, M. MORI, K. YOSHIDA, A. UENO, F. TODA*, and H-J. SCHNEIDER**

Department of Bioengineering, Tokyo Institute of Technology, 4259 Nagatsuta-cho, Midori-ku, Yokohama 226, Japan

**Tokyo Polytechnic College, 2-32-1 Ogawanishi-machi, Kodaira 187, Japan*

***FR Organische Chemie, Universität des Saarlandes, D 66041 Saarbrücken, Germany*

ABSTRACT

Two kinds of anthracene modified β-cyclodextrins having different alkyl chain length between anthracene moiety and β-cyclodextrin were synthesized as DNA intercalators. Guest-induced conformational changes of them were observed by measurements of ^{1}H-NMR spectra and circular dichroism spectra. Their different abilities of intercalation to DNA were observed.

1. INTRODUCTION

Cyclodextrins (CDs) have been extensibly studied as host molecules for molecular recognition and encapsulations of drugs and drug delivery systems. But, CD itself has less specific affinities for biomolecules. For development of molecular recognition using modified CDs with specific and/or selective affinities for biomolecules and also for the interest of DNA intercalation and drug delivery systems, anthryl(alkylamino)-β-CD (**1**) was synthesized. **1** itself does not intercalate to DNA, but **1**-complex with a guest molecule has an ability to intercalate to DNA. [1] The binding ability of **1**-complex to DNA is lower than that of known anthrylamines. [2] To investigate of the effects of the conformation and chain length between anthracene moiety and CD of anthracene modified CDs on the abilities of DNA intercalation, two kinds of anthracene modified β-CDs having different alkylamino chain length were synthesized.

H N N H β-CD

1

J. Szejtli and L. Szente (eds.), Proceedings of the Eighth International Symposium on Cyclodextrons, 233–236.

2. MATERIALS AND METHODS

2.1. Materials

Two kinds of anthracene modified β-CDs (**2** and **3**) were prepared from primary hydroxyl side mono tosylated β-CD as a starting material respectively. For **2**, to attach an anthracene moiety with alkylamino chains, ethylenediamine modified β-CD were synthesized from mono tosylated β-CD. To a solution of anthracene-9-carboxylic acid and DCC in dry DMF was added ethylenediamine modified β-CD at 0 °C. The resulting mixture was stirred for 5 hours at 0 °C and then 24 hours at room temperature. After reprecipitation with acetone, washed with chloroform and purification with ion exchange column chromatography (CM Sephadex C-25), **2** was obtained. For **3**, the same procedure was used except for the use of 1,4-diaminobutane instead of ethylenediamine. The purity of **2** and **3** were confirmed by TLC, elemental analysis and ^{1}H-NMR.

Calf thymus DNA was purchased from Pharmacia and purified by phenol extraction as a conventional method.

2.2. Methods

UV absorption, circular dichroism and fluorescence spectra in pH 7.5 Tris buffer were recorded on a SHIMADZU UV-3100 spectrophotometer, a JASCO-600 spectrophotometer and a HITACHI 850 spectrofluorometer respectively using a 1 x 1 cm quartz cell at 25 °C.

^{1}H-NMR spectra were measured at 500 MHz and 400 MHz on VARIAN model VXR 5000 and UNITY*plus* 400 instruments respectively.

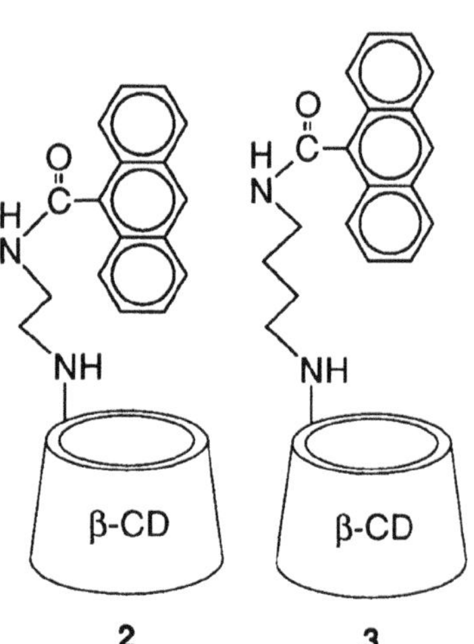

3. RESULTS AND DISCUSSIONS

3.1. Conformations of DNA intercalators

The aromatic ^{1}H-NMR signals of **2** and **3** which are broadening as themselves alone are substantially changed to the similar form of uncomplexed anthracene group after adding 1-adamantanol (**AOH**). (Fig. 1) This signifies the guest molecule (**AOH**) removes the anthryl unit out of the CD cavity. Circular dichroism spectra of **2** alone gives a positive peak indicated that the long axis of the anthracene moiety of **2** is parallel to the axis of the CD cavity. After adding **AOH**, such a positive peak is changed to a negative one. This phenomena also suggested that anthracene moiety of **2** is moved to the out of the CD cavity and located perpendicular to the axis of the CD cavity after making complex with **AOH**. (Fig. 2-a) Circular dichroism spectra of **3** with **AOH** gives also a negative peak, however **3** alone shows no cotton effect. (Fig. 2-b) These data show the anthryl unit of **3** alone is in the equilibrium state between complexed and uncomplexed form by the CD cavity. These differences of conformations of **2** and **3** are caused by the differences of chain length of **2** and **3**.

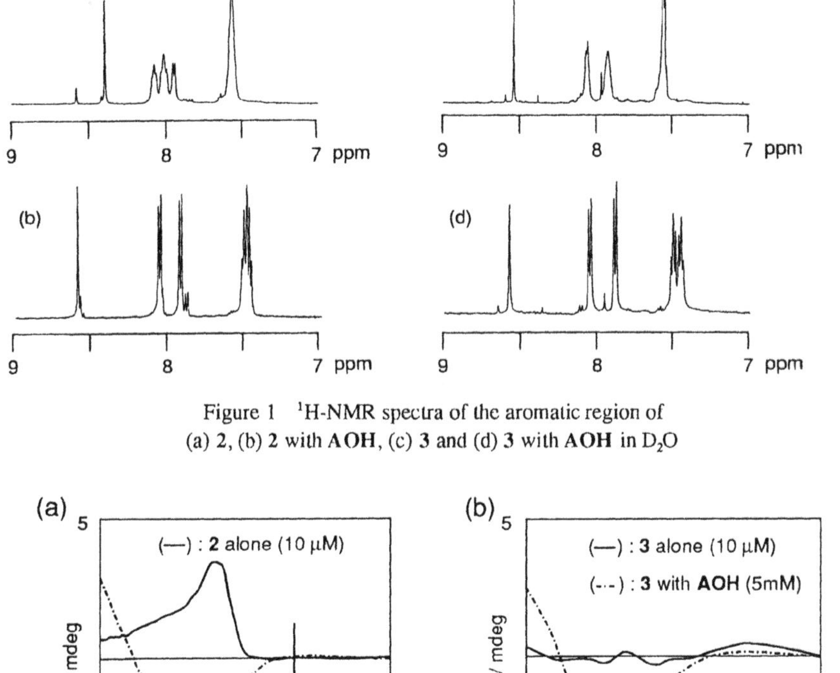

Figure 1 ^{1}H-NMR spectra of the aromatic region of (a) **2**, (b) **2** with **AOH**, (c) **3** and (d) **3** with **AOH** in D_2O

Figure 2 Induced circular dichroism spectra of (a) **2** and (b) **3**

3.2. DNA intercalation

UV absorption and fluorescence spectra of **2** are decreased with increasing concentration of calf thymus DNA only after adding **AOH**. The anthryl unit of **2** is bound to the CD cavity (intramolecular complex form) and an intercalation to DNA is inhibited. After making complex with **AOH**, the anthryl unit of **2** is moved to the outside of the CD cavity and **2** (with **AOH**) can intercalate to DNA.

UV absorption and fluorescence spectra of **3** are slightly decreased with increasing concentration of calf thymus DNA without adding **AOH**. After adding **AOH**, both spectra of **3** are more decreased. **3** itself can slightly intercalate to DNA, because the anthryl unit of **3** is located inside and outside of the CD cavity in the equilibrium state as indicated 3.1.. After making complex with **AOH**, the anthryl unit of **3** is completely removed to the outside of the CD cavity and strongly intercalate to DNA.

From the excitation spectra of **2** and **3**, the energy transfer from DNA to **2** and **3** are observed under the same conditions which DNA intercalation are occurred.

The properties of **1** to DNA intercalation was also determined by ^{1}H-NMR measurements. [1] Same studies for **2** and **3** are now in progress.

4. CONCLUSION

Two kinds of anthracene modified β-CDs (**2** and **3**) were synthesized as DNA-intercalators. **2** can intercalate to DNA only after making complex with a guest molecule, that is, by a guest-induced conformational change, **2**-complex can intercalate to DNA. On the other hand, **3** itself can slightly intercalate to DNA and the ability of intercalation to DNA is enhanced by making complex with a guest molecule. These different abilities of them are caused by different chain length and different conformations of them. The lower DNA intercalative abilities of **1**, **2** and **3** compared with those of known anthrylamines might be caused by the effects of bulky CD moiety. More detailed studies using another guest molecules and the determination of the properties of DNA intercalation by ^{1}H-NMR measurements are now in progress.

5. ACKNOWLEDGMENTS

We are grateful to Nihon Shokuhin Kako Co. LTD. for providing us with samples of CDs.

6. REFERENCES

[1] Ikeda T., Yoshida K., and Schneider H-J., Anthryl(alkylamino)cyclodextrin complexes as chemically switched DNA intercalators, *J. Am. Chem. Soc.*, **117**, 1453-1454, 1995

[2] Kumar C.V. and Asuncion Emma H., Sequence dependent energy transfer from DNA to a simple aromatic chromophore, *J.Chem.Soc.,Chem.Commun.*, 470-472 (1992)
Kumar C.V. and Asuncion Emma H., DNA binding studies and site selective fluorescence sensitization of an anthryl probe, *J.Am.Chem.Soc.*, **115** 8547-8553 (1993)

TIME RESOLVED FLUORESCENT ANALYSIS FOR SEALED HEATING OF DIMETHYL-β-CYCLODEXTRIN AND NAPHTHALENE SYSTEM

Keiji Yamamoto, Takuma Nakao, Etsuo Yonemochi, Toshio Oguchi
Faculty of Pharmaceutical Sciences, Chiba University,
1-33 Yayoicho, Inage-ku, Chiba 263, Japan

ABSTRACT

The inclusion process of naphthalene into heptakis-(2,6-di-*O*-methyl)-β-cyclodextrin (DMβCD) during the sealed heating was investigated by using solid state time resolved fluorescent analysis. Fluorescence lifetimes of naphthalene in naphthalene crystals and in inclusion complex with DMβCD prepared by coprecipitation were determined 59 ns and 88 ns, respectively. Fluorescence lifetime of naphthalene monomer becomes longer after inclusion complex formation. Fluorescence lifetimes of naphthalene monomer and excimer in the sealed heating complex were determined about 80 ns and 100 ns, respectively. By setting the observing wavelength of time resolved fluorescent analysis at 355 nm, the formation and the decay of the excimer state of naphthalene during the sealed heating process were successfully confirmed.

1. INTRODUCTION

For the sealed-heating of naphthalene and heptakis-(2,6-di-*O*-methyl)-β-cyclodextrin (DMβCD), we have already reported that two types of the complexes were obtained, which were different from the inclusion compound obtained by the coprecipitation method.[1] Powder X-ray diffractometry and solid state fluorescence analysis can be used to distinguish between the sealed heated complexes, the coprecipitated complex, and the physical mixture, while they are not enough to the complete distinction.

Time resolved fluorescence analysis is one of the effective techniques to investigate the microenvironment of the molecule. Nelson and Warner used the fluorescence lifetime data to investigate the ability of CD complexation in the presence of alcohols to shield fluorophores from interactions with quenchers.[2]

The aim of this study was to clarify the inclusion process by the sealed heating of naphthalene and DMβCD using time resolved fluorescence analysis, and to contrast the different approaches.

2. MATERIALS AND METHODS

250 mg mixture of amorphous DMβCD and naphthalene (molar ratio = 1:1) was sealed in a 2.0 ml glass ampule, then heated at a definite temperature for 1 min-72 h.

J. Szejtli and L. Szente (eds.), Proceedings of the Eighth International Symposium on Cyclodextrons, 237–240.

Amorphous DMβCD was prepared by grinding. The coprecipitate was prepared by heating the mixture of DMβCD aqueous solution and naphthalene ethyl ether solution. Fluorescence spectra were measured using an FP-770E spectrofluorometer (Japan Spectroscopic Co.) for solid sample. A spectrofluorometer (Horiba NAES-700) was used for solid state time resolved fluorescent analysis.

3. RESULTS AND DISCUSSION

Table 1 shows the fluorescence lifetimes and relative quantum yields of naphthalene in various specimens when the λobs was fixed at 341.0 nm where the emission maximum of naphthalene monomer was observed. In naphthalene crystals, the lifetime of 59.1 ns was observed, even as Hamai evaluated the lifetime of naphthalene monomer in aqueous solution as 40 ns. The lifetime of naphthalene monomer was determined as 88.3 ns in the coprecipitate where the molar ratio of naphthalene to DMβCD was 1.20. In the sealed heated specimens, the similar lifetime was obtained with the coprecipitate, indicating the inclusion of naphthalene molecule to DMβCD cavity, however, the lifetime of coprecipitate was slightly longer than that of sealed heated specimens. Mataga et al. determined the lifetime of naphthalene monomer in organic solvents as 120 ns in cyclohexane and 110 ns in n-hexane, respectively.[3] The elongation of lifetime of naphthalene by the inclusion formation could be related to the environmental changes of naphthalene molecule. The lifetime of 107 ns was obtained in the sealed heated specimen heated at 70°C for 15 min, when λobs was fixed at 390.0 nm. As Mataga et al. reported that the lifetime of naphthalene excimer in cyclohexane was 117 ns, the observed value, 107 ns, seems to be due to naphthalene excimer formed during sealed heating process.

The solid state fluorescence emission spectra of the sealed heated specimens heated at 70°C are shown in Fig. 1. The increase of heating time resulted in the shifting of the emission maximum from 341 nm to 390 nm. Table 2 shows the changes of fluorescence lifetime and relative quantum yield of naphthalene by sealed heating at 70°C when λobs was fixed at 355.0 nm. The shortest lifetime was considered as a result of stray, reflective or scattered light. The sealed heated specimens showed other two components with varied relative quantum yield. While in the physical mixture, only monomer component

Table 1. Fluorescence Lifetimes (τ) and Relative Quantum Yields (ϕ) of Naphthalene in Various Naphthalene-DMβCD Systems (λex = 262.7 nm ; λobs = 341.0 nm)

Sample		τ_1(ns)	ϕ_1(%)	τ_2(ns)	ϕ_2(%)	χ^2
Naphthalene Crystals		5.56	*5.6*	59.1	*94.4*	2.58
Coprecipitate		4.56	*2.6*	88.3	*97.4*	1.19
Sealed-heated at 90°C (Mixing Molar Ratio = 1:1)						
for	3 h	6.06	*5.0*	77.8	*95.0*	1.25
	24 h	4.75	*4.5*	78.3	*95.5*	1.60

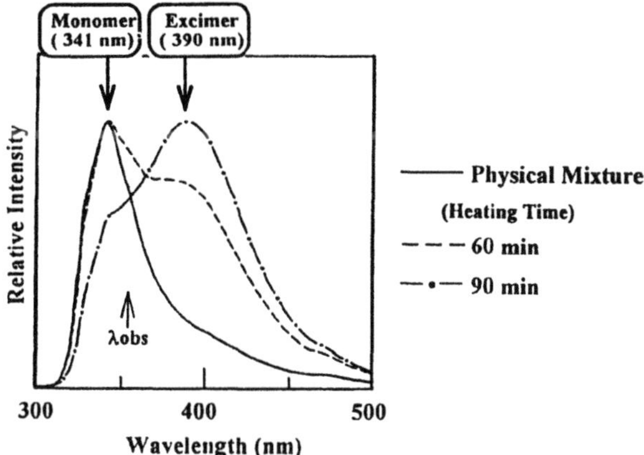

Fig. 1. Changes in Solid State Fluorescence Emission Spectrum of Sealed-heated Sample Heated at 70°C (λex = 262.7 nm)

Table 2. Fluorescence Lifetimes (τ) and Relative Quantum Yields (ϕ) of Naphthalene in Sealed-heated Samples Heated at 70°C for Various Times (λex = 262.7 nm ; λobs = 355.0 nm)

Heating Time	τ_{lamp} (ns)	ϕ_{lamp} (%)	$\tau_{monomer}$ (ns)	$\phi_{monomer}$ (%)	$\tau_{excimer}$ (ns)	$\phi_{excimer}$ (%)	χ^2
Physical Mixture	3.13	3.9	80.8	96.1	—	—	1.19
Sealed-heated for							
60 min	2.25	6.8	77.7	56.6	110	36.6	1.10
70 min	2.45	7.4	78.8	61.3	106	31.3	1.33
80 min	2.49	6.4	71.9	41.9	106	51.7	1.15
90 min	2.81	8.4	81.0	42.3	100	49.3	1.16

was observed. Figure 2 demonstrates the variation of monomer-excimer ratio in sealed heated specimens. The ratios were determined from solid state fluorescence spectra (Fig. 1) and relative quantum yield (Table 2) as well. With the increase of heating time, naphthalene excimer became major component. For the formation of naphthalene excimer, Smith et al. reported that the distance between two naphthalene molecules should be less than 3.7 Å.[4] At the initial step of inclusion compound formation by sealed heating, naphthalene molecules had a certain molecular arrangement in which two naphthalene molecules presented in close distance. As the diameter of DMβCD was not so large to accommodate two naphthalene molecules and the stoichiometry of the inclusion compound was 0.87, two DMβCD and two naphthalene molecules should form the excimer complex. The results of time resolved fluorescence analysis when the specimen was sealed heated at 90°C are shown in Table 3. The amount of excimer complex decreased with the increase of heating time, indicating the formation of stable inclusion complex.

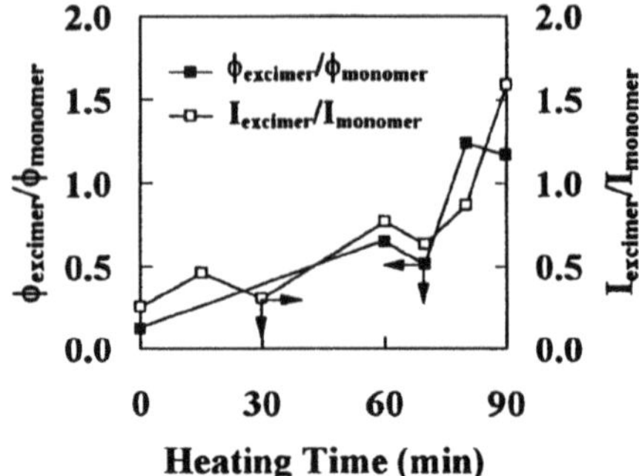

Fig. 2. Changes of $\phi_{excimer}/\phi_{monomer}$ and $I_{excimer}/I_{monomer}$ of Sealed-heated Samples Heated at 70°C

Table 3. Fluorescence Lifetimes (τ) and Relative Quantum Yields (φ) of Naphthalene in Sealed-heated Samples Heated at 90°C for Various Times ($\lambda ex = 262.7$ nm ; $\lambda obs = 355.0$ nm)

Heating Time	τ_{lamp} (ns)	ϕ_{lamp} (%)	$\tau_{monomer}$ (ns)	$\phi_{monomer}$ (%)	$\tau_{excimer}$ (ns)	$\phi_{excimer}$ (%)	χ^2
Physical Mixture	3.13	*3.9*	80.8	*96.1*	–	–	1.19
Sealed-heated for							
30 min	2.88	*3.7*	81.4	*32.5*	95.7	*63.9*	1.07
40 min	2.81	*3.5*	70.7	*39.1*	100	*57.4*	1.04
50 min	2.95	*3.5*	73.8	*39.8*	101	*56.7*	1.09
60 min	3.38	*2.9*	73.9	*56.7*	100	*40.4*	1.21

ACKNOWLEDGEMENTS

This work was supported by a Grant in Aid from the Ministry of Education, Science, Sports and Culture of Japan (07672309)

REFERENCES

1) H. Kawashima, E. Yonemochi, T. Oguchi, and K. Yamamoto, *J. Chem. Soc., Faraday Trans.*, **90**, 3117 (1994)
2) G. Nelson, I. M. Warner, *J. Phys. Chem.*, **94**, 576 (1990)
3) N. Mataga, M. Tomura, H. Nishimura, *Mol. Phys.*, **9**, 367 (1965)
4) F. J. Smith, A. T. Armstrong, S. P. McGlynn, *J. Chem. Phys.*, **44**, 442 (1966)

^{1}H NMR STUDY ON THE INCLUSION OF BICYCLO[3.3.1]NONANES IN CYCLODEXTRINS

E.BUTKUS [1], J. C. MARTINS [2], U. BERG [3]
[1] *Department of Organic chemistry, Vilnius University, Naugarduko 24, 2006 Vilnius, Lithuania,* [2] *High - Resolution NMR Centre, Room 9G621, Vrije Universiteit Brussel,1050-Brussel, Belgium,* [3] *Organic Chemistry 1, Chemical Center, Lund University, P.O. Box 124, S- 2210 Lund, Sweden*

ABSTRACT

α- and ß-cyclodextrins were found to form 1:1 inclusion complexes with 2,6- and 2,9-substituted bicyclo[3.3.1]nonanes. The binding constants and the structure of the complexes were estimated from titration studies and 2D ROESY experiments.

1. INTRODUCTION

The relatively hydrophobic cavities of the doughnut-shaped cyclodextrins (CDs) are known for their ability to bind organic molecules in solution and in the crystalline state. The shape and the internal diameter of the CDs cavity and the size of the guest molecules, i.e. the close match is of primary importance for the complex formation and this may provide useful tools in molecular recognition, substrate-receptor interactions, etc. The interaction of CDs with a number of organic structures has been studied in recent years [1] and among the organic compounds exhibiting the strongest binding to the CDs are substituted adamantanes [2]. The investigation on the interaction of CDs with bicyclo[3.3.1]nonanes which are conformationally flexible structures and have one carbon atom less than adamantane was carried out seeking to gain insight into the bicyclononanes-CDs complex formation and structure.

2. MATERIALS AND METHODS

2.1. Materials

α- and β-Cyclodextrins (Aldrich) were used as received and freeze dried before use. Literature methods were used to prepare bicyclo[3.3.1]nonane-2,6-dione (**1**), endo,endo-bicyclo[3.3.1]nonane-2,6-diol (**2**) and endo-2-anti-9-bicyclo[3.3.1]nonanediol (**4**) [3,4]. Monoacetate **3** was prepared by acetylation of **2** with an equimolar amount of acetic anhydride.

J. Szejtli and L. Szente (eds.), Proceedings of the Eighth International Symposium on Cyclodextrons, 241–244.

2.2. **Proton NMR spectra**

^{1}H NMR spectra were recorded at 303K on a Bruker AMX500 spectrometer equipped with Eurotherm controller for temperature regulation. The complexes were obtained and the titrations were performed by adding variable amounts of the substrates to a cyclodextrin solution in D_2O and *vice versa* at 33 ^oC. The concentration of substrates and the CDs varied between 1-5x10^{-3} M. The solutions were thoroughly mixed and allowed to equilibrate for several minutes in the probe before the spectrum was acquired. The inclusion stability constants K and the estimated associated errors were calculated from ^{1}H NMR titration data using a non-linear least squares fit of the data to a modified version of the Benessi-Hildebrand equation [5] within the program SigmaPlot (Jandel Sci.). For the ROESY spectra the standard pulse sequence was used.

3. RESULTS AND DISCUSSION

The complex formation is expected to induce shifts of the ^{1}H NMR resonances of the cyclodextrin protons, especially of H-3 and H-5, which are directed towards the interior of the cavity. The inclusion of guest bicyclo[3.3.1]nonane molecules **1-4** in the α- and ß-CD cavity in aqueous solution was supported by the changes in the ^{1}H NMR spectra of both partners. The occurrence of chemical shift changes without line broadening allows to characterize the interaction as being fast on the NMR time scale. Furthermore, the fact that significant chemical shift changes occur only for hydrogens on the inner surface (H-3 and H-5) of the CDs, and not for those on the outer surface (H-1, H-2, H-4) allows to establish unambiguously that these changes are the result of inclusion complex formation and not the result of non-specific interactions between both partners.

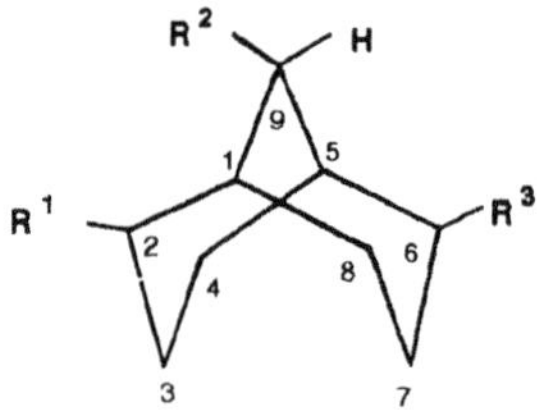

$R^1=R^3=$ -O, $R^2=$H

$R^1=R^3=$OH, $R^2=$H

$R^1=$OH, $R^2=$H, $R^3=OCOCH_3$

$R^1=$H, $R^2=R^3=$OH

An addition of α-CD to the solution of dione **1** resulted in an upfield shift (0.064 ppm) for H-3 proton of the α-CD which is located in the wide opening of the truncated cone. Practically no induced shift was observed for the H-5 proton which is at the narrower side. Downfield shifts for the protons of the included dione **1** were observed (Table I). From these data it follows that substrate molecule does not penetrate deeply into the α-CD cavity. Because of the closer match of the ß-CD cavity size and the shape of bicyclo[3.3.1]nonane (both *ca.*7 Å), a stronger host-guest interaction was expected in this case. Significant chemical shift changes were observed for the substrates **1-4** and ß-CD protons. The upfield shift of the H-5 was larger than of the H-3 proton in dione **1**, i.e. 0.12 and 0.064 ppm, respectively. This observation is consistent with a deeper insertion of the bicyclic framework in the ß-CD cavity than in α-CD cavity. The protons of the guest dione **1** showed also significant downfield shifts of 0.11-0.12 ppm (Table 1). The inclusion of the diol **2** has a larger effect on the H-3 shift than on H-5 shift of ß-CD (0.054 and 0.03 ppm, respectively) showing that presumably the penetration of diol **2** is less deep in the CD cavity compared to the insertion of dione **1**. The proton signals of the bicyclic framework display comparatively large shifts of >0.1 ppm. In the case of 2,9-diol **4** the resonances of the H-3 and H-5 of CD showed small induced shifts (0.045 and 0.01 ppm, respectively), however relatively large shifts were observed for the H_1 proton and some other protons of the bicyclic molecule (Table 1).

TABLE 1. ^{1}H NMR chemical shifts (ppm) displacements ($\Delta\delta_{lim}$) resulting from the inclusion of **1-4** in α- and ß-cyclodextrins and inclusion stability constantsK

	$\Delta\delta_{lim}$				
			Guest+host		
Nucleus	**1**+αCD	**1**+ßCD	**2**+ßCD	**3**+ßCD	**4**+ßCD
Guest					
H_1	0.05	0.12	0.21		0.21
H_4	0.05				0.12
H_9	0.02	0.11	0.105		
H_2				0.07	
H_6				0.08	
CDs					
H-3	-0.64	-0.064	-0.054	-0.07	-0.045
H-5	-0.01	-0.12	-0.03	-0.13	-0.01
K(M^{-1})	60±11	345±12	71±14	195±10	13.5±4

The molar ratios and the stability constants K for the inclusion complexes formed between bicyclononanes and the CDs were determined from ^{1}H NMR titrations. The limiting changes in the chemical shifts of the CD proton resonances upon inclusion of the guest were determined by a titration of CDs with an excess of substrates and *vice versa*. To determine the stoichiometry of the CD-substrate complexes we have plotted the induced shift Δδ values for the ß-CD and bicyclic structures as a function of guest to host molar ratio (R). The changes of Δδ are observed until the value of R=1 for protons of CDs and of substrates when adding the CD to the solution of **1-4**. The variation of the ^{1}H chemical shifts over the range of R values considered is consistent with the formation of a 1:1 inclusion complex. The Δδ data were used to estimate K for the substrate-CD adducts. In the absence of any information about the activity coefficients, only an apparent association constant can be determined [5b] (Table 1).

The inclusion of the guest molecules were further studied by 2D ROESY spectroscopy. No intermolecular NOE's could be detected in 2D NOESY spectra recorded with various mixing times, while intramolecular NOE's were weak and positive and weak and negative for the guest and host molecules, respectively. The observation of weak negative NOE's for CD is mainly the result of the high ß-CD concentration and the degeneracy of the glucopyranose resonances. Therefore, 2D rotating frame NOE spectroscopy was used as ROE is always positive irrespective of τ_C. In the case of the dione **1** the intramolecular cross peaks within the substrate molecule and intermolecular cross peaks with the H-3 and H-5 protons of the CD are observed. The intermolecular NOE's are found between H-3 and $H_{1(5)}$ as well as between H-3 and $H_{9syn(anti)}$. The same NOE's although weaker are found to occur with H-5 of the ß-CD. Thus the protons at carbon atoms 1(5) and 9 are closer to H-3 than they are to H-5. In the case of **2** essentially all ß-CD protons display intermolecular NOE's of more or less equal intensity with the most of the 2,6-diol resonances. Although the ROESY spectra support the formation of inclusion complexes and allow delineate the location of the guest molecules in the host, the pseudo C_7 symmetric cavity of the ß-CD as well as the C_2 symmetry of the guest molecules **1** and **2** does not allow a more quantitative elaboration of the intermolecular distances and lead to the only conclusion that the guest molecules enter into the ß-CD cavity from the large hole to give a symmetrical location of the guest molecule.

4. CONCLUSION

An inclusion complex formation between bicyclo[3.3.1]nonanes and cyclodextrins has been demonstrated on the basis of the ^1H NMR titration experiments. The structure and the binding constants of the complexes have been estimated and the insertion into the cavity relative to one another have been discussed on the basis of chemical shift arguments and the observation of intermolecular NOE's.

REFERENCES

[1] Perli,B., Djedani,F., Berthault,P., New Trends in Cyclodextrins and Derivatives, Editions de Sante, Paris, 1991.

[2] Shortreed, M.E., Wylie, S., Macartney, D.H., Inclusion of (N-Adamantan-1'-ylpyrazinium)pentacyanoferrate(II) ion in α- and β-cyclodextrins, *Inorg. Chem.* **32,** 1824-1829 (1993), and ref. 18-20 cited therein.

[3] Butkus,E., Malinauskiene,J., Kadziauskas,P., Zur Stereochemie der Reduktion von Bicyclo[3.3.1]nonan-2,6-dion, *Z. Chem.*, **20**, 103 (1980).

[4] Berg,U., Butkus,E., Stoncius,A., Stereochemistry of the reduction of bicyclo[3.3.1]nonane-2,9-dione by complex hydrides, *J. Chem. Soc., Perkin Trans. 2*, 97-102 (1995).

[5] (a) Benessi, H.A., Hildebrand, J.H., *J. Am. Chem. Soc.* **71** , 2703 (1949);(b) Bekkers, O., Kettenes-van den Bosch, J.J., van Helden, S.P., Seijkens, D., Beijnen, J.H., Bult, A., Underberg, W.J.M., Inclusion complex formation of antracycline antibiotics with cyclodextrin; a proton nuclear magnetic resonance and molecular modelling study, *J. Incl. Phen.*, **11,** 185-195 (1991).

HYDROXYPROPYL CYCLODEXTRIN COMPLEXES: RELATIONSHIPS BETWEEN THE SUBSTITUTION PATTERN, STABILITY CONSTANTS AND CIRCULAR DICHROISM SPECTRA

ÁGNES BUVÁRI-BARCZA , JUDIT KAJTÁR AND LAJOS BARCZA

Inst. Inorg. Anal. Chem. and Inst. Org. Chem., L.Eötvös Univ., Budapest 112, POB 32, 1518 Hungary

ABSTRACT

The phenomena have been investigated using phenolphthalein because it proved to be an excellent model for inclusion complex formation as a relatively large guest molecule and its interaction with β-cyclodextrin is known rather well. The average degree of substitution (DS) of hydroxypropyl-β-cyclodextrin samples ranged between (0.0) - 2.9 - 16.0 and they can be grouped in three categories as "regular", "less regular" and "irregular" ones. The "regular" samples have been substituted mainly on O(2) (secondary) hydroxy groups, the "irregular" ones have substituents first of all on the primary O(6) side, while the "less regular" ones are hydroxypropylated randomly. The stability constants of phenolphthalein complexes with "regular" samples decrease rather regularly with increasing DS and their circular dichroism spectra are very similar, while the properties of an "irregular" sample (with identical DS) can differ extremely.

1. INTRODUCTION

Hydroxypropylation of the hydroxy groups of β-cyclodextrin (CD) leads to favourable changes in its properties, like in the solubility, solubilizing power, etc. [1]. The average degree of substitution (DS) is the only measure used for the characterization of different products and means the average number of 2-hydroxypropyl (HP: $-CH_2-CH(OH)-CH_3$) units per CD units. Unfortunately, DS can cover rather significant differences in position (2, 3 or 6) and type (primary: 6, secondary: 2, 3) of substitution and can be hardly used for exact prediction. The number of isomers (i.e. the complexity of the sample) increases with increasing DS value [1-3], and the distribution of the different individual substitution numbers is symmetrical around the average DS [1-3].

Our aim was to investigate the connection between the stability of inclusion complexes and the DS value as well as the effect of substitution pattern both on the stability constant

J. Szejtli and L. Szente (eds.), Proceedings of the Eighth International Symposium on Cyclodextrons, 245–249.

and the most probable structure of the inclusion complex (concluded from its circular dichroism spectrum). Phenolphthalein (PP) has been used in these investigations as the most appropriate model for relatively large guest molecules because only one of its phenolic rings has been proved to be included into the CD cavity [4]. This inclusion promotes further interactions resulting in a very stable three-site contact which can be regarded as the molecular recognition of PP by β-CD. The inclusion complex has very characteristic uv-vis and induced circular dichroism (ICD) spectra [4], which seem to be very suitable tools for our purposes.

2. MATERIALS AND METHODS

The hydroxypropyl-β-cyclodextrin (HP-CD) samples were from Cyclolab Ltd. (Hungary), prepared in different ways (e.g. using different alkalinities during the reaction of β-CD with propylene oxide, which influence mostly the pattern of substitution [1,3,5,6]). All the other materials were of analytical grade.

The uv-vis absorption spectra were recorded on a Perkin-Elmer Lambda 15 spectrophotometer and the stability constants of the inclusion complexes were measured and calculated as published elsewhere [7].

ICD spectra were measured using a Jobin-Yvon Dichrograph Mark VI. The concentration of PP was $(1.5\text{-}3.0)\times10^{-5}$ M and the appropriate HP-CD was in relatively large excess to achieve the highest possible complexation.

The temperature was 25.0 + 1.0 °C while the pH was kept constant at pH = 10.5, buffered with sodium carbonate.

3. RESULTS AND DISCUSSION

Like in the parent β-CD complex [4], the very characteristic purple colour of the doubly deprotonated PP (pH=10.5) disappears when complexed by HP-CD of any DS and the uv-vis spectra are practically identical in all cases. All PP - HP-CD inclusion complexes show 1 : 1 stoichiometries. The ICD spectra are different [8], only the complete lack of absorption bands in the visible range is alike. All of these findings point to the existence and importance of the three-site interaction in PP - HP-CD inclusion complexes, too.

However, the result of the complexation can be rather different depending on both the DS values and the substitution pattern. As the data in Table 1 show, regularities can be found only in a definite group of the samples investigated. The samples of this group (referred to as "regular" ones) were prepared at low alkalinities, where mainly the O(2) hydroxy groups of the secondary CD rim are substituted and the formation of oligomeric HP dervatives is less probable. In contrast to this group, the "irregular" samples were produced at high alkalinities which promotes both the HP substitution of the primary rim and the possibility of oligomeric substitution. The third group referred to as "less regular" collects samples from different batches, showing rather randomly mixed substitution pattern.

The stability constants of PP inclusion complexes (Table 1) with "regular" samples decrease rather systematically in function of increasing DS values, while those of the two other groups are even smaller and show no connection to the DS values. Attention must be drawn to the three HP-CDs of DS=8.0, where the ratio of their K values are in proportion to each other as 1.00 : 0.44 : 0.14 [7].

TABLE 1. Stability constants and induced circular dichroism data of phenolphthalein—hydroxypropyl-β-cyclodextrin complexes

Group	DS	K	$\Delta\varepsilon_{270}$	$\Delta\varepsilon_{300}$
"regular"	3.9	1.50×10^4	2.36	-2.19
	6.0	1.23×10^4	2.23	-2.45
	8.0	9.0×10^3	2.22	-2.04
	10.0	7.6×10^3	2.22	-1.80
	12.0	5.6×10^3	2.04	-1.86
	14.0	4.8×10^3	1.86	-1.50
"less regular"	4.0	1.13×10^4	1.84	-2.06
	4.4	9.4×10^3	2.22	-2.10
	5.0	9.9×10^3	1.65	-1.98
	8.0	4.0×10^3	3.29	-0.75
"irregular"	3.1	1.8×10^3	-2.10	1.95
	8.0	1.3×10^3	-2.87	4.04

K=[PP.HP-CD]/[PP][HP-CD]

These rather strange and nearly unexpected differences reflect directly the effects of the change in the substitution pattern, i.e. the differeces in the relative substitution of the primary and secondary rims. It seems that the most sensitive component of the three-site interaction of PP and HP-CD is the H-bonding between the included phenolic ring (more exactly its oxygen) and the hydroxy groups of the CD primary rim.

Table 1 summarizes the important data of the ICD spectra, too: i.e. the $\Delta\varepsilon$ values measured at the maxima (at about $\lambda = 270$ and $\lambda = 300$ nm) of the characteristic ICD bands [4, 9]. It can be seen that the signs of these two $\Delta\varepsilon$ values are identical with those of unsubstituted β-CD [4] except the opposite signs experienced with "irregular" samples.

Like the stability constants, the $\Delta\varepsilon$ values of "regular" samples show also a moderate decreasing trend in function of increasing DS, but those of the two other groups differ extremely. The fluctuation of the values in the case of "less regular" samples could be explained by the perturbed interaction of the phenolic ring rather easily [9], but the change of the whole mechanism might be assumed when the sign is also turned in the case of "irregular" samples. This assumption contradicts both the uv-vis spectra discussed and the similarities of maxima in ICD spectra (only the signs are changed). The discrepancies can be solved using the theory of Kodaka [10] based on the coupled oscillator model. The theory proves that the sign of the ICD is influenced not only by the direction but also by the depth of inclusion of the chromophore. A special pattern of hydroxypropylation may cause increased steric hindrance (and provide modified possibilities for H-bonding). So we can assume, that the PP having been included by one of its phenolic rings is fixed in an inclined position. This inclined position is the cause of the change in the signs while the bands are not shifted.

Comparing the data of Table 1, the connection between the lowest $\underline{K}$ and the most irregular $\Delta\varepsilon$ values is very remarkable. This correlation is valid for a larger collection of data [8], too, proving their common cause as well as the high probability of the explanation proposed.

ACKNOWLEDGEMENT

Financial support of this work from the Hungarian Research Foundation (OTKA Nos. 2239 and 2277) are gratefully acknowledged. We thank Cyclolab Ltd. (Hungary) for supplying the HP-CD samples, Mrs D. Bodnár-Gyarmathy and Dr L. Szente for their helpful collaboration.

REFERENCES

[1] Duchene, D. (Ed.): New trends in cyclodextrins and derivatives, Edition de Sante, Paris (1991)

[2] Pitha, J., Milecki, J., Fales, H., Pannell, L., Uekama, K.: Hydroxypropyl-β-cyclodextrin: preparation and characterization; effects on solubility of drugs, *Int. J. Pharm.* **29**, 73 (1986)

[3] Irie, T., Fukunaga, K., Yoshida, A., Uekama, K., Fales H.M., Pitha, J.: Amorphous water-soluble cyclodextrin derivatives: 2-hydroxyethyl, 3-hydroxypropyl, 2-hydroxy-isobutyl and carboxamido-methyl derivatives of β-cyclodextrin, *Pharm. Res.* **5**, 713 (1988)

[4] Buvári, Á., Barcza, L., Kajtár, M.: Complex formation of phenolphthalein and some related compounds with β-cyclodextrin, *J. Chem. Soc., Perkin Trans. 2*, 1687 (1988)

[5] Rao, C.T., Lindberg, B., Lindberg, J., Pitha, J.: Substitution in β-cyclodextrin directed by basicity: preparation of 2-O- and 6-O-[(R)- and (S)-2-hydroxypropyl] derivatives, *J. Org. Chem.* **56**, 1327 (1991)

[6] Rao, C.T., Pitha, J., Lindberg, B., Lindberg, J.: Distribution of substituents in O-(2-hydroxypropyl) derivatives of cyclomalto-oligosaccharides (cyclodextrins): influence of

increasing substitution, of the base used in the preparation and of macrocyclic size, *Carbohydr. Res.* **223**, 99 (1992)

[7] Buvári-Barcza, Á., Bodnár-Gyarmathy, D., Barcza, L.: Hydroxypropyl-β-cyclodextrins: correlation between the stability of their inclusion complexes with phenolphthalein and the degree of substitution, *J. Incl. Phenom.* **18**, 301 (1994)

[8] Buvári-Barcza, Á., Kajtár, J., Szente, L., Barcza, L.: Hydroxypropyl-β-cyclodextrins: induced circular dichroism spectra of included phenolphthalein as a function of the degree of substitution, *J. Chem. Soc., Perkin Trans. 2,* in press (1996)

[9] Kajtár, M., Horváth-Toró, Cs., Kuthy, É., Szejtli, J.: A simple rule for predicting circular dichroism induced in aromatic guests by cyclodextrin hosts in inclusion complexes, *Acta Chim. Hung.* **110**, 327 (1982)

[10] Kodaka, M.: (a) Sign of circular dichroism induced by β-cyclodextrin, J. Phys. Chem. 95, 2110 (1991) and (b) A general rule for circular dichroism induced by a chiral macrocycle, *J. Am. Chem. Soc.* **115**, 3702 (1993)

FLUORESCENT CYCLODEXTRINS FOR MOLECULE SENSING

MOLECULAR RECOGNITION ABILITIES OF BIS(PYRENE)-MODIFIED β-CYCLODEXTRINS

A. UENO, T. SUZUKI, H. IKEDA

Department of Bioengineering, Faculty of Bioscience and Biotechnology, Tokyo Institute of Technology, 4259 Nagatsuta, Midori-ku, Yokohama 226, Japan

ABSTRACT

Three fluorescent β-cyclodextrins, each having two pyrene moieties at AB, AC, or AD glucose residues, exhibit remarkable changes in the excimer fluorescence intensity associated with guest binding, in spite of the fact that the pyrene moiety is too large to be included in the cavity. The excimer intensity increases for steroids while decreases for sodium dodecyl sulfate, and almost no response was observed for 1-adamantanol which binds strongly with β-cyclodextrin.

1. INTRODUCTION

Cyclodextrins (CDs) are spectroscopically inert, but they can be converted into spectroscopically active hosts by modification with chromophores.[1] We have prepared many fluorophore and dye-modified CDs and demonstrated that they can be used as molecule-recognition sensors[2] on the basis of their guest-induced fluorescence and color-change[3] properties. One example of the fluorescent CDs are pyrene-modified γ-CDs,[4-6] which form excimers by intermolecular or intramolecular interaction between two pyrene moieties. They change their excimer intensities associated with guest binding in aqueous solution and were used for detecting a variety of organic compounds. Here, we report, for the first time, the sensing abilities of β-CD derivatives which have two pyrene moieties.

J. Szejtli and L. Szente (eds.), Proceedings of the Eighth International Symposium on Cyclodextrons, 251–254.

2. MATERIALS AND METHODS

Three isomers of bis(pyrene)-β-CD were prepared. The reaction of each of 6A,6B-, 6A,6C-, and 6A,6D-ditosyl-β-CDs with ethylenediamine afforded a CD derivative bearing two ethylenediamine units at AB, AC, or AD glucose residues, and the product was purifid by column chromatography with CM Sephadex C-25. The reaction of these products with pyreneacetic acid was performed in DMF with DCC and 1-hydroxybenztriazole. Bis(pyrene)-modified β-CDs were identified by 500 MHz-NMR and elemental analysis.

Fluorescent spectra were measured at 25°C in 10% DMSO aqueous solution. The excitation wavelength was 340 nm.

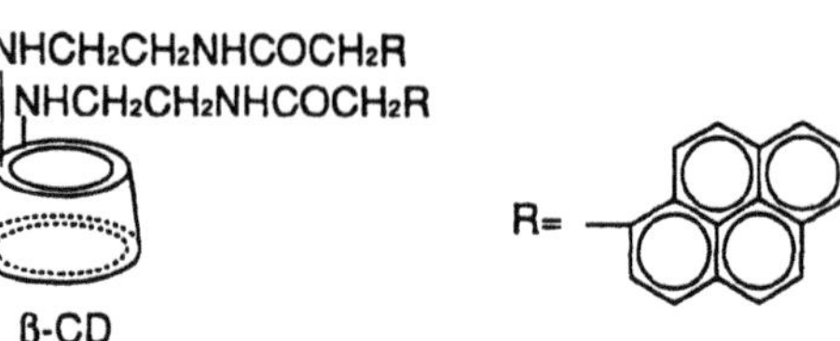

Bis(pyrene)-modified CDs

3. RESULTS AND DISCUSSION

Three isomers of AB, AC, and AD bis(pyrene)-β-CDs exhibit excimer emission around 495 nm in addition to the monomer fluorescence which has peaks around 378 and 398 nm. The excimer emission was enhanced upon addition of chenodeoxycholic acid while it decresed upon addition of sodium dodecylsulfate (SDS) for all samples of the fluorescent hosts (1 μM). However, 1-adamantanol, which is known as a good guest for β-CD, was not effective to change the excimer emission intensity.

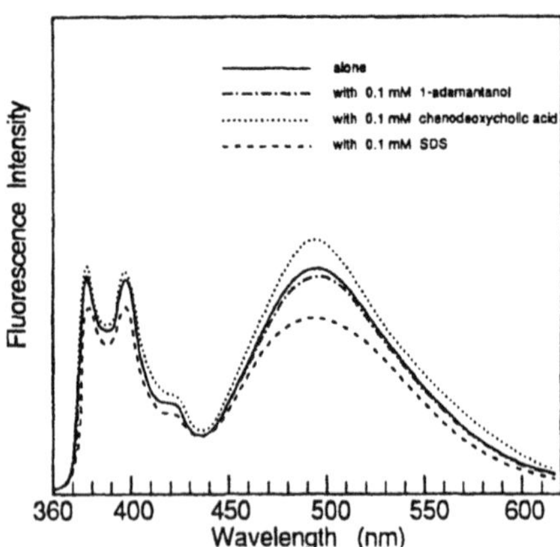

Fig.1 Fluorescence spectra of AC isomer in 10% DMSO aqueous solution

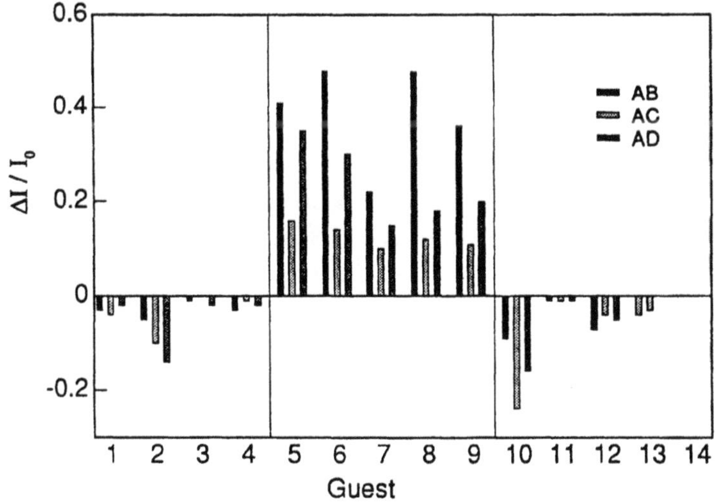

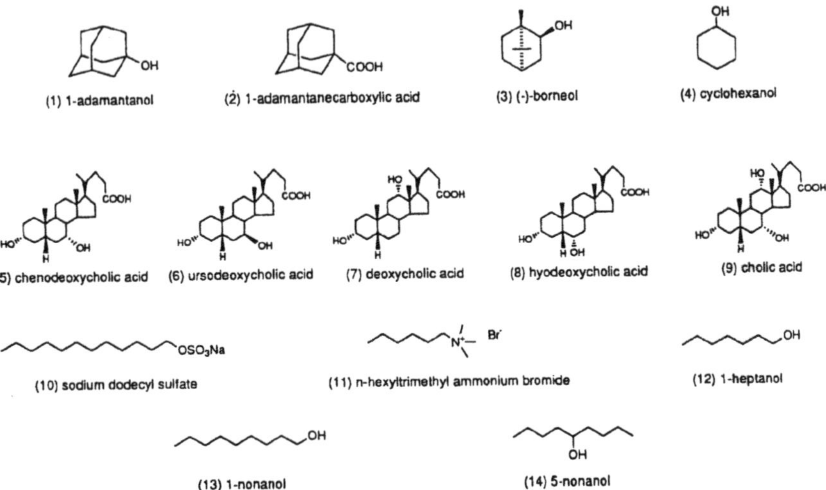

Fig. 2 Variations of the excimer emission intensities of AB, AC, and AD bis(pyrene)-modified CDs (1 μM). The guest concentration is 0.1 mM.

Figure 2 shows that AB, AC, and AD isomers exhibit similar trends in their responses for the guest species examined, for example, increasing and decreasing the excimer emission for the steroid derivatives and others, respectively. Since the pyrene moieties of these fluorescent CDs are unlikely to be included in the β-CD cavity because of the large size of pyrene ring, the excimer formation should occur outside of the cavity. Therefore, the negligible effect of 1-adamantanol seems reasonable if it binds to the β-CD cavity from secondary hydroxyl side. On the other hand, the steroid derivatives and SDS are likely to penetrate the cavity, thus affecting the excimer formation. The different effects between the steroid derivatives and SDS may be due to the steric requirement in the complexes, resulting in facilitated excimer formation for the former while in depressed excimer formation for SDS.

4. CONCLUSION

We constructed a new type of chemosensors as molecule-recognition sensors, which exhibit excimer emission responsive to the shape of guest molecules. This fact demonstrates that fluorescent CDs can be used for sensing molecules even if the fluorophore can not be included in the CD cavity.

REFERENCES

[1] Ueno, A., *Chromophore-modified cyclodextrins for detecting organic compounds with molecular recognition,* in Fluorescent Chemosensors for Ion and Molecule Recognition, (Ed. Czarnik, A. W.) Washington, 1993, Pp. 74-84

[2] Ueno, A., Fluorescent Sensors and color-change indicators for molecules, *Adv. Mater.* 132-134 (1993)

[3] Ueno, A., Kuwabara, T., Nakamura, A., Toda, T., A modified cyclodextrin as a guest-responsive color-change indicator, *Nature*, **356**, 136-137 (1992)

[4] Ueno, A., Suzuki, I., Osa, T., Association dimers, excimers, and inclusion complexes of pyrene-appended γ-cyclodextrins, *J. Am. Chem. Soc.*, **111**, 6391-6397 (1989)

[5] Ueno, A., Suzuki, I., Osa, T., Host-guest sensory systems for detecting organic compounds by pyrene excimer fluorescence, *Anal. Chem.*, 2461-2466 (1990)

[6] Suzuki, I., Ohkubo, M., Ueno, A., Osa, T., Detection of organic compounds by dual fluorescence of bis(1-pyrenecarbonyl)-γ–cyclodextrins, *Chem. Lett.*, 269-272 (1992)

HYDROGEN BONDING INTERACTIONS WITH CYCLODEXTRINS: UTILIZATION OF FLUORENONE AS A NEW PROBE

LÁSZLÓ BICZÓK [1], LÁSZLÓ JICSINSZKY [2], HENRY LINSCHITZ [3]
[1] *Central Research Institute for Chemistry, Hungarian Academy of Sciences, P.O. Box 17. H-1525 Budapest, Hungary*
[2] *CYCLOLAB Research and Development Laboratory Ltd. P.O. Box 435. H-1525 Budapest, Hungary*
[3] *Brandeis University, Waltham, Massachusetts 02254-9110, USA*

ABSTRACT

The interaction of fluorenone with cyclodextrins (CD) was studied by fluorescence lifetime, fluorescence quantum yield and triplet yield measurements. In aqueous solutions, the fast internal conversion of the singlet excited inclusion complexes indicates that fluorenone, while embedded in the CD cavity, still retains hydrogen bonded water. In organic media, the dynamic quenching of fluorescence and singlet-triplet transition was attributed to formation of a short lived, non- or very weakly fluorescent excited fluorenone/CD hydrogen-bonded complex since the methylation of the glycosidic OH-groups significantly decreased the quenching rates.

1. INTRODUCTION

Fluorescence techniques have been widely used to examine the microenvironment of encapsulated molecules, since the fluorescence intensities, lifetimes and emission maxima of some fluorophores are very sensitive to, and reflect the nature of their solvation envelopes. However, the probes which have been used are unable to distinguish specific solvent-solute interactions from the effects of bulk polarity.

We have recently introduced a new molecular probe [1], the singlet excited fluorenone, which is able to selectively detect hydrogen bonding since the rate of its internal conversion markedly increases in the presence of alcohols, while the rate of transition to the triplet is scarcely affected. The singlet excited fluorenone was found to be a good indicator of non-specific interactions with solvents as well because the rate of its triplet formation increases more than two orders of magnitude in going from acetonitrile to hexane.

In the present studies, we exploit these unique photophysical properties to obtain more information on the major characteristics of fluorenone interaction with cyclodextrins and reveal how the methylation of the glycosidic OH-groups influences the structure of fluorenone / ß-cyclodextrin complexes. We also extend these studies to organic solvents and

J. Szejtli and L. Szente (eds.), Proceedings of the Eighth International Symposium on Cyclodextrons, 255–258.

demonstrate that dynamic quenching of the fluorenone singlet excited state in nonaqueous media takes place through short lived hydrogen-bonded complex formation with cyclodextrins.

2. MATERIALS AND METHODS

Fluorenone (FLUKA) was purified by repeated recrystallization from ethanol. Purified ß-cyclodextrin (ßCD), random methylated ß-cyclodextrin (RAMEB), heptakis-(2,6-di-O-methyl)-ß-cyclodextrin (DIMEB) and heptakis-(2,3,6-tri-O-methyl)-ß-cyclodextrin (TRIMEB) are products of CYCLOLAB R&D Lab. Ltd., Hungary. HPLC grade solvents were used as received. The samples were deoxygenated by purging with nitrogen.

Fluorescence lifetimes were measured on an Applied Photophysics SP-3 single photon counting apparatus using a hydrogen lamp operated at 30 kHz. Data were analyzed by a nonlinear least-squares deconvolution method. Corrected fluorescence spectra were recorded on a homemade spectrofluorimeter equipped with a Princeton Applied Research type 1104A/B photon-counting system. Fluorescence quantum yields were determined relative to fluorenone in acetonitrile ($\Phi_F = 0.032$) [1]. Triplet yields were obtained by the "limiting slope" method [2] using fluorenone solution in methylcyclohexane as reference (Φ_{ISC}(MCH)=1.00) [3]. In this method, values of triplet yields are obtained by flash photolysis measurements of the relative slopes of triplet absorbance vs. flash energy plots, in the linear region. A frequency doubled ruby laser or XeF excimer laser was used for excitation and the triplet-triplet absorption of fluorenone was monitored at 440 nm [3]. The molar extinction coefficient of the triplet was taken to be the same for all solvents and in the cyclodextrin complexes.

3. RESULTS AND DISCUSSION

3.1. Fluorenone inclusion complexes in aqueous solutions

In aqueous solution, inclusion of fluorenone in ßCD cavity is demonstrated unambiguously by cyclodextrin-enhanced solubility, absorption spectroscopy and very directly by the appearance of circular dichroism of fluorenone in the presence of ßCD [4]. In all cases, fluorescence decays fit single exponential functions and interference from emission of uncomplexed fluorenone is negligible because of its low solubility and fluorescence yield. Table I. presents the photophysical properties of fluorenone/ßCD complexes.

Table 1.Photophysical parameters of fluorenone

Media	τ_F ns	Φ_F 10^{-3}	Φ_{ISC}	k_{ISC} $10^7 s^{-1}$	k_{IC} $10^7 s^{-1}$
βCD	1.2	2.4	0.04	3.3	79
RAMEB	2.6	3.4	0.06	2.3	36
DIMEB	2.2	4.8	0.08	3.6	41
TRIMEB	3.0	3.6	0.12	3.8	29
THF	2.7	6.4	0.87	32	3
1-octanol	1.2	2.5	0.14	12	71

For comparison, the corresponding data in tetrahydrofuran (THF) and 1-octanol are included as well. It is apparent that the behavior of complexes is much closer to that observed in alcohols than in non-hydroxylic solvents. The short singlet lifetimes (τ_F), the very low values of triplet yields (Φ_{ISC}) and fluorescence yields (Φ_F) arise from the fast internal conversion (k_{IC}) as has been found for uncomplexed fluorenone in alcohols [1]. This indicates that hydrogen bonding interactions with the fluorenone guest are significant. We may exclude, on the basis of the circular dichroism spectrum [4], molecular dimensions and cavity geometry, the possibility that fluorenone is complexed equatorially, with the carbonyl group oriented toward water. The fact that the methylation of the glycosidic OH-groups does not affect significantly the major photophysical characteristics of the complexes suggests that hydrogen bonds do not form between the fluorenone carbonyl and the OH groups of ßCDs. Thus, we conclude that the fast internal conversion of singlet excited fluorenone must be assigned to the H-bonding interaction with water molecules still remaining in the cyclodextrin cavity after fluorenone inclusion.

3.2. Interactions between fluorenone and cyclodextrins in organic solvents

Addition of ßCD to the dimethylformamide (DMF) and dimethylsulfoxide (DMSO) solution of fluorenone leads to ca. 20 nm red-shift of the fluorescence maximum and substantial decrease in the fluorescence intensity but no significant change in the absorption spectra occurs. Fluorescence decaytimes and triplet yields exhibit parallel changes with fluorescence yield in the function of ßCD concentration. All of the three quantities provide linear Stern-Volmer plots. From the slopes $k_q = 9.0 \times 10^8\ M^{-1}\ s^{-1}$ and $3.0 \times 10^8\ M^{-1}\ s^{-1}$ is derived for the rate constant of singlet excited fluorenone quenching by ßCD in DMF and DMSO, respectively. The difference of these values probably reflects the higher hydrogen bonding ability of DMSO.

Table 2. Photophysical processes in the presence of 0.1 M additive in DMF

Additive 0.1 M	τ_F ns	Φ_F	Φ_{ISC}	k_{ISC} $10^7 s^{-1}$	k_q $10^8 M^{-1}s^{-1}$
αCD	7.0	0.013	0.28	4.0	7.7
γCD	6.0	0.011	0.25	4.2	10.1
βCD	6.5	0.012	0.26	4.0	9.0
RAMEB	10.3	0.019	0.41	4.0	3.2
DIMEB	14.2	0.025	0.59	4.2	0.5
TRIMEB	15.2	0.028	0.63	4.1	-
-	15.3	0.028	0.63	4.1	-
MeOH	14.8	0.028	0.63	4.1	0.3

In order to reveal the nature of the quenching process, the structure of cyclodextrins was systematically varied. Table 2. summarizes the photophysical parameters measured in the presence of 0.1 M additives in DMF. It is evident that the photophysical behavior of fluorenone is strikingly insensitive to the cavity size of CDs. In accord with the linearity of Stern-Volmer plots and the single exponential decay of fluorenone fluorescence, this suggests that no inclusion complex formation occurs. However, the fast dynamic quenching caused by CDs shows interaction between singlet excited fluorenone and CDs. Since the methylation of the glycosidic OH-groups significantly decreases the quenching rates, we infer that short lived, non- or very weakly fluorescent hydrogen bonded complex forms between the singlet excited fluorenone and CDs. The interaction is probably facilitated by the enhancement of dipolmoment upon excitation of fluorenone. It is interesting to note that most of the CDs exhibit much higher quenching rate constants than alcohols. This may be attributed to the more efficient vibronic coupling between excited and ground states in the singlet excited fluorenone/CD complexes.

4. CONCLUSION

The interaction of fluorenone with CDs is sharply different in aqueous and organic solutions. In the former case, inclusion complex can be detected both in the ground and excited state. The fast internal conversion of the singlet excited complexes shows that fluorenone, while embedded in the CD cavity, still retains hydrogen bonded water. On the other hand, no intercalation of fluorenone in the CD cavity is observed in organic media. The dynamic quenching of fluorescence and triplet formation is assigned to short lived hydrogen bonded complex between the singlet excited fluorenone and CDs.

ACKNOWLEDGMENT

We much appreciate support of this work by the Division of Chemical Sciences, Office of Basic Energy Sciences, Office of Energy Research, US Department of Energy (Grant No. DE-FG02-89ER14027 to Brandeis University).

REFERENCES

[1] Biczók, L., Jicsinszky, L., Linschitz, H., Solvent dependent radiationless transitions in fluorenone: A probe for hydrogen bonding interactions in the cyclodextrin cavity, *J. Incl. Phen.*, **18**, 237-245 (1994)

[2] Hurley, J. K., Sinai, N., Linschitz, H., Actinometry in monochromatic flash photolysis: The extinction coefficient of triplet benzophenone and quantum yield of triplet zinc tetraphenyl porphyrin, *Photochem. Photobiol.* **38**, 9-14 (1983)

[3] Andrews, L. J., Deroulede, A., Linschitz, H. Photophysical processes in fluorenone, *J. Phys. Chem.* **82**, 2304-2309 (1978)

[4] Yamaguchi, H., Ninomiya, K., Ogata, M., Study of the electronic spectra of 9-fluorenone, Chem. *Phys. Lett.* **75**, 593-595 (1980)

EFFECT OF ALKYL CHAIN LENGTH AND DEGREE OF SUBSTITUTION ON COMPLEXATION OF SULFOALKYL β-CYCLODEXTRINS WITH TESTOSTERONE AND PROGESTERONE

V. Zia[1], E.R. Bornancini[2], E.A. Luna[2], R.A. Rajewski[2], and V.J. Stella[1, 2].
[1] Dept. of Pharmaceutical Chemistry, [2]Center for Drug Delivery Research, The University of Kansas, Lawrence, KS, U.S.A., 66047.

ABSTRACT - This study was designed to test how the sulfoalkyl ether (SAE) modification of β-cyclodextrin (β-CD) enhances the binding capacity of some water insoluble drugs, thereby enhancing their solubility. The SAE-β-CD derivatives contain either sulfopropyl ether (SPE) or sulfobutyl ether (SBE) groups in the 2-, 3-, and 6-hydroxyl positions of the glucose moieties. SAE-β-CD is a mixture of positional and regional isomers containing from one to as many as twelve sulfoalkyl ether (SAE) groups per cyclodextrin. The effect of chain length and the degree of substitution on complexation behavior was investigated using a miniaturized phase solubility method. Unlike the parent β-CD, linear increases in the apparent solubilities of these medicinal agents were observed with binding potentials being comparable to those of β-CD and better. Generally, the binding potential of these derivatives increase with increasing alkyl chain length and reach a maximum with increasing degree of sulfoalkyl substitution.

1 INTRODUCTION

β-CD is a naturally occurring cyclic oligosaccharide consisting of seven α-1,4-linked glucopyranose units. These compounds have long been known to increase the apparent aqueous solubility and/or chemical stability of various medicinal agents through non-covalent inclusion complexation. Pharmaceutical scientists have utilized this property of cyclodextrins to address solubility and stability related formulation problems. Non-modified cyclodextrins, however, have limited use due in part to the low aqueous solubility of β-CD. In an attempt to overcome these limitations of β-CD, a series of sulfoalkyl ether β-CD (SAE-β-CD) derivatives have been developed,

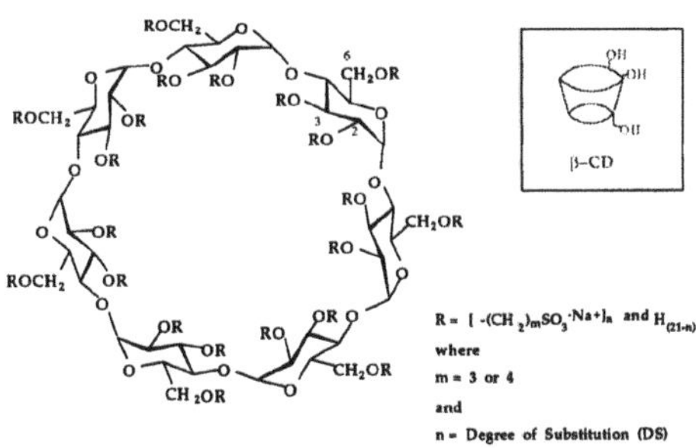

Figure 1. General structure of Sulfoalkyl Ether β-CDs.

which exhibit high aqueous solubility and greater safety [1, 2]. In the present work, the phase-solubility behavior of testosterone and progesterone were investigated in the presence of slufopropyl (SPE-β-CD) and sulfobutyl (SBE-β-CD) ether derivatives of β-cyclodextrin. Therefore, the effect of the degree of sulfoalkyl ether substitution and

J. Szejtli and L. Szente (eds.), Proceedings of the Eighth International Symposium on Cyclodextrons, 259–262.

proximity of the highly hydrated and negatively charged sulfonate group to the cyclodextrin torus on the solubilizing ability of these derivatives is presented.

2 MATERIALS AND METHODS

2.1. Materials

All chemicals were of analytical or reagent grade and used without further purification unless otherwise noted. The preparation and characterization of the sulfoalkyl ether derivatives of β-cyclodextrin have been described elsewhere [1-6]. The sulfoalkyl ether derivatives of β-cyclodextrin described have the general structure depicted in Figure 1. Progesterone and testosterone were obtained from Sigma Chemical Company (St. Louis, Missouri).

2.2. Methods

Characterization of SAE-β-CD derivatives The characterization of SAE-β-CDs have been described else where [3-6]. A specific example is SBE4-β-CD, as depicted in Fig.2, which consists of an average degree of four SBE groups on the cyclodextrin torus, while having from one to as many as ten degrees of substitution. The individual bands designated as SBE1, SBE2, etc. indicate the degree of substitution and hence the magnitude of negative charge of the cyclodextrin. These bands, seperated by ion-exchage chromatography were used in this study as well as the product mixture.

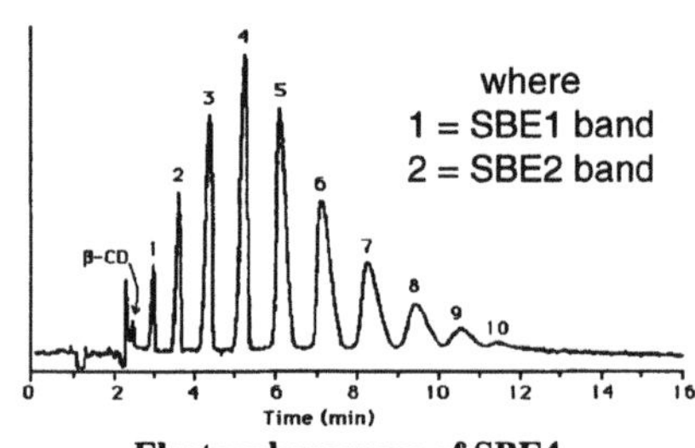

Electropherogram of SBE4 mixture

Figure 2. Representitive electropherogram of SAE-β-CDs.

Phase-Solubility Analysis Aqueous solutions of β-CD and its derivatives were prepared in the range of 0.0 to 50.0 mM in 0.05 M phosphate buffer at pH 6.5 with a constant ionic strength of 0.15. Excess amount of solid drug was placed in 250 ul of each solution. The system was allowed to equilibrate at 25°C for a minimum of 24 hours. Each system was centrifuged and the supernatant was diluted and analyzed by HPLC for the concentration of drug in solution. Phase-solubility diagrams were constructed and binding constants ($K_{1:1}$) were estimated by the method of Higuchi and Conners.

3 RESULTS AND DISCUSSION

3.1. Results

Chemical modification of β-CD to yield more water soluble derivatives resulted in observed A-type behavior for both testosterone and progesterone solubility while the unmodified cyclodextrin displayed B-type behavior.

The effect of the degree of substitution on the ability of the SBE-β-CD derivatives to solubilize testosterone and progesterone is depicted in Figure 4. As can be seen, the binding potential of progesterone is independent of the degree of substitution. In contrast, the binding potential of testosterone increases with an increase in the degree of substitution, indicating the binding potential to be highly sensitive to molecular structure. Although the results of the degree of substitution for testosterone and progesterone to their solubility potentials were not identical, the effect of alky chain lengths were similar.

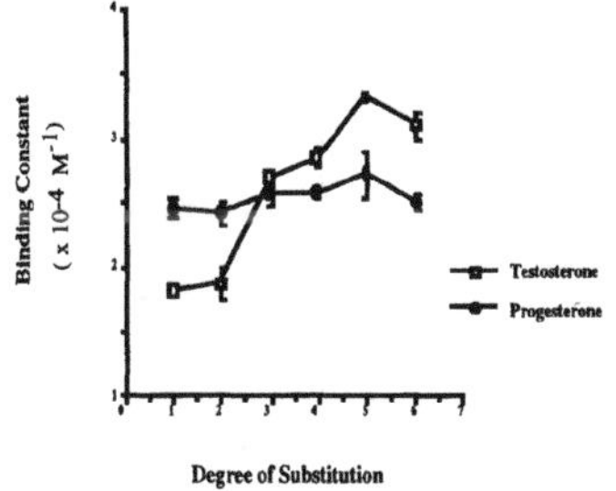

Figure 4. The effect of sulfobutyl ether substitution on the binding of progesterone and testosterone, using the seperated bands from mixtures.

The effect of alkyl chain length on the ability of the derivatives to solubilize testosterone and progesterone are depicted in Figures 5 and 6, respectively. For both testosterone and progesterone, their binding constants in the presence of SPE-β-CD reached a maximum at four degrees of substitution. However, in the presence of SBE-β-CD, both testosterone and progesterone exhibited an increase in binding potential until they reach a plateau at seven degrees of substitution. The relative binding potentials of the SPE-β-CD derivatives with progesterone and testosterone were highly sensitive to the degree of sulfoalkyl substitution, while the SBE-β-CD derivatives were less sensitive. In comparison of SPE-β-CD to SBE-β-CD, in all cases SBE-β-CD exhibited higher binding potential over the SPE-β-CD especially for the highly substituted SBE-β-CD derivatives.

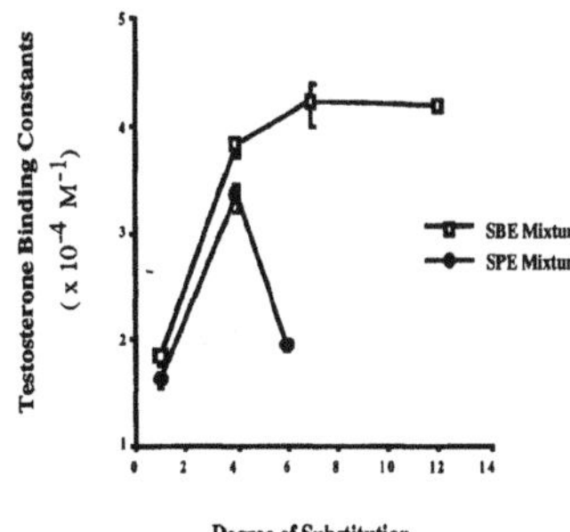

Figure 5. The effect of degree of substitution on testosterone when comparing SBE to SPE-β-CD, using product mixtures.

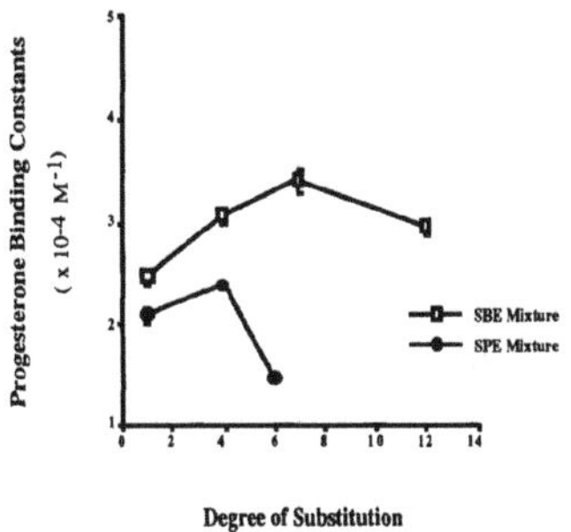

Figure 6. The effect of degree of substitution on progesterone when comparing SBE to SPE-β-CD, using product mixtures.

3.2. Discussion

The SPE-β-CD and SBE-β-CD derivatives exhibit dependence of binding potential on degree of substitution. With these derivatives the effect that the degree of substitution contributes to decreased binding potentials is apparently attenuated when progressing from a sulfopropyl ether spacer to a sulfobutyl ether spacer, illustrated in Figures 5 and 6. The binding constants for the SBE-β-CD derivatives with progesterone and testosterone are significantly larger than those of the SPE-β-CD derivative. These results suggest that if the sulfonic acid moiety and the cyclodextrin torus are separated by a critical distance, the charge density resulting from the sulfoalkyl substitution will have a minimal effect on the relative potential binding capacity of the modified cyclodextrin.

One of the major driving forces for inclusion complexation by cyclodextrins is proposed to be expulsion of "enthalpy rich" water molecules from the cyclodextrin cavity upon inclusion of the guest molecule in the cavity [7]. Chemical modification of cyclodextrins resulting in the attachment of charged substituents being attached directly to the cyclodextrin ring may inhibit the escape of the "enthalpy rich" included water molecules by capping the cavity opening with an aqueous solvation shell associated with the charged substituents. The charged substituents may also provide a hydrogen bonding source for the included water molecules thereby decreasing the energy difference between these water molecules and the bulk solution resulting in a lowering of the driving potential for inclusion complexation. Theoretically, decreases in the binding potential of modified cyclodextrins possessing charged substituents should be attenuated by increasing the distance between the charged substituents and the cyclodextrin torus. The observed binding potentials of the sulfoalkyl ether cyclodextrin derivatives is consistent with the proposed effects of decreasing the relative enthalpy of the included water molecules.

Chemical modification of cyclodextrins may also result in distortion of the geometry of the cyclodextrin torus resulting in a derivative which exhibits compromised potential for inclusion complexation. Distortion of the cyclodextrin torus may result in decreased hydrophobic interactions between the guest molecule and the interior cavity of the cyclodextrin resulting in a decrease in the driving potential for inclusion complexation. Increasing the distance between the charged substituent and the cyclodextrin torus by chemical spacers should decrease this avenue of decreased binding potential due to the relative increase in the conformational degrees of freedom of these molecules in solution. Again the observed binding potentials of the sulfoalkyl ether cyclodextrin derivatives is consistent with the proposed effects of increasing the conformational degrees of freedom of the respective cyclodextrin derivative.

4 CONCLUSION

For both testosterone and progesterone, the binding potential was not only dependant on the degree of substitution, but also the alkyl chain length. Synthesis of modified β-CD with the highly charged and hydrated sulfonate groups near the entrance of the cavity not only decreased the abiltiy of inclusion complexation, but was also magnified with an increase in the number groups placed on the CD torus. However, as the sulfonate group was moved further away from the cavity entrance, the binding potential was increased and was also less sensitive to the degree of substitution.

REFERENCES

[1] R. A. Rajewski. Development and Evaluation of the Usefulness and Parenteral Safety of Modified Cyclodextrins. Ph.D. Dissertation, The University of Kansas, 1990.

[2] R. A. Rajewski and V. J. Stella, Derivatives of Cyclodextrins and Pharmaceutical Uses thereof. in 5,134,127, 1992 U.S. Patent.

[3] R. J. Tait, D. J. Skanchy, D. O. Thompson, N. C. Chetwyn, D. A. Dunshee, R. A. Rajewski, V. J. Stella, and J. F. Stobaugh. Characterization of sulfoalkyl ether derivatives of b-cyclodextrin by capillary electrophoresis. J. Pharm. Biomed. Anal. 10(9) :615-622 (1992).

[4] E. A. Luna, R. J. Tait, D. O. Thompson, V. J. Stella and J. F. Stobaugh. "Evaluation of the utility of capillary electrophoresis (CE) for the analysis of cyclodextrin mixtures", J. Pharm &Biomed. Analysis., submitted.

[5] E. A. Luna, D. Vander Velde, R. J. Tait, D. O. Thompson, V. J. Stella. "Isolation and characterization of sulfobutylether mono derivatives of b-cyclodextrin", submitted to Carbohydrate Research.

[6] E. A. Luna, E. R. Bornancini, D. O. Thompson, V. J. Stella. " Fractionation and characterization of sulfobutylether b-cyclodextrin mixtures", submitted Carbohydrate Research.

MOLECULAR MECHANICS STUDIES ON CYCLODEXTRIN COMPLEXES: INTERACTION OF CROCETIN WITH CYCLODEXTRINS

L. JICSINSZKY[1], H. HASHIMOTO[2], K. MIKUNI[2], I. BAKÓ[3], L. SZENTE[1]

[1] CYCLOLAB R.&D. Lab. Ltd., H-1525 Budapest, P. O. Box: 435, Hungary

[2] Ensuiko Sugar Ref. Co. Ltd., Yokohama, Japan

[3] CRIC-HAS, H-1525 Budapest, P. O. Box: 17, Hungary

ABSTRACT

Crocetin is a light sensitive natural colorant. The aqueous crocetin/cyclodextrin solutions showed increased stability under stressed irradiation conditions. Molecular mechanics geometry optimizations are used to for modeling the interactions between crocetin and cyclodextrins. The energies of obtained geometries suggest that the α-cyclodextrin has the strongest interaction to crocetin among the studied cyclodextrins. These results corresponds to the experimentally observed enhanced photoresistance of crocetin/α-cyclodextrin solutions.

1. INTRODUCTION

The protection of natural colorants from light, heat, and oxygen is essential for the food technology. The low chemical stability of these compounds is usually accompanied by very low aqueous solubility. Their complexation with cyclodextrins is an obvious method to improve these properties. The crocetin/cyclodextrin complexes are chosen to get insight into the probable structure of such group of complexes.

2. MATERIALS AND METHODS

The AM1 semiempirical geometry optimizations of cyclodextrins were performed on an IBM RS6000/m350 computer using MOPAC 6.0 [1-2]. AM1 geometry optimization of crocetin an molecular mechanics calculations and geometry characterizations were performed on a i586-100MHz computer using HyperChem® (Release 4.5) [3]. Molecular lipophylicity potentials and transfer energies were calculated and visualized by WinMGM® 1.0.d [4].

MM calculations were performed by enhanced MM2 method implemented in HyperChem® using the point charge option. Standard (TIP3P) water molecules (340) were added into a 56*56*56 $Å^3$ periodic box containing one β-cyclodextrin and one crocetin molecule. In order to decrease the occurrences of false minima, the energy criteria for the optimizations were chosen enough low for the optimization algorithm (less than 0.001 Kcal/mol/Å) to get sensible minima among the lowest energy states.

J. Szejtli and L. Szente (eds.), Proceedings of the Eighth International Symposium on Cyclodextrons, 263–266.

3. RESULTS AND DISCUSSION

Starting geometries of the components were calculated by AM1 semiempirical method. The electronic properties were similar in all cases but the cyclodextrin geometries calculated by AM1 method were found to be better than PM3 geometries [1] and therefore the AM1 calculated structures were chosen as starting point for the molecular mechanics (MM) calculations.

The Molecular Lypophylic Potentials (MLPs) and Molecular Electrostatic Potentials (MEPs) were calculated for the all optimized starting structures in order to decrease further the sterically and electronically allowed configurations of molecules. The starting more than 400 theoretical arrangements therefore could be reduced to about 200. A detailed analysis of MLPs and MEPs showed that some electronically favored configurations led to unfavorable hydrophobic interactions From 210 reasonable arrangements of the host and guest less than 30 were found to be significantly different and those were chosen for further studies.

In order to compare the "*solution*" and "*in vacuo*" optimizations further optimizations were performed for several structures after the removal of water. The "*solution*" and "*in vacuo*" optimizations showed no significant differences for the chosen configurations. The nature of MM calculations does allow only the comparison of relative energies of the crocetin/cyclodextrin systems so the results are rather qualitative than quantitative.

Preliminary conformational analysis of the resulted best conformations confirmed the presumption of small energy gradient criteria for the MM optimizations.

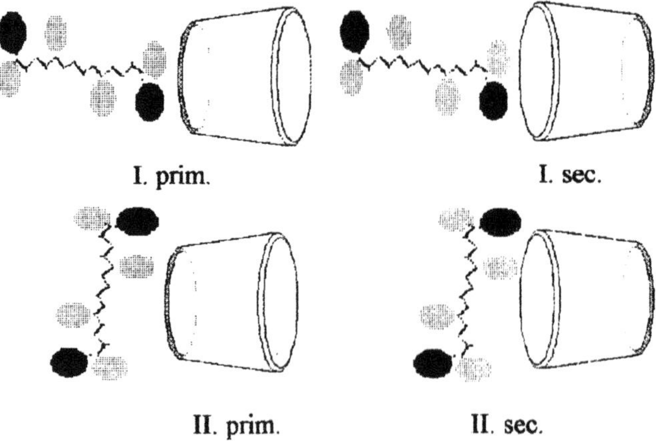

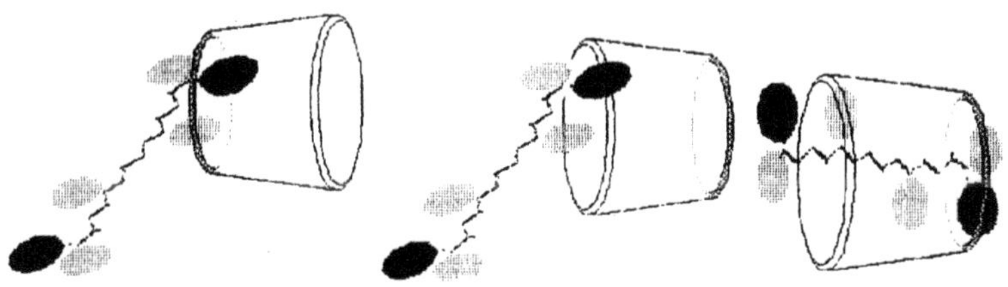

Fig. 1.: Schematic View of the Lowest Energy States of Crocetin/Cyclodexrin Systems in the Aqueous Periodic Box Conditions (Waters are Removed)

Table I.: Relative Energy Differences (Kcal/mol) of the Lowest Energy States (I.-IV.)

	αCD with water	βCD with water	γCD with water	αCD without water	βCD without water	γCD without water
I. prim.	-115.7	-44.4	112.2	37.5	22.1	30.4
I. sec.	-90.4	-110.6	72.4	43.9	17.6	27.7
II. prim.	0.0	0.0	0.0	0.0	0.0	0.0
II. sec.	-17.8	-68.5	98.7	1.3	-0.7	2
III. prim.	-174.6	429.8	527.4	12.7	601.2	576.4
III. sec.	-68.8	-100.2	102.4	39.5	16.4	33.8
IV. (complex)	-64.4	-77.4	119.5	15.6	8.5	25.2

TABLE II.: ENERGY PROFILE OF THE STARTING CONFORMATIONS

	Crocetin	αCD	βCD	γCD
Atom Transfer Energy [Kcal/mol] (Hydrophobic → Hydrophylic)	38.8	6.7	7.9	9.0
Total Solvation Energy [Kcal/mol]	-55.4	-290.5	-336.7	-352.6
Internal Solvation Energy [Kcal/mol]	-53.8	-290.2	-336.2	-349.4
External Solvation Energy [Kcal/mol] (probe: water oxygen)	-1.6	-0.3	-0.6	-3.1
Hydration Energy [Kcal/mol]	-11.6	-58.8	-64.3	-74.0
Molecular Surface Area [$Å^2$]	645	956	1079	1287
Molecular Volume [$Å^3$]	1063	2032	2332	2694
Polarizability [$Å^3$]	38	81	94	107
Refractivity [$Å^3$]	45	195	227	259
lg P	5.7	-7.3	-8.5	-9.7

4. CONCLUSION

Results of our calculation show - in concordance with experimental observations - that a carotene-like colorant, as the crocetin, favors the α-cyclodextrin in aqueous solutions. According to MM calculations a) the energy difference between the "free" and "complexed" crocetin/α-cyclodextrin systems is larger (and negative) than the same for β-, and γ-cyclodextrin systems using "solvated" molecules; b) the "*in vacuo*" energy differences, though they are positive, two order of magnitude higher for β-, and γ-cyclodextrins than that of for α-cyclodextrin; c) only small (almost zero) differences can be found between both the free, solvated crocetin/free, solvated cyclodextrins, as well as the arrangements of molecules where the crocetin is in the proximity of cyclodextrins. The results are relevant to the experimentally observed improved light-stability of crocetin in aqueous αCD solutions in comparison with both the free crocetin and crocetin/β-, and γCD solutions (*see Fig. 1.*, based on [5]).

MLPs of crocetin and cyclodextrins show that the hydrophylic parts dominate in both side of cyclodextrins while crocetin is mainly hydrophobic. For the real inclusion complexation the hydrophobic crocetin, which has only small hydrophylic parts, must go through a strong hydrophylic barrier. The calculated MLP for the optimized geometries by quantum chemical method is in accordance with the published results for crystalline cyclodextrins [6-7].

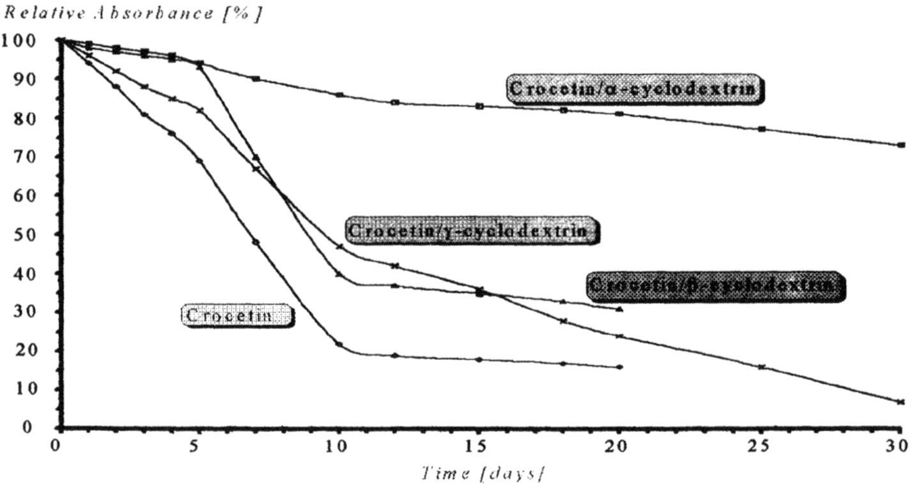

Fig. 1.: Decolorization of Crocetin and Crocetin/Cyclodextrin Complexes under Fluorescent Light

REFERENCES

1. I. Bakó. L. Jicsinszky. Semiempirical Calculations ... *J. Incl. Phenom.* **18**, 275-289 (1993)
2. QCPE:455
3. HyperChem® is a product of Hypercube Inc., Waterloo. Ontario. Canada
4. WinMGM® is a product of Ab Initio Technology (1, Place de l'Etoile. F-67210. France) and written by R. Brasseur, M. Rahman (CBMN-FSA Pass. Des Déportés 2, B-5030 Gembloux, Belgium)
5. H. Hashimoto. Stabilization of .... 1st Int. Symp. Natural Col.. January 1993. Amherst. Mass. USA
6. F. W. Lichtenthaler. S. Immel, Molecular modeling ..., *Tetrahedron: Asymmetry* **5**, 2045-60 (1994)
7. S. Immel, J. Brickmann, F. W. Lichtenthaler. Small-Ring ..., *Liebigs Ann.* **1995** 929-942

CHEMICALLY MODIFIED CYCLODEXTRINS AS CATALYTIC ENZYME MIMICS

HERBERT H. SELTZMAN AND ZDZISLAW M. SZULC
Chemistry and Life Sciences Group
Research Triangle Institute
Research Triangle Park, North Carolina 27709, USA

ABSTRACT

β-Cyclodextrin was chemically modified with both primary rim caps and secondary rim iodosobenzoic acid catalytic moieties to explore the potential cooperativity of binding and catalysis effects to achieve the kind of synergy typical of natural enzymes. The ability of these enzyme mimics to scavenge pinacolyl methylphosphonofluoridate (soman) was tested. Synergism was seen in DIMEB analogs while dominance of the scavenging by the capping effect was seen with the biphenyldisulfonyl capped analogs.

1. INTRODUCTION

In enzymes, the structural components that influence binding and catalysis are separate but proximal features that operate cooperatively to provide the overall efficiency in effecting chemical transformations of substrates. Similarly for β-cyclodextrin (βCD), both the modifications that enhance binding and those that introduce catalysis and higher reactivity can be introduced onto βCD so as to have each contribute its facet cooperatively to the activity of the enzyme mimic. In this paper we report the syntheses of βCDs with both binding and catalysis modifications and their activity in the hydrolysis of the phosphonofluoridate soman, the latter being an irreversible inhibitor of acetylcholinesterase (AChE).

These enzyme mimics evolved from our prior studies which demonstrated that 1) βCD scavenges soman, 2) scavenging can be enhanced by capping, and 3) true catalysis can be introduced into βCD enzyme mimics in the hydrolysis of soman. Thus, βCD scavenges soman in vitro from aqueous media at physiological pH and temperature affording a 72% protection of the activity of AChE at 10^{-3} M.[1,2,3] Scavenging was shown to be due to irreversible[4] phosphonylation of soman to the hydroxyls on the secondary rim of βCD.[1,5]

Dramatic improvement of three orders of magnitude in scavenging was achieved with a βCD that was rigidly capped on the primary rim with the 4,4'-biphenyldisulfonyl group affording scavenging at 10^{-6} M.[1] Other capped βCDs were also shown to have high scavenging activity for soman.[6] Interestingly, the capping group does not have a

J. Szejtli and L. Szente (eds.), Proceedings of the Eighth International Symposium on Cyclodextrons, 267–272.

nucleophilic or catalytic functional group. This meant that the improvement was due to the cap's influence on non-covalent binding which either improved in degree (binding constant)[7] or reoriented[8] the soman within the cavity to facilitate covalent bonding to the native hydroxyls. That such orientation has a marked influence upon the subsequent reaction rate of the substrate with the βCD host has been demonstrated by the results of the acylation of βCD with phenyl acetates[9,10] and the phosphonylation of αCD with sarin isomers.[11] The scavenging affords a mono-phosphonylated capped βCD adduct as demonstrated by HPLC and mass spectral evidence.[6]

The above mimics, while exhibiting high soman scavenging activity, are stoichiometric in their reaction, consuming and being consumed by one molecule of soman. The resulting covalent βCD adduct does not cleave to regenerate the original enzyme mimic. The generation of a catalytic cycle in the function of the enzyme mimic has been achieved by the incorporation of a catalytic moiety onto the secondary rim of the βCD.[12,13,14] The catalytic group employed was iodosobenzoic acid (IBA) which has been shown to function as a true catalyst for the cleavage of organophosphate esters in aqueous micellular solutions at physiological pH.[15] By varying the linker between the IBA and the βCD, as well as its position on the IBA and on the secondary rim of the βCD (C2-OH or C3-OH), a series of enzyme mimics were developed that enhanced the soman scavenging activity of βCD by nearly three orders of magnitude. The ability to hydrolyze a 500 fold molar excess of soman was demonstrated with a turnover rate of 5-13 mole soman/mole mimic/min.[16]

By combining the structural features that contribute to enhanced binding with those that contribute to true catalysis, enzyme mimics can be envisioned that could have additive or even synergistic effects in the scavenging activity in this class of compounds. Thus, it is proposed that primary rim capping can be combined with secondary rim functionalization, employing the most active lead modifications developed in the above earlier work, to afford enzyme mimics with the potential of enhanced activity in scavenging soman by a synergism between binding and catalytic centers typical of natural enzymes (Figure 1).

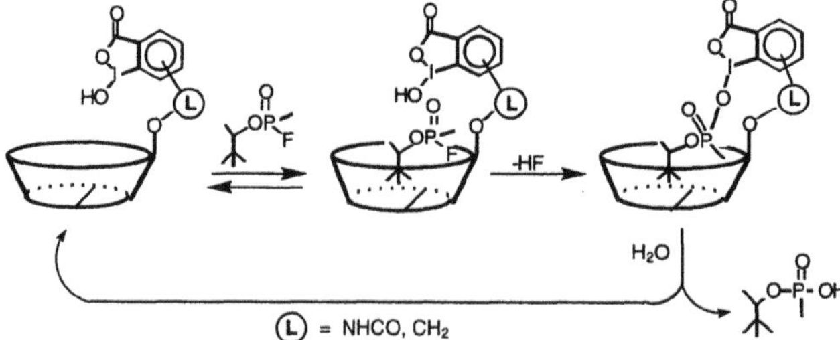

Figure 1. Capped and Iodosobenzoic Acid Conjugated β-Cyclodextrins

2. METHODS

2.1 Synthesis

The target compounds reported here are capped with either a 4,4'-biphenyldisulfonyl rigid cap in the form of 6A,6D-O, O'-(4,4'-biphenyldisulfonyl)-βCD[17] or a flexible cap as a per dimethyl ether in the form of purified heptakis (2,6-di-O-methyl)-βCD (DIMEB).[18] The conjugation of 4- or 5-isocyanato-2-iodobenzoic acid methyl ester to the rigidly capped βCD was achieved with via the tin dioxolane route shown in Figure 2 as developed for the uncapped βCD.[6] Seven regioisomers are possible from 2-O-conjugation (2A-2G conjugates of the 6A,6D-capped βCD). In the case of the 5-carbamoyl-2-iodobenzoic acid methyl ester conjugate, reverse phase chromatography resolved six fractions (one of which contained two isomers). The corresponding 4-carbamoyl conjugates were similarly prepared but isolated as the mixture of isomers. Oxidation of the iodo esters with peracetic acid yielded the corresponding iodosobenzoic acids. The compounds were characterized by their NMR spectra which exhibited a carbamoyl conjugated IBA ring and a negative ion FAB/MS which exhibited the M-H ion for the mono-conjugate as well as M-OH and M-IBA-NCO. An elemental analysis in agreement with theory was obtained for the 4-carbamoyl-IBA conjugate mixture.

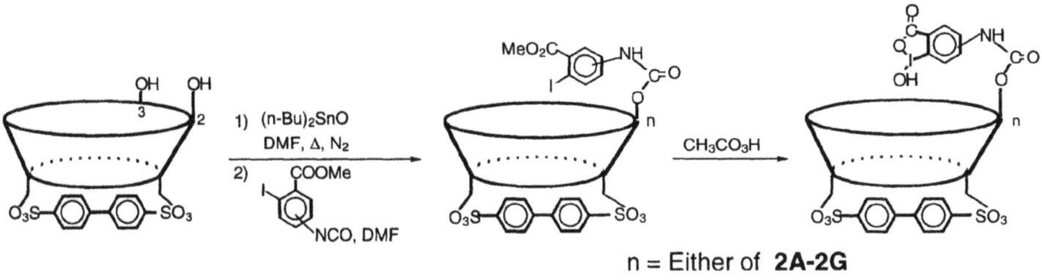

Figure 2. Synthesis of Capped and Conjugated β-Cyclodextrins

The conjugation of the above isocyanates to DIMEB showed resistance to carbamoylation when catalyzed by dimethylaminopyridine typical of some secondary alcohols. The use of the Lewis acid catalyst boron trifluoride etherate[19] afforded mono-conjugates which were oxidized with peracetic acid to yield the IBA-DIMEBs. The latter were identified by NMR which exhibited a down field shifted triplet resonance for the C3-H at the conjugation site and the expected aromatic resonances. Negative ion FAB/MS exhibited an M-H ion as well as M-OH as the base peak.

2.2 Scavenging Assays

Two assays for enzyme-like activity were conducted. The first for scavenging ability and the second for catalytic activity. The scavenging assay determined the extent to which the test cyclodextrin reacted with soman thereby preserving the activity of the subsequently added AChE.[20,21] Thus, soman (5.5×10^{-9} M) and an excess of the test cyclodextrin (10^{-7} to 10^{-3} M) in 0.01 M phosphate buffer (pH 7.4) were incubated at 37 °C for 10 min. AChE (2.2 U) was added and incubated at 37 °C for 10 min. The residual AChE activity was determined by monitoring the rate of hydrolysis of the substrate acetylthiocholine in the presence of the chromogenic reagent DTNB. The data is presented as a plot of percent protection (the percent of residual AChE activity relative to AChE activity in the absence of soman and cyclodextrin) vs concentration of the test cyclodextrin. The percent protection represents the activity of the cyclodextrin in scavenging soman. The more the curve is shifted to low concentration, the more active is the scavenger.

The catalytic activity assay determined the amount of soman consumed by a stoichiometric deficiency of the test cyclodextrin. Thus, soman (5 x 10^{-4} M) and the test cyclodextrin (1 x 10^{-6} M) in phosphate buffer were incubated at 37 °C and aliquots were removed periodically over 60 min. The aliquots were diluted to the concentration range of the above standard assay and treated with AChE and assayed as above. The mU of residual AChE activity was related to the mmoles of soman consumed, by a standard titration curve (mU activity vs mmole soman) using the assay described above. A similar assay for buffer without cyclodextrin was conducted to correct for hydrolysis by buffer. The difference between the test solution and buffer values gives the mmoles of soman scavenged by the cyclodextrin. The assay was conducted with a 500-fold excess of soman on selected 2X-O-(5-carbamoyl-2-iodosobenzoic acid)-6A,6D-O, O'-(4,4'-biphenyldisulfonyl)-βCDs.

3. RESULTS AND DISCUSSION

The results of the scavenging assay with a stoichiometric deficiency of soman, Figure 3, demonstrated that three of the five biphenyldisulfonyl capped-IBA-βCDs[22a] that were tested showed slightly greater activity than either the corresponding IBA-conjugated βCD or the biphenyldisulfonyl capped βCD at 10^{-6} M. Of the two IBA-DIMEBs[22b] that were tested, the 5-carbamoyl analog was of comparable activity to βCD and the 4-carbamoyl analog was nearly two orders of magnitude more active than βCD. In the catalysis assay, the three most active compounds from the above scavenging assay showed no measurable catalysis when a 500 fold excess of soman was employed.

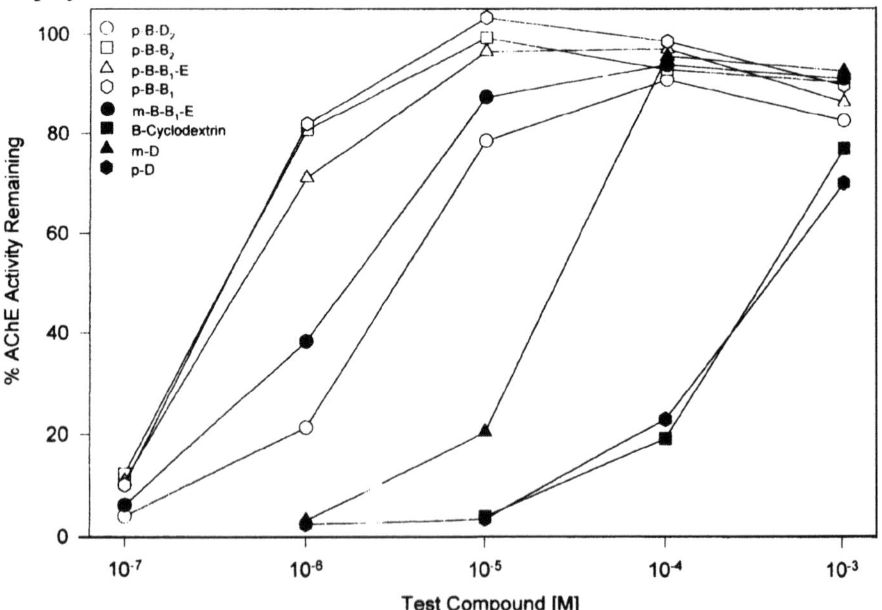

Figure 3. Scavenging of Soman by Cyclodextrin Analogs[22]

Given that the scavenging activity of the biphenyldisulfonyl capped and IBA conjugated test compounds was in the general range of that for the capped analog without the IBA moiety [6A,6D-O, O'-(4,4'-biphenyldisulfonyl)-βCD] (62% at 10^{-6} M), and that no measurable catalysis could be observed, it is concluded that the scavenging demonstrated by these compounds is due to a cap enhanced covalent bonding to the native secondary rim hydroxyls of the βCD without any significant involvement of the IBA moiety. That these particular compounds showed neither an additive nor synergistic enhancement above that of the capped-only-βCD, when these same IBA haptens conjugated to βCD without a capping group induce a two to three order of magnitude scavenging enhancement, suggests that the orienting effect of this cap upon soman is of such a significant magnitude that it fully dominates the course of the reaction to the exclusion of IBA catalysis.

In contrast to the results with the above rigidly capped βCDs, the DIMEB analogs, while not nearly as active versus the corresponding uncapped (e.g. non-methylated) IBA-βCDs, were significantly (infinitely) more active than DIMEB without an IBA moiety (0.3% at 10^{-3} M).[20] This would suggest that IBA-DIMEBs are likely functioning via an IBA mediated (e.g. catalytic) mechanism since DIMEB itself has no significant scavenging activity at the concentrations where these analogs showed 70 to greater than 90% activity. Furthermore, control experiments with IBA conjugated to a non-complex forming alcohol (methanol) showed that in the absence of an enzyme-substrate like complex afforded by βCD, IBA carbamates themselves cannot account for the observed level of scavenging.[6]

4. CONCLUSIONS

The cooperativity of the structural features that contribute to binding of a substrate in an enzyme active site and those that introduce true catalysis can afford a reactivity that exceeds the effect of either feature alone. This was seen in the IBA-DIMEB conjugates. However, a balance and complementarity between these features must be struck lest one feature, such as binding, acts in such a manner as to not complement or to oppose the chemistry and juxtapositional needs of the other. This was seen in the biphenyldisulfonyl capped-IBA conjugated βCDs tested to date where binding effects precluded effective interaction of the IBA moiety. The search for this optimal balance to afford synergism of binding and catalysis is continuing.

ACKNOWLEDGEMENT

This work is supported by the U. S. Army Medical Research and Materiel Command under Contract No. DAMD17-94-C-4012.

REFERENCES

[1] Seltzman, H. H., Koble, D. L., Hendren, R. W., Groblewski, D. B., Cyclodextrins as scavengers of soman, *Proceedings of the Sixth Medical Chemical Defense Bioscience Review*, Columbia, MD, August 4-6, 1987, p. 267-270.

[2] Seltzman, H., Annual Report to the USAMRDC on Contract No. DAMD17-84-C-4101, Dec. 1985.

[3] (a) Saint-Andre, S., Desire, B., *Inactivation du soman par la β-cyclodextrin*, C. R. Acad. Sc. Paris, t. 301, Serie III, No. 3, 67-72 (1985); (b) Desire, B., Saint-Andre, S., Interaction of soman with β-cyclodextrin, *Fund. and Appl. Toxicol.* **7**, 646-657 (1986).

[4] Seltzman, H., Annual Report to the USAMRDC on Contract No. DAMD17-84-C-4101, Aug. 1986.

[5] Voyksner, R. D., Williams, F. P., Smith, C. S., Koble, D. L., Seltzman, H. H., Analysis of cyclodextrins using fast atom bombardment/mass spectrometry and tandem mass spectrometry, *Biomed. Mass Spectrom.*, **18**, 1071-1078 (1989).

[6] Seltzman, H., *Synthesis of soman scavengers*, Final Report to the USAMRDC on Contract No. DAMD17-89-C-9012, Dec. 1992.

[7] Emert, J., Breslow, R., Modification of the cavity of β-cyclodextrin by flexible capping, *J. Am. Chem. Soc.*, **97**, 670-672 (1975).

[8] Breslow, R., Czarniecki, M. F., Emert, J., Hamaguchi, H., Improved acylation rates within cyclodextrin complexes from flexible capping of the cyclodextrin and from adjustment of the substrate geometry, *J. Am. Chem. Soc.*, **102**, 762-770 (1980).

[9] Van Etten, R. L., Sebastian, J. F., Clowes, G. A., Bender, M. L., Acceleration of phenyl ester cleavage by cycloamyloses. A model for enzymatic specificity, *J. Am. Chem. Soc.*, **89**, 3242-3253 (1967).

[10] Van Etten, R. L., Clowes, G. A., Sebastian, J. F., Bender, M. L., The mechanism of the cyclo-amylose-accelerated cleavage of phenyl esters, J. *Am. Chem. Soc.*, **89**, 3253-3262 (1967).

[11] van Hooidonk, C., Breebaart-Hansen, J.C.A.E., Stereospecific reaction of isopropyl methylphos-phonofluoridate (SARIN) with α-cyclodextrin, *Rec. Trav. Chim. Pays-Bas*, **89**, 289 (1970).

[12] Seltzman, H. H., Narula, A. S., Lonikar, M. S., Catalytic cyclodextrin enzyme mimics as soman scavengers, *Proceedings of the 1991 Medical Defense Bioscience Review*, 533-536 (1991).

[13] Seltzman, H. H., Narula, A. S., Lonikar, M. S., Cyclodextrin enzyme mimics as catalytic phos-phonofluoridate hydrolases, *Minutes of the Sixth International Symposium on Cyclodextrins*, Editions DeSante (Ed. Hedges, A. R.), Pp 677-680 (1992).

[14] Seltzman, H. H., Lonikar, M. S., True catalytic hydrolysis of a phosphonofluoridate by modified β-cyclodextrins, *Minutes of the Seventh International Symposium on Cyclodextrins*, Tokyo, Japan, April 1994, Pp 285-288.

[15] Moss, R. A., Alwis, K. W., Bizzigotti, G. O., *o*-Iodosobenzoate: Catalyst for the micellar cleavage of activated esters and phosphates, *J. Am. Chem. Soc.*, **105**, 681-682 (1983); Moss, R. A., Alwis, K. W., Shin, J.-S., Catalytic cleavage of active phosphate and ester substrates by iodoso- and iodoxybenzoates, *J. Am. Chem. Soc.*, **106**, 2651-2655 (1984); Moss, R. A., Chatterjee, S., Wilk, B., Organoiodinane oxyanions as reagents for the cleavage of a reactive phosphate, *J. Org. Chem.*, **51**, 4303-4307 (1986).

[16] Seltzman, H. H., Lonikar, M. S., Catalytic β-cyclodextrin enzyme mimics as soman hydrolases, *Proceedings of the 1993 Medical Defense Bioscience Review*, **3**,1075-1083 (1993).

[17] Tabushi, I., Shimokawa, K., Fujita, K., Specific bifunctionalization on cyclodextrin, *Tetrahedron Lett.*, 18, 1527-1530 (1977).

[18] Spencer, C. M., Stoddard, J. F., Zarzycki, R., Structure mapping of an unsymmetrical chemically modified cyclodextrin by high-field nuclear magnetic resonance spectroscopy, *J. Chem. Soc., Perkin Trans.*, **2**, 1323-1336 (1987).

[19] Ibuka, T., Chu, G.-N., Aoyagi, T., Kitada, K., Tsukida, T., Yoneda, F., A convenient Lewis acid catalyzed preparation of carbamates from secondary alcohols and isocyanates, *Chem. Pharm. Bull.*, **33**, 451-453 (1985).

[20] Seltzman H., Final Report to the USAMRDC on Contract No. DAMD17-84-C-4101 (1988).

[21] Schoene K., Titration of acetylcholinesterase with soman, *Biochem. Pharmacol.*, **20**, 2527-2528 (1971).

[22] a) p-B-fraction #, m-B-mixed isomers; p(m) = 5(4)-carbamoyl; b) m-D, p-D

CATALYTIC ESTEROLYSIS OF p-NITROPHENYL ACETATE BY β–CYCLODEXTRIN ASSOCIATED TO POLY(VINYLAMINE) AND A BENZYLATED DERIVATIVE

B. MARTEL[1] , M. MORCELLET[1], G. CRINI[2]
[1]Laboratoire de chimie macromoléculaire UA-CNRS351 Université des Sciences et Technologies de Lille 59650 Villeneuve d'Ascq France ; [2]Istituto scientifico di chimica e biochimica «G. Ronzoni » via G Colombo 81 20133 Milano Italia

ABSTRACT

We first studied poly(vinylamine)/β–cyclodextrin systems referring to their catalytic ability in p-nitrophenylacetate hydrolysis. The catalytic effect of cyclodextrin and polymer is almost equal to the addition of their separate effects when they are mixed. On the other hand, covalently bound systems were more efficient than the equivalent mixture and we deduced the occurance of a cooperative effect between the hydrophobic cavity of cyclodextrin and the nucleophilic amino groups. In a second part, we observed that mixtures of benzylated polyvinylamine and β–cyclodextrin result in an inhibition of the catalytic effect.

1. INTRODUCTION

Cyclodextrins and some of their derivatives have been reported to have an enzyme like action in ester cleavage [1] ; an apolar cavity surrounded by hydroxyl groups playing the role of the binding site and the active site, respectively. The hydrolysis of p-nitrophenylacetate (PNPA) by β–cyclodextrin (CD) follows a Michaelis-Menten kinetic [1,2]. Alkylated and arylated polyamines are reported to have a better catalytic activity in nitrophenyl ester hydrolysis than the parent polymers [3-7]. The kinetic obeys a second order mechanism. Their apolar side groups have the tendency to assemble in aqueous solutions ; the consequence is that the polymer takes a coiled conformation. This micelle like structure has the property to solubilize the substrate (PNPA) in the vicinity of the catalytic sites (amino groups) [8]. In this paper, we compared the catalytic activity of CD mixed with or covalently bound to poly(vinylamine) (PVAm) or mixed with benzylated poly(vinylamine) (PVAB).

2. MATERIALS AND METHODS

J. Szejtli and L. Szente (eds.), Proceedings of the Eighth International Symposium on Cyclodextrons, 273–276.

Kinetic measurements : The kinetic of the hydrolysis of PNPA was monitored by following the appearance of the p-nitrophenolate ion at 400nm versus time. Conditions of the reaction : pH = 8.74 ; tris buffer μ = 0.05 ; T° = 25°C [2,3]
The mechanism of Michaelis-Menten type reaction is :

$$C + S \underset{k_{-1}}{\overset{k_1}{\rightleftharpoons}} CS \xrightarrow{k_2} CS' + P$$

where C = cyclodextrin, S = Substrate ;

k_2 and the Michaelis constant $K_m = (k_{-1}+k_2)/k_1$ were determined by means of the Lineweaver-Burk equation :

$$\frac{1}{(k_{obs} - k_{un})} = \frac{K_m}{k_2.[C_o]} + \frac{1}{k_2}$$

For the PVAm or PVAB catalysed reactions, the pseudo first order constant k_{obs} followed the equation $k_{obs} = k\ [cat] + k_{un}$ where [cat] is the polyamine concentration and k_{un} is the uncatalysed reaction constant rate.
Synthesis of **PVACD1** *to* **5** : We applied a method reported previously [9]; monotosylated cyclodextrin was reacted with PVAm to yield copolymers of different degrees of substitution (DS) as reported in table 1.
Synthesis of **PVAB** : Benzyl chloride was reacted with PVAm in methanol as reported in [3]. A copolymer with 54% of N-benzylated groups was obtained.

3. RESULTS AND DISCUSSION

Mixture of CD and PVAm : The catalytic power of CD was investigated in the presence of PVAm ; a Michaelis-Menten type kinetic was observed (figure1). The values of the ratio k_2 / K_m (table 1) showed that the polyamine slightly reduces the catalytic activity of CD. The increase of the k_2 constant is cancelled by the destabilisation of the intermediate complex (K_m increases). In other words, the polymer partly hinders the formation of the inclusion complex.

Catalyst	DS(mol%)	$k_2(x10^3\ s^{-1})$	$K_m(x\ 10^3M)$	$k_2/K_m(M^{-1}.s^{-1})$	$NH_2/OH(CD)$
PVACD1	0.3	24	10.5	2.30	16
PVACD2	0.9	11	8.3	1.30	5
PVACD3	1.5	9	11.5	0. 80	3
PVACD4	2	6	4.5	1.30	2
PVACD5	5	1.6	7.1	0.20	1
CD	-	0.5	8.5	0.060	-
CD + PVAm[a]	-	0.75	18.2	0.040	-

Table 1 Kinetic constants from the Michaelis-Menten equation. (a) : [PVAm] = 10^{-2} M

Covalent system PVACD : Figure 3 reports the apparent reaction rate k_{obs} against the concentration of CD for a covalent and a non covalent system where the ratio

CD/NH_2 is 1/50 in both cases (PVACD4 here used). From the general aspect of the plots, a different mechanism can be observed. The reaction catalysed by the non covalently bound system proceeds according to a second order kinetic (straight line), indicating that PVAm is the main catalyst. On the other hand, a Michaelis-Menten type reaction occurs when PVACD4 is the catalyst, meaning that the reaction occurs through a complex intermediate. The Lineweaver-Burk plots were obtained for the different PVACD systems and the resulting calculated constants were reported in table 1. The results report that the catalytic power (or the ratio k_2/K_m) is inversely proportional to DS and proportional to the ratio NH_2/OH. Table 1 shows that the catalytic power of the PVACD systems differ by the difference of the value of k_2 whilst K_m remains constant. The results confirm that amino groups are involved in the cleavage of the ester function of the encapsulated substrate (k_2). Furthermore, if the values of K_m are compared between the covalent and the non covalent systems (table 1) the stability of the intermediate inclusion compound is two fold that measured for the covalent system and remains unchanged compared to native CD. At last, when considering both plots in figure 2, in the CD concentration range studied, the catalytic power of the grafted system is more important than that of the mixture CD-PVAm. So, we propose that the mechanism occurs through a cooperative action of the cavity of CD and the amino groups of the polymer. This is in agreement with what was reported by Seo et al. [10,11] who studied a covalent Poly(allylamine)-CD system.

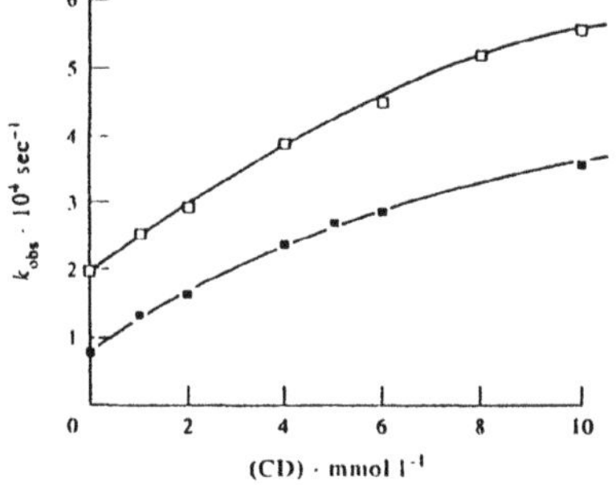

Fig. 1 k_{obs} against CD concentration (■) β–CD alone ; (□) β–CD+ PVAm (=10^{-2} M)

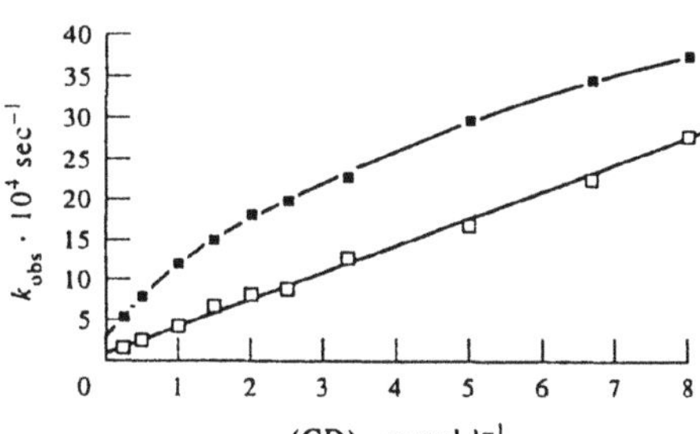

Fig. 2 k_{obs} for a covalent system PVACD4 (■) and for a mixture of PVACD and PVAm (□)

Mixture of PVAB and CD : We show in figure 4 that the partly benzylated PVAm (DS = 54%) has an important catalytic activity in the hydrolysis of PNPA. The second order constant k_{obs} for PVAB is 20 fold that measured for the PVAm catalysed reaction. Figure 3 reports the variation of the first order constant without and with CD. CD obviously inhibits the reaction. Capillar viscosimetric measurements showed that the viscosity of a solution of PVAB sharply decreased when CD is added. So, CD involves a more compact conformation of the polymer. The addition of urea into the PVAB/CD solution makes the viscosity increase. So, urea, by the destruction of the hydrogen bonds, involves the stretching of the polymer backbone. As a matter of fact, the initial compact conformation is supposed to consist of benzyl groups included into CD moieties bound to each other by a dense hydrogen bonding network. This hypothesis has been partly confirmed by fluorimetric measurements : opposite to what was observed in PVAB solutions, the fluorescent probe is not solubilized into the polymer hydrophobic microenvironment neither included into the CD cavity when they are mixed [3]. So, there is a reciprocal interaction that destroys the solubilizing power of PVAB and the inclusion ability of the CD so that the respective catalytic effect of both compounds is quenched.

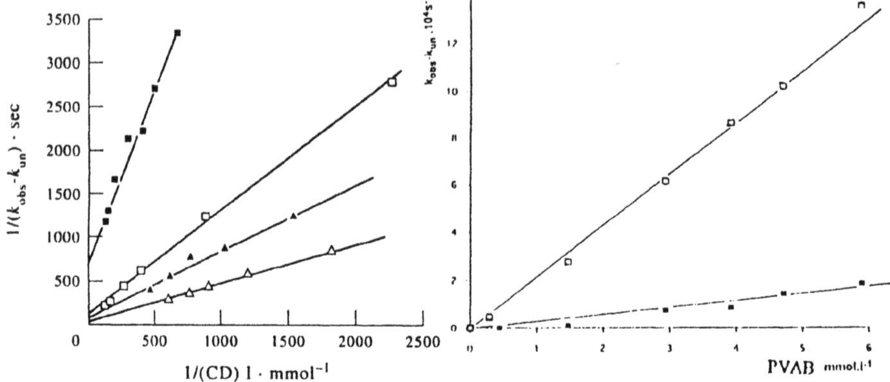

Fig. 3 Lineweaver-Burk plots obtained from the reactions catalyzed by (△) PVACD1 ; (▲) PVACD2 ; (□) PVACD3 ; (■) PVACD4

Fig. 4 k_{obs} for the reaction catalysed by (□) PVAB ; (■) PVAB+CD

CONCLUSION

In this paper, we studied the behaviour of mixed and covalent systems in the catalytic hydrolysis of PNPA. CD, PVAm and PVAB have a separate catalytic activity. For PVAm-CD systems a cooperative action was observed when they are covalently bound. Mixtures of PVAB and CD result in an inibition of their respective catalytic effect.

REFERENCES

[1] Bender M.L. , Komiyama M., *Cyclodextrin Chemistry*,New York, 1977

[2] Martel B., Morcellet M., *Eur. Pol. J.* **31**, 1089-1093 (1995)

[3] Martel B., Pollet A., Morcellet M., *Macromolecules*, **27**, 5258-5262 (1994)

[4] Everaerts A., Samyn C., Smets G., *Makromol. Chem.*, **185**, 1881 (1984)

[5] Koltz I.M., Strykcr V.H., *J. Am. Chem. Soc.* **90**, 2717 (1968)

[6] Seo T., Kajihara T., Iijima T., *Makromol. Chem.* **191**, 1645 (1990)

[7] Pshezhetskii V.S., Lukyanova A.P., Kabanov V.A., *J. Molec. Catal.*, **2**, 49 (1977)

[8] Seo T, Take S., Miwa K., Hamada K., Iijima T. *Macromolecules* **24**, 4255 (1991)

[9] Martel B., Leckchiri Y., Pollet A., Morcellet M., *Eur. Pol. J.* **31**, 1083-1088 (1995)

[10] Seo T, Kajihara, Iijima T., *Makromol.Chem.* **188**, 2071 (1987)

[11] Seo T, Kajihara, Iijima T., Miwa K., *Makromol.Chem*, 192, 2357 (1991)

CYCLODEXTRIN HOMO- AND HETERO-DIMERS AS ENZYME MODELS

H. IKEDA,* S. NISHIKAWA, J. TAKAOKA, T. AKIIKE, Y. YAMAMOTO, A. UENO, and F. TODA†

Department of Bioengineering, Tokyo Institute of Technology, Nagatsuta-cho, Midori-ku, Yokohama 226, Japan
† Polytechnic College, 2-32-1, Ogawanisi, Kodaira 187, Japan

ABSTRACT

Three kinds of dimers (one homo-dimer of β-CD and two kinds of hetero-dimers of α-CD and β-CD) were synthesized as artificial hydrolases. The dimers were prepared by the condensation of 6-deoxy-6-(L-histidylamino)-cyclodextrin and 6-(carboxymethylthio)-6-deoxycyclodextrin with dicyclohexylcarbodiimide. The enzyme-like activities were studied by measuring the rates for the cleavage reaction of some kinds of nitrophenyl alkanoates. They showed large acceleration ability and substrate specificity for the acyl chain length of the substrate. As the acyl chain length of the substrate was longer, the transition state was more stabilized by the dimers. The only homo-dimer showed allosteric behavior with 1.8 Hill constant, when *p*-nitrophenyl methoxyethoxyethoxyacetate was used for the substrate.

1. INTRODUCTION

Cyclodextrins (CD's) have been extensively studied as enzyme models and as molecular receptors due to their abilities to bind various hydrophobic compounds into their hydrophobic cavities. We have been studied imidazole-modified cyclodextrins, which show maximum activity for ester cleavage reactions in neutral pH region and the turnover property. The most successful enzyme model in our studies is imidazole-appended 2,6-dimethyl-β-cyclodextrin [1]. This model caused about 1000-fold acceleration of the hydrolysis of *p*-nitrophenyl acetate, and k_{cat} for this reaction of this model is 2.67 x 10^{-2} s^{-1}, which is over twice as much as that of α-chymotrypsin.

J. Szejtli and L. Szente (eds.), Proceedings of the Eighth International Symposium on Cyclodextrons, 277–280.

However K_m of the former is much bigger than that of the latter. Therefore, the total reaction activity k_{cat}/K_m of the former is still smaller than that of the latter. This means that the binding ability of the enzyme model is much smaller than that of the natural enzyme, whereas the catalitic activity of the former after making inclusion complex is larger than that of the latter. So, if the binding ability of an enzyme model can be improved, total catalitic acceleration of the model will be larger than that of the natural enzyme. Recently O. S. Tee reported that 2 : 1 (CD/substrate) complex played an important role for the cleavage of aryl alkanoates. Some CD dimers have very strong binding ability. So, imidazole-appended CD dimers would be expected to be better artificial hydrolases. In this symposium, we present the syntheses of CD homo- and hetero-dimers and these enzyme-like activities.

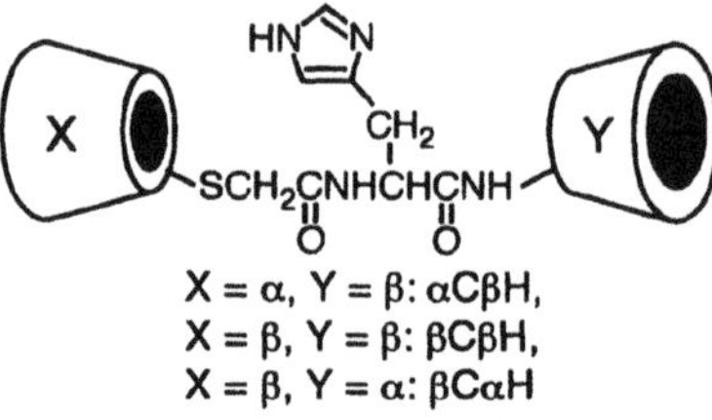

2. MATERIALS AND METHODS

2.1 Materials

Imidazole-appended cyclodextrin dimers were prepared by the condensation of 6-deoxy-6-(L-histidylamino)-cyclodextrin and 6-(carboxymethylthio)-6-deoxy-cyclodextrin with dicyclohexylcarbodiimide (DCC) and 1-hydroxybenzotriazole (HOBt). 6-Deoxy-6-(L-histidylamino)-cyclodextrin was obtained by the reaction between 6-amino-6-deoxy-cyclodextrin and N^{α}-*tert*-butyloxycarbonyl-N^{im}-tosyl-L-histidine in the presence of DCC and HOBt followed by deprotection. 6-(Carboxymethylthio)-6-deoxy-cyclodextrin was synthesized from 6-deoxy-6-iodo-cyclodextrin and methyl sodium sulfidoacetate followed by demethylation. These compounds were identified by elemental analysis, some kinds of NMR spectra including 2D NMR, and mass spectra.

2.2 Methods

Ester cleavage reaction was followed by monitoring the appearance of nitrophenolate ion using a Simadzu UV-3100 spectrophotometer. The reaction was conduced in a quartz cell in the water-jacketed cell holder of the UV-3100. The temperature was maintained at 25 °C by a HAAKE F3 circulating water bath. The reaction was initiated by adding a stock solution of the ester in acetonitrile or DMSO to a buffer solution in the quartz cell. The rates used in the calculation of rate constants were averages of at least three determinations which agreed within 3 %.

3. RESULTS AND DISCUSSION

3.1 Cleavage Reaction of *p*-Nitrophenyl Alkanoates

The enzyme-like activities of three kinds of dimers were studied by measuring the rates for the cleavage reaction of some kinds of nitrophenyl alkanoates in a 16.1 % DMSO aqueous phosphate buffer solution of pH 7.8 at 25 ° C. The reaction proceeded by a Michaelis-Menten mechanism, showing saturation behavior with increasing substrate concentration. Table 1 shows the dependence of kinetic parameters on the acyl length

TABLE 1 Kinetic Parameters for the Cleavage of *p*-Nitrophenyl Alkanoates by Cyclodextrin Homo- and Hetero-dimers

catalysis	substrate	k_{cat} / 10^{-4} s^{-1}	K_m / 10^{-4} M	k_{cat}/K_m / $M^{-1}s^{-1}$	K_{TS} / 10^{-5} M	k_{cat}/k_{un} / -
αCβH	C2	2.54	9.14	0.278	4.89	18.7
	C4	1.13	3.31	0.341	1.77	18.7
	C6	1.73	1.70	1.02	0.726	23.4
βCβH	C2	8.04	38.7	0.208	6.98	55.4
	C4	2.32	5.53	0.420	1.54	35.9
	C6	4.76	5.78	0.824	0.939	61.6
βCαH	C2	1.47	8.32	0.177	8.21	10.1
	C4	0.995	1.83	0.544	1.19	15.4
	C6	0.815	1.19	0.685	1.13	10.5

$C_nH_{2n+1}-C(=O)-O-C_6H_4-NO_2$ C2: n=1 C4: n=3 C6: n=5

Substrate (**1**)

of the esters.

The transition state stabilization (K_{TS}) of the cleavage reaction was evaluated using the Kurz approach as presented by O. S. Tee. As the acyl chain length of the substrate was longer, the transition state was more stabilized by the dimers.

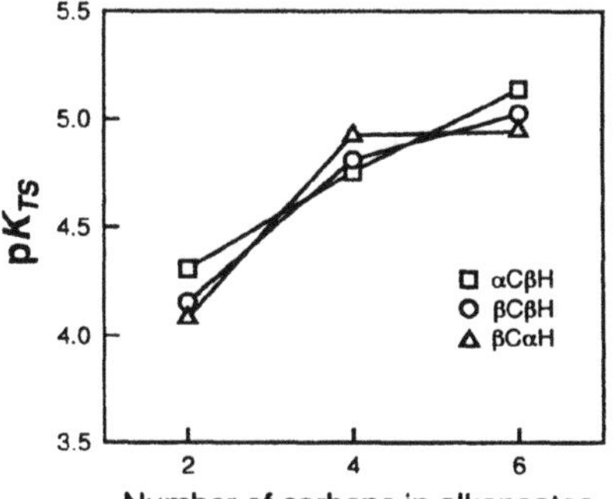

Fig. 1 Dependence of pK_{TS} on acyl length of the substrates for the cleavage of *p*-nitropheyl alkanoates by the cyclodextrin dimers.

3.2 Allostery of Homo-Dimer

The presence of some cooperativity between two CD cavity of the CD dimer is suggested by the result of dependency on the acyl chain length of the substrates in the reaction, but it could not confirmed by the reaction with the alkanoate esters, because of low solubility of the ester in water. So, we synthesized a new ester that can be more soluble in water and the cleavage ability of three kinds of dimers were studied in high substrate concentration condition. The only homo dimer showed sigmoid behavior with increasing substrate concentration as shown in Fig. 2, whereas two kinds of hetero-dimers showed simple saturation behavior. This sigmoidal behavior was fitted to the Hill equation and the Hill constant was 1.8. This means that high cooperativity exits between two β-CD cavity of the homo dimer but the hetero dimers have no allostery. The MWC model can explain the allosteric interactions of natural allosteric proteins. The most important assumption of the MWC model is symmetry of the structure of the protein. So, our result, that the only homo-dimer showed allostery, would be supported by the MWC model.

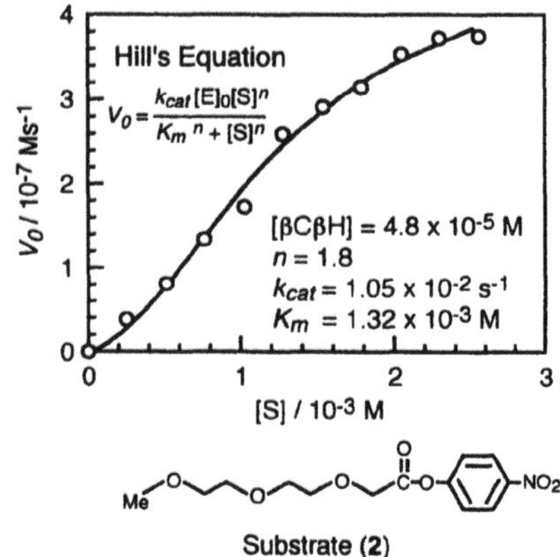

Fig. 2 Dependence of the initial rate on the substrate concentration for the cleavage of the substrate (**2**) by βCβH in pH 7.4 phosphate buffer (0.1M) / acetonitrile (15 : 1) at 25 °C.

4. CONCLUSION

Three kinds of CD dimers were synthesized as artificial hydrolases. As the acyl chain length of the substrates were longer, the transition state was more stabilized by the dimers. We constructed a new artificial allosteric system with 1.8 Hill constant by the use of CD homo-dimer and *p*-nitrophenyl methoxyethoxyethoxyacetate as the artificial enzyme and the substrate, respectively.

Acknowledgments

We are grateful to Nihon Shokuhin Kako Co. LTD. for providing us with cyclodextrins.

Reference

[1] Ikeda, H., Kojin, R., Yoon, C-j., Ikeda, T., Toda, F., Artificial Hydrolase Using Modified Dimethyl-β-Cyclodextrin, *J. Incl. Phenom.*, **7**, 117 (1989)

[2] Akiike, T., Nagano, Y., Yamamoto, Y., Nakamura, A., Ikeda, H., Ueno, A., Toda, F., Coupled Cyclodextrin Appending Imidazole as an Enzyme Model, *Chem. Lett.*, 1089-1092 (1994).

HYDROLYSIS OF N - BENZOYL - L - TYROSINE ETHYL ESTER ANCHORED IN CD BY α-CHYMOTRYPSIN AND SUBTILISIN

K. SANDEEP PRABHU AND **C. S. RAMADOSS**
Division of Biological Sciences, Vittal Mallya Scientific Research Foundation, P. B. # 406, K. R. Road, Bangalore 560 004, INDIA.

ABSTRACT

N-Benzoyl-L-tyrosine ethyl ester (BTEE), a water insoluble substrate, was rendered soluble upon complexation with β-cyclodextrin (or its methyl derivative) and but not with γ-CD. Kinetics of the hydrolysis of such CD anchored BTEE by subtilisin Carlsberg showed a 2 to 4 fold increase in both *kcat* and *Km* values, over the methanol solubilised substrate. Whereas, α-chymotrypsin hydrolyzed the CD anchored BTEE with only a marginally higher reaction rate than the methanol solubilised BTEE.

1. INTRODUCTION

The enzymatic catalyses of hydrophobic compounds is rather slow due to their insoluble nature in an aqueous environment. Water miscible organic solvents used for this purpose do infact strip the essential water layer from the surface of the enzymes making them inactive or with a altered specificity [1]. Also, detergents that are employed, in some cases, inhibit the enzyme activity [2] and at higher temperatures, the micellar structure is disrupted leading to turbidity [3]. The cyclodextrin solubilised hydrophobic substrates have been used for enzyme catalysis [4, 5, 6]. We have investigated the catalysis of subtilisin and chymotrypsin on the CD anchored BTEE hydrolysis.

2. MATERIALS AND METHODS

2.1. Materials

N-Benzoyl-L-tyrosine ethyl ester (BTEE), β-cyclodextrin (βCD), γ-cyclodextrin (γCD), and subtilisin Carlsberg (EC 3.4.21.14) were purchased from Sigma Chemical Co., USA. β-methyl-cyclodextrin (βmCD) was a gift from Dr. Kolossa, Wacker-Chemie GmbH, Germany.

J. Szejtli and L. Szente (eds.), Proceedings of the Eighth International Symposium on Cyclodextrons, 281–284.

2.2. Solubilization of BTEE

BTEE was dissolved in 100μl of methanol prior to its addition to a warm solution of 2.5 molar excess CD, the final volume was made up to 10ml with assay buffer (0.05M Tris-HCl, 8.00 + 0.01M $CaCl_2$). All assays were done at 25^oC (unless otherwise mentioned). The protein was estimated by the method of Lowry *et al* [7] using bovine serum albumin as standard.

2.3. Enzyme Assay

Enzyme activity was calculated from the initial phase of the increase in absorbance at 256nm after addition of a suitable aliquot of enzyme to the buffered substrate solution. A molar extinction coefficient of 810 was used to calculate the specific activity [8].

3. RESULTS AND DISCUSSION

3.1. Solubilities of BTEE in βCD, βmCD and γCD solutions

The formation of inclusion complex between CD and BTEE was monitored by examining the effect of increasing CD concentration on the BTEE solubility. βCD and βmCD formed a totally soluble complex at a concentration 2.5 fold excess over BTEE. Whereas, γCD failed to solubilise BTEE (Figure 1). As reported earlier, αCD failed to solubilise BTEE [3].

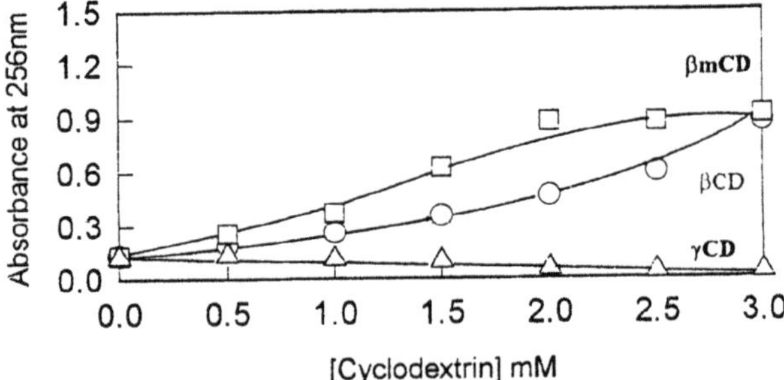

Figure 1: Solubility curves of BTEE in βCD, βmCD and γCD solutions.

3.2. Comparative kinetics of subtilisin Carlsberg and α-chymotrypsin catalyzed hydrolysis of CD anchored BTEE

The CD anchored BTEE was hydrolyzed by both subtilisin and α-chymotrypsin. The pH optima for the hydrolysis of BTEE either anchored in CD or solubilised in methanol were essentially identical. The hydrolysis of CD anchored BTEE by subtilisin and α-chymotrypsin followed typical Michaelis-Menten kinetics.

With subtilisin, the *Km* value was 2 to3 fold higher for the CD anchored BTEE than the methanol solubilised substrate. The *Vmax* was also higher by 3 to 4 fold. However, since both these parameters were increased, the specificity constant remained unaltered. To rule out the effect of sugar moieties on the kinetic parameters of subtilisin, two sugars were tested in the assay system. It was observed that the kinetic parameters remained essentially unchanged in the presence of either 0.1M α-D-glucose or 0.1M D-sorbitol, suggesting that the observed increase in *Vmax* was not due to the polyol effect of the sugar moieties present in the cyclodextrin molecule. The inclusion of βCD (3mM) either in the preincubation mixture or in the assay system had no effect on the hydrolysis of BTEE showing that βCD *per se* had no effect on enzyme activity.
Further, experiments carried out at lower pH (7.00), higher temperature (37°C) and with higher concentration of the enzyme showed similar increase in the kinetic parameters for subtilisin suggesting that the observation on the increase in *Km* and *Vmax* is a true effect of anchoring the substrate in CD. Although changes in *Km* and *Vmax* were noted, the inhibition profile at higher substrate concentrations was similar for methanol solubilised as well as CD anchored substrates. In contrast, an increase in *Km* was not observed with α-chymotrypsin but a marginal increase in the *Vmax* value was observed (Table 1). Therefore, the specificity constant for the βCD anchored substrate was higher.

TABLE 1. Comparative kinetic parameters for CD anchored BTEE hydrolysis by subtilisin Carlsberg and α-chymotrypsin.

Substrate in	**Subtilisin Carlsberg**			**α-Chymotrypsin**		
	Km^{*}	$Vmax^{+}$	$kcat/Km^{**}$	Km^{*}	$Vmax^{+}$	$kcat/Km^{**}$
Methanol	156.2	25.0	3.0 x 10^4	40.5	101.1	1.4 x 10^6
βCD	454.6	91.9	4.1 x 10^4	38.5	134.2	2.1 x 10^6
βmCD	352.9	59.3	3.6 x 10^4			-ND-

* *Km* in μM ; + *Vmax* in μmol/min/mg ; ** *kcat/Km* in $Sec^{-1}.M^{-1}$; ND = not determined.

This process of increase in *kcat and Km* values observed with subtilisin may be due to more productive encounter of the substrate with the enzyme and some structural changes in the enzyme as a result of the binding of the CD-complex.

3.4. Effect of temperature on the hydrolysis of CD anchored BTEE

The Ea values for the hydrolysis of methanol solubilised, βCD-BTEE and βmCD-BTEE by subtilisin were 5.47, 4.04 and 3.36 kcal/mol respectively. A marginal decrease in Ea with CD-BTEE suggested a possible better transition state stabilization. In contrast, the Ea values for α-chymotrypsin catalyzed hydrolysis of BTEE showed an increase in the case of βCD-BTEE when compared to methanol solubilised BTEE [3].

4. CONCLUSION

Our study shows that both subtilisin and chymotrypsin could access the CD anchored substrate. It is apparent that CD anchoring of substrate brings about varied changes in kinetic parameters depending on the enzyme system. Literature on the use of CD in enzymology reveals other advantages of CDs such as, thermostability [9] and renaturing capacity [10] of the protein apart from solubilizing hydrophobic substrates.

ACKNOWLEDGMENTS

KSP thanks the CSIR for the award of Senior Research Fellowship and the Foreign Travel Grant. KSP is also grateful to Professors J. Szejtli, P. V. Subba Rao and P. R. Krishnaswamy for all their help that made it possible for him to attend the symposium.

REFERENCES

[1] Zaks, A. and Klibanov, A. M., Enzymatic catalysis in nonaqueous solvents. *J. Biol. Chem.*, **263**(7), 3194-3201 (1986)

[2] Cook, H. W. and Lands, W. E. M., Further studies of the kinetics of oxgenation of arachidonic acid by soybean lipoxygenase. *Can. J. Biochem.*, **53**, 1220-1231 (1975)

[3] Sandeep Prabhu, K. and Ramadoss, C. S., α-Chymotrypsin catalyzed hydrolysis of N-benzoyl-L-tyrosine ethyl ester anchored in β-cyclodextrin. *Biocatal. Biotransform.*, **12**, 281-291 (1995)

[4] Jyh-Ping Chen., Enhancement of enzymatic hydrolysis rate of olive oil in water by dimethyl-β-cyclodextrin. *Biotechnol. Lett.*, **2**(9), 633-636 (1989)

[5] Otero, C., Cruzado,C and Ballesteros, A., Use of cyclodextrins in enzymology to enhance the solubility of hydrophobic compounds in water. *Appl. Biochem. Biotechnol.*, **27**, 185-194 (1991)

[6] Jyothirmayi, N. and Ramadoss, C. S., Soybean lipoxygenase catalyzed oxygenation of unsaturated fatty acid encapsulated in cyclodextrin. *Biochim. Biophys. Acta.*, **1083**, 193-200 (1991)

[7] Lowry, O.H., Rosenbrough, N.J., Farr, A.L. and Randall, R.J., Protein measurement with the folin phenol reagent. *J. Biol. Chem.*, **193**, 265-275 (1951)

[8] Corey, D.R. and Margaret, P.A., Cyclic peptides as proteases: a reevaluation. *Proc. Natl. Acad. Sci.* (USA), **91**, 4106-4109 (1994)

[9] Ezure, Y., Maruo, S., Kojima, M., Yamashita, H. and Sugiyama, M. Effect of α-cyclodextrin on thermostability of glucoamylase. *Agric. Biol. Chem.*, **52**(4), 1073-1074 (1988)

[10] Karrupiah, N and Sharma, A., Cyclodextrins as protein folding aids. *Biochem. Biophys. Res. Commun.*, **211**(1), 60-66 (1995)

Chapter 3.
CYCLODEXTRINS IN PHARMACEUTICALS

PIROXICAM SODIUM : EFFECT OF INCLUSION COMPLEXATION WITH β-CYCLODEXTRIN

B.D. GLASS, M.S. WORTHINGTON, A. KRALLIS, C. STUBBS*

School of Pharmaceutical Sciences, Rhodes University, Grahamstown, 6140, R.S.A.

**Lennon Limited, P.O. Box 4002, Korsten, Port Elizabeth, 6014, R.S.A.*

ABSTRACT

The photostability of piroxicam sodium (PNa) complexed with β-cyclodextrin (β-CD) was investigated in solution by comparative evaluation of PNa and the 1:1, 1:2, and 2:1 PNa-β-CD complexes and physical mixtures. The order of stability of PNa in the complexes/physical mixtures is reported as the 1:2 complex being the most unstable followed by 2:1>1:1 with the uncomplexed PNa being the most stable.

1. INTRODUCTION

Piroxicam (4-hydroxy-2-methyl-N-(2-pyridyl)-2H-1,2-benzothiazine-3-carboxamide-1,1-dioxide) is a frequently prescribed non-steroidal anti-inflammatory drug (NSAID) displaying low aqueous solubility with improvement of the absorption and gastrointestinal tolerability providing the reason for the preparation of the PNa-β-CD complex. Since piroxicam has been implicated in photosensitivity reactions [1] and the mechanism of photodegradation reported [2], this study aims to develop a protocol to determine the photostability of PNa in the presence of the complexing agent β-CD.

2. MATERIALS AND METHODS

2.1 Materials

Piroxicam was kindly donated by Lennon Ltd (Port Elizabeth, South Africa) and β-CD was purchased from Aldrich Chem. Co. Inc. (Milwaukee). Water for chromatography was obtained using a Milli-Q® water purification system (Waters Assoc., USA). All other materials were of analytical reagent grade.

J. Szejtli and L. Szente (eds.), Proceedings of the Eighth International Symposium on Cyclodextrons, 287–290.

2.2 Methods

Piroxicam was converted into the sodium salt by titration with NaOH in water and the 1:1, 1:2 and 2:1 complexes (freeze drying) and physical mixtures prepared. Verification of inclusion complexation was achieved by DSC, IR and NMR spectroscopy. Photostudies were undertaken at a concentration of 1.2mM of PNa (at a concentration of less than 1.0mM the complex is completely dissociated) in an immersion-well photoreactor/400W medium pressure mercury lamp with samples withdrawn at 10 minute intervals and analyzed by HPLC.

3. RESULTS AND DISCUSSION

Disappearance of the NH/OH stretching band at 3330cm^{-1} confirmed the presence of the sodium salt of piroxicam while shifts in the carbonyl stretching band for PNa were observed in the IR spectra of all three complexes. Significant shifts observed for the H_3 and H_5 protons in the NMR spectra and a decrease in the height of the endothermic peak at 219.47°C seen for all complexes as compared to the corresponding physical mixtures confirmed inclusion complexation. Table 1 indicates the %PNa remaining (HPLC analysis) for PNa and the various complexes/physical mixtures on irradiation for 60 minutes sampling at 10 minute intervals.

TABLE 1. %PNa versus time

time	1:1		2:1		1:2		PNa
(min)	comp./phys.mix		comp./phys.mix		comp./phys.mix		
0	100.00	100.00	100.00	100.00	100.00	100.00	100.00
10	96.42	94.05	98.83	82.89	85.42	81.91	98.50
20	84.46	79.59	82.19	48.54	61.14	32.16	82.53
30	60.22	58.15	63.99	13.69	29.61	6.12	67.01
40	36.38	41.34	33.56	2.20	4.85	5.04	54.94
50	23.45	24.10	10.08	2.20	4.06	3.70	41.50
60	14.43	11.59	3.23	2.20	3.60	2.35	29.31

Graph 1 provides a comparative evaluation of the photodegradation of PNa and the various complexes.

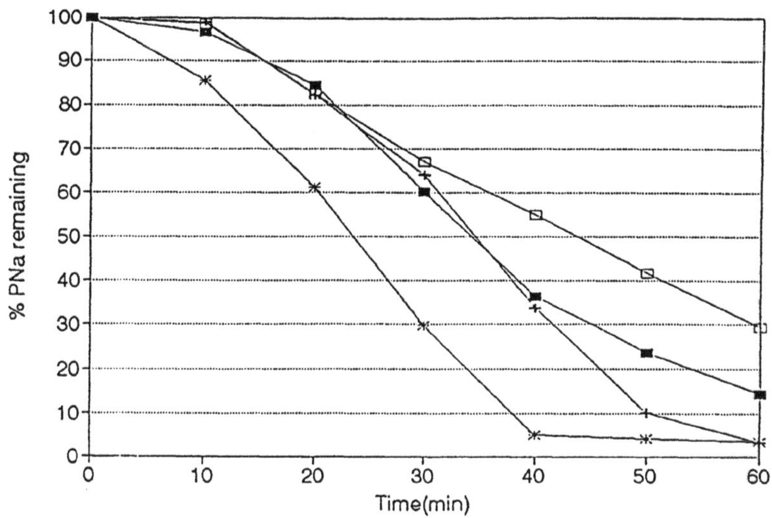

Graph 1. %PNa versus time

Although inclusion complexation appears to afford photostability for PNa as compared to the physical mixtures its degradation is enhanced when compared to PNa. From the graph it can be seen that the 1:2 complex degraded the fastest followed by the 1:1 and 1:2 complexes.

4. CONCLUSION

These results concur with those reported by Backensfeld et al [3] where all cyclodextrins studied accelerated the rate of decomposition of piroxicam when stored in solutions containing a 10-fold excess of the respective cyclodextrins at temperatures ranging from 21-71°C. The destabilizing effect was evident in that the piroxicam concentration at 61°C/475 days without CD was 81.6% as compared to 61.2% in the presence of β-CD. Our study reports 41.50% PNa remaining after 50 minutes irradiation as opposed to 23.45% in the 1:1 complex. These results could be explained in terms of the proposed model of the inclusion equilibria for the PNa-β-CD system where it can be seen that potential sites for hydrolysis and photolytic degradation are not included in the β-CD cavity [4].

5. REFERENCES

[1] Brooks, G., Drug-induced photosensitivity, *Pharm. J.*, **245**, 19-21 (1990)

[2] Miranda, M.A., Vargas, F., Photodegradation of piroxicam under aerobic conditions. The photochemical keys of the piroxicam enigma?, *J. Photochem. Photobiol.*, **8**, 199-202 (1991)

[3] Backensfeld, T., Muller, B.W., Kalter, K., Interaction of NSA with cyclodextrins and hydroxypropyl cyclodextrin derivatives, *Int. J. Pharm.*, **74**, 85-93 (1991)

[4] Fronza, G., Mele, A., Redenti, E., Ventura, P., Proton nuclear magnetic resonance spectroscopy studies of the inclusion complex of piroxicam with β-cyclodextrin, *J. Pharm. Sci.*, **81**(12), 1162-1165 (1992)

INTERACTION OF PIROXICAM POLYMORPHS WITH β-CYCLODEXTRIN IN GROUND MIXTURES

SATIT PUTTIPIPATKHACHORN and KANOKWON PRASITPORN

Faculty of Pharmacy, Mahidol University, Sri-ayudhya Road, Bangkok 10400 Thailand

ABSTRACT

The interactions of two piroxicam polymorphs, viz., needle-form and cubic-form, with β-cyclodextrin (βCD) in the ground mixtures were investigated by differential scanning calorimetry, Fourier-transformed infrared spectroscopy and powder X-ray diffractometry. The results indicated that the cubic-form piroxicam might form the inclusion complex and/or the hydrogen bonding with βCD in the ground mixtures whereas the needle-form piroxicam interacted with βCD through the intermediate transformation to the cubic form. It was confirmed by the evidence of the crystallisation of the inclusion complex after the ground mixtures were exposed to 75%RH at 45°C. It was demonstrated that the solid-state inclusion complex formation of piroxicam with βCD in the ground mixtures depened on drug polymorphism

1. INTRODUCTION

Co-grinding is a process for preparing inclusion complexes of drugs with cyclodextrin (CD) in which no organic solvent is used [1]. Piroxicam, a non-steroidal anti-inflammatory drug, has two polymorphic form, namely cubic form and needle form [2]. In this study, the molecular interaction of different piroxicam polymorphs with βCD under grinding treatment were investigated by differential scanning calorimetry (DSC), Fourier-transformed infrared (FTIR) spectroscopy and powder X-ray diffractometry. The dissolution behavior of the ground mixtures was also studied.

J. Szejtli and L. Szente (eds.), Proceedings of the Eighth International Symposium on Cyclodextrons, 291–295.

2. MATERIALS AND METHODS

2.1. Materials

The cubic-form piroxicam was used as received (Secifarma, Italy). The needle-form piroxicam was prepared by fast crystallizing from the methanolic solution [2]. The βCD (Kleptose®, Roquette, France) was used.

2.2. Preparation of physical mixtures and ground mixtures

The physical mixtures and the ground mixtures of each piroxicam polymorph with βCD at 1:1 and 1:2 molar ratio were prepared by gentle mixing in a motar and by grinding for 1, 3, 5 and 24 h in a ball mill respectively.

2.3. DSC study

The DSC thermograms were measured by Perkin Elmer model DSC7 differential scanning calorimeter at heating rate of 10 °C/min under a 50 ml/min nitrogen gas flow. The open aluminum pan was used.

2.4. FTIR study

The FTIR spectra were carried out with Nicolet model 52DX FTIR spectrophotometer using KBr disc method.

2.5. Powder X-ray diffraction study

The powder X-ray diffraction patterns were measured using Phillips model PW1730/10 diffractometer using Cu-Kα radiation.

2.6. Dissolution study

The dissolution behavior was studied using a USP dissolution tester apparatus II and simulated gastric fluid without pepsin was used as the dissolution medium. The apparatus was operated at the rate of 75 rpm and 37 °C. The amount of drug dissolved were determined spectrophotometrically at 333 nm.

3. RESULTS AND DISCUSSION

3.1. Changes in DSC thermograms

The DSC thermograms of the physical and ground mixtures of cubic-form piroxicam with βCD were shown in Figure 1A. As the grinding time increased, the intensity of an endothermic peak decreased. Two endothermic peaks were observed in DSC thermogram of 5 h ground mixture. In the 24 h ground mixture, the endothermic peak at higher temperature disappeared and only one small broad peak around 187 °C remained. It was suggested that the endothermic peak at lower temperature might be that of the inclusion complex between piroxicam and βCD. Changes in DSC thermogram of the physical mixture of needle-form piroxicam with βCD by grinding were shown in Figure 1B. After grinding for 1 h, two endothermic peaks were observed at 196.3 and 200.6°C. The endothermic peak at higher temperature was attributable to the melting peak of cubic-form piroxicam. These two endothermic peaks were clearly resolved when grinding for 3 h. It was indicated that grinding caused a polymorphic transformation of piroxicam from needle form to cubic form. After grinding for 5 h, a broad endothermic peak with low intensity was observed at 186.1°C. This small broad peak was still observed in the 24 h ground mixture. This result indicated that piroxicam molecules in the ground mixture were further transformed to another state which could be attributed to inclusion complex formation between piroxicam and βCD.

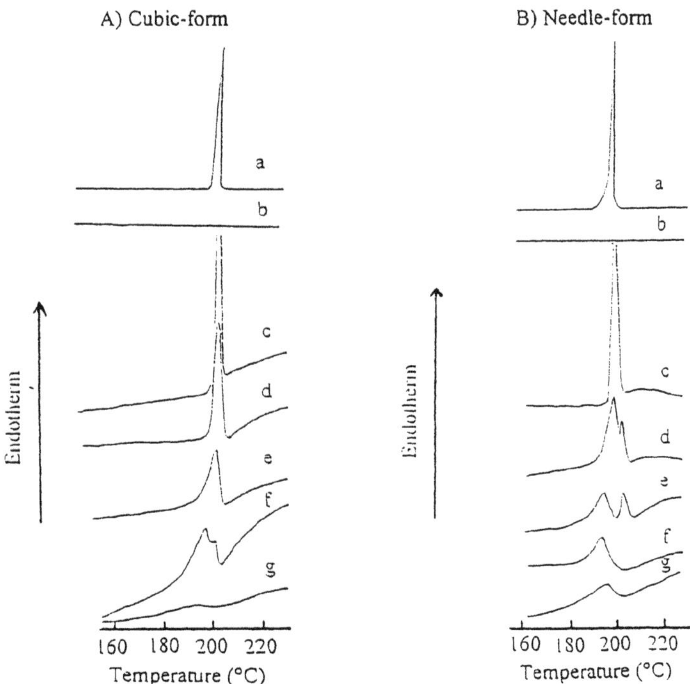

Fig. 1 DSC Thermograms of Physical Mixtures and Ground Mixtures of Piroxicam with βCD at 1:1 Molar Ratio

a) piroxicam, b) βCD, c) physical mixture, d) 1 h ground mixture, e) 3 h ground mixture, f) 5 h ground mixture, g) 24 h ground mixture

3.2. Changes in FTIR spectra

The two piroxicam polymorphs showed slightly different fingerprint regions in the IR spectrum [2]. The FTIR peak intensity at 3337 cm^{-1} which was attributed to -NH and -OH stretching of piroxicam decreased with increased grinding time. The FTIR peak at 1630 and 1529 cm^{-1} were due to the stretching vibration of carbonyl group (the amide I band) and the amide II band of piroxicam respectively. These two peaks were shifted to 1646 and 1533 cm^{-1} respectively when grinding proceeded for 24 h. In the piroxicam crystal, the carbonyl group showing IR peak at 1630 cm^{-1} formed the intramolecular hydrogen bond with hydroxyl group [3]. The amide II band at 1529 cm^{-1} of piroxicam crystal was attributed to mainly the bending vibration of -NH group mixed with stretching vibration of -CN group. Kojic-Prodic and Ruzic-Toros [3] reported that the -NH group formed an intermolecular hydrogen bond to an O atom bonded to S atom. Therefore, the shift in IR peaks of the amide I band and the amide II band to higher wavenumber indicated the decrease in strength of both intermolecular and intramolecular hydrogen bonding. In addition, the amide III band (mainly νC-N mixed with δNH) at 1298 cm^{-1} also slightly shifted to 1301 cm^{-1} after grinding for 24 h. The FTIR peak at 1149 cm^{-1} which was due to symmetric vibration of $-SO_2-$ group also shifted to higher wavenumber. These results also confirmed the change in extent of both intermolecular and intramolecular hydrogen bonding of piroxicam molecules. The changes in IR patterns by grinding as mentioned above could be attributed to the inclusion complex formation and/or the hydrogen bonding formation between piroxicam and βCD. The similar results of the shift in FTIR spectra were observed as in the case of cubic-form piroxicam.

3.3. Changes in powder X-ray diffraction patterns

The crystallinity of piroxicam decreased with increased grinding time and grinding for 24 h almost changed the state of piroxicam to be an amorphous form. Similar results were also obtained in the case of the needle-form piroxicam and βCD mixture. After storage at 45°C and 75%RH for 90 days, the 24 h ground mixture of both piroxicam polymorphs with βCD showed new characteristic diffraction peaks, indicating the inclusion complex between piroxicam and βCD in the ground mixture was changed from amorphous state to crystalline state by adsorbing water vapor.

3.4. Dissolution behavior

The dissolution rates of both piroxicam polymorphs were markedly enhanced when ground with βCD. This implied a solubility enhancing effect of βCD by forming inclusion complex. No change in dissolution rates of the ground mixtures was observed after storage at 75%RH and 45°C for 90 days.

4. CONCLUSION

It was demonstrated that the cubic-form piroxicam could form hydrogen-bond and/or inclusion complex with βCD directly by grinding whereas the needle-form piroxicam interacted with βCD through the intermediate transformation to the cubic-form piroxicam.

REFERENCES

[1] Nakai, Y., Molecular behavior of medicinals in ground mixtures with microcrystalline cellulose and cyclodextrins, *Drug Dev Ind Pharm*, 12, 1017-1039 (1986)

[2] Mihalic, M., Hofman, H., Kufinec, J., Krile, B., Caplar, V., Kajfez, F., Blazevic, N., *Piroxicam*, in Analytical Profiles of Drug Substances, Vol. 15, (Ed. Florey,K.), Florida: Academic Press Inc, 1986, Pp. 508-531

[3] Kojic-Prodic, B., Ruzic-Toros, Z., Structure of the anti-inflammatory drug 4-hydroxy-2-methyl-N-2-pyridyl-2H-1λ^6,2-benzothiazine-3-carboxamide1,1-dioxide (piroxicam), *Acta Cryst* B38, 2948-2951(1982)

STUDY OF THE INFLUENCE OF γ-CYCLODEXTRIN ON THE MOLSIDOMINE PHOTOSTABILITY.

G. PIEL[1], L. POCHET[2], L. DELATTRE[1], J. DELARGE[2]

Laboratory of Pharmaceutical Technology[1] and Laboratory of Pharmaceutical Chemistry[2], Institute of Pharmacy, University of Liège, rue Fusch 3-5, 4000 Liège, Belgium.

ABSTRACT

Molsidomine, a very useful therapeutic agent for the treatment of coronary heart disease, is extremely light sensitive. The aim of this work is to study the influence of cyclodextrins on the molsidomine photostability. Different molsidomine-γ-cyclodextrin (CD) complexes were prepared by spray-drying and the inclusion could be demonstrated by different methods as the DSC and the FTIR spectroscopy. The photostability was tested at 4500 lux. We could show that γ-CD would not have a better protective effect on molsidomine than lactose, mannitol or NaCl, in the same weight ratio and that there is a linear relationship between the half life time and the percentage of molsidomine contained in the mixtures or the complexes.

1. INTRODUCTION

Molsidomine (N-ethoxycarbonyl-3-morpholino-sydnonimine) (Figure 1) is an established therapeutic agent for the treatment of coronary heart disease. Molsidomine is effective in the treatment of angina pectoris, congestive heart failure, acute myocardial infarction and treatment of pulmonary hypertension.

Figure 1: Molsidomine structure.

The problem encountered with molsidomine is its extremely high unstability. As a matter of fact, the sydnonimine derivatives are extremely light sensitive. Following irradiation

J. Szejtli and L. Szente (eds.), Proceedings of the Eighth International Symposium on Cyclodextrons, 297–300.

with artificial or natural light, nitrogen, morpholine, ammonia and other degradation products are obtained [3].
The CD's (cyclic oligosaccharides containing 6, 7 or 8 D-glucopyranose units) can be used to improve the light stability of active ingredients [4].
The aim of this work is to study the influence of CD's on the molsidomine photostability.

2. MATERIALS AND METHODS

2.1 Materials.

γ-CD was obtained from Wacker Chemie GmbH (München, Germany), lactose from DMV (Netherlands), mannitol from Chemurgie GmbH (Hamburg, Germany) and NaCl from Merck Belgolabo (Belgium). All other materials were of analytical grade.
The complexes and the physical mixtures were prepared by the spray-drying method.

2.2 Methods.

Photostability studies: Each product was submitted to a constant light exposure of 4500 lux with neon lamps up to attaining 50% of degradation.
Molsidomine assay: Intact molsidomine was assayed using a HPLC method.

3. RESULTS ANS DISCUSSION

Different molsidomine-γ-CD complexes were prepared by spray-drying, in different stoichiometric ratios. The inclusion could be demonstrated for the (1:2), (1:1) and (2:1) complexes by different methods as the DSC and the FTIR spectroscopy.
Figure 2 shows that the (2:1) complex (cont. ± 21% of molsidomine) is more stable than the (1:1) complex (cont. ± 14.5% of molsidomine), itself more stable than the (1:2) complex (cont. ± 7.7% of molsidomine). This figure also shows that the spray-dried molsidomine (spray-dried in the same conditions as the complexes) is as stable as the (1:2) complex and that the crystalline molsidomine is the most stable.
We could observe that the molsidomine degradation follows first order kinetics; therefore it is possible to linearise the degradation and to express the stability by the half-life time of the product.
To better understand the influence of the cyclodextrins, we spray-dryied physical mixtures of molsidomine with lactose, mannitol or NaCl in the same weight ratios as the corresponding complexes.
We could show that there is no significant difference of stability between the (1:2) γ-CD complex and the physical mixtures with lactose, mannitol or NaCl. The same results were obtained for the (1:1) and (2:1) complexes.

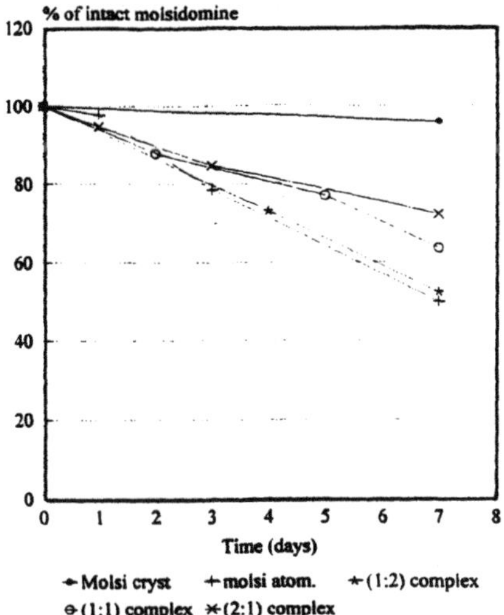

Fig.2 Photostability study of the crystalline molsidomine, the spray-dried molsidomine and the (2:1), (1:1) and (1:2) molsidomine-γ-CD complexes.

Subsequently, we wanted to better understand why the (2:1) complex is more stable than the (1:2) so why the most concentrated complex is the most stable. We carried out a spray-drying on physical mixtures of molsidomine with γ-CD, lactose, mannitol or NaCl containing 1, 45, 60, 70, 80 and 90% of molsidomine.
We could observe that the degradation rate increases with the dilution. Table 1 and figure 3 show that there is a linear relationship ($r = 0.995$) between the half-life time and the concentration of molsidomine.

TABLE 1: Influence of the molsidomine concentration on the half-life time of the complex and the physical mixtures.

Concentration of molsidomine* (%)	Half-life time* (days)
1.00 ± 0.03	3.33 ± 0.57
7.70 ± 0.22	8.34 ± 0.68
14.52 ± 0.30	13.13 ± 4.70
21.32 ± 3.36	16.80 ± 3.55
43.92 ± 2.14	40.88 ± 5.13
58.44 ± 1.91	52.09 ± 2.07
68.10 ± 2.99	58.21 ± 5.89
77.57 ± 1.31	61.43 ± 1.10
86.37 ± 2.82	67.88 ± 6.14

* mean of the complexes or mixtures with γ-CD, lactose, mannitol and NaCl.

This tends to show that the γ-CD would not be protective against the degradation. Molecular modeling studies and RMNC[13][5] studies showed that the ethoxycarbonyl part of the molsidomine molecule is located into the cavity of the CD. So, the most sensitive part (sydnonimine part) would be exposed to light. The complexation would not have any protective effect in this case.

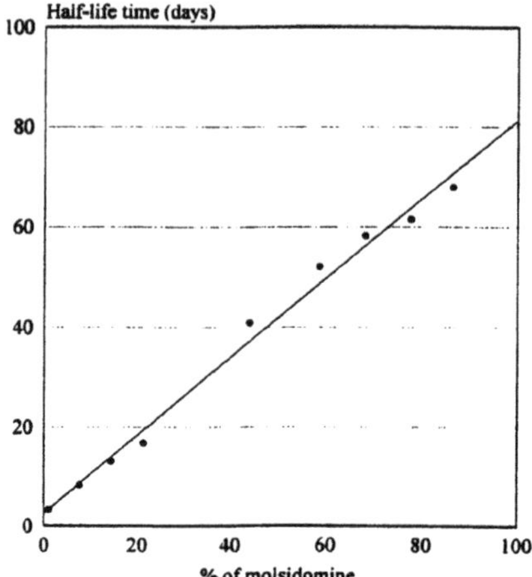

Fig. 3 Influence of the molsidomine concentration on the half-life time of the complexes or the mixtures.

4. CONCLUSION.

As a conclusion we can say that γ-CD would not have a better protective effect on molsidomine than lactose, mannitol or NaCl, in the same weight ratio. There is a linear relationship between the half life time and the percentage of molsidomine contained in the mixtures or the complexes. Further studies will investigate the influence of different CD's derivatives on the molsidomine stability

REFERENCES.

[1] Reden, J., Molsidomine, *Blood Vessels* **27**, 282-294 (1990)

[2] Scheen, A.J., L'oxyde nitrique (NO), nouvelle cible thérapeutique, *Rev.Med.Liège*, **50(1)**, 22-24 (1995)

[3] Asahi, Y., Shinozaki, K., Nagaoka, M.,Chemical and kinetic study on stabilities of 3-morpholinosydnonimine and its N-ethoxycarbonyl derivative, *Chem.Pharm.Bull.*, **19(6)**, 1079-1088 (1971)

[4] Fromming, K.H. and Szejtli, J., Cyclodextrins in Pharmacy, Kluwer academic publishers, Dordrecht, 1994

[5] Vikmon, M., Inclusion complexes of N-ethoxycarbonyl-3-morpholino-sydnonimine or salts formed with cyclodextrin derivatives, preparation thereof and pharmaceutical compositions containing the same, US Patent 07 807 852 (1992).

PREPARATION AND INVESTIGATION OF PRODUCTS CONTAINING TOLPERISONE·HCl AND CDs

L. ANTAL[1], GY. DOMBI[2], CS. NOVÁK[3] and M. KATA[1]
[1]Department of Pharmaceutical Technology, [2]Department of Pharmaceutical Chemistry, Albert Szent-Györgyi Medical University, H-6720 Szeged, [3]Institute of General and Analytical Chemistry, Technical University, H-1521 Budapest, Hungary

ABSTRACT

Tolperisone·HCl (**T**) is an original Hungarian drug molecule. It is well soluble in water [1]. From a scientific aspect, it may be very interesting to investigate how a compound that is readily soluble in water forms inclusion complexes with cyclodextrins (**CDs**) [2]. The aim was to study the conditions of complex formation with different **Cds**, such as **α-CD**, **β-CD**, **γ-CD**, **DIMEB** and **RAMEB**, by using various preparation technniques and investigation methods.

1. INTRODUCTION

T is a β-aminoketone with antinicotine activity (skeletal muscle relaxant). It was discovered by **Pórszász and Nádor** (1956, G. Richter Pharm. Factory, Budapest). Mydeton® and Mydocalm® are its well-known registered preparations and a further 44 products are marketed worldwide.
The publications in the surveyed literature dealt only with its stability and pharmacological features. No experiments on the complex formation of **T** with **Cds** have been reported, because the attention of researchers naturally concentrates on the study of the conditions of complex formation of drug molecules that are only slightly soluble in water, in order to increase their solubility and to ensure a better bioavailability of pharmacons with fewer side-effects, that are more favourable from an economic standpoint.

2. MATERIALS AND METHODS

2.1. Materials

Tolperisone·HCl (G. Richter Pharm. Factory, Budapest), mp 176-177 °C, mol wt 281.8; and different **CDs** (Cyclolab Ltd., Budapest).

2.2. Methods

Methods for preparation of products: mixing, kneading and spray-drying. Molar ratios of products (**T:CD**): 1:1 and 1:2.

J. Szejtli and L. Szente (eds.), Proceedings of the Eighth International Symposium on Cyclodextrons, 301–303.

Investigation methods: solubility and dissolution tests (USP XXII and Ph. Hg. VII, rotating basket method), determination of surface tension (Krüss Interfacialtensiometer K 8600), n-octanol/water distribution ratio measurements, stability tests and stability constant determinatiom, IR spectroscopy and X-ray diffraction, in vitro diffusion (Sartorius Resorptionsmodell), thin-layer chromatography and thermal analysis (in a separate poster).
Analysis: spectrophotometrically at 265 nm in the concentration range 2-15 μg/mL (Specord UV-VIS, C. Zeiss, Jena, Germany, and Spektromom 195, MOM, Budapest).

2. RESULTS AND DISCUSSION

Pure **T** is well soluble in water and dissolved during 5 minutes in water, but **T** Mydeton® tablets required 30 minutes.
The *dissolution time* of **T** from 1:1 kneaded products prepared with the different **CDs** was generally less than 10 min, while that from 1:2 kneaded products was not more than 5 min for **α-CD** and **DIMEB**, 10 min for **γ-CD** and 15 min for **β-CD** and **RAMEB**. The standard deviation was at most 4.26; it was generally (more than 85%) less than 1.34.
The values of partition coefficients (**P**) differed interestingly.

TABLE 1. Concentrations in n-octanol and water (mg/100 mL), and partition coefficients (P)

Product	c_o/c_w	P ± S.D.
T + DIMEB	435.4 / 52927.9	0.01 ± 0.35
T + RAMEB	8239.5 / 55180.2	0.15 ± 0.56
T	13138.1 / 56118.6	0.23 ± 0.34
Mydeton	315.3 / 287.5	1.09 ± 0.23
T + β-CD	780.8 / 405.4	1.93 ± 0.67
T + α-CD	1186.2 / 438.45	2.70 ± 1.23
T + γ-CD	1486.4 / 421.9	3.52 ± 1.12

The solubilities in water of products containing **α-CD**, **β-CD** and **γ-CD** or **DIMEB** and **RAMEB** are similar, but their concentrations in octanol are very different, and this strongly influences the values of **P**.
The surface tension of 0.1% **T** solution is 53.20±0.50. The values of surface tension for the products: 69.00 > 63.30 > 54.00 > 50.70 > 47.05 > 41.75 (**T+β-CD, T+α-CD, T+γ-CD, T+DIMEB, T+RAMEB** and Mydeton® tablet), and for **CDs**: 66.90 > 65.15 > 61.45 > 61.35 > 54.50 (**β-CD, γ-CD, α-CD, DIMEB** and **RAMEB**).
Results of in vitro diffusion experiments. Averages of at least 3 determinations with artificial gastric juice containing 0.1% **T** or 0.1% **T** + 0.9% **CD**.
Pure **T** has a low diffusion ability. The diffusion coefficient was lowered by **β-CD** and **γ-CD** complexation and increased by **α-CD**, **DIMEB** and **RAMEB**. This means that complexation between **T** and **β-CD** or **γ-CD** decreases the diffusion process, while other **CDs** increase it. In consequence of the relatively small diameter of **β-CD**, it may form a stronger complex with **T**. These results were confirmed by IR spectroscopy, X-ray diffraction and thermal analysis.

TABLE 2. In vitro diffusion constants

	$/k_d$ ($\cdot 10^{-5}$) [cm / min]/
T	4.62 ± 0.25
Mydeton	5.65 ± 1.27
T + α-CD	4.86 ± 0.12
T + β-CD	2.41 ± 0.03
T + γ-CD	3.88 ± 0.35
T + DIMEB	5.98 ± 0.52
T + RAMEB	8.24 ± 0.15

The cavity of **DIMEB** is irregular and more flexible than the ring of **β-CD**. Further, **DIMEB** has surface activity. Consequently, these conditions help **T** molecules to pass through the membrane of the Sartorius apparatus into the artificial plasma.
RAMEB decreases the surface tension of its solutions even more and its complex stability constant is also the smallest; therefore, more **T** may diffuse than in the case of **DIMEB**. The Mydetone® dragee contains gum Arabic, its solutions are surface active, and thus its diffusion ability is better.
Pure **T** also has surface activity. It increases the surface tension of **α-CD** and **β-CD**, but decreases that of other **CD**s.
The amount of uncomplexed **T** increases in the sequence **α-CD**, **β-CD**, **γ-CD** complexes. This is due to the larger diameter of the given **CD**. The IR measurements prove the host-guest interactions in the **α-CD**, **β-CD** and **γ-CD** complexes, but in the products formulated with **DIMEB** and **RAMEB** another kind of interactions is observed.

4. CONCLUSIONS

The aim of the experiments was to study whether a pharmacon well soluble in water may form inclusion complexes with **CD**s or not. The authors made products 1:1 and 1:2 **T:CD** ratios with different **CD**s. The products partially contain inclusion complexes, as proved by solubility and dissolution tests, surface tension and partition ratio measurements, stability tests and stability constant determinations, IR spectroscopy and X-ray diffraction, thin-layer chromatography and thermal analysis.

ACKNOWLEDGEMENT

The authors gratefully thank **Prof. J. Szejtli and his co-workers** for providing CD products and advice.

REFERENCES

[1] Antal, L., Bodnár, E. and Kata, M.: In vitro diffusion study of **T** and its products (Hungarian). Gyógyszerészet 38, 295-299 (1994)

[2] Szejtli, J.: Cyclodextrin Technology. Kluwer Academic Publ., Dordrecht, 1988

IMPROVEMENT IN THE DISSOLUTION PROPERTIES OF KETOCONAZOLE THROUGH MULTICOMPONENT β-CYCLODEXTRIN COMPLEXATION

Esclusa-Díaz M.T., Pérez-Marcos, M.B., Vila-Jato, J.L. and Torres-Labandeira J.J.

Pharmaceutical Technology Department. School of Pharmacy. University of Santiago de Compostela. Campus Universitario Sur. 15706 Santiago de Compostela. Spain.

ABSTRACT

Increases the solubility of ketoconazole was studied. Two systems were used: binary complexes prepared with β-cyclodextrin and multicomponent systems (β-cyclodextrin and an acid compound), obtained by spray-drying. X-ray diffractometry and differential scanning calorimetry showed differences between ketoconazole/cyclodextrin complexes and their corresponding physical mixtures and individual components. The solubility of ketoconazole increased significant with the cyclodextrin complexes. However, enhancement was better from the multicomponent systems.

1. INTRODUCTION

Ketoconazole (KET) is a weak base and has a very low aqueous solubility because of its hydrophobic structure. It can be solubilized only under extremely acidic conditions and the improvement of its oral bioavailability has been problematic because the absorption is markedly inhibited by agents that increase gastric pH or in achloridric patients (1). Cyclodextrin has been used to improved the poor aqueous solubility of drugs from their formulation as inclusion compounds (2). Nevertheless, the usefulness of natural cyclodextrins has been limited by relatively low aqueous solubility particularly, β-cyclodextrin (BCD). Salt formation with different acids and the cyclodextrin in the multicomponent complex has been studied to improve the solubility of these base-type drugs (3).
The aim of this study was to investigate the influence of the complexation of KET with BCD (both binary and multicomponent complexes) on the solubility of the solubility the the drug in aqueous solutions pH 5 and 6. For this purpose, different molar ratio and preparation techniques were employed. In order to analyze the obtained products, X-ray diffractometry and DSC were used. Solubility diagrams and dissolution studies were carried out in aqueous media pH 5 and 6.

2. MATERIALS AND METHODS

2.1. Materials

Ketoconazole was supplied by Guinama (Valencia, Spain) and β-cyclodextrin by Cyclolab (Budapest, Hungary). All other materials and solvents were of analytical reagent grade.

J. Szejtli and L. Szente (eds.), Proceedings of the Eighth International Symposium on Cyclodextrons, 305–308.

2.2. Phase solubility studies

Solubility diagrams were obtained according to Higuchi and Connors (4) in phosphate buffer solutions of pH 5 and 6. The apparent stability constant of the KET-BCD complex, assuming 1:1 stoichiometry, were calculated from the slope of the initial straight portion of the solubility diagram.

2.3. Preparation and characterization of solid inclusion complexes and physical mixtures

The solid complexes of KET-BCD (1:1, 1:2, molar ratio) and KET-BCD with hydrochloric (1:2:2 molar ratio) or citric acid (1:2:1, molar ratio), were prepared by spray-dried. Physical mixtures of an appropriate amount of KET and BCD were obtained by pulverizing and thereafter mixing both solids in a Turbula T2C mixer.

Powder X-ray diffraction patterns were obtained using a Philips X-ray diffractometer (PW 1710 BASED) using Cu-K_α radiation. The differential scanning calorimetry (DSC) patterns were taken by a Shimadzu DSC-50 system equipped with a computerized data station.

2.4. Dissolution studies

Dissolution rates were determined according to Nogami et al. (5), in phosphate buffer solutions of pH 5 and 6 as the dissolution medium, at 37 °C for 180 minutes. The concentration of KET was determined by UV spectrophotometry at 225 nm. All samples were analyzed in triplicate. Dissolution efficiencies after 180 min (DE_{180}) were calculated according to Khan (6). The effects of drug formulation on dissolution efficiency at each pH were investigated by one-way analysis of variance with the Scheffé test for multiple comparisons.

3. RESULTS AND DISCUSSION

3.1. Interaction between KET and BCD in aqueous medium

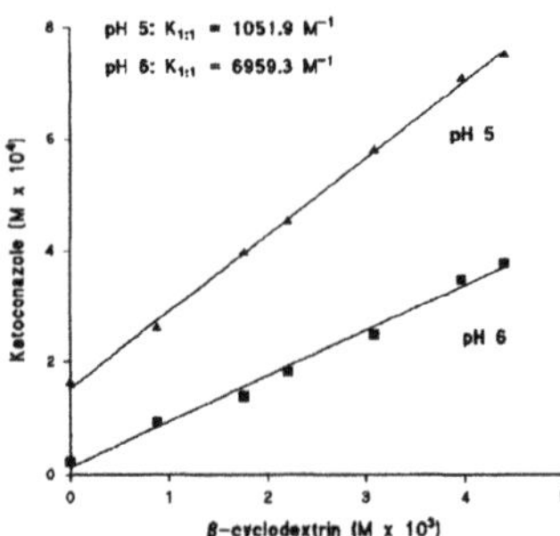

Fig. 1. Phase solubility diagrams of KET and BCD in buffer pH 5 and 6 at 37°C.

Phase solubility profiles of KET with BCD are shown in figure 1. Both diagrams can be classiffied as A_L type according to Higuchi and Connors (4).

This indicates that, within the BCD concentration range tested, a soluble complex is formed. From the calculated values of the stability constant (see figure 1) a different interaction in both medium between the drug and the cyclodextrin can be deduced .

The reason for that is the higher ionization of ketoconazole at pH5.

3.2. Characterization of solid complexes

The X-ray diffractograms of formulations are shown in the figure 2 (left) . The diffraction patterns of the physical mixtures correspond to the superimposed diffractograms of KET and BCD, while those of the complexes show fewer and less intense peaks. Figure 2 (right) shows the DSC thermograms of the preparations. The physical mixtures show two endothermic peaks at 100 and 149°C corresponding respectively to BCD (water) and KET (fusion). These peaks disappeared in the case of the all complexes.

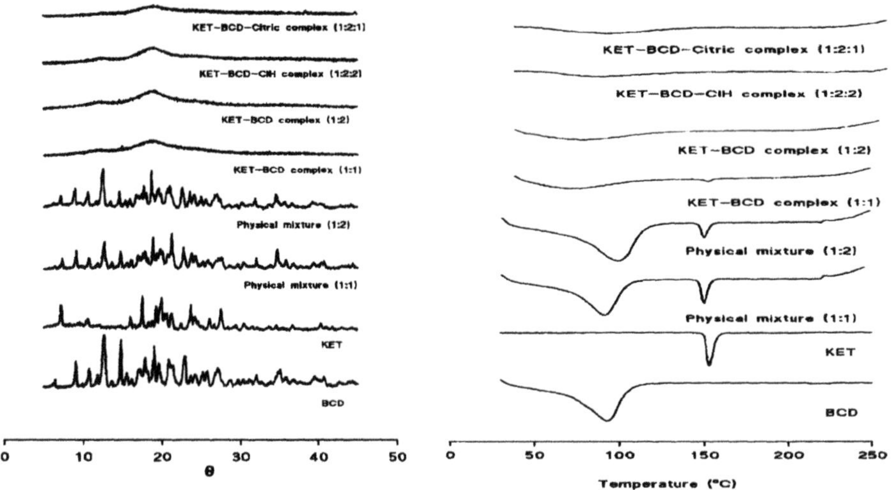

Fig. 2. Power X-ray diffraction patterns (left) and differential scanning calorimetry (right) of the indicated formulations.

These results indicates the formation fo amorphous complexes between the cyclodextrin and the drug, either bynary or multicomponent systems.

3.3. Effects of complexation on dissolution of the drug

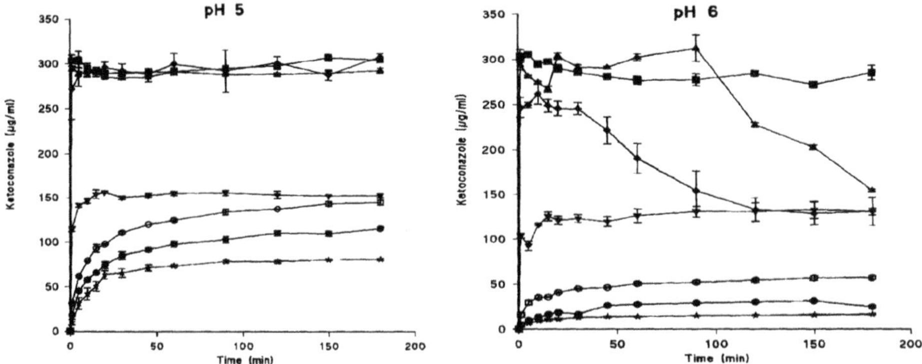

Fig. 3 Dissolution profiles in buffer pH 5 (left) and 6 (right).

Key: ★ KET; ● Physical mixture 1:1 M:M; ○ Physical mixture 1:2 M:M; ▼ KET-BCD 1:1 inclusion complex; ♦ KET-BCD 1:1 inclusion complex; ■ KET-BCD-ClH inclusion complex; ▲ KET-BCD-Citric acid inclusion complex

Figure 3 (left) illustrates the dissolution profiles obtained in buffer pH5. One-way ANOVA in dissolution efficiency (0-180 min) reveals significant differences between the different formulations ($F_{6,14}$ = 2379.708, α<0.01).

The Sheffé test grouped the formulations as follows:

KET PM 1:1 PM 1:2 KET-BCD 1:1 KET-BCD 1:2 KET-BCD-HCl KET-BCD-CITRIC ACID

Physical mixtures of KET with BCD showed a higher dissolution rate than pure KET and, this effect increases with the amount of BCD in the mixture. In the binary systems as the amount of BCD raises in the complexes, a better dissolution profile is shown.
Nevertheless, no differences were found between the binary system 1:2 and the multicomponent complexes - prepared with the same ratio of BCD. This fact is probably due to the ionization of the drug in this medium.
In medium pH 6, the behaviour of the prepared systems defers from the above at pH 5 (figure 3, right). The solubility of KET is lower because ionization decreases. For this reason, in general, total amount of KET dissolved from all the preparations is lower than at pH 5. Therefore, in spite of the higher stability constant calculated at pH 6, the effect of cyclodextrins on the drug solubility is lower than in the more acidic medium.
Analysis of variance indicates that the "formulation" has a significant effect on 180-min dissolution efficiency ($F_{6,14}$=754.308, α <0.01) and the Sheffé test grouped the systems as follows:

KET PM 1:1 PM 1:2 KET-BCD 1:1 KET-BCD 1:2 KET-BCD-HCl KET-BCD-CITRIC ACID

The dissolution behaviour of KET from the physical mixtures, and 1:1 complex is similar to that obtained at pH 5. From the binary system 1:2, it was not possible to dissolve all the drug (around 80%) and after 30 minutes, the concentration decreased due to the recrystallization of KET in the medium (8).
Both multicomponent complexes show different dissolution profiles. The one prepared with citric acid reached a percentage close to 100% but after 90 minutes the concentration decreases as that above, whereas with the one prepared with HCl 100% of the dose dissolved and remains in the solution throughout.

4. CONCLUSIONS

Binary and multicomponent inclusion complexes of ketoconazole and β-cyclodextrin can be obtained by the spray-drying method. This provides evidence of improved drug solubility in both media studied. No differences were found between binary and multicomponent complexes prepared with the same molar ratio (1:2) at pH 5.
However, in medium pH 6, inclusion complex formed in the presence of hydrochloric acid resulted in higher solubility enhancement for the drug.

ACKNOWLEDGEMENTS

This work was supported by a research project from Xunta de Galicia XUGA20302A93.

REFERENCES

1. Van der Meer J.W.M., Keuning J.J., Scheijgrond H.W., Heykants J., Van Cutsem J., Brugmans J. J. Antimicrob. Chemother. 6, 552 (1982).
2. Torres Labandeira J.J., Echezarreta López M., Santana Penin L., Vila Jato J.L. Eur. J. Pharm. Biopharm. 39, 255 (1993).
3. Szente L., Szejtli J., Vikmon M., Szemán J., Fenyvesi e., Pasini M., Redenti E., Ventura P. Proceed. 1st World Meeting APGI/APV, pg.579, Budapest (1995).
4. Higuchi T., Connors K.A. Adv. Anal. Chem. Instrum. 4, 117 (1965).
5. Nogami H., Nagai T., and Yotsuyanagi, T. Chem. Pharm. Bull., 17, 499-509 (1969).
6. Khan K.A. J. Pharm. Pharmacol. 27, 48-49 (1975).
7. Vromans H., Eissens A.C. and Lerk C.F., Acta Pharm. Technol., 35, 250-255 (1989).
8. Szejtli, J. Pharm. Technol. Int., 3, 15-22, (1991).

STUDY OF COMPLEXATION OF GLICLAZIDE WITH ß-CYCLODEXTRIN IN SOLUTION BY NMR TECHNIQUES

J.R. MOYANO, J.M. GINES, M.J. ARIAS, J.I. PEREZ-MARTINEZ, G. BETTINETTI# and F. GIORDANO@
Department of Pharmacy and Pharmaceutical Technology. Faculty of Pharmacy. University of Seville. 41080 Seville (SPAIN)
 # Department of Pharmaceutical Chemistry. Faculty of Pharmacy. University of Pavia. Via Taramelli 12. 27100 Pavia (ITALY)
@ Pharmaceutical Department. Faculty of Pharmacy. University of Parma. Viale delle Scienze. 43100 Parma (ITALY)

ABSTRACT

The study of complexation between GL and ß-CD in liquid medium has been carried out by phase-solubility, ^{1}H and ^{13}C NMR studies. A formation complex is observed from the phase solubility diagram, being the average association constant of 1094 M^{-1}. The NMR studies revealed the preferent complexation of the aliphatic moiety of GL. The aromatic moiety is also entrapped, but in minor extent, by the CD molecules.

1. INTRODUCTION

Gliclazide (GL) (see Figure 1 for its structure) is an orally active hypoglycemic agent, included in the second generation sulphonylurea group. It is characterized by a poor solubility in water and gastrointestinal fluids, which yields an absorption process limited by its dissolution rate and interindividual variability on its bioavailability [1].The aim of this work is the study of complexation of GL with the ß-CD in liquid medium, in order to evaluate this CD as an adequate complexant agent for our objectives. Phase-solubility and ^{1}H and ^{13}C NMR studies were used to evidence the inclusion process.

2. MATERIALS AND METHODS

GL was kindly supplied by Servier (E-Madrid) and β-CD was purchased from Roquette (F-Lestrem). D_2O and deuterated sodium hydroxide were purchased from Merck (E-Barcelona). The solubility studies have been carried out by the Higuchi and Connors technique at 298 K. ^{1}H and ^{13}C NMR spectra recorded at 310K using a Bruker ACF 200 spectrometer operating at 200.13 and 50.3 MHz, respectively. The chemical shifts

J. Szejtli and L. Szente (eds.), Proceedings of the Eighth International Symposium on Cyclodextrins, 309–312.

were referred to an external sodium trimethylsilylpropionate (TSP) at 0 ppm. The solvent employed was a 0.2 N solution of NaOD in D_2O. The nOe measurements were made during *steady-state* experiments, by irradiation of the H3 signal of the β-CD, at a temperature of 310 K.

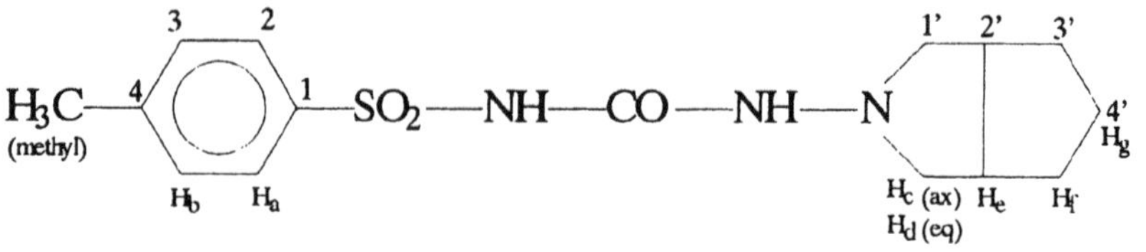

3 2 1' 2' 3'
H_3C—4 (ring) 1—SO_2—NH—CO—NH—N (bicycle) 4' H_g
(methyl)
H_b H_a H_c (ax) H_e H_f
H_d (eq)

Figure 1. Structure of GL.

3. RESULTS AND DISCUSSION

3.1. Phase-solubility studies

The studies revealed that GL shows a typical Bs solubility curve in presence of ß-CD (Figure 1), with the apparition of a precipitate, corresponding to a complex GL-ß-CD. The apparent 1:1 stability constant value, calculated from the first straight line portion of the solubility diagram, appears to be 1094 M^{-1}. By the other hand, the stoichiometry of the complex formed, calculated from the plateau region of the diagram, was of 1:2 drug:CD.

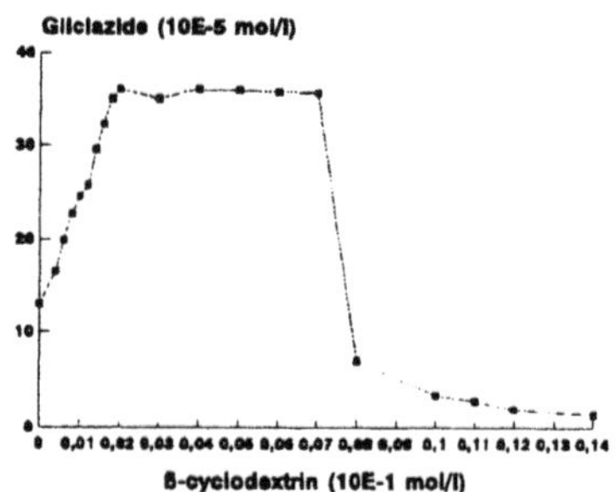

Figure 2. Phase-solubility diagram GL-β-CD.

3.2 ^{1}H NMR spectroscopy

The chemical shifts for the GL and β-CD protons are summarized in Tables 1 and 2, respectively. The GL is characterized by the presence of two groups (azabicyclooctyl and tolyl) which potentially may interact with the CD cavity. For them, it is observed that the protons of the azabicyclooctyl group showed the higher chemical shifts variations, indicating the preferential complexation of this radical compared with the aromatic one.

For the β-CD, the presence of GL is related to an upfield shift of its H3 proton and the apparition of the H5 proton signal, which is overlapped with the H6 signal, indicating also an upfield shift of this one, which is ascribed to the GL complexation. In our case, they are not observed stronger changes of the chemical shifts of H3 and H5 signals. This fact also indicates a preferential complexation of the aliphatic moiety, which does not induces an strong anisotropic effect such as a ring current of the π electrons of an aromatic group.

TABLE 1. ^{1}H chemical shifts corresponding to the Gl, in absence and presence of β-CD.

Gl protons	δ_{Free}	$\delta_{Complex}$	$\Delta\delta$ (ppm)
Ha	7.700	7.710	0.010
Hb	7.360	7.360	0
Hc (eq)	3.100	3.220	0.120
Hd (ax)	2.190	2.060	-0.130
He	2.530	2.600	0.070
Hf	1.550	1.620	0.070
Hg	1.402	1.500	0.098
Methyl	2.390	2.390	0

TABLE 2. ^{1}H chemical shifts corresponding to the β-CD, in presence and absence of Gl.

β-CD protons	δ_{Free}	$\delta_{Complex}$	$\Delta\delta$ (ppm)
H1	4.940	4.940	0
H2	3.490	3.495	0.005
H3	3.860	3.840	-0.020
H4	3.420	3.420	0
H5	-	3.780	-
H6	3.820	-	-

3.3. NOE studies

For these experiments, a significant nOe effect is observed between the protons of the azabicyclooctyl group and the H3 protons, principally for the Hc (1.17 %) and He (2.08 %), indicating the complexation of this one on the CD cavity. For the tolyl group, is also registered a significant signal enhancement of their protons signals (0.78 %), as well indicating the complexation of this group, but in minor extent that the aliphatic ring. All the above mentioned observations may conclude in two posibilities: a) the presence of a bimodal complexation [2] (i.e., the existence of two 1:1 complexes), with a formation of a 1:2 complex at high concentrations of β-CD or b) the direct formation of an 1:2 inclusion compound. This matter will be treated in future studies.

3.4. ^{13}C NMR spectroscopy

The ^{13}C NMR asignments signals for the pure components and the complex are reported in Tables 3 and 4. The most shifted ^{13}C NMR signals for the drug molecule corresponded to the azabicyclooctyl moiety, being these results in good agreeement with the ^{1}H NMR studies. By the other hand, the GL-induced large shifts of the C3 and C5 carbons of the β-CD indicate the formation of an inclusion complex, because these carbons are situated in the CD cavity.

TABLE 3. ^{13}C chemical shifts corresponding to the GL, in absence and presence of β-CD.

Gl carbons	δ_{Free}	$\delta_{Complex}$	Δδ (ppm)
C1	145.540	145.483	-0.057
C2	128.623	128.651	0.028
C3	132.022	131.948	-0.074
C4	142.480	142.605	0.125
Methyl	23.193	23.258	0.065
C1'	64.825	65.444	0.619
C2'	42.540	42.751	0.211
C3'	33.712	34.062	0.350
C4'	27.053	27.077	0.024

TABLE 4. ^{13}C chemical shifts corresponding to the β-CD, in presence and absence of Gl.

β-CD protons	δ_{Free}	$\delta_{Complex}$	Δδ (ppm)
C1	105.673	105.750	0.077
C2	75.935	76.009	0.071
C3	76.742	76.834	-0.092
C4	84.598	84.608	0.010
C5	74.583	74.684	0.101
C6	63.187	63.104	-0.083

CONCLUSIONS

From these studies, it is clearly observed the complexation of GL with β-CD in aqueous medium. Under these conditions (0.2 N in NaOD), the NMR studies evidenced the preferential complexation of the azabicyclooctyl moiety, but the phase solubility diagram revealed the formation of a 1:2 compound, indicating also the participitation of the tolyl moiety.

REFERENCES

[1] Palmer, K.J., Brogden, R.N., Gliclazide. An Update of its Pharmacological Properties and Therapeutic Efficacy in Non-Insulin-Dependent Diabetes Mellitus. *Drugs*, **46**, 92-125 (1993)

[2] Redenti, E., Pasini, M., Ventura, P., The Terfenadine/β-Cyclodextrin Inclusion Complex. *J. Incl. Phenom.*, **15**, 281-292 (1993)

X-RAY STRUCTURES AND THERMAL ANALYSES OF NEW CD/DRUG INCLUSION COMPOUNDS

M.R.CAIRA, V.J.GRIFFITH, G.R.BROWN AND L.R.NASSIMBENI
Department of Chemistry, University of Cape Town, Rondebosch, 7700, South Africa.

ABSTRACT

Single crystal X-ray methods, thermogravimetry and differential scanning calorimetry have been used to characterize several cyclodextrin/drug complexes in order to reconcile crystal packing features with behaviour on heating.

1. INTRODUCTION

Despite the widespread use of cyclodextrins (CDs) as solubilizers for poorly soluble drug substances [1], there are relatively few studies describing comprehensive investigation of CD/drug complexes by thermal analysis and single crystal X-ray diffraction. Used in combination, thermogravimetry (TG), differential scanning calorimetry (DSC) and single crystal X-ray analysis can elucidate the nature of host-guest interactions as well as the relation between structure and thermal decomposition. With non-volatile drug guests, thermal decomposition involves crystal water loss as the first stage. Dehydration may occur in one step or in a series of steps. If the crystal structure of the complex is known, sequential water loss on heating may be correlated with the known crystal sites of hydration in favourable cases.

2. MATERIALS AND METHODS

2.1 Materials

All cyclodextrins were obtained from Cyclolab, Hungary. Diclofenac sodium was purchased from Syntex, USA, and meclofenamate sodium and (*L*)-menthol from Sigma Chemical Company, USA.

2.2 Methods

Single crystals of the CD-complexes were obtained by slow evaporation or slow cooling of aqueous solutions containing host and guest in known molar ratios. Single crystal X-ray structures were determined by direct phasing or isomorphous replacement techniques. Thermal analysis was performed on a Perkin Elmer PC7-Series Thermal Analysis system with 5-10 mg samples at a scanning rate of

J. Szejtli and L. Szente (eds.), Proceedings of the Eighth International Symposium on Cyclodextrons, 313–316.

10°C min^{-1} under constant N_2-purge. Full details of crystallization conditions, X-ray structures and thermal analyses appear in the references cited.

3. RESULTS AND DISCUSSION

3.1 Crystal structures and thermal analyses

The β-CD complexes of diclofenac sodium [2] and meclofenamic acid sodium salts [3] are monomeric species crystallizing in the space groups $P6_1$ and $P2_12_12_1$ respectively with 11 and 16 H_2O molecules in the respective formula units. Structural analyses revealed that in each case the phenyl ring associated with the -COOH function is included in the host cavity from the primary hydroxyl side. In the former case, all H_2O molecules are ordered and three of them are coordinated to a Na^+ ion whose octahedral coordination sphere is completed by three O atoms from neighbouring CD molecules. Fig.1(a) shows the combined TG and DSC traces for this species. Three points of inflection are observed in the TG trace, representing successive loss of 6, 2.2, 1.8 and 1.0 H_2O molecules. The temperature range for H_2O loss (30-244°C) is considerably larger than that observed for complexes containing neutral guests and is attributed to the fact that some H_2O molecules are strongly coordinated to the Na^+ ion (O•••Na^+ 2.3(1)-2.46(1)Å). In addition to being bound to Na^+, one H_2O molecule is strongly hydrogen bonded to three other O atoms and it is suggested that loss of this molecule is reflected in the DSC endotherm B of Fig.1(a). It appears likely that just prior to this event, the other two H_2O molecules bound to Na^+ are lost (shoulder in complex endotherm A). Analogous behaviour was observed for β-CD complexes with the K^+ and Cs^+ salts of diclofenac, and for the γ-CD/diclofenac sodium complex. TG and DSC curves for the β-CD/meclofenamic acid Na^+ salt complex are shown in Fig.1(b). A point of inflection in the TG trace indicates stepwise loss of approximately 11.2 and 3.8 H_2O molecules and these events are reflected as overlapping endotherms A, B in the DSC trace. Correlation with structural data is difficult in this case due to crystallographic disorder of both Na^+ and H_2O molecules. The former is disordered over two sites 1.7Å apart with s.o.f.s 0.6 and 0.4. These sites are respectively octahedrally and tetrahedrally coordinated by O atoms. Nevertheless, we infer, by analogy with the data for the diclofenac complexes, that the H_2O molecules most strongly retained are those bound to Na^+.

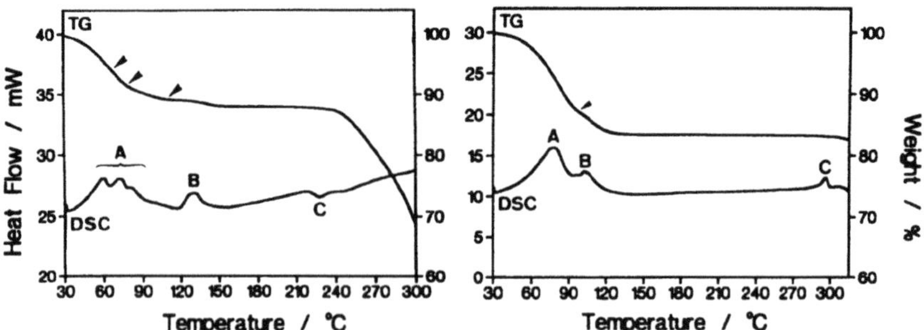

Fig.1 TG/DSC traces for β-CD complexes of (a) diclofenac sodium (left) and (b) meclofenamic acid sodium salt (right)

The β-CD/ibuprofen complex is dimeric (space group C2). X-ray analysis [4] showed that the CD molecules pack to form infinite linear channels. Severe disorder of the guests in the channel prevented their modelling by X-ray methods. However, the H_2O molecules are relatively ordered and of the 13.3 H_2O molecules per host CD (from TG analysis), nine were located outside the channels. These different environments of the H_2O molecules were again reflected as two-step H_2O losses in both TG and DSC traces. We have also determined the crystal structure of the TRIMEB/ibuprofen complex [5]. This complex contains ordered drug guest and is anhydrous. Consequently, the DSC trace shows only one sharp fusion endotherm for the complex.

(*L*)-menthol forms 1:1 inclusion compounds with β-CD ($P2_1$), TRIMEB ($P2_12_12_1$) and DIMEB ($P2_12_12_1$). X-ray analysis showed that the β-CD complex is dimeric, with the guest adopting distinctly different orientations in the cavities of symmetry-independent hosts. These orientations in turn differed from that of the guest within the cavity of TRIMEB [6]. The guest -OH group does not engage in hydrogen bonding with host O atoms in either of these complexes and the guest molecules are retained by their hosts by purely hydrophobic interactions. The DSC onset temperature for water loss from the β-CD complex is 47°C, but for the TRIMEB complex an unusually high value of 124°C was measured. The crystal structure data are consistent with these observations, revealing that in the former crystal, H_2O molecules occupy channel-like regions between columns of complex units, allowing their relatively easy escape on heating, whereas with TRIMEB as host, the H_2O molecules are enclosed in cavities formed by closely packed complex molecules.

The crystal structure of the DIMEB/(*L*)-menthol complex was not determined but the measured unit cell data are similar to those for e.g. the DIMEB/*p*-iodophenol complex [7], except that the length c is approximately doubled. The TG/DSC trace is shown in Fig.2. It differs from the traces for the other (*L*)-menthol complexes in that two weight losses, with corresponding endotherms (A, D), are observed. By chemical and ^{1}H-NMR analysis, we ascertained that these weight losses correspond to loss of H_2O followed by loss of one molecule of menthol. The traces were highly reproducible, the DSC trace showing, in addition, exothermic peaks B and C attributed respectively to a phase change for the dehydrated complex and a second phase change which possibly initiates loss of menthol.

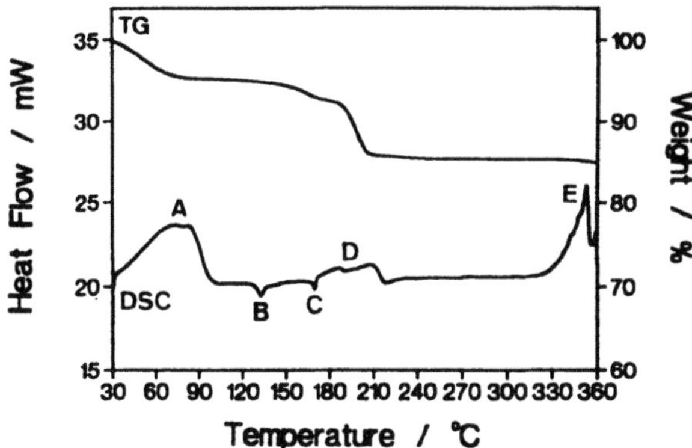

Fig. 2 TG/DSC trace for the complex between DIMEB and (L)-menthol

By analogy with the structure of the DIMEB/*p*-iodophenol complex, in which the DIMEB cavity contains H_2O molecules and the guest resides outside, it is possible that the guest (*L*)-menthol in the DIMEB complex is similarly located. This may account for the distinctly different behaviour on heating compared with that for the *β*-CD and TRIMEB complexes, but single crystal X-ray analysis of the DIMEB/(*L*)-menthol complex is necessary to support this explanation.

3.2 Importance of C-H···O bonds in TRIMEB

Examination of the crystal structures of TRIMEB complexes we have prepared invariably reveals the presence of host intramolecular $C(6G_n)$-H···$O(5G_{n-1})$ hydrogen bonds with C···O in the range 3.0-3.4Å. These interactions stabilise the host conformations not only in the complexes but also in TRIMEB monohydrate [8].

4. CONCLUSIONS

Interpretation of detailed features in TG and DSC curves of CD-complexes is facilitated when three-dimensional X-ray crystal structural data are available. This has been demonstrated for several CD/drug complexes.

ACKNOWLEDGEMENTS

We thank the University of Cape Town, the Foundation for Research Development, and South African Druggists for financial support.

REFERENCES

[1] Frömming, K.-H, Szejtli, J. Cyclodextrins in Pharmacy, Kluwer Academic Publishers, Dordrecht, 1994.

[2] Caira, M.R., Griffith, V.J., Nassimbeni, L.R., van Oudtshoorn, B., Synthesis and X-ray crystal structure of *β*-cyclodextrin diclofenac sodium undecahydrate, a *β*-CD complex with a unique crystal packing arrangement, *J.Chem.Soc., Chem.Commun.*, 1061-1062, (1994).

[3] Griffith, V.J., Physicochemical characterisation of cyclodextrin-drug complexes, Ph.D. Thesis, University of Cape Town, 1996.

[4] Brown, G.R., Caira, M.R., Nassimbeni, L.R., in preparation, 1996.

[5] Brown, G.R., Caira, M.R., Nassimbeni, L.R., Inclusion of ibuprofen by heptakis(2,3,6-tri-O-methyl)-*β*-cyclodextrin: an X-ray diffraction and thermal analysis study, *J. Incl. Phen.*, submitted (1995).

[6] Caira, M.R., Griffith, V.J., Nassimbeni, L.R., van Oudtshoorn, B., X-ray structures of 1:1 complexes of (*L*)-menthol with *β*-cyclodextrin and permethylated *β*-cyclodextrin, *Supramol. Chem.*, in press (1996).

[7] Harata, K., The structure of the cyclodextrin complex. XXI. Crystal structures of heptakis(2,6-di-O-methyl)-*β*-cyclodextrin complexes with *p*-iodophenol and *p*-nitrophenol, *Bull.Chem.Soc.Jpn.*, **61**, 1939-1944, (1988).

[8] Caira, M.R., Griffith, V.J., Nassimbeni, L.R., van Oudtshoorn, B., Unusual $^1C_4$ conformation of a methylglucose residue in crystalline permethyl-*β*-cyclodextrin monohydrate, *J.Chem.Soc., Perkin Trans.2*, 2071-2072 (1994).

INCREASING THE SOLUBILITY CHARACTERISTICS OF D-NORGESTREL WITH CYCLODEXTRINS

Z. AIGNER[1], GY. DOMBI[2] and M. KATA[1]
[1]*Department of Pharmaceutical Technology,* [2]*Department of Pharmaceutical Chemistry, Albert Szent-Györgyi Medical University H-6701 Szeged, P.O.Box 121, Hungary*

ABSTRACT

Levonorgestrel dissolves only slightly in water. Attempts were made to increase the solubility properties of this drug by complexing with cyclodextrins. The products were investigated with a dissolution tester and a Sartorius resorption model. X-ray and NMR spectra of the inclusion complexes were analysed.

1. INTRODUCTION

Levonorgestrel (**Lev**) is one of the most popular contraceptives. It is used in combination with estrogen. It was discovered by *Hughes et al.* (1963). It is a white or nearly white powder, which dissolves only slightly in water.
Cyclodextrins (**CDs**) form inclusion complexes with numerous guest drug molecules and this complexation increases the solubility and rate of dissolution of these drugs.
Our aim was to increase the solubility of **Lev** by using **CDs** and to investigate these products (dissolution rate, in vitro diffusion properties, X-ray, NMR, etc.).

2. MATERIALS AND METHODS

2.1. Materials

Levonorgestel (D-norgestrel, **Lev**), 13-ethyl-17-hydroxy-18,19-dinorpregn-4-en-20-yn-3-one (Chemical Works of G. Richter Ltd., Budapest, Hungary) [1].
α-, β-, γ-CD, methyl-β-CD, dimethyl-β-CD, hydroxyethyl-β-CD, hydroxypropyl-β-CD, RAMEB (Cyclolab Ltd., Budapest, Hungary) [2].

2.2. Apparatus

USP rotating-basket dissolution apparatus, type DT; kneading mixer, type LK5 (Erweka Apparatebau GmbH., Heusenstamm, Germany); Sartorius resorption model (Germany); Spektromom 195 (MOM, Budapest, Hungary); Specord UV-VIS (C. Zeiss, Jena, Germany); DRON UM-1 X-ray apparatus; BRUKER Avance Drx 400 NMR spectrometer.

J. Szejtli and L. Szente (eds.), Proceedings of the Eighth International Symposium on Cyclodextrons, 317–320.

2.3. Preparation of products

Products were prepared in four different mole ratios (**Lev:CD** mole ratio = 2:1, 1:1, 1:2 and 1:3).
Physical mixtures: The ground components were mixed in a mortar and sieved through a DIN 0.315 mm sieve.
Kneaded products: Physical mixtures of **Lev** and **γ-CD** were mixed (Erweka LK5) in the same quantity of ethanol + water (1:1). They were kneaded until the bulk of the solvent mixture had evaporated. After this, they were dried at room temperature and then at 105 °C, and were next pulverised and sieved (DIN 0.315 mm).
Products were stored under normal conditions at room temperature in closed glass containers.

2.4. Dissolution of drug

In the USP rotating-basket dissolution apparatus, 20 mg of pure **Lev**, or product containing 20 mg of **Lev**, was examined in 900.0 g of distilled water. The basket was rotated at 100 rpm. Sampling was performed after 5, 10, 15, 30, 60 and 90 min. The sample volume was 5.0 mL. The **Lev** contents of the samples were determined spectrophotometrically.

2.5. Membrane diffusion

Measurements were performed on 100.0 mL of artificial gastric juice (pH = 1.1 ± 0.1) or artificial intestinal juice (pH = 7.0 ± 0.1) and artificial plasma (pH = 7.5 ± 0.1). 20 mg of active agent, or product containing 20 mg of active agent, was in the donor phase in all cases. The temperature was 37.5 ± 1.5 °C. During the examination, 5.0 mL samples were taken five times (after 30, 60, 90, 120 and 150 min) and their active agent contents were determined spectrophotometrically. The amount of diffused active agent was calculated.

2.6. NMR spectra

The high-resolution NMR spectra were measured on a BRUKER Avance DRX 400 Fourier transform NMR spectrometer at 400 MHz ^{1}H frequency. The samples were dissolved in $CDCl_3$ and the deuterium signal of the solvent was used to lock the spectrometer. The spectra were recorded at room temperature; 32 K data points were used with a digital resolution of 0.26 Hz/pt.

3. RESULTS AND DISCUSSION

3.1. Preliminary experiments

A mixture of 0.03 g of **Lev** and 0.50 g of **CD** derivative was diluted to 20.0 g with water and stirred for 15 min. Suspension systems were filtered and the UV spectra were recorded. A system without **CD** was used as a control. As **γ-CD** had the highest influence on the solubility (by a factor of 310), this derivative was used for the further examinations.
The absorption maximum was 243 nm. The absorption obeyed the Bouguer-Lambert-Beer law in the concentration interval 0-15 µg/mL.

3.2. Dissolution studies

The amount of **Lev** that dissolved in distilled water during 90 min was less than 2.64%.

The physical mixtures yielded a higher dissolution of active agent as compared to **Lev**. The highest value for physical mixtures was obtained for the 1:3 composition (10.85 mg/900 mL), which is more than a 20-fold solubility increase relative to the pure active agent. Maximum dissolution was attained at 30 min, and this value did not change significantly later (saturation).
On dissolution of the kneaded products, similarly as for the physical mixtures, the best results were obtained for the 1:3 composition (17.34 mg/900 mL). Dissolution was better and faster than for the physical mixtures. The maximum was reached at about 5-15 min.
Summarizing: increasing CD ratio increased the amount of dissolved material. The rate of dissolution and the amount of dissolved material depended on the preparation methods. There were significant differences in the amount of dissolved drug between analogous compound products made by different preparation methods.

3.3. Membrane diffusion examinations

The results of these examinations can be seen in **Tables 1 and 2**.

TABLE 1. Membrane diffusion examinations in artificial gastric juice (mg /100 mL)

		Physical mixtures				Kneaded products			
Time	**Lev**	2:1	1:1	1:2	1:3	2:1	1:1	1:2	1:3
30'	0.0664	0.0987	0.0483	0.0806	0.0894	0.0763	0.0922	0.0786	0.0998
60'	0.0971	0.1020	0.0521	0.0899	0.0993	0.0976	0.1041	0.0930	0.1190
90'	0.0878	0.1152	0.0818	0.1015	0.1059	0.1069	0.1226	0.1012	0.1349
120'	0.0747	0.1053	0.0796	0.1179	0.1020	0.1174	0.1259	0.1078	0.1503
150'	0.0894	0.1119	0.0916	0.1201	0.1097	0.1267	0.1346	0.1127	0.1487

TABLE 2. Membrane diffusion examinations in artificial intestinal juice (mg/100 mL)

		Physical mixtures				Kneaded products			
Time	**Lev**	2:1	1:1	1:2	1:3	2:1	1:1	1:2	1:3
30'	0.1168	0.0905	0.1300	0.1695	0.1185	0.1168	0.1284	0.1531	0.1432
60'	0.1234	0.1020	0.1201	0.1695	0.1331	0.1284	0.1333	0.1744	0.1794
90'	0.1152	0.1119	0.1366	0.1794	0.1421	0.1514	0.1580	0.1827	0.2041
120'	0.1168	0.1201	0.1333	0.1810	0.1399	0.1547	0.1563	0.1893	0.1991
150'	0.1366	0.1185	0.1300	0.1909	0.1564	0.1563	0.1432	0.1909	0.2041

Summarizing: most of the products showed a slight increase in diffusion as compared to **Lev**. There was no significant difference between the different compositions. The amount of **Lev** that diffused from the kneaded product was higher.

3.4. X-Ray investigations

Inclusion complex formation was investigated by X-ray techniques. As can be seen in **Fig. 1, Lev** and **γ-CD** have crystalline structures, while their 1:3 kneaded product is amorphous: the spectrum contains no characteristic peaks of **Lev** or **γ-CD**. The change in this spectrum during 1 month is not important.

3.5. ^{1}H NMR investigation

Comparison of the ^{1}H NMR spectra of norgestrel and its complex formed with CD revealed the following changes:

- the chemical shift of the olefinic proton at position 4 moved from 5.83 ppm to 5.97 ppm;

- the chemical shift of the ethinyl proton at position 17 showed a paramagnetic shift from 2.59 ppm to 3.11 ppm;
- the chemical shifts of the angular ethyl group at position 13 exhibited a paramagnetic shift.

From the paramagnetic shifts summarized above, it can be seen that, during the complexation, a H-bond is formed between the hydroxy groups of **CD** and the 3-oxo and 17-OH groups of **Lev**. The paramagnetic shielding effect of the oxygens of **CD** is best seen on the neighbouring protons, e.g. at positions 4 and 17 (**Fig. 2**).

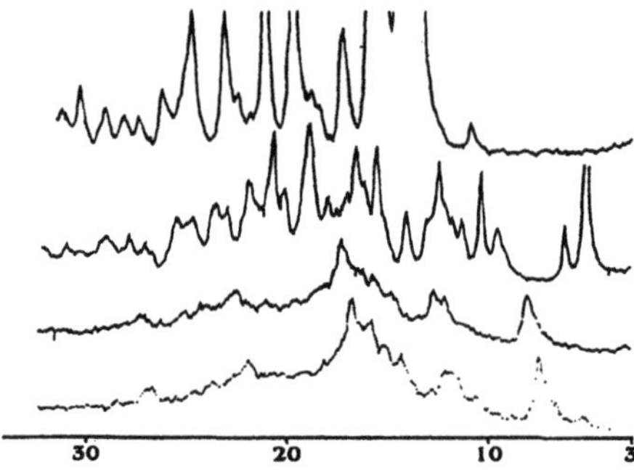

Fig. 1. X-ray spectra of **Lev**, **γ-CD**, 1:3 kneaded product

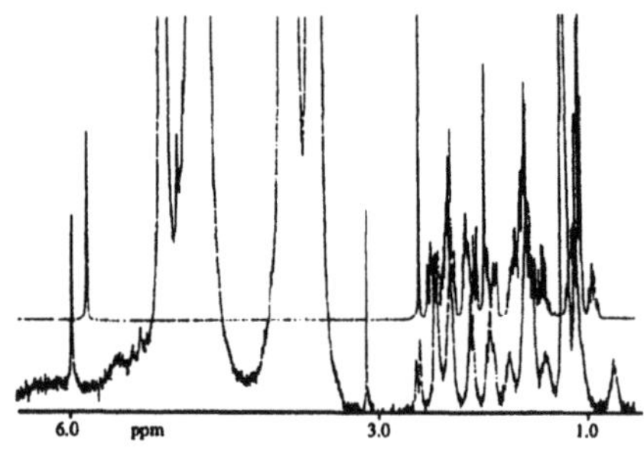

Fig. 2. ^{1}H NMR spectra of **Lev** and 1:3 kneaded product

4. CONCLUSIONS

- **CD** derivatives increase the solubility of **Lev**; the best solubility increase was found for γ-CD.
- Dissolution of the active agent increases on increase of the **CD** content; the best dissolution results were obtained for the 1:3 **Lev**:**γ-CD** composition.
- The dissolution rate is more than 20 times for the physical mixture, and more than 33 times for the kneaded product.
- Influence of the preparation method: kneading better increases the dissolution properties.
- The **CDs** slightly infuence the diffusion rate of **Lev**; the increase is not significant.
- The X-ray and NMR spectra of the products confirmed the solubility and diffusion results and showed the presence of complex formation.

The use of **CDs** affords possibilities of increasing the solubility characteristics of **Lev**, and by this means the quantity of drug in the dosage forms may be decreased.

References

[1] USP XXIII and NF XVIII. US Pharmacopeial Convention, Inc., 1994
[2] Szejtli, J., Cyclodextrin Technology, Kluwer Academic Publishers, Dordrecht, 1988

ENHANCED WATER-SOLUBILITY OF ALBENDAZOLE BY HYDROXY-PROPYL-β-CYCLODEXTRIN COMPLEXATION

V.L. BASSANI[1], D. KRIEGER[1], D. DUCHENE[2] and D.WOUESSIDJEWE[2]

[1] *Faculty of Pharmacy,. Av. Ipiranga, 2752, 90610-000 Porto Alegre, UFRGS, Brazil*

[2] *Laboratoire de Physico-chimie, Pharmacotechnie et Biopharmacie, URA CNRS 1218, Faculté de Pharmacie, Université de Paris-Sud, 92290 Châtenay Malabry, France*

ABSTRACT

The inclusion complexation of methyl (5-(propylthio)-1*H*-benzimidazol-2-yl) carbamate, albendazole (ABZ) with 2-hydroxypropyl-β-cyclodextrin (HPβCD) in water was investigated with a view to improving the low aqueous solubility of the drug. The combination of albendazole and HPβCD in a molar ratio of 1/10 resulted in a significant increase in the aqeous solublity of the drug, up to 3500 times. Albendazole/HPβCD complexes could be recommended as a parenterally administered formulation because of its good solubility properties and the safety of the cyclodextrin used.

1. INTRODUCTION

Albendazole (Figure 1) belongs to a group of benzimidazol derivatives with a broad spectrum of activity against human and animal helminthe parasites such as nematodes, metacestodes and hydatoses [1, 2]. These potential anthelmintic effects of ABZ, its relatively good tolerance and its low cost explain its wide use against veterinary and human parasites for more than two decades. However, the low water solubility of the drug (0.2 μg/ml at pH 7.4) is one of the limiting factors for its bioavalability. In some cases, high-dose therapy by the oral route then becomes necessary ,which could lead to adverse reactions such as gastro-intestinal disturbances and liver impairment [3, 4]. Futhermore the correct intake of doses is critical especially in veterinary medicine.

The improvement of water solubilty of ABZ by utilizing hydrophilic cyclodextrin appears to be one solution to these problems. Some authors carried out a complexation of ABZ with dimethyl-β-cyclodextrin [5]. In this study, we report the combination of ABZ and hydroxypropyl-β–cyclodextrin to achieve inclusion complexes which could be used in injectable formulations.

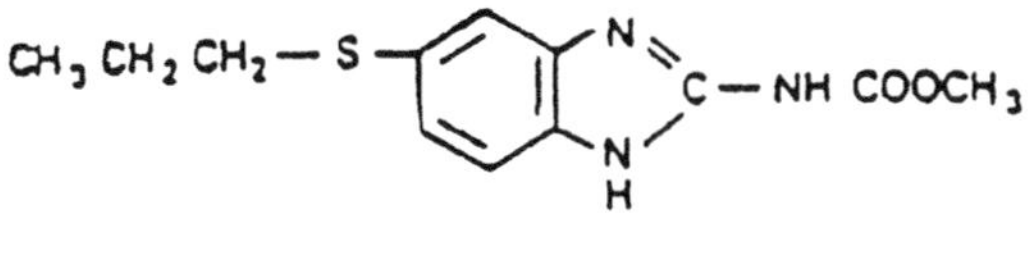

Fig. 1 Chemical structure of albendazole

J. Szejtli and L. Szente (eds.), Proceedings of the Eighth International Symposium on Cyclodextrons, 321–324.

2. MATERIALS AND METHODS

2.1. Materials

Albendazole was supplied by Smith, Kline & French (France). 2-hydroxypropyl-β-cyclodextrin (BETA W 7 HP 0,9) was purchased from Wacker Chimie SA (Lyon, France). All other materials were of analytical grade.

2.2. Methods

2.2.1. Solubility studies and complex preparation

Solubilities of ABZ were determined by adding an excess amount of the drug (66 mg) to 25 ml of aqueous solutions containing increasing concentrations of HPβCD. Typically, seven samples of the following molar ratios ABZ/HPβCD were prepared: 1/2, 1/4, 1/6, 1/8 and 1/10. The suspensions formed were stirred at 37 °C for at least 6 days, after which equilibrium is reached. After cooling to 25 °C, the suspensions were filtered through a 0.45 μm membrane filter (Millipore HVLP). The supernatants were suitably diluted in water (HCl 0.1 M) and analyzed spectrophotometrically at 230 nm.

Preparation of the complex was carried out in the same way, with a molar ratio of 1/10 for ABZ and HPβCD respectively. 0.25 mmol of ABZ was suspended in 25 ml of aqueous solution containing 2.5 mmol of HPβCD. The suspension was also stirred for 6 days, but at a higher temperature (55 °C) than for the solubility studies. The supernatant obtained as described above was freeze-dried in order to give a solid state complex.

2.2.2. Characterization of the complex

The complex was studied by UV-spectrophotometry using a Varian UV-VIS spectrophotometer (Cary/1E). Circular dichroism spectra were observed using a Jobin Yvon Mark V spectrometer. An appropriate quantity of the solid inclusion compound of ABZ/HPβCD (6.55 mg ABZ per gramme of complex) was dissolved in distilled water to obtain a mother solution of 1.5 mg ABZ per ml. This solution was filtered through a 0.45 μm membrane filter and the filtrate was diluted 10 times with distilled water prior the UV and circular dichroism studies. An equivalent quantity of physical mixture ABZ/HPβCD was treated in the same way.

3. RESULTS AND DISCUSSION

3. 1. Solubility

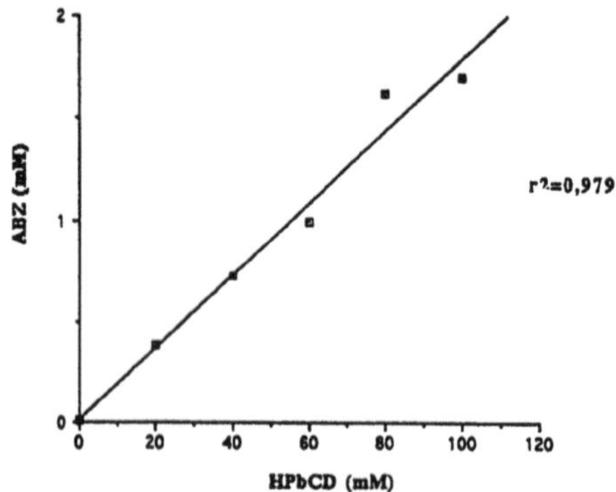

Fig. 2 The phase-solubility diagram of ABZ in aqueous HPβCD at 37 °C

The phase-solubility diagram of ABZ (Figure 2) appeared to be linear and could correpond to the type A, which mean a soluble inclusion is probably formed.

3.2. Characterization of ABZ/HPβCD complex

The interaction of ABZ with HPβCD in the liquid phase was first examined by UV-spectrophotometry. The spectra displayed in Figure 3 show a strong peak at about 220 nm and one very weak peak at 240 nm, which probably correspond to the absorbance of ABZ. The intensity of absorbance is higher for the inclusion compound than for the physical mixture, indicating a possible interaction between ABZ and HPβCD.

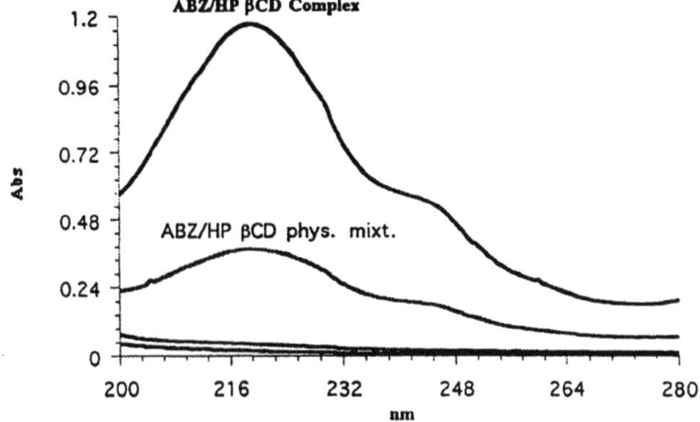

Fig. 3 UV-spectrum of ABZ/HPβCD complex and physical mixture

In the circular dichroism spectrum (Figure 4), a relatively strong broad negative peak is detected at about 220 nm. As in the UV spectum, the modification of dichroic spectrum is more important in the case of the complex.

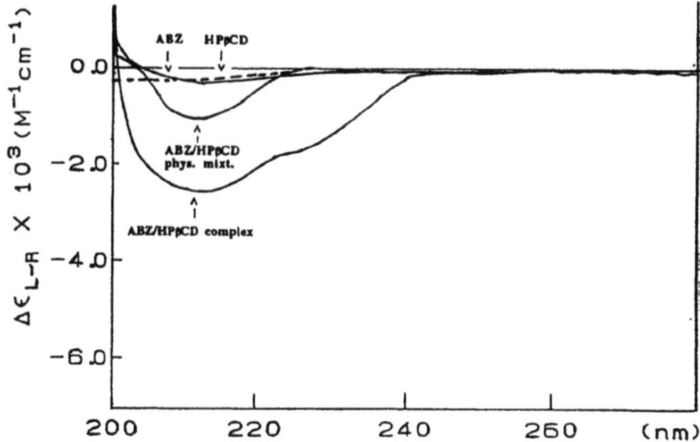

Fig. 4 Circular dichroism spectra of ABZ in the presence and absence of HPβCD

The increase in apparent solubility of ABZ, and the results of UV-spectrophotometry and circular dichroism obviously suggest the possible interaction of ABZ with the cavity of HPβCD.

The solubility study of ABZ from the solid complex (molar ratio of 1/10 ABZ/HPβCD) performed at 25 °C showed that 745 μg was rapidly dissolved in 1 ml of distilled water. Owing to the low water solubility of this drug, 0.2 μg/ml phosphate buffer, pH 7.4 [6], it could be assumed that the complex affords a significant solubility enhancement of ABZ, up to a 3500 times increase in solubility.

Recently, a pharmacokinetic study of albendazole was reported in swine following the oral administration of the drug suspended in water. One of their main conclusions was that albendazole displays very large differences on absorption, metabolism and elimination [7]. It seems obvious that the poor water solubility of the drug plays a role on the discrepancy observed. The combination of albendazole with hydrophilic cyclodextrins to give an aqueous solution of the drug could be beneficial for both oral and parenteral formulations.

CONCLUSION

The noticeable water solubility enhancement for albendazole could be obtained by its complexation with 2-HPβCD in the appropriate molar ratio 1/10. The good solubility of the complex makes possible the preparation of solutions for both oral and parenteral administration of the drug.

REFERENCES

[1] Gyurik, R.J., Chow, A.W., Zaber, B., Brunner, E.L., Miller, J.A., Villani, J.A., Petka, L.A., Parish, R.C., Metabolism of albendazole in cattle, sheep, rats, and mice, *Drug Met. Disp.*, **9**, 503-508 (1981).

[2] Prieto, J.G., Alonso, M.L., Justel, A., Santos, L., Tissue levels of albendazole after *in vivo* intestinal and gastric absorption in rats, *J. Pharm. Biomed. Anal.*, **6**, 1059-1063, (1988).

[3] Fourestié, V., Bougnoux, M.E., Ancelle, T., Liance, M., Roudot-Thoraval, F., Naga, H., Pairon-Pennachioni, M., Rauss, A., Lejonc, J.L., Randomized trial of albendazole versus tiabendazole plus flubendazole during an outbreak of human trichinellosis, *Parasitol Res.*, **75**, 36-41, (1988).

[4] Davis, A., Multicentre clinical trials of bendimidazole carbamate in human echinococcosis, *Bull. WHO.*, **64**, 384-388, (1986).

[5] Khata, M., Schauer, M., Increasing of the solubility characteristics of albendazole with dimethyl-β-cyclodextrin, *Acta Pharm. Hung.*, **61**, 23-31, (1991).

[6] Bogan, J.A., Marriner, S.E., *Pharmacodynamic and toxicological aspects of albendazole in man and human*, in Proc. Int. Congr. Ser R. Soc. Med., **61**, 13-21, (1984).

[7] Zhangliu, C., Zhenling, Z., Zonghui, Y., Fung, K.-F., *Pharmacokinetic study of albendazole in swine*, in Proc. 5th. EAVPT. Congr. Copenhagen, 1991, Pp. 385-386.

COMPARATIVE STUDY OF THE INCLUSION PROPERTIES OF ß-CYCLODEXTRINS FOR KETOPROFEN AND IBUPROFEN IN SOLUTION AND IN THE SOLID STATE

P. MURA[1], G. P. BETTINETTI[2], A. MANDERIOLI[1], M. T. FAUCCI[1], G. BRAMANTI[1], M. SETTI[3]

[1]*Dipartim. Scienze Farmaceutiche - Università di Firenze, Via Capponi 9, 50121 Firenze (I)*

[2]*Dipartim. Chimica Farmaceutica - Università di Pavia, Viale Taramelli 12, 27100 Pavia (I)*

[3]*Dipartim. Scienze della Terra - Università di Pavia, Via Abbiategrasso 209, 27100 Pavia (I)*

ABSTRACT

The interactions of ketoprofen and ibuprofen with some randomly alkylated ß-cyclodextrins (methyl, hydroxypropyl and hydroxyethyl derivatives) were investigated both in aqueous solution (using phase-solubility analysis) and in the solid state (using differential scanning calorimetry and X-ray diffractometry). The molecular features of the guest play a role in such interactions which were particularly strong in the case of methyl ß-cyclodextrin, the best complexing, solubilizing and amorphizing carrier for both drugs.

1. INTRODUCTION

Amorphous randomly alkylated ß-cyclodextrins generally show higher water solubility and better solubilizing efficacy and complexing properties than crystalline, native ß-cyclodextrin [1]. The nature of the substituent may play a role in their performance, as we have demonstrated in the case of naproxen (NAP). Actually methyl ß-cyclodextrin was a better partner than native ß-cyclodextrin and some hydroxypropyl and hydroxyethyl derivatives, and thus the optimal carrier for this drug [2-4]. We have extended our studies to other non-steroidal antiinflammatory drugs (NSAIDs) with low water solubility with the aim of investigating possible relationships between solubilizing properties of the host and molecular features of the guest. In the present work, the interactions of three randomly alkylated ß-cyclodextrins, i.e. methyl, hydroxypropyl and hydroxyethyl ß-cyclodextrins with ketoprofen and ibuprofen, which are aryl derivatives of propionic acid (as NAP), are investigated using phase-solubility analysis to study inclusion complexation in aqueous solution and differential scanning calorimetry (DSC) supported by X-ray powder diffractometry to characterize the drug-cyclodextrin solid systems.

2. MATERIALS AND METHODS

2.1. Materials

Ketoprofen[1] (KETO), ibuprofen[1] (IBU), ß-cyclodextrin[1] (ßCd) and ß-cyclodextrin derivatives[2] : Methyl (MeßCd), hydroxyethyl (HEßCd) and hydroxypropyl (HPßCd),

J. Szejtli and L. Szente (eds.), Proceedings of the Eighth International Symposium on Cyclodextrons, 325–328.

with an average substitution degree per anhydroglucose unit of 1.8, 1.6, and 0.9 respectively, were used as received. Physical mixtures in the drug to carrier 1:1 (mol/mol) ratio were prepared by gently and smoothly blending suitable amounts of KETO (or IBU) and cyclodextrin (Cd) powders (75-150 µm sieve granulometric fraction) in an agate mortar with a pestle.

2.2. Solubility studies

Excess amounts of KETO or IBU were added to water or Cd aqueous solution (in the 5 to 100 mmol L^{-1} concentration range) in sealed glass containers, and stirred at constant temperature (25, 37 or 45 °C) up to equilibrium. Aliquots were withdrawn and filtered (pore size 0.45 µm), and the drug concentration was determined with a derivative spectroscopic method in the range 268-242 nm (KETO) and 274-270 nm (IBU).

2.3. Thermal analysis and X-ray analysis

DSC was performed with a Mettler TA4000 apparatus equipped with a DSC 25 cell. Samples were weighed (Mettler M3 microbalance) in Al pans (5-10 mg) pierced with a perforated lid, and scanned at 10 K min^{-1} between 30 and 200 °C under static air.
X-ray diffraction patterns were collected with a computer-controlled Philips PW 1800 apparatus in the 2-50° 2 θ interval (scan rate 1° min^{-1}), using a CuKα radiation monochromatized with a graphite crystal.

3. RESULTS AND DISCUSSION

The equilibrium phase solubility diagrams were all linear (A_L-type), except for the IBU-ßCd system which showed a typical B_S-type shape (Fig. 1).

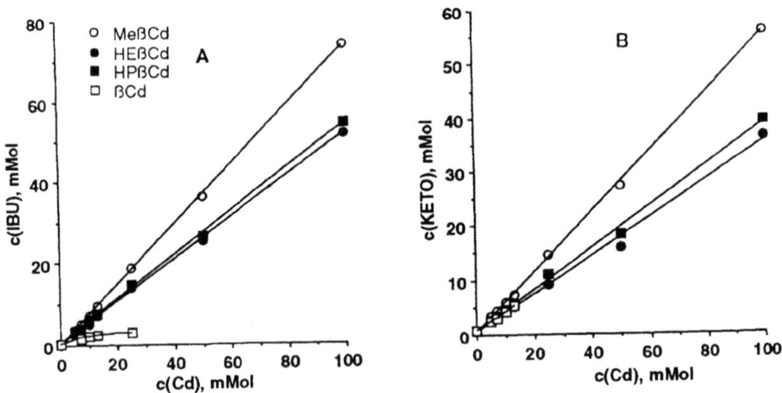

Fig. 1. Phase-solubility diagrams of IBU (A) and KETO (B) in aqueous solution at 37°C.

The apparent stability constants, calculated from the linear portion of the solubility diagrams, are listed in Table 1, with those of the corresponding NAP-ßCds inclusion complexes [3] for comparison purposes.

TABLE 1. Stability constants of ibuprofen, ketoprofen and naproxen with ß-cyclodextrins

Cd	IBUPROFEN Stability constant $K_{1:1}$ (M^{-1})			KETOPROFEN Stability constant $K_{1:1}$ (M^{-1})			NAPROXEN Stability constant $K_{1:1}$ (M^{-1})		
	25°C	37°C	45°C	25°C	37°C	45°C	25°C	37°C	45°C
ßCd	10560	10461	10162	806	770	704	1702	1388	----
MeßCd	12080	10180	7750	2002	1748	1695	6892	5855	5135
HEßCd	4270	3895	3528	777	767	752	2145	1800	1729
HPßCd	5378	4244	3815	970	875	826	2581	1973	----

The decrease in $K_{1:1}$ values with increasing temperature indicated the exothermic nature of inclusion complexation. As for NAP [3], MeßCd was the best solubilizing agent also for KETO and IBU, and formed the most stable inclusion complex. The "strength" of the complexes of KETO with ßCd, HEßCd and HPßCd was similar ($K_{1:1,25°C}$≈1000 M^{-1}), and close to that found for the analogous complexes with NAP ($K_{1:1,25°C}$≈2000 M^{-1}) [3]. In the IBU-Cd system, instead, the strength of the IBU-ßCd complex was about twice that of the IBU-HEßCd and IBU-HPßCd complexes and very close to that of the IBU-MeßCd complex. The solubilizing ability of ßCd was however significantly lower than that of MeßCd, probably due to the low aqueous solubility of the inclusion complex (see Fig. 1A). The best performance of MeßCd might be attributed to the substituent methyl groups that expand the hydrophobic region of the Cd by capping the cavity and increase substrate binding through the hydrophobic effect [5].

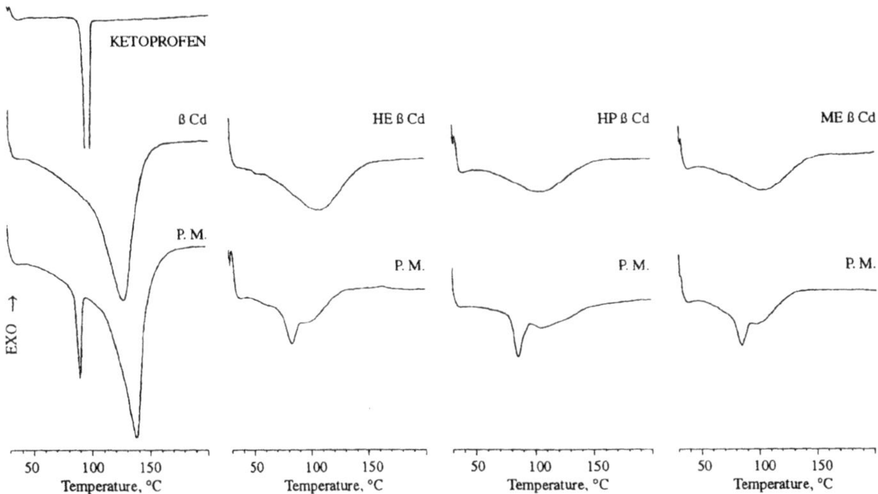

Fig. 2 DSC curves of KETO and its equimolar physical mixtures (P.M.) with ß-cyclodextrins.

DSC analysis (as shown in Fig. 2 for the KETO-ßCds systems) showed that only in the systems with ßCd derivatives the fusion endotherm of IBU (T_{peak} = 76.4±0.3°C, $\Delta_{fus}H$= 111.7±1.9 J/g) and KETO (T_{peak}= 95.8±0.2°C, $\Delta_{fus}H$ = 98.5±0.7 J/g) broadened, with a lower peak temperature and a marked decrease in the associated enthalpy. This indicated a strong interaction in the solid state of both drugs with amorphous Cds, in particular with MeßCd .

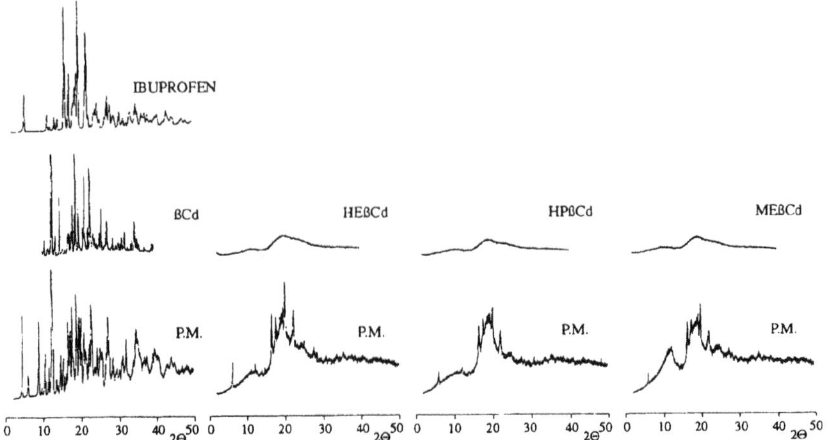

Fig. 3 X-ray diffraction patterns of IBU and its equimolar physical mixtures (P.M.) with ß-cyclodextrins

X-ray diffraction patterns of the physical mixtures of both drugs with ßCd (as shown in Fig. 3 for the IBU-ßCds systems) were the superimposition of those of the single components. In systems with amorphous Cds a decrease in drug crystallinity similar to that found for NAP [3] was instead evident, in agreement with DSC results.

4. CONCLUSION

The analogies observed for the interactions of IBU, KETO and NAP with ßCds suggest similar complexation mechanisms and inclusion modes of the guest molecules in the host cavities. MeßCd gives the most stable inclusion complex and is the most effective derivative in improving the drug solubility properties. This derivative displays also the highest efficiency in drug amorphization in the solid state. The molecular features of the guest play a role in the inclusion (drug-Cd affinity degree decrease in the order IBU>NAP >KETO). Unlike KETO and NAP, IBU displays a marked affinity for native ßCd.

NOTES [1]Sigma Chemical Co.; [2]Kindly donated by Wacker-Chemie GmbH.

REFERENCES

[1] Müller B.W., Brauns U., Backensfeld T., Cyclodextrin derivatives for solubilisation, stabilisation and absorption of drugs, *Proc. IV Int. Symp. Cyclodex.*, O. Huber and J. Szejtli (Eds), Kluwer Academic Publishers, pp. 369-382, 1988.

[2] Bettinetti G.P., Melani F., Mura P., Monnanni M., Giordano F., Carbon-13 NMR study of naproxen interaction with cyclodextrins, *J. Pharm. Sci.*, **80,** 1162-1170 (1991).

[3] Bettinetti G.P., Gazzaniga A., Mura P., Giordano F., Setti M., Thermal behaviour and dissolution properties of naproxen in combinations with chemically modified ß-cyclodextrins, *Drug Dev. Ind. Pharm.*, **18,** 39-53 (1992).

[4] G.P. Bettinetti, P. Mura, F. Melani, F. Giordano, M. Setti (S)-(+)-6-Methoxy-α-methyl-2-naphthaleneacetic acid and ß-cyclodextrin derivatives: inclusion in aqueous media and solid phase interaction, *Proc. 5th Int. Symp. Cyclodex.*, pp. 239-242, Ed. D. Duchene, Eds. de Santé, Paris, 1990.

[5] Green A.R., Guillory J.K., Heptakis (2,6-di-O-methyl)-ß-cyclodextrin complexation with antitumor agent chlorambucil, *J. Pharm. Sci.*, **78** 427-431 (1989).

PHASE SOLUBILITY ANALYSIS IN STUDYING THE INTERACTION OF NIFEDIPINE WITH SELECTED CYCLODEXTRINS IN AQUEOUS SOLUTION

M.S. WORTHINGTON, B.D. GLASS, L.J. PENKLER*

School of Pharmaceutical Sciences, Rhodes University, Grahamstown, 6140, R.S.A.

* *Druggists Group Research, South African Druggists,Ltd, P.O. Box 4002, Korsten, Port Elizabeth, 6014, R.S.A.*

ABSTRACT

The solubilizing potential and complexing tendencies of six cyclodextrins (CyD) with nifedipine in aqueous solution were evaluated using phase solubility methods. Solubility curves of nifedipine with β-CyD, 2-hydroxypropyl-β-CyD (2HP-β-CyD) and 2-hydroxypropyl-γ-cyclodextrin (2HP-γ-CyD) were classified as type A_L, while for heptakis (2,6-dimethyl)-β-CyD (DIMEB), randomly methylated-β-CyD (RAMEB) and γ-CyD, A_P type phase behaviour was observed. Stability constants, calculated from phase solubility diagrams, decreased in the order: DIMEB > RAMEB > β-CyD > 2HP-β-CyD > γ-CyD > 2HP-γ-CyD.

1. INTRODUCTION

Nifedipine (2,6-dimethyl-3,5-dicarboxyl-4-(2'-nitrophenyl)-1,4-dihydropyridine) is a calcium channel antagonist displaying very poor aqueous solubility and hence inferior dissolution and oral bioavailability. Recent studies have shown that the interaction of nifedipine with β-CyD,[1-3] 2HP-β-CyD[4] and some branched cyclodextrins[5,6] in aqueous solution and the solid state, have enhanced the above-mentioned physico-chemical properties. This study continues to investigate the solubilizing potential and complexing tendencies of γ-CyD, 2HP-γ-CyD, randomly methylated-β-CyD (RAMEB) and heptakis(2,6-di-O-methyl)-β-CyD (DIMEB) in aqueous solution using phase solubility techniques.

J. Szejtli and L. Szente (eds.), Proceedings of the Eighth International Symposium on Cyclodextrons, 329–332.

2. MATERIALS AND METHODS

2.1. Materials

Nifedipine, β-CyD, γ-CyD, 2HP-β-CyD (Average Degree of Substitution, D.S. 4.8) and 2HP-γ-CyD (D.S. 5.3) were kindly donated by South African Druggists, Ltd (Port Elizabeth, South Africa). DIMEB and RAMEB (D.S. 12.4) were purchased from Cyclolab (Budapest, Hungary). Water for chromatography was obtained using a Milli-Q® water purification system (Waters Assoc., U.S.A.). All other materials were of analytical reagent grade.

2.2. Methods

Solubility measurements were carried out according to the method described by Higuchi and Connors.[7] Excess amounts of nifedipine were added to fixed volumes of 0.05M potassium phosphate buffer pH 5.8 containing various concentrations of CyDs. Nitrogen was passed through the preparations to avoid hydrolysis or photodegradation during equilibration. The solutions were shaken in a water-bath at 25°C ± 0.1°C for 24 hours, within which time equilibrium solubility of nifedipine was achieved. Each sample was centrifuged prior to analysis by HPLC. Phase solubility studies were performed in triplicate; from which apparent 1:1 and 1:2 stability constants (K)[7,8] were calculated. All phase solubility studies were performed in a dark-room under red light.

3. RESULTS AND DISCUSSION

Soluble substrate:ligand complexes were formed under the present experimental conditions with all CyDs studied (Figure 1). Nifedipine solubility increased linearly as a function of β-CyD, 2HP-β-CyD and 2HP-γ-CyD concentrations and thus the solubility curves were classified as type A_L,[7] indicating the formation of 1:1 substrate:ligand complexes. For DIMEB, RAMEB and γ-CyD positive deviations from linearity were observed at higher CyD concentrations, therefore displaying A_P-type[7] phase behaviour. This indicates the formation of higher order complexes, eg. 1:2 nifedipine : CyD, at high CyD concentrations in solution. The magnitude of the interactions between host and guest molecules, reflected by the stability constants (Table 1), decreased in the order DIMEB > RAMEB > β-CyD ≈ 2HP-β-CyD > γ-CyD ≥ 2HP-γ-CyD.

Generally, weak interactions were observed in solution between nifedipine and the CyDs, with only the highly hydrophobic derivatives, DIMEB and RAMEB, showing significantly high stability constants. The superior solubilizing potential of DIMEB and RAMEB has been frequently observed for many highly hydrophobic drugs.[9,10] The low aqueous solubility of β-CyD, and to a lessor extent γ-CyD, has been attributed to intramolecular hydrogen bonding between the secondary hydroxyl groups, thus imparting a rigidity to the macrocycle and preventing hydration of Cyd by water. Selective or random methylation of β-CyD prevents the formation of these hydrogen bonds and consequently hydration of the CyD is made possible, therefore substantially increasing their solubility.[9,11] In addition, methylation of the hydroxyl groups expands the hydrophobic region of the CyD cavity, thus enhancing substrate binding via a hydrophobic effect.[12,13]

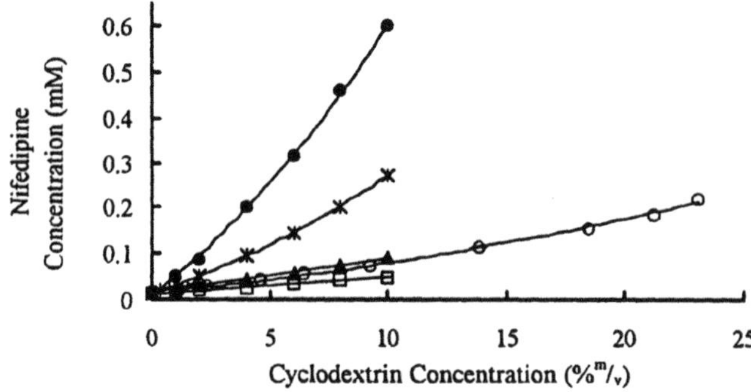

Figure 1. Phase solubility diagrams of nifedipine-cyclodextrin systems in 0.05M potassium phosphate buffer pH 5.8 at 25°C. Key: (•) DIMEB; (✱) RAMEB; (+) β-CyD; (▲) 2HP-β-CyD; (O) γ-CyD; (□) 2HP-γ-CyD.

The hydroxypropyl derivatives of γ- and β-CyD showed roughly equivalent or even slightly poorer solubilizing capacities when compared to their respective parent CyDs. This observation, however, is not uncommon and there are numerous drugs for which the solubilizing potential of these derivatives are reportedly similar or weaker than the parent CyDs.[10,14]

TABLE 1. Solubility enhancement (S.E.*), types of phase diagrams and stability constants (K) for nifedipine-cyclodextrin systems in 0.05M potassium phosphate buffer pH 5.8 at 25°C.

	Solubility enhancement	Type of diagram	Stability Constants (M^{-1})	
			$K_{1:1}$	$K_{1:2}$
β-CyD	2.1	A_L	77.9	-
γ-CyD	14.7	A_P	53.1	3.5
2HP-β-CyD	5.1	A_L	77.2	-
2HP-γ-CyD	3.3	A_L	49.3	-
RAMEB	19.5	A_P	184.9	5.8
DIMEB	37.5	A_P	283.1	11.3

* S.E.: Solubility enhancement in 10% m/v aqueous solutions of respective cyclodextrin derivatives; in the case of β- and γ-CyD, maximum obtained solubility, namely 18.5 and 23.2 %m/v, respectively.

4. CONCLUSION

Comparison of stability constants obtained from the phase solubility studies using β-CyD, γ-CyD, 2HP-β-CyD, 2HP-γ-Cyd, RAMEB and DIMEB indicates that the affinity of nifedipine is greater for the β-CyDs than for the γ-CyDs. It is proposed that the complex-forming moiety of nifedipine may be the 2'-nitrophenyl group; the dimensions of which would be more geometrically compatible for a closer and stronger interaction with the β-CyD cavity than for the larger γ-CyD cavity.

5. ACKNOWLEDGEMENTS

Financial support from South African Druggists, Ltd and the Foundation for Research Development is gratefully acknowledged.

6. REFERENCES

[1] Mielcarek, J., Grodzka, J., Inclusion compounds of nifedipine and other 1,4-dihydropyridine derivatives with cyclodextrins I. Complexation of nifedipine with β-cyclodextrin, *Acta Pol. Pharm.*, **51**, 15-20 (1994)

[2] Nozawa, Y., Yamamoto, A., Effects of roll mixing with β-cyclodextrin on enhancing solubilities of water insoluble drugs, *Pharm. Acta Helv.*, **64**, 24-29 (1989)

[3] Acatürk, F., Kişlal, Ö., Çelebi, N., The effect of some natural polymers on the solubility and dissolution characteristics of nifedipine, *Int. J. Pharm.*, **85**, 1-6 (1992)

[4] Uekama, K., Ikegami, K., Wang, Z., Horiuchi, Y., Hirayama, F., Inhibitory effect of 2-hydroxypropyl-β-cyclodextrin on crystal-growth of nifedipine during storage: Superior dissolution and oral bioavailability compared with polyvinylpyrrolidone K-30, *J. Pharm. Pharmacol.*, **44**, 73-78 (1992)

[5] Yamamoto, M.., Yoshida, A., Hirayama, F., Uekama, K., Some physicochemical properties of β-cyclodextrins and their inclusion characteristics, *Int. J. Pharm.*, **49**, 163-171 (1989)

[6] Suzuki, H., Uede, H., Kobayashi, S., Nagai, T., Panosyl-β-cyclodextrin and its solubilization ability for poorly water soluble drugs, *Eur. J. Pharm. Sci.*, **1**, 159-164 (1993)

[7] Higuchi, T., Connors, K.A., Phase-solubility techniques, *Adv. Anal. Inst.*, **4**, 117-212 (1965)

[8] Higuchi, T., Kristiansen, H., *J.*, Binding specificity between small organic solutes in aqueous solution: Classification of some solutes into two groups according to binding tendencies, *Pharm. Sci.*, **59**, 1601-1608 (1970)

[9] Uekama, K., Pharmaceutical applications of methylated cyclodextrins, *Pharm. Int.*, **6**, 61-65 (1985)

[10] Szejtli, J., Cyclodextrins in drug formulations, *Pharm. Technol.*, **15**, 36-44 (1991)

[11] Green, A.R., Guillory, J.K., Heptakis(2,6-di-O-methyl)-β-cyclodextrin complexation with the antitumor agent chlorambucil, *J. Pharm. Sci.*, **78**, 427-431 (1989)

[12] Szejtli, J., Cyclodextrins and their inclusion complexes, Akadémia Kiado, Budapest, 1982

[13] El-Gendy, G.A., El-Gendy, M.A., Heptakis (2,6-di-O-methyl)-β-cyclodextrin complexation, with glutethimide, *Eur. J. Pharm. Biopharm.*, **39**, 249-254 (1993)

[14] Haskins, N.J., Saunders, M.R., Camilleri, P., The complexation and chiral selectivity of 2-hydroxypropyl-β-cyclodextrin with guest molecules as studied by electrospray mass spectrometry, *Rapid Commun. Mass Spectrom.*, **8**, 423-426 (1994)

PHASE SOLUBILITY ANALYSIS IN STUDYING THE INTERACTION OF NIFEDIPINE WITH SELECTED CYCLODEXTRINS IN AQUEOUS SOLUTION

M.S. WORTHINGTON, B.D. GLASS, L.J. PENKLER*

School of Pharmaceutical Sciences, Rhodes University, Grahamstown, 6140, R.S.A.

* *Druggists Group Research, South African Druggists,Ltd, P.O. Box 4002, Korsten, Port Elizabeth, 6014, R.S.A.*

ABSTRACT

The solubilizing potential and complexing tendencies of six cyclodextrins (CyD) with nifedipine in aqueous solution were evaluated using phase solubility methods. Solubility curves of nifedipine with β-CyD, 2-hydroxypropyl-β-CyD (2HP-β-CyD) and 2-hydroxypropyl-γ-cyclodextrin (2HP-γ-CyD) were classified as type A_L, while for heptakis (2,6-dimethyl)-β-CyD (DIMEB), randomly methylated-β-CyD (RAMEB) and γ-CyD, A_P type phase behaviour was observed. Stability constants, calculated from phase solubility diagrams, decreased in the order: DIMEB > RAMEB > β-CyD > 2HP-β-CyD > γ-CyD > 2HP-γ-CyD.

1. INTRODUCTION

Nifedipine (2,6-dimethyl-3,5-dicarboxyl-4-(2'-nitrophenyl)-1,4-dihydropyridine) is a calcium channel antagonist displaying very poor aqueous solubility and hence inferior dissolution and oral bioavailability. Recent studies have shown that the interaction of nifedipine with β-CyD,[1-3] 2HP-β-CyD[4] and some branched cyclodextrins[5,6] in aqueous solution and the solid state, have enhanced the above-mentioned physico-chemical properties. This study continues to investigate the solubilizing potential and complexing tendencies of γ-CyD, 2HP-γ-CyD, randomly methylated-β-CyD (RAMEB) and heptakis(2,6-di-O-methyl)-β-CyD (DIMEB) in aqueous solution using phase solubility techniques.

J. Szejtli and L. Szente (eds.), Proceedings of the Eighth International Symposium on Cyclodextrons, 329–332.

Data were collected on the BW7 beamline of EMBL. Usage of a bigger image plate in comparison with the room temperature data collection allowed to collection of data to a maximum resolution of 0.9 A. R_{sym} for this dataset was 5.8 %.

2.3 NMR Spectroscopy

The structure of the complex in aqueous solution was also studied by NMR Spectroscopy. Spectra were recorded at 200 MHz, using D_2O as solvent, at room temperature. Gaussian enchancement was used.

3. RESULTS AND DISCUSSION

3.1. Structure solution

The structure was solved using the room temperature data. The position of the Br atom was located from both the Patterson synthesis and the anomalous difference Patterson synthesis, utilizing the program SHELXS86. Attempts to get a good quality map using either Patteron expansion methods or direct methods were unsuccesfull.

A novel protocole was then applied to obtain an interpretable map. The program suite ARP, primarly used in protein crystallography for improvement and completion of models, was used. Using only the Br atom as a starting model, 40 cycles of sparse matrix unrestrained least squares refinement were applied. After each cycle 2 oxygen atoms were added to the model automatically on the basis of the density in the difference Fourier synthesis map. Another 60 cycles followed, in which the model was updated after each cycle more extensively, 6 atoms were allowed to be added and 6 to be removed in accordance with density and geometrical criteria. Atoms were added where density above 4 rms in the difference Fourier synthesis was present and if an atom already existed in a distance between 1.1 and 3.0 A, which covers covalent and hydrogen bonding interactions. After that, the model, consisting at that stage of only oxygen atoms, showed clearly all atoms of the cyclodextrin, the ligand and 10 water molecules.

Only 4 additional atoms did not make chemical sense (two in each side of the Br atom to compensate apparently for the high anisotropy of that atom along that direction and two to compensate for well defined hydrogen atoms of the structure) and thus were removed.

It must be noted, that as it could be judged after complete refinement of the structure, the initial phase error which was 72°, dropped to only 14° with respect to the final model, just by applying this protocole.

3.2. Refinement

After identifying and assigning the correct atom types the model was subjected to 10 cycles of full matrix least squares refinement using SHELXL-93. Hydrogens were added and an additional 20 cycles of refinement were performed in which each non-hydrogen atom was assigned an anisotropic temperature factor. After adjustment of the weighting scheme the final R factor converged to 7.9 % indicating a structure of very good quality.

For the low temperature data, the hydrogens and anisotropic temperature factors of the room temperature model were removed and the model was subjected to 10 cycles of refinement.

A similar protocole like for room temperature data was then applied. The final R factor converged to 6.4 %, which is exceptionally low for structures of cyclodextrin complexes.

3.3. Description of the structure

The hydrophobic adamantane group is burried within the hydrophobic cavity of one cyclodextrin molecule. The cyclodextrin lies at an angle of 20° with respect to the plane formed by the crystallographic a and b axes. The aliphatic tail extends away from the cyclodextrin in a direction almost parallel to the a axis of the cell. After the tertiary nitrogen, the direction of the chain is changed by 90° and becomes effectively parallel to the long c axis, getting gradually disordered. Towards its end it regains order and comes to the proximity of the cyclodextrin ring of the molecule related by the 2_1 symmetry along the long c axis. Thus, the end of the chain interacts from the opposite side of the cyclodextrin ring than the side where the adamantane group lies.
The Br atom is coordinated by the O10 atom of the adamantane moiety and three more water molecules. All three water molecules form additional hydrogen bonds both to other waters and cyclodextrin sugar ring oxygens.
A rather complex hydrogen bonding network between the water molecules and the cyclodextrin and ligand is present. Every water molecule is involved in at least one hydrogen bond with mainly the O2, O3 and O6 atoms of the seven sugars comprising the cyclodextrin. All O2, O3 and O6 atoms are involved in a hydrogen bond with at least one water from the same or another asymmetric unit.
Only one water molecule forms a hydrogen bond to the oxygen atached to the adamantane moiety. This water in turns interacts with another water attached to the O3 atom of one of the sugar rings. This is the only hydrogen bonding interaction between the ligand and the cyclodextrin molecule in which the hydrophobic part of the ligand is burried in. Interestingly this oxygen atom is hydrogen bonded to the O3 atom of the sugar ring of a cyclodextrin molecule related by the 2_1 symmetry along the a axis.

3.4 ^{1}H NMR Spectroscopy

The complex formation induces chemical shift changes in the resonance of the CD protons, especially of protons directed towards the interior of the βCD cavity (H-C3,H-C5) and of the drug protons. (Tale I). The proton NMR spectrum of the pure molecule in D_2O consists of from six different groups of peaks.
The largest Δδ values were observed for the 4 axial and 9 axial protons of adamantane ring in the range of 0.16 ppm. A slight modification was observed to the signals corresponding to the 4 equatorial and 9 equatorial protons of the ring. The significant displacement of the signals corresponding to the rest of adamantane ring protons were expected because of the existing slope of ring axis in the βCD cavity.

4. CONCLUSION

In the prepared I-1-10 βCD complex the hydrophobic adamantane group is burried within the hydrophobic cavity of one βCD molecule. The aliphatic tail is extended away from the βCD and becames gradually more disordered. Towards its end the tail becames more ordered and interacts with the βCD of a neighboring assymetric unit. The Br ion comes to 3.2 A distance with the O atom attached to the C10 of the adamantane ring.
The observed resonance modifications of both molecules revealed interactions between CD and the adamantane moiety of the active molecule, which shields this part from the solvent. Therefore it appears that the structure of the complex is quit similar both in solution and in the solid state.

TABLE I Chemical Shifts δ(ppm) of ADM-10 and β-CD in the Free and Complex State.

Proton*	δ_o (free)	δ_c (complex)	$\Delta\delta$ (δ_c-δ_o)
-$(CH_2)_9\underline{CH_3}$(t,2H)			
-$(CH_2)_7$-(m,14H)	1,2768	1,2802	+0,003
	1,3537	1,3696	+0,002
Ad(d,2H,4eq,9eq)	1,5668	1,6266	+0,071
	1,6128	1,6840	+0,071
Ad(m,12H)	1,7374	1,7923	+0,055
	1,7772	1,8561	+0,079
Ad(d,2H,4ax,9ax)	2,0860	2,2477	+0,162
	2,1462	2,3075	+0,163
$N(CH_3)_2$(s,6H)	3,0581	3,0726	+0,014
β-CD			
H3	4,043	3,926	-0,117
	3,878	3,987	-0,102
	3,949	3,800	-0,147
H5	3,540	3,855	-0,287
	3,586	3,873	-0,315
Anomeric	5,0800	5,0652	-0,015
	5,0976	5,0482	-0,050

*Only the protons that show chemical shift changes

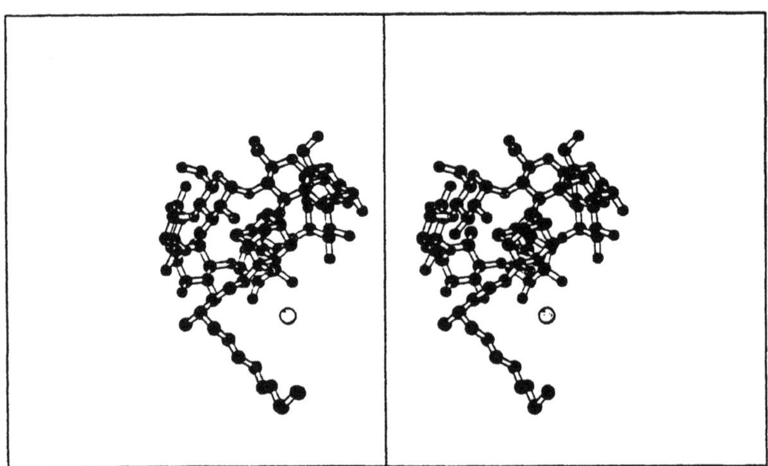

Stereo-pair showing the structure of the complex of I-1-10 with βCD

REFERENCES

[1]Antoniadou-Vyza, E., Tsitsa, P., Tsantili-Kakoulidou, A., Hitiroglou, E.: New adamantan-2-ol and adamantan-1-methanol derivatives as potent antibacterials. Synthesis, antibacterial activity and lipohpilicity studies. *J. Eur. Med. Chem.* 31, 105-110 (1996)

[2]. Krishnamurthy, V., Pradeep,S., Olah, A. : Study of Substituent Effects on One-Bond NMR Coupling Constants in adamantane Derivatives. *J. Org. Chem.* 48, 3373-78 (1983).

[3]. Arndt, U.W.; Wonacott, A.J. : *The rotation method in crystallography*, North Holland Publishing company 75-103 (1977)

[4]. Cosier, J.Glazer, A.M. : A nitrogen-gas-stream cryostat for general Xray diffraction studies. *J. Appl. Cryst.* 105-107 (1986)

[5]. Otwinowski, Z. *Denzo: An oscillation data processing program for macromolecular crystallography*, Yale University, New Haven, USA.(1993)

[6]Sheldrick, G.M. :SHELXL-93, *Program for crystal structure refinement.* University of Gottingen, Germany(1993)

[7]Lamzin, V. : Automated refinement of protein models. *Acta Crystal.* D 49, 129-147 (1993)

COMPLEXATION OF NEW ACTIVE ANTIBACTERIAL ADAMANTAN DERIVATIVES WITH β CD: PREPARATION AND CHARACTERIZATION OF COMPLEXES STUDY OF THE THERMOTROPIC PROPERTIES OF PURE AND COMPLEX FORM WITH DIPALMITOYL PHOSPHATIDYLCHOLINE BILAYERS

ANTONIADOU-VYZA E.[1], TSITSA P.[1], THEODOROPOULOU E.[2], MAVROMOUSTAKOS T.[2],

1-Department of Pharmaceutical Chemistry, University of Athens 15771, Greece.

2-Institute of Organic and Pharmaceutical Chemistry, The National Hellenic Research Foundation, 48 Vas. Constantinou, Athens 11523, Greece

ABSTRACT

β-Cyclodextrin (βCD) complexes of I1-8, I1-10, I1-12, were prepared, isolated and characterized in solid and liquid form. Their thermotropic effects were studied by inserting them in DPPC bilayers both in pure and complexed forms. The results have shown that the presence of I1-8 causes spliting, broadening and lowering of the phase transition of DPPC bilayers. Their effects are more significant when I1-10 and I1-12 are inserted. These differential effects are eliminated when the above studied three molecules are incorporated in a complex form with βCD in DPPC bilayers. The obtained results suggest that the bromine salts in the bilayer are likely to remain in the complex form rather than released in the membrane.

INTRODUCTION

The microbiology of skin infections is changing. Thus, there is an increasing incidance of infections caused by strains, that are resistant to many antibiotics. The development of new potent topical preparations may change the traditional preference for systemic drugs in managing these disorders. The octyl, decyl and dodecylbromide salts of 2-(3-dimethylaminopropyl)-tricyclo [3.3.1.13,7]decyl-2-ol (I-1-8, I-1-10, I-1-12), were synthesized in our laboratory and showed an activity of MIC values ranging from $1x10^{-5}$ to $1x10^{-3}$ and limited aqueous solubility.
Although in recent literature a plethora of drug CD complexes have been prepared and studied for their solubility chemical stability and other physicochemical properties, studies on the interaction of this supramolecular assemblies with phospholipidic bilayers is rare.
The effectiveness of the quaternary ammonium salts appears to be a result of disturbance caused to the microbial cell membrane leading to leakage of intracellular compounds. At the same time, some microorganism possess impermeable cell membranes that prevent influx of the drug. Ohers lack the both case hydrophilic drug carrier such as CDs may facilitate the transversation of

J. Szejtli and L. Szente (eds.), Proceedings of the Eighth International Symposium on Cyclodextrons, 337–340.

transport system that isequired for entrance of the drug into the bacterial cell. In this kind of outer membranes. The study of thermotropic properties of various drugs in a complex form with cyclodextrins in phospholipid bilayers is very useful because they give valuable information of their potential use as drug releasing agents.

2. MATERIALS and METHODS

2.1 Materials

The active compounds were synthesised and purified in our laboratory.. All other chemicals, were reagent grade and were used without further purification. βCD complexes of the active compounds has been prepared by using the precipitation method. Dipalmitoylphosphatidylcholine (DPPC) was obtained from Avanti Polar Lipids, Inc. AL, USA.

2.2 Methods

Drug-cyclodextrin interactions in aqueous solution were studied by Proton Nuclear Magnetic Resonance Spectroscopy (^{1}H NMR) and in solid state by Thermal Analysis (DSC). Appropriate amount of phospholipid and the bromine salts of quaternary dimethylamino adamantanol (or their complex with β-CD) were dissolved in spectroscopic grade chloroform and the two solutions mixed. The solvent was then evaporated by passing a stream of O_2-free nitrogen over the solution at 50 °C and the residue was placed under vacuum (0.1 mmHg) for 12 h.
Differential Scanning Calorimetry: The containing phospholipid prepared sample was hydrated (50% w/w) and was transferred to a stainless steel capsule (Perkin Elmer) and sealed hermetically. Thermograms were obtained on Perkin-Elmer DSC-7 instrument. All samples were scanned at least twice until identical thermograms were obtained using a scanning rate of 2.5 °C/min. Samples containing drug alone and mixed or complexed with β-CD were scanned using a scanning rate of 2,5 °C/min.

3. RESUTS AND DISCUSSION

3.1 Characterization of the complexes

The different thermotropic properties between the mixture and complexes is a result of their complexation.

3.2 Phosphatidylcholine bilayers

Certain hydrated phospholipids spontaneously form bilayers which share many of the conformational and dynamic properties of the natural membranes. Studies with these fully hydrated phospholipids are, therefore, useful since they allow us to gain insight into the physical chemistry of lipid interactions in natural membranes in which phospholipids are believed to exist largely as liquid crystalline bilayers. Among the membrane phospholipids, phosphatidylcholines are a major component and their phase properties have received a great deal of attention. The calorimetric measurements for a hydrated DPPC preparation show two endothermic transitions in the temperature range usually used to study membranes; a broad low-enthalpy pretransition (T'c=35.3 °C) and a main transition (Tc=41.2 °C) (Fig. 3, top). Below the pretransition, the phospholipid molecules are arranged in a one-dimensional lamellar gel phase ($L_{\beta'}$), while above the main transition they exist in the liquid crystalline phase (L_{α}). At temperatures between T'c and Tc, there is a ripple phase

($P_{\beta'}$) which on the basis of solid-state NMR evidence, has been shown to be composed of coexisting gel and liquid crystalline components.

3.3 Effects of drugs in phosphatidylcholine bilayers

The thermograms of DPPC with or without the active molecules are shown in Fig. 3. The presence of 0.10 molar ratio (x=0.10) I-1-8 in DPPC bilayers results in splitting and broadening of the phase transition as well as lower of the phase transition temperature of DPPC bilayers. The other two molecules I-1-10 and I-1-12 exert similar thermotropic effects in DPPC bilayers. Thus, the presence of either of the above two molecules lowers the phase transition and causes more significant broadening in the main phase transition temperature. A splitting was not observed in the presence of these two molecules depicting that they do not create any heterogeneity (domains) in the membrane bilayer as it happens with I-1-8 .

The thermograms of DPPC/(x=0.10)β-CD bilayers with or without either of the studied molecules are shown in Fig. 4. The presence of β-CD affects only marginally DPPC bilayers by lowering (ca. 1.0 °C) and splitting the top of the main phase transition. The presence of either comlexes with β-CD of the three molecules in the DPPC bilayers lowers the phase transition temperature and broadens the phase transition in a similar way. It appears, therefore, that the differential effects of the three studied compounds are eliminated when are incorporated in a complex form with β-CD in DPPC bilayers. The obtained results can be interpreted that the bromine salts in the bilayers are likely to remain in the complex form rather than released in the membrane.

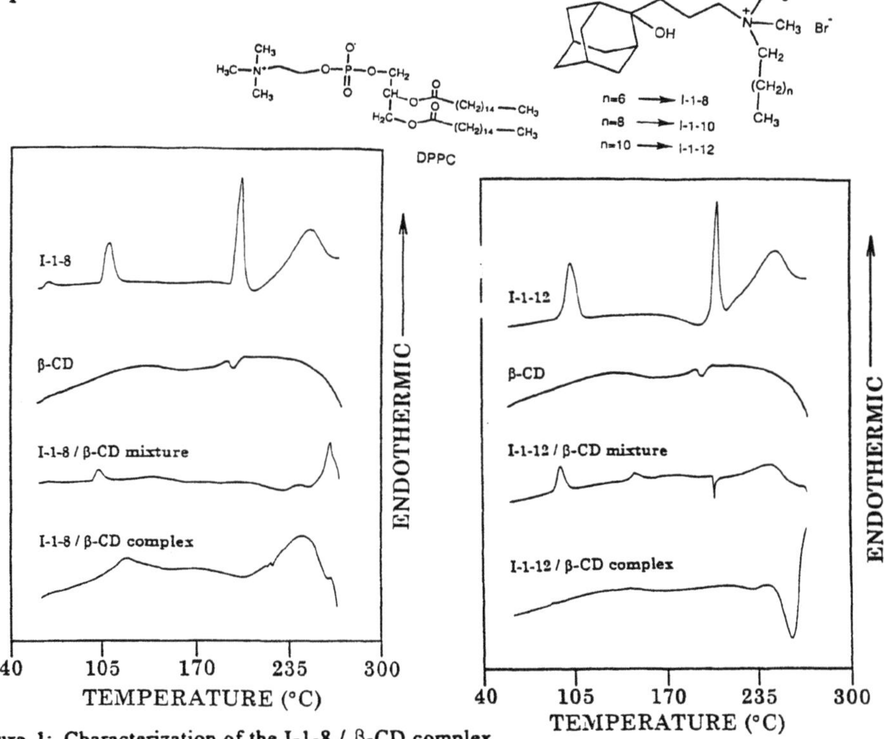

Figure 1: Characterization of the I-1-8 / β-CD complex

Figure 2: Characterization of the I-1-12 / β-CD complex

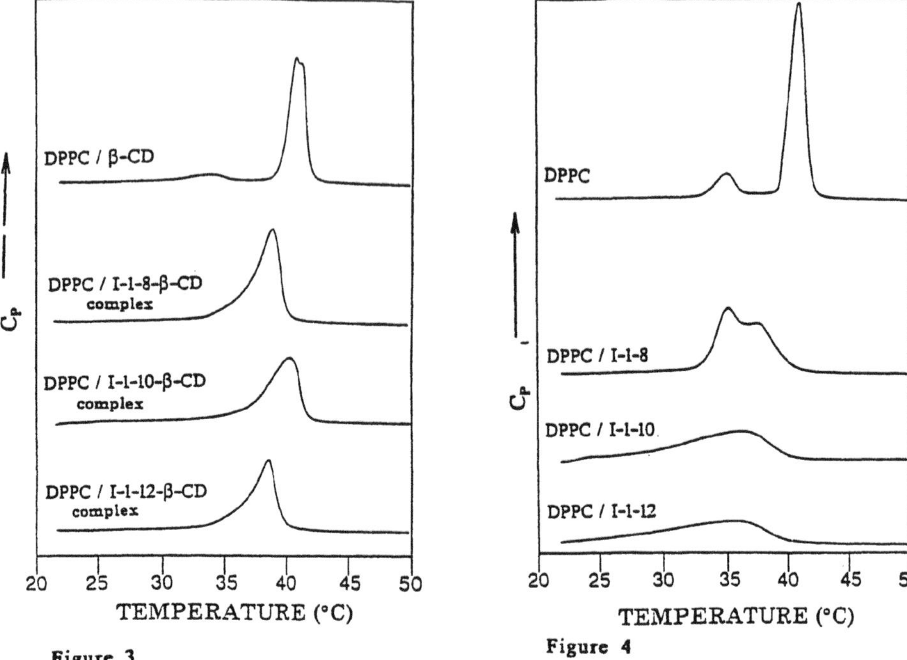

Figure 3: Normalized thermograms of DPPC, DPPC / (x=0.10) I-1-8, DPPC / (x=0.10) I-1-10 and DPPC (x=0.10) I-1-12.

Figure 4: Normalized thermograms of DPPC / β-CD, DPPC / (x=0.10) I-1-8-β-CD, DPPC / (x=0.10) I-1-10-β-CD and DPPC / (x=0.10) I-1-12-β-CD.

REFERENCES

[1]. Antoniadou-Vyza E., Tsisa p., Tsantili-Kakoulidou A., Hitiroglou E.: New adamantan-2-ol and adamantan-1-methanol derivatives as potent antobacterials. Synthesis, antibacterial activity and lipophilicity studies. *J. Eur. Med chem.* 31, 105-110 (1996)

[2]. Bruggemann, E.P., Melchior, D.L. : Alterations in the organization of phosphatidylcholine cholesterol bilayers by tetrahydrocannabinol. *J. Biol. Chem.* 258 : 8298 (1983).

[3]. Carlson, J.C., Gruber, M.Y., Thompson J.E.: A study of the Interaction between Progesterone and Membrane Lipids. *Endocrinology* 113 , 190 (1983).

[4]. Saenger W.: Cyclodextrin Inclusion Compounds in Research and Industry *Angew. Chem. Int. Ed. Engl.* 19: 344 (1980).

[5]. Szejtli, J., Cserhàti, T., Szögyi, M.: Interactions between Cyclodextrins and Cell-Membrane Phospholipids.: *Carbohydrate Polymers* 6: 35, (1986).

[6]. Castelli, F., Puglisi, G., Pignatello, R., Gurrieri, S. : Calorimetric studies of the interaction of 4-biphenylacetic acid and its β-cyclodextrin inclusion compound with lipid model membrane. *Int. J. Pharm.* 52, 115 (1989).

[7]. Castelli, F., Giammona, G., Puglisi, G., Carlisi, B., Currieri.S.: Interaction of macromolecular pro-drugs with lipid model membrane: calorimetric study of 4-biphenylacetic acid linked to α,β-poly(N-hydroxyethyl)-DL-aspartamide interacting with phosphatidylcholine vesicles. *Int. J. Pharm.* 59: 19 (1990).

[8]. Castelli, F., Giammona, G., Raudino, A., Puglisi, G.: Macromolecular prodrugs interaction with mixed lipid membrane. A calorimetric study of naproxen linked to polyaspartamide interacting with phosphatidylcholine and phosphatidylcholine-phosphatidic acid vesicles. *Int. J. Pharm.* 70: 43 (1991).

[9]. Castelli, F., Puglisi, G., Giammona, G., Ventura, C.A.: Effect of the complexation of some nonsteroidal anti-inflammatory drugs with β-cyclodextrin on the interaction with phophatidylcholine liposomes. *Int. J. Pharm.* 88: 1: 1, (1992).

INCREASING THE SOLUBILITY OF FUROSEMIDE WITH CYCLODEXTRINS

R. M. KREAZ[1], GY. DOMBI[2] and M. KATA[1]
[1]Department of Pharmaceutical Technology and [2]Department of Pharmaceutical Chemistry, Albert Szent-Györgyi Medical University, H-6701 Szeged/Hungary, P.O.Box 121

ABSTRACT

In the past 20 years, cyclodextrin (**CD**) research has achieved considerable results, as indicated by the seven earlier symposia on **CDs**, and the new dosage forms prepared with them. Furosemide (**F**) is slightly soluble in water (10.26 mg/100 g) [1-3]. Different **CDs** increase the solubility of **F** to various degrees, depending on their ratios and the methods of preparation. The parameters of the solutions are changed and the inclusion complex formation was analysed.

1. INTRODUCTION

F, a frequently used diuretic and antihypertensive pharmacon, was discovered by **Sturn et al.** in 1962 (Hoechst). It is slightly soluble in water, although its sodium salt is very soluble. Its registered tablets contain 40 mg and its injections 20 mg of **F** in 2 ml. Its structural formula is as follows:

COOH
NH—CH_2
O
NH_2SO_2
Cl

The aims of the experiments were to improve the bioavailability of **F** by increasing its solubility and dissolution properties by complexation with **CDs** in various ratios, using different methods of preparation, and to investigate the products by various techniques [4-6] such as X-ray analysis, etc.

J. Szejtli and L. Szente (eds.), Proceedings of the Eighth International Symposium on Cyclodextrons, 341–344.

2. MATERIALS AND METHODS

2.1. Materials

F and **F-Na** (Chinoin, Budapest); **α-CD, β-CD, γ-CD**, hydroxypropyl-β-CD, dimethyl-β-CD and random-methylated β-CD (= **HP-β-CD, DIMEB** and **RAMEB**; Cyclolab Ltd., Budapest).

2.2. Methods of preparation

Powder mixing, kneading, precipitation, spray-drying and freeze-drying in 1:1/2, 1:1 and 1:2 molar ratios (**F**:**CD**).

TABLE 1. Composition of investigated products in %

Ratios	β-CD	HP-β-CD	RAMEB	DIMEB
1:2	12.72	12.17	11.15	12.45
1:1	22.60	21.71	20.06	22.14
1:1/2	33.39	35.67	33.41	36.26

2.3. Analysis

The products were investigated by spectrophotometry at 282 nm in the concentration range 2-15 μg/g (Spektromom 195, MOM, Budapest), where the presence of the different CDs in the usual concentrations does not influence the absorbance of the solutions. The dissolution rate was determined at 100 rpm according to USP XXIII [1, 3]. The X-ray spectra were recorded with a DRON UM-1 instrument and the IR spectra with Perkin-Elmer Paragon FT-IR apparatus.

3. RESULTS AND DISCUSSION

Determination of solubility. In a pre-experiment, the **CDs** increased the solubility of **F** in the sequence **F** < **γ-CD** < **α-CD** < **β-CD** < **RAMEB** < **HP-β-CD** < **DIMEB** (under the usual conditions, i.e. at room temperature, during 1 week). In preformulation experiments, it was found that 300 mM **DIMEB** increases the solubility 137-fold, **RAMEB** increases it 75-fold.

The extent of dissolution of pure **F** during 60 min is 39.96 mg/900 mL distilled water. The effects of **CDs** on the extent of dissolution: **DIMEB** increases the solubility properties of **F** to the largest extent: the quantity of **F** dissolved was in all cases more than that for pure **F**.

TABLE 2. Extent of dissolution of products, in mg/900 mL during 60 min

Methods	Ratios	β-CD	HP-β-CD	RAMEB
Mixing	1:2	28.13	61.21	62.01
	1:1	37.88	62.50	56.57
	1:1/2	20.30	54.34	60.58
Kneading	1:2	66.65	63.77	78.31
	1:1	64.34	63.13	73.52
	1:1/2	57.06	55.78	49.23
Spray-drying	1:2	62.82	33.33	92.27
	1:1	70.17	51.62	79.12
	1:1/2	69.85	61.00	42.52

The dissolution properties of products containing **RAMEB** and **HP-β-CD** exhibit good results, while the products containing **β-CD** are the poorest. For **HP-β-CD**, there is practically no difference in the extent of dissolution at the different ratios.

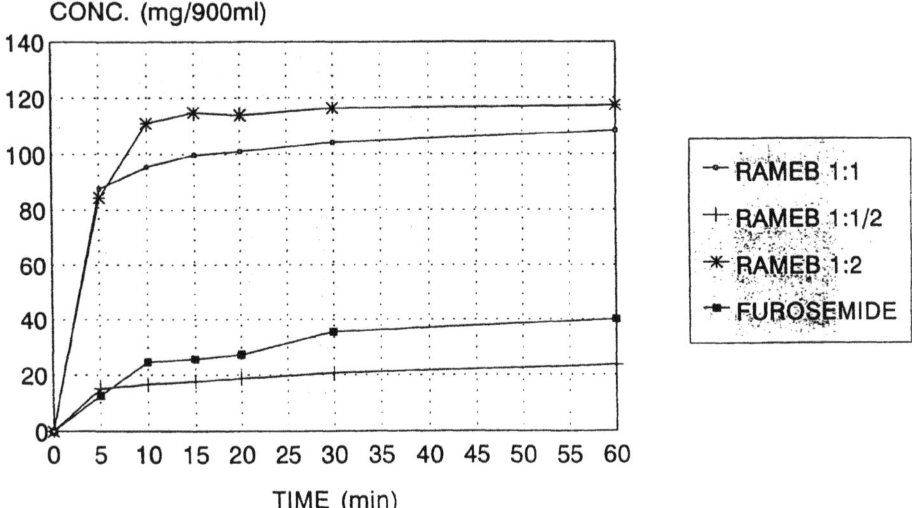

Fig.1. Influence of RAMEB and its ratio on the dissolution of **F** from freeze-dried products

As concerns the methods of preparation, the **RAMEB** and **HP-β-CD** kneaded products are the best, while **β-CD** gives similar results for the spray-dried and kneaded products. Inclusion complex formation was investigated by X-ray and IR techniques. It was established that **F** and **β-CD** have crystalline structures, whereas **HP-β-CD**, **DIMEB** and **RAMEB** have amorphous structures. Consequently, all physical mixes contain characteristic peaks of **F** and, of course, powder mixtures and precipitated products of **F** and **β-CD** also give characteristic peaks of both components.
All kneaded products exhibit such peaks, their intensities being in direct proportion to their **F** content. Finally, all freeze-dried and spray-dried products were completely amorphous.
ΔH_{sol} of **F** is 16.37 kJ/mol at 20-40 °C and 33.98 kJ/mol at 40-60 °C.
The IR spectra confirm the earlier findings.

4. CONCLUSIONS

The different **CDs** increase the solubility of **F** to various degrees, depending on their ratios and the methods of preparation. **DIMEB** increases the solubility properties of **F** to the largest extent (at 300 mM, the increase is 137-fold). An **F**:**CD** ratio of 1:2 was generally the best, and the dissolution of spray-dried and kneaded products was the highest. In parallel with this, the parameters of the solutions also changed. The findings of these investigations are confirmed by the X-ray and IR spectral results on **F**, **CDs** and their products.

ACKNOWLEDGEMENT

The authors gratefully thank **Prof. J. Szejtli** for providing the CD products and advice.

REFERENCES

[1] Pharmacopoeia Hungarica, 7th Edition. Medicina, Budapest, 1986, p. 996
[2] The Merck Index, 11th Edition. Merck and & Co., Inc., Rahway, N.J., USA, 1989, p. 674
[3] USP, 23rd Edition. US Pharm. Conv., Inc., 1994, p. 696
[4] Sejtli, J.: Cyclodextrins and their Inclusion Complexes. Academia Press, Budapest, 1982
[5] Duchêne, D.: Cyclodextrins and their Industrial Uses. Ed. de Santé, Paris, 1987
[6] Szejtli, J.: Cyclodextrin Technology. Kluwer Academic Publ., Dordrecht, 1988
[7] Kata, M. and Kedvessy, G.: Increasing the Solubility Characteristics of Pharmaca with CDs. Pharm. Industry 49, 98-100 (1987)
[8] Kata, M., Giordano, F., Hadi, I.A. and Selmeczi, B.: Conditions of CD Complexation. Acta Pharm. Hung. 63, 285-289 (1993)

STUDY OF THE SOLUBILITY AND PHOTOSTABILITY OF NIFEDIPINE AND ITS DERIVATIVES WITH CDs

Á. GYÉRESI[1], B. TŐKÉS[1], G. REGDON[2], M. KATA[2] and G. NAGY[3]
[1]Department of Pharmaceutical Chemistry, Univ. of Medicine and Pharmacy
4300 Târgu-Mures, Romania
[2]Department of Pharmaceutical Technology and [3]Department of Pharmacognosy
Albert Szent-Györgyi Medical University, H-6701, Szeged/Hungary, P.O.Box 121

ABSTRACT

Nifedipine (**Nif**) and its derivatives (**Nifs**), such as nimodipine (**Nim**), nisoldipine (**Nis**) and nitrendipine (**Nit**), are frequently used antianginal, anti-hypertensive or cerebral vasodilator pharmacons. They have two main disadvantages: they are practically insoluble in water and they are light-sensitive in solution [1-3]. Cyclodextrin derivatives (**CDs**) may be used as inclusion complex-forming agents to increase the solubility of drugs and different dyes are also available to prevent the photo-degradation of these molecules [4-5].

1. INTRODUCTION

The above-mentioned pharmacons were discovered in 1968, 1972 and 1977. They are very important in human medicine therapy. They are slightly soluble in water (e.g. **Nif** is nearly 10 mg/mL and others < 10 mg/mL). On the other hand, **CDs** may be used to increase the solubility properties of **Nifs** [2, 4, 5] which are very easily decomposed in solution. Different dyes are available to hinder these processes [3, 5]. According to the experimental results, **CDs** may also increase the photostability effect of the dyes.
Structural formula of **Nif** is as follows:

M_W of **Nifs**:
Nif 346.34
Nim 418.45
Nis 388.42
Nit 360.37

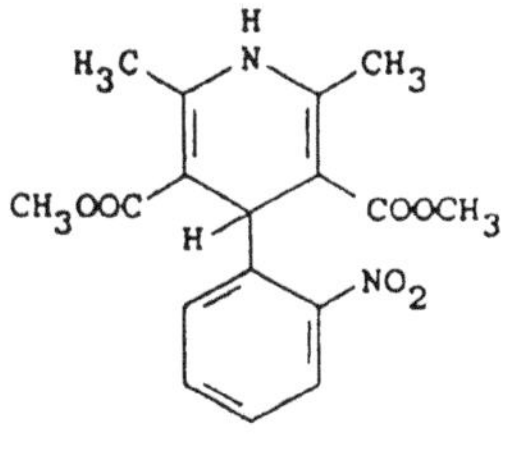

2. MATERIALS AND METHODS

2.1. Materials

Nit (Alkaloida, Tiszavasvári, Hungary), other **Nifs** (Bayer, Germany), **α-CD**, **β-CD**, **γ-CD**, **Me-β-CD**, **HE-β-CD**, **HP-β-CD**, **DIMEB** and **RAMEB** (Cyclolab Ltd., Budapest), and dyes: Chinolingelb 70 E (BASF, Germany), tartrazine (Hopkins & Williams, Essex, England) further Chrysoin S, Echtgelb and Orange Gelb (= **CH**, **T**, **CG**, **EG** and **OG**).

2.2. Methods

Determination of phase-solubility and spectrophotometric analysis in UV light with Specord UV-VIS Spectrophotometer (C. Zeiss, Jena, Germany) and Spektromom 195 (MOM, Budapest).

J. Szejtli and L. Szente (eds.), Proceedings of the Eighth International Symposium on Cyclodextrons, 345–348.

In vitro dissolution test according to USP XXIII and Ph. Hg. VII rotating basket method. For the determination of the stability of **Nifs**, a polarography was used (Polarograph LP 7E, Line recorder TZ 213S, Laboratorne Pristroje, Praha).
Determination of the stability of **Nifs** in methanol + water solutions in the presence of dyes and **CDs** with UV-VIS Scanning Spectrophotometer, UV-2101 PC (Shimadzu, Japan).

3. RESULTS AND DISCUSSION

The solubility of **Nifs** in distilled water at λ = 239 nm is about 10 mg/mL for **Nif**, 3.7 mg/mL for **Nim**, 3.2 mg/mL for **Nis** and 2.2 mg/mL for **Nit**.
All **CDs** as **γ-CD, β-CD, HP-β-CD** and **RAMEB** improved the solubility of **Nifs.**

TABLE 1. Influence of **CDs** on the solubility of **Nifs** in μg/mL

	in water	γ-CD	β-CD	HP-β-CD	RAMEB
Nif	10.0	16.7	13.3	19.2	52.5
Nim	3.7	9.2	32.5	30.8	83.3
Nis	3.2	15.8	21.7	34.6	85.0
Nit	2.2	8.3	9.2	6.7	29.2

It may be established that **RAMEB** increases to the greatest degree the solubility of **Nifs**, e.g. it is 26 times in the case of **Nis**, and **CDs** hardly improve the solubility properties of **Nit**.
Different dyes influence the stability of **Nifs** to dissimilar degrees (**Tables 2 and 3**).

TABLE 2. Effect of dyes on the stability of **Nifs**

Nif	CG	+ 19.3%
Nif	T	+ 19.3%
Nim	EG	0.0 %
Nis	CH	+ 16.9 %
Nit	OG	- 15.1%

TABLE 3. Effect of **CDs** on the stability of **Nit**

Nit	initial conc.	100.0%
Nit	β-CD	75.0 %
Nit	HP-β-CD	71.9 %
Nit	DIMEB	50.0 %

The periods of action of sunlight were: 0, 1, 6, 24 and 96 hours, or 30 min of UV light (λ = 254 nm).

We have considered the increase or decrease of absorbance of the sunlight treated solutions at 279.5 nm for **Nif**. During a period of 1 hr, the increase of absorbance was 96.6% and during 96 hrs, it was 111.2%. In the presence of Chinolingelb (**CH**) and tartrazine (**T**), the increase of absorbance was 19.3%; they were the best dyes. The increase of absorption in the presence of other dyes: 28.4% in the presence of Echtgelb (**EG**), 37.2% Orange Gelb (**OG**) and 59% Chrysoingelb (**CG**).
The change of absorbance was measured at 360.5 nm for **Nim**. During a 96 hrs sunlight treatment, decrease of absorbance was 26.7%, while a 30 min UV light treatment, resulted in a 22.7%. Different dyes stabilized **Nim** well, the largest decomposition was only 7.6% (in the presence of **OG**) and the **EG** was the best dye. In the presence of **CG** and **EG**, the absorbance gently increased in the first hours, later (after 24 hrs), it decreased. The order of „goodness of stability" of dyes against 30 min UV treatment is: **CG** > **OG** > **EG** > **CH** > **T**.

The change of absorbance of **Nis** was measured at 279 nm. The absorbance increased by 93.3% after a 96 hrs sunlight treatment, and it also increased by 28.8% after a 30 min UV treatment. The **CH** was the best dye: increase of absorbance was 16.9% after a 96 h sunlight treatment. Others are in order: **T** = 24.7% > **EG** = 25.1% > **OG** = 31.6% > **CG** = 49.3%.

We also checked the change of absorbance of **Nit** at 357.5 nm. During a 4 days sunlight treatment its absorbance decreased by 83.3%, while during a 30 min UV light treatment, it decreased by 14.4% only. **EG** is the best dye against sunlight effect (decrease is 4.1%). Other dyes are in order: **CG** > **OG** > **CH** > **T**.

TABLE 4. Decomposition of **Nif** in solution in the presence of **Cds** and sunlight (concentration of **Nif** at preparation = 100%)

β-CD	37.5 %
RAMEB	68.4 %
DIMEB	81.3 %
HP-β-CD	83.6 %

Different solutions were prepared of **Nif** and **CG**, **Nim** and **EG**, **Nis** and **CH**, and all with **CDs**. **RAMEB** was the best. Influence of **RAMEB** and a dye on stability of **Nifs** can be seen in **Table 5.** (initial concentration 100.0%).

TABLE 5. Influence of **RAMEB** and a dye on stability of **Nifs**

Composition	sunlight, 60 min	UV light, 30 min
Nif + CG + RAMEB	+ 17%	+ 10 %
Nim + EG + RAMEB	- 1 %	+ 1 %
Nis + CG + RAMEB	+ 23%	+ 12 %
Nit + OG + RAMEB	- 10 %	0 %

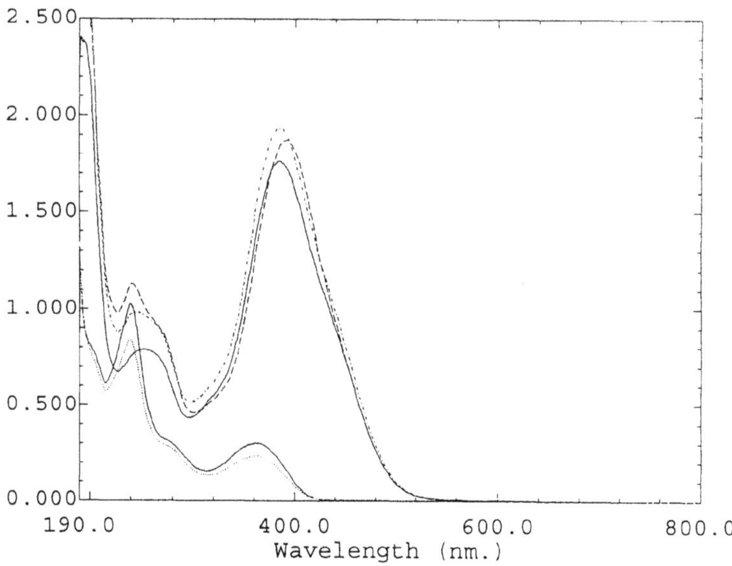

Fig. 1. RAMEB and CG influence on decomposition of sunlight treated Nisoldipine solution

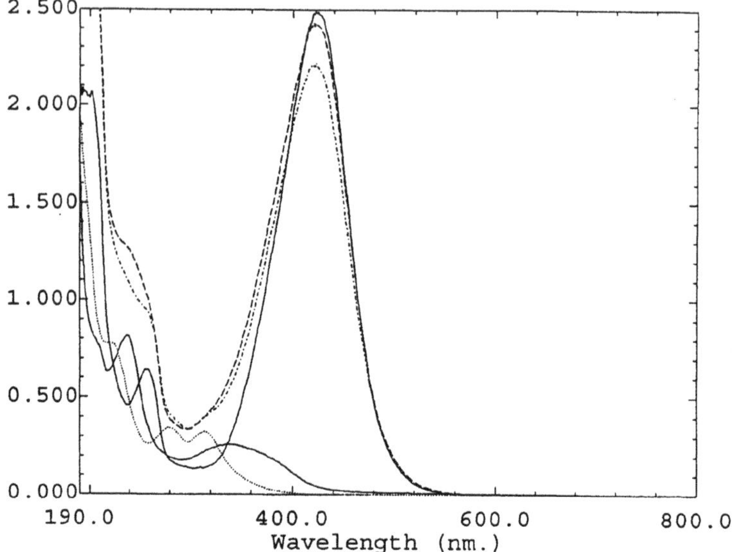

Fig. 2. RAMEB and EG influence on decomposition of UV light treated Nimodipine solution

4. CONCLUSIONS

- Nifedipine derivatives are water-insoluble pharmacons, at the same time, they are very light-sensitive compounds: both sunlight and UV light promote the process of photo-decomposition having them in solution.
- Different **CDs** increase the solubility of **Nifs** to various degrees.
- Various kinds of dyes hinder the decomposition of these drugs. Finally, **CDs** and dyes in combination may prevent it much better than one by one.

ACKNOWLEDGEMENTS

The authors wish to thank **Prof. J. Szejtli** for providing the CDs and advice.

REFERENCES

[1] The Merck Index, 11th Edition. Merck & Co., Inc., Rahway, N.J., USA, 1989, pp. 1031, 1036, 1038 and 1040

[2] Müller, B. W. and Albers, E.: Complexation of dihydropyridine derivatives with CDs. Int. J. Pharm. 79, 273-288 (1992)

[3] Thoma, K. und Klimek, R.: Untersuchungen zur Photoinstabilität von Nifedipin. Pharm. Industry 53, 388-396 (1991)

[4] Szejtli, J.: Cyclodextrin Technology. Kluwer Academic Publ., Dordrecht, 1988

[5] Gyéresi, Á., T_kés, B., Nagy, G., Regdon, G. and Kata, M.: Proc. 1st World Meeting APGI/APV, Budapest, 9/11 May 1995

EFFECT OF TEMPERATURE AND HUMIDITY ON THE STABILITY OF TOLBUTAMIDE/ß-CYCLODEXTRIN INCLUSION COMPLEXES

F. VEIGA and A. SOUSA
Department of Pharmaceutical Technology, Faculty of Pharmacy
University of Coimbra, 3000 Coimbra (Portugal).

ABSTRACT

Complexes of tolbutamide with ß-cyclodextrin were prepared by kneading and coprecipitation. The inclusion complexes were stored under controlled conditions at 5°C, 40°C, 50°C and 40°C/75% relative humidity. The influence of temperature and the humidity was examined by performing DSC thermal analysis, X-ray diffractometry, Raman spectroscopy and dissolution rate.

The results indicated that the physicochemical properties and dissolution rate of TBM/ßCD inclusion complexes were unchanged after storage at 5°, 40°C and 50°C over six months.

1. INTRODUCTION

Cyclodextrins are well know to form inclusion complexes with various molecules in solution and in the solid state. This phenomenon has been widely used to improve the solubility, dissolution rate, bioavailability and stability of poorly water soluble drugs (1-4).

Tolbutamide (TBM), an oral hypoglycaemic agent, is practically insoluble in water.

The aim of this work was to investigate the influence on the physicochemical characteristics and the dissolution rate of TBM/CDs inclusion complexes prepared by two methods, after storage at different conditions of temperature and humidity.

The effect of the storage on such complex was determined by differential scanning calorimetry (DSC), X-ray diffraction, Raman spectroscopy and dissolution studies.

2. MATERIALS AND METHODS

2.1. Materials

Tolbutamide (Sigma) and ß-cyclodextrin (Roquette) were used as received.

2.2. Preparation of solid complexes

The kneaded and coprecipitated complexes were prepared as described in our paper (5).

J. Szejtli and L. Szente (eds.), Proceedings of the Eighth International Symposium on Cyclodextrons, 349–352.

2.3. Storage Conditions

Samples of the inclusion complexes were stored sealed in vials at 5°C, 40°C and 50°C over 6 months. Under the other storage conditions, 40°C/75% relative humidity (RH) over 6 months, the powder was spread on glass disks in order to expose a greater surface area to the moisturised atmosphere.

2.4. Differential scanning calorimetry (DSC)

Thermal analysis were carried out using a Shimadzu, model 50. The measurements were done using aluminium sample pans at a scanning speed of 10°C/min. under a nitrogen stream from 25 to 250°C.

2.5. X-ray diffraction

X-ray diffraction patterns of different samples were obtained using an Enraf-Norius powder diffractometer equipped with an horizontal mounted INEL CPS120 curved position-sensitive detector. Cu-Kα_1 radiation was selected by a bent quartz-crystal monochromator.

2.6. Raman Spectroscopy

The Raman spectra were recorded on a Spex 1403 double spectrometer. The 514.5 nm line of an argon ion laser (Spectra Physics, model 2020-03) was used as Raman excitation.

2.7. Dissolution studies

The dissolution tests were performed according to the USP XXIII paddle method. The dissolution rates were measured in an apparatus (Hanson Research) connected to the spectrophotometer by a peristaltic pump, so that the absorbance was monitored automatically at 229 nm. All samples were analysed at least six times.

3. RESULTS AND DISCUSSION

The results, with exception for dissolution profiles, are not presented because of the excessive amount of information which would not fit the available space.

TBM/β-CD kneaded system

Concerning inclusion complexes of TBM/ß-CD prepared by kneading, the thermograms after storage under different conditions of temperature and/or humidity display the same thermogram as at time 0. The two endothermic peaks, one at ca 126°C and other at ca 155°C were attributed to the presence of a few TBM crystals in the compound and to the shift of melting point of TBM to a higher temperature, respectively. This last occurrence can provide an indication of the inclusion complex formation (6).

The X-ray diffraction patterns of these inclusion complexes, after stored at 5°C, 40°C, 50°C and 40°C/75% RH showed exactly the same at time 0. It was concluded that the storage conditions did not change the interactions between the TBM and ß-CD.

The results obtained with Raman studies were in agreement with those of DSC and X-ray patterns.

TBM/β-CD coprecipitated system

A comparison of the thermograms of TBM/ß-CD coprecipitated inclusion complexes at time 0 and after storage, showed that all display two endothermic peaks, one at ca 126°C and other at ca 155°C. These findings were similar with those of TBM/ß-CD kneaded system.

Also, the X-ray and Raman spectroscopic of TBM/ß-CD coprecipitated system were similar with those of TBM/ß-CD kneaded compound, which indicated that under above mentioned storage conditions no change had occurred relatively at time 0.

Dissolution Studies

Storage at 5°C, 40°C, 50°C and 40°C/75% RH did not have any marked effect on the dissolution rate of none of studied compound, Figs. 1 and 2. These results are in agreement with the previous findings obtained by others authors (7), namely with the TBM/ß-CD coprecipitated inclusion complex.

4. CONCLUSION

The results of this work have shown that the physicochemical properties and dissolution rate of TBM/ß-CD compounds were unchanged after the storage at 5°C, 40°C, 50°C and 40°C/75% RH over six months, which there are in good agreement with previous studies also using TBM as model drug (7-8).

Therefore, we can conclude that the systems studied are stable even when in presence of humidity. Thus, the systems are potential candidates for further incorporation into dosage forms.

REFERENCES

[1] Duchêne, D., Glomot, F. and Vaution, C., Pharmaceutical applications of cyclodextrins, in: Cyclodextrins and Their Industrial Uses, Ed. D. Duchêne, Editions de Santé Publishers, Paris, pp.211-257, 1987.

[2] Szejtli, J., Cyclodextrins in Pharmaceuticals, in: Cyclodextrin Technology, Ed. J.E.D. Davies, Kluwer Academic Publishers, Dordrecht, pp.186-306, 1988.

[3] Saenger, W., Cyclodextrin inclusion compounds in research and laboratory. *Angew. Chem. Int. Ed. Engl.*, **19**, 344-362 (1980).

[4] Uekama, K., Otagiri, M., Cyclodextrins in drug carrier systems. CRC Critical Reviews in *Therapeutic Drug Carrier Systems*, **3**, 1-40 (1987).

[5] Veiga, F., Teixeira-Dias, J.J.C., Kedzierewicz, F.; Sousa. A. and Maincent, P., Inclusion complexation of tolbutamide with ß-cyclodextrin and hydroxypropyl-ß-cyclodextrin, *Int.J.Pharm.*, in press.

[6] Kedzierewicz, F., Hoffman, M. and Maincent, P., Comparison of tolbutamide ß-cyclodextrin inclusion compounds and solid dispersions. *Int. J. Pharm.*, **58**, 221-227 (1990).

[7] Kedzierewicz, F., Villieras, F., Zinutti, C., Hoffman, M. and Maincent, P. A 3 year stability study of tolbutamide solid dispersions and ß-cyclodextrin complex. *Int. J. Pharm.*, **117**, 247-251 (1995).

[8] Torres-Labandeira, J.J., Otero-Espinar, F., Anguiano-Igea, S., Blanco-Méndez, J. and Vila-Jato, J.L., Influence of storage on the biopharmaceutical properties of the tolbutamide/ß-cyclodextrin inclusion complex. *S.T.P. Pharma. Sci.* **5**, 326-330 (1991).

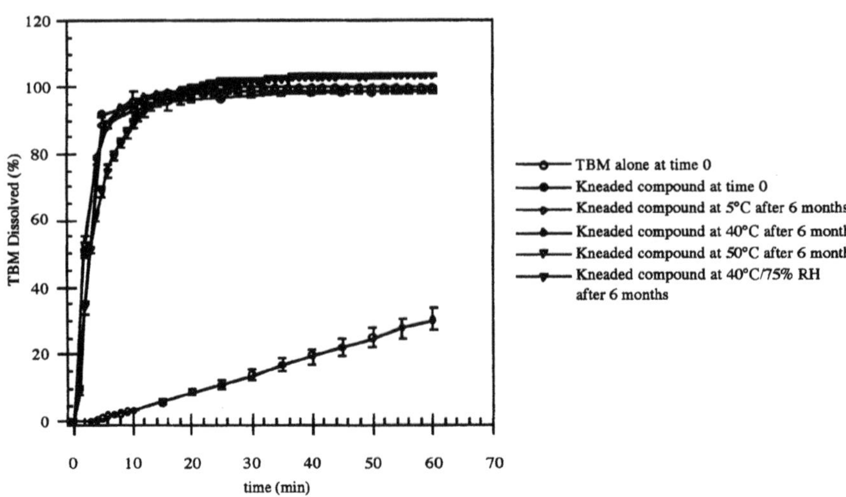

Fig. 1 - Dissolution profiles of TBM and its ßCD kneaded systems. Each point: Mean±S.D.

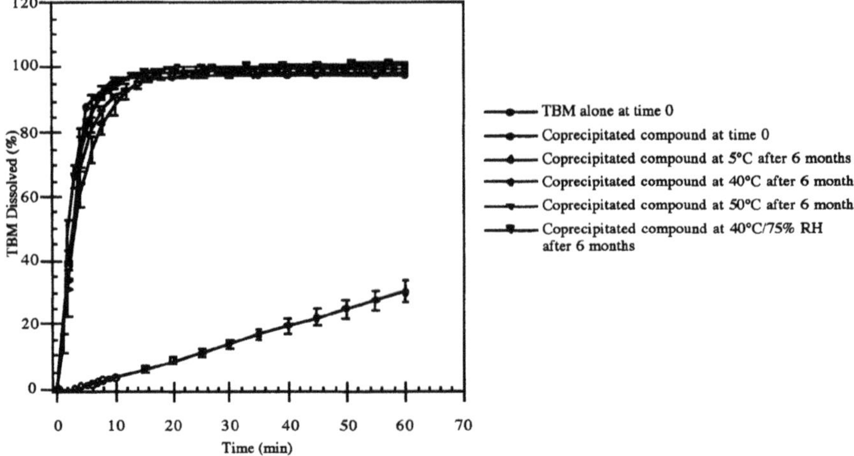

Fig. 2 - Dissolution profiles of TBM and its ßCD coprecipitated systems. Each point: Mean±S.D.

CRYSTALLIZATION AND POLYMORPHIC TRANSITION BEHAVIOR OF CHLORAMPHENICOL PALMITATE IN 2-HYDROXYPROPYL-β-CYCLODEXTRIN MATRIX

F. Hirayama, M. Usami, K. Kimura and K. Uekama

Faculty of Pharmaceutical Sciences, Kumamoto University, 5-1 Oe-honmachi, Kumamoto 862, Japan

ABSTRACT

Chloramphenicol palmitate (CPP) was converted to an amorphous complex when spray-dried with 2-hydroxypropyl-β-cyclodextrin (HP-β-CyD), and no crystallization of CPP was observed for at least 2 months under the storage condition of 50 ℃ and 50% relative humidity. The dissolution rate of CPP/HP-β-CyD complex in aqueous HCO-60 solution was much faster than CPP polymorphs (complex > metastable forms (B and subB) > stable form (A)), which was reflected in the *in-vivo* absorption behavior of CPP following oral administration in dogs.

1. INTRODUCTION

It is important to control the crystallization and polymorphic transition of solid drugs, since crystal modifications affect various pharmaceutical properties, such as stability, solubility, dissolution rate and bioavailability of drugs. Since cyclodextrins (CyDs) can improve various pharmaceutical properties of drug molecules, they can serve as multi-functional drug carriers. From a practical point of view, amorphous CyDs such as 2-hydroxypropyl-CyDs are useful for the control of solid properties of poorly water-soluble drugs [1], because they convert crystalline drugs to amorphous complexes which are usually water-soluble. In this study, the crystallization and polymorphic transition behavior of chloramphenicol palmitate (CPP) in HP-β-CyD matrix was investigated.

J. Szejtli and L. Szente (eds.), Proceedings of the Eighth International Symposium on Cyclodextrons, 353–356.

2. MATERIALS AND METHODS

Materials: CPP and HP-β-CyD (degree of substitution: 4.8) were donated by Sankyo Co. Ltd. (Tokyo, Japan) and Nihon Shokuhin Kako Co. (Tokyo, Japan), respectively. Chloramphenicol (CP) was purchased from Nakalai Tesqe (Tokyo, Japan).

Preparation of CPP polymorphs and CPP/HP-β-CyD complex:: CPP polymorphs (Forms A, B and C) were prepared according to the methods reported [2]. The complex of CPP with HP-β-CyD was prepared by the spray-drying method, *i.e.*, CPP and HP-β-CyD in a 1:2 molar ratio were dissolved in CH_2Cl_2-EtOH (1:1.2 %v/v) and subjected to spray-drying, using a Pulvis GA32 Yamato spray-dryer (Tokyo, Japan) under the following conditions: air flow rate, 0.45 cm^3 min^{-1}; air pressure, 1.0 kgf cm^{-2}; inlet and outlet temperatures, 85 and 55 ℃, respectively.

***In-vivo* absorption studies:** The sample (equivalent to 125 mg CPP) wrapped in a wafer was orally administered along with water (100 ml) to fasted beagle dogs (weight: 9-11 kg). Blood samples (1 ml) was collected from vengular vein using an injection syringe with heparin, and centrifuged (1000 g) for 10 min. CP in plasma was determined by high-performance liquid chromatography.

3. RESULTS AND DISCUSSION

The CPP/HP-β-CyD system showed a typical Ap-type [3] phase solubility diagram, whereas the parent β-CyD system showed a mixed Ap/Bs-type diagram with an ascending curvature at low β-CyD concentration (< about 9 x 10^{-3} M). The stoichiometry of the solid CPP/parent β-CyD complex was 1:2 (guest:host) molar ratio, and the stability constants of 1:1 and 1:2 complexes were 1900 M^{-1} and 4500 M^{-1} for the parent β-CyD complex and 1200 M^{-1} and 3400 M^{-1} for the HP-β-CyD complex, respectively. When CPP was spray-dried in the absence of additives, a metastable CPP (Form subB) was exclusively formed, which has exothermic and endothermic peaks at 64 ℃ and 88 ℃ due to the transition to the other metastable CPP (Form B) and the melting of Form B, respectively, in DSC curves. Form subB was easily converted to Form B with a half-life of 30 min at 50 ℃ and 50 % relative humidity (R.H.). On the other hand, CPP was converted to an amorphous complex when spray-dried with HP-β-CyD and no crystallization of CPP was observed for at least 2 months under the above storage condition, and only 7.2

% crystallization to Form B after 6 months. When the complex was stored for 2 weeks even at a severe condition of 80 ℃, 75 %R.H., it converted to a stable CPP (Form A) only in small amounts (0.7 %).

Figure 1 shows the dissolution profiles of the CPP/HP-β-CyD complex and CPP polymorphs (Forms subB, B and A) in 50 %v/v isopropanol/water and 0.01 %w/v HCO-60/water solutions. The initial dissolution rate of the complex in the isopropanol solution was rather slower than those of CPP polymorphs (rate: Form subB > Form B ~ complex > Form A), although the total CPP released from the complex was larger after 1 h. On the other hand, the complex dissolved rapidly in the aqueous HCO-60 solution, and the amount of dissolved CPP was much larger than those of the polymorphs (rate and amount: complex >> Form B > Form subB > Form A). The rapid dissolution of the complex in the aqueous HCO-60 solution may be due to its high wettability to water, compared with Form subB. Figure 2 shows the plasma level of CP *vs* time curves obtained after oral administration of the complex or CPP polymorphs to dogs. The plasma level of CP was enhanced by the administration of CPP/HP-β-CyD complex, and no ageing effect on the absorption of CPP was observed for the HP-β-CyD complex, even after 6 months at 50 ℃, 50 %.R.H.

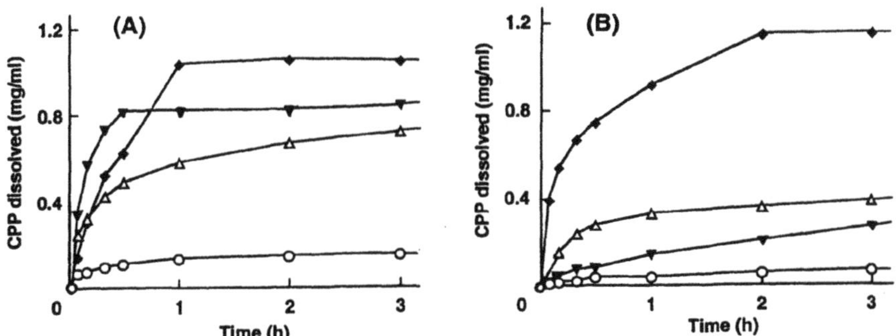

Fig. 1 Dissolution Profiles of CPP from Various Preparations (equivalent to 62.5 mg CPP) in 50 %v/v Isopropanol/Water Solution at 25 ℃ (A) and 0.01%w/v HCO-60/Water Solution at 37 ℃ (B), Measured by Dispersed Amount Method at 91 rpm
○ : Form A, △ : Form B, ▼ : Form subB, ◆ : HP-β-CyD complex.
Each point represents the mean of 2-4 experiments.

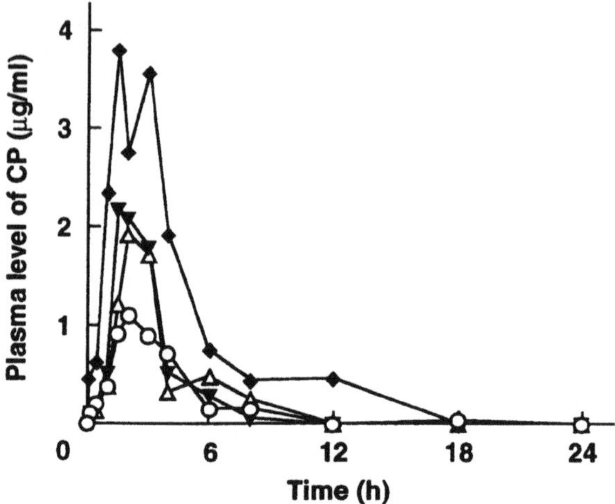

Fig. 2 Plasma Levels of CP after Oral Administration of CPP/HP-β-CyD Complex (equivalent to 125 mg/body CP) or CPP Polymorphs in Dogs
○ : Form A, △ : Form B, ▼ : Form subB, ◆ : complex.
Each point represents the mean of 3 dogs.

4. CONCLUSION

The present results suggest that HP-β-CyD is useful for the preparation and stabilization of amorphous CPP, and will provide a rational basis for the design of formulation and storage conditions in solid dosage forms of poorly water-soluble drugs.

REFERENCES

[1] Hirayama,F., Wang,Z., Uekama,K., Effect of 2-hydroxypropyl-β-cyclodextrin on crystallization and polymeric transition of nifedipine in solid state, *Pharm. Res.*, 11, 1766 (1994).

[2] Kaneniwa,N, Otsuka,M., Effect of Grinding on the transformation of polymorphs of chloramphenicol palmitate, *Chem. Pharm. Bull*, 33, 1660 (1985).

[3] Higuchi,T., Connors, K.A., Phase-solubility techniques, *Adv. Anal. Chem. Instr.*, 4, 117 (1965).

DIFFERENTIAL EFFECTS OF CYCLODEXTRIN DERIVATIVES ON AGGREGATION AND THERMAL BEHAVIOR OF INSULIN

KEIICHI TOKIHIRO, TETSUMI IRIE AND KANETO UEKAMA

Faculty of Pharmaceutical Sciences, Kumamoto University, 5-1, Oe-honmachi, Kumamoto 862, Japan

ABSTRACT

Maltosyl-β-cyclodextrin (G2-β-CyD) suppressed the aggregation of insulin in a neutral solution, while the sulfate of α-CyD (S-α-CyD) accelerated the aggregation. On the other hand, the sulfobutyl ether of β-CyD (SBE-β-CyD) showed differential effects on the insulin aggregation, depending on the degree of substitution; *i.e.* the inhibition at relatively low substitution and acceleration at higher substitution. Differential scanning calorimetric (DSC) studies indicate that the self-association of insulin stabilized the native conformation of the peptide, as indicated by an increase in the mean unfolding temperature (Tm). G2-β-CyD and SBE-β-CyD decreased the Tm value of insulin oligomers, while S-α-CyD increased the Tm value. These results suggest that a proper use of the CyD derivatives is effective in designing rapid and long-acting insulin preparations.

1. INTRODUCTION

Our previous studies have shown that hydrophilic CyDs including G2-β-CyD and 2-hydroxypropyl-β-CyD significantly inhibited the adsorption of bovine insulin to hydrophobic surfaces of containers and its aggregation by interacting with hydrophobic amino acid residues of the peptide [1]. Recently, the sulfates and sulfoalkyl ethers of CyDs have been evaluated as a new class of parenteral drug carriers, because they are highly hydrophilic and less hemolytic than the parent CyDs [2]. In the present study, we examined the effects of the CyD derivatives on the aggregation and thermal behavior of insulin in both acidic and neutral solutions by means of the ultrafiltration method, DSC and liquid chromatography-mass (LC/MS) spectrometry.

2. MATERIALS AND METHODS

2.1. Materials

Bovine insulin (27.5 IU/mg) was obtained from Sigma Chemical Co. (St. Louis, MO, USA). G2-β-CyD, S-α-CyD with an average degree of substitution of 11.4 and SBE-β-

J. Szejtli and L. Szente (eds.), Proceedings of the Eighth International Symposium on Cyclodextrons, 357–360.

CyD with average degrees of substitution of 4.0 and 7.0 (SBE4-β- and SBE7-β-CyDs) were donated by Ensuiko Sugar Refining Co. Ltd. (Yokohama, Japan), Kokusan Chemical Co. Ltd. (Tokyo, Japan), and CyDex L.C. (Overland Park, KS, USA), respectively.

2.2. Methods

The aggregation of insulin was evaluated by measuring the remaining insulin in the filtrate (filter: DISMIC-13CP045AN; Advantec Co., Tokyo, Japan) after standing freshly prepared insulin solutions (0.15 mM, in pH 6.8 phosphate buffer) in a silicone-coated glass tube at 25°C. The DSC thermograms of insulin solutions were recorded by a MC-2 microcalorimeter (MicroCal, Inc., Amherst, MA, USA). The LC/MS spectra of insulin solutions were measured by a M-1200H LC/MS system (Hitachi, Tokyo, Japan) equipped with an electrospray ionization (ESI) source.

3. RESULTS AND DISCUSSION

3.1. Effects of CyD Derivatives on Aggregation of Insulin

Insulin in a neutral solution (pH 6.8) was mostly assembled as zinc-containing hexamers, eventually leading to the precipitation of higher order aggregates of the peptide in a concentration- and time-dependent manners. As shown in Fig. 1, G_2-β-CyD significantly suppressed the aggregation of insulin, while S-α-CyD accelerated the insulin aggregation. G_2-β-CyD may interact with hydrophobic amino acid residues of insulin, and thus prevent the aggregation by eliminating intermolecular hydrophobic contacts [1]. Since S-α-CyD has highly concentrated negative charges located near the entrance of the cavity and shows limited inclusion ability, the neutralization of cationic charges in insulin by S-α-CyD may contribute to the accelerated aggregation of the peptide.

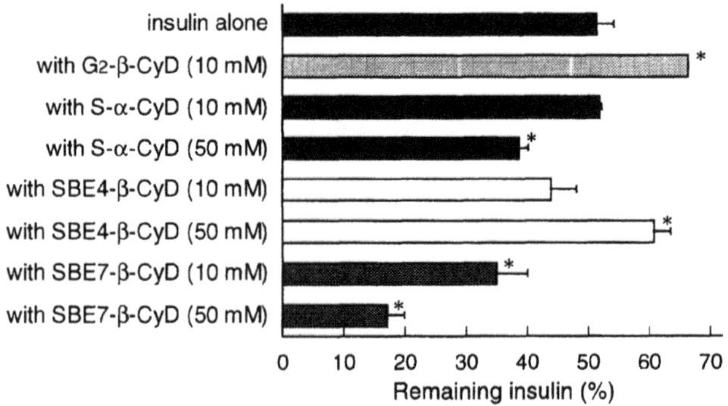

Fig. 1 Effects of CyD Derivatives on Aggregation of Insulin 24 h after Preparation of Insulin Solution (0.15 mM) in Phosphate Buffer (pH 6.8) at 25°C

Each point represents the mean±S.E. of 2-14 experiments.
*$p<0.05$ *versus* insulin alone.

On the other hand, SBE-β-CyD showed differential effects on insulin aggregation, depending on the degree of substitution; *i.e.* inhibition at relatively low substitution and acceleration at higher substitution. Since the sulfonate groups in SBE-β-CyD are appropriately spaced from the CyD cavity with an butyl chain and do not interfere with the inclusion process, SBE4-β-CyD may inhibit the insulin aggregation in a manner similar to G2-β-CyD. In case of SBE7-β-CyD, the electric effects seem to be more of a factor than the inclusion effects, eventually leading to the acceleration of the insulin aggregation. The differential effects of the CyD derivatives on the insulin aggregation were confirmed by the ultrafiltration experiments, in which G2-β-CyD and SBE4-β-CyD facilitated the permeation of insulin through the ultrafiltration membranes, while S-α-CyD and SBE7-β-CyD reduced the membrane permeation of insulin. Furthermore, G2-β-CyD facilitated the permeation of insulin through the membranes in an acidic solution (pH 2.0), in which insulin existed mainly as a zinc-free dimer in this condition, indicating that G2-β-CyD affects the equilibrium between the dimer and the monomer.

3.2. Effects of CyD Derivatives on Thermal Behavior of Insulin

The DSC thermograms of insulin solutions showed that self-association of insulin stabilized the native conformation of the peptide, as indicated by an increase in Tm. Fig. 2 shows the effects of the CyD derivatives on DSC thermograms of insulin in phosphate buffer (pH 6.8). G2-β-CyD and SBE-β-CyD significantly reduced the Tm value of insulin oligomers, the former being more effective. These CyD derivatives may shift the equilibrium in favor of the unfolded insulin by dissociating the oligomers and/or binding to hydrophobic side chains exposed on the unfolded peptide. On the other hand, S-α-CyD increased the Tm value of insulin, reflecting on the higher degree of association of the peptide.

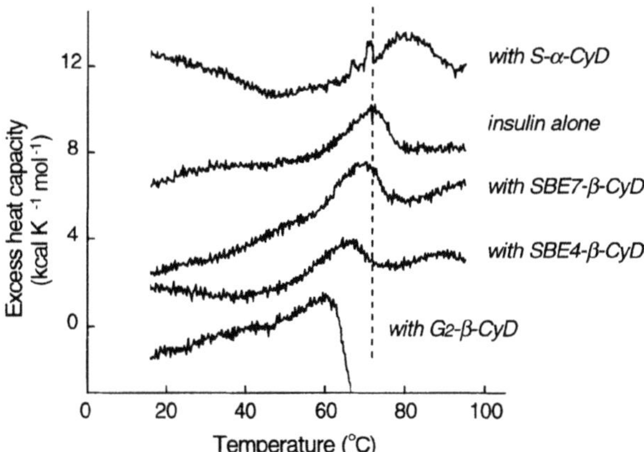

Fig. 2 Effects of CyD Derivatives (0.1 M) on DSC Thermogram of Insulin (0.1 mM) in Phosphate Buffer (pH 6.8)

The complexation of insulin with G2-β-CyD was further confirmed by the LC/MS analysis. Fig. 3 shows the positive ion ESI mass spectra of insulin in the absence and presence of G2-β-CyD in acidified mixtures of water and methanol. In the absence of G2-β-CyD, insulin gave a bell-shaped multiple charge state distribution to the $(M+6H)^{6+}$ multiple-protonated species. In the presence of G2-β-CyD, a peak corresponding to the complex or electrostatic adduct of the charged insulin with G2-β-CyD at a molar ratio of 1:1 was observed.

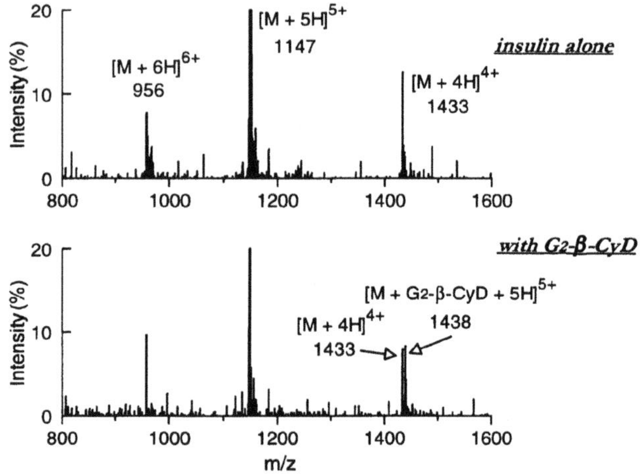

Fig. 3 ESI Mass Spectra of Insulin (0.1 mM) in Positive Ion Mode in the Absence and Presence of G2-β-CyD (2 mM) in Water/Methanol/Acetic Acid (47 / 47 / 6) Solution

4 CONCLUSION

The present results indicate that the CyD derivatives interact with insulin in a differential manner and hence a proper use of the CyD derivatives is effective in designing rapid and long-acting insulin preparations.

ACKNOWLEDGEMENTS

This work is partly supported by the Sasakawa Scientific Research Grant from the Japan Science Society.

REFERENCES

[1] Tokihiro, K., Irie, T., Uekama, K., Pitha, J., Potential use of maltosyl-β-cyclodextrin for inhibition of insulin self-association in aqueous solution, *Pharm. Sci.*, **1**, 49-53 (1995)

[2] Shiotani, K., Uehata, K., Irie, T., Uekama, K., Thompson, D. O., Stella, V. J., Differential effects of sulfate and sulfobutyl ether of β-cyclodextrin on erythrocyte membranes *in vitro*, *Pharm. Res.*, **12**, 78-84 (1995)

ENERGETICS OF CYCLODEXTRIN-INDUCED DISSOCIATION OF INSULIN

MICHELLE LOVATT[1], ALAN COOPER[1] & PATRICK CAMILLERI[2]

[1]Department of Chemistry, Glasgow University, Glasgow G12 8QQ, Scotland; and [2]SmithKline Beecham Pharmaceuticals, The Frythe, Welwyn, Herts AL6 9AR, UK.

ABSTRACT

The energetics of dissociation of bovine insulin oligomers in aqueous solution under various conditions have been investigated by dilution microcalorimetry. Addition of cyclodextrins increases dissociation of insulin oligomers in solution in a manner consistent with interaction of these cyclic polysaccharides with protein side chains. For example, assuming monomer-dimer equilibrium, in the absence of cyclodextrins dilution data (25 °C, pH 2.5) are consistent with a dimer dissociation constant (K_{diss}) of about 12 μM and a dimer dissociation enthalpy (ΔH_{diss}) of +41 kJ mol^{-1}. Addition of methyl-β-cyclodextrin (up to 200 mM) makes dissociation significantly more endothermic (ΔH_{diss} = 79 kJ mol^{-1}) and reduces the apparent dimer dissociation constant by more than two orders of magnitude ($K_{diss} \approx$ 1.7 mM). Qualitatively similar results are observed with α-cyclodextrin and other β-cyclodextrin derivatives.

1. INTRODUCTION

Insulins are known to occur in a variety of aggregation (oligomer) states in solution depending on concentration, pH, temperature, Zn^{2+} concentration and other ionic conditions [1], and their aggregation states can potentially affect their use in therapeutic situations. Based on previous observations on interaction of cyclodextrins with globular proteins [2-4] we predicted that complexation of these cyclic polysaccharides with surface protein residues might significantly affect the state of aggregation of protein in solution. The non-polar cavities of toroidal cyclodextrin molecules have a particular affinity for small non-polar groups and can both enhance the solubility and improve stability of such molecules in water. Cyclodextrins are finding increasing use as solubilizing agents and stabilizing excipients for protein drugs, including insulins [5]. We show here by calorimetric measurement of heats of dilution that the dissociation of bovine insulin in solution is significantly enhanced in the presence of various cyclodextrins. The energetics of this process are consistent with association of

J. Szejtli and L. Szente (eds.), Proceedings of the Eighth International Symposium on Cyclodextrons, 361–364.

cyclodextrin molecules with insulin surface residues. Such a complexation might also bring about conformational changes in insulin which might contribute indirectly to the disaggregation process.

2. MATERIALS AND METHODS

Bovine insulin concentrations were verified by UV absorbance measurements on diluted aliquots assuming a molar extinction coefficient (ε_{280}) of 5734 M^{-1} [6]. Buffers used were 0.1M glycine/HCl pH 2.5 or 0.1M Na-phosphate pH 7.4, containing appropriate concentrations of cyclodextrins where required. Calorimetric dilution experiments were done using a Microcal OMEGA titration microcalorimeter following standard instrumental procedures at 25°C [7,8]. In a typical dilution experiment small aliquots (10-20μl) of concentrated insulin, dissolved in buffer or buffer/cyclodextrin mix, were injected into the calorimeter reaction vessel containing the identical buffer mixture. Integrated heat pulse data, after correction for mixing controls done separately under identical conditions, were analysed by non-linear regression in terms of a simple monomer-dimer equilibrium model to give the apparent equilibrium constant (K_{diss}) and enthalpy of dissociation (ΔH_{diss} per mole dimer).

3. RESULTS AND DISCUSSION

Dilution of a series of small aliquots of insulin into a larger volume of buffer in the microcalorimeter gives a sequence of endothermic heat pulses characteristic of molecular dissociation (Fig.1A). A typical heat of dilution curve which can be fitted in terms of a monomer-dimer equilibrium model yielding K_{diss} and ΔH_{diss} is shown in Fig.1B. Addition of cyclodextrins to the buffer mixture gives rise to two significant effects (Fig.1C): (i) insulin dilution heats become more endothermic, and (ii) the dilution curves become more attenuated, indicating greater dimer dissociation in the presence of cyclodextrins.

Previous studies have shown that, at low pH, insulin is predominantly dimeric in solution at high concentrations [1]. In the absence of cyclodextrins at pH 2.5 the insulin dimer dissociation constant (K_{diss}), obtained from non-linear regression analysis of dilution data, is around 12μM and in agreement with previous determinations by other techniques [1]. At pH 7.4 the oligomeric state of insulin is less clear, and hexamers or higher oligomers almost certainly exist under the conditions used here. Nevertheless, the dilution data fit reasonably to a dimer model.

Non-linear regression analysis of experimental data for the variation of apparent dimer dissociation constant with cyclodextrin concentration (Fig.2) indicates that two sequential binding sites are adequate for describing the data satisfactorily over the accessible concentration range.

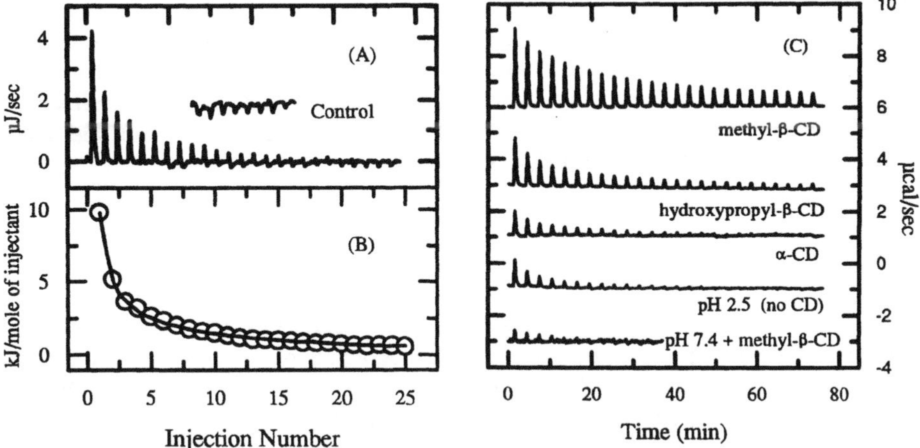

Fig. 1 Calorimetric data for the endothermic dissociation of insulin dimers at pH 2.5: (A) Raw data for injection of insulin, 1.53mM, 25 × 10 μl injections, into buffer at 25 °C, with control data; (B) Integrated injection heats, corrected for control heats and fit (solid line) to a dimer dissociation model with K_{diss} = 12μM and ΔH_{diss} = 41 kJ mol^{-1}; (C) Examples of raw calorimetric dilution data showing the effects of different cyclodextrins on insulin dissociation, all at pH 2.5 except where indicated. For comparison, cyclodextrin concentrations (when present) are all approximately 100 mM in this case.

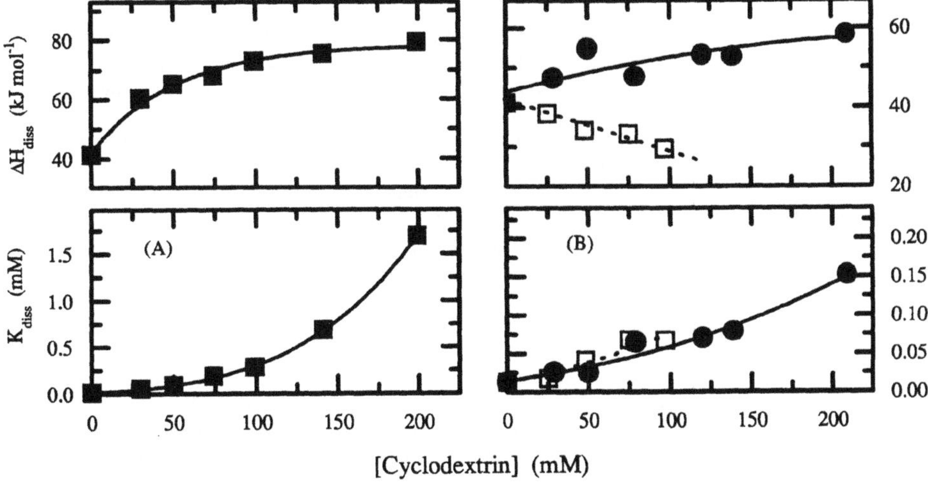

Fig. 2 Change in apparent dissociation constants (K_{diss}, lower panels) and enthalpies of dissociation of insulin (ΔH_{diss}, upper panels) at pH 2.5 with increasing cyclodextrin concentrations. In the lower panels the curves show the theoretical fits to a simple sequential binding model. (A) Methyl-β-cyclodextrin (filled squares). (B) Hydroxypropyl-β-cyclodextrin (filled circles) and α-cyclodextrin (open squares).

The numerical values obtained for the site-binding constants (K_1, K_2) are consistent with the relatively weak affinities expected on the basis of previous observations of interaction between cyclodextrins in solution and aromatic amino acid side chains and similar groups [2,9,10]. For methyl-β-cyclodextrin, which shows the biggest effect, K_1 and K_2 are estimated to be about 20 and 6.5 M^{-1}, respectively. For the other cyclodextrins $K_1 \approx$ 10-15 M^{-1}, with $K_2 < 5\ M^{-1}$.

4. CONCLUSION

In the absence of cyclodextrins the dissociation of insulin oligomers is endothermic. Addition of α-cyclodextrin, in addition to encouraging oligomer dissociation, also makes this dissociation less endothermic in a manner consistent with the exothermic binding of α-cyclodextrins to exposed groups on insulin monomers after dissociation. In contrast, although methyl- and hydroxypropyl-β-cyclodextrins similarly induce oligomer dissociation, this dissociation is observed to be more endothermic. This suggests that the binding of these modified β-cyclodextrins to exposed insulin residues, although thermodynamically favourable, is endothermic and consequently entropy-driven.

REFERENCES

[1] Blundell, T.L., Dodson, G., Hodgson, D. and Mercola, D., Insulin: The structure in the crystal and its reflection in chemistry and biology, *Adv.Prot.Chem.*, **26**, 279-402 (1972).

[2] Cooper, A., Effect of cyclodextrins in the thermal stability of globular proteins, *J.Amer.Chem.Soc.*, **114**, 9208-9209 (1992).

[3] Cooper, A. & McAuley-Hecht, K.E., Microcalorimetry and the molecular recognition of peptides and proteins, *Phil.Trans.R.Soc. Lond. A*, **345**, 23-35 (1993).

[4] Camilleri, P., Haskins, N.J. & Howlett, D.R., Beta-cyclodextrin interacts with the Alzheimer amyloid beta-A4 peptide, *FEBS Lett.*, **341**, 256-258 (1994).

[5] Brewster, M.E., Hora, M.S., Simpkins, J.W. and Bodor, N., Use of 2-hydroxypropyl-beta-cyclodextrin as a solubilizing and stabilizing excipient for protein drugs, *Pharm.Res.*, **8**, 792-795 (1991).

[6] Porter, R.R., Partition chromatography of insulin and other proteins, *Biochem.J.*, **53**, 320-328 (1953).

[7] Wiseman T., Williston, S., Brandts, J.F. & Lin, L.-N., Rapid measurement of binding constants and heats of binding using a new titration calorimeter, *Anal.Biochem.*, **179**, 131-137 (1989).

[8] Cooper, A. & Johnson, C.M., *Isothermal Titration Microcalorimetry*, in Microscopy, Optical Spectroscopy, and Macroscopic Techniques Methods in Molecular Biology, (Jones, C., Mulloy, B., and Thomas, A.H., eds.), Humana Press, Totowa, N.J.,1994, Vol.22, pp. 137-150.

[9] Cooper, A. & MacNicol, D.D., Chiral host-guest complexes: interaction of α-cyclodextrin with optically active benzene derivatives, *J.Chem.Soc.Perkin II*, 761-763 (1978).

[10] Horsky, J. & Pitha, J., Inclusion complexes of proteins, *J.Inclusion Phenomena and Mol. Recognition in Chemistry*, **18**, 291-300 (1994).

COMPLEXATION OF DRUG COMPOUNDS WITH IONIC AND NON-IONIC CYCLODEXTRINS.

MÁR MÁSSON, THORSTEINN LOFTSSON, HAFRÚN FRIÐRIKSDÓTTIR, DORTE SEIR PETERSEN AND SIGRÍÐUR JÓNSDÓTTIR§.

Department of Pharmacy, University of Iceland, IS-107 Reykjavik, Iceland.
§*Science Institute, University of Iceland, IS-107 Reykjavik, Iceland.*

ABSTRACT

The complexation and stabilisation of ionic drugs with ionic and non-ionic cyclodextrins was studied by measuring the degradation rate in cyclodextrin solutions. The complexation constants for the anionic chlorambucil and cationic atropine increased up to 90% with counter ionic cyclodextrin as compared to non-ionic cyclodextrins but was reduced by 50 to 100%, when the drug and cyclodextrin molecules carried the same type of charge. The degradation rate of the drug within the cyclodextrin cavity was not affected by the ionisation.

1. INTRODUCTION

The use of cyclodextrins in drug formulations is already well documented and has resulted in adoption of ß-cyclodextrin into the pharmacopoeia of many countries. The cyclodextrins are used to increase both solubility and stability of drugs, and to promote up-take of drugs through biological membranes.
Some of the natural parent cyclodextrins have the disadvantage of low aqueous solubility, which can also lead to some toxicity. Therefor there is much interest in cyclodextrin derivatives with larger aqueous solubility, such as the non-ionic 2-hydroxypropyl-ß-cyclodextrin (HP-CD).
Recently ionic cyclodextrins have become available. Ionic cyclodextrins can have certain advantage over other cyclodextrin derivatives, like high solubility and low up-take of the cyclodextrin through biological membranes. Many drugs are ionic in solution and thus we were interested to know if there existd any co-operative effect between the lypophilic binding of the drug molecule in the cyclodextrin cavity, and the ionic interaction. Further more we wanted to see whether the charge would have any influence on the drug stability.
Three drugs were selected for the present study. Chlorambucil, a anti-cancer drug, atropine, a anticholinergic drug, and acetyl salicyclic acid, a analgesic drug. The charge and the rate determining step in the degradation is shown in Figure 1. The

J. Szejtli and L. Szente (eds.), Proceedings of the Eighth International Symposium on Cyclodextrons, 365–368.

complexation and stability of the drugs were investigated with two negatively charged cyclodextrin, one positively charged and three neutral cyclodextrins as reference.

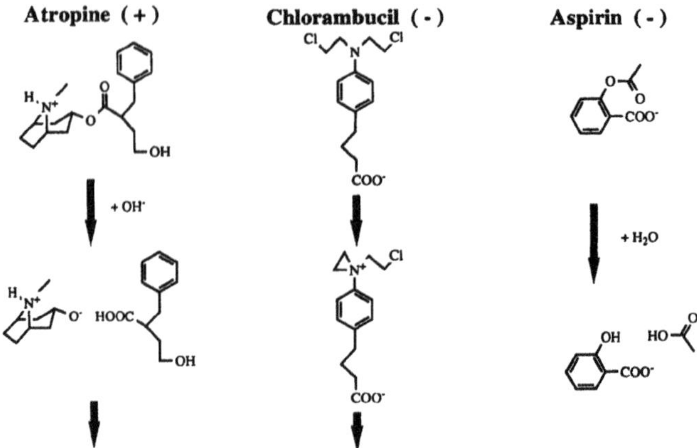

Figure 1. The rate determining step in the breakdown of the drugs [1].

2. MATERIALS AND METHODS

2.1 Materials

The following cyclodextrin derivatives were used: carboxymethyl-ß-cyclodextrin (CM-CD; MS 0.5) trimethylamoniumpropyl-ß-cyclodextrins (TMA-CD; MS 0.5) and hydroxypopyl-ß-cyclodextrin from Wacker (HP-CD; MS 0.6), acetyl-ß-cyclodextrin (A-CD; MS 1.0) and methyl-ß-cyclodextrin from Consortium (M-CD; MS 0.6), and sulfobutyl-ß-CD form CyDex (SB-CD, Mw ~ 2160). Chlorambucil was supplied by the courtesy of Wellcome Foundation Ltd. (UK). All other chemicals were commercially available chemicals of reagent or analytical grade.

2.2 Kinetic studies and chromatography conditions

The degradation media consisted of 10 mM aqueous phosphate buffer containing from 0 to15% cyclodextrin. The stock solutions of atropine, acetyl salicyclic acid, and chlorambucil were made in water, ethanol, and methanol respectively. Between 10 and 7.5 µl of the drug stock solution was added to 1.5 ml of the cyclodextrin solutions, which were kept on a temperature-controlled-sample-rack, and the break-down was monitored by HPLC analysis. The initial drug concentration was 36 $\times 10^{-5}$ M for atropine, 5.7 $\times 10^{-5}$ M for aspirin and 3.4 $\times 10^{-5}$ M for chlorambucil. The chromatographic conditions for aspirin [2] and chlorambucil [3] analysis have been reported previously . For the analysis of atropine a C_{18} reverse-phase, 30 cm, column was used with a acetonitrile : 0.25 M aqueous acetic acid (80:20) mobile phase containing 6 mM tetradecyltrimethylammoniumbromide, pH 6.9. The retention time with 1.5 ml/min flow rate was 3.2 min.

2.3 Interpretation of the kinetic data.

The first order rate constant for the for degradation of the drug in the cyclodextrin solutions, k_{obs}, and k_o for pure buffer, was obtained from liner regression of the natural logarithm of the drug peak height in HPLC, against the time. The complexation constant K_c; and the degradation rate constant k_c, for drug in the complex, was obtained from a Linweaver-Burk type of plot for the different cyclodextrin concentrations, as previously described [4].

3. RESULTS AND DISCUSSION

3.1. Complexes of chlorambucil and atropine.

The cyclodextrin concentration had no effect on the degradation rate of aspirin (acetyl salicyclic acid) at pH 7.0, for all the cyclodextrins tested. NMR measurement of sodium salicyclic acid, 5 mM in 0.75% cyclodextrin solutions, also showed that the complextion of the ionized compound was insignificant at this pH. Previously [2] we have shown that both aspirin and salicyclic acid form a good complex with ß-CD at pH 1, were they are protonated and therefore neutral.

TABLE 1. K_c and k_c constants for the two drugs

Cyclodextrin	charge	$K_c \times 10^{-3}$ M^{-1}	$k_c \times 10^{3}$ min^{-1}	k_c/k_0
Chlorambucil (40 °C, pH 7.35)	-			
CM-CD	-	0.7	2.8	0.043
SB-CD	-	1.4	0.3	0.005
HP-CD	0	3.1	3.4	0.031
A-CD	0	2.6	2.1	0.051
M-CD	0	4.7	3.7	0.032
TMA-CD	+	4.9	2.0	0.056
Atropine (78 °C, pH 7.56)	+			
CM-CD	-	0.15	0.38	0.19
SB-CD	-	0.10	$k_c << k_0$	-
HP-CD	0	0.05	0.35	0.18
M-CD	0	0.28	0.54	0.27
TMA-CD	+	-	-	-

The K_c and k_c values for atropine and chlorambucil are shown in Table 1. When the anionic chlorambucil was complexed with the cationic TMA-CD the K_c value increased 5-90%, as compared to the non-ionic cyclodextrins, and with the anionic cyclodextrin it was reduced 45-85%. The degradation rate of the cationic atropine did not change

when it was dissolved with TMA-CD, indicating that a complex was not formed. The K_c of atropine with the anionic cyclodextrins was two to three times higher than with HP-CD, but only about one-half that for M-CD. The K_c of chlorambucil with M-CD was also relatively high. M-CD is thought to be more hydrophobic than other cyclodextrin derivatives. Stable pH could not be maintained in aqueous A-CD solutions at 78 °C, and therefore the measurements were not reliable for atropine.
The k_c values for SB-CD were much lower than for the other cyclodextrins, for both drugs tested, but otherwise the k_c varied less than 40% of the median value and there appeared to be no systematic difference between the ionic and non-ionic cyclodextrins.

3.3. NMR studies of chlorambucil complexes.

The change in the NMR chemical shifts of the chlorambucil protons, when complexed with cyclodextrin, showed that the aromatic group and partially the amine alkyl chains were inside the cyclodextrin cavity. There was very little change in the chemical shifts of the protons on the alkyl chain bearing the carboxyl group, indicating that this chain extended out into the solution. The relative shift changes were the same except for TMA-CD were the relative shift changes of the aromatic protons were different. In this case the aromatic group was probably accommodated differently inside the cyclodextrin so that a better ionic pair could form between the drug anion and the cyclodextrin cation.

4. CONCLUSION

The ionic cyclodextrins form stronger complexes with counter-ionic drugs than non-ionic cyclodextrins, and weaker complexes with cyclodextrins carrying the same type of charge. The charge of the cyclodextrin did not appear to have any significant effect on the degradation rate of the drug inside the cyclodextrin cavity.

ACKNOWLEDGEMENTS

The work was supported by grant from the University of Iceland Research Fund.

REFERENCES

[1] Connors, K. A., Amidon, G. L., Stella, V. J., Chemical Stability of Pharmaceuticals, John Wiley & Sons, New York, 1986.

[2 Loffssson, T., Olafsdottir, B. J., Fridriksdottir, H., Jonsdottir, S., Cyclodextrin complexation of NSAIDs: physicochemical characteristics, *Eur. J. Pharm. Sci.* , **1**, 95-101, (1993)

[3] Loftsson, T., Bjornsdottir, S., Palsdottir, G., Bodor, N., The effects of 2-hydroxuypropyl-ß-cyclodextrin on the solubility and stability of chlorambucil and melphalan in aqueous solution, *Int. J. Pharm.*, **57**, 63-72, (1989)

[4] Loftssson, T., Effects of cyclodextrins on the chemical stability of drugs in aqueous solutions, *Drug Stability*, **1**, 22-33, (1995).

THE APPLICATION OF EQUILIBRIUM DIALYSIS TO THE DETERMINATION OF DRUG-CYCLODEXTRIN STABILITY CONSTANTS

SYDNEY O. UGWU, MARCOS J. ALCALA, RENU BHARDWAJ and JAMES BLANCHARD
Department of Pharmacology and Toxicology
University of Arizona, Tucson, AZ 85721 U.S.A.

ABSTRACT

The equilibrium dialysis method was applied to the determination of drug-cyclodextrin stability constants using diflunisal and 2-hydroxypropyl-β-cyclodextrin (HPBCD) as a model system. Analysis of data showed the existence of a linear Scatchard plot, which is indicative of the formation of a 1:1 diflunisal:HPBCD complex. The stoichiometry of the complex was verified using the appropriate mass action law equation. The complexation constant (K_c) was 3801 ± 541 M^{-1}. The K_c obtained using the equilibrium dialysis method was comparable to that obtained using a potentiometric method (5564 ± 56 M^{-1}).

1. INTRODUCTION

The equilibrium dialysis method has been employed in several studies investigating binding of drugs to proteins and other macromolecules (1). However, this method has not been widely applied to the characterization of drug binding to cyclodextrins. Equilibrium dialysis has several advantages over the commonly used phase-solubility method. Saturated drug solutions are not required for this method, therefore, only relatively small amounts of drug are needed. The control of pH and ionic strength of solutions is relatively easy compared to the saturated solutions used in phase-solubility studies. It is also possible to study the effects on complexation of varying either the drug or ligand concentration. This method also avoids some of the problems associated with the accurate determination of S_o (the solubility of the drug in the absence of ligand) reported elsewhere (2). Furthermore, it is possible to measure free drug, bound drug, and ligand concentrations. Since the concentrations of free drug, bound drug and ligand are known, the stability constant can be calculated directly without resorting to data transformation.

In the present study, the equilibrium dialysis method was applied to the determination of the stability constant of HPBCD:diflunisal complexes since the binding parameters for this

J. Szejtli and L. Szente (eds.), Proceedings of the Eighth International Symposium on Cyclodextrons, 369–372.

complex were recently evaluated using the potentiometric method and an ion-selective electrode (3), and thus, could be compared to parameters obtained in this study. The results obtained by the equilibrium dialysis and potentiometric titration methods were then compared.

2. MATERIALS AND METHODS

2.1 Materials

The materials used in this study were as follows: potassium phosphate, monobasic crystals, and potassium phosphate, dibasic crystals (J.T. Baker, Phillipsburg, NJ), diflunisal (5[2,4-Difluorophenyl] salicylic acid), Sigma Chemical Co., St. Louis, MO), Encapsin HPB (2-Hydroxypropyl-ß-cyclodextrin; HPBCD) (D.S. = 4.1, MW = 1,372.8, American Maize Products Company, Hammond, IN), sodium hydroxide (MCB Manufacturing Chemists, Inc., Cincinnati, OH), methanol, and acetonitrile (Burdick & Jackson, Muskegon, MI), potassium chloride (J.T. Baker, Phillipsburg, NJ), nylon membrane, 0.6-mil, acrylic plastic dialysis cells (Model 289, Bel-Art Products, Pequannock, NJ), filters, 0.45-μm (Millipore Corp., Bedford, MA). All solutions were prepared using deionized, distilled water.

2.2 HPLC Assay

The HPLC system consisted of an Altex (Altex Scientific Inc., Berkeley, CA) Model 110A pump, a Rheodyne (Cotati, CA), Model 7125 injector with a 50 μL loop and a Hitachi/Spectra-Physics (Fremont, CA) Model 100-30 variable-wavelength UV detector set at 262 nm. The analytical column was a Phenomenex (Torrance, CA) C-18 column (10 μm, 300 X 3.9 mm i.d.) fitted with a Whatman (Clifton, NJ) C_{18} (30 μm) guard column (10 X 4.6 mm). The mobile phase consisted of 58% v/v of 0.01 M phosphate buffer, pH 7.0: 26.3% v/v acetonitrile: 15.7% v/v methanol. A flow rate of 1.0 mL per min was utilized. Duplicate 50-μL injections were made for each sample.

2.3 Equilibrium dialysis studies

The nylon membranes were washed for 1 hr in deionized, distilled water to remove any contaminants. The dialysis cells were assembled with the membrane acting as a semipermeable barrier between the two compartments. Then, 0.8 mL of buffer solution was added to one cell compartment (the aqueous compartment) and 0.8 mL of solutions consisting of varying volume ratios of diflunisal (0.01 M) in HPBCD (0.009 M) and HPBCD (0.009 M) was added to the other cell compartment (binder compartment). The cells were placed on a water bath shaker (GCA/Precision Scientific, Chicago, IL) and agitated at 100 oscillations/min for 21 hours at room temperature (23-25°C) until equilibrium was achieved. Samples were removed from both cell compartments and analyzed for diflunisal using the above HPLC method.

The impermeability of the nylon membrane to HPBCD was verified by placing 0.8 mL of buffer solution in one cell compartment and 0.8 mL of 0.009 M HPBCD solution in the other compartment. After equilibration (21 hr), both compartments were assayed for HPBCD using a previously reported method (4). Possible changes in the volume of the two compartments, due to an osmotic pressure gradient (5), were insignificant during the period of the study. Binding of diflunisal to the nylon membrane was also determined to be insignificant (< 0.5%).

2.4 Data Analysis

In the equilibrium dialysis studies, the concentration of diflunisal measured on the buffer side of the cell compartment represents the free (unbound) concentration, and the difference in diflunisal concentration on the buffer and binder sides of the cell compartments represents the bound diflunisal concentration. The binding of diflunisal to HPBCD and the calculation of the associated binding parameters were analyzed by the methods of Scatchard (6) and Plumbridge et al. (7).

Assuming that only one class of sites exists, and that there are n independent and equivalent binding sites per molecule of cyclodextrin, each having a complexation constant (Kc), the following equation may be written:

$$r/D_f = K_c(n-r) \qquad \text{(Eq. 1)},$$

where r is the average number of moles of diflunisal bound per mole of HPBCD; D_f is the concentration of free (unbound) diflunisal, n is the number of binding sites per molecule of HPBCD and K_c is the complexation constant for the binding of diflunisal to HPBCD. In accordance with the recommendation of Plumbridge et al. (7) the binding parameters (n, K_c) were obtained by regressing D_f (the free or unbound drug concentration) on D_t (the total drug concentration) according to the following equation:

$$D_f = D_t(1 + K_cD_f)/(1 + K_cD_f + nK_cC_{Lt}) \qquad \text{(Eq. 2)},$$

where C_{Lt} is equal to the total ligand (HPBCD) concentration.

3. RESULTS AND DISCUSSION

The data obtained from the equilibrium dialysis experiments (N=4) were subjected to a Scatchard analysis. The regression of r/D_f vs. r produced highly significant linear correlations ($p < 0.0005$). The observed linearity of the data plots is indicative of the presence of one class of binding sites and the adherence of the binding data to Eq. 1. The mean values for K_c and n, which were calculated from the regression of D_f on D_t are 3801 ± 541 M^{-1} and 0.906 ± 0.059, respectively. The %CV for Kc and n were 14.20 and 6.58, respectively.

The stability constant for the diflunisal-HPBCD interaction determined by the equilibrium dialysis method was comparable to the Kc determined by the potentiometric method (3), reported to be 5564 ± 56 M^{-1}. This indicates that the equilibrium dialysis method described here can be used reliably for studying drug binding to cyclodextrins. The potentiometric method is also a relatively simple method; however, it requires the use of ion-selective electrodes which are not always easy to construct. As previously discussed, there are several advantages of equilibrium dialysis over phase solubility methods which can often make it the preferred method.

4. CONCLUSIONS

The major factor limiting the widespread use of the equilibrium dialysis method is the availability of a dialysis membrane with an appropriate selectivity for the cyclodextrin and drug molecules to be studied. Ideally, the membrane should totally restrict the passage of the cyclodextrin but allow free movement (equilibration) of drug molecules. At the present time, there is a limited commercial availability of dialysis membranes with a low molecular weight cutoff (≤ 500 Daltons). The molecular weight cutoff of the nylon membrane is not known, however this membrane proved to be selectively permeable to HPBCD and diflunisal in this study. The molecular weight of diflunisal (MW = 250) is in the same MW range as many other typical drug molecules, therefore the nylon membrane should be applicable to the study of binding of many other drugs to cyclodextrins.

REFERENCES

1. M.C. Meyer and D.E. Guttman. The binding of drugs by plasma proteins. J. Pharm. Sci., 57:895-918 (1968).
2. K.A. Connors. Binding Constants: The Measurement of Molecular Complex Stability, John Wiley & Sons, New York, NY, 1987, p. 266.
3. E.E. Sideris, M.A. Koupparis and P.E. Macheras. Effect of cyclodextrins on protein binding of drugs: the diflunisal/hydroxypropyl-β-cyclodextrin model case. Pharm. Res. 11:90-95 (1994).
4. M. Vikmon. Rapid and simple spectrophotometric method for determination of micro-amounts of cyclodextrins. 1st Int. Sympos. Cyclodextrins, 1981, Budapest, Hungary, pp. 69-74.
5. G.F. Lockwood and J.G. Wagner. Plasma volume changes as a result of equilibrium dialysis. J. Pharm. Pharmacol. 35:387-399 (1983).
6. G. Scatchard. The attractions of proteins for small molecules and ions. Ann. N.Y. Acad. Sci., 51:660-672 (1949).
7. T.W. Plumbridge, L.J. Aarons and J.R. Brown. Problems associated with analysis and interpretation of small molecule/macromolecule binding data. J. Pharm. Pharmacol. 30:69-74 (1977).

SOLUBILIZATION OF β-CYCLODEXTRIN

The effect of polymers and various drugs on the solubility of β-cyclodextrin

HAFRÚN FRIÐRIKSDÓTTIR AND THORSTEINN LOFTSSON

Department of Pharmacy, University of Iceland
P.O.Box 7210, IS-127 Reykjavík, Iceland

ABSTRACT

The effect of polymers and various drugs on the solubility of β-cyclodextrin (βCD) was determined. The apparent stability constant (Kc) of the drug-βCD complex was calculated from the phase-solubility diagrams. The water solubility of βCD was determined to be 1.9% (w/v), but addition of 0.25-1% (w/v) of polyvinylpyrrolidone (PVP) increased the solubility to about 2.1% (w/v). The solubility of βCD in 10% (w/v) sodium-salicylate water solution was 29% (w/v), but 38%(w/v) if 0.25%(w/v) PVP was present. Comparable effects were observed when the effects of lipophilic drugs on the solubility of βCD were studied. The solubility of βCD in a supersaturated aqueous solution of carbamazepine was 2.8% (w/v), but 5.5% (w/v) when 0.1% (w/v) hydroxypropyl methylcellulose (HPMC) was added to the solution.

1. INTRODUCTION

The aqueous solubility of βCD is relatively low (about 1.9% (w/v)) and this limits the use of βCD in the pharmaceutical preparatives [1]. Water-soluble cellulose derivatives and other common macromolecular pharmaceutical excipients, such as polyvinylpyrrolidone (PVP), have been shown to form complexes with cyclodextrins (CDs) and such polymer-cyclodextrin complexes possess physicochemical properties, such as solubilizing properties, different from those of individual cyclodextrin molecules [2-4]. We have discovered that the water-soluble polymer have a synergistic effect on the capacity of CDs to solubilize lipophilic compounds in aqueous solutions. The polymer form a three-way complex with the drug and CD molecules. These three-way complexes have larger apparent stability constants than the simple drug cyclodextrin complexes [2-4]. The purpose of this study was to investigate the effects of both various polymers and various drugs (or other substances) on the solubility of βCD in aqueous solution.

J. Szejtli and L. Szente (eds.), Proceedings of the Eighth International Symposium on Cyclodextrons, 373–376.

2. MATERIALS AND METHODS

2.1 Materials

βCD was obtained from CELDEX (Japan), carboxymethyl-cellulose sodium salt (CMC) and hydrocortisone from Norsk Medisinaldepot (Norway), HPMC 4000 and PVP of molecular weight 40,000 from Mecobenzon (Denmark), sodium salicylate and sulfamethoxazole from Icelandic Pharmaceuticals (Iceland), acetazolamide from Agar (Italy), carbamazepine from Aldrich Chemical Company (USA). All other chemicals used were of pharmaceutical or special analytical grade.

2.2 Quantitative determination

The quantitative determination of βCD was performed on high-performance liquid chromatographic (HPLC) system composed of a Model 501 pump from Waters, a Rheodyne 7125 injector operated at 1,0 ml/min flow rate and PAD-2 pulsed amperometric detector from Dionex with a gold working electrode and a silver-silver chloride reference electrode. The column was a CarboPac PA1 Analytical Column (4x250mm) from Dionex. The eluent contained 150 mM sodiumhydroxide and 300 mM sodium acetate in water. Duration times for detection were: E_1= 100 mV, (t_1=120 ms), E_2=600mV, (t_2=120ms) and E_3=-800mV, (t_3=300ms). The PAD response time was set at 1 s. Quantitative determinations of the drugs were performed on a reversed-phase HPLC component system [4].

2.3 Solubility studies

An excess amount of βCD and/or the drug to be tested was added to water or aqueous polymer solutions. The suspensions formed were heated in an autoclave (M7 Speed Clave from Midmark Corporation, USA) in sealed containers (120°C for 20 min). After equilibration at room temperature (23°C) for 4-7 days, the suspensions were filtered through a 0,45μm membrane filter (Gelman), diluted with water (determination of the βCD) or with 70%(v/v) aqueous methanol solution (determination of the drugs). The stability constants (Kc) were determined from the phase-solubility diagrams according to the method of Higuchi and Connors [5].

2.4 Diffusion through cellophane membrane

The effect of heating the polymer βCD solutions on the diffusion of βCD through cellophane membrane (SPECTROPOR membrane tubing, m.w. cut off 12,000-14,000) was investigated in Franz diffusion cells. The aqueous solutions tested (as a donor phase) contained 1%(w/v) βCD with or without polymer, half of each solutions was heated in an autoclave as decribed above. Water was used as receptor phase. Samples (200μl) were withdrawn from the receptor phase every 15 min for 90 min. The βCD concentration was determined by HPLC. Each experiment was repeated at least three times and the results reported are mean values ± standard error of the mean.

3. RESULTS AND DISCUSSION

3.1 Solubility

Sodium salicylate formed an inclusion complex with βCD, increasing the solubility of βCD from 1.9 %(v/w) in water to 20% (w/v) in 5% aqueous sodium salicylate solution (at pH about 6). The solubility of βCD was increased somewhat if various polymers were added to the sodium salicylate solutions. The solubility of βCD in 5%(w/v) sodium salicylate solution containing CMC, HPMC or PVP are 21, 24, and 24 % (w/v), respectively. The solubility of βCD increased linearly as a function of sodium salicylate concentration and, thus, 1:1 complex formation could be assumed. The apparent stability constants (Kc) of the 1:1 complex are shown in Table 1. The effect of various polymers on the solubility of βCD and various drugs (saturated solutions of βCD and drug) are shown in Table 2.

TABLE 1. The effect of three way complex formation on the apparent stability constant of βCD and sodium salicylate.

	Kc (M^{-1})	Ratio of Kc polymer/Kc water
no polymer	51	1.0
CMC (0,25%(w/v))	64	1.2
PVP (0,25%(w/v))	87	1.7
HPMC (0,1%(w/v))	93	1.8
HPMC(0,25% (w/v))	80	1.6

TABLE 2. The effect of polymers on the aqueous solubility of βCD and on the solubility of various drug in saturated βCD solutions. The solubility of βCD in pure water was determinated to be 19 mg/ml. (* 0.25%(w/v) of the polymer, # 0.1%(w/v) of the polymer).

TABLE 2a.

Polymer	Carbama-zepine (mg/ml)	βCD (mg/ml)
water	2.16	28
PVP*	2.27	45
CMC*	2.58	29
HPMC#	6.46	55
HPMC*	5.53	43

TABLE 2b.

Polymer	Sulfa-methoxazol (mg/ml)	βCD (mg/ml)
water	3.25	39
PVP*	5.75	48
CMC*	5.11	42
HPMC#	5.09	46
HPMC*	4.12	53

TABLE 2c.

Polymer	metha-zolamide (mg/ml)	βCD (mg/ml)
water	1,12	26
PVP*	1.75	29
CMC*	1.20	51
HPMC#	1.77	32
HPMC*		

3.2 Diffusion through cellofan membrane

The effect of heating various polymer and βCD on the diffusion of βCD through a cellofan membrane (as a flux) is shown in Table 3.

TABLE 3.
The effect of heating βCD polymer solution on the diffusion of βCD through cellofan membrane.

donor phase	Flux x 10^{-7} (M $min^{-1}cm^{-2}$) ± S.E.*	
	three way complexes	simple complex
1% βCDin 0,25%(w/v) HPMC aqueous solution	68±11	70±13
1% βCDin 0,25%(w/v) PVP aqueous solution	46±4	56±4
1% βCDin 0,25%(w/v) CMC aqueous solution	53±7	100

(* standard error of the mean)

The results show that addition of various polymer to βCD solutions and heating the solutions in an autoclave enhances the complex formation between βCD and sodium salicylate, as shown in Table 1. Kc was increased from 20-80 % when the polymers were present in the aqueous complexation media. Similar results were obtained when the effect of lipophilic drugs on the solubility of βCD was studied (Table 2). The solubility of βCD increased from 1.9% in water to 2.8% when the solution was saturated with carbamazepine and to 5.5% when both carbamazepine and HPMC were present in the solution, (an increase from 51% to 300%). Comparable results were obtained when acetazolamide (an increase from 40% to 250% with HPMC) or sulfamethoxazole (an increase from 56% to 250% with HPMC) were in the solution. By comparing the solubility of the βCD in water with the solubility in aqueous polymer solutions, it can be seen that the polymers increased the solubilizing effect of βCD (about 10%). When the lipophilic drug and polymer are mixed together, a much greater solubilization than when polymer and lipophilic drugs are used separately (Table 2) is obtained. This solubilization is more than additive, it is synergistic.

Conclusion

The results show that addidion of water-soluble polymers to aqueous solutions and heating the solutions in an autoclave both enhances the solubility of βCD and the solubility of lipophilic drugs.

ACKNOLEDGMENT

This work was supported by a grant from Icelandic Research Fund

REFERENCES

[1] Amdidouche, D., Darrouzet, H., Dominique, D., Poelman, M.C., Inclusion of retinoic acid in β-cyclodextrin. *Int. J. Pharm.* **54,** 175-179 (1989).

[2] Haldon, T., Cwiertina, B., Physical and chemical interaction between cellulose ethers and β-cyclodextrins. *Pharmazie* , **49**, 497-500 (1994),

[3] Sigurðardóttir, A.M., Loftsson T., The effect of polyvinylpyrrolidone on cyclodextrin complexation of hydrochortisone and its diffusion through hairless mouse skin. *Int. J. Pharm.*, **126**, 73-78 (1995)

[4] Loftsson,T., Friðriksdóttir, H., Sigurðardóttir, A. M., Ueda, H., The effect of water-soluble polymers on drug-cyclodextrin complexation. *Int. J. Pharm.*, **110**, 169-177 (1994)).

[5] Higuchi, T., Connors, K.A., Phase-solubility techniques. *Adv. Anal. Chem. Instrum.*, **4**, 117-212 (1965)

DICLOFENAC - ß - CYCLODEXTRIN INCLUSION IN SOLUTION PROTON MAGNETIC RESONANCE AND MOLECULAR MODELLING STUDIES

DARRYL V. WHITTAKER, LAWRENCE J. PENKLER, LUÉTA A. GLINTENKAMP, M.C. BOSCH VAN OUDTSHOORN AND PHILLIPUS L. WESSELS†

Druggists Group research, South African Druggists Limited, P.O. Box 4002, Korsten, 6014, Port Elizabeth, South Africa
†Department of Chemistry, University of Pretoria, 0002, South Africa

ABSTRACT

The nature of the inclusion complex of the NSAID [2-[(2,6-dichlorophenyl) amino]phenyl]acetic acid (diclofenac) with ß-cyclodextrin and 2-hydroxypropyl-ß-cyclodextrin was studied by selected 1D and 2D NMR experiments and by molecular modelling studies using Hyperchem® software. The results indicate that simultaneous inclusion of both rings occurs even at low drug/CD ratios giving rise to multiple equilibria in solution (two isomeric 1:1 complexes and a 1:2 complex). Continuous variation plots are supportive of 1:1 stoichiometry indicating that isomeric 1:1 complexes predominate in the range of concentrations studied.

1. INTRODUCTION

The aqueous solubility of diclofenac is pH dependant increasing substantially from < 0.02 mg/mL at pH 2.0 to >16 mg/mL at pH 7.0 due to ionization of the acetate functionality (pKa ~ 4). At neutral pH in 2-hydroxypropyl-ß-cyclodextrin solutions, the solubility of diclofenac is enhanced by the combined effect of ionization and inclusion complexation permitting preparation of stable aqueous solutions containing 25mg/mL, suitable for intravenous and intramuscular administration[1]. The net solubility enhancement of the drug is therefore largely a consequence of increased saturation solubility of the drug due to ionization, with a lesser contribution from inclusion complexation of the ionized (and unionized) species. However, the complex stability constant decreases substantially with increasing pH (from greater than 1500 M^{-1} at pH 2.0 to less than 100 M^{-1} at pH 7.0)[2], suggesting that the ionized drug is less complexable than the free acid.

J. Szejtli and L. Szente (eds.), Proceedings of the Eighth International Symposium on Cyclodextrons, 377–380.

Thus, in order to better understand the nature of inclusion complexation in aqueous solution, selected proton magnetic resonance experiments combined with energy minimization calculations were performed at neutral pH.

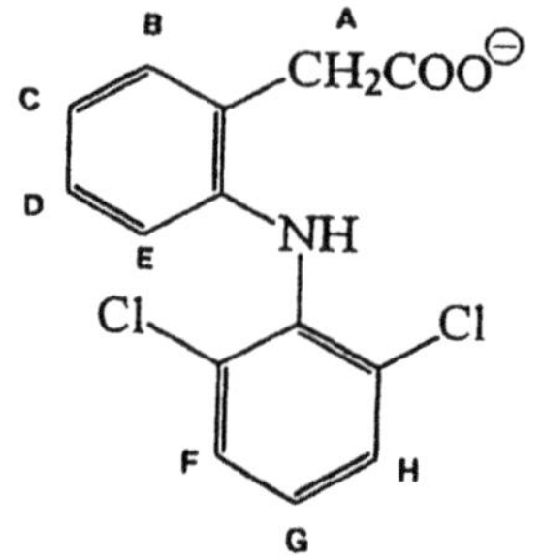

Figure 1. Structure and proton notation of Diclofenac

2. MATERIALS AND METHODS

2.1 NMR spectroscopy

Proton NMR spectra were recorded on a Bruker AMX 500 spectrometer operating at 500.13 Mhz (303K). Chemical shifts are given relative to external TMS. 25 mM solutions of diclofenac in BCD and HPBCD and a lyophilised (1:1) complex were prepared in D_2O (pD = 7.00). Resonance assignments of diclofenac, BCD and HPBCD were obtained from the 2D experiments and correlated with published data[3]. ROESY experiments were performed using a 150 ms spin-locking time and were transformed to a 1K x 1K data matrix. Continuous variation plots were obtained from solutions of the drug and CD where the total concentration was kept at 10 mM and the ratio of the two components varied between 0 and 1.

2.2 Computational Methodology

Calculations were performed using the Hyperchem® program running on an IBM PC. The molecular structures of diclofenac and BCD were generated by manual techniques. In the case of BCD, improper torsional constraints were imposed using the atoms $(C5 - C4)_n$ $(C1 - O5)_{n-1}$, where the value of the improper torsion was constrained about 0° with a force constant of 16 kcal.deg^{-1}to induce C7 symmetry.

Geometry optimisation was performed *in vacuo* using the MM+ force field with the Polak Ribierre conjugate gradient method and termination condition of 0.01 kcal.mol^{-1} In all calculations, charges were omitted to prevent exaggerated electrostatic interactions resulting from lack of solvent shielding effect. The optimised CD structure was in agreement with experimental data derived from X-ray analysis[(4)].

3. RESULTS AND DISCUSSION

As expected, 1D NMR spectra of solutions of diclofenac and CD contained only one set of resonances for each proton or group of equivalent protons confirming that the fast exchange process prevails. Significant *upfield* shifts for the internally oriented H3' and H5' protons of the CD, attributable to the inclusion of an aromatic substituent were observed. However, *downfield shifts* for aromatic protons of *both* rings (TABLE 1) are suggestive of bimodal inclusion. Further evidence of bimodal inclusion complexation was obtained from ROESY spectra of solutions of the complex. Intermolecular cross-peaks were observed between certain aromatic protons from both rings of the drug and the 3' and 5' protons of BCD and HPBCD, even at low drug/CD ratios, indicating that three different complexes are possible in solution, namely two isomeric 1:1 complexes and a 1:2 complex.

TABLE 1. Proton NMR chemical shift data (500 MHz) at 303 K (D2O, ext TMS) for Diclofenac and an equimolar 1:1 solution of Diclofenac (5mM) and ß-CD (5mM)

Diclofenac	Free	Complex	Δδ	Nature of shift
Ha	2.989	3.027	-0.038	*Downfield*
Hb	6.597	6.609	-0.012	*Downfield*
Hc	6.299	6.306	-0.007	*Downfield*
Hd	6.448	6.475	-0.027	*Downfield*
He	5.811	5.790	+0.021	*Upfield*
Hf,h	6.817	6.848	-0.031	*Downfield*
Hg	6.485	6.475	+0.010	Upfield
ß-Cyclodextrin				
H1'	4.385	4.361	+0.024	*Upfield*
H2'	2.965	2.939	+0.026	*Upfield*
H3'	3.275	3.208	+0.067	*Upfield*
H4'	2.897	2.869	+0.028	*Upfield*
H5'	3.175	3.100	+0.075	*Upfield*
H6,6'	3.192	3.183	+0.009	*Upfield*

In the solid state, the diclofenac/BCD complex crystallises in a unique hexagonal crystal system with the phenylacetate ring preferentially included due to the bulkiness of the dichlorophenyl substituent[5]. However, in solution, the NMR data clearly shows that the dichlorophenyl substituent is also complexed. The equivalent aromatic protons Hf,h on the more hydrophobic dichlorophenyl ring showed slightly more intense ROE cross peaks to H3' and H5' and slightly larger downfield shifts in the 1D spectra than the protons Hd and He on the phenylacetate ring, suggestive that it may be preferrentially included. No cross peaks were observed between Hb and Hc on the phenylacetate ring. and the H3' and H5' protons of BCD. Refer to Figure 1.

Continuous variation plots constructed from proton chemical shift data indicate that complex stoichiometry is predominantly 1:1, however slight skewing of the curves for the diclofenac protons to the left of $r = 0.5$ suggests the presence of 1:2 complexes ($r \sim 0.4-0.5$).

Further insight into the geometry of the proposed complex(es) was obtained from molecular modelling studies. Several rigid body docking experiments were carried out to position diclofenac in the BCD cavity ensuring non-violation of van der Waals contacts. Four different binary complexes were geometry optimized with inclusion of phenylacetate and chlorophenyl rings respectively, and seven different ternary complexes with simultaneous inclusion of both aromatic rings were similarly treated. Optimized conformations were in close agreement with NMR chemical shift and ROE data.

4. CONCLUSION

There is evidence that complexation of both aromatic substituents of diclofenac occurs in solution at neutral pH despite the low complex stability constant attributable to the ionizable acetate functionality.

REFERENCES

[1] Penkler, L.J. *et al* European Patent. 94308690.0, South African Druggists Ltd.

[2] Orienti, I., Fini, A., Bertasi, V., and Zecchi, V, Inclusion Complexes between Non Steroidal Antiinflammatory Drugs and ß-cyclodextrin, *Eur J Pharm. Biopharm* 37 (1991) 110-112.

[3] Florey, K, *Analytical Profiles of Drug Substances*, Volume 19, Academic Press Inc, 1990.

[4] Lindner, K. and Saenger, W. *Carbohydr. Res.* 99 (1982), 103-115.

[5] M.R. Caira, V.J. Griffith, L.R. Nassimbeni and M.B. van Oudtshoorn, Synthesis and X-Ray Crystal Structure of ß-cyclodextrin. Diclofenac Sodium Undecahydrate, a ß-CD Complex with a Unique Crystal Packing Arrangement, *J. Chem. Soc. Chem. Commun* (1994) 1061-1062.

CYCLODEXTRINS IN NASAL DRUG DELIVERY: TRENDS AND PERSPECTIVES

E. MARTTIN, J.C. VERHOEF, S.G. ROMEIJN, F.W.H.M. MERKUS
Leiden/Amsterdam Center for Drug Research, Leiden University
P.O. Box 9502, 2300 RA Leiden, The Netherlands

ABSTRACT

Cyclodextrins (CDs) are used as excipients in nasal drug formulations because they act as solubilisers and/or absorption enhancers. With several CDs very efficient nasal drug absorption has been reported, but large interspecies differences have been found. Studies concerning the safety of CDs in nasal drug formulations demonstrate the non-toxicity of the CDs and also clinical data show no adverse effects. Therefore, some CDs can be expected to become effective and safe excipients in nasal drug delivery.

1. Introduction

Nasal drug delivery is an attractive approach for the systemic delivery of high potency drugs with a low oral bioavailability due to extensive gastro-intestinal breakdown and high hepatic first-pass effect. For lipophilic drugs nasal drug delivery is possible if they can be dissolved in the dosage form. Peptide and protein drugs often have a low nasal bioavailability because of their large size and hydrophilicity, resulting in poor transport properties across the nasal mucosa. CDs are used to improve the nasal absorption of these drugs by increasing their aqueous solubility and by enhancing their nasal absorption.

2. CDs as absorption enhancers of lipophilic drugs

For drugs with a low aqueous solubility CDs are used to increase their solubility in various nasal drug formulations. Especially methylated ßCDs are excellent excipients, because they are very good solubilizers in a low concentration. The steroid hormones estradiol and progesterone were delivered intranasally in rats, rabbits and humans [1-4]. Nasal administration of estradiol makes it possible to decrease the dose administered relative to oral administration, avoiding high blood levels of the metabolites of estradiol and thus providing a physiological estrone/estradiol ratio [4]. Estradiol was administered as an inclusion complex with dimethyl-ß-cyclodextrin (DMßCD) in the molar ratio 1:2 to rats and rabbits resulting in mean absolute bioavailabilities of 94.6% and 67.2%, respectively [1]. In oophorectomized women the same ratio of DMßCD and estradiol was administered, leading

J. Szejtli and L. Szente (eds.), Proceedings of the Eighth International Symposium on Cyclodextrons, 381–386.

to a rapid absorption of estradiol [3]. During a 6-month trial successful estradiol replacement therapy was achieved without side effects [3]. Progesterone and estradiol were administered simultaneously in rats and humans, resulting in absorption comparable to separate administration of both steroids [2, 4]. Absorption of progesterone in rabbits administered as a powder containing lactose and ßCD as excipients resulted in 34% bioavailability, but the influence of ßCD on the absorption of progesterone in this formulation was not clarified [5]. The antiviral lipophilic drug pirodavir was given intranasally to humans with 10% (w/v) hydroxypropyl-ß-cyclodextrin (HPßCD), and was effective in preventing clinical cold [6].

3. CDs as absorption enhancers of oligopeptide drugs

Oligopeptide drugs can be delivered nasally without the use of absorption enhancers, but their bioavailability is generally low. With CDs their biovailability can be increased considerably. The ACTH (4-9) analogue, Org 2766, was administered to rats with α-cyclodextrin (αCD) and DMßCD in concentrations of 5% (w/v), resulting in absolute bioavailabilities of 76% for both CDs. At a lower concentration of DMßCD, 2% (w/v), the bioavailability was 70%, which is almost as high as for 5% DMßCD [7]. In the case of leuprolide, a LHRH agonist, the nasal absorption was 35.5% for 5% (w/v) αCD, and this concentration was optimal [8]. However, in humans the highest bioavailability of leuprolide with 5% αCD was only 10%. The nasal absorption of another LHRH antagonist, buserelin, was studied in rats using several CDs [9]. The highest bioavailability was obtained with DMßCD, resulting in 60.1% bioavailability, whereas with αCD and dimethyl-α-cyclodextrin (DMαCD) they were 37.9% and 37.1%, respectively. Other CDs, such as ßCD and the hydroxypropyl- and carboxymethyl- derivatives of αCD and ßCD, were ineffective [9].

4. CDs as absorption enhancers of polypeptide and protein drugs

For nasal delivery of polypeptide and protein drugs, absorption enhancers are mandatory, because the size and hydrophilicity of these molecules limit their nasal bioavailability to less than 1% [10]. Nasal administration of salmon calcitonin with methylated ßCDs resulted in significant reduction in calcium blood levels. In rats nasal administration of calcitonin with randomly methylated ß-cyclodextrin (RAMEB) or DMßCD in concentrations of 5% (w/v) gave hypocalcaemic responses of 23.9% and 21.5%, respectively. Trimethyl-ß-cyclodextrin (TMßCD) in similar concentrations of 5% resulted in a smaller reduction in calcium levels, only 15.1% (see Table I) [11].
A 100% bioavailability of insulin in rats was achieved with 5% DMßCD, which was the best absorption enhancing CD [12]. At lower DMßCD concentrations the mean bioavailability was also high: 91% for 3% DMßCD and 63% for 2% DMßCD [13]. Studies with the same aqueous formulation in rabbits and man showed differences in nasal insulin absorption, because the bioavailability in rabbits and man was very low [14-16]. Nevertheless, administration of insulin with DMßCD to rabbits as a powder formulation resulted in an absolute bioavailability of 12.9% [16]. Following nasal administration of an insulin/DMßCD

TABLE 1. Intranasal calcitonin [11]

			Decrement in serum calcium (%)
Rat	i.v.	-	24.0 ±3.7
13.4 IU/kg	i.n.	-	3.1 ± 6.5
	i.n.	5% TMßCD	15.1 ± 3.3
	i.n.	5% RAMEB	23.9 ± 2.4
	i.n.	1% DMßCD	11.4 ± 6.5
	i.n.	2% DMßCD	19.0 ± 4.0
	i.n.	3% DMßCD	23.1 ± 5.1
	i.n.	5% DMßCD	21.5 ± 3.6
	i.n.	10% DMßCD	23.4 ± 3.3
Rabbit	i.v.	-	11.6 ± 6.3
12.6 IU/kg	i.n.	-	2.1 ± 1.1
	i.n.	5% DMßCD	9.5 ± 3.9

Data are given as mean ± SD of five animals

powder to human volunteers and diabetes mellitus patients, a bioavailability of only 5.1% was achieved, which is too low for therapeutic use by diabetes patients [17]. From these insulin results and those of the oligopeptide leuprolide it can be concluded that high nasal absorption of peptide drugs in rats or rabbits is not always indicative of high absorption in humans [8, 14-17].

Recombinant human granulocyte colony-stimulating factor (rhG-CSF) (molecular weight 19 kDa) was also administered intranasally to rabbits with CDs in a concentration of 20% (w/v). With DMßCD a bioavailability of 16% was obtained, whereas αCD resulted in bioavailabilities of about 11% [18]. After nasal administration of rhG-CSF with αCD substantial biological effects of the protein drug were observed in rabbits [19].

In animal studies with polypeptide and protein drugs the same rank order of effectiveness of CDs was found as for oligopeptides, with DMßCD and RAMEB as the optimal absorption enhancers followed by αCD [7, 11, 12, 15, 18, 20]. For DMßCD an optimum in absorption enhancing effect can be found depending on the concentration. This optimum is reached at the relatively low concentrations of 2-3% [7, 11, 13].

5. CDs in combination with other absorption enhancers

Cyclodextrins are also used in combination with other absorption enhancers to increase the solubility of the absorption enhancer, or to protect the nasal mucosa from damaging effects of the nasal absorption enhancer. HPßCD protected the rat nasal mucosa from damage by laureth-9 while a substantial absorption of insulin was achieved [21]. The combination of αCD and didecanoyl-L-α-phosphatidylcholine resulted in a bioavailabity of 23% for human growth hormone in rabbits, but the formulation was very damaging to the nasal epithelium [22]. HPßCD was used in rats to solubilise 1-[-(decylthio)ethyl]azacyclopentane-2-one, to increase the absorption of buserelin, resulting in a nasal bioavailability of 71% [23].

6. Mechanisms of nasal absorption enhancement by CDs

Several mechanisms have been postulated to explain the nasal absorption enhancement by CDs. The absorption enhancement by CDs is probably explained by a combination of these factors. CDs can have solubilizing and stabilizing effects on peptide and protein drugs, and prevent the aggregation and enzymatic breakdown of these drugs [24-26]. The nasal absorption enhancement with CDs can also be caused by the interaction of CDs with the nasal mucosa. Comparing the release of membrane components from brush border membrane vesicles and intestinal and nasal membranes it was found that DMßCD and ßCD both release cholesterol, whereas αCD releases phospholipids [27, 28]. In a study on the release of marker compounds from the rat nasal cavity 15 min after a single administration of DMßCD (2%) and RAMEB (2%) cholesterol was released but no intracellular enzymes, showing that no cell damage occurred. In addition, a high correlation was observed between the increase of permeability of the nasal mucosa after DMßCD and RAMEB administration and the release of proteins from the rat nasal cavity *in vivo* [29]. Nasal administration of CDs can result in moderate mucus extrusion from goblet cells, changing mucus conditions in the nasal cavity [30]. The increase of nasal absorption of large molecules by CDs might also be caused by transient opening of the tight junctions [26]. Our recent confocal laser scanning microscopy studies also suggest that CDs can open the tight junctions, since the transport of large dextran molecules in the presence of RAMEB was observed to be paracellular (to be published).

7. Safety of CDs as nasal absorption enhancers

The successful application of CDs as nasal absorption enhancers is not only dependent on their efficacy but also on their safety. The use of CDs should be free of irritation of the nasal mucosa and of hampering the nasal mucociliary clearance [31]. This protective mechanism can be damaged by substances that decrease the ciliary beating. The effects of DMßCD and RAMEB on the ciliary beating of new-born chicken tracheal tissue and human adenoid tissue *in vitro* show that these substances exert only a mild effect on the cilia [13, 32]. The decrease in ciliary beat frequency by RAMEB or DMßCD, both at concentrations of 2%, is comparable to the effect of a solution of physiological saline (0.9% NaCl) (see Fig. 1) [32]. Histological studies of DMßCD (2%) and RAMEB (2%) show that they cause minor effects on rat nasal epithelium, comparable to physiological saline and the preservative benzalkonium chloride (0.01%) [30]. These results are confirmed by clinical studies, demonstrating that the tolerability of a spray containing steroids and DMßCD was used for more than 6 months without nasal irritation or any other nasal side effect [3, 4].

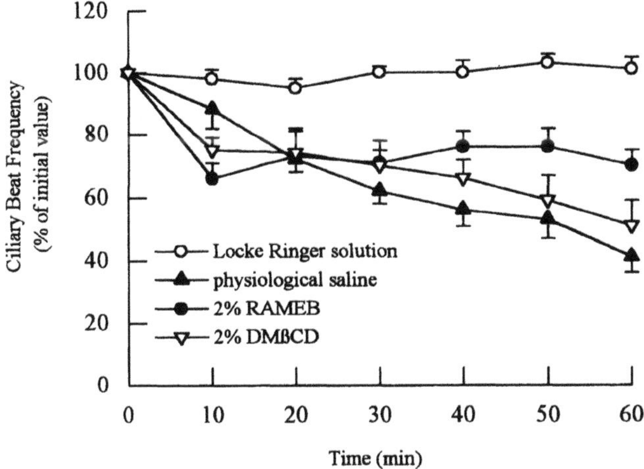

Figure 1. Effect of Locke-Ringer solution, physiological saline, 2% RAMEB and 2% DMßCD on ciliary beat frequency of chicken embryo trachea *in vitro*. Data represent means of ± SEM of 7-10 experiments [32]

REFERENCES

[1] Hermens W.A.J.J., Deurloo M.J.M., Romeijn S.G., Verhoef J.C., Merkus F.W.H.M., Nasal absorption enhancement of 17-β-oestradiol by dimethyl-β-cyclodextrin in rabbits and rats, *Pharm. Res.*, **7**, 500-503 (1990)

[2] Schipper N.G.M., Hermens W.A.J.J., Romeijn S.G., Verhoef J., Merkus F.W.H.M., Nasal absorption of 17-beta-estradiol and progesterone from a dimethyl-ß-cyclodextrin inclusion formulation in rats, *Int. J. Pharm.*, **64**, 61-66 (1990)

[3] Hermens W.A.J.J., Belder C.W.J., Merkus J.M.W.M., Hooymans P.M., Verhoef J., Merkus F.W.H.M., Intranasal estradiol administration to oophorectomized women, *Eur. J. Obs. Gynecol. Reprod. Biol.*, **40**, 35-41 (1991)

[4] Hermens W.A.J.J., Belder C.W.J., Merkus J.M.W.M., Hooymans P.M., Verhoef J., Merkus F.W.H.M., Intranasal administration of estradiol in combination with progesterone to oophorectomized women: a pilot study, *Eur. J. Obs. Gynecol. Reprod. Biol.*, **43**, 65-70 (1992)

[5] Provasi D., De Ascentiis A., Minutell A., Colombo P., Catellani P.L., Nasal powders of progesterone: manufacturing and bioavailability, *Eur. J. Pharm. Biopharm.*, **40**, 223-227 (1994)

[6] Hayden F.G., Andries K., Jansen P.A.J., Safety and efficacy of intranasal pirodavir (R77975) in experimental rhinovirus infection, *Antimicrob. Agents Chemother.*, **36**, 727-732 (1992)

[7] Schipper N.G.M., Verhoef J.C., De Lannoy L.M., Romeijn S.G., Brakkee J.H., Wiegant V.M., Gispen W.H., Merkus F.W.H.M., Nasal administration of an ACTH(4-9) peptide analog with dimethyl-β-cyclodextrin as an absorption enhancer: pharmacokinetics and dynamics, *Br. J. Pharmacol.*, **10**, 335-340 (1993)

[8] Adjei A., Sundberg D., Miller J., Chun A., Bioavailability of leuprolide acetate following nasal and inhalation delivery to rats and healthy humans, *Pharm. Res.*, **9**, 244-249 (1992)

[9] Matsubara K., Abe K., Irie T., Uekama K., Improvement of nasal bioavailability of luteinizing hormone-releasing hormone agonist, buserelin, by cyclodextrin derivatives in rats, *J. Pharm. Scie.*, **84**, 1295-1300 (1995)

[10] Lee W.A., Permeation enhancers for the nasal delivery of protein and peptide therapeutics, *Biopharm. Drug Disp.*, **Nov/Dec**, 22-25 (1990)

[11] Schipper N.G.M., Verhoef J.C., Romeijn S.G., Merkus F.W.H.M., Methylated β-cyclodextrins are able to improve the nasal absorption of salmon calcitonin, *Calcif. Tissue Int.*, **56**, 280-282 (1995)

[12] Merkus F.W.H.M., Verhoef J., Romeijn S.G., Schipper N.G.M., Absorption enhancing effect of cyclodextrins on intranasally administered insulin in rats, *Pharm. Res.*, **8**, 588-592 (1991)

[13] Schipper N.G.M., Verhoef J., Romeijn S.G., Merkus F.W.H.M., Absorption enhancers in nasal insulin delivery and their influence on nasal ciliary functioning, *J. Control. Rel.*, **21**, 173-186 (1992)

[14] Merkus F.W.H.M., Verhoef J., Romeijn S.G., Schipper N.G.M., Interspecies differences in the nasal absorption of insulin, *Pharm. Res.*, **8**, 1343 (1991)

[15] Watanabe Y., Matsumoto Y., Kawamoto K., Yazawa S., Matsumoto M., Enhancing effects of cyclodextrins on nasal absorption of insulin and its duration in rabbits, *Chem. Pharm. Bull.*, **40**, 3100-3104 (1992)

[16] Schipper N.G.M., Romeijn S.G., Verhoef J.C., Merkus F.W.H.M., Nasal insulin delivery with dimethyl-β-cyclodextrin as an absorption enhancer in rabbits: powder more effective than liquid formulations, *Pharm. Res.*, **10**, 682-686 (1993)

[17] Merkus F.W.H.M., Schipper N.G.M., Verhoef J.C., The influence of absorption enhancers on the intranasal insulin absorption in normal and diabetic subjects, *J. Control. Rel.*, in press

[18] Watanabe Y., Matsumoto Y., Yamaguchi M., Kikuchi R., Takayama K., Nomura H., Maruyama K., Matsumoto M., Absorption of recombinant human granulocyte colony-stimulating factor (rhG-CSF) and blood leukocyte dynamics following intranasal administration in rabbits, *Biol. Pharm. Bull.*, **16**, 93-95 (1993)

[19] Watanabe Y., Kikuchi R., Kiriyama M., Nakagawe K., Oe J., Nomura H., Muruyama K., Matsumoto M., Increase in total blood leukocyte count following intranasal administration of recombinant human granulocyte colony-stimulating factor (rhG-CSF) in rabbits with cyclophosphamide-induced leukopenia, *Biol. Pharm. Bull.*, **18**, 1084 (1995)

[20] Irie T., Wakamatsu K., Arima H., Aritomi H., Uekama K., Enhancing effects of cyclodextrins on nasal absorption of insulin in rats, *Int. J. Pharm.*, **84**, 129-139 (1992)

[21] Jabbal Gill I., Fisher A.N., Hinchcliffe M., Whetstone J., Farraj N., De Ponti R., Illum L., Cyclodextrins as protection agents against enhancer damage in nasal delivery systems II. Effect on in vivo absorption of insulin and histopathology of nasal membrane, *Eur. J. Pharm. Sci.*, **1**, 237-248 (1994)

[22] Agerholm C., Bastholm L., Johansen P.B., Nielsen M.H., Elling F., Epithelial transport and bioavailability of intranasally administered human growth hormone formulated with the absorption enhancers didecanoyl-L-α-phosphatidylcholine and α-cyclodextrin in rabbits, *J. Pharm. Sci.*, **83**, 618-663 (1994)

[23] Abe K., Irie T., Adachi H., Uekama K., Combined use of 2-hydroxypropyl-beta-cyclodextrin and a lipophilic absorption enhancer in nasal delivery of the LHRH agonist, buserelin acetate, in rats, *Int. J. Pharm.*, **123**, 103-112 (1995)

[24] Irwin W.J., Dwivedi A.K., Holbrook P.A., Dey M.J., The effect of cyclodextrins on the stability of peptides in nasal enzymic systems, *Pharm. Res.*, **11**, 1698-1703 (1994)

[25] Shao Z., Krishnamoorthy R., Mitra A.K., Cyclodextrins as nasal absorption promoters of insulin - mechanistic evaluations, *Pharm. Res.*, **9**, 1157-1163 (1992)

[26] Hovgaard L., Bronsted H., Drug delivery studies in Caco-2 monolayers. IV. Absorption enhancers effects of cyclodextrins, *Pharm. Res.*, **12**, 1328-1332 (1995)

[27] Nakanishi K., Nadai T., Masada M., Miyajima K., Effect of Cyclodextrins on Biological Membrane .2. Mechanism of Enhancement on the Intestinal Absorption of Non-Absorbable Drug by Cyclodextrins, *Chem. Pharm. Bull.*, **40**, 1252-1256 (1992)

[28] Nakanishi K., Masukawa T., Miyajima K., *The interaction with cyclodextrin and biological membrane on the drug absorption*, in Proc. 7th Int. Cyclodex. Symp., Tokyo 1994, 377-381

[29] Marttin E., Verhoef J.C., Romeijn S.G., Merkus F.W.H.M., Effects of absorption enhancers on rat nasal epithelium in vivo: release of marker compounds in the nasal cavity, *Pharm. Res.*, **12**, 1151-1157 (1995)

[30] Marttin E., Verhoef J.C., Romeijn S.G., Zwart P., Merkus F.W.H.M., Acute effects of absorption enhancers on rat nasal epithelium in vivo: histopathology, submitted

[31] Merkus F.W.H.M., Schipper N.G.M., Hermens W.A.J.J., Romeijn S.G., Verhoef J.C., Absorption enhancers in nasal drug delivery: efficacy and safety, *J. Control. Rel.*, **24**, 201-208 (1993)

[32] Romeijn S.G., Verhoef J.C., Marttin E., Merkus F.W.H.M., The effect of nasal drug formulations on ciliary beating in vitro, *Int. J. Pharm.*, in press

EFFECTS OF CYCLODEXTRINS ON NASAL ABSORPTION AND ANALGESIC ACTIVITY OF OPIOIDS IN RATS

T. KONDO, K. NISHIMURA, M. HIRATA, T. IRIE and K. UEKAMA

Faculty of Pharmaceutical Sciences, Kumamoto University, 5-1 Oe-honmachi, Kumamoto 862, Japan

ABSTRACT

Heptakis(2,6-di-*O*-methyl)-β-cyclodextrin (DM-β-CyD) enhanced the nasal absorption rate of morphine and its entry into the cerebrospinal fluid, while 2-hydroxypropyl-γ-CyD (HP-γ-CyD) sustained the plasma and cerebrospinal fluid levels of the opioid. When fentanyl was administered nasally to rats, HP-β-CyD reduced the respiratory depression of fentanyl without drastic loss of the analgesic activity. DM-β-CyD inhibited the degradation of leucine-enkephalin in rat nasal mucosal homogenates, and increased the disappearance rate of leucine-enkephalin and its fragments from the nasal cavity under *in situ* recirculating perfusion. These results suggest that a proper use of CyDs in nasal opioid preparations may improve the pharmacokinetic and pharmacodynamic behaviors of the opioids tested.

1. INTRODUCTION

We have previously reported that the nasal cavity is a potential route of administration for morphine in order to bypass hepatic first-pass metabolism and is able to deliver the opioid more effectively into the central nervous system. Furthermore, CyDs were found to modify the nasal absorption of morphine [1]. In this study, we compared the effects of CyDs on the nasal absorption and analgesic activity of three kinds of narcotic analgesics with different physicochemical and biopharmaceutical properties (morphine, fentanyl and leucine-enkephalin) in rats.

2. MATERIALS AND METHODS

2.1. Materials

Morphine hydrochloride (Sankyo Co., Ltd., Tokyo, Japan), fentanyl citrate (Sankyo Co.,

J. Szejtli and L. Szente (eds.), Proceedings of the Eighth International Symposium on Cyclodextrons, 387–390.

Ltd.) and leucine-enkephalin acetate (Sigma Chemical Co., St. Louis, MO, USA) were used as supplied. DM-β-CyD and HP-CyDs with an average degree of substitution of 4.8 were donated from Nihon Shokuhin Kako Co. (Tokyo). Maltosyl-β-CyD (G_2-β-CyD) was donated from Ensuiko Sugar Refining Co., Ltd. (Yokohama, Japan).

2.2. *In vivo* nasal absorption studies

The nasal absorption studies using male Wistar rats (200-300 g) were performed according to the procedure described previously [1]. The blood and cerebrospinal fluid samples were taken from the jugular vein and the cisterna, respectively. The concentrations of morphine and its metabolites in the plasma and cerebrospinal fluid were assayed by high performance liquid chromatography method [2].

2.3. Evaluation of antinociceptive activity

A cutaneous thermal (hot-plate) test was used for assessing nociceptive responses. The rat was placed on a metal surface maintained at 52.5±0.3℃. The test measure was the latency between the time of placing the animal on the surface and the behavioral endpoint (licking the hind paw or jumping).

2.4. Stability of leucine-enkephalin in nasal homogenates

The rat nasal mucosa was homogenated in a 10-fold volume of cold isotonic phosphate buffer (pH 7.4). The homogenates were centrifuged at 3020 x g for 10 min at 4℃ and the supernatant (0.2 ml) was added to isotonic phosphate buffer (0.8 ml) containing 0.125 mM leucine-enkephalin and 12.5 mM CyDs at 37℃. The samples (0.1 ml) from the mixture were removed and a citrate buffer (pH 2.3, 0.2 ml) was spiked to terminate the enzymatic reaction at 0℃. The concentrations of leucine-enkephalin and its fragments were assayed by high performance liquid chromatography method.

3. RESULTS AND DISCUSSION

Fluorescence and nuclear magnetic resonance spectroscopic studies indicated that the larger cavity of γ-CyDs can accommodate the bulky morphine molecule more deeply, while the smaller cavity of α- and β-CyDs partially includes the hydrophobic moieties of fentanyl and leucine-enkephalin molecules, reflecting the steric complementarity between host and guest molecules.

When morphine in solution was administered nasally to rats, co-administration of DM-β-CyD significantly enhanced the rate of nasal absorption of morphine by facilitating the nasal epithelial permeability and consequently enhanced the entry of the opioid into the cerebrospinal fluid (Fig. 1). DM-β-CyD did not affect the cerebrospinal fluid-to-plasma level ratio of morphine, suggesting that the direct entry of the opioid into the cerebrospinal

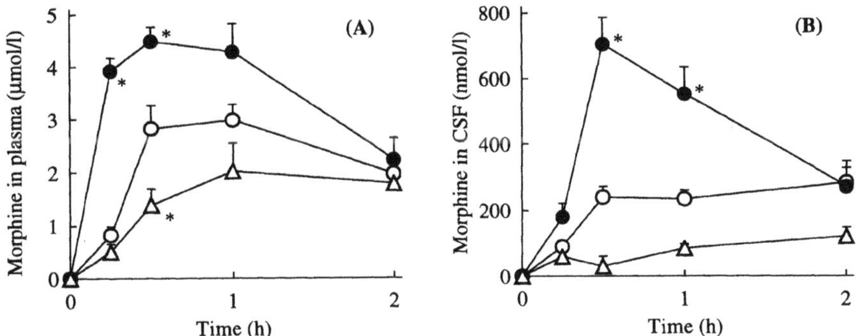

Fig. 1. Plasma (A) and cerebrospinal fluid (CSF, B) levels of morphine after nasal administrations of morphine hydrochloride (5 mg/kg) with CyDs in rats
O: morphine hydrochloride alone, ●: with DM-β-CyD (50 mM), Δ: with HP-γ-CyD (200 mM). Each point represents the mean±S.E. of 3-4 rats. * $p < 0.05$ *versus* morphine hydrochloride alone.

fluid is insignificant. In contrast, HP-γ-CyD tended to sustain the plasma and cerebrospinal fluid levels of morphine, probably due to the formation of a less permeable complex through membranes (Fig. 1). The nasal administration of morphine produced a dose-dependent analgesic response. The efficacy of morphine to block the nociceptive response increased in the order: oral < nasal < subcutaneous route. DM-β-CyD enhanced the analgesia induced with morphine administered nasally, while HP-γ-CyD reduced it.

The analgesic effect of a lipophilic opioid, fentanyl, administered nasally was almost identical to that obtained after subcutaneous administration. When fentanyl was administered nasally to rats, HP-β-CyD reduced the analgesic activity and the respiratory depression due to the supraspinal distribution of the opioid in a dose-dependent manner, probably through the formation of a less membrane-permeable complex. There was an optimal concentration of HP-β-CyD (~50 mM) to reduce the supraspinal side effects of fentanyl without drastic loss of the analgesic activity (Fig. 2).

The *in vitro* degradation of an opioid pentapeptide, leucine-enkephalin, predominately occurred by removal of the tyrosine residue at the N-terminal end to yield *des*-tyrosine leucine-enkephalin. Of the CyDs tested, G_2-β-CyD showed the most prominent inhibitory effect on the enzymatic degradation of leucine-enkephalin (Fig. 3). Under the *in situ* recirculating perfusion, DM-β-CyD increased the rate of disappearance of leucine-enkephalin and its major fragments from the nasal cavity, indicating that DM-β-CyD facilitates the nasal absorption of the opioid peptide.

The present results suggested that a proper use of CyDs in nasal opioid preparations may improve the pharmacokinetic and pharmacodynamic behaviors of opioids.

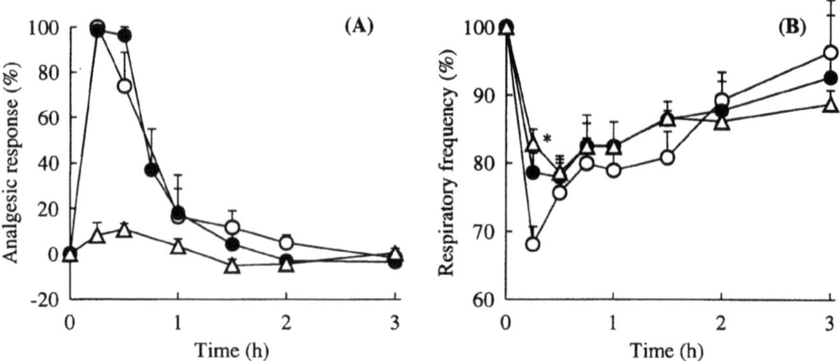

Fig. 2. Analgesic responses (A) and respiratory frequencies (B) after nasal administrations of fentanyl citrate (A; 100, B; 50 μg/kg), HP-β-CyD and hydroxypropylcellulose-H (1 % w/v) in rats
O: fentanyl citrate alone, ●: with HP-β-CyD (50 mM), Δ: with HP-β-CyD (100 mM).
Each point represents the mean±S.E. of at least 3 rats. * $p < 0.05$ *versus* fentanyl citrate alone.

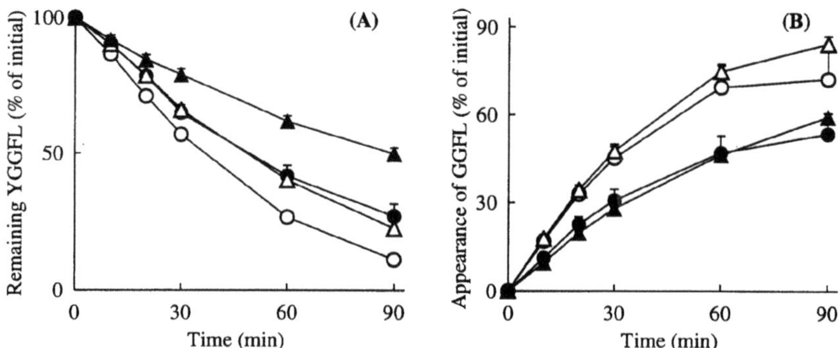

Fig. 3. Time courses for degradation of leucine-enkephalin (YGGFL, A) and appearance of *des*-tyrosine YGGFL (GGFL, B) after incubation of YGGFL (0.1 mM) with β-CyDs (10 mM) in rat nasal mucosal homogenates at 37°C
O: leucine-enkephalin alone, ●: with DM-β-CyD, Δ: with HP-β-CyD, ▲: with G_2-β-CyD.
Each point represents the mean±S.E. of 3-5 rats.

REFERENCES

[1] Kondo, T., Nishimura, K., Irie, T., Uekama, K., Cyclodextrin derivatives that modify nasal absorption of morphine and its entry into cerebrospinal fluid in the rat, *Pharm. Sci.*, **1**, 163-166 (1995)

[2] Kondo, T., Irie, T., Uekama, K., Combination effects of α-cyclodextrin and xanthan gum on rectal absorption and metabolism of morphine from hollow-type suppositories in rabbits, *Biol. Pharm. Bull.*, **19**, 280-286 (1996)

2-HYDROXYPROPYL-ß-CYCLODEXTRIN IN EYE DROPS. EVALUATION OF ARTIFICIAL TEAR-DROPS IN HUMAN PATIENTS

EINAR STEFÁNSSON, SIGRÍÐUR THÓRISDÓTTIR, ÓLAFUR GRÉTAR GUÐMUNDSSON, THORSTEINN LOFTSSON, HAFRÚN FRIÐRIKSDÓTTIR, JÓHANNES KÁRI KRISTINSSON.

University of Iceland, Reykjavík, Iceland.

ABSTRACT:

Relatively little is known about the effects of cyclodextrins (CDs) on the human eye, especially in patients with dry eyes. In an effort to gain more knowledge on the effects of CDs on the human eye we designed a 2-hydroxypropyl-ß-cyclodextrin (HPßCD) containing eye drop preparation with cholesterol, which is a natural tear ingredient. A pilot study in rabbits and human patients was followed by an open trial and then a double masked study in patients with mild dryness. All these patients reported that they felt better in both eyes. One patient complained of crusts on the eyelid margin and this was related to the HPßCD containing artificial tears. No changes were observed in visual acuity or intraocular pressure, nor on the ocular surface or anterior eye structure.

1. INTRODUCTION

Human tears contain three main components, an aqueous portion, a lipid and a mucus component. However, treatment of tear deficiency, i.e. dry eyes, depends mainly on saline tear substitutes, and disregards the lipid and mucus components. In this study we develop and test a tear substitute that attempts to mimic human tears and contains saline, and cholesterol dissolved with the aid of 2-hydroxypropyl-ß-cyclodextrin.
Cyclodextrins are cyclic oligosaccharides with a hydrophilic outer surface, and therefore they are usually soluble in water. The molecule has a central lipophilic cavity and cyclodextrins are capable of forming inclusion complexes with a wide variety of hydrophobic drug molecules (1). ß-Cyclodextrin, is the one most practical to use in pharmaceutical formulations but it has a relatively low aqueous solubility, haemolytic activity and nephrotoxicity which limits its use (2,3). A ß-cyclodextrin derivative, 2-hydroxypropyl-ß-cyclodextrin also forms inclusion complexes with various hydrophobic drugs (4.5.6). This results in the improvement of solubility, dissolution rate, bioavailability and unlike ß-cyclodextrin and its methyl derivatives, it is without toxicity (7,8,9,10) such as to the corneal epithelium. The properties of this ß-cyclodextrin derivative makes it a desirable vehicle in ophthalmic eye-drop formulations of poorly water soluble drugs. It has previously been reported that this compound is non-toxic to the rabbit eye (8,11).

J. Szejtli and L. Szente (eds.), Proceedings of the Eighth International Symposium on Cyclodextrons, 391–394.

2. MATERIALS AND METHODS

Artificial tear drops containing 2-hydroxypropyl-ß-cyclodextrin and cholesterol were formed. A short term study (single application, one week) and an intermediate-term study (up to one month) were performed in people with dry eyes. The effect of this artificial tear solution on the structures of the eyes was evaluated with slit lamp biomicroscopy and a measurement of the visual acuity and intraocular pressure. The study was approved by the Landakot hospital ethics board and, before entering the study, all patients were informed of its nature and risks.
Cholesterol was obtained from Sigma Chemical company (USA), 2-hydroxypropyl-ß-cyclodextrin (HPßCD) of molar substitution 0.9 from Wacker-Chemie (Germany), and hydroxypropyl methylcellulose from Mecobenzon (Denmark). A commercially available tear substitute, Isoptonaturale® was obtained from Alcon Laboratories (USA). All other chemicals used were of pharmaceutical or special analytical grade. Concentration of chemicals in the formulation is given in per cent (%) to indicate weight-in-volume (% w/v).
An aqueous solution was created with HPßCD (20%), cholesterol (0.05%), hydroxypropyl methylcellulose (0.1%), benzalkonium chloride (0.01%), sodium edetate (0.05%) and sodium chloride (0.14%). The solution was sterilised in an autoclave (120°C for 20 min). The isotonicity was monitored with a automatic osmometer from Knauer (Germany) and the pH of the final solution was about 5.

2.1. Mild dry eye syndrome

Human patients were eligible for the trial if they had mild dryness of eyes, and fulfilled all of the four following criteria: (1) complained of dryness in one or both eyes;
(2) had not previously been treated for dry eyes; (3) had Schirmer's test, under topical anaesthesia, less than 5 mm in 5 minutes in one or both eyes; and (4) a slit lamp examination revealed no pathological changes on the surface of the eye other than mild punctate staining with fluorescein dye. Four patients entered the single application part of the study. The visual acuity and the intraocular pressure were measured, as was done in all study groups, and the surface of the eyes and cornea examined with a slitlamp microscope, before one drop of the cyclodextrin-cholesterol teardrop was instilled to one eye, the other eye serving as untreated control. Identical examinations were repeated 30 and 60 minutes after the installation and patients were asked to report any discomfort in their eyes. In the next phase patients were examined as above and given a 15 ml bottle containing cyclodextrin-cholesterol tear drops and asked to administered the drops three times a day in one eye. Six patients participated in this part of the study and they returned for examination after one week.
A group of 8 patients, who complained of mild dryness and met the same criteria as the patients above were given two identical bottles, one containing the cyclodextrin-cholesterol eye drop solution and the other Isoptonaturale®. The bottles were labelled 1-R, 1-L, 2-R and 2-L and so on for use in right (R) and left (L) eye. The patients administered the eye drop solution three times a day for four weeks and returned for examination (as above) one and four weeks from the start of the trial.

2.2. Severe dry eye syndrome

Patients suffering from severe dryness were recruited for this part of the study. All the patients had been treated for dry eyes, but this treatment was discontinued before the trial. The Schirmer´s test, under anaesthesia, gave less than 5 mm after 5 minutes and there were pathological changes including marked staining with fluorescein or Rose Bengal dye on the surface of the eyes. Patients were given the standard information sheet and two eye drop solutions labelled as before, and instructed to use the eye drops in right and left eye as frequently as they wished. They returned for evaluation after a week and were then asked to cross over and use bottle R in left eye and L in right eye. The endpoint evaluation was after another week of treatment.

3. RESULTS

In the single application and one week unmasked parts of the mild dryness study, no subjective complaints were reported. All patients indicated that their eyes felt better . The visual acuity and intraocular pressure did not change and the surface of the eye and cornea were normal in all study groups. In the double masked study in patients with mild dryness, all the patients reported they felt better in both eyes. One patient complained of crusts on the eyelid margin and it was related to cyclodextrin solution. In the patients with severe dryness, one stopped using both solutions after two days of treatment because of severe discomfort in both eyes. Three complained of crusting on the eyelid margin which was related to a cyclodextrin residue.

8 patients with mild dryness

Sex M	Sex F	Age (years) (Average ± SD)	Initial visual acuity (Average ± SD) R	Initial visual acuity (Average ± SD) L	Final visual acuity (Average ± SD) R	Final visual acuity (Average ± SD) L
2	6	71.1 ± 6.0	0.8 ± 0.3	0.8 ± 0.4	0.8 ± 0.3	0.8 ± 0.4

Schirmer's test (Average, mm ± SD) R	Schirmer's test (Average, mm ± SD) L	Intraocular pressure (Average, torr ± SD) R	Intraocular pressure (Average, torr ± SD) L	IOP change (n=3) (Average, torr ± SD) R	IOP change (n=3) (Average, torr ± SD) L
2.8 ± 1.3	3.8 ± 2.0	15.9 ± 3.3	16.7 ± 3.1	0.7 ± 5.0	16.7 ± 3.1

4 patients with severe dry eye syndrome

Sex M	Sex F	Age (years) (Average ± SD)	Initial visual acuity (Average ± SD) R	Initial visual acuity (Average ± SD) L	Final visual acuity (Average ± SD) R	Final visual acuity (Average ± SD) L
0	4	58.3 ± 14.7	0.9 ± 0.3	0.7 ± 0.4	1.0 ± 0.2	0.7 ± 0.4

Schirmer's test (Average, mm ± SD) R	Schirmer's test (Average, mm ± SD) L	Intraocular pressure (Average, torr ± SD) R	Intraocular pressure (Average, torr ± SD) L
5.5 ± 3.5	5.0 ± 3.2	9.8 ± 1.7	10.5 ± 2.5

4. DISCUSSION

The tear substitute containing 2-hydroxypropyl-ß-cyclodextrin, cholesterol and saline was well tolerated by human patients with mild dryness of eyes, but not by those with severe dryness. We believe that evaporation of the water component leaves a hypertonic cyclodextrin solution that may be irritating to the eye and also form crusts as cyclodextrin is deposited from the saturated solution. Apart from these crusts, no damage was seen on the eyes, and visual acuity, intraocular pressure and the surface of the eyes remained undamaged. Previous studies on the toxicity of cyclodextrins both for oral, parenteral and ocular administration have been reported (4,9). It has been shown that dimethyl-ß-cyclodextrin has a toxic effect on the corneal epithelium in rabbits but 2-hydroxypropyl-ß-cyclodextrin had no such effect and is therefore a more suitable vehicle in ophthalmic formulations (8). The possible toxic effect of 2-hydroxypropyl-ß-cyclodextrin in the human eye has not previously been reported.
The study indicates that 2-hydroxypropyl-ß-cyclodextrin is a non-toxic excipient for ophthalmic formulations in humans. The cyclodextrin-cholesterol artificial tear solution was found to be free of any toxic effects, except crusting of the eyelids, which can probably be improved by reducing the cyclodextrin concentration.

REFERENCES

1. J. Szejtli. Cyclodextrins in Drug Formulations: Part I. Pharm.Tech. Int. 3(2), 15-22,)1991).
2. A. Yoshida, H. Arima, K. Uekama, and J. Pitha. Pharmaceutical evaluation of hydroxyalkyl ethers of ß-cyclodextrin. Int. J. Pharm: 46 : 217-222 (1988).
3 J. Pitha and J. Pitha, Amorphous water-soluble derivatives of cyclodextrin: nontoxic dissolution enhancing excipients. J. Pharm. Sci., 74:987-990 (1985)
4 D. Duchene, D. Wouessidjewe. Physicochemical characteristics and pharmaceutical uses of cyclodextrin derivatives, Part I. Pharm. Tech., June 1990, 26-34.
5. D. Duchene, D. Wouessidjewe. Physicochemical characteristics and pharmaceutical uses of cyclodextrin derivatives, Part II. Pharm. Tech., August 1990, 24-26.
6. T. Loftsson, M.E. Brewster, H. Derendorf, and N. Bodor. 2-Hydroxypropyl-ß-cyclodextrin: properties and usage in pharmaceutical formulations. Pharm. Ztg. Wiss., 4/136, 5-10. (1991)
7 J. Pitha, T. Irie, P. Sklar, and J. Nye. Drug solubilizers to aid pharmacologists: Amorphous cyclodextrin derivatives. Life Sci. 43 : 493 (1988).
8. T. Jansen, B. Xhonneux, J. Mesens, and M. Borgers. Beta- Cyclodextrins as vehicles in eye-drop formulations: An evaluation of their effects on rabbit corneal epithelium. Lens and Eye Toxicity Research, 7(3&4), 459-468 (1990).
9. A. Usayapant, A.H. Karara, and M.M. Narurkar. Effect of 2-Hydroxypropyl-ß-cyclodextrin on the Ocular Absorption of Dexamethasone and Dexamethasone Acetate. Pharm. Res., 8, No. 12, 1495-1499, (1991).
10. H.W. Frijlink, A.C. Eissns, A.J.M. Schoonen and C.F. Lerk. The effects of cyclodextrins on drug absorption. II. In vivo observations. Int. J. Pharm., 64 195-205, (1990) .
11. T. Loftsson, H. Fridriksdóttir, S Thórisdóttir, E. Stefánsson, A.M. Sigurdardóttir, Ö. Gudmundsson, T. Sigthórsson. 2-Hydroxypropyl-ß-cyclodextrin in topical carbonic anhydrase inhibitor formulations. European J. Pharmaceutical Sci. 1, 175-180, (1994).

THE EFFECTS OF HP-β-CD ON AQUEOUS SOLUBILITY, STABILITY AND IN VITRO CORNEAL PENETRATION OF ANANDAMIDE

P. JARHO[1], A. URTTI[2], D. W. PATE[1,3], P. SUHONEN[2] and T. JÄRVINEN[1]
[1]*Department of Pharmaceutical Chemistry*, [2]*Department of Pharmaceutics, University of Kuopio, P.O.Box 1627, FIN-70211 Kuopio, Finland*
[3]*HortaPharm B.V., Amsterdam, The Netherlands*

ABSTRACT

Anandamide (arachidonylethanolamide; AEA), an endogenous ligand for the cannabinoid receptor, has a low aqueous solubility and an instability which hinders its use in aqueous formulations. In the present study, AEA was found to form an inclusion complex with hydroxypropyl-β-cyclodextrin (HP-β-CD) resulting greater aqueous solubility and stability of the AEA. The effect of HP-β-CD on corneal penetration of AEA was investigated *in vitro* by using isolated corneas of rabbits. The complexation of AEA with HP-β-CD increased corneal penetration of AEA compared to a suspension of the compound. Maximum permeability was achieved with the lowest HP-β-CD concentration that dissolved the AEA completely. The corneal permeability of AEA correlated well with the concentration of free AEA in solution.

1. INTRODUCTION

Arachidonyl ethanolamide (AEA) was the first anandamide to be identified as an endogenous ligand for the cannabinoid receptor [1]. AEA possesses cannabimimetic pharmacological activity, including the ability to decrease intraocular pressure in rabbits [2]. Anandamides are unsaturated fatty acid derivatives and show very poor aqueous solubility. So far, the low aqueous solubility of AEA has been overcome by using nonaqueous solvents (e.g., ethanol, dimethylsulfoxide and propyleneglycol), emulsifiers or mixtures of these agents which are, however, not suitable for ophthalmic formulations. In addition, AEA shows low aqueous stability which hinders its use in aqueous solutions [3].

The aims of the present study were to increase the aqueous solubility and stability of AEA with HP-β-CD and to investigate the effect of this complexation on the *in vitro* corneal penetration of AEA.

2. MATERIALS AND METHODS

Arachidonylethanolamide (Fig. 1) was obtained from Cayman Chemical (Ann Arbor, MA

J. Szejtli and L. Szente (eds.), Proceedings of the Eighth International Symposium on Cyclodextrons, 395–398.

USA). HP-β-CD (Encapsin®; mw=1297.4, degree of molar substitution 0.40) was purchased from Janssen Biotech (Olen, Belgium). All other reagents were obtained commercially and were of analytical reagent grade. Liquid chromatography (HPLC) was performed with a system consisting of the Beckman solvent module, UV detector and System Gold data module (Beckman Instruments Inc., San Ramon, CA, USA). A deactivated Supelcosil LC8-DB (15 cm x 4.6 mm i.d., 5 μm) reversed-phase column (Supelco, Bellefonte, PA, USA) was used for the separations.

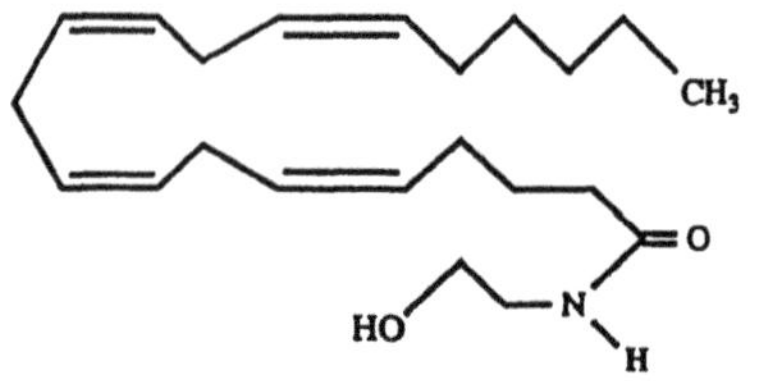

Fig. 1 Chemical structure of arachidonylethanolamide (AEA)

The complexation of AEA with HP-β-CD was determined by using the phase-solubility method of Higuchi and Connors [4]. An excess of AEA was added to phosphate buffer solutions (0.16 M, pH 7.4, ionic strength of 0.5) containing HP-β-CD (0 - 145 mM). After equilibration (24 hours), the suspensions were filtered through 0.45 μm membrane filters and analysed by HPLC.

The stability of AEA in phosphate buffer (0.16 M, ionic strength of 0.5, pH 7.4) was studied in 2.5 % and 10.0% HP-β-CD solution at 50 °C, 60 °C and 70 °C. AEA solutions were prepared by dissolving 2.5 mg of AEA in 25.0 mL of aqueous buffer containing the desired HP-β-CD concentration. The solutions were placed in a constant temperature environment and samples were taken at the appropriate intervals. The remaining AEA was determined with the HPLC method described above.

The effect of HP-β-CD concentration on the permeability of AEA through isolated cornea of pigmented rabbits was studied using glass diffusion cells. The solution containing AEA (0.5 mg/mL) was added to the donor cell (epithelial side of the cornea) and a similar solution, without AEA, placed in the receiver cell (endothelial side). However, the HP-β-CD concentration on the endothelial side was always at least 5% to control the solubility of penetrated AEA. Samples (200 μL) were withdrawn from the receiver side for a period of 4h. The samples were analysed by HPLC. The methods used in these penetration studies have been previously described in detail [5].

3. RESULTS AND DISCUSSION

In the solubility studies HP-β-CD markedly increased the aqueous solubility of AEA. Fig. 2 shows the phase-solubility diagram of AEA with HP-β-CD at pH 7.4. The phase-solubility diagram of AEA with HP-β-CD is of the A_p-type, indicating formation of 1:1 and 1:2 AEA/CD-complexes. The stability constants for the 1:1 and 1:2 complexes were calculated to be 39 419 M^{-1} and 12 M^{-1} [4].

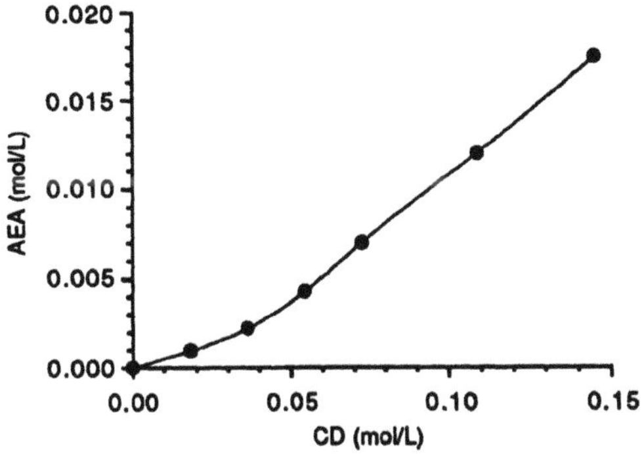

Fig. 1 The phase-solubility diagram of AEA with HP-β-CD.

In the stability studies overall degradation of AEA followed first-order kinetics. Table 1 shows the observed half-lives ($t_{1/2}$) for overall degradation of AEA at the temperatures studied. The stability of AEA in the absence of HP-β-CD was not determined due to the very poor solubility of AEA (0.413 μg/mL), which was below the analytical sensitivity of the HPLC detector employed. However, the $t_{1/2}$ of AEA without CD has been reported to be about 12 h in aqueous solution [3]. Based on the the linear relationship between log k_{obs} and 1/T, $t_{1/2}$ of AEA was calculated with the Arrhenius method to be 62 years in 10% HP-β-CD solution at 25 °C. However, it must be pointed out that the Arrhenius method can overestimate the stability of the drug in a CD containing solution if the complexation of drug with CD is dependent on temperature. In any case, the results show that the stability of AEA can be increased dramatically with CDs, which may have a practical value for future AEA experiments.

TABLE 1: Observed $t_{1/2}$ for overall degradation of AEA in 2.5% and 10.0% HP-β-CD solutions at various temperatures

Solution	Temperature (°C)	$t_{1/2}$ (days)
HP-β-CD (2.5%)	70	74.4
	60	507.8
	50	911.2
HP-β-CD (10.0%)	70	130.3
	60	349.2
	50	1088.9

In permeability studies complexation of AEA with HP-β-CD increased the corneal penetration of AEA compared to a suspension of the compound. The apparent corneal permeability coefficients of AEA as a function of HP-β-CD concentration are shown in Figure 3. The results show also that increased complexation of AEA with HP-β-CD at high concentrations of HP-β-CD decreases the penetration of AEA through the cornea. In addition, Figure 2 shows the calculated concentration of free AEA on the epithelial side at different HP-β-CD concentrations [6]. The linear part of the diagram represents AEA vehicles where CD concentration is too low to dissolve the AEA completely (i.e., the concentration of free AEA molecules is theoretically equal to intrinsic solubility, S_0, of AEA). The lowest HP-β-CD concentration able to dissolve AEA completely was calculated to be 23.9 mM (31.0 mg/mL). The penetration of AEA through rabbit cornea is clearly dependent on HP-β-CD concentration and correlates well with the calculated concentrations of free AEA on the epithelial side

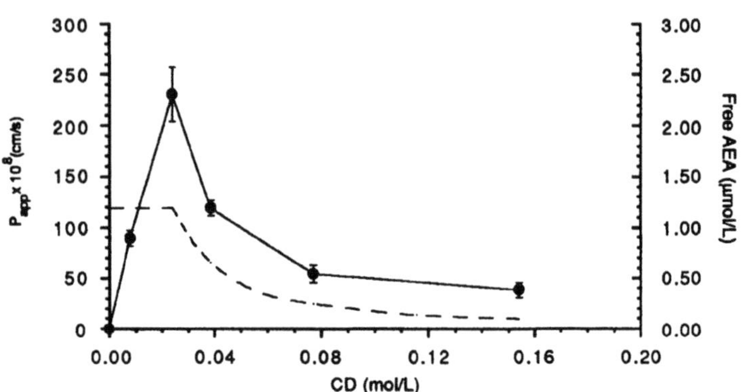

Fig 2: The determined permeability coefficient (mean ± S.E.) of AEA through the cornea of pigmented rabbits (n=2-6) as a function of HP-β-CD concentration (solid line), and calculated concentration of free AEA on the epithelial side as a function of HP-β-CD concentration (broken line)

Acknowledgements

We wish to thank the Pharmacal Research Foundation (P.J.), the Emil Aaltonen Foundation (P.J.), the Finnish Cultural Foundation (P.J.) the Academy of Finland (T.J. and A.U) and The Technology Development Centre (Finland) for financial support.

References

[1] Devane, W.A., Hanus, L., Breuer, A., Pertwee, R.G., Stevenson, L.A., Griffin, G., Gibson, D., Mandelbaum, A., Etinger, A. and Mechoulam, R., Isolation and structure of rat brain constituent that binds to the cannabinoid receptor. *Sci.*, **258**, 1946-1949 (1992).

[2] Pate, D.W., Järvinen, K., Urtti, A., Jarho, P. and Järvinen T., Ophthalmic arachidonylethanolamide decreases intraocular pressure in normotensive rabbits. *Curr. Eye Res.*, **14**, 791-797 (1995).

[3] Cabral, G.A., Toney, D.M., Fischer-Stenger, K., Harrison, M.P. and Marciano-Cabral, F., Anandamide inhibits macrophage-mediated killing of tumor necrosis factor-sensitive cells. *Life Sci.*, **56**, 2065-2072 (1995).

[4] Higuchi, T. and Connors, K.A., Phase-solubility techniques. *Adv. Anal. Chem. Instr.*, **4**, 117-212 (1965).

[5] Suhonen, P., Järvinen, T., Peura, P. and Urtti, A. Permeability of pilocarpic acid diesters across albino rabbit cornea in vitro. *Int. J. Pharm.*, **74**, 221- 228 (1991).

[6] Jarho, P., Urtti, A and Järvinen T. Hydroxypropyl-β-cyclodextrin inreases the aqueous solubility and stability of pilocarpine prodrugs. *Pharm. Res.*, **12**, 1371-1375 (1995).

THE EFFECT OF SBE7-β-CD WITH VISCOUS VEHICLE ON OCULAR DELIVERY OF PILOCARPINE PRODRUG IN RABBITS

P. JARHO[1], K. JÄRVINEN[2], A. URTTI[2], V. J. STELLA[3] and T. JÄRVINEN[1]

[1]Dept. of Pharm. Chem., [2]Dept. of Pharmaceutics, University of Kuopio, P.O.B. 1627, 70211 Kuopio, Finland.[3]Center for Drug Delivery Research and Dept. of Pharm. Chem., University of Kansas, Lawrence, KS 66047, USA

ABSTRACT

The purpose of this study was to compare the effects of viscous and non-viscous pilocarpine prodrug/SBE7-β-CD solutions on apparent ocular absorption and tolerability of the pilocarpine prodrug. The pilocarpine prodrug formed 1:1 inclusion complexes with SBE4-β-CD and SBE7-β-CD, and 1:1 and 1:2 complexes with HP-β-CD at pH 7.4. Coadministered SBE7-β-CD eliminated the eye irritation due to the pilocarpine prodrug, but also decreased the miotic response. Apparent ocular absorption of the prodrug was improved by increasing the viscosity of prodrug/SBE7-β-CD solution with PVA without inducing any eye irritation. In conclusion, administration of pilocarpine prodrug in viscous SBE7-β-CD solution decreases substantially eye irritation while ocular absorption is not affected compared to plain prodrug solution.

1. INTRODUCTION

Bispilocarpic acid diesters are dimeric pilocarpine double prodrugs. These prodrugs show promising biopharmaceutical properties (improved ocular absorption and prolonged duration of action) but their clinical acceptability is mostly limited due to the eye irritation [1].

The eye-irritation due to bispilocarpic acid diesters has been decreased with SBE4-β-CD [2] and HP-β-CD [3], but the ocular absorption of the studied prodrug decreased with increased CD concentrations. These results are consistent with the general assumption that only the free drug, not CD/drug complex, penetrates through biological membranes [4].

J. Szejtli and L. Szente (eds.), Proceedings of the Eighth International Symposium on Cyclodextrons, 399–402.

2. MATERIALS AND METHODS

2.1. Chemicals

The pilocarpine prodrug, O,O'-dipropionyl-(1,4-xylylene) bispilocarpate was synthesized and identified according to the previously described method [5]. Sulfobutylether β-CDs (SBE7-β-CD and SBE4-β-CD) were kindly supplied by CyDex, L.C. (Kansas, USA). HP-β-CD (Encapsin®) was purchased from Janssen Biotech (Olen, Belgium) and poly(vinyl alcohol) (PVA) (mw = 124 000 - 186 000 g/mol) from Aldrich Chemicals Company, Inc. (Milwaukee, USA). All the other chemicals used were of analytical grade.

2.2. Solubility studies

The stability constants for inclusion complex formation between the O,O'-dipropionyl-(1,4-xylylene) bispilocarpate and CDs were determined at pH 7.4 using the phase-solubility method.

2.3. Miotic studies

Pilocarpine prodrug solutions in the presence and absence of SBE7-β-CD were prepared by dissolving the required amount of pilocarpine prodrug and SBE7-β-CD in 5 mL of distilled water or distilled water containing 30 mg/mL of PVA. The pH of the solutions was adjusted to 5.0 or 6.0 and the solutions were made isotonic by NaCl.

The test solution (25 μL) was instilled on the upper corneoscleral limbus of the pigmented rabbit. To measure the pupil diameter, the eyes were photographed from a constant distance before and after drug administration. The studies were set up in a 6 x 6 masked cross-over design. At least 72 h wash-out time was allowed for each rabbit between dosings.

The magnitude of miotic responses (%) was calculated as $((PD_0 - PD_t) / PD_0) \times 100\%$, where PD_0 is a baseline pupillary diameter and PD_t is pupillary diameter at the time point. Areas under the miotic response (%) versus time curves ($AUC_{0\text{-}6h}$) were calculated using the trapezoidal method. All the values are expressed as mean ± S.E. (standard error).

2.4. Evaluation of eye irritation

To estimate the discomfort caused by an instilled eyedrop, the extent of eyelid closure (closed or half closed) after unilateral eyedrop administration (25 μL) was observed and recorded. The amount of mucoidal discharge at 1 h after eyedrop administration was also recorded. Mucoidal discharge was scored from 0 to 2 as follows: 0 = normal, no lacrimation. 1 = slight discharge (any amount different from normal) and 2= strong discharge (moisten the lids and hair just adjacent to the lids).

2.5. Statistical analysis of data

A one-factor analysis of variance (ANOVA for repeated measurements) was used to test the statistical significance of differences between groups; significance in the differences in the means was tested using Fishers's Protected Least Significant Difference (PLSD) at 95% confidence.

3. RESULTS AND DISCUSSION

3.1. Solubility studies

The phase-solubility diagrams of prodrug with SBE4-β-CD ($K_{1:1}$ = 14 777 M^{-1}) and SBE7-β-CD ($K_{1:1}$ = 9 844 M^{-1}) are A_L-type indicating formation 1:1 prodrug/CD complexes at this pH and CD concentration range. The intrinsic solubility (S_o) of prodrug in phosphate buffer solution at 25 °C was 15.3 ± 0.7 μg/mL (mean ± S.E., n = 5).

The phase-solubility diagram of prodrug with HP-β-CD is A_P-type indicating formation of 1:1 and 1:2 prodrug/CD-complexes ($K_{1:1}$ = 6 065 M^{-1} and $K_{1:2}$ = 9 M^{-1}).

Coadministered PVA (10 mg/mL) decreased slightly the complexation of prodrug with SBE7-β-CD.

3.2. Miotic response and eye irritation

The results for miosis and eye irritation studies are summarized in Table 1. Coadministered SBE7-β-CD eliminated the eye irritation due to prodrug, but also decreased significantly the apparent ocular absorption. Increasing the viscosity of the prodrug/SBE7-β-CD eyedrops increased the apparent ocular absorption of prodrug but did not affect the ocular irritation.

The results show that the eye irritation of ophthalmic pilocarpine prodrug is eliminated with viscous SBE7-β-CD solution without impairing the ocular absorption of prodrug.

TABLE 1. The $AUC_{0\text{-}6h}$ and eye irritation data after ocular administration (25μl) of 12 mM O,O'-dipropionyl-(1,4-xylylene) bispilocarpate (in the presence or absence of SBE7-β-CD and PVA (mean ± S.E., n = 6)

Solution	pH	Eyelid closure[a] (min)	Drainage[b] (range 0-2)	$AUC_{0\text{-}6\,h}$ (% x h)
Isotonic NaCl	5.0	0.0 ± 0.0#	0.0 ± 0.0#	23.2 ± 4.1#
12 mM Prodrug	5.0	2.3 ± 1.2*	1.7 ± 0.2*	85.5 ± 14.5*
12 mM Prodrug + 24 mM SBE7-β-CD	5.0	0.0 ± 0.0#	0.0 ± 0.0#	64.1 ± 11.5*
12 mM Prodrug + 36 mM SBE7-β-CD	5.0	0.0 ± 0.0#	0.0 ± 0.0#	40.7 ± 13.4#
12 mM Prodrug + 24 mM SBE7-β-CD + 3% PVA	5.0	0.0 ± 0.0#	0.0 ± 0.0#	87.3 ± 15.3*
12 mM Prodrug + 24 mM SBE7-β-CD + 3%PVA	6.0	0.0 ± 0.0#	0.0 ± 0.0#	97.5 ± 15.5*
36 mM SBE7-β-CD + 3% PVA, control	5.0	0.0 ± 0.0#	0.0 ± 0.0#	15.6 ± 7.6#

[a] = Period of time when eye was closed or half-closed; [b] = The amount of mucoidal discharge 1h after eyedrop administration; * = Significantly different from value for the control ($p<0.05$, by Fisher's PLSD test); # = Significantly different from value for 12 mM prodrug ($p<0.05$, by Fisher's PLSD test)

ACKNOWLEDGEMENTS

The author wish to thank the Academy of Finland and Technology Development Centre (Finland) for financial support and CyDex, L.C. for supplying SBE-β-CD material.

REFERENCES

[1] Suhonen, P., Järvinen, T., Lehmussaari, K., Reunamäki, T. and Urtti, A. Rate control of ocular pilocarpine delivery with bispilocarpic acid diesters. *Int. J. Pharm.* 127: 85-94 (1996)

[2] Järvinen, T., Järvinen, K., Urtti, A., Thompson, D. and Stella, V.J. (1995b) Sulfobutyl ether β-cyclodextrin (SBE-β-CD) in eyedrops improves the tolerability of a topically applied pilocarpine prodrug in rabbits. *J. Ocul. Pharmacol. Ther.* 11: 95-106 (1995)

[3] Suhonen, P., Järvinen, T., Lehmussaari, K., Reunamäki, T. and Urtti, A. Ocular absorption and irritation of pilocarpine prodrug is modified with buffer, polymer, and cyclodextrin in eyedrop. *Pharm. Res.* 12: 529 - 533 (1995).

[4] Frijlink, H.W. , Eissens , A.C., Schoonen, A.J.M. and Lerk, C.F. The effects of cyclodextrins on the drug absorption II. In vivo observation. *Int J. Pharm.* 64: 195-205 (1990).

[5] Järvinen, T., Suhonen, P., Auriola, S., Vepsäläinen, J., Urtti, A. and Peura, P. O,O'- (1,4-xylylene) bispilocarpic acid esters as new potential double prodrugs of pilocarpine for improved ocular delivery. I. Synthesis and analysis. *Int. J. Pharm.* 75: 249-258 (1991).

CYCLODEXTRINS AS SKIN PENETRATION ENHANCERS

Effects of polymers on cyclodextrin complexation and transdermal drug delivery

THORSTEINN LOFTSSON AND ANNA M. SIGURÐARDÓTTIR

Department of Pharmacy, University of Iceland
PO Box 7210, IS-127 Reykjavik, Iceland

ABSTRACT

The flux of hydrocortisone and enalaprilat was determined from aqueous vehicles containing various amounts of 2-hydroxypropylβCD, carboxymethylβCD, randomly methylated βCD or maltosylβCD through hairless mouse skin. When the drug was in suspension, the flux was increased as the cyclodextrin (CD) concentration was increased. The flux decreased at higher CD concentrations, when all the drug was in solution. Maximum flux through the skin was obtained when just enough CD was used to keep all the drug in solution. Addition of small amount of a water-soluble polymer, such as hydroxypropyl methylcellulose or polyvinylpyrrolidone, to the aqueous complexation medium, and heating in sealed container to 120-140°C for 20-40 minutes, resulted in up to 200% larger drug permeability compared to preparations containing no polymer.

1. INTRODUCTION

The main barrier to dermal and transdermal drug delivery is the thin outermost layer of the skin, the stratum corneum. A drug must first penetrate this barrier in order to reach its site of action and the clinical usefulness of a drug is frequently limited by its inability to pass. It is well known that various vehicle additives, so called penetration enhancers, can temporary alter or damage the skin barrier and thereby increase the permeability of drugs into or through the skin. CDs are relatively large hydrophilic molecules which, under normal conditions, do not penetrate into the skin and, thus, hydrophilic CDs do not affect the skin barrier. The drug permeation enhancement obtained from hydrophilic drug-CD complexes is probably mainly due to the fast equilibrium of the drug molecules bound in the complex with free drug molecules out in the solution (Fig. 1). This results in rapid delivery of the drug to the skin surface and enhanced drug permeability through the skin without effecting the skin barrier [1,2]. It has been shown that polymers, such as water-soluble cellulose derivatives and other rheological agents, can form complexes with CDs and that such complexes possess physicochemical properties different from those of individual CD molecules [3-5]. In aqueous solutions water-soluble polymers increase the solubilizing effect of CDs on various hydrophobic drugs by increasing the apparent stability constants of the drug-CD complexes. Water-soluble polymers are also capable to increase aqueous solubilities of the parent CDs without decreasing their complexing abilities, thus, making them more attractive as pharmaceutical excipients.

J. Szejtli and L. Szente (eds.), Proceedings of the Eighth International Symposium on Cyclodextrons, 403–406.

The purpose of this study was to investigate the effect of drug-CD complexes and drug-CD-polymer ternary complexes on the permeability of drugs through hairless mouse skin.

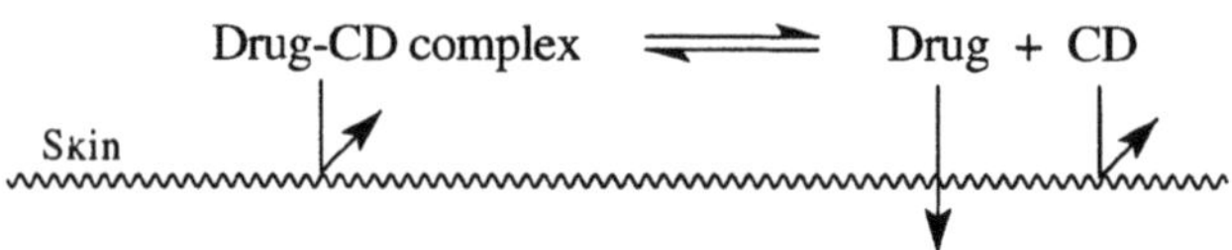

Fig. 1 The effect of cyclodextrins on transdermal drug delivery.

2. MATERIALS AND METHODS

2.1 Materials

2-HydroxypropylβCD MS 0.6 (HPβCD), carboxymethylβCD Na-salt MS 0.6 and randomly methylated βCD MS 1.8 were obtained from Wacker-Chemie (Germany) and maltosylβCD from Ensuiko Sugar Refining Co. (Japan). Hydrocortisone was obtained from Norsk Medisinaldepot (Norway). Enalapril maleate from Chemische Fabrik Schweizerhall (Switzerland). Enalaprilat was formed by hydrolysis of enalapril. Polyvinylpyrrolidone of MW 360,000 (PVP) and hydroxypropyl methylcellulose 4000 (HPMC) were obtained from Sigma Chemical Company (USA). All other chemicals were of commercially available products of special reagent grade. Hairless mice were obtained from Bommice (Denmark).

2.2 Solubility studies

Solubilities were determined by adding excess amount of the drug to be tested to aqueous solutions containing various amounts of different kinds of CDs with or without the polymer. The suspensions formed were heated in an autoclave in sealed containers (120-140°C for 20-40 minutes) and then allowed to equilibrate for at least three days at 23°C. After equilibration was attained, aliquots of the suspensions were filtered through 0.45 μm membrane filters and analyzed by HPLC. The stability constant of the drug-CD (1:1) complex was determined by the solubility method [5].

2.3 Skin permeation studies

The skin permeability study was previously described [5]. Briefly, the drug flux through hairless mouse skin was determined in Franz diffusion cells. The donor phase consisted of suspension or solution of the drug in aqueous CD solution, or aqueous CD solution containing 0.25% (w/v) PVP or 0.10% (w/v) HPMC, which had been heated in an autoclave as described above. Aqueous CD pH 5.0 buffer solution was used as a donor phase when enalaprilat was tested. After equilibration for three days at room temperature, 2.0 ml of the donor phase was applied to the skin surface. Samples (150 μl) were withdrawn from the receptor phase every 12 h for three days and replaced with fresh buffer solution. The samples were kept frozen until analyzed by HPLC. Each experiment was repeated at least three times and the results reported are the mean values ± standard error of the mean (s.e.m.). Then the total amount of dissolved hydrocortisone was determined by HPLC and, based on the total solubility and the value of K_C, the concentration of the free drug was calculated. Finally, based on the total hydrocortisone solubility in the vehicle the permeability coefficient (P) at the maximum permeability, i.e. when just enough CD was used to dissolve all hydrocortisone in the vehicle, was determined [6].

3. RESULTS AND DISCUSSION

The effect of CD concentration on the flux of hydrocortisone from aqueous vehicles containing hydrocortisone-CD complexes through hairless mouse skin are shown in Fig. 2. The hydrocortisone concentration was kept constant but the CD concentration was increased from 1 to 20% (w/v) and this was done both when no polymer was present in the vehicle and when 0.25% (w/v) PVP was present. Since only the free drug molecules are able to permeate the skin the total flux of the drug through the skin (I_t) can be described by:

$$I_t = I_D f_D \quad \text{and} \quad f_D = \frac{1}{1 + K_C([CD]_t - [D]_t)} \quad \text{and} \quad K_C = \frac{[D\text{-}CD]}{[D]\,[D\text{-}CD]} \quad \text{if } [CD] >> [D]$$

where I_D is the flux of the free drug through skin, f_D is the fraction of free drug in solution, K_C is the equilibrium constant for the 1:1 drug-CD complex (D-CD), $[CD]_t$ is the total CD concentration and $[D]_t$ the total drug concentration in [1]. When hydrocortisone was in suspension increasing the CD concentration resulted in increasing amount of dissolved drug and, since the rate of drug release from the drug-CD complex was much faster than the rate of drug dissolution, this increase in solubility lead to larger flux through the membrane. On the other hand, when all hydrocortisone was in solution increasing the CD concentration lead to an increased CD complexation of the drug molecules and since the hydrated hydrocortisone-CD complex is unable to permeate the skin this lead to a decrease in the flux. Maximum flux through the membrane was obtained when just enough CD was used to keep all hydrocortisone in solution. Addition of the polymers and heating in an autoclave did not only increase the complexation but also enhanced transdermal drug delivery. Apparently the polymers interacted with the hydrocortisone-CDs complex resulting in better contact with the skin surface. At maximum flux addition of PVP resulted in about 40% increase in the hydrocortisone flux from aqueous 2-hydroxypropylβCD (HPβCD) vehicle (Fig. 2). Similar results have been obtained with other CD derivatives [5]. As shown in Table 1 it is not enough to add the polymers to the complexation medium. Addition of polymers to the unheated vehicles did not enhance the

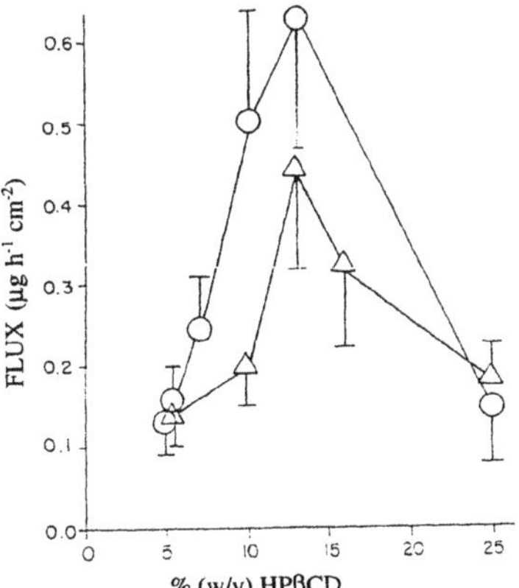

Fig. 2 The effect of HPβCD concentration on the flux of hydrocortisone through skin in vitro. Aqueous HPβCD solution (Δ); HPβCD solution containing 0.25% PVP (O).

TABLE 1. The effect of heating on transdermal delivery of enalaprilat from 10% (w/v) HPβCD solutions at pH 5.0 containing 2.5% enalaprilat in a suspension. The concentration of dissolved enalaprilat was between 2.0 and 2.3% (w/v).

Donor phase	Flux (mg h^{-1} cm^{-2})		Ratio
(w/v per cents)	Un-heated	Heated	
HPβCD	18±2	-	-
HPβCD, 0.25% PVP	16±6	23±7	1.4
HPβCD, 0.10% HPMC	14±3	37±12	2.6

transdermal delivery of enalaprilat. However, heating the vehicles after addition of the polymers resulted in significant enhancement.

TABLE 2. The effect of addition of 0.25% (w/v) PVP and heating to 120-140°C for 20-40 minutes on the stability constant (Kc) of the 1:1 hydrocortisone-CD complex and the permeability coefficient (P) of hydrocortisone through hairless mouse skin at maximum permeability.

Cyclodextrin	Kc (M^{-1})			P x 10^5 (cm/h)		
	no PVP	PVP	Ratio	no PVP	PVP	Ratio
2-HydroxypropylβCD	900	1100	1.2	2.9	4.0	1.4
CarboxymethylβCD	3200	9000	2.8	8.0	24	3.0
Methylated βCD	1700	2000	1.2	15	31	2.1
MaltosylβCD	2400	2700	1.1	6.1	8.0	1.3

An other way to look at this is to calculate the permeability coefficient of the drug based on the concentration of free drug in solution (Table 2). The stability constant of the drug-CD complex was increased from 10 to 180% (the K_C-ratio was from 1.1 to 2.8) when PVP was added complexation media. In a saturated solution, or drug suspension, the free drug concentration is constant (equal to drug solubility) but in unsaturated solution this increase in K_C will decrease the amount of free drug in the aqueous solution, and, since only the free drug and not the drug bound within the complex is able to permeate the skin, this should result in decreased drug permeability through the skin. However, both the total flux (see ref. 5) and the permeability coefficient were increased (Table 2).

ACKNOWLEDGMENTS

This work was supported by a grant from the Icelandic Science Foundation.

REFERENCES

[1] Loftsson, T., Bodor, N., *Effect of cyclodextrins on percutaneous transport of drugs*, in Percutaneous Penetration Enhancers (Eds. Smith, E,W., Maibach, H.I.) Boca Raton, 1995. Pp. 335-342

[2] Loftsson, T., Sigurðardóttir, A.M., The effect of polyvinylpyrrolidone and hydroxypropyl methylcellulose on HPβCD complexation of hydrocortisone and its permeability through hairless mouse skin, *Eur. J. Pharm. Sci.*, **2**, 297-301 (1994)

[3] Hladon, T., Cwiertnia, B., Physical and chemical interactions between cellulose ethers and β-cyclodextrins, *Pharmazie*, **49**, 497-500 (1994)

[4] Loftsson, T., Friðriksdóttir, H., Sigurðardóttir, A.M., Ueda, H., The effect of water-soluble polymers on drug-cyclodextrin complexation, *Int. J. Pharm.*, **110**, 169-177 (1994)

[5] Sigurðardóttir, A.M., Loftsson, T., The effect of polyvinylpyrrolidone on cyclodextrin complexation of hydrocortisone and its diffusion through hairless mouse skin, *Int. J. Pharm.*, **126**, 73-78 (1995)

[6] Loftsson, T., Experimental and theoretical model for studying simultaneous transport and metabolism of drugs in excised skin, *Arch. Pharm. Chemi, Sci. Ed.*, **10**, 104-110 (1982)

NOVEL CD-BASED DRUG DELIVERY SYSTEM

Effects of polymers on cyclodextrin complexation and drug bioavailability

T. LOFTSSON[1], E. STEFÁNSSON[2], H. FRIÐRIKSDÓTTIR[1]
AND J.K. KRISTINSSON[2]

Departments of [1]Pharmacy and [2]Ophthalmology, Faculty of Medicine, University of Iceland, PO Box 7210, IS-127 Reykjavik, Iceland

ABSTRACT

Enhanced cyclodextrin (CD) complexation can be obtained by adding small amount of some water-soluble polymer to an aqueous complexation medium followed by heating, for example in an autoclave, to 120 to 140°C for 20 to 40 minutes. The water-soluble polymers participate directly in the complex formation, through formation of ternary complexes (or co-complexes), increasing the apparent stability constants of the drug-CD complexes. The ternary complexes possess physicochemical properties that are different from those of regular CD complexes. First, enhanced complexation increases the solubilizing properties of the CDs which reduces the total amount of CD needed to solubilize a given amount of drug. Second, each polymer molecule is able to form complexes with many drug and CD molecules. These large ternary complexes appear to be adsorbed to the surfaces of biological membranes such as to the surface of the skin or the eye, ensuring better contact between the drug molecules and the biological membranes. This results in enhanced drug delivery to the membranes and, consequently, enhanced drug bioavailability. Third, in some cases drug dissolution from solid ternary complexes is faster than from simple CD complexes. Less CD is needed in the ternary complexes and this can, at least partly, explain the faster drug dissolution observed. Finally, through polymer-CD complexation the water-soluble polymers increase the aqueous solubilities of the parent CDs without decreasing their abilities to form complexes with drugs. This will make the parent CDs, such as βCD, more attractive as pharmaceutical excipients.

1. INTRODUCTION

Water-soluble polymers have been used in pharmaceutical formulations for many years and although the polymers are usually considered to be chemically inert, they are known to form complexes with small molecules in aqueous solutions [1]. Thus, water-soluble polymers can, through complex formation, enhance aqueous solubility of lipophilic water-insoluble drugs [2]. Furthermore, addition of water-soluble polymers to aqueous surfactant solutions have been found to increase the solubilizing effects of the surfactants [3]. The surfactant-polymer interactions are either between individual surfactant molecules and the polymer chain (i.e. simple adsorption), or in the form of polymer-aggregate complexes (i.e. complex formation between the micelles and the polymer chain). The formation of such structures in surfactant-polymer systems is often illustrated as

J. Szejtli and L. Szente (eds.), Proceedings of the Eighth International Symposium on Cyclodextrons, 407–412.

formation of such structures in surfactant-polymer systems is often illustrated as resembling a string of pearls or water droplets on a spider's web [4]. Polymer-surfactant complexes, e.g. the PVP-sodium laurylsulphate complex, have a larger solubilizing effect than the sum of the individual solubilizing effects of the polymer and the surfactant. That is, the polymer has a synergistic effect on the capacity of the surfactant to solubilize water-insoluble compounds. For example, addition of 0.10% polyvinylpyrrolidone (PVP) about doubles the solubilizing effect of the non-ionic surfactant dodecyl-(oxyethylene)-ether [3]. It has been shown that water-soluble polymers have similar effect on solubilizing abilities of CDs [5]. Thus, the solubilizing effect of 2-hydroxypropyl-β-CD (HPβCD) was improved by 12 to 129% when 0.25% PVP was present in the solution, on average the solubilizing effect was improved by 49%. Similar results were obtained with other CDs, such as hydroxyethyl-β-CD (HEβCD), randomly methylated β-CD (MβCD) and glucosyl-α-CD, glucosyl-β-CD, maltosyl-α-CD and maltosyl-β-CD, and other polymers, such as carboxymethylcellulose (CMC) and hydroxypropyl methylcellulose (HPMC) [5,6]. Various studies (such as NMR studies, thermodynamic studies, solubility studies and bioavailability studies) have shown that the polymers participate directly in the complex formation by forming ternary drug-CD-polymer complexes and that these ternary complexes have physicochemical properties which are different from those of simple binary drug-CD complexes. The purpose of this present investigation was to investigate further the physicochemical properties of these ternary complexes.

2. MATERIALS AND METHODS

2.1 Materials

Acetazolamide (Agrar, Italy), hydrocortisone (Norsk Medicinaldepot, Norway), dexamethasone (Sigma Chemical Co., USA), carbamazepine (Aldrich Chemical Co., USA), βCD (Celdex, Japan)), HPβCD with molar substitution of 0.6 (Wacker-Chemie, Germany), γCD (Wacker-Chemie), CMC sodium (Norsk Medicinaldepot, Norway), PVP MW 40,000 (Mecobenzon, Denmark) and HPMC 4000 (Mecobenzon). All other chemicals were commercially available products of special reagent grade.

2.2 Solubility studies

The solubility studies and determination of the apparent stability constants of the drug-CD (1:1) complexes have previously been described [5]. Briefly, excess amount of the drug to be tested was added to aqueous CD solution containing no polymer or small amount (0.1 to 0.4% w/v) of a water-soluble polymer. The suspension formed was heated in an autoclave in a sealed container (120-140°C for 20-40 minutes) and then allowed to equilibrate for at least three days at 23°C. After equilibration was attained, an aliquot of the suspension was filtered through 0.45 μm membrane filter and analyzed by HPLC. The stability constants were determined from the phase-solubility diagrams.

2.3 Testing of eye drops in humans

The animal experiments have previously been described [7]. Each patient gave informed consent prior to entry into the studies.

Dexamethasone eye drop solutions contained the drug (0.67% w/v) dissolved in a vehicle consisting of HPβCD (10% w/v (heated) or 11.5% w/v (unheated)), HPMC (0.10% w/v), benzalkonium chloride (0.01% w/v), sodium edetate (0.05% w/v) and sodium chloride

in an autoclave (121°C for 20 minutes). Both eye drop solutions were filtered through a 0.22 μm membrane filter. To prevent possible drug precipitation in the eye drop solutions during storage about 10% more HPβCD was used than needed to dissolve the drug, based on the solubility studies. The final dexamethasone concentration in the aqueous eye drop solutions, after heating and filtration, was determined by HPLC. Patients scheduled to undergo cataract surgery were recruited to the dexamethasone study. One drop (50 μl) was administered into the lower conjunctival fornix of the eye prepared for surgery at predetermined time points pre-operatively. Intra-operatively, 0.1 ml of aqueous humor was withdrawn with a small needle and syringe, and analyzed by HPLC.
Acetazolamide eye drop solutions contained the drug (1.0% w/v) dissolved in a vehicle consisting of HPβCD (16% w/v), HPMC (0.10% w/v), benzalkonium chloride (0.01% w/v), sodium edetate (0.05% w/v) and sodium chloride (0.36% w/v) in water. The eye drop solution was heated in an autoclave (121°C for 20 minutes). A baseline curve was obtained by measuring intraocular pressure (IOP) with an applanation tonometer prior to the administration of the first drop. Those with maximal IOP under 21 torr in either eye were excluded. The eye with higher mean IOP was selected as the study eye. Each patient was instructed to administer the eye drops three times a day, with the first drop administered exactly 2 hours before the first examination at 9 AM. On study days 3, 7, 14 and 28 the IOP was measured at 9 AM, 12 AM and 3 PM.

3. RESULTS AND DISCUSSION

3.1 Complexation and solubilization

The effects of HPβCD and polymers on the aqueous solubility of acetazolamide are shown in Table 1, and the effects of polymers on the aqueous solubility of acetazolamide and βCD are shown in Table 2. The solubility of acetazolamide in pure water is only 0.70 mg/ml. Addition of 10% HPβCD resulted in 3.6-fold increase in the solubility when no polymer was included in the formulation but 8.4-fold increase when 0.10% HPMC was included. That is, for acetazolamide the ternary drug-HPβCD-HPMC complex was 130% more

TABLE 1. The effect of polymers on the solubility of acetazolamide in aqueous 10% (w/v) HPβCD solutions at 23°C (solubility ratio in brackets).

Polymer (% w/v)	Solubility (mg/ml)
No polymer	2.52 (1.00)
0.25% CMC	3.11 (1.24)
0.25% PVP	3.92 (1.56)
0.10% HPMC	5.86 (2.33)

TABLE 2. The effect of polymers on the aqueous solubility of βCD and the solubility of acetazolamide in saturated βCD solutions. The solubility of βCD in pure water was determined to be 1.9% (w/v).

Polymer (% w/v)	Solubility βCD (% w/v)	Solubility of acetazolamide (mg/ml)
No polymer	2.53 (1.3)*	1.44 (1.00)
0.25% CMC	2.86 (1.5)	1.54 (1.06)
0.25% PVP	3.14 (1.7)	1.77 (1.23)
0.10% HPMC	4.52 (2.4)	3.17 (2.20)

*Solubility ratio compared to pure water.

powerful solubilizer than the simple drug-HPβCD complex (Table 1). The solubility of pure βCD is 1.9% in water but the total solubility of βCD increased to 2.53% when aqueous solution was saturated with both βCD and acetazolamide, i.e. the total βCD solubility is the sum of the solubility of βCD and that of the acetazolamide-βCD complex (Table 2). Since the aqueous solubility of the ternary complex was greater than that of the simple complex, the total βCD solubility increases when a water-soluble polymer was added to the complexation media. For example, when 0.10% HPMC was added the total aqueous solubility of βCD was determined to be 4.52% and that of acetazolamide 3.17 mg/ml. Thus, under these conditions the solubility of acetazolamide in 4.52% βCD solution is about equal to that in pure aqueous 14% HPβCD solution. The solubility of carbamazepine in pure water was determined to be 0.11 mg/ml but 0.49 mg/ml (or 4.5-fold larger) in aqueous 5% γCD solution but about 1.11 mg/ml (or 10-fold larger) when both 5% γCD and 0.10% HPMC were present in the complexation medium (Table 3). Somewhat less enhancement was obtained when HPMC was replaced by PVP or CMC (Table 3). Thus, for carbamazepine the γCD ternary complexes were up to 127% more powerful solubilizers than the simple γCD complexes. In aqueous 10% HPβCD solution the solubility was determined to be 7.00 mg/ml increasing to 9.20 mg/ml when 0.25% CMC was present and 8.50 mg/ml when 0.25% PVP was present, indicating that the ternary complexes were 20 to 30% more powerful solubilizers than the simple binary HPβCD complexes. However, the solubility of carbamazepine in 5.5% βCD-HPMC solution was comparable with that in pure 9.4% HPβCD solution (Table 4). Phase solubility studies have shown that the water-soluble polymers not only increase the aqueous solubility of βCD but that they also enhance βCD's complexing abilities resulting in better overall solubilization. In other words, the polymers do not only increase the aqueous solubility of βCD but, through formation of ternary complexes, they also increase the apparent stability constants of the drug-βCD complexes. Drug ionization resulted in even much larger solubilization enhancement [8]. During preformulation of the aqueous eye drop solutions the effect of addition of 0.10% (w/v) HPMC on the apparent stability constant of the drug-HPβCD (1:1) complex was tested. For the dexamethasone-HPβCD complex it increased from 1230 M^{-1} to 1550 M^{-1}, that of the acetazolamide-HPβCD complex increased from 76 M^{-1} to 150 M^{-1} and that of the

TABLE 3. The effect of polymers on the solubility of carbamazepine in aqueous 5% (w/v) γCD solutions at 23°C (solubility ratio in brackets).

Polymer (% w/v)	Solubility (mg/ml)
No polymer	0.49 (1.00)
0.25% CMC	0.65 (1.33)
0.25% PVP	1.07 (2.18)
0.10% HPMC	1.11 (2.27)

TABLE 4. The effect of polymers on the aqueous solubility of βCD and the solubility of carbamazepine in saturated βCD solutions. The solubility of βCD in pure water was determined to be 1.9% (w/v).

Polymer (% w/v)	Solubility βCD (% w/v)	Solubility of carbamazepine (mg/ml)
No polymer	2.84 (1.5)*	2.16 (1.00)
0.25% CMC	2.92 (1.5)	2.58 (1.19)
0.25% PVP	4.54 (2.4)	2.27 (1.05)
0.10% HPMC	5.50 (2.9)	6.46 (2.99)

*Solubility ratio compared to pure water.

methazolamide-HPβCD complex from 28 M^{-1} to 69 M^{-1}. This enhanced complexation reduced the amounts of HPβCD needed to solubilize the drugs in the aqueous formulations.

3.2 Permeability and bioavailability

Fig. 1 shows the concentration of dexamethasone in aqueous humor after application of two almost identical solutions, one was heated in an autoclave but the other not. Both solutions contained 0.67% (w/v) dexamethasone as determined by HPLC after heating and filtration. The AUC from 0 to 9 hours was over 4-fold larger for the heated solution than for the unheated one and the drug levels were always higher after application of the heated solution; significantly higher at extended time points. Thus, the bioavailability of dexamethasone from the ternary dexamethasone-HPβCD-HPMC complex, formed during the heating process, was much greater than from the simple dexamethasone-HPβCD complex.

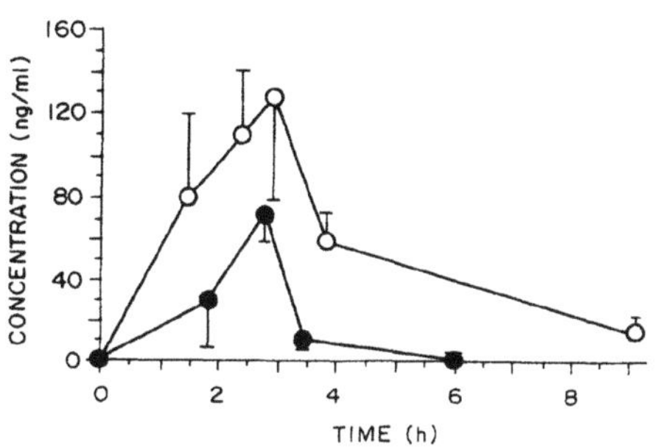

Fig. 1 The effect of HPMC and heating on the mean dexamethasone concentration (±s.e.m.) in the aqueous humor after administration of 0.67% dexamethasone in aqueous HPβCD eye drop solutions. O: heated, ●: not heated.

TABLE 5. Effect of one drop (50 μl) of aqueous HPβCD eye drop solution containing timolol maleat, equivalent to 0.5% (w/v) timolol, 1% (w/v) acetazolamide or 1% methazolamide. Each value represents the mean IOP lowering of ten experiments±s.e.m. in rabbits

Time (h)	Timolol	Acetazolamide	Methazolamide
0	0	0	0
1	-3.94±0.73	-2.11±0.57	-1.18±0.40
2	-3.29±0.56	-2.39±0.76	-3.10±0.36
3	-3.77±0.60	-1.69±0.66	-2.19±0.28
4	-3.15±0.44	+0.10±0.44	-2.08±0.41
5	-1.80±0.60		-1.76±0.52
6	-1.01±0.49		-0.47±0.61

The mean peak IOP lowering activity of 1% (w/v) acetazolamide eye drops in ocular hypertensive human subjects was 15.6%, 2 hours after installation of the eye drops, 3, 7, 14 and 28 days after beginning of treatment. This is less than after topical application of 0.5% timolol but not much less than what has been observed after application of 2% dorzolamide eye drops (about 18%). The IOP lowering effects of acetazolamide and methazolamide eye drop solutions were compared in rabbits (Table 5). The rabbit studies show that methazolamide is more active than acetazolamide although both are less active than timolol. Methazolamide eye drops are currently being tested in humans.
Both the dexamethasone and the acetazolamide eye drop solutions consisted of clear solutions of low viscosity (between 2.5 and 3.0 mPa s) which were well tolerated. No side effects or irritation was observed.

3.3 Other drug delivery forms

Previously, we have shown how the ternary drug-CD-polymer complexes increase drug flux through hairless mouse skin [9,10]. The ternary complexes have also been used to enhance buccal absorption of 17β-estradiol from sublingual tablets [11] and to formulate hydrocortisone solution for treatment of oral diseases such as widespread mucosal ulceration.

4. CONCLUSION

Our results show that the ternary drug-CD-polymer complexes possess several advantages over the simple binary drug-CD complexes. First, they are more powerful solubilizers than the simple CD complexes. Second, they are able to solubilize CDs which have limited aqueous solubility, making for example βCD more attractive as a pharmaceutical excipient. Third, they can enhance drug bioavailability from CD complexes. Fourth, they can result in extended drug release, for example after topical application to the eye. Finally, they can sometimes enhance drug dissolution from solid CD complexes.

ACKNOWLEDGMENTS

This work was supported by a grant from the Icelandic Research Council.

REFERENCES

[1] Takeru Higuchi, a Memorial Tribute, Vol. 3 - Equlibria and Thermodynamics (Eds. Riley, C.M., Rytting, J.H., Kral, M.A.), Allen Press, Lawerence (Kansas), 1991, pp. 220-272

[2] Loftsson, T., Friðriksdóttir, H., Guðmundsdóttir, T.K., The effect of water-soluble polymers on aqueous solubility of drugs, *Int. J. Pharm.*, in press (1996)

[3] Attwood, D.; Florence, A.T. *Surfactant Systems. Their Chemistry, Pharmacy and Biology*; Chapman and Hall, London, 1983, pp. 361-365.

[4] Myers, D. *Surfactant Science and Technology*; VCH Publishers, New York, 1988, pp. 142-145.

[5] Loftsson, T., Friðriksdóttir, H., Sigurðardóttir, A.M., Ueda, H., The effect of water-soluble polymers on drug-cyclodextrin complexation, *Int. J. Pharm.*, **110**, 169-177 (1994)

[6] Loftsson, T., Friðriksdóttir, H., Sigurðardóttir, A.M., *The effect of polymers on cyclodextrin complexation*, in Proc. 7th International Cyclodextrins Symposium (Eds. Osa, T., et al.) Tokyo, 25/28 April 1994. Pp. 218-221

[7] Loftsson, T., Friðriksdóttir, H., Stefánsson, E., Thórisdóttir, S., Guðmundsson, Ö., Sigthórsson, T., Topically effective ocular hypotensive acetazolamide and ethoxyzolamide formulations in rabbits, *J. Pharm. Pharmacol.*, **46**, 503-4-504 (1994)

[8] Loftsson, T., Guðmundsdóttir, T.K., Friðriksdóttir, H., The influence of water-soluble polymers and pH on hydroxypropyl-β-cyclodextrin complexation of drugs, *Drug Devel. Ind. Pharm.*, **22**, 403-407 (1996).

[9] Sigurðardóttir, A.M., Loftsson, T., The effect of polyvinylpyrrolidone on cyclodextrin complexation of hydrocortisone and its diffusion through hairless mouse skin, *Int. J. Pharm.*, **126**, 73-78 (1995)

[10] Loftsson, T., Sigurðardóttir, A.M., *Cyclodextrins as skin penetration enhancers*, in Proc. 8th International Cyclodextrins Symposium (Eds. Szejtli, J., et al.) Budapest, 31 March/2 April 1996.

[11] Friðriksdóttir, H., Loftsson, T., Guðmundsson, J.A., Bjarnason, G.J., Kjeld, M., Thorsteinsson, T., Design and in vivo testing of 17β-estradiol-HPβCD sublingual tablets, *Pharmazie*, **51**, 39-42 (1996)

NEW FUNCTIONS OF PERACYLATED β-CYCLODEXTRINS AS SUSTAINED-RELEASE DRUG CARRIERS

K. Uekama, F. Hirayama, and T. Irie

Faculty of Pharmaceutical Sciences, Kumamoto University, 5-1, Oe-honmachi, Kumamoto 862, Japan

ABSTRACT

A series of peracylated β-CyDs with different alkyl chains (acetyl to octanoyl) were prepared, and their bioadhesive, biodegradable, and film-forming properties were evaluated with particular attention to their potential as possible novel sustained-release carriers for water-soluble drugs in parenteral, oral and transdermal formulations.

1. INTRODUCTION

Various cyclodextrin (CyD) derivatives have been prepared in order to extend the bioadaptability and multi-functional characteristics of parent CyD as novel drug carriers [1, 2]. Among the chemically modified CyDs, the hydrophilic CyDs have been mainly focused on the modification of pharmaceutical properties of hydrophobic drug molecules [3, 4]. On the other hand, the hydrophobic CyDs deserve special attention since they can control the release rate of hydrophilic drugs [5, 6]. Therefore, we attempted to demonstrate a new strategy of application of the hydrophobic peracylated β-CyD derivatives for the design of prolonged release and improved bioavailability of water-soluble drugs in various routes of administration.

2. MATERIALS AND METHODS

Peracylated β-CyDs with different alkyl chains (acetyl - octanoyl) were prepared by acylating all hydroxyl groups of β-CyD, using corresponding acid anhydrides in pyridine solution. A single component of each derivative was isolated by silica gel column chromatography, and characterized by NMR and fast-atom bombardment mass spectrometry, and elemental analysis, as reported [7]. The TV-β-CyD films were mainly prepared by the casting method, spreading the ethanol solution (100 ml) containing TV-β-CyD and drug in a 1:1 molar ratio with or without surfactant and/or oleic acid on a backing membrane (for example: polyethylene terephthalate film). The solutions were allowed to stand at 25 ℃ for 24 hr to evaporate the solvent. The adhesive property of TV-β-CyD film was evaluated by a probe-tack tester, measuring the detachment force between the probe and samples.

J. Szejtli and L. Szente (eds.), Proceedings of the Eighth International Symposium on Cyclodextrons, 413–418.

3. RESULTS AND DISCUSSION

3.1. Some characteristics of peracylated β-CyDs

When hydroxyl groups of CyDs were substituted using acetyl or longer acyl groups, the solubility of these CyDs in water decreased proportionally to their degree of substitution or the length of the alkyl chains (Table 1). The concentrated solutions of peracylated β-CyDs in organic solvents were highly viscous and sticky and gelation took place upon evaporation of the solvents. These properties were thought to be particularly useful for a slow-release carrier of water-soluble drugs. TB-β-CyD, for example, formed, an amorphous complex with molsidomine, a peripheral vasodilator, in 1:2 molar ratio, where the drug molecules are monomolecularly dispersed both in the cavity and the interspace of the alkyl side chains of TB-β-CyD.

Table 1. Some physicochemical properties of peracylated β-CyDs at 25 ℃

β-CyD derivative	Acyl group	Molecular weight	Melting point (℃)	$[M]_D$ [a]	Solubility [b] (mg/dl)
β-CyD		1135	280	+1850 [c]	119.0
TA-β-CyD	$COCH_3$	2018	201-202	+2522	823.0
TP-β-CyD	COC_2H_5	2312	168-169	+2450	423.5
TB-β-CyD	COC_3H_7	2607	126-127	+2607	199.6
TV-β-CyD	COC_4H_9	2901	54-56	+2640	283.0
TH-β-CyD	COC_5H_{11}	3196	- [d]	+2620	3.7
TO-β-CyD	COC_7H_{15}	3785	- [d]	+2763	- [e]
TD-β-CyD	COC_9H_{19}	4374	- [d]	+2668	- [e]
TL-β-CyD	$COC_{11}H_{23}$	4963	- [d]	+2829	- [e]

a) Molar rotation in $CHCl_3$. b) In 80% (v/v) ethanol. c) In water.
d) Oily substance. e) Could not be determined because of the low solubility.

3.2. Parenteral application

Buserelin acetate (BLA), a leuteinizing hormone-releasing hormone (LHRH) agonist, reduces the plasma levels of endogenous sex hormones to a castrate level through down-regulation, when administered continuously or daily. These paradoxical and pharmacological effects are utilized in the treatment of endocrine-dependent diseases such as endometriosis, and both precocious and uterine leiomyoma. We have previously reported that the oily injection of BLA with a sustained-release feature can be achieved by ethylated β-CyDs [8]. The present study deals with the possible use of peracetylated CyDs (TA-CyDs) as sustained release carriers, since the release of BLA from the peanut-oil suspension into the aqueous phase was significantly retarded by the addition of TA-CyDs [9]. A single subcutaneous injection of the oily suspension of BLA containing TA-β-CyD or TA-γ-CyD in rats enhanced the retardation of plasma BLA levels, giving 25 and 30 times longer mean residence times, respectively, than that with BLA alone. Simultaneously, as shown in Fig. 1, the suppression of plasma testosterone levels to induce castration, the pharmacological effect of BLA, continued for 1 to 2 weeks and significant weight reduction in genital organs was observed due to the antigonadal effect. Both TA-CyDs were found to be degraded enzymatically in the rat skin homogenates. For example, the residual amounts of TA-β- and TA-γ-CyDs were 72.4 and 59.9 % after 8 hr incubation, respectively. Thus, both TA-CyDs have potential for use as bioabsorbable sustained-release carriers for water-soluble peptides following subcutaneous injection of oily suspension.

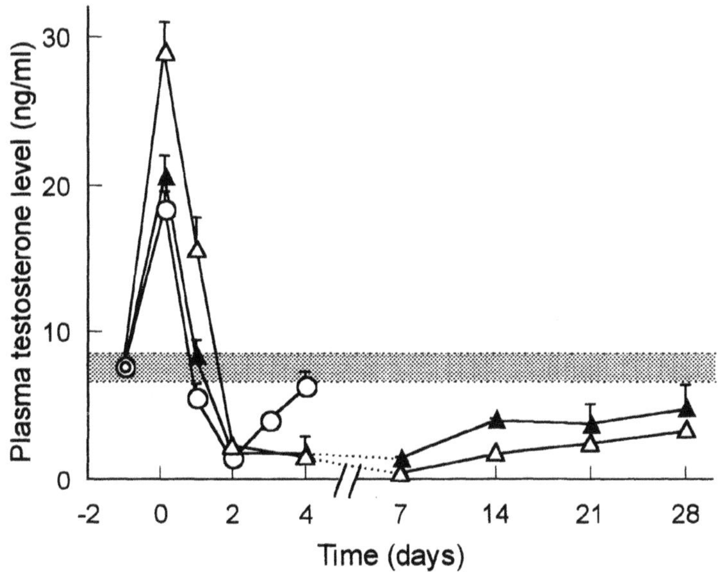

Fig. 1. Plasma levels of testosterone following the subcutaneous administrations of BLA and its TA-CyD complexes in oily suspension (1 mg/kg) in rats.
(○)BLA alone; (▲)TA-β-CyD complex; (△)TA-γ-CyD complex.
Each value represents the mean ± S.E. of 3- 5 rats.

3.3. Oral application

As shown in Fig. 2, the release rate of molsidomine, a water-soluble drug, was markedly retarded by the complexation with peracylated β-CyDs in the decreasing order of the solubility of the host molecules. Since the peracylated derivatives with substituents longer than the hexanoyl moiety released the drug at very slow rates, peracetyl- (TA-), perbutanoyl- (TB-) and perhexanoyl- (TH-) β-CyDs were used in the *in vivo* studies following oral administration of the complexes to dogs. Among them, TB-β-CyD suppressed a peak plasma level of molsidomine and maintained a sufficient drug level for long periods, while other peracylated β-CyDs with shorter or longer chains were ineffective in controlling the *in vivo* release behavior of molsidomine [7, 10]. It was also noteworthy that TB-β-CyD significantly improved the bioavailability of salbutamol, a short acting bronchodilator, after oral administration to dogs, probably due to the reduction of the metabolism in GI tracts [11]. The prominent retarding effect of TB-β-CyD was ascribable to its mucoadhesive property and hydrophobicity compared with other peracylated β-CyDs. These facts suggest that TB-β-CyD is particularly useful as a multi-functional carrier to modify the release rate of water-soluble drugs in oral preparations.

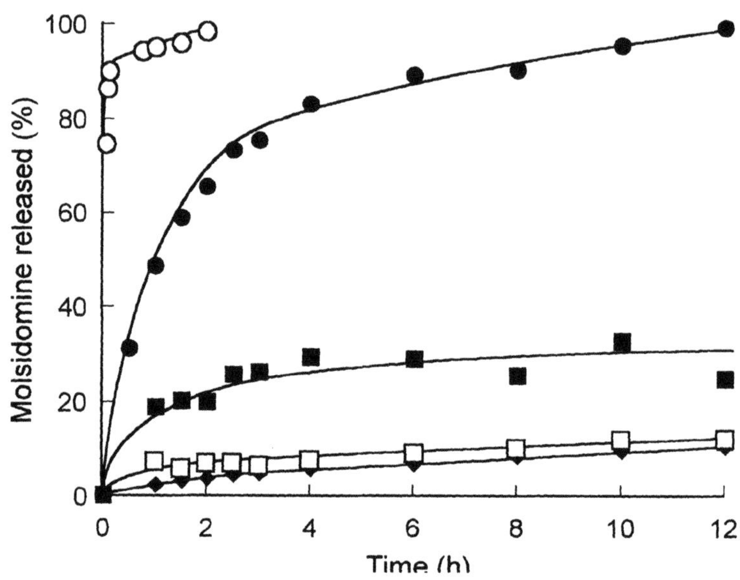

Fig. 2. Release profiles of molsidomine from capsules containing molsidomine or its peracylated β-CyD complexes (equivalent to 10 mg drug) in JP XII 2nd fluid (pH 6.8) at 37 ℃ .

(○) drug alone; (●) TA-β-CyD complex; (■) TP-β-CyD complex; (□) TB-β-CyD complex; (◆) TH-β-CyD complex.

3.4. Transdermal application

In order to examine one of the most potentially useful applications of the TV-β-CyD film, we attempted to design a prolonged release system for some water-soluble drugs such as molsidomine and isosorbide dinitrate (ISDN), peripheral vasodilators with a short biological half-life, using a transdermal formulation in rat models. When a 100 ml benzene solution of TV-β-CyD (15 mg) was spread on the water surface, a transparent film (about 1.5 cm diameter) was preferentially formed among the peracylated β-CyDs, resembling a wrap film or a contact lens. From the practical point of view, an adhesive TV-β-CyD film was also obtainable by means of the ordinary casting method, when an ethanol solution containing TV-β-CyD and drug was spread on a polyethylene backing sheet. This film seemed to be a candidate for a functional drug reservoir, because the water-soluble drugs were slowly released from the film. The film properties such as surface area, thickness, and release rate of the incorporated drug were modified by the addition of a small amount of surfactant. For example, the diameter expanded with an increase in concentration of the non-ionic surfactant, polyoxyethylene hydrogenated castor oil (HCO-60), whereas the addition of HCO-60 higher than the critical miceller concentration (c.m.c.: 0.02 %) made the size of the film smaller, more irregular in shape and also very uneven in thickness. As a tentative measure of adhesion *in vitro*, the force of detachment of TV-β-CyD film was compared with some adhesive polymers. The detaching force of TV-β-CyD film (3.7 N) was much higher than those of the adhesive polymers (0.7, 1.0 and 0.5 N for hydroxypropylcellulose (HPC), hydroxypropylmethylcellulose (HPMC) and poly-2-hydroxyethyl-methacrylate (PEHMA), respectively), suggesting that TV-β-CyD film is preferable for use as an adhesive drug carrier. Fig. 3 shows a typical example of the plasma concentration-time profile of molsidomine after the topical application of TV-β-CyD film to the abdominal skin of rats *in vivo*. The plasma level increased up to 40-50 ng/ml within about 30 min, and then constant drug level (> 10 ng/ml) was maintained for at least 3 days, suggesting a superior sustained release property of the TV-β-CyD film. In addition, neither the film detachment from the skin nor local skin irritation was observed under the experimental conditions, probably due to both the adhesive and bioadaptable property of TV-β-CyD. The present formulation has many advantages in reducing the frequency of dosing, prolonging the drug efficacy and avoiding the local irritancy

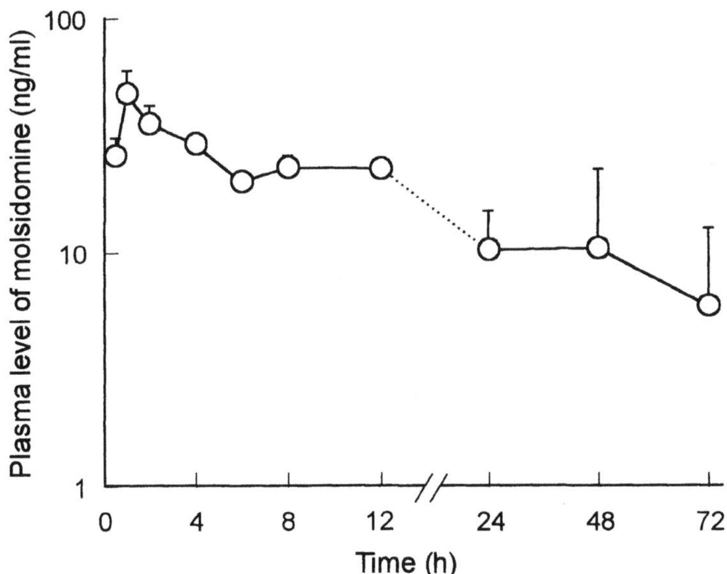

Fig. 3. Plasma levels of molsidomine after topical application of the TV-β-CyD film (4.5 cm diameter) containing the drug (550 μg) onto the abdominal skin of rats. Each point represents the mean ± S.E. of 5 rats.

associated with the long-term therapy. Moreover, TV-β-CyD film may be in preference to other synthetic polymer-films from the viewpoints of safety and environmental pollution, because of its possible degradation to non-toxic oligosaccharides and pentanoic acid.

4. CONCLUSION

The presented results suggest that the peracylated β-CyDs are useful as novel hydrophobic carriers with multi-functional properties which are available to modify the release rate of water-soluble drugs in a variety of administration routes. TA-β-CyD may be useful as a parenteral drug carrier, owing to the biodegradation in subcutaneous administration. Bioadhesive property of TB-β-CyD can be widely used in oral and transmucosal formulations. The TV-β-CyD film may be a new candidate in transdermal preparations, because of its superior sustained release property, as well as its adhesiveness and bioadaptability.

REFERENCES

[1] Szejtli, J. *Cyclodextrin Technology*, Kluwer Academic Publishers, Dordrecht, 1988
[2] Uekama, K., Otagiri, M., Cyclodextrins in drug carrier systems. *CRC Criti. Rev. Ther. Drug Carrier Systems*, **3**, 1-40 (1987)
[3] Duchene, D., *New Trends in Cyclodextrins and Derivatives*, Edition de Sante, Paris, 1991
[4] Uekama, K., Hirayama, F., Irie, T., Application of Cyclodextrins, in *Drug Absorption Enhancement*, (Ed., de Boer, A.G.), Harwood Academic Publishers, Chur, Switzerland, 1994, pp. 411-456
[5] Szychers, M., *High Performance Biomaterials: A Comprehensive Guide to Medical and Pharmaceutical Applications*, Technomic Publishing Co., Lancaster, PA, 1991
[6] Hirayama, F., Yamanaka, M., Horikawa, T., Uekama, K., Characterization of peracylated β-cyclodextrins with different chain lengths as a novel sustained release carrier for water-soluble drugs. *Chem. Pharm. Bull.*, **43**, 130-136 (1995)
[7] Uekama, K., Horikawa, T., Yamanaka, M., Hirayama, F., Peracylated β-cyclodextrins as novel sustained-release carriers for a water-soluble drug, molsidomine. *J. Pharm. Pharmacol.*, **46**, 714-717 (1994)
[8] Uekama, K., Arima, H., Irie, T., Matsubara, K., Kuriki, T., Sustained release of buserelin acetate, a luteinizing hormone-releasing hormone agonist, from an injectable oily preparation utilizing ethylated β-cyclodextrin. *J. Pharm. Pharmacol.*, **41**, 874-876 (1989)
[9] Matsubara, K., Irie, T., Uekama, K., Controlled release of the LHRH agonist buserelin acetate from injectable suspensions containing triacetylated cyclodextrins in an oil vehicle. *J. Controlled Release*, **31**, 173-180 (1994)
[10] Uekama, K., Hirayama, F., Improvement of Drug Properties by Cyclodextrins, in "*The Practice of Medicinal Chemistry*," (Ed. by C.G. Wermuth), Academic Press., London, 1996, pp. 793-825
[11] Hirayama, F., Horikawa, T., Yamanaka, M., Uekama, K., Enhanced bioavailability and reduced metabolism of salbutamol by perbutanoyl-β-cyclodextrin after oral administration in dogs. *Pharm. Sci.*, **1**, 517-520 (1995)

PROLONGED RELEASE OF DRUG FROM TRIACETYL-β-CyD COMPLEX FOR ORAL AND RECTAL ADMINISTRATION

K. NAKANISHI[1], T. MASUKAWA[1], T. NADAI[1], K. YOSHII[2], S. OKADA[2] and K. MIYAJIMA[3]
Faculty of Pharmaceutical Sciences, Setsunan University,[1] Nagaotoge-cho Hirakata, Osaka 573-01, Japan, National Institute of Hygienic Sciences Osaka Branch,[2] and Faculty of Pharmaceutical Sciences, Kyoto University,[3]

ABSTRACT

Triacetyl-β-cyclodextrin (TA-β–CyD), a hydrophobic cyclodextrin derivative, that is insoluble in water, was used to form a complex with flufenamic acid (FA). FA-TA-β–CyD complex formation was demonstrated by differential scanning calorimetry and powder X-ray diffractometry. The release rate of FA from the FA-TA-β–CyD complexes in phosphate buffer pH 6.8 was significantly retarded compared to that of FA from the FA and glucose mixture. When the FA-TA-β–CyD complexes were administered directly into the intraduodenal lumen, the plasma concentration of FA remained at a plateau level (10-18 μg/ml) for 6-8 h. An increased mean residence time of FA following FA-TA-β–CyD complexes administration was observed. These results indicate that TA-β–CyD may serve as a hydrophobic carrier in prolonged-release preparations of FA.

1. INTRODUCTION

Cyclodextrins (CyD), which can include many kinds of molecules within their cavities, are now being widely applied to pharmaceutical formulations to achieve improvements in low solubility, dissolution rate, stability, and the bioavailability of drugs.[1)] Hydrophobic CyD derivatives, such as ethylated and peracylated β–CyDs, have been employed in attempts as slow release carriers for water-soluble drugs to improve the release rate of the drugs.[2)] FA used as a model drug is poorly soluble in water and a short biological half-life in rats. The aim of this study was to evaluate the pharmaceutical applications of TA-β–CyD as a slow-release carrier for FA, compared to the use of glucose as a non-complex excipient.

2. MATERIALS AND METHODS

2.1 Materials TA-β–CyD (molar substitution 3, purity > 99.9%) was donated by Ensuiko Sugar Refining Co. (Japan). FA was purchased from Wako Pure Chemicals Co. (Japan). Scheme I is shown the structure TA-β–CyD and FA.

Scheme I Structure of TA-β-CyD and FA used in this study

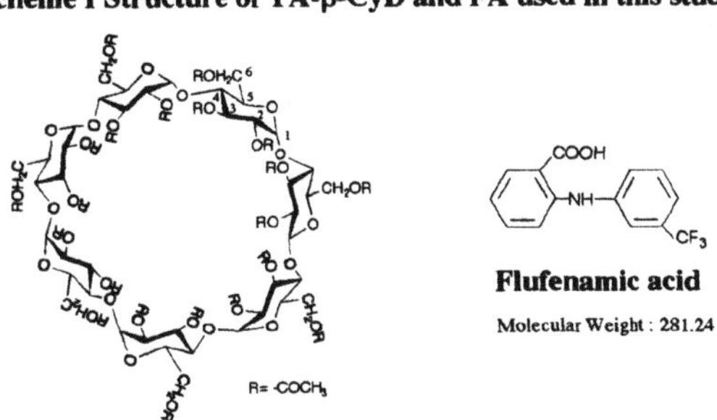

Triacetyl-β-Cyclodextrin Molecular Weight : 2018

J. Szejtli and L. Szente (eds.), Proceedings of the Eighth International Symposium on Cyclodextrons, 419–422.

2.2 Preparation of Solid Complexes Complexes of FA with TA-β-CyD at various molar ratios (1:1, 1:2, 1:3; FA-TA-β–CyD) were prepared by a kneading method using ethanol as a solvent.[3]
2.3 Differential Scanning Calorimetry (DSC) and Powder X-Ray Diffraction Complex formation was studied by DSC, using a Shimadzu DSC-50 (Shimadzu Corp.) with DSC crimp cell (sample size 2 mg, heating rate 5 °C/min). The powder X-ray diffraction pattern was determined with a MXP 3VA diffractometer (MAC Science Co.Ltd.).
2.4 *In Vitro* Release Study The release rate of drug from the FA-TA-β–CyD complexes in isotonic phosphate buffers pH 6.8 was measured by the dispersed amount method.
2.5 Absorption Experiment The absorption experiments were carried out with male Wistar rats (240-260g). The small intestine was exposed, and the duodenal segment was cut, and silicon tubing was inserted. The FA and TA-β–CyD mixture or the FA-TA-β–CyD complexes (equivalent to 2.81 mg FA) were administered directly into the intraduodenal lumen with a syringe through the silicone tubing and the remaining drug in the syringe was washed out twice with 0.5 ml isotonic saline. Rectal absorption of FA was carried out after insertion a suppository including the FA and TA-β-CyD mixture and the FA-TA-β-CyD complexes. At appropriate intervals for up to 10 h, blood samples were collected from the femoral artery.
2.6 Analytical Method FA *in vitro* release was determined spectrophotometrically at 235 nm. FA in plasma was measured by HPLC according to the method of Dusci and Hacket.[4]

3. RESULTS AND DISCUSSION

3.1 Interaction Behavior of TA-β–CyD in the Solid State The interaction behavior in the solid state was investigated by DSC. The molar ratio of the complex between FA and TA-β-CyD obtained by DSC was found to be 1:1. Figure 1 shows a typical example of the DSC thermogram curves of the FA and TA-β–CyD mixture and the FA-TA-β–CyD complexes. The DSC thermogram of the FA and TA-β–CyD mixtures showed an endothermic melting peak around 127°C, corresponding to the melting peak of the free FA, and the mixture (1:2) was broadened. The FA-TA-β-CyD complex(1:1) showed no endothermic peaks, due to the melting of both components. Further, the diffraction peaks of FA also disappeared on complex formation (1:1).

Fig.1 Differential scanning calorimetry of FA, TA-β-CyD, FA and TA-β-CyD mixture and FA-TA-β-CyD complexes

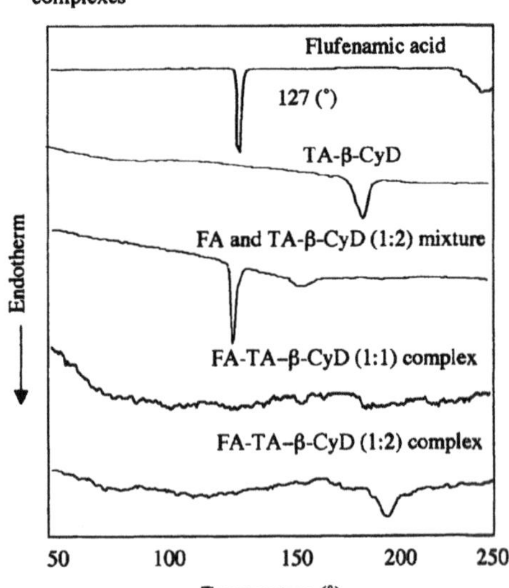

3.2 Release Rate of FA from TA-β–CyD Complexes The release rate of FA from the drug-glucose mixture was very fast, due to the high solubility in phosphate buffer pH 6.8, as shown in Fig.2. The FA-TA-β-CyD complexes released the drug very slowly and the release profiles of the drug from the complexes consisted of faster and very slow phase. The faster release in the initial stage depended on the free FA in the preparations, while the slower release was due to the release of FA from the complexes.
3.3 Absorption Experiments The

plasma concentration of FA versus time curves obtained after the intraduodenal administration of powder containing either the FA and TA-β-CyD mixture or the FA-TA-β-CyD complexes (equivalent to 2.81 mg of FA) to rats is shown in Fig. 3. When the equivalent doses of FA were administered the FA and TA-β-CyD mixture and the complexes, the intestinal absorption of FA in FA and TA-β-CyD mixture was very fast. On the other hand, the complexes did not show a sharp peak plasma concentration, but produced a prolonged plateau plasma level of FA for 6-8 h. Furthermore, the plasma levels of FA after administration of suppository including FA-TA-β-CyD complex (1:2) was similar behavior with that of FA intraduodenal administration of the complex.

Fig. 2 Dissolution profiles of FA and its TA-β-CyD complex in pH 7.4 isotonic phosphate buffer at 37℃

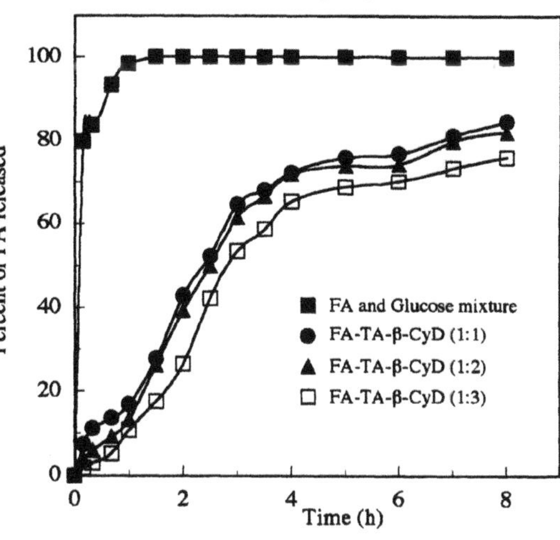

The pharmacokinetic parameters obtained from the data , where the area under the plasma concentration-time curves (AUC_{0-10}), the mean residence time (MRT) in the systemic circulation, and the variance of residence time (VRT). The AUC_{0-10} values after administration of the complexes in the powder form were slightly lower than those of the FA solution and the glucose mixture following intraduodenal administration. However, there was no significant difference in absolute bioavailability between i.v. administration (100 %) and the drug TA-β-CyD mixture (90 %) or the complexes (1:1; 86 %, 1:2; 92 %, 1:3; 84 %). The MRT values after administration of the complex (1:2) were 2.2-fold and 2.0-fold those seen with the i.v. administration and with intraduodenal administration, respectively of the FA solution.

Scheme II shows the MRT, the mean absorption time (MAT), the mean dissolution time (MDT), and the mean release time calculated by moment analysis.[5] The MRT of the FA-TA-β-CyD com-

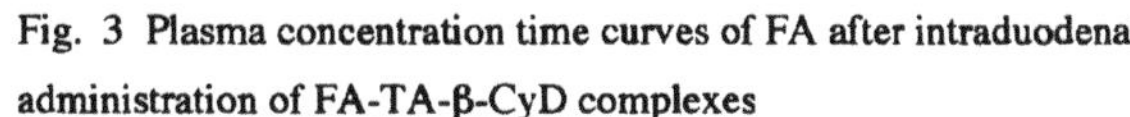
Fig. 3 Plasma concentration time curves of FA after intraduodenal administration of FA-TA-β-CyD complexes

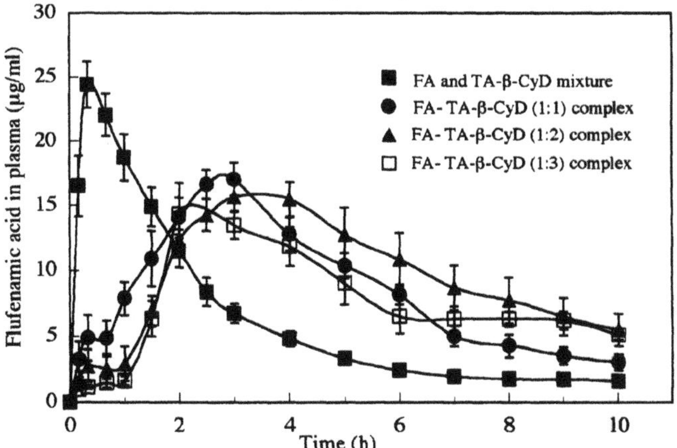

plex ($MRT_{complex}$) was 4.91 h, and it consisted of $T_1+T_2+T_3+T_4$, where T_1 is the mean release time from the complex, T_2 is the mean time for the dissolution of drug, T_3 is the mean time for the absorption of the dissolved drug, and T_4 is the mean time for the elimination of drug from the body, shown as MRT_{iv}. The mean release time of the complex, 2.12 h *in vivo* experiments, was estimated by subtracting the values for the mean MDT_{powder}, $MAT_{solution}$, and MRT_{iv} from the mean $MRT_{complex(1:2)}$ value. The mean release time of the complex in *in vitro* experiments was 1.81 h, similar to the value in the *in vivo* experiments. Thus, the increase in MRTcomplex value after the intraduodenal administration of the FA-TA-β–CyD complex (1:2) was due to the retarded release of the drug from the complex in the intestinal lumen, indicating that the release of FA from the complex was the rate-limiting step.

Scheme II Illustration of the meaning of the MRT, MAT, MDT and Mean Release Time

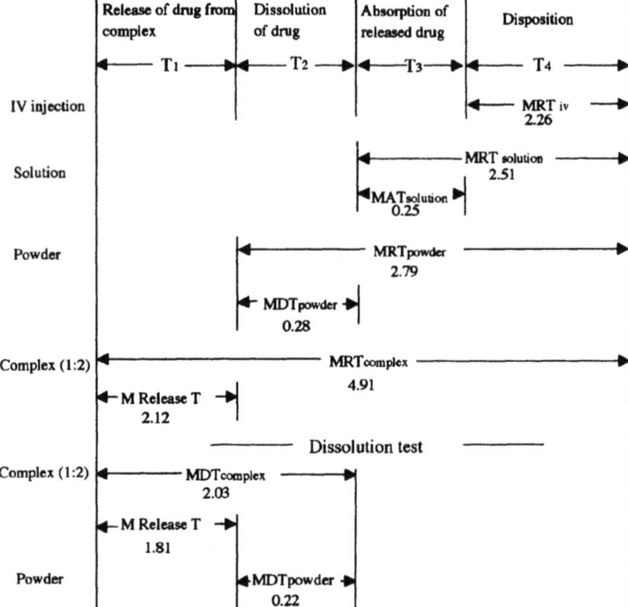

4. CONCLUSION

The present study show that an initial high plasma peak concentration does not occur after the administration of FA-TA-β–CyD complexes, suggesting that the side effects may also be reduced. The form of the FA-TA-β-CyD complex used here appears to be appropriate for practical clinical applications that would result in reduced side effects of the drug and prolonged action.

REFERENCES

[1] Chow D.D. and Karara A.H., Characterzation, dissolution and bioavailability in rats of ibuprofen-β-cyclodextrin complex system, *Int.J.Pharm.*, **28**, 95-101 (1986).

[2] Uekama K., Horikawa T., Yamanaka M., Hirayama F., Peraclated β-cyclodextrins as novel sustained-release carrier for water-soluble drug, molsidomine, *J.Pharm.Pharmacol.* **46**, 714 -717 (1994).

[3] Tsuruoka M., Hashimoto T., Seo H., Ichimasa S., Ueno O., Fujinaga T., Otagiri M., Uekama K., Enhanced Bioavailability of phenytoin by β-cyclodextrin complexation,*Yakugaku Zasshi*, **101**, 360-367 (1981).

[4] Dusci L.J. and Hackett L.P., Determination of some antiinflammatory drugs in serum by high-performance liquid chromatography, *J.Chromatogr.*, **172**, 516-519 (1979).

[5] Tanigawara Y., Yamaoka K., Nakagawa T., Uno T., Momeny analysis for the separation of mean in vivo disintegration, dissolution, absorption, and disposition time of ampicillin products, *J.Pharm.Sci.*, **71**, 1129-1133 (1982)

AMPHIPHILIC CYCLODEXTRINS AND TARGETING OF DRUGS

D. DUCHENE and D. WOUESSIDJEWE

Physico-chimie, Pharmacotechnie, Biopharmacie, URA CNRS 1218
Faculté de Pharmacie, Université de Paris-Sud

Rue Jean Baptiste Clément, 92290 Châtenay Malabry, France

ABSTRACT

For some years now, new cyclodextrin derivatives have been synthesized and work is still continuing. Among these derivatives, amphiphilic cyclodextrins seem to have a very promising future. *Skirt-shaped* cyclodextrins obtained by grafting alkyl chains on the secondary hydroxyl groups are described here, as well as some of their physical characteristics. They have the remarkable characteristic of being able to form nanoparticles easily, either nanocapsules or nanospheres. Nanocapsules can protect the gastro-intestinal tract against the ulcerogenic effect of non-steroidal anti-inflammatory drugs, such as indomethacin. Nanoparticles can be loaded either with hydrophilic or lipophilic drugs, and, in the latter case, they are capable of a fast release of the drug with a possible increase in bioavailability.

1. INTRODUCTION

For some fifteen to twenty years, cyclodextrins have been studied for their inclusion ability. In pharmacy, one of the most interesting consequences of this property is the increase in solubility possibly resulting in an increase in biovailability of poorly water-soluble guest products. However, due to certain drawbacks of natural cyclodextrins, a number of derivatives have been prepared. Most of them are more water soluble than the mother cyclodextrins.

More recently, researchers have been interested in derivatives presenting a relative external hydrophoby in order to improve the affinity of cyclodextrins for the biological membranes. These derivatives have been prepared by grafting hydrocarbon chains on the hydroxyl groups of either α, β or γ-cyclodextrin. They appear to constitute a very promising tool for pharmaceutical technology.

2. MAIN AMPHIPHILIC CYCLODEXTRIN DERIVATIVES

Series of amphiphilic cyclodextrins have been prepared, with a more or less complicated structure.

Lollipops result from the grafting of only one aliphatic acid chain on a 6-amino-β-cyclodextrin [1]. Because of the possibility for the aliphatic chain to enter the

J. Szejtli and L. Szente (eds.), Proceedings of the Eighth International Symposium on Cyclodextrons, 423–430.

cyclodextrin cavity, hampering the inclusion of a guest compound, voluminous groups (tertbutyl) have been linked to the end of the aliphatic chain, preventing this drawback; such cyclodextrins have been named *cup-and-ball* [2]. *Medusa-like* cyclodextrins are obtained by grafting of hydrophobic groups (alkyls with chain lengths from C_{10} to C_{16}) on all the primary hydroxyl groups (C6) of a β-cyclodextrin [3, 4, 5]. *Skirt-shaped* cyclodextrins correspond to the esterification of all the secondary hydroxyl groups (C2 and C3) [6, 7]. *Bouquet-shaped* cyclodextrins result from the grafting of 14 polymethylene chains on 3-monomethylated β-cyclodextrin, such a technique leads to obtaining an equal number (7) of chains on the side of primary (C6) and the side of secondary (C2) hydroxyl groups [8]. Belonging to the *bouquet* family are also the per(2,6-di-*O*-alkyl)-cyclodextrins in which the alkyl chain is a propyl, butyl, pentyl, 3-methylbutyl or dodecyl chain [9].

These various cyclodextrins differ not only by the position and length of the substituents, but also by the nature of the chemical bond. In fact, this can be sulphinyl-deoxy [3], amino-deoxy [4, 10, 11, 12, 13], ether [9, 14], ester [5, 6, 7].

This latter bond, used in the synthesis of *skirt-shaped* cyclodextrins, is particularly interesting because it can be degraded *in vivo* by the natural esterases in the organism. This series of amphiphilic cyclodextrins appears to have a great potential interest in pharmaceutical technology, and the remainder of the present paper will focus on them.

3. SYNTHESIS OF *SKIRT-SHAPED* CYCLODEXTRINS

This synthesis is a complete esterification of the secondary hydroxyl groups of a β-cyclodextrin after protection of the primary hydroxyl groups [6]. The same procedure can be applied to α and γ-cyclodextrin [6, 7, 15].

The first stage of the synthesis is protection of the primary hydroxyl groups by tert-butyldimethyl chlorosilane in anhydrous pyridine. The second stage is esterification of the secondary hydroxyl groups by acyl chloride in anhydrous pyridine (or chloroform) in the presence of dimethylaminopyridine. Finally, in the third stage, primary hydroxyl groups are deprotected by treatment with boron trifluoride in ethanol-free ethylether. At each stage, the persubstituted fraction is separated from the others by silica gel chromatography and undergoes a series of crystallizations and dryings.

4. MAIN CHARACTERISTICS OF AMPHIPHILIC CYCLODEXTRINS

Amphiphilic cyclodextrins, resulting from either substitution on the primary hydroxyl groups (*medusa-like* cyclodextrins) or on the secondary hydroxyl groups (*skirt-shaped* cyclodextrins) and whatever the nature of the bond between the lipohilic chain and the cyclodextrin, are water insoluble and present self-organizing properties [16]. Because of the great similarity in the behaviour of both kind of amphiphilic cyclodextrin, we shall not limit our review only to the *skirt-shaped* derivatives, but we shall also include the *medusa-like* molecules.

4.1. Formation of Langmuir-Blodgett monolayers

Monolayers of amphiphilic cyclodextrins can be obtained by spreading their chloroform solutions on pure water, the hydrophilic groups (-OH or -NH_2) are directed towards the water, the lipophilic hydrocarbon chains being towards the air [3, 7, 10, 11, 17, 18].

By studying monolayers formed of α-, β- or γ-cyclodextrins with C_{14} hydrocarbon chains (esterification) at positions 2 and 3, it has been demonstrated that the density increases slightly with the surface pressure for α- and γ-cyclodextrins, but that in the case of β-cyclodextrin the density which is very low at low surface pressure increases strongly at high surface pressure. Furthermore, the external (film/air interface) and internal (substrate/film interface) roughnesses are equal for α- and γ-cyclodextrins, whereas the external roughness is higher than the internal one in the case of β-cyclodextrin [18]. This seems to be the consequence of the very good lateral packing of α- and γ-cyclodextrins due to their 6- and 8-fold molecular symmetries. In the case of β-cyclodetxrin, the film has holes related to the 7-fold symmetry of the cyclodextrin, which induces disorder at high surface pressures resulting in an increase in the external roughness [18].

A comparison of β-cyclodextrin diesters (C2 and C3) with chain length varying from 2 to 14 carbons (C_2, C_6, C_8, $C1_0$, C_{12} and C_{14}) reveals that the derivative with a C_6 chain is surprisingly the most surface-active molecule of the series [19].

4.2. Solubility in organic solvents and formation of vesicular systems

β-cyclodextrin diesters (C2 and C3) having hydrocarbon chains in C_6, C_{12} and C_{14}, which are water-insoluble, when dissolved in tetrahydrofuran [7] present a solubility increasing slowly with temperature up to 50 °C, temperature at which they have almost the same solubility (500 g/l). Above this value, a sharp increase is observed. At low temperature, the solubility is inversely related to the chain length ($C_6 >> C_{12} > C_{14}$). At high temperature, the contrary is observed: at 60 °C solubilities are 850, 940 and 1410 g/l respectively for the diesters in C_6, C_{12} and C_{14}[7].

With an increase in concentration, the surface tension of the β-cyclodextrin diesters is subjected to a small decrease with an inflection point resembling that observed for surfactants in aqueous solutions at the critical micellar concentration. This suggests the formation of vesicles, such as in the case of phospholipids. For diesters in C_{12} and C_{14}, similar aggregates were observed in pyridine. The size of the vesicle is around 500 nm in tetrahydrofuran, and 4000 nm in pyridine [7].

The temperature dependence of solubility in tetrahydrofuran and ^1H NMR studies lead the authors to propose a geometry of the aggregates with the polar head groups interacting and the hydrophobic chains oriented towards the solvent. In pyridine, the proposed structure is bilayer with the polar head group oriented towards the solvent [7].

Similar results were obtained with β-cyclodextrin monoester (C_{14} chains grafted in C6) dissolved in dimethyl sulphoxide, the solvent being bound to the hydroxyl groups, and the hydrophobic chains interacting [4].

4.3. Molecular associations with amphiphilic cyclodextrins

The possibility of inclusion in amphiphilic cyclodextrins has been investigated on *medusa-like* cyclodextrins.

It has been shown that heptakis(6-dodecylamino-6-deoxy)-β-cyclodextrin in chloroform can include in its cavity 4-nitrophenol and give a salt ion pair stabilized by additional hydrogen bonding involving phenoxide anion and the neighbouring NH groups [12].

The monolayer method has been frequently used to investigate the possible interactions between chemical entities and amphiphilic cyclodextrins. This technique evidences the inclusion of azobenzenes, and especially 4-(4-dimethyl aminophenol)azobenzoic acid, in heptakis(6-dodecylamino-6-deoxy)-β-cyclodextrin [13].

With the same kind of α-, β- or γ-cyclodextrin derivatives, but with C_{16} chains, it appears that cholesterol can be included only in the γ-cyclodextrin derivative. In the case of the α- and β-cyclodextrin derivatives, with smaller cavities, cholesterol is accommodated in the intermolecular cavity formed by the alkyl chains at low pressure, and at high pressure a mixed monolayer is formed [11]. Similarly, a phospholipid (dipalmitoylphosphatidylcholine) cannot be included in these C_{16} amphiphilic α-, β- and γ-cyclodextrin derivatives, but is intercalated between the cyclodextrins with interactions between their hydrocarbon chains which are of identical length [10].

β-cyclodextrin with C_{14} chains on the C6 position, can be incorporated in phospholipid vesicles. The *medusa* molecules are in the phospholipid phase, the secondary face of the cyclodextrin molecules (hydroxyl groups) being exposed towards the aqueous compartment [4].

5. NANOPARTICLES OF *SKIRT-SHAPED* CYCLODEXTRINS

Skirt-shaped cyclodextrins have the remarkable property of producing nanoparticles easily, either nanocapsules or nanospheres.

5.1. Preparation of nanoparticles

5.1.1. Nanospheres

Nanospheres can be prepared either from amphiphilic β- or γ-cyclodextrins esterified in C2 and C3, and bearing hydrocarbon chains from C_6 to C_{14} [20]. Preparation is very simple and consists in dissolving the amphiphilic cyclodextrin in an organic solvent (acetone) and pouring this solution into water under stirring. Nanospheres precipitate spontaneously. The presence of a surfactant in the organic phase is not necessary for the obtention of blank nanopsheres.

Loading of nanospheres can be achieved either concomitantly with the preparation of nanospheres, by dissolving the active ingredient (hydrophilic or lipophilic) in the phase corresponding to its solubility, or by adsorption on to the surface of the already prepared nanospheres.

Nanospheres can also be prepared by the emulsion/evaporation method. The cyclodextrin is dissolved in an organic solvent inmiscible in water, this poured in water under stirring, and finally the organic solvent is evaporated [21].

Nanospheres, which have a diameter of about 90 to 300 nm according to the preparation process, have been characterized by freeze-fracture scanning microscopy [22], photon scanning tunnelling microscopy, scanning force microscopy and no-contact scanning force microscopy [23].

The structure of nanoparticles has not yet been determined, and more especially the arrangement of the amphiphilic cyclodextrin molecules inside the particle is not known. However, because of the ability of these amphiphilic moleules to form monolayers, and bilayer vesicles, it is assumed that, at least, the amphiphilic cyclodextrins form an external layer in which the primary hydroxyl groups are oriented towards the aqueous solution.

5.1.2. Nanocapsules

The preparation method is very similar to the previous one. The amphiphilic cyclodextrin and an oil are dissolved in an organic solvent, the solution is poured in water under stirring, and nanocapsules are formed spontaneously. As above, the presence of surfactant is not necessary for the obtention of blank nanocapsules, but recommended in the presence of an active ingredient. According to the preparation process, the diameter is of about 100 to 900 nm. Loading of the nanocapsules can be achieved by dissolving the active ingredient in the phase corresponding to its solubility [24].

5.2. **Application of *skirt-shaped* amphiphilic cyclodextrin nanocapsules**

Amphiphilic β-cyclodextrin (C_6 chains) nanocapsules loaded with indomethacin have been investigated by oral administration to the rat [25]. It appeared that not only the nanocapsules protect against both gastric and intestinal ulceration, when compared with an aqueous solution of indomethacin administered orally, but they also lead to an increase in bioavailability.

5.3. **Application of *skirt-shaped* amphiphilic cyclodextrin nanospheres**

5.3.1. Nanospheres loaded with hydrophilic drug

Nanospheres of amphiphilic β- and γ-cyclodextrin (C_6 chains) have been loaded with a hydrophilic active ingredient, doxorubicin hydrochloride, either during the nanoprecipitation or after preparation of blank nanoparticles. Best results are obtained in the presence of surfactants such as sodium dodecyl sulphate associated with poloxamer 188 [15, 26, 27].

5.3.2. Nanospheres loaded with lipophilic drugs

Nanospheres of amphiphilic γ-cyclodextrin (C_6 chains) have been loaded, during nanoprecipitation, with lipophilic active ingredients: hydrocortisone, testosterone and progesterone in the presence of pluronic F68. The amount of loaded active ingredient is specially high in the case of progesterone (7% w/w) [15, 28]. DSC thermograms and X-ray diffraction patterns showed that progesterone is in an amorphous state, and probably molecularly dispersed in the nanoparticles [28]. The release of the active ingredient occurs very rapidly with dilution of the nanosphere suspension [29]. This result is very interesting because it demonstrates the possibility of administering high amount of water insoluble active ingredients with the possibility of a high bioavailability.

Very similar results, concerning the loading capacity and the release rate, were obtained when the nanospheres are loaded with progesterone after their preparation. This suggests that the progesterone is at the nanoparticle surface. However this does not exclude the possibility of inclusion in the cyclodextrin cavities, progesterone being able to enter the γ-cyclodextrin through the narrowest side (primary hydroxyl groups).

5.3.3. Cytotoxicity of nanoparticles

Cytotoxicity studies were carried out on P388 murine leukaemic cells [15]. It appears that suspensions of blank nanospheres of amphiphilic β- or γ-cyclodextrin (C_6 chains) are not cytotoxic.

6. CONCLUSION

Amphiphilic cyclodextrins are very attractive molecules because of the possibility of giving vesicular systems and of retaining their ability to include guest molecules.

Further knowledge on these new target systems will result from a collaboration between chemistry, physics, pharmaceutical technology, pharmacology and toxicology.

ACKNOWLEDGEMENTS

The authors wish to express their most sincere thanks to A. Baszkin, A. W. Coleman, J. P. Devissaguet, H. Fessi, H. Galons, A. Gulik, E. Lemos-Senna, F. Leroy-Lechat, S. Lesieur, C.-C. Ling, H. Parrot-Lopez, B. Perly, C. de Rango, M. Skiba, P. Tchoreloff and P. Zhang, for their works on *skirt-shaped* cyclodextrins, and their contribution to the study of nanoparticles.

REFERENCES

[1] Bellanger, N., Perly, B., NMR investigations of the conformation of new cyclodextrin-based amphiphilic transporters for hydrophobic drugs: molecular lollipops, *J. Mol. Struct.*, **15**, 215-226 (1992)

[2] Lin, J., Synthèse des cyclodextrines amphiphiles et étude de leur incorporation dans des phases phospholipidiques, Thesis, Université de Paris VI, 1995

[3] Kawabata, Y., Matsumoto, M., Tanaka, M., Takahashi, H., Irinatsu, Y., Tamura, S., Tagaki, W., Nakahara, H., Fukuda, K., Formation and deposition of monolayers of amphiphilic β-cyclodextrin derivatives, *Chem. Lett.*, 1933-1934 (1986)

[4] Djedaïni, F., Coleman, A.W., Perly, B., *New cyclodextrin-based media for vectorization of hydrophilic drug, Mixed vesicles composed of phospholipids and lipophilic cyclodextrins*, in Minutes of the 5th International Symposium on Cyclodextrins, (Ed. Duchêne D.), Editions de Santé, Paris, 1990, 328-331

[5] Liu, F.-Y., Kildsig, D. O., Mitra A. K., Complexation of 6-acyl*O*-β-cyclodextrin derivatives with steroids, Effects of chain length and substitution degree, *Drug Dev. Ind. Pharm.*, **18**, 1599-1612 (1992)

[6] Zhang, P., Ling, C.-C., Coleman, A. W., Parrot-Lopez, H., Galons, H., Formation of amphiphilic cyclodextrins via hydrophobic esterification at the secondary hydroxyl face, *Tetradron Lett.*, **32**, 2769-2770 (1991)

[7] Zhang, P., Parrot-Lopez, H., Tchoreloff, P., Baszkin, A., Ling, C.-C., de Rango, C., Coleman, A. W., Self-organizing systems based on amphiphilic cyclodextrin diesters, *J. Phys. Org. Chem.*, **5**, 518-528 (1992)

[8] Canceill, J., Jullien, L., Lacombe, L., Lehn J.-M., Channel-type molecular structures, Part 2: Synthesis of bouquet-shaped molecules based on a β-cyclodextrin core, *Helv. Chim. Acta*, **75**, 791-812 (1992)

[9] Wenz, G., Synthesis and characterization of some lipophilic per(2,6-di-*O*-alkyl)cyclomalto-oligosaccharides, *Carbohydr. Res.*, **214**, 257-265 (1991)

[10] Taneva, S., Ariga, K., Tagaki, W., Okahata, Y., Association of amphiphilic cyclodextrins with dipalmitoylphosphatidylcholine in mixed insoluble monolayers at the air-water interface, *J. Coll. Interf. Sci.*,**131**, 561-566 (1989)

[11] Taneva, S., Ariga, K., Okahata, Y., Association between amphiphilic cyclodextrins and cholesterol in mixed insoluble monolayers at the air-water interface, *Langmuir*, **5**, 111-113 (1989)

[12] Takahashi, H., Irinatsu, Y., Kozuka, S., Tagaki, W., Host-guest complexation of a lipophilic heptakis(6dodecylamino-6-deoxy)-β-cyclodextrin with nitrophenols in chloroform, *Mem. Fac. Eng. Osaka City Univ.*, **26**, 93-99 (1985)

[13] Tanaka, M., Ishizuka, U., Matsumoto, M., Nakamura, T., Yabe, A., Nakanishi, H., Kawabata, Y., Takahashi, H., Tamura, S., Tagaki, W., Nakahara, H., Fukuda, K., Host-guest complexes of an amphiphilic β-cyclodextrin and azobenzene derivatives in Langmuir-Blodgett films, *Chem. Lett.*, 1307-1310 (1987)

[14] Takeo, K., Mitoh, H., Uemura, K., Selective chemical modification of cyclomalto-oligosaccharides via tert-butyldimethylsilylation, *Carbohydr. Res.*, **187**, 203-221 (1989)

[15] Lechat-Leroy, F., Investigation de la cytotoxicité et de la capacité de transporteur d'un nouveau système colloïdal à base de cyclodextrines amphiphiles, Application à la vectorisation d'un principe actif anticancéreux: la doxorubicine, Thesis N°422, Université de Paris-Sud, 1995

[16] Wenz, G., Cyclodextrins as building blocks for supramolecular structures and functional units, *Angew. Chem. Int. Ed. Engl.*, **33**, 803-822 (1994)

[17] Parrot-Lopez, H., Ling, C.-C., Zhang, P., Baszkin, A., Albrecht, G., de Rango, C., Coleman, A. W., Self-assembling systems of the amphiphilic cationic per-6-amino-β-cyclodextrin 2,3-di-*O*-alkyl ethers, *J.Am. Chem. Soc.*, **114**, 5479-5480 (1992)

[18] Schalchli, A., Benattar, J. J., Tchoreloff, P., Zhang; P., Coleman, A. W., Structure of a monolecular layer of amphiphilic cyclodextrins, *Lagmuir*, **9**, 1968-1970 (1993)

[19] Tchoreloff, P. C., Boissonnade, M. M., Coleman, A. W., Baszkin, A., Amphiphilic monolayers of insoluble cyclodextrins at the water/air interface, Surface pressure and surface potential studies, *Lagmuir*, **11**, 191-196 (1995)

[20] Skiba, M., Wouessidjewe, D., Coleman, A. W., Fessi, H., Devissaguet, J. P., Duchêne, D., Puisieux, F., Préparation et utilisations de nouveaux systèmes colloïdaux dispersibles à base de cyclodextrines, sous forme de nanosphères, French Patent, 92 07287 (1992)

[21] Lemos-Senna, E., unpublished results

[22] Skiba, M., Wouessidjewe, D., Coleman, A. W., Fessi, H., Devissaguet, J. P., Duchêne, D., Puisieux, F., *Development of new submicronic particles from chemically-modified cyclodextrins, Part1: Nanospheres,* 7th Intern. Cyclodextrin Symp., Tokyo, 25 /28 April 1994, pp. 400-404

[23] Skiba, M., Puisieux, F., Duchêne, D., Wouessidjewe, D., Direct imaging of modified β-cyclodextrin nanospheres by photon scanning tunneling and scanning force microscopy, *Int. J. Pharm.*, **120**, 1-11 (1995)

[24] Skiba, M., Wouessidjewe, D., Fessi, H., Devissaguet, J. P., Duchêne, D., Puisieux, F., Préparation et applications de nouveaux systèmes colloïdaux nanovésiculaires dispersibles à base de cyclodextrines, sous forme de nanocapsules, French Patent, 92 07285 (1992)

[25] Skiba, M., Morvan, C., Duchêne, D., Puisieux, F., Wouessidjewe, D., Evaluation of gastrointestinal behaviour in the rat of amphiphilic β-cyclodextrin nanocapsules loaded with indomethacin, *Int. J. Pharm.*, **126**, 275-279 (1995)

[26] Leroy-Lechat, F., Wouessidjewe, D., Puisieux, F., Duchêne, D., *Optimization of entrapment of doxorubicin in nanospheres of a modified β-cyclodextrin*, 7th International Cyclodextrin Symposium, Tokyo, 25 /28 April 1994, pp. 463-466

[27] Leroy-Lechat, F., Wouessidjewe, D., Duchêne, D., Puisieux, F., *Stability studies of the doxorubicin association to new colloidal carriers made of amphiphilic cyclodextrins*, APGI/APV 1st World Meeting Pharmaceutics, Biopharmaceutics, Pharmaceutical technology, Budapest, 9/11 May 1995, pp. 499-500.

[28] Lemos-Senna, E., Lesieur, S., Wouessidjewe, D., Duchêne, D., Progesterone loaded in modified cyclodextrin nanoparticles, Evaluation of the physical state of the drug, APV 42nd Annual Congress, Mainz, 7/9 March 1995, *Eur. J. Pharm. Biopharm.*, **42**, 34S (1996)

[29] Lemos-Senna, E., Wouessidjewe, D., Duchêne, D., *Release profiles of a hydrophobic drug incorporated in modified cyclodextrin nanospheres*, 8th International Cyclodextrin Symposium, Budapest, 30 March-2 April 1996

RELEASE PROFILES OF A HYDROPHOBIC DRUG INCORPORATED IN MODIFIED CYCLODEXTRIN NANOSPHERES

E. LEMOS-SENNA, D. WOUESSIDJEWE, D. DUCHÊNE
Laboratoire de Physico-chimie, Pharmacotechnie et Biopharmacie, URA CNRS 1218, Université de Paris-Sud, Chatênay-Malabry, 92296, FRANCE

ABSTRACT

The release profiles of a hydrophobic drug, progesterone, from amphiphilic cyclodextrin nanospheres were investigated in sink conditions. The results demonstrated that the release was determined by instantaneous partition of the drug between the medium and the carrier. This property was found to be interesting when an improvement in the bioavailability of the drug is desired.

1. INTRODUCTION

The modification of the hydrophilic secondary face of cyclodextrins by grafting fatty acid chains leads to amphiphilic molecules which have been used to prepare nanospheres. These colloidal suspensions can be considered as potential delivery systems for poorly soluble drugs, particularly for parenteral administration. The capacity of the nanospheres prepared from hexyl-γ-cyclodextrin ester (γ-CDC6) to carry a hydrophobic model drug, progesterone (P), has been evaluated. It was shown that the drug is molecularly dispersed in the carrier (1). In this paper, the release profiles of progesterone from amphiphilic cyclodextrin nanospheres are described. The influence of the formulation parameters, such as the surfactant used and the size of the particle, on the progesterone release was evaluated.

2. MATERIALS AND METHODS

2.1. Materials

The γ-CDC_6 was obtained by a synthetic route described by Zhang *et al.* (2). P was supplied by Sigma Chemical Co. (St Louis, MO, USA.). The surfactants Pluronic F68 and Tween 80 were obtained from ICI (Clamart, France). Solvents for HPLC and other

J. Szejtli and L. Szente (eds.), Proceedings of the Eighth International Symposium on Cyclodextrons, 431–434.

analytical procedures were obtained from Prolabo (Paris, France).

2.2. Methods

2.2.1. Preparation of cyclodextrin nanospheres

Cyclodextrin nanospheres were prepared by a nanoprecipitation method (3). This method consists in injecting an organic solution of both γ-CDC_6 (30 mg) and P (2.5 mg) into an aqueous phase under magnetic stirring containing the Pluronic F68 or the Tween 80 surfactant at several concentrations. The organic solvent is totally removed under reduced pressure, and the colloidal suspensions are concentrated to the desired final volume in the same way. In these conditions, the loading of the drug ranged between 60 and 80 µg/mg γ-CDC_6, which is equivalent to 200 to 300 mg/ml of P in the suspension. The particle size of nanospheres, measured by means of a laser light scattering method (Nanosizer N4MD, Coultronics, Margency, France) was between 100 and 300 nm.

2.2.2. Progesterone release from nanospheres

Progesterone release from cyclodextrins nanopheres was carried out at 37 °C under mechanical stirring after dilution of the colloidal suspensions. These dilutions were performed in an isotonic phosphate buffer solution pH 7.4 or in a water/PEG mixture (60/40). In order to separate the particles from the medium, ultracentrifugation or centrifugal ultrafiltration techniques were employed. In the first case, at given time intervals, 2 ml of the diluted suspensions was subjected to ultracentrifugation at 120,000 g for 1 h. In the second case, 400 µl of the diluted suspensions was deposited in the Ultra-free MC Unit (100,000 NMWL, Polysulfone membrane type) and subjected to centrifugation at 5000 g for 5 min. The released progesterone was determined either in the supernatants or in the ultrafiltrates by HPLC. In order to evaluate the effect of dilution on the initial progesterone release, the loaded nanospheres were diluted to different volumes of the two continuous phases mentioned above. The diluted suspensions were subjected to ultracentrifugation at 120,000 g for 1 h. The supernatants were assayed for progesterone by HPLC. The P associated in the nanopheres was expressed in µg/mg γ-CDC_6 .

3. RESULTS AND DISCUSSION

In order to carry out the *in vitro* release kinetics experiments under perfect sink conditions, the colloidal suspensions were diluted 60 to 70-times or 10 to 20-times, respectively, in phosphate buffer solution pH 7.4 or water/PEG mixture (60/40). Figure 1 shows the release profiles of progesterone from the cyclodextrin nanospheres. It can be noted that the release of the drug into the sink solution is very fast. Between 70 and 90% of the P loaded in the nanospheres was released within 1 h rapidly reaching equilibrium.

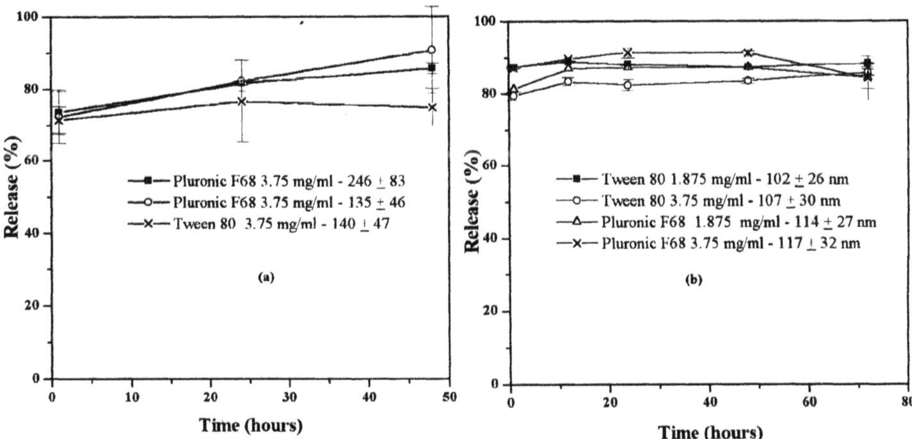

Fig. 1. Progesterone release profiles from cyclodextrin nanospheres (%) (n=3). (a) phosphate buffer solution pH 7.4, (b) water/PEG (60/40).

To confirm these results, the centrifugal ultrafiltration separation method was also assayed. In this case, the release performed in the water/PEG mixture (60/40) (Figure 2) shows an extremely fast release of the drug in the first 5 minutes after the dilution. The results obtained in both approaches indicated that the release of the hydrophobic drug is determined by a partition of the drug between the medium and carrier and the equilibrium is rapidly reached. In both cases, no influence of the formulation parameters on the P release was observed.

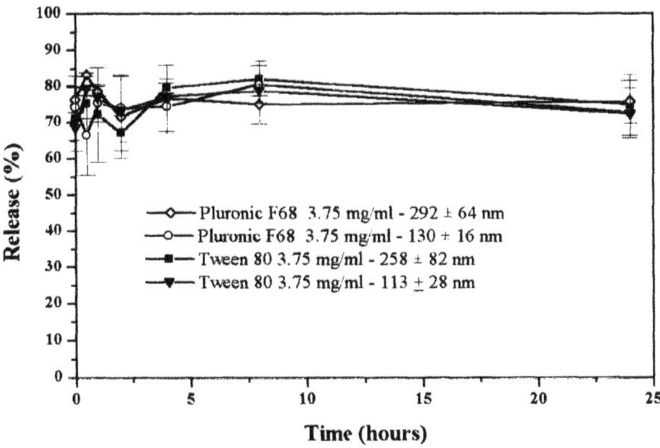

Fig. 2. Progesterone release profiles obtained after separation of particles by the centrifugal ultrafiltration technique (%) (n = 3).

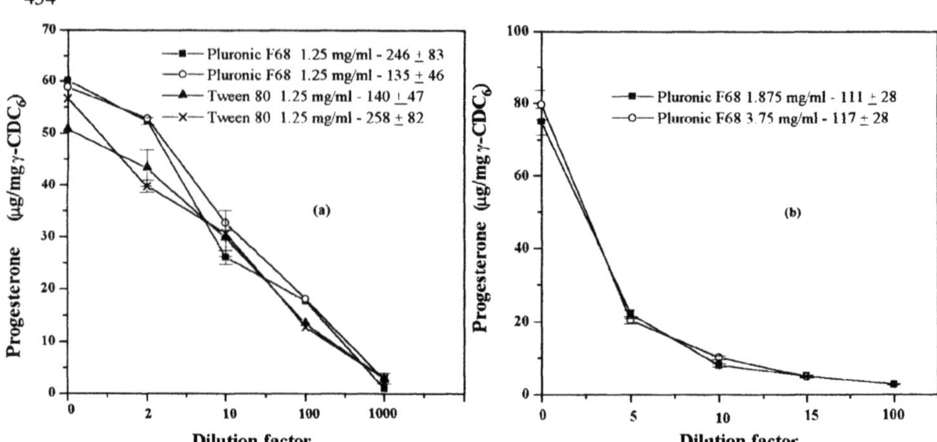

Fig 3. Effect of dilution on progesterone release from cyclodextrin nanospheres (n = 3). (a) phosphate buffer solution pH 7.4, (b) water/PEG (60/40).

In the study of the effect of dilution on progesterone release (Figure 3), it may be observed that the drug loaded in the nanospheres decreases with the dilution. Almost the totality of the drug was released in the phosphate buffer solution pH 7.4 or water/PEG mixture (60/40) when the suspensions were diluted , respectively, 1000 or 100-times.

4. CONCLUSIONS

In this paper, we demonstrated that a molecularly dispersed hydrophobic drug incorporated in cyclodextrin nanospheres is released rapidly after dilution. The release was governed by a partition phenomenon and depends only of the solubility on the drug in the medium. This release profile can be interesting when an improvement in bioavailability of the drugs is desired.

REFERENCES

[1] Lemos-Senna, E., Lesieur, S., Wouessidjewe, D., Duchene, D. Progesterone-Loaded in Modified Cyclodextrin Nanoparticles: Evaluation of the Physical State of the Drug. *Eur. J. Pharm. Biopharm.*, **42** (Suppl.), 34S (1996).

[2] Zhang P., Ling C.-C., Coleman A. W., Parrot-Lopez, H., Galons, H. *Tethahedrom Letters*, **32**, 2769-2770 (1991).

[3] Fessi H., Devissaguet J.-P., Puisieux F. Thies C. Process for the preparation of dispersible colloidal systems of a substance in the form of nanoparticles. U.S. patent 5,118,528 (1992).

INFLUENCE OF CYCLODEXTRINS ON THE IN VITRO DRUG LIBERATION OF RECTAL SUPPOSITORIES CONTAINING BENZODIAZEPINE DERIVATES

G. REGDON jr., I. BÁCSKAY, Á. GERGELY, K. HÓDI,
G. REGDON sen., M. KATA

Albert Szent-Györgyi Medical University
Department of Pharmaceutical Technology
H-6701 Szeged, P.O.Box. 121. Hungary

ABSTRACT

In addition to the oral and parenteral use of benzodiazepine derivatives, there is also need for their rectal administration. Due to shortage of space, the extensive in vitro and in vivo examination of diazepam and nitrazepam cannot be detailed now (see References). Both pharmacons proved to be analogous in character, thus it is similarly possible to formulate them in rectal suppositories.

1. INTRODUCTION

Benzodiazepines are minor tranquillizers widely used in psychotherapy.

There is need for the therapeutic rectal use of benzodiazepines for anxyolitic, antiepileptic and hypnotic purposes.

The more extensive rectal use of diazepam and nitrazepam is hindered partly by its poor solubility.

The authors' aim was

- ☞ to investigate the increase in the solubility of the pharmacons caused by cyclodextrin derivatives,
- ☞ to perform the electronmicroscopic morphological study of the preparations formulated in different ways, and
- ☞ to use these products in medicated suppositories and to carry out their in vitro study in order to determine probable advantages offered by their use as medicines.

J. Szejtli and L. Szente (eds.), Proceedings of the Eighth International Symposium on Cyclodextrons, 435–438.

2. MATERIALS AND METHODS

2.1. Materials

2.1.1. Pharmacons

10 mg of diazepam and nitrazepam each / 2.0 g suppository

Only the pharmacon, without additives

Physical mixes

Physical mixes were prepared from the calculated quantity of the pharmacon (Gedeon Richter Ltd. Budapest, Hungary) and the DIMEB (Cyclolab Ltd. Budapest, Hungary) in the proportion of 2:1, 1:1 and 1:2.

Kneaded products

The calculated quantity of the pharmacons and DIMEB was mixed with 50% water-ethanol mixture. The solvents were evaporated at 105°C.

Spray-embedded products

Spray-embedded products were prepared with a laboratory-sized NIRO MINOR ATOMIZER (Denmark) apparatus. The pharmacons dissolved in 96% alcohol and DIMEB dissolved in distilled water were mixed and spray-embedded at 105°C.

2.1.2. Vehicles, additives

- Lipophilic glyceride derivates of various hydroxyl-values (Witepsol, Estarinum, Suppocire, Butyrum cacao) and hydrophilic macrogolum derivates of various molecular masses were used.
- Tensides (Span, Tween) served as moistening agents and as agents increasing solubility.
- Various cyclodextrin derivates were used to increase solubility.

2.2. Methods

2.2.1. Electron microscopic study

Morphological studies were performed with a Hitachi Model S-2400 type Japanese scanning electron microscope. The sample surfaces were made electrically conductive by evaporating silver on them in a Polaron (U.K.) cathode spray.

2.2.2. In vitro examination method

- The diffusion of powders containing diazepam (nitrazepam) and DIMEB was measured in vitro with a Hungarian-made vibrostate (Manufacturer: KUTESZ-Budapest, Hungarian Academy of Sciences) at a temperature of 37±0.5°C, on the basis of the principle of dynamic diffusion.

- 20.0 mL of distilled water was used as the acceptor phase, the phase was changed after each sampling (30, 60, 120, 240 minutes).
- A kidney dialyzing membrane with a surface of 18 cm^2 and a pore diameter of 2-8 nm was used as a dialyzing membrane (Union Carbide Co. Chicago, USA).
- The quantity of the diffused pharmacon was determined spectrophotometrically at λ = 230 nm (diazepam) and λ = 258 nm (nitrazepam).
- The liberation of the pharmacon from rectal suppositories was similary studied in the above described way. The suppository bases and DIMEB did not interfere with the determination of diazepam and nitrazepam.

3. RESULTS AND DISCUSSION

The results of the experiments confirm (see Table 1) that the properties of the selected suppository base exert a considerable influence both on the quantity of the diffused pharmacon and on the values of in vitro relative availability. These vehicles are all harmless physiologically, but they are not indifferent with respect to pharmacology.

TABLE 1. Diffusion data of nitrazepam-DIMEB spray-embedded product from various suppository bases (Pharmacon content: 10 mg / 2.0 g suppository)

Vehicle	Diffused pharmacon (mg)		In vitro relative availability	
	240 min	S_{SEM}	%	S_{SEM}
Massa Estarinum 299	1.030	3.08	428.1	5.28
Witepsol W 35	0.732	6.16	302.7	8.60
Butyrum cacao	0.660	8.99	357.1	14.54
Suppocire AM	1.258	6.68	400.3	11.94
Witepsol H 15	1.394	6.75	444.2	12.94
Witepsol H 35	1.154	5.35	368.1	9.04
Witepsol S 55	1.113	8.14	349.4	14.49
Macrogolum 1540	4.382	0.95	1112.4	4.95
10% Macrogolum 400 90% Macrogolum 1540	5.245	3.00	1257.4	54.46
5% Polysorbatum 20 10% Macrogolum 400 85% Macrogolum 1540	5.728	3.17	1461.2	20.37

It is clear from Fig.1 that the DIMEB additive can be used effectively to increase the solubility of diazepam (and similarly that of nitrazepam). The three hydrophilic bases all showed a considerably better liberation than the lipophilic bases. It is best to use DIMEB as a spray-embedded product.

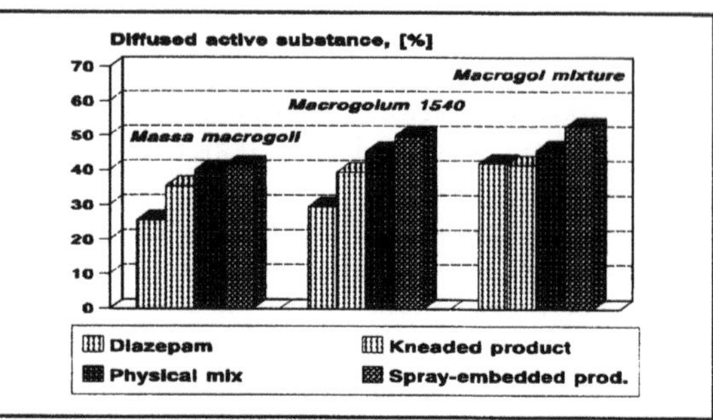

Fig. 1. Drug liberation of diazepam-DIMEB rectal suppositories from hydrophilic vehicles

4. CONCLUSION

☞ The dimethyl-beta-cyclodextrine additive proved to be the best with respect to increasing the solubility of both diazepam and nitrazepam.

☞ The greatest extent of drug liberation, that is the highest percentage of in vitro availability measured with membrane diffusion was also seen from hydrophilic macrogol bases in both cases.

☞ The formulation of the two benzodiazepine derivatives (diazepam - Seduxen® and nitrazepam - Eunoctin®) in rectal suppositories seems to be possible. Similar experiences were obtained both with respect to formulation and as to in vitro drug liberation and optimal in vitro availablity.

☞ The solubility of the pharmacons was increased gradually by the CD-derivates in the following order: pharmacon $< \alpha$-CD $< \gamma$-CD $< \beta$-CD $<$ DIMEB

☞ Our in vitro results are in correlation with the in vivo experiments, which were reported previously.

☞ Our experimental work was successful: a possibility arose to widen the choice of products.

REFERENCES

[1] **Regdon, G., Bácskay, I., Kata, M., Selmeczi, B., Szikszay, M., Sánta, A., Bálint G.S.:**
Formulation of diazepam-containing rectal suppositories and experiences of their biopharmaceutical study.
Pharmazie 49, 346-349 (1994)

[2] **Kata, M., Bácskay, I., Hódi, K., Regdon, G.:**
Formulation of inclusion complexes containing diazepam and dimethyl-β-cyclodextrin and their pharmaceutical study.
Boll. Chim. Farmaceutico (Milano) 134, 557-563 (1995)

PREPARATION AND PHARMACEUTICAL EVALUATION OF ESTER-TYPE CONJUGATES OF BIPHENYLYL ACETIC ACID AND CYCLODEXTRINS AS COLON-TARGETING PRODRUGS

K. Minami, F. Hirayama, and K. Uekama

Faculty of Pharmaceutical Sciences, Kumamoto University, 5-1 Oe-honmachi, Kumamoto 862, Japan

ABSTRACT

Biphenylyl acetic acid (BPAA) was conjugated with one of primary hydroxyl groups of α-, β-, and γ-cyclodextrins (α-, β- and γ-CyDs) through an ester-linkage. The serum levels of BPAA after oral administration of the α- and γ-CyD conjugates to rats were significantly delayed, accompanying a marked increase in the serum levels, whereas the β-CyD conjugate resulted in little increase to the serum levels. The *in vitro* drug release behavior also suggests that the ester-type drug conjugates of α- and γ-CyD serve as colon-targeting prodrugs providing a delayed drug action.

1. INTRODUCTION

The physicochemical and bio-pharmaceutical properties of cyclodextrin (CyD) conjugate in which a drug is covalently bound to CyDs, may differ greatly from those of the inclusion complex. CyDs are known to be absorbed only in small quantities from the small intestine whereas they are fermented in colon microflora at different rates depending on the cavity size, and absorbed as small saccharides from the gastrointestinal (GI) tract. This property of CyD may be particularly useful as a colon-targeting carrier and thus CyD conjugates may serve as a source of colon-targeting prodrug. In a previous paper [1], we reported that the ester-type conjugate, rather than the amide-type conjugate, of an anti-inflammatory drug, biphenylyl acetic acid (BPAA), with β-CyD may serve as a colon-targeting prodrug. In this study, we prepared the ester-type α- and γ-CyD conjugates of BPAA and their *in vitro* drug release and *in vivo* absorption behaviour in the rat model following oral administration were investigated and compared with those of β-CyD conjugate.

2. MATERIALS AND METHODS

Preparation of conjugates: 6^A-O-{(4-biphenylyl)acetyl}-α, β-, and γ-CyDs (α-, β- and γ-CyD ester conjugates) were prepared according to the method of Coates et al. [2]. Sodium 4-biphenylyl acetate was added to 6^A-O-(*p*-toluenesulfonyl)-α-, or β-CyDs or 6^A-O-(β-naphthalenesulfonyl)-γ-CyD/DMF, and the mixture was stirred at 100 °C for

J. Szejtli and L. Szente (eds.), Proceedings of the Eighth International Symposium on Cyclodextrons, 439–442.

30 h. The reaction solution was concentrated under reduced pressure, a large amount of acetone was added, and the resulting precipitates were filtered. The β-CyD conjugate was further purified by a preparative thin-layer chromatography (Silica gel 60F (Merck F265), eluant: acetonitrile/water 3:7 v/v). The α- and γ-CyD conjugates were purified by a column chromatography (DIAION HP-20 (Mitubishi Kasei Co.), mobile phase 0-60 %v/v methanol/water solution).

Hydrolysis of conjugates in rat biological media: Male Wistar rats weighing 300-400 g were used for hydrolysis studies with GI tract contents and biological media. The experimental procedures were the same as those reported previously [1].

***In vivo* absorption studies:** Male Wistar rats weighing 180-200 g were fasted for 18 h prior to drug administration, while water was allowed ad libitum. BPAA or the ester conjugates (10 mg/5 ml water of BPAA per kg of rat body weight) were administered orally as suspension (BPAA alone, β- and γ-CyD conjugates) or solution (α-CyD conjugate), because of different solubilities in water (1.3×10^{-4} M, 1.2×10^{-2} M, 1.3×10^{-5} M and 4.3×10^{-4} M for BPAA alone, α-CyD conjugate, β-CyD conjugate and γ-CyD conjugate, respectively, at 25 °C). Blood samples were taken periodically from the jugular vein after administration, and the serum was assayed for BPAA by high-pressure liquid chromatography under the same condition as those reported previously [1].

3. RESULTS AND DISCUSSION

The BPAA/α-, β- and γ-CyD conjugates were confirmed, by means of NMR and Mass spectroscopies, to be a single component, in which BPAA was introduced to one of the

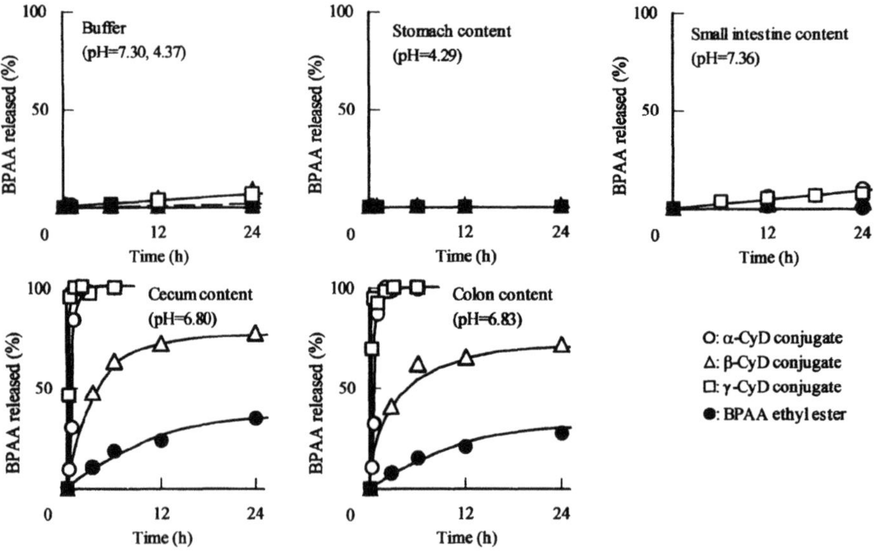

Fig. 1 Releases of BPAA during Incubation of CyD Ester Conjugates (1×10^{-5} M) in Rat GI Tract Contents (10 %w/v) or Isotonic Buffer/DMF (1.0 %v/v) Solution at 37 °C

primary hydroxyl groups of each CyD through an ester linkage. The solubility of conjugates unexpectedly differ from each other, the aqueous solubility decreasing in the order of α-CyD conjugate (1.2×10^{-2} M) > γ-CyD conjugate (4.3×10^{-4} M) > BPAA alone (1.3×10^{-4} M) > β-CyD (1.3×10^{-5} M). The low solubility of the β-CyD conjugate may be at least partly ascribed to the strong intermolecular association between the BPAA moiety and the neighbouring CyD cavity, whereas the high solubility of the α-CyD conjugate may be due to its amorphous form because the α-CyD cavity is too small to accommodate the BPAA moiety.

The ester conjugates were hydrolyzed to BPAA and parent CyDs in aqueous solution, with half-lives of about 7 h (γ-CyD conjugate), 8 h (β-CyD conjugate), 11 h (α-CyD conjugate) and 22 h (BPAA ethyl ester) at pH 9.0 at 37 °C, and only in small amounts (< 10 %) under the physiological pH condition (about 7).

Figure 1 shows the release profiles of BPAA from the three conjugates and the ethyl ester in rat GI contents. In the contents of stomach and small intestine, the conjugate released the drug only in small amounts (< 10 %) which could be accounted for by spontaneous hydrolysis in aqueous solution. On the other hand, the conjugate release the drug significantly in the rat cecal and colonic contents, in the order of γ-CyD conjugate > α-CyD conjugate >> β-CyD conjugate: 100 % release from the α- and γ-CyD conjugates within 2-3 h and more than 50 % from the β-CyD conjugate within 24 h.

Figure 2 shows the *in vitro* release profiles of BPAA in rat intestine homogenates without contents, liver homogenates and blood. No appreciable drug release from the conjugates was observed for 24 h, whereas BPAA ethyl ester hydrolyzed rapidly to BPAA, because of the high esterase activity in the homogenates and blood.

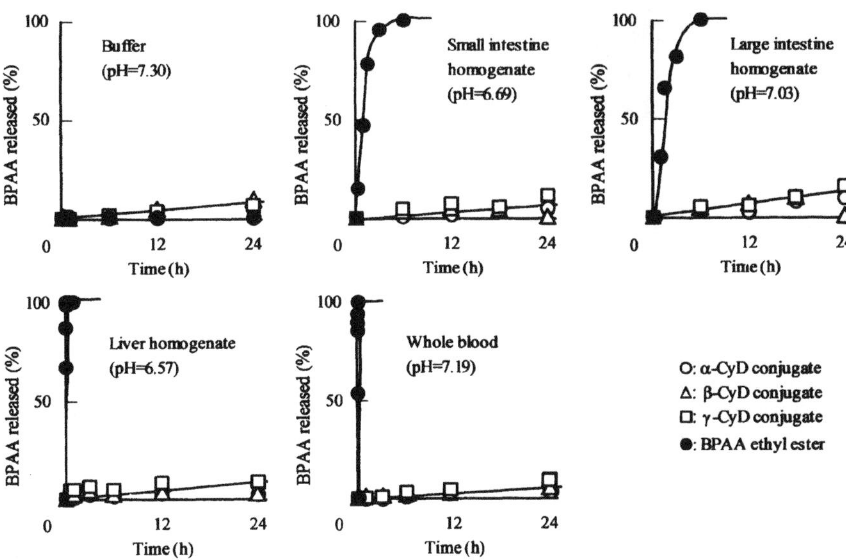

Fig. 2 Releases of BPAA during Incubation of CyD Ester Conjugates (1×10^{-5} M) in Homogenates of Rat GI Tract and Liver and in Whole Blood or Isotonic Buffer/DMF (1.0 %v/v) Solution at 37 °C

Figure 3 shows the serum levels of BPAA after oral administration of the conjugates and BPAA alone to rats. The β- and γ-CyD conjugates and BPAA alone were administered as suspensions in water because of the low solubilities, whereas the α-CyD conjugate was in solution. The BPAA alone system showed rapid increase and decrease in the serum level of the drug. In the case of the α- and γ-CyD conjugates, however, the serum levels of BPAA increased after a lag time of 2-3 h, and reached to the maximum levels after about 9 and 8 h, respectively, accompanying a significant increase in the serum levels. The β-CyD conjugate gave little increase in serum levels under the experimental conditions, probably because of its slower drug release compared to those of the α- and γ-CyD conjugates, as is apparent from Figure 1. The above results suggest that BPAA activation took place site-specifically in the rat cecum and colon, although the hydrolysis mechanism, particularly for the slower release of the β-CyD conjugate, should be further studied.

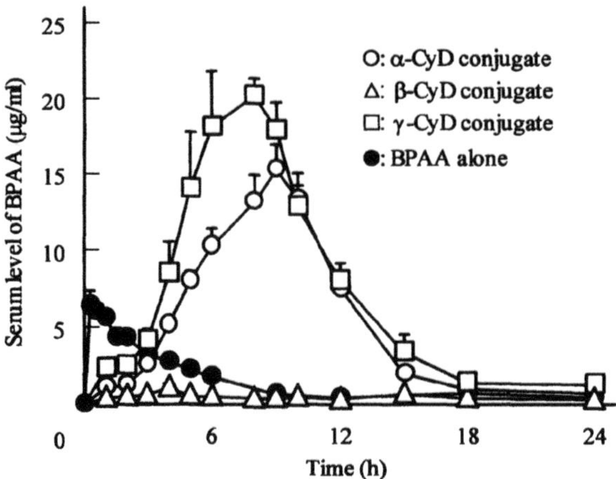

Fig. 3 Serum levels of BPAA after Oral Administration of Suspensions Containing BPAA or CyD Ester Conjugates (equivalent to 10 mg/kg BPAA) to Rats.

4. CONCLUSION

The present results suggest that ester-type drug conjugates of α- and γ-CyDs can function as delayed action-type prodrugs which release a parent drug selectively in cecum and colon. Thus, the CyD conjugation approach may provide a versatile means for constructing colon-specific drug delivery system.

REFERENCES

[1] Hirayama, F., Minami, K., Uekama, K., *In-vitro* evaluation of biphenylyl acetic acid-β-cyclodextrin conjugates as colon-targeting prodrugs: drug release behaviour in rat biological media, *J. Pharm. Pharmacol.*, in press.

[2] Coates, J., Easton, C., Fryer, N.L., Lincoln, S.F., Complementary diastereoselectivity in the synthesis and hydrolysis of acylated cyclodextrins, *Chem. Lett.*, 1994, 1153-1156.

CONTROLLED RELEASE OF DRUGS FROM CD POLYMERS SUBSTITUTED WITH IONIC GROUPS

FENYVESI, É.[1], UJHÁZY, A.[1], SZEJTLI, J.[1], PÜTTER, S.[2], GAN, T. G.[2]
[1]*Cyclolab R&D Lab., Ltd., H-1525 Budapest P.O. Box 435, Hungary*
[2]*Medice Chem.-pharm. Fabrik, Pütter GmBH, D-58638 Iserlohn P.O. Box 2063, Germany*

ABSTRACT

Various disinfecting drugs (ethacridine lactate, methylene blue, gentian violet, brilliant green, fuchsin acid, cetylpyridinium chloride) were incorporated into CD bead polymers substituted with carboxymethyl groups and a retarded release rate was measured. These polymers were successfully used as sustained release wound powders as well as in chewing gum formulations.

1. INTRODUCTION

The insoluble cyclodextrin polymers (CDP) usually prepared in the form of tiny beads are special sorbents for binding certain components from aqueous solutions. Beside the physical adsorption on the surface and inside the pores of the gel structure inclusion complex formation in the cyclodextrin cavities also takes place. When ionic groups are linked covalently in the polymer the interactions between the matrix and the substance are supplemented with salt formation. Cationic polymers are the suitable sorbents for anionic substances, like ethacridine lactate (EAL), brilliant green, methylene blue, fuchsin acid, cetylpyridinium chloride (CPC), etc. The effect of carboxymethylation of the polymer on the release rate of these drugs has been studied.
Controlled release wound powder and chewing gum formulations have been developed applying drug/carboxymethyl cyclodextrin polymer (CMCDP) complexes [1, 2].

J. Szejtli and L. Szente (eds.), Proceedings of the Eighth International Symposium on Cyclodextrons, 443–447.

2. EXPERIMENTAL

2.1. Materials

βCD polymer prepared by crosslinking with epichlorohydrin, ethyleneglycol bis(epoxypropyl) ether mixed crosslinking agent in the presence of polyvinyl alcohol (swelling: 5 mL/g, CD content 55 %) was used.
EAL and CPC (Fluka) as well as the other dyes (Reanal) were of analytical grade.

2.2. Methods

Carboxymethylation of the polymer
The polymer was swollen in a chloroacetic acid solution. Concentrated NaOH solution was added and the suspension was stirred gently or shaken in a shaking machine at room temp. for 24 h. After that it was filtered, neutralized, dehydrated and dried. The COOH-content was measured by titration.
Preparation of polymer complexes
The polymer was swollen in the aqueous ethanolic (50 v/v %) solution containing the proper amount of the active ingredient, then shaken for 2 h and dried.
Measurement of the dissolution rate:
The sample of the polymer complex (0.2 g) was stirred in 20 mL phosphate buffer solution at 37 ^{o}C and aliquots of the supernatant were measured spectrophotometrically.
Multilayer model-wound experiments
For modeling a wet wound 4 sheets of 4x4 cm filter paper were put on each other and wetted with 1.5 mL of 0.05 M pH=7.0 phosphate buffer and 50 mg dry EAL/CDP complex beads were spread on the top sheet. After a given time the beads were removed and the EAL diffunded into the filter paper layers was washed out with 4x20 mL 50 % (v/v) aqueous ethanol by stirring at room temp. for 1 h. The dissolved EAL was measured spectrophotometrically after filtration.
Preparation and investigation of chewing gums
The chewing gums of the following composition were prepared by warming the gum base to about 70 °C in a kneader, adding glycerol, sorbitol solution and the active ingredient either in itself or complexed with CDP. The composition was kneaded after each addition until homogeneous mass was achieved.

300 g base
20 g glycerol
230 g 70 % sorbitol solution
ad.450 g sorbitol powder
+ 5 g CPC with or without CDP

The formulations were subjected to test panel studies with multiple panelists. The CPC released during mastication was measured by assaying the gum bolus after chewing it for an appropriate time (5 and 30 min).

Assay of the gums

A 3 g piece of chewing gum is weighed into a round flask. 20 mL chloroform, 10 mL 0.004 M sodium lauryl sulfate, 5 mL 2 N sulfuric acid solution and 35 mL distilled water were added and boiled for 20 min at 60 °C under reflux. The emulsion is cooled, 1 mL butter yellow indicator solution is added and titrated with 0.004 M CPC solution till orange color.

3. RESULTS AND DISCUSSION

The dissolution rate of the cationic disinfectant, EAL decreased by using carboxymethyl polymer supports. The higher the carboxymethyl content of the polymer the lower is the rate of dissolution of the drug (Fig. 1). On the basis of these results the release of EAL can be controlled by introducing the proper amount of COOH groups. The EAL/CMCDP polymer complex is suitable to develop a controlled release wound powder formulation [1].

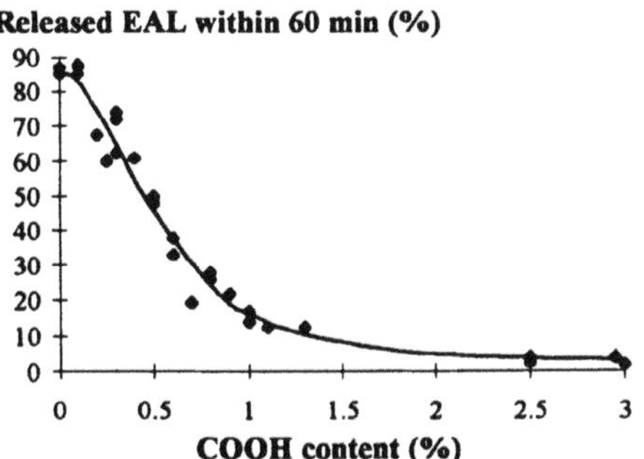

Fig. 1 Release of Ethacridine Lactate from carboxymethyl cyclodextrin polymers as a function of the COOH content of the polymer

Three formulations with identical (2%) EAL content were selected for clinical trial as wound powder for the healing of ulcus cruris and other oozing wounds. The results of the dissolution rate measurement are not characteristic because of the high excess of dissolution medium. The wound powder acts in the low volume of the wound exudate. The release rate of the drug in the multilayer model-wound experiment can be seen in Fig. 2.

The salt formation with the carboxymethyl polymer resulted in decreased release rate in case of some other disinfectants, such as cetylpyridinium chloride, acridine derivatives, methylene blue, triphenyl methane dyes, too (Table I).

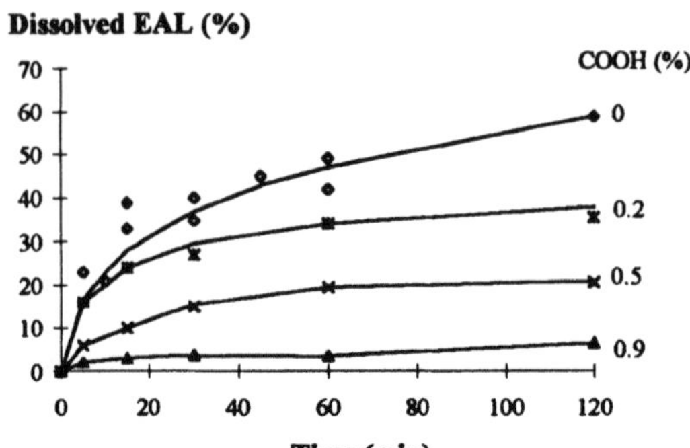

Fig. 2 Dissolution of Ethacridine lactate from CDP-25 in the multilayer model-wound experiment

Table I Effect of carboxymethylation on the drug release from cyclodextrin polymer

disinfectant	carboxymethyl group content of the polymer (mmole/g)	molar ratio of drug/CD	dissolved drug into water in 60 min (%)
CPC	0	1:1	45
	1.0	1:1	23
EAL	0	0.4:1	86
	0.2	0.4:1	60
methylene blue	0	0.1:1	34
	0.6	0.1:1	9.2
brilliant green	0	0.1:1	23
	0.2	0.1:1	1.7
fuchsin acid	0	0.1:1	17
	0.2	0.1:1	1.2

Cetylpyridinium chloride (CPC) was used as model drug, a possible candidate for the development of a chewing gum formulation against sore throat and mouth infections.
Using CPC/CDP or CPC/CMA (carboxymethyl amylose of about the same COOH content: 2 %) complex a higher percentage of CPC is released from the gum but only in the first 5 min. Thus, no retardation effect was observed. This means that neither the complex formation nor the salt formation alone can retard the drug release from the gum. The formulation with CPC/CMCDP, which combines the salt and complex formation has more favorable properties: both the release rate and the amount of unused drug is lower

than in case of the other formulations (Fig. 3). The carboxymethyl cyclodextrin polymer has a significant retardation effect.

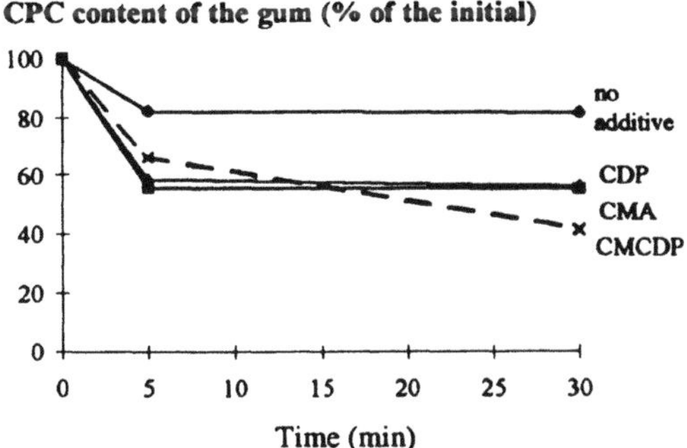

Fig. 3 Release of CPC from chewing gums during mastication

4. CONCLUSIONS

Carboxymethyl polymers are good sorbents for cationic drugs. The release rate of the drug can be controlled by the degree of the carboxymethylation.
By combining the effect of cationic disintegrants with the effect of the swelling carboxymethyl polymers controlled release wound powders have been developed for the treatment of infected wounds, ulcus cruris, traumatic injuries and burns.
The disinfectant/carboxymethyl polymer combination can be successfully used to obtain sustained release chewing gum formulations against sore throat and mucosal infections.

ACKNOWLEDGMENT

The preparation of the chewing gums is greatly acknowledged to S. Molnár, head of the Research Lab. of Zamat Chocolate Factory. The analytical work of Zs. Nagy is highly appreciated.

REFERENCES

[1] Szejtli, J.; Puetter, S., Inclusion complexes of polymerized cyclodextrins with pharmaceutical drugs, Eur. Pat. Appl. 575976 (1993)

[2] Szejtli, J.; Puetter, S., Chewing gum compositions, Eur. Pat. Appl. 575977 (1993)

EFFECT OF CYCLODEXTRINS ON THE PHYSICAL PROPERTIES AND DISSOLUTION RATE OF CARBAMAZEPINE TABLETS

Lj. Tasić[1], J. Milić–Aškrabić[1], D. Rajić[2]
[1]Faculty of Pharmacy, Belgrade University, 11221 Belgrade
[2]Military–Technical Institute, 11000 Belgrade, Katanićeva 15

ABSTRACT

The effect of β-cyclodextrin (βCD) and 2-hydroxipropyl β-cyclodextrin (HPβCD) on physical properties and dissolution rate of Carbamazepine (CBZ) tablet samples in comparison with conventional formulation (without CDs) and commercial tablet (Tegretol®) was studied. Weight variation, crushing force and disintegration time were acceptable for all samples. The friability values of examined CBZ tablets were higher than commercially CBZ tablet. Both CDs (βCD and HPβCD) were showed some increasement in dissolution rate of CBZ from tablets, regarding conventional tablet formulation, but not regarding Tegretol®.

1. INTRODUCTION

Carbamazepine (CBZ) is presently the most widely used antiepileptic agent in the world. Limited water solubility and poor wettability properties of CBZ contribute to slow, erratic and variable absorption from the commercially available immediate–release tablets [1]. These properties of CBZ can be improved via complexation with various modified β–cyclodextrins, especially 2–hydroxypropyl cyclodextrin (HPβCD) [2,3]. Otherwise, the formation of CDs inclusion complexes with high dosage drugs (>100 mg) showed some practical limitations, since the usually molar ratio in CD–drug complexes was 1:1 this resulting in pretty high weight of solid dosage forms (≥1 g). Practically the design of tablet formulation with CBZ/HPβCD complex was difficult, since for usually 200 mg CBZ doses tablet needs 1431 mg (if molar ratio 1:1) or 3280 mg of complex (if molar ratio 1:2.6) [3]. The possibility of using βCD as a complexation agent with intention of drugs solubilising was limited by their low water solubility, but relatively new described possibility of using βCD as a filler–binder in tablet formulation [4] offers an attractive approach to design tablets with CDs.

J. Szejtli and L. Szente (eds.), Proceedings of the Eighth International Symposium on Cyclodextrons, 449–452.

Based on all these reasons the present work is connected with the effect of βCD and HPβCD on physical characteristics and dissolution rate of some CBZ fast–release tablet formulations. Comparison was done by the CBZ conventional tablet formulation without CDs and Tegretol® tablets, as well.

2. MATERIALS AND METHODS

2.1. Materials

Carbamazepine (supplied by Hemofarm DD Vršac, Yu); β–cyclodextrin and 2 hydroxypropyl β–cyclodextrin (Roquette Freres, France); Avicel® PH 101 (FMC Co., USA); Ac–Di–Sol® (FMC Co., USA); polyvinylpyrrolidone K–25 (Kollidon® 25, BASF, Germany); Tegretol® 200 mg tablets (lot:148400, Ciba–Geigy, Switzerland) were used as received. Talc, starch, lactose, magnesium stearate and all other materials were of pharmaceutical purity grade.

2.2. Methods

2.2.1. Preparations of tablet samples.
Three different CBZ powder mixtures were used for preparation of granulate: *CBZ conventional*, containing CBZ (200 mg), lactose (171.5 mg) and starch (73.5 mg); *CBZ/βCD mixture*, containing CBZ (200 mg) and β–cyclodextrine (245 mg); *CBZ/HPβCD*, containing CBZ (160 mg) and CBZ/HPβCD complex (286 mg, corresponding to 40 mg of CBZ). CBZ/HPβCD complex (molar ratio 1:1) was prepared by following procedure: HPβCD was dissolved in water at 60–70°C. CBZ was added to solution of HPβCD and mixed for 24h. Sample was filtered (pore size 10 μm), concentrated by evaporation under vacuum and dried under IR lamp (60°C) until constant weight. Dry powder mass was sieved through 0.15 mm sieve. The inclusion complex CBZ/HPβCD was checked by DSC.
Tablet samples were prepared by wet granulation of three different CBZ powder mixtures by 5% (w/w) ethanol solution of polyvinylpyrrolidone. The mass was sieved (1.2 mm), dried at 60°C; the granules sieved (1 mm) again. Standard tablet excipients: Avicel® PH 101 (9.57%), talc (1.91%), Ac–Di–Sol® (1.91%) and magnesium stearate (0.48%) were added, mixed and compressed directly on an excenter tablet machine (Korsh EKO,E.Korsch,GmbH, Berlin,Germany) fitted with 12 mm flat faced punches. Target tablet weight was 522.5mg.

2.2.2. Physico- pharmaceutical methods
Weight variation was determined according to the Ph.Yug.IV. *Disintegration time* was measured at 37±0.5°C in distilled water, using disintegration test apparatus (Disintegration test apparatus, type 2T3 Erweka, Hausenstamm, Germany).

Crushing strength was determined by using an Erweka tester (Erweka type TH 24, Hausenstamm, Germany). *Friability* was determined after 10 min. of rotating in the friabilator (Friability tester type TA 3, Erweka, Hausenstamm, Germany) at 20 rpm.
Dissolution rate of CBZ from different tablet samples were determined in 900 ml water containing 1% sodium lauryl suplphate at 37±0.5^{0}C and at a rotational speed of 100 rpm; by the basket method (Dissolution test apparatus, type DT 6, Erweka, Hausenstamm, Germany) according to the USP 23. Suitable aliquots were removed at the specific times filtered and spectrophotometrically quantified at 285 nm (UV/VIS 914 Spectrophotometer GBS, Dandendong, Australia).

3. RESULTS AND DISCUSSION

Results of physical characteristics of examined CBZ samples are shown in Table I. The values obtained for weight variation, tablet disintegration time and crushing strength were acceptable for CBZ conventional tablet formulation. The friability value was above 1% with all samples (3.25% for CBZ conventional, 7.06% for CBZ/βCD, and 296% for CBZ/HPβCD) and close to 1% with Tegretol® tablets (1.15%). The upper level of acceptability for friability value is

Table 1: Physical characteristics of CBZ tablets

sample	A	B	C	D
CBZ conventional	0.5041 0.86	4.45	3.25	169
CBZ/ CD	0.5257 0.22	4.64	7.06	177
CBZ/HP CD	0.5064 0.086	5.30	2.96	415
Tegretol®	0.2783 0.23	6.19	1.15	104

symbols: A mean weight ±SD(g)
B crushing force (kg)
C friability (%)
D disintegration time (s)

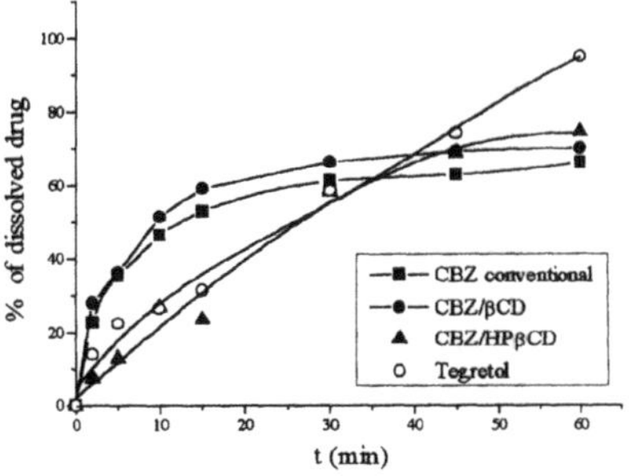

Figure 1.Dissolution profiles of CBZ from tablets

1% [5]. These results may originate from low pressure force during compression, but especially high friability of CBZ/βCD tablet samples is indicative. This means that the βCD is relatively good filler, but is not so good binder. This conclusion is in agreement with our previous results [5].
The dissolution profiles of CBZ samples and Tegretol® are shown in Fig. 1. Generally, the dissolution rate of drug from tablets is a function of method of preparation the tablets and many production and formulation factors. Basically, the powder

characteristics of drug material (particle size, shape, crystal structure) has important influence on dissolution. According to the USP 23 proposition for CBZ tablets (Q not less than 75% in 60 min.) the dissolution rate of CBZ/HPβCD and Tegretol® tablets were confirmed. The profile of dissolution curves of CBZ conventional formulation and CBZ/βCD formulation were similar to each other, and were not confirmed on USP proposition. During the first 15 min. the high dissolution of CBZ occurred from CBZ/βCD tablet (60% of dissolved drug) as a consequence of wettability and surfactant like properties of βCD [6]. CBZ/HPβCD formulation showed a good profile as an outcome of presence of inclusion complexes of CBZ/HPβCD, which significantly improves the solubility of CBZ [2]. In this tablet samples only 20% of CBZ amount was included as a complex formation, and it seemed that was not enough, regarding Tegretol® dissolution profile.

4. CONCLUSION

The experimental results show that the CDs used in CBZ tablet formulation has some influence on physical characteristics of tablet, particularly the βCD. The βCD conformed as a good filler in observed CBZ tablets. Both CDs (βCD and HPβCD) bringinged some increasement in dissolution rate of CBZ from tablets, regarding conventional tablet formulation.CBZ/HPβ CD complex which included in tablet composition only at 20% (drug contain) showed higher effect on dissolution rate, regarding tablet composition of CBZ/βCD mixture. Used CDs has influence on CBZ dissolution profile, but not enough to overtake Tegretol®.

REFERENCES

[1] Mechichan, J.J., Rutt, H., Individualizing Drug Therapy: Practical Applications of Drug Monitoring 2, ed. by Talylor W.J., Finn, A.A., Gross Townssend Frank Inc., New York, 1981, pp. 6–8,

[2] Choudhury, S., Nelson, K.F., Improvement of oral bioavailability carbamazepine by inclusion in 2– hydroxypropyl–β–cyclodextrine, Int. J. Pharm., 85 , 175–180 (1992)

[3 Brewster, M., Anderson, W., Pop, E., Sawchuk, R., Cheung, B., and Webb, A., Improved oral bioavailability of carbamazepine through the use of chemically modified cyclodextrins, Proc. 1st World Meeting APGI/APV, Budapest, 9/11 May 1995, pp. 627–628

[4] Pande, G.S., Shanfraw, R.F., Characterization of β-cyclodextrin for direct compression tabletting, Int. J. Pharm., 101, 71-80 (1994)

[5] Tasić, Lj., Milić–Aškrabić, J., Rajić, D., Kilibarda, V., Maksimović, M., Effects of β–cyclodextrin on physico–pharmaceutical characteristics and dissolution rate of Etodolac in tablets, Eur. J. Pharm. Biopharm, 1996, in press

[6] Duchene, D. and Wouessidjewe, D., Physicochemical Characteristics and Pharmaceutical Uses of Cyclodextrin Derivates, Pharm. Tech. Inter. 2 (N°6), 21–29 (1990)

THE EFFECT OF 2-HYDROXYPROPYL-β-CYCLODEXTRIN ON RELEASE OF NAPROXEN FROM PLURONIC F-127 GEL

N. TARIMCI, T. KILINÇ, D. ERMİŞ
Department of Pharmaceutical Technology, Faculty of Pharmacy, Ankara University, 06100, Tandoğan-Ankara / TURKEY

ABSTRACT

In order to improve the solubility and stability in aqueous topical dosage forms, inclusion complexes of naproxen with 2-hydroxypropyl-β-cyclodextrin were prepared by two methods kneading and ground mixture. IR spectroscopy, x-ray powder diffraction and differential scanning calorimetry were used to characterize the systems obtained. The solubility of naproxen were improved by inclusion in 2-hydroxypropyl-β-cyclodextrin and an aqueous gel (Pluronic F-127). The results indicate that , the apparent release rate and permeation of naproxen was significantly increased by the naproxen/2-hydroxypropyl-β-cyclodextrin complexations.

1. INTRODUCTION

2-Hydroxypropyl-β-cyclodextrin (HPBCD) is a new derivative of β-cyclodextrin (BCD) which exhibits high water solubility. The nonsteroid antiinflammatory (NSAI) drugs, which are generally slightly soluble in water. They usually cause some from of gastrointestinal ulceration and bleeding. So most of them have been used widely in topical systems.
Pluronic F-127 (P.F-127) is a non-toxic polyoxy propylene-polyoxyethylene surface active block polymer with an avarage molecular weight of 11500 (1). Reverse thermal gelation is one of the characteristics of aqueous solutions of this compound, eg. 20-25% P.F-127 gels are fluid at refrigerator temperature (4-5°C), but are highly viscous gels at room temperature. These aqueous clear gels appear to have a potential as topical drug delivery systems.
The major objective of the study was to increase the aqueous solubility of the Naproxen (NAP) which is a NSAI drug with HPBCD and to investigate the release of NAP from the complexes in P.F-127 gel.

J. Szejtli and L. Szente (eds.), Proceedings of the Eighth International Symposium on Cyclodextrons, 453–456.

2. MATERIALS AND METHODS

2.1. Materials

Naproxen and 2-hydroxypropyl-β-cyclodextrin were kindly supplied by Abdi İbrahim Pharm. Co. Ltd. (Turkey) and Chinoin Pharmaceutical and Works (Hungary) , respectively. Pluronic F-127 (BASF).

2.2. Methods

2.2.1. Solubility studies

Solubility measurements were carried out according to the method of Higuchi and Conners (2). Excess amount of NAP was added to aqueous solutions containing various concentrations of HPBCD, and were vigorously shaken at 37±0.5°C for 24h to attain the equilibrium. Aliquots were withdrawn, filtered suitably diluted and analyzed spectrophotometrically for their NAP content.

2.2.2. Preparation of physical mixture

NAP/HPBCD physical mixture was prepared in 1:1 molar ratio, mixing thoroughly both components until a homogeneous mixture was obtained.

2.2.3. Preparation of inclusion compounds

Inclusion compounds between NAP and HPBCD in 1:1 molar ratio were prepared by two methods kneading and ground mixture.

2.2.4. Characterization of the systems

Infrared spectroscopy (KBr disk method) , differential scanning calorimetry (DSC) and x-ray powder diffraction were used to characterize the systems obtained.

2.2.5. Preparation of the dermal form

The P.F-127 gel 20% was prepared by the cold process (3): An appropriate amount of P.F-127 was slowly added to cold distilled water (5-10°C) under constant agitation, after which the dispersion was stored overnight in a refrigerator. With time a clear, viscous solution formed appropriate amount of NAP or NAP/HPBCD complexes were then added to the cold solutions, and the system incubated in the gel form at 30°C until the clarity was formed. The formulations contained 1% NAP.

2.2.6. Release studies

The release of NAP or its complexes was determined using modified Franz cell method with a cellophane membrane that was previously allowed to hydrate in distilled water for at least 2h. The receptor compartment consisted in an aqueous solution buffered at pH 7.4. At pH 7.4 of the receptor phase, sink conditions were assured. It was stirred with a magnetic stirring bar in a constant temperature water bath maintained at 37+0.5°C. The samples were withdrawn at appropriate time intervals. Each sample was assayed spectrophotometrically for their NAP content at 331 nm. All experiments were carried out in triplicate.

3. *RESULT AND DISCUSSION*

3.1. *Solubility Studies*

The solubility of naproxen increased with increasing concentration of CD, showing an A_L type solubility curve. Similar results had previously been reported by Loftsson and co-workers (4).The solubility results are seen in Table 1. However, there does not seem to be any significant difference in the improvement caused by the complex compared with that resulting from the physical mixture.

Table 1. Solubility of Naproxen and their complexes with HPBCD at 37°C after 48h agitation (n:3)

Samples	*Solubility (mg/ml)*
NAP	0.055
Physical mixture.HPBCD	0.380
Complex. HPBCD (Kneading)	0.432
Complex. HPBCD (Ground Mixture)	0.379

3.2. *Naproxen / CD complex characteristics*

The IR spectras of solid complexes and its physical mixture did not show any significant difference between the main naproxen absorption bands on the one hand and the complexes and physical mixture on the other.

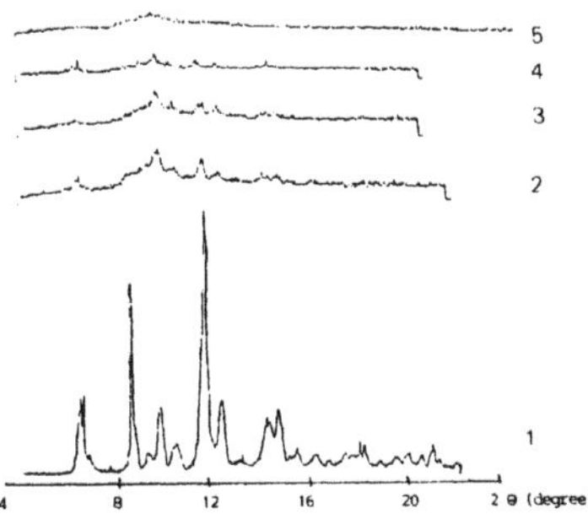

Fig. 1. Powder x-ray diffraction NAP and NAP/HPBCD complexes (1) NAP, (2) Phy. mixt., (3) Kneading, (4) Ground mixt. (5) HPBCD

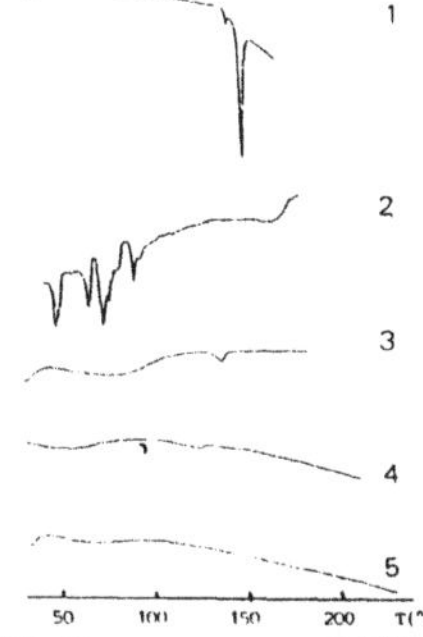

Figure 2. DSC Thermogram of NAP and its HPBCD preparations (1) NAP (2) Phy. mixt., (3) Kneading, (4) Ground mixt. (5) HPBCD

Fig. 1 shows the powder x-ray diffraction patterns of naproxen/CD complexes in comparison with the corresponding physical mixtures at the same molar ratio. As shown in the same figure, the x-ray diffraction patterns of the NAP / HPBCD complexes showed that the degree of crystallinity of NAP was apparently changed.

These data are indicative of the transformation of NAP from the crystalline to the amorphous state by formation of an inclusion complex with HPBCD.
The thermograms concerning the complexes of naproxen and HPBCD are shown in Fig. 2. DSC curve of NAP exhibited an endothermic peak at 151°C (melting). These two peaks disappeared in the case of the complex prepared by kneading and ground mixture which indicates the formation of an amorphous complex between HPBCD and NAP.

3.3. Analysis of dermal form

The release behaviour of CD complexes from P.F-127 gel was examined and compared with that of NAP alone. Fig. 3 shows the amount of NAP released from gel bases containing NAP and its CD complexes as a function of the square root of time. It is evident that the release rate of NAP was significantly improved by inclusion complexation, particularly in kneading method. The linearities of the plots may indicate that the release of NAP is diffusion controlled (5).
The enhanced release of NAP from the complexes may be due to the faster dissolution of the complexes and lower binding affinity of the complexes in the gel base.

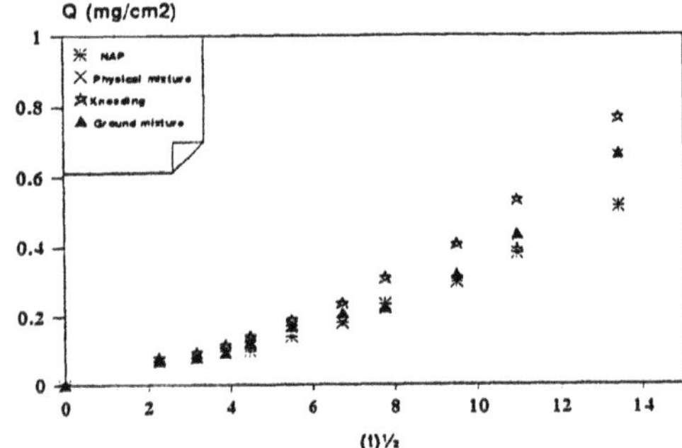

Fig. 3. The release profiles of naproxen from Pluronic F-127 gel

4. CONCLUSION

The preliminary results suggest that HPBCD complexation may provide good topical bioavailability of NAP. However, these results must be confirmed by tests of cutaneous penetration and invivo availability.

REFERENCES

1- BASF Wyandotte Corp., Industrial Chemicals Group, Pluronic polyols - toxicity and irritation data. Publications No. OS-3012 (765), Wyandotte, Mich.
2- Higuchi, T. and Connors, K., Phase solubility techniques, *Anal. Chem. Instr.*, *4*, 117-212 (1965)
3- Schmolka, I. R., Artificial Skin I. Preparation and properties of Pluronic F-127 gels for treatment of burns. *J. Biomed. Mater. Res.*, *6*, 571-582 (1972)
4- Loftsson, t., Olafsdottir, B.J., Frioriksdottir, H., Jonsdottir, S., Cyclodextrin complexation of NSAIDs : physicochemical characteristics, *European J. Pharm. Sci.*, *1*, 95-101 (1993)
5- Higuchi, W.I., Analysis of data on the medicament release from ointments, *J. Pharm. Sci.*, *51*, 802-84 (1962)

THE EFFECTS OF CYCLODEXTRINS ON THE INCOMPATIBILITY DEXAMETHASONE-POLYMYXIN B IN SOLUTION

Echezarreta-López, M.; Vila-Jato, J.L.; Torres-Labandeira, J.J.
Pharmaceutical Technology Department. School of Pharmacy. University of Santiago de Compostela. Spain

ABSTRACT

The inclusion complex formation of dexamethasone sodium phosphate with dimethyl-ß-cyclodextrin and hydroxypropyl-ß-cyclodextrin was studied by nuclear magnetic resonance technique. The structure of both inclusion complexes was found to be similar. Polymyxin B sulphate does not affect the formation of these inclusion complexes. This fact was found useful to in order avoid the interaction produced between dexamethasone and polymyxin B in ophthalmical preparations.

1. INTRODUCTION

Water-soluble derivative of anti-inflamatory steroids such as dexamethasone sodium phosphate (DX) exist as anions and often form water-insoluble complexes with basic antibiotics (1). During formulation of an ophthalmic solution containing DX and polymyxin B sulphate (PB) a precipitate is formed, due to the interaction between both drugs (2).
Cyclodextrins are well-known for their ability to form inclusion complexes with lipophilic compounds. Inclusion complex formation with drugs may result in an increase of solubility, stability, and/or bioavailability of drugs (3,4,5).
The objective of this work was to examine the influence of the formulation of DX as inclusion complexes with cyclodextrins - dimethyl-ß-cyclodextrin and hydroxypropyl-ß-cyclodextrin (d.s 2.7) - on its interaction with PB. The effect of PB on DX-CD interaction was analyze by the proton nuclear magnetic resonance (^{1}H-NMR) technique. Evaluation of the effect of inclusion complex formation on the incompatibility DX-PB in two dissolution media - water and phosphate buffer pH 7.4 - was carried out by turbidity measurements.

2. MATERIALS AND METHODS

2.1 Materials

Dexamethasone sodium phosphate (DX) and polymyxin B sulphate (PB) were donated by Laboratorios Cusí, Barcelona (Spain) and dimethyl-ß-cyclodextrin (DIMEB) and hydroxypropyl-ß-cyclodextrin (d.s.2,7) (HPBCD) were purchased from Cyclolab, Budapest (Hungary).

2.2 ^{1}H-NMR studies

Nuclear magnetic resonance spectra were obtained with a Bruker WN 300 spectrometer at 25°C. Samples were dissolved in deuterated water. The internal reference was the peak due to small amounts of DHO and H_2O present as impurity assigned a value δ= 4.8 ppm.

J. Szejtli and L. Szente (eds.), Proceedings of the Eighth International Symposium on Cyclodextrons, 457–460.

2.3 Turbidimetry studies:

2.3.1 Interaction DX-PB

The intensity of turbidity produced by mixing varying volumes of DX (0,1%) and PB (0,1%) was measured spectrophotometrically at 420 nm at 25°C. Measurements were made after mixing at one minute intervals and the highest reading was recorded.

2.3.2 Evaluation of the effect of cyclodextrins on incompatibility DX-PB.

Increasing volumes of a cyclodextrin solution (0,5%) were added to a solution of DX and PB (1:1) (w/w). Turbidity measurements were carried out as above.

3. RESULTS AND DISCUSSION

The insertion of a guest molecule into the CD cavity is clearly reflected in changes in ^{1}H-NMR chemical shift values (6,7).

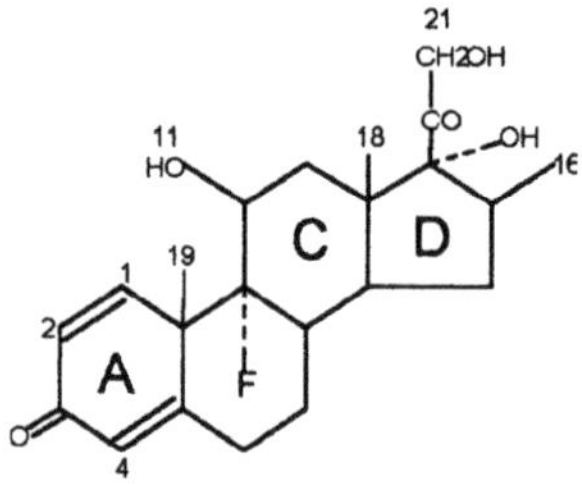

Figure 1. Structure molecular of dexamethasone

Figure 2 shows the chemical shifts of the protons corresponding to the ring A of DX (C_1-H, C_2-H, C_4-H) in the presence of different concentrations of DIMEB and HPBCD. The signals of all protons of DX molecule vary until the 1:1 molar ratio in the DX/CD complex is obtained.(8).

The changes undergone by these protons are shifts upfield. As described elsewhere (9,10) these changes suggest that in the part of the molecule of DX that interacts most strongly with, and is encapsulated by both CDs molecules.

In spite of the fact that above results indicate a similar structure in both inclusion compounds, chemical shifts of other protrons show different interactions between both components.
In fact, methyl groups of DX molecule shifted downfield with increasing HPBCD concentration. These changes indicated a different inclusion because of the different cavity size (10,11).

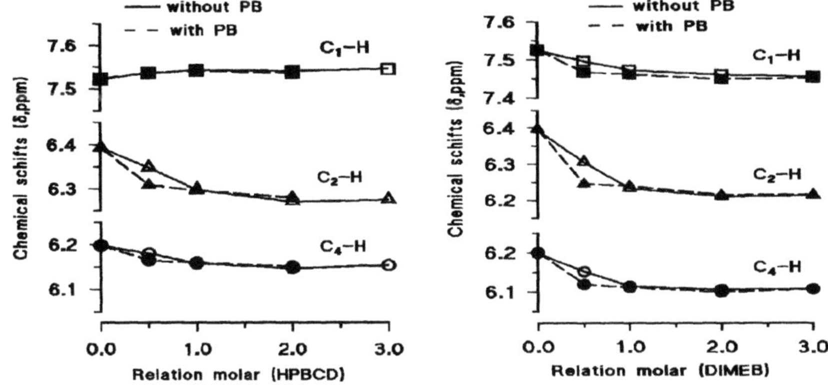

Figure 2. Representation of changes of chemical shifts of protons of dexamethasone molecule in presence of increase amounts of cyclodextrin

Once the structure was characterize, the influence of PB on the DX inclusion complex formation was checked. In order to do this, ^{1}H-NMR spectrum were repeated introducing a constant amount of PB in the solutions.
Figure 2 shows the chemical shifts of DX ring A protons. It is evident that the presence of PB in the solution does not affect significantly the DX complex.
In order to study the incompatibility of DX and PB in solution, turbidimetric measurements produced by the insoluble complex DX/PB were carried out (figure 4 y 5).

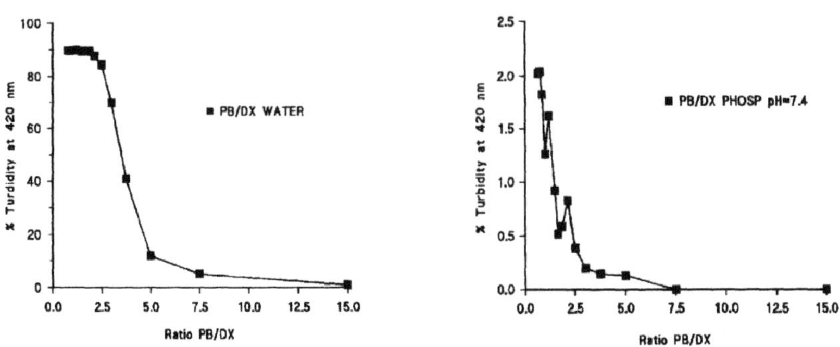

Figure 3. Results of turbidity measurements at 420 nm of dexamethasone sodium phosphate-polymyxin B sulphate system

These studies were made in two media, water and phosphate buffer solution pH 7.4, because the complex formation was found to be affected by the composition of the dissolution (2). The effect of the medium is clear because in water 90% turbidity can be obtained with a small ratio of PB/DX, however in the buffer solution the highest percentage is 2%. In both cases, complex formation decreases with the ratio of PB/DX.

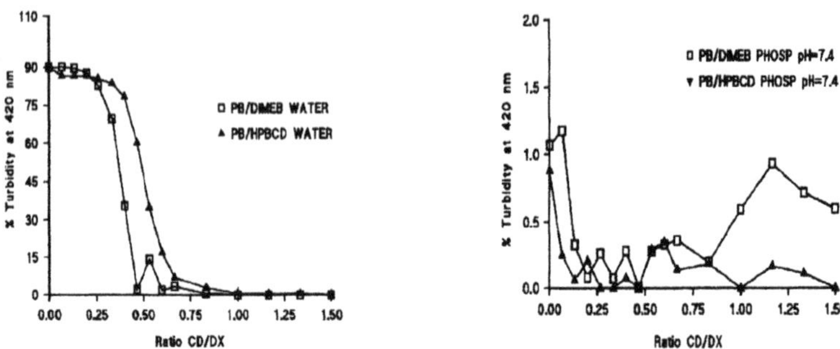

Figure 4. Results of turbidity measurements at 420 nm of dexamethasone sodium sulphate-polymyxin B sulphate system in presence of increases amounts of cyclodextrin.

Figure 4 shows the effect of DIMEB and HPBCD on complex formation. In these studies, a ratio PB/DX equal to 1:1 (w/w) was used because its turbidity it has been found to be much higher. In water, turbidity disappeared with a 1:1 w/w ratio CD/DX. The effect is more variable in a buffer solution, but a decrease in the turbidity was found.

4. CONCLUSION

* Dexamethasone form inclusion complexes with DIMEB and HPBCD in dissolution. In both cases, it is ring A of DX molecule structure which is included in cyclodextrin cavity.
* The DX inclusion complex formation on CDs is not affected by the presence in the dissolution of polymyxin B sulphate.
* DX inclusion complex formation protects this drug from the interaction with PB. This effect is produced from a ratio 1:1 w/w CD/DX, and it could be useful in the preparation of ophthalmic formulations.

ACKNOWLEDGEMENTS

This work was supported by laboratories Cusí, S. A. (Spain)

REFERENCES

(1) McGynity, Brown, R.L.; J.Pharm. Sci., **64**, 1528, (1975).
(2) Aggag, M,; Khahil,S.A.H.. Manufacdurating Chemist and Aerosol News, pp. 43, (1977).
(3) Duchêne,D.; Debruères, B; Vaution, C.. S.T.P. Pharma, **1**, 37,(1985).
(4) Racz, I; Plachy, J.; Szabo-Tabori, M.; Stadler-Szöke, A.; Wikmon, M.; Szjtli, J Acta Pharm. Hung., **54**,154, (1984).
(5) Frömming, K.H.; Szejtli,J. Cyclodextrins in pharmacy. Ed. Kluwer. Academic Publishers, pg.134, (1994).
(6) Ueda, H; Nagai, T. Chem. Pharm. Bull., **28**, 1415-1421, (1980).
(7)Takkar, A. L.; Demarco, P. V.. J. Pharm. Sci. **60**, 652-653, (1971).
(8) Djedaïni, F.; Lin, S-Z.; Perly, B.; Wouessidjewe, D.. J, Pharm. Sci., **79**, 643-646, (1990).
(9) Uekama, K.; Fijinaga, T.; Hirayama, F; Otagiri, M. Yamasaki, M. Int. J. Pharm., **10**, 1-15, (1982).
(10) Torres-Labandeira, J.J. Echezarreta-López, M., Santana-Penin, L.; Vila-Jato, J.L. Eur. J. Pharm. Biopharm., 39, 255-259,(1993).
(11) Frömming,K.H.; Fridich, R.; Medinert, K.E.. Eur. J. Pharm. Biopharm., **39**, 148-152,(1993).

INVESTIGATION OF THE EFFECT OF β-CD ON IN VITRO RELEASE OF KETOCANOZOLE FROM DIFFERENT GEL BASES

N.ÇELEBİ[1], Z.İ. GÜL[1], F.OCAK[1] S.YILDIZ[2] and F.ACARTÜRK[1]

[1]Dept. of Pharm. Tech., Fac.of Pharm.,Univ.of Gazi, 06330 Etiler-Ankara-TURKEY
[2]Dept. of Pharm.Microbiol., Fac.of Pharm.,Univ. of Ankara, 06100 Tandoğan-Ankara-TURKEY

ABSTRACT

The solid complex of ketocanozole(KET) with ß-CD in a molar ratio of 1:1 was prepared by kneading method.The kneaded mixture of KET with ß-CD in solid state was confirmed by differential scanning calorimetry (DSC) and X-ray diffractometry techniques.
The release of KET and its kneaded and physical mixtures from gel bases was studied using a modified Franz diffusion cell with a cellophane membrane in pH 5.0 buffer solution at 37±0.5°C. In addition, release charecteristics were compared with commercial products. On the other hand, antimycotic activity of KET and its mixtures was investigated by inhibition zone measurements of Candida albicans.
It was found that the release of KET from gel bases was significantly increased by the KET/ß-CD complexation by kneading method. Microbiological tests showed that KET/ß-CD complex was much more effective than that of physical mixture, ketocanozole and ß-CD alone against C.albicans ATCC 10231.

1. INTRODUCTION

Cyclodextrins are cyclic oligosaccharides which are known to form inclusion complexes with many lipophilic drugs, thus changing the physicochemical and biopharmaceutical properties of the drugs. Antimycotic imidazole derivatives such as econazole and miconazole are lipophilic drugs, and complexation of these drugs with cyclodextrins has been reported [1-3]. Recently, ß -CD and its derivatives were studied for penetration enhancement of the drugs [4,5].

The objective of the present study was to examine the effect of ß-CD on in vitro release of ketocanozole (KET) which is an antifungal imidazole derivative from different gel bases. In addition, the influence of complexation with ß-CD on antimycotic activity of KET was investigated.

2. MATERIALS AND METHODS

2.1. Materials

Ketocanozole (KET) and ß -CD were kindly supplied by İlsan-İltaş Pharmaceutical Co. Ltd (Turkey) and Cyclolab (Hungary), respectively. Carbopol 940,971P,Noveon AA1 and Polyvinyl alcohol(Maurol 40-80) were the gifts from B.F.Goodrich and Hoechst, respectively. Commercial products (1,2) were kindly provided by Eczacıbaşı and İlsan-İltaş. Candida albicans strains (ATCC 10231) were used for microbiological assays. Sabouraud Dextrose Agar (SDA)(Difco) was used for cultivation of the microorganisms and Yeast Morphology Agar (YMA)(Difco) was employed for the inhibition zone measurements.

J. Szejtli and L. Szente (eds.), Proceedings of the Eighth International Symposium on Cyclodextrons, 461–464.

2.2. Methods

2.2.1. Phase -solubility Studies

The phase-solubility studies were carried out according to the method described by Higuchi and Connors [6] in buffer solution in pH 5.0 at 37 ± 0.5 °C .

2.2.2. Preparation of KET / ß -CD Complex

The solid complex of KET with ß-CD in 1:1 molar ratio was prepared by kneading method. The physical mixture was also prepared in the same molar ratio by simple mixing.

2.2.3. Characterization of Complex

The solid complex of KET/ ß -CD was confirmed by differential scanning calorimetry (DSC-Shimadzu DT 40; temparature speed 10° C /min; scanning speed 10 mm/min) and X-ray powder diffractometry (Philips, Ni-filtered Cu-K_α ; radiation voltage 40 kV; current 30 mA; time constant 2 s; scanning speed 2 ° / cm ; Bragg angle 2 θ).

2.2.4. Gel Formulations

KET and its physical mixture and solid complex was incorparated with different gel bases (gel I, gel II, gel III). The formulations are given in Table 1.

Table1. The Content of Gel Formulations

Ingredients (g)	Gel I	Gel II	Gel III
Ketocanozole	2	2	2
Carbopol 934P	2	-	-
Carbopol 971P	-	0.50	-
Noveon AA-1	-	0.50	-
Polyvinyl alcohol (Maurol 40-80)	-	-	20
Triethanolamine	1.7	0.65	-
Glycerol	-	50	-
Propylene glycol	-	-	8
Distilled water	94.4	46.4	70

2.2.5. In vitro Release of Ketocanozole From Gel Bases

The in vitro release of KET and its kneaded and physical mixtures from gel bases was determined using a modified Franz diffusion cell with a cellophane membrane in pH 5.0 buffer solution at 37±0.5 °C . At appropriate intervals, samples were withdrawn and replaced by fresh test fluid. Samples were assayed spectrophotometrically for KET content at 224 nm. The experiment was carried out in triplicate.

2.2.6. Microbiological Assay

Solutions of KET (4mg/ml), KET/ß-CD complex (13.75 mg/ml), KET/ß-CD physical mixture (12.58 mg/ml) and ß-CD (9.75 mg/ml) in sterile distilled water were prepared.
Strains were cultured for 24 h on SDA. Fungal suspensions were prepared to contain approximately 10^5 cells per ml in sterile distilled water. One ml of the suspension were mixed with 45 ml melted (45°C) YMA and poured into petri dishes (∅12cm). Four holes with a diameter of 6mm were punched on each agar plates and the holes were filled with 50 µl of the solutions to be tested. Inhibition zones were read after 24 hours incubation at 35 °C.

3. RESULTS AND DISCUSSION

The solubility of drug increase linearly as a function of the ß-CD concentration and the solubility curve can be generally classified as type A_L. The apparent stability constant of KET/ß-CD complex, as calculated from the linear portion of the solubility plot, amounted to 581 M^{-1}. It was assumed that KET formed a complex with ß-CD in a molar ratio of 1:1.
The in vitro release profiles of KET from different gel bases are shown in Fig.1-3. The general rank order for in vitro release of KET was found to be: gel I> gel II> gel III .

Table 2. Assessment of Zero Order Kinetics

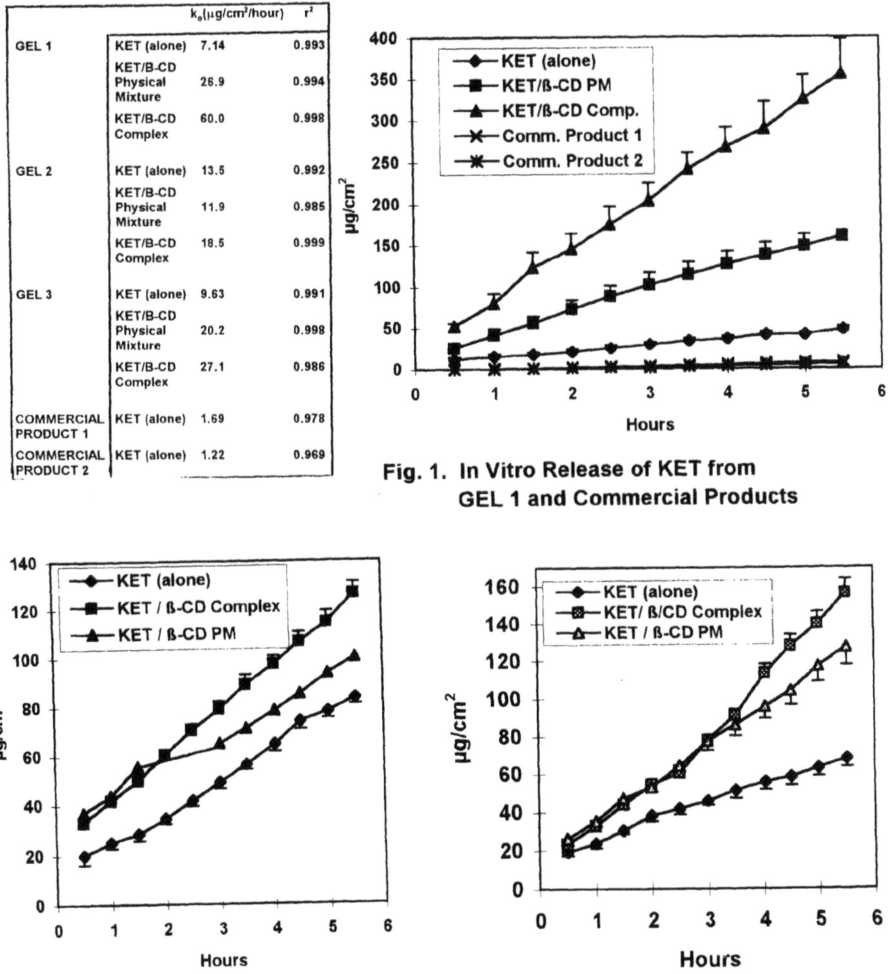

		k_o(µg/cm²/hour)	r^2
GEL 1	KET (alone)	7.14	0.993
	KET/B-CD Physical Mixture	26.9	0.994
	KET/B-CD Complex	60.0	0.998
GEL 2	KET (alone)	13.5	0.992
	KET/B-CD Physical Mixture	11.9	0.985
	KET/B-CD Complex	18.5	0.999
GEL 3	KET (alone)	9.63	0.991
	KET/B-CD Physical Mixture	20.2	0.998
	KET/B-CD Complex	27.1	0.986
COMMERCIAL PRODUCT 1	KET (alone)	1.69	0.978
COMMERCIAL PRODUCT 2	KET (alone)	1.22	0.969

Fig. 1. In Vitro Release of KET from GEL 1 and Commercial Products

Fig. 2. In Vitro Release of KET from GEL 2

Fig. 3. In Vitro Release of KET from GEL 2

The release of KET was increased by ß -CD addition as compared with that of physical mixture and drug alone for all gel formulations. In addition, in vitro release of KET was compared with the commercial products (1,2). As shown in Fig.1, the release of KET from gel I was significantly higher than those of commercial products. The increase in drug release may be explained increasing solubility of KET with ß -CD.

The kinetic assessment of release data showed that the best kinetic fit was zero order kinetics (Table 2). The release rate constant of KET/ ß-CD complex in gel I was the highest (kr_o = $\mu g.cm^{-2}.h^{-1}$).

Differantial scanning calorimetry (DSC)and X-ray diffractometry analysis were performed to investigate the interaction between KET and ß-CD. According to the DSC thermograms, the intensity of the endothermic peak at 146 ° C decreased in the kneaded mixture as compared with the physical mixture and KET alone; however, the peak did not disappear completely. This may be due to the energy level of the kneaded mixtures being near to that of the physical mixture. The intensities of the diffraction peaks of the kneaded mixture was reduced vs that of KET alone. However, a weak interaction did occur between drug and ß-CD and the drug powder was dispersed as separate crystals in the kneaded mixtures.

The antimycotic activity of the complex was superior to the activity of ketocanozole, the physical mixture of KET and ß-CD.

4. CONCLUSION

In conclusion, the release of KET from gel bases increased by complexation. But, differantial scanning calorimetry and X-ray diffractometry data indicate that there was a weak interaction between KET and ß-CD. Therefore the kneading method was not suitable for the preparation of KET/ß-CD inclusion complex.

ACKNOWLEDGEMENTS

The authors would like to thank CYCLOLAB(Hungary) for supplying ß-cyclodextrin . Thanks also go to Ilsan-Iltas Pharmaceutical Company for ketocanozole. B.F Goodrich and Hoechst are also acknowledged.

REFERENCES

[1] Mura,P., Liguori,A., Bramanti,G., Bettinetti,G., Campisi,E., Faggi,E., Improvement of dissolution properties and microbiological activity of miconazole and econazole by cyclodextrin complexation, ***Eur. J. Pharm. Biopharm.***, **38,** 119-123 (1992)

[2] Pedersen,M.,Edelsten,M., Nielsen,V.F., Scarpellini,A., Skytte,S., Slot,C., Formation and antimycotic effect of cyclodextrin inclusion complexes of econazole and miconazole, ***Int. J. Pharm.***, **90**, 247-254 (1993)

[3] Pedersen,M., Effect of hydrotropic substances on the complexation of clotrimazole with ß-cyclodextrin, ***Drug Dev. Ind. Pharm.***, **19**, 439-448, (1993)

[4] Vollmer, U., Müller,B.W., Peeters, J., Mesens, J., Wilffert,B., Peters,T., A study of the percutaneous absorption-enhancing effects of cyclodextrin derivatives in rats, ***J. Pharm. Pharmacol.***, **46**, 19-22 (1994)

[5] Çelebi, N., Kışlal,Ö., Tarımcı,N., The effect of β-cyclodextrin and penetration additives on the release of naproxen from ointment bases, ***Pharmazie,*** **48**, 914-917 (1993)

[6] Higuchi, T., Connors, K.A., Phase-solubility techniques,., ***Adv. Anal. Inst.,*** **4**, 117-212 (1965)

DRUGS IN CYCLODEXTRINS IN LIPOSOMES: A NOVEL APPROACH TO DRUG STABILITY AGAINST PHOTOCHEMICAL OXIDATION.

Y.L. LOUKAS*, V. VRAKA* & G. GREGORIADIS
Centre for Drug Delivery Research, School of Pharmacy, University of London, 29-39 Brunswick Square, London WC1N 1AX, UK
*Present address: Riga Ferreou 21, Ano Ilioupolis, 163 43 Athens, GREECE. E-mail: ylloukas@compulink.gr

ABSTRACT

Sodium ascorbate (SA) is oxidized in aqueous solutions by reaction with the dissolved oxygen and this process is accelerated by the presence of light (photochemical oxidation). In the present study we employed a novel system based on the combination of dehydration-rehydration liposomes, cyclodextrins and sunscreen agents in order to improve the stability of the drug. Analysis of various formulations revealed that a DRV liposomal formulation containing the α-cyclodextrin inclusion complex of the vitamin and incorporating the water soluble light absorber sulisobenzone and the lipid soluble light absorber oil red O provided maximum protection, increasing the half-life of SA from 56 min to 112.13 h.

1. INTRODUCTION

Drugs are usually sensitive to a number of external factors, such as oxygen, light, temperature e.t.c. This results in difficulties either in the application of drugs or their storage. Although there have been made attempts to stabilize sensitive drugs, researchers are still seeking ways to improve further the stability of such drugs. Data from previous work[1,2,3], indicated that photodegradation of cyclodextrin-included riboflavin by UV light was greatly delayed when the vitamin was entrapped in dehydration-rehydration multilamellar liposomes incorporating in their bilayers the light absorbers oil red O, oxybenzone and dioxybenzone and the antioxidant β-carotene.

In the present study multilamellar liposomes prepared by the dehydration-rehydration method[4] were composed of equimolar egg phosphatidylcholine (PC) and cholesterol and contained sodium ascorbate[6,7] (SA) as such or in the form of an α-cyclodextrin (αCD)

J. Szejtli and L. Szente (eds.), Proceedings of the Eighth International Symposium on Cyclodextrons, 465–470.

complex together with and incorporating a number of light absorbers in the aqueous phase and/or the lipid bilayer. Analysis of the various formulations revealed that a DRV liposomal formulation containing the α-cyclodextrin inclusion complex of the vitamin and incorporating the water soluble light absorber sulisobenzone and the lipid soluble oil red O provided maximum protection, increasing the half-life of SA from 56 min to 112.13 h.

2. MATERIALS AND METHODS

2.1 Materials

SA, oil red O, sulisobenzone and cholesterol were obtained from Sigma Chemical Company (Poole, Dorset, UK); αCD was purchased from Wacker Chemie (GmbH, Munich). PC was from Lipid Products (Nuthill, Surrey, UK). All other reagents were of analytical grade. Double distilled water was used throughout.

2.2 Methods

Measurement of SA in various preparations, the kinetics of its degradation and the assay of components used for photoprotection were carried out in a Compuspec UV/visible spectrophotometer (Wallac) connected to a personal computer.

2.3 Preparation of the SA:αCD inclusion complex and entrapment of agents into liposomes

The inclusion complex of SA with αCD was prepared by the freeze-drying method.
Entrapment of SA as such or as an αCD inclusion complex into liposomes (aqueous phase) composed of PC and equimolar cholesterol was carried out (by a procedure which produces multilamellar dehydration-rehydration vesicles DRV[4,5]) in the absence or presence of the lipid soluble UV absorber oil red O (entrapped in the lipid phase). In some experiments, the latter was supplemented with the water soluble light absorber sulisobenzone, co-entrapped with the SA:αCD in the liposomal aqueous phase.

2.4 Exposure to UV radiation

A number of SA formulations were tested for their effectiveness in protecting the vitamin against air oxygen and UV light. Solutions or liposomal suspensions of SA (3 ml) in cuvette cells (1x1x4 cm^3; 1 cm path length) were exposed sideways at 24 °C ± 1 to UV light from a distance of 3 cm using a Black-Ray (365 nm) UVB lamp (San Gabriel, USA) with 6W rating and 460 $\mu W\ cm^{-2}\ dm^{-1}$ intensity. Liposome-entrapped SA (or free SA or SA:αCD complex) samples from the cuvettes removed at time intervals, were first solubilised with isopropanol and then appropriately diluted with distilled water and assayed for SA spectrophotometrically.

3. RESULTS AND DISCUSSION

3.1 Determination of entrapped materials in DRV liposomes and yield of entrapment

The determination of materials incorporated in the various DRV liposomes (Table 1) was performed by derivative UV spectrophotometry. The zero order spectra of the materials exhibited extensive overlaps (Fig. 1) when materials were present in the same preparation. Therefore, the use of the second order derivative (D_2) was deemed necessary, as it also provided good resolution and good S/N ratio.

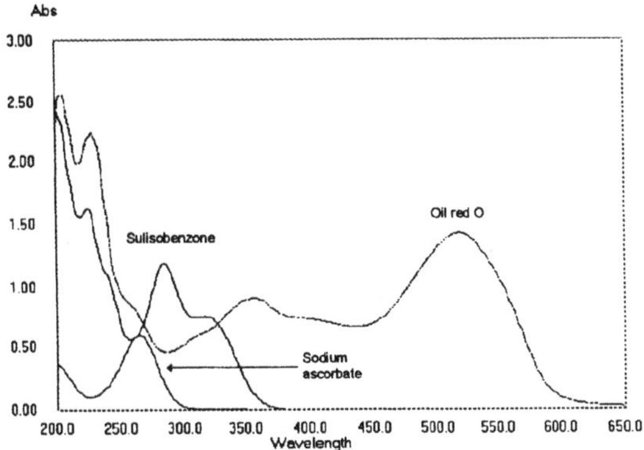

Fig. 1 Zero order spectra of sodium ascorbate, sulisobenzone and oil red O

In the preparations where the water soluble light absorber (sulisobenzone) was present together with the SA (or its inclusion complex) (Table 1, DRV 2 and 6) the second order derivative of both spectra (Fig. 2) gave at 222 nm a clear maximum for SA and a zero reading for sulisobenzone. Furthermore, it was also observed that at 353 nm the D_2 spectrum of sulisobenzone exhibited a clear maximum while the D_2 spectrum of SA showed zero absorbance. Thus, by using these two maxima at the two wavelengths we were able to determine the % entrapment values shown on Table 1 for DRV 2 and 6. The spectra of SA in both pure and complexed form were almost identical.

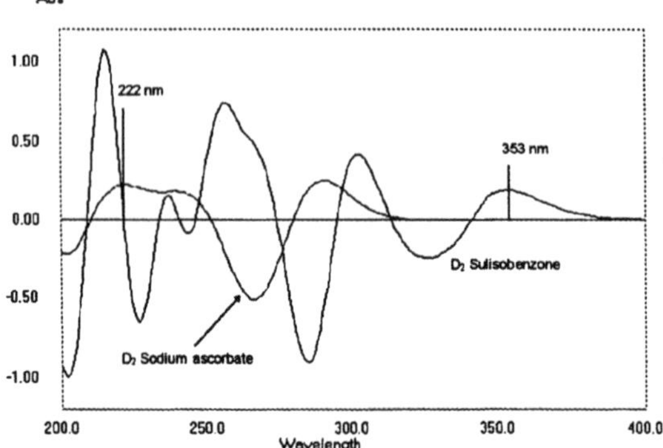

Fig. 2 Second order derivative spectra of SA and sulisobenzone.

Similarly, in the formulations in which the vitamin was present together with the lipid soluble light absorber (oil red O) (Table 1, DRV 3 and 7), the second order derivative of both spectra (Fig. 3) gave a clear maximum reading at 583 nm for oil red O (at this wavelength SA does not absorb). For the calculation of the pure or complexed SA, the maximum at 223 nm was used because at this wavelength the D_2 of oil red O spectrum is zero.

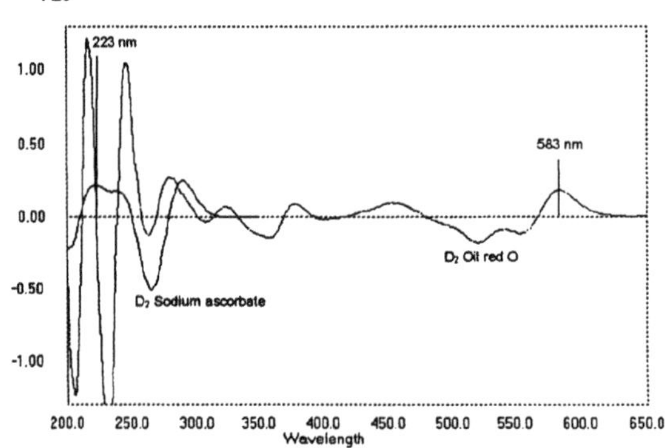

Fig. 3 Second order derivative spectra of SA and oil red O.

Finally, for the preparations in which all materials were present in the same formulation (Table 1, DRV 4 and 8), the second order derivative for all three spectra did not provide clear maxima for any of the materials whilst the absorbancies of the other two compounds were zero. Also, higher order of derivatization did not provide good S/N ratio. Therefore, in this case the lipid soluble oil red O was extracted from the other DRV components by using chloroform. Oil red O was then calculated by the zero order spectra in the organic phase and the pure or complexed SA remaining together with sulisobenzone in the aqueous phase were calculated as described above.

As judged by UV spectrophotometry of a control "empty" DRV preparation, interference by the DRV components PC and cholesterol in the measurement of the entrapped materials was nil.

Entrapment values (Table 1) for materials in the DRV preparations were calculated according to the equation % entrapment = $[(A_0 - A)/A_0] \times 100$, where A_0 is the total absorbance of the materials in the pellets and pooled supernatants and A is the absorbance of the non-entrapped materials in the pooled supernatants.

3.2 Kinetics studies

Degradation of SA as such or liposome-entrapped followed apparent first-order kinetics expressed as: $C=C_oe^{-kt}$ (eq. 1). This equation (in exponential form) can be transformed logarithmically to a linear form: $\ln C=\ln C_o-kt$ (eq. 2), where C_o is the initial concentration of SA, C is the concentration of SA after t min irradiation and k is the first order degradation rate constant. The graphic representation of eq. 2 is linear with the slope of the line giving the degradation rate constant k from which the corresponding half life ($t_{50\%}$) of SA exposed to light can be then calculated. The ability of the various liposomal preparations to protect SA from being oxidized was defined as the stabilization ratio (k_F/k_L), where k_F and k_L are, respectively, the photodegradation rate constants of the free SA in aqueous solution and of SA in the various liposomal formulations studied.

3.3 Stabilization of liposome-entrapped SA

A number of SA formulations described in Table 1 were tested for their effectiveness in protecting the vitamin against UV light. Results in Table 2 indicate that entrapment of SA into DRVs in the absence of other materials, increases its half-life from about 1 h (observed for SA in solution) to 9 h. Half-life was increased to 16 h by the addition of sulisobenzone in the aqueous phase of the DRV and to 60 h when oil red O alone was present (in the bilayers). There was a further increase in the half-life of SA (to 81 h) when oil red O in the bilayers was supplemented with sulisobenzone in the aqueous phase.
In experiments with αCD-included SA, a half-life of 2 h for SA was found when the complex was exposed to UV light (results not shown). This was improved to 17 h when the SA:αCD complex was entrapped into DRV (Table 2; DRV 5). A much greater half-life (91.6 h) was achieved when oil red O was also present in the bilayers (DRV 7). As with DRV containing free SA, stabilization was increased even more (to 112 h) in the presence of sulisobenzone in the aqueous phase (Table 2; DRV 8). Thus, by a combination of an αCD inclusion complex entrapped in DRV liposomes and the use of light absorbers in the lipid bilayer and the aqueous phase, the half-life of SA was increased 120 fold.

TABLE 1. Entrapment values (% of material used) for various DRV preparations consisting of 1:1 PC:Cholesterol and the amounts of materials used for the entrapped materials.

DRV Preparations	SA (1.9 mmoles)	SA:αCD (1.9 mmoles)	Oil Red O (3.8 mmoles)	Sulisobenzone (1.9 mmoles)
DRV 1	71	-	-	-
DRV 2	30	-	-	35
DRV 3	62	-	55	-
DRV 4	28	-	51	31
DRV 5	-	35	-	-
DRV 6	-	15	-	20
DRV 7	-	31	49	-
DRV 8	-	14.5	45	18

TABLE 2. Degradation rates of SA in DRV preparations exposed to UV light

Sample	$t_{1/2}$ (h)	k (x 10^4 min^{-1})	Stabilization Ratio
SA in solution	1	124	-
DRV 1	9	12.6	10
DRV 2	16	7.19	17
DRV 3	60	1.91	65
DRV 4	81	1.42	87
DRV 5	17	6.52	19
DRV 6	37	3.12	40
DRV 7	91	1.26	98
DRV 8	112	1.03	120

4. CONCLUSION

In conclusion, sensitive agents may be protected from light induced oxidation (photochemical oxidation) by incorporating these into a cyclodextrin-liposome-based system. Our data indicate that optimal protection of SA is provided by DRV liposomes containing the αCD inclusion complex of the vitamin within their aqueous phase and the light absorbers oil red O and sulisobenzone in the lipid and aqueous phase respectively.

REFERENCES

[1] Loukas, Y., Jayasekera, P., Gregoriadis, G., Characterization and photoprotection of a model γ-cyclodextrin-included photolabile drug entrapped in liposomes incorporating light absorbers, *J. Phys. Chem.*, **99**, 11035-11040 (1995)

[2] Loukas, Y., Jayasekera, P., Gregoriadis, G., Novel liposome-based multicomponent systems for the photoprotection of photolabile agents, *Int. J. Pharm.*, **117**, 85-94 (1995)

[3] Gregoriadis, G., Loukas, Y., Stabilization of photosensitive materials, British patent GB94.1111.50 (1994)

[4] Kirby, C., Gregoriadis, G., Dehydration-rehydration vesicles: A simple method for high yield entrapment in liposomes, *Biotechnology*, **2**, 979-984 (1984)

[5] Ho, A.H.L., Puri, A., Sugden, J.K., Effect of sweetening agents on the light stability of aqueous solutions of L-ascorbic acid, *Int. J. Pharm.*, **107**, 199-203 (1994)

[6] Asker, A.F., Canady, D., Cobb, C., Influence of DL-methionine on the photostability of ascorbic acid solutions, *Drug Dev. Ind. Pharm.*, **11(12)**, 2109-2125 (1985)

SBE7-β-CD, A NEW, NOVEL AND SAFE POLYANIONIC β-CYCLODEXTRIN DERIVATIVE: CHARACTERIZATION AND BIOMEDICAL APPLICATIONS.

Valentino J. Stella

The Center for Drug Delivery Research and the Department of Pharmaceutical Chemistry, The University of Kansas, Lawrence, Kansas, 66047.

1. INTRODUCTION

β-Cyclodextrin (β-CyD) and some other CyDs have utility for solubilizing and stabilizing drug entities, however, they are nephrotoxic when administered by injection (parenterally) to animals. Therefore, the objective of our work was to identify, prepare and evaluate various CyD derivatives with superior inclusion complexation ability and maximal *in vivo* safety for various biomedical uses. A systematic study led to SBE7-β-CD, a polyanionic 2-, 3-, and 6-variably substituted sulfobutyl ether (average degree of substitution of 7) of β-CyD as a non-nephrotoxic derivative with superior properties over HP-β-CD, a competitor product [1,2].

This mini-review paper will focus on the issues that led to the identification of SBE7-β-CD as the candidate of choice [1,2], its safety features [3,4,5], its characterization by CE [6,7,8,9,10,11] and NMR, its various applications including its use as a solubilizer and stabilizer of pharmaceuticals [1,2,12,13,14,15], and its ability to improve drug delivery [16,17,18,19,20,21,22].

2. WHY SULFOBUTYL ETHERS?

The genesis for our interest in cyclodextrins goes back to the mid- to early-1980s when the need for new solubilizing agents was triggered by clinical concerns with the use of surfactants in parenteral products. Although the nephrotoxicity of parenteral cyclodextrins was previously reported, we decided to explore the use of charged cyclodextrins as potentially less renally toxic materials. We noted in initial studies that placing a charge very close to the cyclodextrin nucleus inhibited inclusion complexation [1] but by moving the charge further from the cyclodextrin nucleus, optimally in our hands by a butyl group, the binding capacity of the cyclodextrin was retained. The sulfonate group as chosen as the charged group of choice because it is metabolically more stable than other charged entities, is ionized at all relevant physiological pH values and the synthesis of the alkyl sulfonates could be performed economically. Two sulfobutyl ether derivatives of cyclodextrins were chosen for further study, SBE4-β-CD and SBE7-β-CD, sulfobutyl ether, sodium salt, derivatives variably substituted on the 2-, 3- and the 6-positions of β-cyclodextrin. The numbers four and seven refer to the average degree of substitution per cyclodextrin molecule. Earlier papers focused on the use of SBE4-β-CD, however, the more commercially viable derivative is SBE7-β-CD. SBE7-β-CD was chosen because of its

J. Szejtli and L. Szente (eds.), Proceedings of the Eighth International Symposium on Cyclodextrons, 471–476.

greater safety, high binding capacity and because it can be prepared free of unreacted β-cyclodextrin. Similar derivatives were prepared for α- and γ-cyclodextrins.

3. WHY GREATER SAFETY?

The acute safety of sulfoalkyl ether derivatives (SAE-β-CDs) of β-cyclodextrin, especially SBE4-β-CD and SBE7-β-CD have recently been documented [2,3,4,5]. Specifically, the parenteral safety of these derivatives was determined by survival of male mice post intraperitoneal (IP) injection, kidney histopathology and plasma urea nitrogen levels of mice determined 24 hours after injection, relative *in vitro* hemolytic potential and activated partial thromboplastin times (APTT). In addition, the 24-hour renal excretion behavior of the derivatives was measured. Where appropriate, the results obtained with these cyclodextrin derivatives were compared with results obtained for β-CyD and HP-β-CD. The SAE-β-CD derivatives did not produce mortality in mice following IP injection at doses exceeding 5.45 mmol/kg. No significant histological lesions were observed in the kidney tissue of mice receiving the cyclodextrin derivatives. The SAE-β-CD derivatives were excreted faster and to a greater extent than β-CyD and at rates comparable to HP-β-CD. The hemolytic potential of these derivatives was less than that of β-CD and comparable (for lower degrees of substitution) or better than HP-β-CD. The SAE-β-CD derivatives did not increase APTT clotting times indicating that these derivatives have no significant anticoagulant activity. It has been hypothesized that the greater safety of SBE4-β-CD and SBE7-β-CD could be due to their inability to form 1:2 complexes with cholesterol, except at very high concentrations of the SBE4-β-CD and SBE7-β-CD. Presumably this is due to coulombic repulsion between the polyanionic cyclodextrins. SBE7-β-CD as Captisol™ has recently undergone extensive safety studies and is currently being used in a number of clinical and preclinical studies. Other safety studies are ongoing.

4. CHARACTERIZATION OF SBE4-β-CD AND SBE7-β-CD.

Characterization of the sulfoalkyl ether derivatives of cyclodextrins, which had been reported earlier by Parmeter et al. [23] had primarily involved elemental analyses. We found that the materials defined by Parmeter were highly contaminated by unreacted β-CyD and hydroxybutane sulfonic acid. By modifying the synthetic procedure and cleaning the samples by ultrafiltration, we were able to isolate better defined materials that had low to negligible contamination with unreacted β-CyD. Furthermore, we were able to develop CE based indirect methods of analyses [6,8,9] that were able to check batches of SAE-β-CDs for purity and reproducibility. Recently, we isolated the mono-band, SBE1-β-CD, and using ion exchange chromatography, the pure mono-2-,3- and 6-sulfobutyl ethers [10]. These isomers provided valuable insight into the NMR spectra of the more complex mixtures (ongoing work) such as SBE4-β-CD and SBE7-β-CD as well as newer materials being developed in our laboratories.

5. SOLUBILIZATION AND STABILIZATION BY SBE4-β-CD AND SBE7-β-CD.

Numerous phase solubility studies have been performed in our laboratory comparing the complexation ability of various SAE-β-CDs to other modified and unmodified CyDs. We have generally found that SBE4-β-CD and SBE7-β-CD have inclusion complexation ability superior to HP-β-CD on a molar basis and comparable to the β-CyD with the added advantage of being more water soluble than β-CyD. For example, we performed a study to determine the role that charge might play in the interaction of charged and uncharged drugs with HP-β-CD and the anionically charged SBE7-β-CD [15]. The binding of the acidic drugs, indomethacin, naproxen and warfarin and the basic drugs, papaverine, thiabendazole, miconazole and cinnarizine with the two cyclodextrins was determined at 25°C as a function of pH and cyclodextrin concentration by the phase-solubility method.

Except for miconazole and cinnarizine (A_P-type diagrams), all other materials studied displayed A_L-type diagrams. By comparing the binding constants of both the charged and uncharged forms of the same drugs to both HP-β-CD and SBE7-β-CD, the following conclusions could be drawn. The binding constants for the neutral forms of the drugs were always greater with SBE7-β-CD than with HP-β-CD. A plot of the binding constants for the SBE7-β-CD versus those for HP-β-CD gave a slope of four. For the anionic agents, the binding constants between SBE7-β-CD and HP-β-CD were similar while the binding constants for the cationic agents with SBE7-β-CD were superior to those of HP-β-CD, especially when compared with the neutral form of the same drug. We concluded that a clear charge effect on complexation, attraction in the case of cationic drugs and perhaps inhibition in the case of anionic drugs, was seen with the SBE7-β-CD.

SBE-β-CDs have also been found to stabilize drugs against chemical degradation [12,13]. SBE4-β-CD was found to stabilize pilocarpine against chemical degradation [12]. Additionally, the effect of SBE4-β-CD on the aqueous hydrolysis kinetics of the antitumor drug O^6-benzylguanine (BG) was studied [13]. BG has poor aqueous solubility and undergoes rapid hydrolysis to the poorly water soluble guanine. The stability of a parenteral BG formulation was studied after storage at 25°C, 37°C and 50°C. Compared to the intrinsic solubility of BG (0.14 mg/ml, 25°C), 0.05 M SBE4-β-CD enhanced its solubility to 2.9 mg/ml at 25°C and 3.9 mg/ml at 50°C. Solubility data yielded binding constants of 565 M^{-1} at 25°C and 342 M^{-1} at 50°C. The solubility of guanine was only slightly enhanced by SBE4-β-CD. Hydrolysis kinetics of BG were studied at 50°C over a pH range of 1-9 and the maximum stability was observed at pH 8-8.5. In the presence of 0.05 M SBE4-β-CD, hydrolysis was about 9.5 times slower at pH 1, 14.6 times slower at pH 6 and 10 times slower at pH 8. The effect of varying SBE4-β-CD concentration was studied at pH 2.2 and 4.8 at 50°C. Hydrolysis rate constants decreased with increasing SBE4-β-CD concentrations. A non-linear regression analysis of this data yielded binding constant values of 311 M^{-1} and 270 M^{-1} at pH 2.2 and 4.8 respectively. A formulation containing 2.5 mg/ml of BG and 0.05 M SBE4-β-CD in a pH 8 phosphate buffer was stored in ampoules at 25°C, 37°C and 50°C. Guanine production in the samples was measured since its low solubility (2.5 μg/ml) imposed a limitation on the shelf life. Guanine levels exceeded its apparent solubility after 1-2 months of storage at 50°C. At 37°C guanine levels were only 1.6 μg/ml after 343 days of storage whereas those at 25°C were negligible and below the limit of quantitation (approximately 0.1 μg/ml). The greater stability at room temperature was attributed to the higher binding constant value observed and greater intrinsic stability of BG in the complex.

Recently, we have studied the effect of SBE7-β-CD on the chemical stability of melphalan and various nitrosoureas (unpublished results). Very encouraging results have been obtained. SBE7-β-CD was shown to be superior to HP-β-CD on a molar basis in all cases.

6. *IN VIVO* APPLICATIONS OF SBE4-β-CD AND SBE7-β-CD

The *in vivo* applications of cyclodextrins has been reviewed recently [24]. The emphasis in that review was on the more pharmaceutically acceptable cyclodextrins including the use of SBE-β-CDs.

The IV pharmacokinetics of methylprednisolone (20 mg/Kg) in six rats were studied after administration in a co-solvent (60:12:28 PEG 400:ethanol:water) mixture, a 0.075 M SBE4-β-CD solution and as its two water soluble prodrugs, the 21-phosphate ester, disodium salt and the 21-hemisuccinate ester, monosodium salt [16,17]. The aim of the work was to assess what effect the SBE4-β-CD would have on methylprednisolone pharmacokinetics while the comparison to the prodrugs would provide some insight into a formulation versus a chemical approach to the parenteral delivery of a sparingly water soluble drug. The plasma concentration time curves and the pharmacokinetic parameters of methylprednisolone from the SBE4-β-CD solution and co-solvent mixture were not significantly different. For example, the AUC ± S.E. values from zero to infinity of methylprednisolone from the co-solvent and the SBE4-β-CD solutions were 326.7 ± 20.6 and 317.4 ± 15.4 μgmin/ml, respectively. The AUC values of methylprednisolone from its 21-phosphate and 21-hemisuccinate esters were 59.2 ± 4.4 and 33.17 ± 5.3 % of that from the co-solvent, respectively. These results confirm that IV administered drugs such as methylprednisolone appear to be rapidly and quantitatively released from SBE4-β-CD inclusion complexes.

The IM pharmacokinetics of prednisolone (5 mg/kg) in eight rabbits after its administration in a co-solvent (40:10:50; PEG 400:ethanol:water) mixture, in a slightly hypertonic 0.09 M SBE4-β-CD solution and from a water soluble prodrug, the 21-phosphate ester, disodium salt were studied [18,19]. Muscle damage as measured by changes in plasma creatine kinase (CK) levels caused by the administration of the three solutions was also assessed. The prednisolone plasma AUC values over 24 hours from the SBE4-β-CD formulation and the phosphate ester were 87.0±12.6 and 78.0±14.1% of that from co-solvent, respectively. The apparent bioavailability of prednisolone over 24 hours from the SBE4-β-CD formulation and its prodrug was not significantly different from that of the co-solvent. The changes in CK levels from the SBE4-β-CD were identical to those from normal saline, however, the co-solvent mixture caused significantly elevated CK levels. The presence or absence of prednisolone had no effect on the relative CK levels for the cyclodextrin solution and the normal saline. There was a small effect noted for the co-solvent, with and without prednisolone. These results confirm that IM administered drugs, such as prednisolone, appear to be rapidly, quantitatively and safely released from SBE4-β-CD inclusion complexes.

The absolute bioavailabilities (Fabs) of cinnarizine after oral administration as two modified β-CyDs (SBE4-β-CD or HP-β-CD) solutions, an aqueous suspension and two capsules in fasted beagle dogs were determined [20]. Cinnarizine was administered orally (25.0 mg) and intravenously (12.5 mg) to four dogs. Blood samples were drawn for 24.5 h post-dosing and cinnarizine levels in plasma were determined by HPLC with spectrofluorometric detection. Cinnarizine pharmacokinetics after IV administration as a 1.25 mg/ml SBE4-β-CD solution followed triexponential behavior ($t_{1/2}$ = 12.6 ± 0.4 h and Cl = 1.4 ± 0.17 L/h/kg). A very low bioavailability of cinnarizine with a wide inter-subject variation was observed after oral administration as a suspension (Fabs = 8 ± 4%) or capsule containing only cinnarizine (Fabs = 0.8 ± 0.4%). Administration of cinnarizine as a CyD-complex either as a solution (Fabs = 55 - 60%) or in the capsule (Fabs = 38 ± 12%) significantly enhanced the bioavailability. Since the solutions showed excellent bioavailability, the logical conclusion is that once presented as a solution, cinnarizine is well absorbed and that cinnarizine rapidly dissociates from its inclusion complexes. Presumably, the elevated bioavailability from the SBE4-β-CD containing capsule was due to rapid dissolution and release of cinnarizine.

SBE-β-CDs have also been explored for their ophthalmic use [12,21,22]. For example, SBE4-β-CD did not damage the corneal epithelium *in vitro* and was well-tolerated by the rabbit eye *in vivo*. Coadministered SBE4-β-CD did not significantly affect the miotic response of pilocarpine solutions at pH values of 4.5 or 7.0 when the molar ratio of SBE4-

β-CD to pilocarpine was between 0.2:1 - 7:1. The effect of the coadministered SBE4-β-CD on the miotic response of pilocarpine solutions was also compared to that HP-β-CD which had been suggested to increase ocular bioavailability of pilocarpine in rabbits. Our studies could not confirm the earlier observations[12]. The effects of SBE4-β-CD on eye irritation and miotic response of an ophthalmically applied pilocarpine prodrug, O,O'-dipropionyl-(1,4-xylylene) bispilocarpate, in albino rabbits was also studied [21]. Compared to the commercial pilocarpine eyedrop solution (163 mM, equivalent to 3.4% pilocarpine), 12 - 24 mM pilocarpine prodrug solutions (equivalent to 0.5 - 1.0% pilocarpine, respectively) decreased peak miotic intensity (I_{max}) and increased the time to reach peak (t_{max}), but did not significantly affect values for the area under the miosis versus time curves (AUC), i.e. 12 - 24 mM pilocarpine prodrug appeared to be equivalent to 163 mM pilocarpine. Ocularly applied 12 - 24 mM pilocarpine prodrug solutions, however, were more irritating than a commercial pilocarpine eyedrop solution. Coadministered SBE4-β-CD significantly decreased the eye irritation of the pilocarpine prodrug solutions. Coadministered SBE4-β-CD did not affect the miotic response of prodrug solution when the molar ratio of SBE4-β-CD to prodrug was low. However, increasing the molar ratio of SBE4-β-CD to prodrug decreased the I_{max} and AUC values. The results show that eye irritation of the pilocarpine prodrug is prevented by levels of SBE4-β-CD that do not affect the apparent ocular absorption of the prodrug.

A subsequent study attempted to analyze whether a viscous SBE7-β-CD solution could also be used to prevent the irritation of these bispilocarpine prodrugs without compromising their efficacy [22]. Therefore, the effects of coadministered SBE7-β-CD with and without poly(vinyl alcohol) (PVA) on the miotic response and eye irritation of the prodrug were investigated in pigmented rabbits. Coadministered SBE7-β-CD eliminated the eye irritation due to the pilocarpine prodrug, but also decreased the miotic response. Ocular absorption of the prodrug was improved by increasing the viscosity of prodrug/SBE7-β-CD solution with PVA without inducing any eye irritation. Eye irritation due to viscous prodrug/SBE7-β-CD solutions was comparable to isotonic NaCl-solution.

7. CONCLUSIONS.

The SBE-β-CDs, especially SBE7-β-CD appear to be novel safe and effective complexation agents for biomedical use.

ACKNOWLEDGMENTS

This work was supported CyDex L.C. and by grants from the Kansas Technological Enterprise Corporation.

REFERENCES

[1] Rajewski, R. A., Development and Evaluation of the Usefulness and Parenteral Safety of Modified Cyclodextrins, Ph.D. dissertation to the University of Kansas, 1990.
[2] Stella, V. J., Rajewski, R. A., Derivatives of Cyclodextrins Exhibiting Enhanced Aqueous Solubility and the Pharmaceutical Use Thereof. *U.S. Patent* 5,134,127, July 18, 1992.
[3] Rajewski, R., Traiger, G., Bresnahan, J., Stella, V. J., Thompson,D. O., Preliminary Toxicology and Histopathology of Sulfoalkyl Ether β-Cyclodextrin Derivatives, *J. Pharm. Sci.*, **84**, 927-932 (1995).
[4] Shiotani, K., Uehata, K., Ninomiya, K., Irie, T., Uekama, K., Thompson, D. O. Stella,V. J., Different Mode of Interaction of Chlorpromazine with Sulphated and Sulphoalkylated Cyclodextrins and Effects on Erythrocyte Membranes, *Proc. 7th Int. Cyclodextrin Symp.*, 492-495 (1994).
[5] Shiotani, K., Uehata, K., Irie, T., Uekama, K., Thompson, D. O., Stella,V. J., Differential Effects of Sulfate and Sulfobutyl Ether of β-Cyclodextrin on Erythrocycte Membranes *In Vitro*, *Pharm. Res.*, **12**, 78-84 (1995).

[6] Tait, R. J., Skanchy, D. J., Thompson, D. O., Chetwyn, N. C., Dunshee, D. A., Rajewski, R. A., Stella, V. J., Stobaugh, J. F., Characterization of Sulphoalkyl Ether Derivatives of β-Cyclodextrin by Capillary Electrophoresis with Indirect UV Detection, *J. Pharm. Biomed. Anal.*, **9**, 615-622 (1992).
[7] Tait, R. J., Thompson, D. O., Stella, V. J., Stobaugh, J. F., Sulfobutyl Ether β-Cyclodextrin as a Chiral Discriminator for Use with Capillary Electrophoresis, *Anal. Chem.*, **66**, 4013-4018 (1994).
[8] Stella, V. J., Luna, E., Tait, R. J., Stobaugh, J. F., Thompson, D. O., The Analysis of Anionically Modified Cyclodextrins by Capillary Electrophoresis, *Proc. 7th Int. Cyclodextrin Symp.*, 198-201 (1994).
[9] Luna, E. A., Bornancini, E. R. N., Tait, R. J., Thompson, D. O., Stobaugh, J. F., Rajewski, R. A., Stella, V. J., Evaluation of the Utility of Capillary Electrophoresis (CE) for the Analysis of Modified Sulfobutylether Cyclodextrin Mixtures, *J. Pharm. Biomed. Anal.*, in press.
[10] Luna, E. A., Vander Velde, D., Tait, R. J., Thompson, D. O., Rajewski, R. A., Stella, V. J., Isolation and Magnetic Resonance Characterization of Monosubstituted Sulfobutyl ether β-Cyclodextrin, *Carbohyd. Res.* submitted.
[11] Luna, E. A., Bornancini, E. R. N., Thompson, D. O., Rajewski, R. A., Stella, V. J., Fractionation and Characterization of Sulfobutyl ether β-Cyclodextrin Mixtures, *Carbohyd. Res.* submitted.
[12] Järvinen, T., Järvinen, K., Thompson, D., Stella, V. J., The Effect of a Modified β-Cyclodextrin, SBE4-β-CD, on the Aqueous Stability and Ocular Absorption of Pilocarpine, *Cur. Eye Res.*, **13**, 897-905 (1994).
[13] Gorecka, B. A., Bindra, D. S., Sanzgiri, Y. D., Stella, V. J., Effect of SBE4-β-CD, a Sufobutyl Ether β-Cyclodextrin, on the Stability and Solubility of O^6-Benzylguanine (NSC-637037) in Aqueous Solutions, *Int. J. Pharm.*, **125**, 55-61 (1995).
[14] Johnson, M. D., Hoestrey, B. L., Anderson, B. D., Solubilization of a tripeptide HIV protease inhibitor using a combination of ionization and complexation with chemically modified cyclodextrins, *J. Pharm. Sci.*, **83**, 1142-1146 (1994).
[15] Okimoto, K., Rajewski, R. A., Uekama, K., Jona, J., Stella, V. J., The Interaction of Charged and Uncharged Drugs with Neutral (HP-β-CD) and Anionically Charged (SBE7-β-CD) β-Cyclodextrins, *Pharm. Res.*, **13**, 256-264 (1996).
[16] Stella, V. J., Lee, H. K., Thompson, D. O., The Effect of SBE4-β-CD on I.V. Methylprednisolone Pharmacokinetics in Rats: Comparison to a Co-Solvent Solution and Two Water Soluble Prodrugs, *Int. J. Pharm.*, **120**, 189-195 (1995).
[17] Stella, V. J., Lee, H. K., Thompson, D. O., The Effect of a Parenterally Safe, Anionic β-Cyclodextrin Derivative, Variably Substituted Alkylsulfonates (SBE4-β-CD), on I.V. Methylprednisolone Pharmacokinetics in Rats, *Proc. 7th Int. Cyclodextrin Symp.*, 365-368 (1994).
[18] Stella, V. J., Lee, H. K., Thompson, D. O., The Effect of SBE4-β-CD on I.M. Prednisolone Pharmacokinetics and Tissue Damage in Rabbits: Comparison to a Co-Solvent Solution and a Water Soluble Prodrug, *Int. J. Pharm.*, **120**, 197-204 (1995).
[19] Stella, V. J., Lee, H. K., Thompson, D. O., The Effect of a Parenterally Safe, Anionic β-Cyclodextrin Derivative, SBE4-β-CD, on I.M. Tissue Damage and Prednisolone Pharmacokinetics in Rabbits, *Proc. 7th Int. Cyclodextrin Symp.*, 369-372 (1994).
[20] Järvinen, T., Järvinen, K., Schwarting, N., Stella, V. J., β-Cyclodextrin Derivatives, SBE4-β-CD and HP-β-CD, Increase the Oral Bioavailability of Cinnarizine in Beagle Dogs, *J. Pharm. Sci.*, **84**, 295-299 (1995).
[21] Järvinen, T., Järvinen, Urtti, A., Thompson, D., Stella, V. J., Sulfobutyl Ether β-Cyclodextrin (SBE-β-CD) in Eyedrops Improves the Tolerability of a Topically Applied Pilocarpine Prodrug in Rabbits, *J. Ocul. Pharmacol. Ther.*, **11**, 95-106 (1995).
[22] Jarho, P., Järvinen, K., Urtti, A., Stella, V. J., Järvinen, T., Modified β-Cyclodextrin (SBE7-β-CD) with Viscous Vehicle Improves the Ocular Delivery and Tolerability of Pilocarpine Prodrug in Rabbits. *J. Pharm. Pharmacol.*, in press.
[23] Parmeter, S. M., Allen, E. E. Jr., Hull, G. A., Cyclodextrins with Anionic Properties, *U.S. Patent* 3,426,011, .
[24] Rajewski, R. A., Stella, V.J. Pharmaceutical Applications of Cyclodextrins. II. *In Vivo* Drug Delivery, *J. Pharm. Sci.* in press.

SELECTIVE SYNTHESIS OF NEGATIVELY CHARGED DERIVATIVES OF β-CYCLO-DEXTRIN AND THEIR BIOLOGICAL EFFECTS

Ruth Baumann, Paul Rys, Georg K. Uhlschmid* and Zhining Wu

Chemical Engineering & Industrial Chemistry Laboratory,
Swiss Federal Institute of Technology (ETH), CH-8092 Zürich, Switzerland
*Research Div., Dept. Surgery, Univ. Hospital, CH-8091 Zürich, Switzerland

ABSTRACT

It was possible to selectively introduce sulfate and carboxyl groups into defined positions of β-cyclodextrin. In addition, the distance between the sulfate groups could be increased with the aid of alkyl spacers in order to lower the negative charge density on the cyclodextrin. Investigations of the biological activity of these derivatives yield more information about the angiogenesis-inhibition mechanism. Thus, the modified cyclodextrins were tested for biological activity on the chick chorioallantoic membrane (CAM).

1 INTRODUCTION

Heparin shows a variety of biological effects, *e. g.*, it is widely used as an anticoagulant. It has been shown that for a large number of these biological effects heparin fragments as small as a hexasaccharide can act as heparin mimics [1]. But even simpler molecular analogues might have similar properties: Thus, sulfated cyclodextrins, compared with heparin, were found to show the same or even better effects [2]. Hydrocortisone, for instance, can be converted into a potent angiogenesis inhibitor by coadministration with a sulfated cyclodextrin. This effect can only be observed if the cyclodextrin bears at least ten sulfate groups. Since very little is known about the influence of the position of the sulfate groups or the charge density on the activity of the sulfated cyclodextrin, our attention was drawn to the synthesis of selectively sulfated and sulfatoalkylated ß-cyclodextrins. Also part of our research has been focusing on the controlled and regioselective synthesis of *carboxylated* cyclodextrins, another possibility to introduce a certain negative charge density.

J. Szejtli and L. Szente (eds.), Proceedings of the Eighth International Symposium on Cyclodextrons, 477–480.

2 MATERIALS AND METHODS

2.1 Materials

The trimethylammoniumsulfate-complex was obtained from Aldrich. Cyclodextrin, reagents, and solvents were purchased from Fluka Chemie AG. Pruducts were purified by column chromatography on Sephadex G-25-80 from Fluka and eluted with deionized water.

2.2 Methods

A typical procedure for the sulfation of the fully protected and hydroxyalkylated β-cyclodextrins: The modified cyclodextrin and a fivefold excess of trimethylammoniumsulfate were dissolved in dimethylformamide and stirred for six days at 60 °C. After three days another portion of the sulfating agent was added to the reaction mixture. After six days the solution was added to 0.5*M* NaOH and the resulting mixture was stirred for 1 h, concentrated to dryness. The residue was dissolved in little water, and the salts removed by column chromatography (Sephadex, water).

Propenylation of β-cyclodextrin was conducted by a two-step procedure: The first pentenylation (NaOH, DMSO, 5-bromo-1-pentene, and 20 °C) yielded the heptakis(2,6-O-pentenyl)-β-cyclodextrin as the major product; further pentenylation was carried out under more rigorous conditons (NaH, THF, 60 °C and 3 d).

Hydroboration-oxidation of alkenylated cyclodextrins: The alkenylate was allowed to react with a 0.5*M* solution of 9-borabicyclo[3,3,1]nonane (1.5 fold excess) in THF for 5 h at room temperature. The reaction mixture was then heated to reflux overnight. The organoboranes were oxidized by adding successively 25 mL of methanol, 20 mL of 6*M* NaOH and dropwisely 33 ml of 30 % hydrogen peroxide. The mixture was heated for 1 h at 50 °C, then cooled to room temperature and concentrated.

A typical preparation of carboxymethylated β-cyclodextrins: (A) To a solution of cyclodextrin derivative in DMSO were added finely powdered NaOH and 2-chloro-*N*,*N*-dimethylacetamide (a threefold excess). The mixture was stirred for 4 d at room temperature. The reaction was quenched with water and the reaction product was extracted with dichloromethane and purified over Sephadex. (B) A heterogenious mixture of cyclodextrin derivative, potassium tert-butoxide (6 fold excess), diethyl ether and water were stirred vigorously for 3 d at room temperature. The reaction mixture was neutralised with 5*M* hydrochloric acid and concentrated to 5 mL. The residue was purified over Sephadex.

Leghorn eggs with 4-5 days old embryos were prepared the standard way and kept under standard conditions in a special incubator (37 °C, 85 % humidity, turned all 2-3 h). On day 9 the cyclodextrins were administrated to the chorio allantoic membrane. Depending on their solubility, the cyclodextrins were solved in physiological saline or Ringer-lactate solution.

3 RESULTS AND DISCUSSION

Direct selective sulfation of β-cyclodextrin (**1**) is difficult because the activity differences between the three kinds of hydroxy groups are too small. To introduce the sulfate groups into defined positions of the cyclodextrin, part of the hydroxy groups had to be protected.

The intermediates **2,4,6** & **8** could be synthesized by described methods [3,4,5]. Sulfation with the trimethylammoniumsulfate complex in dimethylformamide led to cyclodextrin sulfates with the sulfate groups in defined positions (**3,5**,**7** & **9**).

These products (**3,5**,**7** and **9**) were tested for metachromatic activity with the cationic dye Azur A. Table 1 shows the metachromatic activity of these compounds, which was measured by titration of a solution of the compound in water against an Azur A solution. The value for the metachromatic activity is the negative slope of the curve of absorbance at 632 nm *vs.* the cyclodextrin sulfate concentration in mol/L [6].

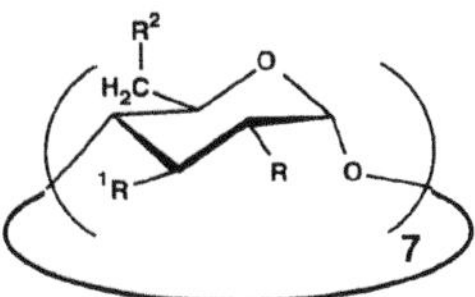

	R	R1	R2		R	R1	R2
1	OH	OH	OH	**14**	$O(CH_2)_5OH$	$O(CH_2)_5OH$	$O(CH_2)_5OH$
2	OH	OMe	OMe	**15**	$O(CH_2)_5OSO_3^-Na^+$	$O(CH_2)_5OSO_3^-Na^+$	$O(CH_2)_5OSO_3^-Na^+$
3	$OSO_3^-Na^+$	OMe	OMe	**16**	OCH_2CONMe_2	OCH_2CONMe_2	OCH_2CONMe_2
4	OBz	OH	OH	**17**	$OCH_2COO^-Na^+$	$OCH_2COO^-Na^+$	$OCH_2COO^-Na^+$
5	OBz	$OSO_3^-Na^+$	$OSO_3^-Na^+$	**18**	OH	OH	$OSiMe_2$t-Bu
6	OAc	OAc	OH	**19**	OCH_2CONMe_2	OCH_2CONMe_2	$OSiMe_2$t-Bu
7	OAc	OAc	$OSO_3^-Na^+$	**20**	$OCH_2COO^-Na^+$	$OCH_2COO^-Na^+$	OH
8	OH	OMe	OH	**21**	OH	OH	Cl
9	$OSO_3^-Na^+$	OMe	$OSO_3^-Na^+$	**22**	OCH_2CONMe_2	OCH_2CONMe_2	Cl
10	OAll	OAll	OAll	**23**	$OCH_2COO^-Na^+$	$OCH_2COO^-Na^+$	Cl
11	$O(CH_2)_3OH$	$O(CH_2)_3OH$	$O(CH_2)_3OH$	**24**	OCH_2CONMe_2	OCH_2CONMe_2	SH
12	$O(CH_2)_3OSO_3^-Na^+$	$O(CH_2)_3OSO_3^-Na^+$	$O(CH_2)_3OSO_3^-Na^+$	**25**	$OCH_2COO^-Na^+$	$OCH_2COO^-Na^+$	SH
13	$O(CH_2)_3CH{=}CH_2$	$O(CH_2)_3CH{=}CH_2$	$O(CH_2)_3CH{=}CH_2$	**26**	$OCH_2COO^-Na^+$	$OCH_2COO^-Na^+$	SO_3H

Fig.1: Intermediates and products

Table 1: Metachromatic activity of sulfated β-cyclodextrins

Compound number	Metachromatic activity
2	$2.50 \cdot 10^4$
4	$1.80 \cdot 10^5$
6	$1.34 \cdot 10^5$
8	$6.02 \cdot 10^5$

Allylation of **1** was carried out according to a method described by Takeo et al [8]. For the pentenylation reaction a method similar to the preparation of peracetylated cyclodextrin [9] was applied. Derivatives **10** and **13** were obtained by complete allylation and pentenylation, and were subsequently converted into the perhydroxyallylated cyclodextrins **11** and **14** by a hydroboration-oxidation sequence at the unsaturated terminals of the alkenyl functions. Further sulfation of **11** and **14** afforded the cyclodextrin sulfates

12 and **15**, resp. with allyl chains of different lengths functioning as spacers for the sulfate groups within the cyclodextrin skeleton.

Compound **21** and its sulfhydral and sulfonate derivatives were prepared in analogy to a method described by Rajewski [10].

To introduce carboxylate groups into defined positions of β-cyclodextrin and to get percarboxmethylated cyclodextrins, *N,N*-dimethylamidomethylation of cyclodextrin proved to be a more effective pathway compared to the direct reaction of cyclodextrin with sodium-monochloroacetate in aqueous NaOH, which usually gave the carboxymethylated product with a rather low degree of substitution [11].

4 CONCLUSION

Metachromatic activity shows a clear dependence on the charge density of the sulfated β-cyclodextrin. The comparatively low activity of the 3,6-sulfated-2-benzylated β-cyclodextrin can be explained by the benzyl groups shielding the sulfate groups from each other.
Investigations of the activity on the growth of blood capillary vessels led to non-uniform, very dose-dependent results.

REFERENCES

[1] Folkman J., Weisz P.B., Joullié M.M., Li W.W., Erwing W.R., Control of Angiogenesis with Synthetic Heparin Substitutes, *Science*, **243**, 1490-1493 (1989)

[2] Weisz P.B., Hermann H.C., Joullié M.M., Kumor K., Levine E.M., Macarak E.J., Weiner D.B., Angiogenesis and heparin mimics, Angiogenesis: Key Principles-Science-Technology-Medicine, (ed. by R. Steiner, P.B. Weisz & R. Langer) Basel, 1992, Pp. 107-117

[3] Fügedi P., Synthesis of heptakis(6-O-tert-butyldimethylsilyl)cyclomaltoheptaose and octakis(6-O-tert-butyldimethylsilyl) cyclomalto-octaose, *Carbohydr. Res.*, **192**, 366-369 (1989)

[4] Bergeron R.J., Meeley M.P., Machida Y., Selective Alkylation of Cycloheptaamylose, *Bioorganic Chemistry*, **5**, 121-126 (1976)

[5] Takeo K., Mitho H., Uemura K., Selective chemical modification of cyclomalto-oligo-saccharides via tert--butyldimethylsilylation, *Carbohydr. Res.*, **187**, 203-221 (1989)

[6] Grant A.C., Linhardt R.J., Fitzgerald G.L., Park J.J., Langer R., Metachromatic Activity of Heparin and Fragments, *Anal. Biochem.*, **137**, 25-32 (1984)

[7] Steiner R., Angiostatic activity of anticancer agents in the chick embryo chorioallantoic membrane (CHE-CAM) assay, Angiogenesis: Key Principles-Science-Technology-Medicine, (ed. by R. Steiner, P.B. Weisz & R. Langer) Basel, 1992, 449-454

[8] Takeo, K., Uemura, K., Mitoh, H., Derivatives of α-cyclodextrin and the synthesis of 6-O-α-D-glucopyranosyl-α-cyclodextrin, *J. Carbohydr. Chem.*, **7**, 293-308 (1988)

[9] Bates, P. Parker, D., A chiral sensor based on a peroctylated α-cyclodextrin, *J. Chem. Soc., Chem. Commun.*, ,153-155 (1992)

[10] Rajewski, R. A., Development and evaluation of the usefulness and parenteral safety of modified cyclodextrins, PhD Thesis, University of Kansas, 1990

[11] Reuben, J., Rao, C. T., Pitha, J., Distribution of substituents in carboxmethyl ethers of cyclomaltoheptaose, *Carbohydr. Res.*, **258**, 281-285 (1994)

ENHANCED PHARMACOKINETIC PROPERTIES OF ORAL AND PARENTERAL DICLOFENAC-CYCLODEXTRIN DELIVERY SYSTEMS

LAWRENCE J. PENKLER, DARRYL V. WHITTAKER,
LUÉTA A. GLINTENKAMP and M.C. BOSCH VAN OUDTSHOORN

Druggists Group Research, South African Druggists Limited
PO Box 4002, Korsten 6014, Port Elizabeth, South Africa

ABSTRACT

The application of beta cyclodextrins to enhance the oral and parenteral delivery of the extensively used NSAID diclofenac sodium is described. Tablets containing microcrystalline 1:1 diclofenac sodium/beta-cyclodextrin complex prepared by kneading in a one pot ROTO granulator were formulated with an alkali agent. In a single dose cross-over study (n=6) the test product yielded mean T_{max} after 20 min compared with 40 min for the reference commercial non-enteric coated tablets, and showed more complete absorption. A stable 75mg/3mL aqueous solution of diclofenac sodium/2-hydroxypropyl-beta-cyclodextrin (1:2 mol/mol) was administered via *im* and *iv* injection to 6 volunteers. Relative to the commercial *im* reference product, the test showed significantly enhanced *im* and interesting *iv* pharmacokinetics.

1. INTRODUCTION

Diclofenac (*Voltaren*®) has been used for over two decades as a non-steroidal anti-inflammatory drug (NSAID) with distinct anti-inflammatory, analgesic and anti-pyretic properties. The drug is available in a variety of dosage forms, including among others, enteric coated and modified release tablets and ampoules for intramuscular (*im*) injection. More recently, a rapid release oral preparation (*Voltarol Rapid*® / *Cataflam*®) has been registered in several countries and the *im* product has been adapted for use in *iv* infusions. These latest developments relate to the need for rapid absorption and distribution of diclofenac to improve therapeutic efficiency in pain management. The time to reach peak plasma concentrations (T_{max}) for commercial enteric coated tablets, rapid release and *im* diclofenac formulations are 1-4 hr, 30-60 min and 20 - 30 min respectively. [1] We report on the successful application of cyclodextrin technology to the development of patented oral[2,3] and parenteral[4]

J. Szejtli and L. Szente (eds.), Proceedings of the Eighth International Symposium on Cyclodextrons, 481–486.

delivery systems of diclofenac aimed at enhancing the oral pharmacokinetics and extending parenteral administration to a well tolerated *iv/im* bolus injection of this significant therapeutic agent.

2. MATERIALS AND METHODS

2.1. Materials

Diclofenac sodium (DIC) was from Labochim, Italy; beta-cyclodextrin (BCD) was from Wacker Chemie, Germany; and 2-hydroxypropyl-beta-cyclodextrin (HPBCD, *Encapsin^{TM}*) was from Janssen Biotech, Belgium. All materials were of commercial pharmaceutical grade. The average degree of substitution of the HPBCD was 4.6 2-hydroxypropyl groups per cyclodextrin molecule. *Cataflam®* tablets and *Voltaren® im* injection were from Ciba-Geigy, South Africa.

2.2 Methods

Kneaded complexes were prepared in a ROTO P50 granulator, (Zanchetta C.s.r.l., Italy) by blending 1:1 mol/mol stoichiometric quantities of DIC and BCD at 120 rpm for 10 min, then spraying purified water with mixing at 150 rpm until a uniform paste was obtained. Mixing at 120 rpm was continued for 2 hr at 50 °C and then drying in vacuo until a coarse, flowing granular product was obtained. The product was characterised by powder X-ray diffraction and diffuse-reflectance infrared spectroscopy. The complex was blended with and without sodium bicarbonate together with microcrystalline cellulose, sodium starch glycollate and magnesium stearate. Tablets containing the equivalent of 50 mg DIC were compressed at 100N and coated with Opadry-OYB II, Colorcon, France. Dissolution of tablets was tested in purified distilled water and in a flow through dissolution apparatus under simulated conditions (pH 1.2, 45 min; pH 6.8, 90 min). The oral bioavailability of the DIC/BCD tablets containing sodium bicarbonate was assessed in a single dose cross-over double blind study in six healthy fasted volunteers against *Cataflam®* tablets as reference.[5]
Proton magnetic resonance at 500 MHz and molecular mechanics studies of the interactions of DIC with BCD and HPBCD respectively are reported elsewhere in these proceedings.[6] Solutions containing 25mg/mL DIC were prepared in water for injections containing a 2 mol equivalent HPBCD relative to the guest. Osmolality was adjusted to 300 mOsm/kg using sorbitol and the solution sterilised by filtration before aseptic filling into amber ampoules. The parenteral bioavailability via *im* and *iv* routes was determined in a single dose cross-over double blind study in six healthy fasted volunteers against *Voltaren® im* as reference.[7]

3. RESULTS AND DISCUSSION

3.1 DIC/BCD complex characterisation

The kneaded 1:1 DIC/BCD complex was characterised by a fine microcrystalline morphological structure with good flow and compaction properties suitable for direct compression. The equilibrium solubility of DIC from the complex in unbuffered water at pH 7 was 38 mg/mL compared with 7 mg/mL for uncomplexed DIC. The X-ray powder diffractogram of the kneaded complex surprisingly corresponded exactly with line positions in the calculated diffraction pattern obtained from single crystal X-ray crystallographic analysis of DIC/BCD.(8) This finding indicates that the kneading method produces a well characterised inclusion complex with highly conserved structure.(2) Diffuse reflectance IR-spectroscopy revealed reduced intensity of carboxylate and aromatic bands indicative of inclusion phenomena.

3.2 Oral Formulation Considerations

The positive influence of cyclodextrins on dissolution and rate and extent of absorption of sparingly soluble drugs is well documented.(9) In the case of oral NSAID's, cyclodextrin complexation is particularly advantageous since rapid dissolution and absorption may reduce local gastric irritation associated with NSAID's.

The solubility and stability constant of the DIC/BCD complex is strongly pH dependent.(10) At low pH the stability constant is high ($>1000M^{-1}$) whereas solubility is low, and the converse applies at alkaline pH. Simulated flow through dissolution studies with DIC/BCD tablets indicated incomplete dissolution in acidic media due to a gelling effect at the tablet surface preventing tablet wetting. The latter effect further delayed dissolution in alkaline media. In order to enhance the rate of dissolution of the complex in gastric fluid, sodium bicarbonate was included in the formulation to create an alkaline diffusion layer around the disintegrating tablet as well as to aid disintegration at low pH. With the cyclodextrin complex alkali system,(3) 100% disintegration could be obtained within 5 minutes at pH 1.2, and 100% dissolution within 5-10 minutes in unbuffered water.

3.3 Pharmacokinetics of oral DIC/BCD

The summary of pharmacokinetic data for DIC/BCD and commercial reference product is given in Table 1 and graphically depicted in Figure 1. The results indicate that the DIC/BCD formulation is within the conventional bioequivalence range (80 - 125%) relative to the reference product with respect to the extent of absorption. The cyclodextrin product is however absorbed at a significantly faster rate as evidenced by the 50% shorter mean T_{max}. The fact that T_{max} occurs within 20 minutes suggests that absorption, at least in part, is already taking place in the stomach.

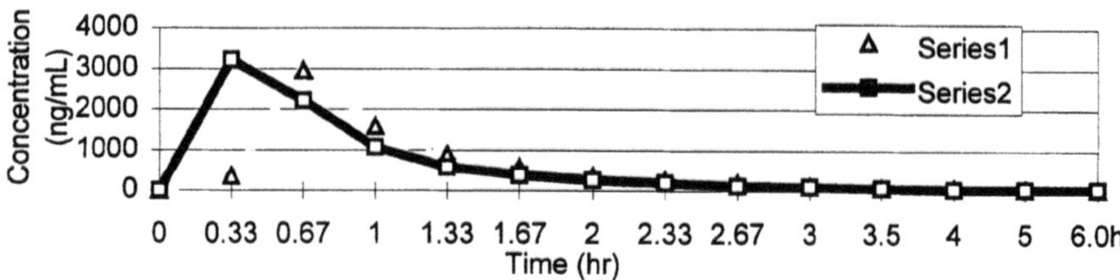

Figure 1. Mean Plasma Diclofenac Concentrations (n=6):
Dose 2 x [50 mg diclofenac potassium tablet, equivalent to 44.3 mg diclofenac] - *Cataflam*® (Series 1);
Dose 2 x [50 mg DIC/BCD tablet, equivalent to 46.6 mg diclofenac] *DIC/BCD* (Series 2).

The geometric mean for the variable C_{max}, calculated from individual values, was 30% higher for DIC/BCD (see Table 1), which, taken together with the ratio over AUDC, suggests more complete absorption from the cyclodextrin formulated product. The increased rate of absorption leads to lower contact time of diclofenac with the gastric mucosa, which, together with the neutralizing effect of the alkali agent, should significantly reduce local gastric irritation associated with non-enteric coated diclofenac.

TABLE 1. Summary Of Pharmacokinetic Data For Oral Diclofenac
Dose: 2 x 50 mg tablets, (n = 6)

		Diclofenac potassium (*Cataflam*®) - reference			**Diclofenac sodium/BCD (DIC/BCD)**				
Variable	UNIT	GEOMETRIC MEAN	SD	Range	GEOMETRIC MEAN	SD	Range	Mean Ratio (%)*	90% CI (%)**
C_{max}	ng/mL	3050	1.22	2253-3825	3992	1.46	2440-6260	131	102-168
T_{max}#	h	0.67		0.33-0.67	0.33		0.33-0.67	49%	
AUDC	ng.h/mL	2827	1.17	2434-3631	3047	1.24	2230-3898	108	95.6-122
C_{max}/AUDC	L/h	1.08	1.12	0.92-1.27	1.31	1.22	0.93-1.61	121	99.3-149
$CL_{tot/f}$	mL/min/kg	7.41	1.15		6.87	1.25			

* : Geometric mean of individual "test/reference" ratios.
** : 90% Conventional confidence interval for the "test/reference" mean ratio after logarithmic transformation of the data.
\# : Medians, ranges, non parametric point estimate of "test-reference" median difference, and corresponding confidence interval.

3.4 DIC/HPBCD Complex Characterisation

Bimodal inclusion of DIC is afforded by the two aromatic rings, giving rise to a number of possible isomeric 1:1 complexes in addition to a 1:2 DIC/HPBCD complex. The interaction between DIC and HPBCD was investigated with high field proton magnetic resonance and molecular mechanical calculations from which evidence for the bimodal inclusion in solution was obtained.[6]

3.5 Parenteral Formulation Considerations

Commercial diclofenac preparations for parenteral administration contain several solubilizing adjuvants such as propylene glycol and benzyl alcohol, limiting the route of administration to strict intramuscular (*im*) use. Stable aqueous solutions containing 25 mg/mL DIC at pH 7.4 were prepared with a 1:2 DIC/HPBCD stoichiometry using sorbitol to adjust osmolality to 300 mOsm/kg.[4] Use of buffers was avoided due to interference of cations with complex equilibria leading to slow precipitation of DIC. Interestingly the 1:2 stoichiometry is critical to prevent slow crystallization of DIC tetrahydrate which occurs only after 6 months in samples with 1:1.5 stoichiometry.

3.6 Pharmacokinetics of parenteral DIC/HPBCD

The summary of pharmacokinetic data for DIC/HPBCD and commercial reference product is given in Table 2 and graphically depicted in Figs 2 and 3. The results indicate that the two products are bioequivalent with respect to diclofenac absorbed after *im* administration. DIC/HPBCD however reached higher maximum concentrations than the reference product (see Table 2 and Fig. 2). The absolute biovailability of *im* DIC/HPBCD is almost 100% relative to the *iv* route by comparison of the AUDC values of 3696 vs 3852 ng.h/ml respectively (see Fig. 3). No clinically significant adverse effects were observed. The preparation was well tolerated in all subjects by both routes and less pain after *im* administration was noted.

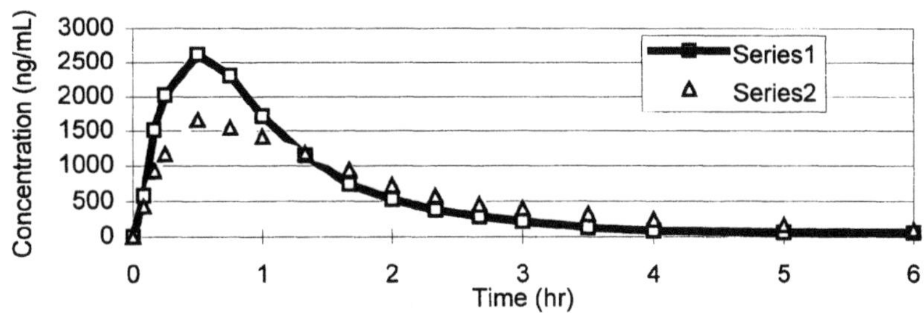

Figure 2. Mean Plasma Diclofenac Concentrations after *im* administration of 75 mg diclofenac sodium (n=6). Series 1, DIC/HPBCD; Series 2, reference product - *Voltaren*®

TABLE 2. Summary Of Pharmacokinetic Data For Intramuscular Diclofenac (n = 6; Dose; 75 mg diclofenac sodium)

		Diclofenac sodium (*Voltaren®*) *im*			DIC/HPBCD *im*				
Variable	UNIT	GEOMETRIC MEAN	SD	Range	GEOMETRIC MEAN	SD	Range	Mean Ratio (%) *	90% CI (%) **
C_{max}	(ng/mL)	1682	1.26	1357-2504	2761	1.39	1662-4336	164	119-226
T_{max}#	(h)	0.50		0.50-0.75	0.50		0.25-0.75	-0.13	
AUDC	(ng.h/mL)	3861	1.13	3342-4522	3696	1.09	3271-4243	95.7	87.0-105
C_{max}/AUDC	(L/h)	0.44	1.21	0.32-0.55	0.75	1.32	0.46-1.02	172	132-223
$Cl_{tot/f}$	mL/min/kg	4.12	1.09		4.30	1.06			

* : Geometric mean of individual "test/reference" ratios.

** : 90% Conventional confidence interval for the "test/reference" mean ratio after logarithmic transformation of the data.

\# : Medians, ranges, non parametric point estimate of "test-reference" median difference, and corresponding confidence interval.

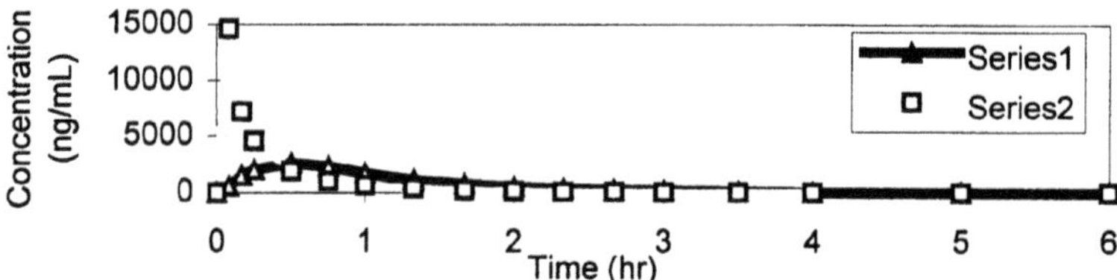

Figure 3. Mean Plasma Diclofenac Concentrations after *iv* (Series 1) and *im* (Series 2) administration of DIC/HPBCD (n=6).

4. CONCLUSION

The pharmacokinetic results, viewed against a background of small sample size, indicate that the rational application of cyclodextrin technology leads to improved oral and parenteral diclofenac products with significantly enhanced pharmacokinetics.

REFERENCES

[1] Voltaren® (Diclofenac): Twenty years of clinical experience, an update. Ciba-Geigy Ltd., Basle Switzerland. 1994

[2] Penkler, L.J., Glintenkamp, L-A., et al. European Pat. Appl. 94307380.9, South African Druggists Ltd.

[3] Penkler, L.J., Glintenkamp, L-A., et al . PCT Appl. PCTBG95/2679, South African Druggists Ltd.

[4] Penkler, L.J., Glintenkamp, L-A.,. et al . Europe Pat. Appl. 94308690.0, South African Druggists Ltd.

[5] FARMOVS Institute for Clin. Pharmacol. & Drug Dev., Univ. of the Orange Free State, Bloemfontein, 20/94

[6] Whittaker, D.V., Penkler L.J., Glintenkamp, L-A., et al. Proc. 8th Int. Cyclodextrin Symposium, Budapest, (1996)

[7] FARMOVS Institute for Clin. Pharmacol. & Drug Dev., Univ. of the Orange Free State, Bloemfontein. 19/94

[8] Caira, M.R., Griffiths, V., Nassimbeni, L.R. *J. Chem. Soc. Chem. Commun.*, 1061-1062, (1994)

[9] Frömming, K-H. & Szejtli, J. *Cyclodextrins in Pharmacy*, Kluwer Academic Publishers, Dordrecht, (1994)

[10] Orienti, I., Fini, A., Bertasi, V. et al *Eur J. Pharm. Biopharm.* **37**, 110-112, (1991)

IN VITRO AND IN VIVO STUDIES ON SODIUM NIMESULIDE-β-CYCLODEXTRIN INCLUSION COMPLEX .

G. PIEL, I. DELNEUVILLE AND L. DELATTRE

Laboratory of Pharmaceutical Technology, Institute of Pharmacy, University of Liège, 5 rue Fusch, 4000 Liège, Belgium.

ABSTRACT

Nimesulide (NI), a non steroidal anti-inflammatory drug, is very poorly soluble in water. The aim of this work is to study the in vitro and in vivo characteristics of a NI Na-β-cyclodextrin (CD) complex in the stoichiometric ratio 1:1. The complex was prepared by spray-drying and the inclusion could be demonstrated by DSC and solubility method. This study has shown that the NI Na-β-CD complex can increase the aqueous solubility of NI (about 5000 times) as well as the solubility in acidic medium (7 times) and at pH 6.8 (34 times). A preliminary study in 3 healthy volunteers has shown promising results for further galenic developments.

1. INTRODUCTION

NI (4'-nitro-2'-phenoxymethane sulphonanilide) is a non steroidal anti-inflammatory drug (NSAID) useful in the treatment of inflammatory and pain states [1] . Like many NSAIDs, NI is very poorly soluble in water (≈ 0.01 mg/ml) and this gives rise to problems for the preparation of pharmaceutical formulations with good release and non variable bioavailability [2].
The CD complexation with NSAIDs frequently results in more rapid absorption of the NSAIDs after oral administration [3].
Patents applications WO 91/17774 [2] and WO 94/02177 [4] describe complexation of NI with CD's, preferably with β-CD, with improved water solubility and faster absorption. Until now, the best way to obtain this complex seems to be the freeze drying or spray drying from an aqueous ammonium hydroxide solution. The alkali salts of NI were subjected to complexation to give NI alkali-CD complexes whereby solubility was considerably increased [6].

J. Szejtli and L. Szente (eds.), Proceedings of the Eighth International Symposium on Cyclodextrons, 487–490.

The objective of this work is to study the in vitro and in vivo characteristics of a NI Na-β -CD complex in the stoichiometric ratio 1:1.

2. MATERIALS AND METHODS

2.1 Materials

β-CD was obtained from CNI (Neuilly sur Seine, France) and γ-CD from Wacker Chemie GmbH (München, Germany).
All other materials were of analytical grade.
NI Na, NI Na-β-CD and NI Na-γ-CD complexes were prepared by the spray drying method [6].

2.2 Methods

Differential Scanning Calorimetry (DSC): DSC was carried out with a differential scanning calorimeter (DuPont instrument 910) at a scanning speed of 10°C /min under N_2 stream.
Determination of acetone solubility: Excess of NI Na-β-CD physical mixture or complex were shaken with extra pure acetone. After filtration, the NI Na was assayed at 397 nm.
Solubility studies: The solubility of the complexes and the physical mixtures was measured in purified water and in buffers (μ=0.2) of pH 1.4 and 6.8.
Preliminary in vivo studies: 3 healthy volunteers received granules (dispersed in a few water) and capsules containing NI Na-β-CD complex corresponding to 100 mg of NI. The study was designed as a cross over and the order of galenic form administration was completely randomized. Blood samples were drawn during 24 hours following drug intake for the assay of NI and OH-NI by HPLC.

3. RESULTS AND DISCUSSION

The inclusion of the NI Na salt into β-CD could be proved by different methods.
Because of the free solubility of the NI Na in acetone, we could show significant differences between the physical mixture and the complex. From the physical mixture, 100% of the NI Na can be dissolved in acetone whereas less than 1% of the NI Na can be dissolved from the complex. This shows that NI Na is included in the β-CD.
The DSC pattern confirms the inclusion of NI Na. As can be seen in figure 1, the complexation of NI Na in β-CD is shown by the disappearance of the exothermic peak corresponding to a probable recrystallization of the NI Na. A minor loss of water at a lower temperature can also prove the rearrangement of the water during complexation. The endothermic peak corresponding to the melting point of the NI Na is masked

because of a reaction between the NI and the CD, in both the physical mixture and the complex.

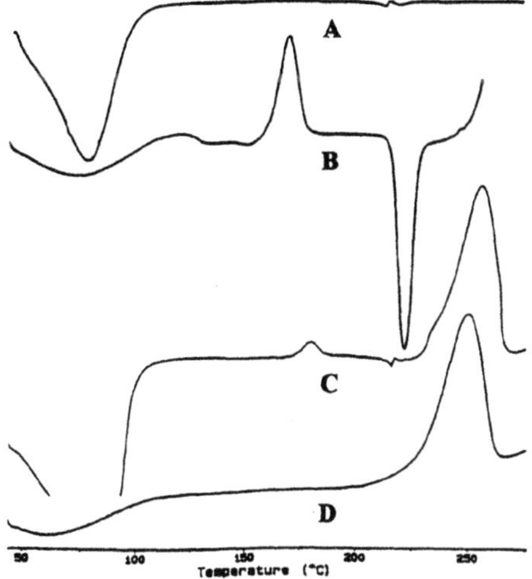

Figure 1: DSC thermograms of the β-CD (**A**), the NI Na (**B**), the NI Na-β-CD physical mixture (1:1) (**C**) and the NI Na-β-CD complex (**D**).

As can be seen in table 1, the water solubility of the complex is significantly higher than that of NI but not so high as that of NI Na. The CD's seem to be a limiting factor. Otherwise, the β-CD complex is more soluble than the γ-CD complex.

TABLE 1: Solubility values expressed as Nimesulide dissolved (mg/ml)

	water	pH 1.4	pH 6.8
Nimesulide	0.014	0.005	0.015
Nimesulide Na	109.500[1]	0.008	0.042
Nimesulide Na β-CD complex	74.050[2]	0.035	0.508
Nimesulide Na γ-CD complex	57.050	0.011	0.147
Nimesulide Na β-CD mixture		0.035	0.496
Nimesulide Na γ-CD mixture		0.011	0.121

[1] pH rises to about 12 [2] pH rises to about 9

Table 1 also shows the values of solubility at different pH values. CD's improve the solubility at lower pH values and at pH close to the neutrality. At pH 1.4, β-CD increases the solubility of NI about 7 times whereas γ-CD increases it 2.5 times. At pH 6.8, β-CD increases the solubility 34 times and γ-CD only 10 times. β-CD is better than γ-CD to increase the solubility of NI at different pH values. It is noteworthy that the solubility of physical mixtures of NI Na with β- and γ-CD, at pH 1.4 and 6.8 is the same as the solubility of the corresponding complexes. This can be explained by the fact that in solution, the complexes are dissociated and NI Na is totally (pH 1.4) or partially (pH

6.8) transformed in NI . The increasing solubility effect of the CD is due to the formation of a NI-CD complex in situ.
Three healthy volunteers received granules of NI Na-β-CD complex in a quantity equivalent to 100 mg of NI as an oral suspension and as an oral capsule. Table 2 gives the pharmacokinetic parameters in comparison with the pharmacokinetic parameters of a NI granule and a NI oral tablet (n=12) [5]. As can be seen, we obtained promising results as Cmax and AUC are higher and Tmax are lower. Further studies with appropriate galenic form and with more volunteers will be performed to confirm these results.

TABLE 2: Pharmacokinetic parameters of NI Na-β-CD complex (eq. to 100 mg of NI) granule and capsule and NI (100 mg) granule and tablet after administration of a single oral dose to healthy volunteers.

	Cmax ($\mu g.ml^{-1}$)	Tmax (h)	AUC_{0-24} ($\mu g.h.ml^{-1}$)
1.Complex granule	4.95	1.00	38.42
2.Nimesulide granule*	4.69	2.17	17.32
ratio 1./2.	**1.16**	**0.56**	**2.22**
3.Complex capsule	4.26	1.78	27.39
4.Nimesulide tablet*	2.86	2.63	22.69
ratio 3./4.	**1.64**	**0.82**	**1.21**

*From reference [5]

4. CONCLUSION

This study has shown that the NI Na-β-CD complex can increase the aqueous solubility of NI (about 5000 times) as well as the solubility in acidic medium (7 times) and at pH 6.8 (34 times). A preliminary study in 3 healthy volunteers has shown promising results for further galenic developments.

REFERENCES

[1] Biscarini, L., Patoia, L., Del Favero, A., Nimesulide - A new non steroidal inflammatory agent, *Drugs today*, **24**, 23-27 (1988)

[2] Boehringer Ingelheim Italia, Inclusion compounds of nimesulide with cyclodextrins, WO 91/17774 (1991)

[3] Loftsson, T., Olafsdottir, B.J., Fridriksdottir, H., Jonsdottir, S., Cyclodextrin complexation with non steroidal antiinflammatory drugs: physicochemical characteristics, *Eur.J.Pharm*, **1**, 95-101, (1993)

[4] Boehringer Ingelheim Italia, Methods of preparation of inclusion compounds of nimesulide with cyclodextrins, WO 94/02177 (1993)

[5] Bernareggi, A., The pharmacokinetic profile of Nimesulide in healthy volunteers, *Drugs,* **46**, (supp 1), 64-72 (1993)

[6] Géczy J., New Nimesulide Salt Cyclodextrin Inclusion Complexes, WO 94/28031 (1994)

EFFECT OF CYCLODEXTRIN BEAD POLYMER ON WOUND HEALING

I. FELMÉRAY[1], É. FENYVESI[2], T. NEUMARK[3], J. TAKÁCS[3], A. GERLÓCZY[2], J. SZEJTLI[2]
[1]*Dermatological Dept., FlórFerenc Hospital, H-1137 Budapest, Jászai M. tér 5. Hungary*
[2]*Cyclolab R&D Lab., Ltd., H-1525 Budapest P.O.Box 435, Hungary*
[3]*Electronmicroscopic Lab. of Natl. Inst. for Rheumatism and Physiotherapy, H-1525 Budapest 114, P.O.Box 54 Hungary*

ABSTRACT

Cyclodextrin bead polymer, a swellable carbohydrate derivetive proved to be effective in healing the wounds inflicted on the back of rats and in healing venous leg ulcers of human patients.
According to microscopic studies, polymer beads implanted into the muscular tissue of rats did not cause inflammation cell reaction for up to 6-week observation period.

1. INTRODUCTION

The topical treatment of different oozing wounds like burns, venal leg ulcers, etc. means a drainage of the wound area. An ideal wound healing agent not only absorbs the exudate, but removes bacteria from the wounded area, allows some evaporation of the fluids, is not incorporated into the scar tissue, can be removed from the wounded area without damage to the wound itself, and does not cause hypersensitivity.
βCD bead polymer (CDP) is prepared by crosslinking of βCD with epichlorohydrin in the presence of polyvinyl alcohol (1, 2). It consists of regular spherical beads of 0.1-0.3 mm diameter which absorb 4-5 fold of their weight from water and aqueous solutions (e.g. biological liquids). Beside absorbing the dissolved components of an aqueous solution, the smaller molecules and the side chains or groups of the larger molecules (e.g. amino acids, polypeptides, lipids etc.) are bound in the CD rings in the form of inclusion complexes. The CD-polymer, due to its physical-chemical qualities is an appropriate candidate for the treatment of crural ulcers.

J. Szejtli and L. Szente (eds.), Proceedings of the Eighth International Symposium on Cyclodextrons, 491–494.

2. MATERIALS AND METHODS

2.1. Wound healing and tissue compatibility studies on rats

The fur on the back of Wistar rats was shaved and wounds were inflicted by a copper rod of 250 °C temperature on the back of anesthetized rats. CDP was sprinkled on to the wounds and healing process was controlled daily. Samples were taken on the 2nd, 4th and 8th days for histopathological examinations.
Tissue reaction to CDP was studied following i.m. injection of 0.5 ml suspension of gas-sterilized polymer beads into the femoral muscle of rats. Beads were suspended in physiological solution of NaCl. Tissue samples were taken 3 and 6 weeks later. All the samples were fixed in formaline, embedded in paraffine and sectioned. Segments were stained with haematoxylin eosine or with Van Gieson's staining.

2.2. Treatment of venous leg ulcers

The CDP powder was spread onto the wound surface in the thickness required to the depth of the ulcer. The wound was covered with Folpack foil then with dry covering dressing. Treatment was applied once daily; the CDP beads were removed by washing with physiological saline solution.

3. RESULTS

3.1. Wound healing effect of CDP on rats

Histological studies showed that CDP mixed with the exudatum and cellular debris formed an inhomogeneous protective layer connected with the wound surface. New skin formation began from the wound edges on the 2nd day. On the 4th day a newly formed, continuous epithelium covered the wound under the protective polymer layer. No incidence of trapping of the beads into the epithelial tissue could be observed. Wounds healed up within 7 to 8 days. No inflammation could be observed around the wounds and in the deeper layer of subcutis either. No exudation was observed.

3.2. Tissue compatibility of CDP in rats

According to histological studies performed on the 3rd week, a characteristic foreign body granuloma covered by a connective tissue capsule could be observed in the muscular tissue. There was no inflammation cell reaction in the tissues surrounding the granuloma.

The great enlargement photos show that histiocytes, foreign body giant cells and other cells of connective tissue origin are involved in the granuloma. The sequestration of the foreign body granuloma began on the 6th week (Fig.1.). The collagen fibers and fibrocytes are organized into a layered structure. There is already a thin capsule of connective tissue origin around the single beads, and the contraction of beads may be observed. There is no inflammation even on the 6th week.

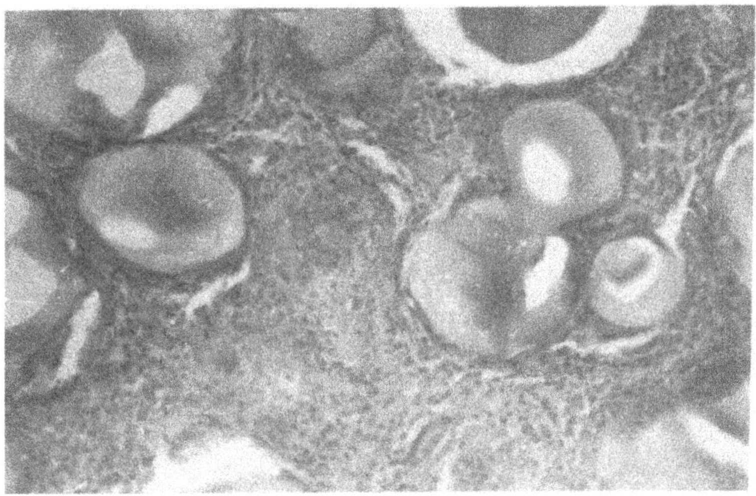

Fig.1. A characteristic foreign body granuloma (Haematoxylin eosin, x210)

3.3. Effect of CDP on the healing of venous leg ulcers

CDP has been applied for ulcers existing for long years and being therapy-resistant (postthrombotical ulcer, scleroderma, rheumatoid arthritis, ulcus trophicum).

The therapeutical effect - tested on 9 patients - was extraordinarily good. The overinfected, coated wounds became clear on the average after 5 days. On the formed granulation tissue the epithelization began under essentially more favourable condition. Both the depth and the basic-area of the ulcers was reduced quickly.

In no case did the patients complain of pain during the application. No allergenic process was observed.

The ulcer-healing effect of CDP is illustrated by the following 4 examples.

1st example: The patient, a 60-year old man, had thrombophlebitis several times, venous thrombosis one time. Ulcus cruris and elephantiasis have developed gradually. Due to

poor venous circulation and lymphoedema, the ulcus became exudative, strongly overinfected. The ulcus showed only minimal improvement during usual local treatments. Following a 14-day treatment with CDP, the exudation of the wound decreased considerably. On the 23th day of treatment, the ulcus was completely cleansed. Epithelization began not only from the wound edges but also from the central parts.
2nd example: The patient, a 60-year old man, had ulcus cruris developed in consequence of postthrombotic syndrome. The ulcus was deep, necrotic at the edges, and overinfected. Following a 15-day CDP treatment, the ulcus was intensively healing, epithelization has begun.
3rd example: The patient, a 70-year old man, had ulcus cruris developed in consequence of postthrombotic syndrome. The ulcus was large, deep and strongly overinfected.
Following a 12-day CDP treatment, the surface of the wound cleansed, and the size of the ulcer decreased considerably.
4th example: The patient, a 55-year old woman, had ulcus cruris developed in consequence of postthrombotic syndrome. Following a 14-day CDP treatment, the wound was cleansed, epithelization has began and the size of the ulcer decreased.

4. CONCLUSION

CDP proved to be effective in healing the wounds on the back of rats. No inflammation or crust formation could be observed.
The lack of any signs of the inflammation cell reactions for up to 6 weeks after i.m. injection of CDP beads to rats suggests that CDP has no local irritating effect.
According to some preliminary trials the therapeutic effect of CDP in humans is very good, too. In case of a large number of patients suffering from Ulcus cruris for years the overinfected wounds were cleansed in 10-12 days. In consequence of the fast healing administration of antibiotics was not necessary and in this way resistence problems were avoided. Not any allergic process, eczematisation, or other toxic effects were observed.

REFERENCES

1. Szejtli, J.; Fenyvesi, É.; Zoltán, S.; Zsadon, B.; Tüdős, F.: Cyclodextrin-polyvinyl alcohol polymers and a process for the preparation thereof in a pearl, foil fiber or block form. U.S. Patent 4,274,985 (1980), HU Appl. CI-1845 (1978)
2. Szejtli, J.: Cyclodextrin technology (Reidel, Dordrecht, 1988) p. 63-65

TOPICAL APPLICATION OF CYCLODEXTRIN ETHERS IN THE CONTROL OF PAIN

M.EVEREST-TODD

Willow House Research Centre

Ings Farm, North Kelsey, Lincolnshire, England, LN7 6LG

ABSTRACT

The mediators of pain and inflammation are generated from membrane lipids with arachidonic acid as the central molecule. It has been found that β- cyclodextrin and dimethyl β-cyclodextrin will clathrate arachidonic acid and the complex is resistant to oxidation. Topical application of β-cyclodextrin and the ether has been found to alleviate erythema and associated pain when applied to open skin lesions using a variety of techniques.

1. INTRODUCTION

The implications of the arachidonic cascade in perturbed tissue produced via phospholipase followed by cyclo-oxygenase or lipoxygenase result in a stream of extremely potent molecules. These ecosanoids may be prevented by cyclo-oxygenase inhibitors such as aspirin, lipoxygenase inhibitors such as benoxaprofen or by inhibiting the release of arachidonic acid with steroidal agents. It was anticipated that β-cyclodextrin could bind the arachidonic acid as it was produced thereby rendering the molecule resistant to oxidation. In order to investigate this hypothesis a range of ecosanoids were introduced to β-cyclodextrin and the dimethyl derivative and subjected to evaluation for resistance to oxidation.

J. Szejtli and L. Szente (eds.), Proceedings of the Eighth International Symposium on Cyclodextrons, 495–498.

Topical application of polysaccharides had been practiced by the ancients as stimulators to the healing processes. This is illustrated by the placing of spider webs over wounds known in the British Isles for over 1000 years.

The decision was made to follow three investigations:

- biochemical investigation of β-cyclodextrin and the ecosanoids;
- clinical investigation of the effect of topical application of β-cyclodextrins and
- clinical investigation of the effect of topical application of β-cyclodextrin together with glucosamine.

2. MATERIALS AND METHODS

2.1. Materials

Preparation of the β-cyclodextrin and dimethyl β-cyclodextrin complexes with arachidonic acid was based on the method described by Cramer and Heuglein (1958). Formulations of β-cyclodextrin for topical application were produced by the addition of a 96% w/w emulsifiable concentrate of coconut oil to a 5.0% w/w aqueous solution of the β-cyclodextrin to give a final concentration of 2.5% w/w cyclodextrin.

A similar method was used to produce the dimethyl β-cyclodextrin equivalent and a comparative group incorporating glucosamine as the hydrochloride at 2.0% w/w.

Surgical dressings were prepared non-woven polyester fabric containing 6 μg cm^2 β-cyclodextrin used in conjunction with a similar fabric containing 12 μg cm^2 glucosamine in conjunction with a water retaining polyacrylate fibre. The design permitted the cyclodextrin to be in direct contact with wound.

2.2. Method

The effectivity of the clathration of arachidonic acid by the cyclodextrins was evaluated by a modified primary irritation test as set out in the Federal Register Method for the testing of primary irritant substances No. 187.

This method evaluates skin reactions as follows:

Erythema and eschar formation	Value
No erythema	0
Very slight erythema (barely perceptible)	1
Well defined erythema	2
Moderate to severe erythema	3
Severe erythema (beet redness) to slight eschar formation (injuries in depth)	4

Applications were made to intact skin using a patch test technique and an exposure time of 6 hours. Readings were taken immediately after exposure and 24 hours later.

Clinical studies were performed on male and female volunteers in accident and emergency situations.

3. RESULTS AND DISCUSSION

3.1. Results

The investigation into effective protection of the arachidonic acid by the cyclodextrins employed a standard biological test as used in the safety evaluation of pharmaceuticals. Mammalian reaction to the arachidonic acid appears to be common to all species but variations will occur in range of ecosanoids subsequently generated.

Results from the patch tests are given in Table I and indicate effective blocking of the lipoxygenase and cyclo-oxygenase pathway.

TABLE I

Chemical applied as test	Value	
	patch + 0 hours	patch + 24 hours
β-cyclodextrin	0	0
β-cyclodextrin + A.A.	1	0
DIMEB	1	0
DIMEB + A.A.	1	0
A.A. (Arachidonic acid)	4	4

The formulation of β-cyclodextrin as a 2.5% w/w concentration in a coconut oil cream was applied to 5 patients suffering from eczema. All reported relief from pruritis within ten minutes of application. In 5 similar cases the formulation used had the addition of 2.0% w/w glucosamine. The results were similar but there was significant regeneration of dermal tissue by the fourth day as opposed to similar tissue regeneration by the seventh day in the cases of β-cyclodextrin alone.

Surgical dressings incorporating β-cyclodextrin or dimethyl β-cyclodextrin in the outer layer of fabric backed with a fabric containing glucosamine, enclosing a non-woven pad of water retaining acrylic material were saturated with sterile water and applied to 2nd degree burns. All such dressings gave relief from pain within minutes of application but the most therapeutic effect was demonstrated by the combinations with glucosamine.

3.2. Discussion

These initial studies have indicated direct benefit from the use of cyclodextrins to clathrate the molecule from which the main mediators of pain and inflammation are derived. The principle mediator of inflammation in the epidermis is the leukotriene LTB_4 which forms a clathrate with cyclodextrin.

It would appear that one use of dimethyl β-cyclodextrin could be as a non-steroidal anti inflammatory agent.

REFERENCES

Cramer, F., Heuglein, F.M., Chem. Ber., 91, 308 (1958)

ANESTHETIC ACTIVITY AND PHARMACOKINETICS OF THE NEUROSTEROID ALFAXALONE FORMULATED IN 2-HYDROXYPROPYL-β-CYCLODEXTRIN IN THE RAT

M. E. BREWSTER[a,b,*], W. R. ANDERSON[a,b], A. WEBB[c], N. BODOR[b] AND E. POP[a,b].

[a]Pharmos, Corp., Alachua, FL, 32615 and Rehovot, Israel, [b]Center for Drug Discovery, College of Pharmacy, University of Florida, Gainesville, FL 32610 and the [c]College of Veterinary Medicine, University of Florida, Gainesville, FL 32610

ABSTRACT

Alfaxalone (3α-hydroxy-5α-pregnan-11,20-dione) is a potent steroid hypnotic agent whose poor water solubility has prompted the development of a 2-hydroxypropyl-β-cyclodextrin (HPβCD) formulation. The anesthetic potency of the complex was examined in a rat model using righting reflex as a measure of sleeping time. The data suggested a sex difference in activity with females proving to be more sensitive than male to the effects of the drugs especially at higher doses. Pharmacokinetic properties of alfaxalone were then determined using a novel analytical method in which the steroid was derivatized with dinitrophenylhydrazine to confer a potent and selective chromophore. Analysis of the time-concentrations profiles indicated that alfaxalone was more readily eliminated from males than female suggesting that the difference in activity between the sexes was related to differential clearance.

1. INTRODUCTION

Alfaxalone is a steroid anesthetic chemically related to progesterone. The compound is reported to possess superior clinical properties regarding induction, maintenance and recovery from anesthesia as compared to other commonly used intravenous agents [1]. While this drug was extensively used and highly considered in the anesthesiologist's armamentarium, its poor water solubility required that it be formulated in the electroneutral detergent, Cremophor [2]. This excipient is associated with anaphylactoid reactions in sensitive individuals prompting its withdrawal from the market in the mid-1980's [3]. In an attempt to resurrect this otherwise useful agent, the steroid was reformulated using 2-hydroxypropyl-β-cyclodextrin [3,4]. The complex formed was found to be highly water soluble generating aqueous alfaxalone solutions well in excess of those attainable using Cremophor and was also stable in solutions under a variety of storage conditions for well over two years. Furthermore, while the Alfaxalone-Cremophor formulation in highly toxic to sensitive species such as the dog, alfaxalone can be safely and conveniently administered to Cremophor sensitive animals with no side-effects or elevation in blood histamine [4]. In continuing our development of a cyclodextrin-based formulation for alfaxalone, we have looked in detail at its

J. Szejtli and L. Szente (eds.), Proceedings of the Eighth International Symposium on Cyclodextrons, 499–502.

pharmacological profile in the rat and developed sensitive and selective analytical procedures to allow for pharmacokinetic assessment. These two data sets then provided a means of correlating pharmacodynamic and pharmacokinetic information.

2. MATERIALS AND METHODS

2.1 Materials

2-Hydroxypropyl-β-cyclodextrin (HPβCD) was obtained from Roquette (France) and was characterized by a degree of substitution of 4.2 by FAB-mass spectrometry. Alfaxalone was obtained from Gideon-Richter (Hungary). Solid complexes were prepared by adding an excess of the steroid to a 43.5% aqueous solution of HPβCD. After equilibrating for 5 days at room temperature, the suspensions were filtered through 0.45 μm membranes and freeze-dried using a Labconco Model 18 lyophilizer. The solid complexes were then milled through a 60 mesh sieve (particle size ≤ 250 μm). The degree of drug incorporation for alfaxalone•HPβCD (Alf•CD) was then determined by HPLC, as previously described [5], and was found to be 154.2 mg/g (15.4 % w/w).

2.2 Animal Studies

Anesthetic and pharmacokinetic assessments were completed using male and female adult rats (HSD:CD, BW = 175-250 g) procured from Harlan Sprague-Dawley, Inc. (Prattville, AL). All animals were housed in standard hanging wire cages in a light- (12h:12h, lights on at 07:00), temperature- (22 °C) and humidity- (65%) controlled vivarium. Animals were provided Teklad 4% rodent chow and tap water ad libitum.

For sleeping time assessments, alfaxalone was administered in doses between 3-132 mg/kg (9-397 μmol/kg) i.v. Rats were fasted 18 h before treatment and 6-8 animals were included in each group. Animals were restrained in a "Broome" holder and the drug injected via the lateral tail vein. The dose of HPβCD was maintained at 435 mg/kg. Parameters examined included loss and re-establishment of righting response, time to anesthetic induction and total sleeping time. The significance of differences in mean sleeping times was analyzed using analysis of variance (ANOVA) with post-hoc Tukey's comparisons. For all tests, the level of probability was $p < 0.05$.

For pharmacokinetic evaluation, groups of male and female Sprague-Dawley rats (n=6) were administered an i.v. dose of alfaxalone (50 mg/kg, 150 μmol/kg) in 22.5% w/v aqueous solution of HPβCD. The injection volume was 1.69 mL/kg and the HPβCD dose, 380 mg/kg. At various times after administration of the drug (0.25 to 24 h), animals were sacrificed and blood and organs collected for alfaxalone determinations. Plasma was separated from heparinized blood by centrifugation was stored as such.

2.3 Analytical Methodology

Tissue (1.0 g) was prepared for HPLC analysis by homogenization in 3.0 mL of deionized water using a Polytron PT 1200C tissue homogenizer (Kinematica AG; Bern, Switzerland). A sample (400 μL) of either the aqueous homogenate or plasma was collected in a 1.5 mL conical microcentrofuge tube, mixed with 800 μL of ice-cold acetonitrile and centrifuged in a Beckman Microfuge 12 at 10,000xg for 10 min. The deproteinized supernatant (500 μL) was then removed and treated with 500 μL of a solution of dinitrophenylhydrazine in acidified aqueous acetonitrile (1.0 mg/mL). The mixture was allowed to react for 30 min at 40 °C.

The derivatized samples were than analyzed by HPLC using an Alltima C18 (5 μm particle size, 150 x 4.6 mm i.d.) analytical column fitted with an Alltima C8 guard column. The mobile phase contained acetonitrile:acetate buffer pH4 (75:25), the flow rate was 1 mL/min and all determinations were completed at ambient temperature. The alfaxalone-dinitrophenylhydrazone was quantitated at 372 nm using a Kratos Spectroflow 757 UV/VIS variable wavelength detector with a retention time of 11 min. The limits of detection were 100 mg/mL for plasma and 200 ng/mL for other tissues. Pharmacokinetic evaluation was completed using the RSTRIP software ver. 2.02 (MicroMath, Inc., Utah).

3. RESULTS AND DISCUSSION

3.1 Anesthetic Potency

The anesthetic potency of alfaxalone complexed with HPβCD was assessed in a rat model using sleeping time as an end point. As illustrated in Figure 1, males and females were equally affected by alfaxalone at dose below 100 μmol/kg (33 mg/kg). As the dose is increased, however, females appear to be markedly more sensitive to the effects of the drug than males. At 60 mg/kg (181 μmol/kg) for example, female rats sleep 2.6-times longer than male rats and at 72 mg/kg (217 μmol/kg) almost three-fold longer. These data are in contrast to previously published data which suggested that while alfaxalone did manifest a sex difference in activity after i.p. dosing that this effect was not apparent after i.v. administration. On the other hand, these early studies used maximal doses of 48 mg/kg suggesting that sex differences may have been present but the abbreviated dose range precluded their observation [6]. We have also observed sex differences in the action of alfaxalone after i.p. administration.

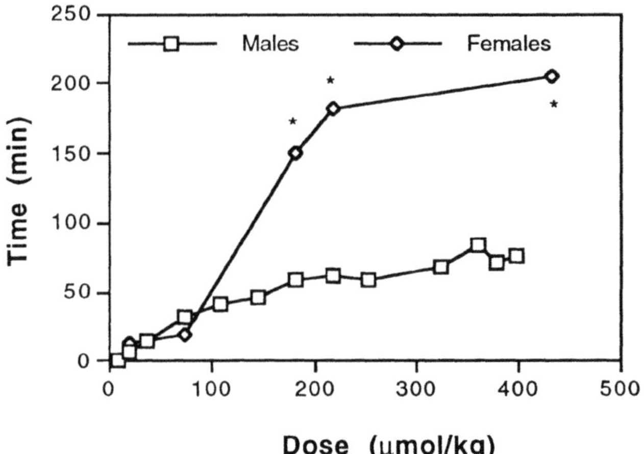

Figure 1. Effect of Rising Doses of Alfaxalone in HPβCD on Sleeping time (in min) in Male and Female Sprague-Dawley Rats (* $p < 0.05$).

3.2 Pharmacokinetic Studies

The disposition of alfaxalone (at a dose of 50 mg/kg, 150 μmol/kg) in male and female Sprague-Dawley rats was assessed using a novel HPLC-based analytical method. The approach employed pre-column derivatization of alfaxalone to generate the alfaxalone dinitrophenylhydrazone. The added chromophore provided for both increased sensitivity and selectivity. The method proved to be precise, accurate and rugged. Analysis of the plasma and organ concentration data by pharmacokinetic modeling was completed with results given in Table I.

Table I. Pharmacokinetic Parameters (Area under the Concentration Curves (AUC_∞) and Terminal Half-Lives ($\beta\ t_{1/2}$) for Alfaxalone in Male and Female Rats.

	AUC (mg•min/L)		$\beta\ t_{1/2}$ (min)	
Tissue	Male	Female	Male	Female
Plasma	906	1812	12.5	26.4
Brain	532	1268	10.1	20.3
Liver	27.8	3039	16.6	33.3
Kidney	615	2367	7.5	15.8
Heart	503	1166	11.3	21.9
Lung	474	1495	10.7	16.6
Fat	1156	3701	9.2	36.1

The data indicate a sex difference in alfaxalone disposition after i.v. administration at this dose examined with terminal half-lives twice as short in male as compared to females. In addition total body clearance was also approximately twice as fast in the case of the males. Interestingly, liver concentrations of alfaxalone were 100-times lower in the male suggesting altered hepatic metabolism or distribution. In any case, the more rapid elimination of alfaxalone in the male is consistent with its shorter duration of action. Interestingly, a previous study came to the opposite conclusion. Fink et al., found that male and female rats dosed with 12 mg/kg alfaxalone (in Cremophor/ethanol/saline) disposed the steroid with equal efficacy [7]. Since we observed pharmacological sex difference only at doses above 30 mg/kg, it is possible that pharmacokinetic differences may only be manifested at high doses.

REFERENCES

[1] Morgan, M., Whitwam, J., Editorial:Althesin, *Anaethesia*, **40**, 121-123 (1985).

[2] Wright, P., Dundee, J., Attitudes to intravenous anaesthesia, *Anaethesia*, **37**, 1209-1213 (1982).

[3] Brewster, M., Estes, K., Bodor, N., Development of a non-surfactant formulation for alfaxalone through the use of chemically modified cyclodextrins, *J. Parent. Sci. Technol.*, **43**, 262-265 (1989).

[4] Estes, K., Brewster, M., Webb, A., Bodor, N., A non-surfactant formulation for alfaxalone based on an amorphous cyclodextrin: activity studies in rats and dogs, *Int. J. Pharm.*, **65**, 101-107 (1990).

[5] Brewster, M., Anderson, W., Loftsson, T., Huang, M., Bodor, N., Pop, E., Preparation, characterization and anesthetic properties of 2-hydroxypropyl-β-cyclodextrin complexes of pregnanolone and pregnenolone in rat and mouse, *J. Pharm. Sci.*, **84**, 1154-1159 (1995).

[6] Child, K., Currie, J., Davis, B., Dobbs, M., Pearce, D., Twissell, D., The pharmacological properties in animals of CT1341 - a new steroid anaesthetic agent, *Brit. J. Anaesth.*, **43**, 2-13 (1971).

[7] Fink, G., Sarkar, D., Dow, R., Dick, H., Borthwick, N., Malnick, S., Twine, M., Sex difference in response to alphaxalone anaesthesia may be oestrogen dependent, *Nature*, **298**, 270-273 (1982).

IMPROVED ORAL BIOAVAILABILITY OF THE BRAIN-TARGETING ESTROGEN, E2-CDS, THROUGH THE USE OF CARBOXYMETHYLETHYL-β-CYCLODEXTRIN

M. E. BREWSTER[a,b,*], T. MURAKAMI[c], W. R. ANDERSON[a,b], N. BODOR[b], and E. POP[a,b].

[a]Pharmos, Corp., Alachua, FL, 32615 and Rehovot, Israel, [b]Center for Drug Discovery, University of Florida, Gainesville, FL 32610 and [c]Department of Pharmacy, School of Medicine, Hiroshima University, Hiroshima, Japan

ABSTRACT

Improved oral bioavailability for the poorly water soluble, acid labile drug, E2-CDS was attempted through the use of carboxymethylethyl-β-cyclodextrin (CMEβCD). E2-CDS could be readily included into CMEβCD with the complex containing 25 mg E2-CDS/g powder. Oral administration of the complex to rats generated significant brain and blood levels of the E2-CDS metabolite, E2-Q^+ indicative of gastrointestinal absorption. In addition, the CMEβCD formulation almost doubled brain and blood levels of E2-Q^+ relative to the analogous HPβCD dosage form.

1. INTRODUCTION

Brain-targeted delivery of estrogens may be highly beneficial in the treatment of not only hormonal deficiencies such as post-menopausal vasomotor symptoms but also cognitive decrements such as Alzheimer's disease [1,2]. The specificity associated with a brain-selective modality include decreased peripheral toxicities most notably cancer of the breast and uterus. One approach for achieving this goal is the chemical delivery (CDS) wherein a molecular targetor is covalently attached to the drug of interest, i.e., estradiol [3,4]. A drug designed along these lines (E2-CDS, estradiol 17β-(1-methyl-1,4-dihydronicotinate) has been tested extensively in animals and appears to function in a manner consistent with the CDS concept [5]. That is a single dose of the drug administered i.v. to animals results in rapid peripheral elimination of the parent hormone while brain levels were elevated for a prolonged period. These increased CNS concentrations resulted in potent behavioral (re-establishment of copulatory behavior in castrate male rats, decreased rate of weight gain) and physiological (increased high affinity choline uptake, decreased prostate weight) effects [3,5].

One problem with E2-CDS is its poor water solubility which effectively prevents convenient i.v. dosing. This shortcoming was resolved through the use of 2-hydroxypropyl-β-cyclodextrin (HPβCD) [6]. This chemically modified β-cyclodextrin increased the aqueous solubility of E2-CDS by more than 100,000-fold (62 ng/mL in aqua pura to 40 mg/mL in 40% w/v HPβCD) and allowed for the configuration of

J. Szejtli and L. Szente (eds.), Proceedings of the Eighth International Symposium on Cyclodextrons, 503–506.

parenteral dosage forms. These formulations have been evaluated in two human clinical trials using both i.v. and buccal administration routes [7,8]. While parenteral administration is important for E2-CDS, the development of the compound would be assisted by the availability of an oral preparation. Unfortunately, E2-CDS is labile to acid and would not be expected to survive passage through the stomach without significant degradation. One approach to this conundrum is carboxymethylethyl-β-cyclodextrin (CMEβCD) [9]. This derivative, similar to enteric coating materials such as cellulose acetate phthalate (CAP), is highly insoluble in acid conditions but is readily soluble in slightly basic conditions. Thus, including a drug into CMEβCD may offer some protection of the guest to the stomach environment after which it may be released in the intestine where it can be absorbed.

2. MATERIALS AND METHODS

2.1 Materials

2-Hydroxypropyl-β-cyclodextrin (HPβCD) was obtained from Roquette (France) and was characterized by a degree of substitution of 4.2 by FAB-mass spectrometry. Carboxymethylethyl-β-cyclodextrin (CMEβCD) was kindly provided by Wako Pure Chemical Industry, Ltd., Japan. E2-CDS was prepared as previously described as were complexes of E2-CDS and HPβCD. For complexes of E2-CDS and CMEβCD, two procedures were followed. In the first, the method of Pitha was followed in which 100 mg of E2-CDS and 500 mg of CMEβCD were dissolved in 10 mL of ethanol and sonicated for 1h [10]. The solvent was then removed, the residue reconstituted with water, filtered and lyophilized. The resulting powder was analyzed by HPLC and was found to contain 25 mg E2-CDS/g. Prototype formulations for animal evaluation were prepared in a different manner. Two grams of either HPβCD or CMEβCD were dissolved in 20 mL of a 0.10 M, pH 9.0 borate buffer. In the case of the CMEβCD system, the pH had to be adjusted with 1N sodium hydroxide. E2-CDS (150 mg) was then dissolved in 2 mL of ethanol and added to each cyclodextrin solution. After stirring for 3 h at 0 °C under argon, the solvent was removed in vacuo, reconstituted with pH 9 borate buffer and lyophilized. One gram of the respective complexes was then suspended in 15 mL of a 1% solution of carboxymethylcellulose (3000-6000 cps) for oral administration to rats. The final suspension was designed to contain 5 mg E2-CDS/mL of vehicle.

2.2 Animal Studies

Male rats (HSD:CD, BW = 220-250 g) were procured from Harlan Sprague-Dawley, Inc. (Prattville, AL). All animals were housed in standard hanging wire cages in a light- (12h:12h, lights on at 07:00), temperature- (22 °C) and humidity- (65%) controlled vivarium. Animals were provided Teklad 4% rodent chow and tap water ad libitum. Animals were fasted 18 h prior to drug administration and were anesthetized (50:5 mg/kg - ketamine:xylazine, i.p) for the procedure. Suspensions (4 mL/Kg giving an E2-CDS dose of 20 mg/kg) were administered through a calibrated intubation tube and animals (n=5) were sacrificed 4 h later. Blood and brain were removed, weighed and frozen until assayed. In preparing samples for HPLC analysis, 1.0 mL of 0.05M ammonium acetate buffer was added to either 1.0 mL of whole blood or whole brain. The systems were then thoroughly homogenized using a Polytron PT-1200C homogenizer, followed by addition of 4 mL of acetonitrile and 1.0 mL of saturated saline. The samples were vortexed and stored at -20 °C for 1 h. The organic phase was then separated and filtered through 0.45 µm membranes.

2.3 Analytical Methodology

E2-Q+ was assayed using an HPLC method. The system included a Waters 510 pump, a Shimadzu SIL-9A autoinjector, a Kratos Spectroflow 757 (UV/VIS) variable wavelength detector and a SpectraPhysics ChromJet integrator. Compounds were eluted on a Spherisorb C8 analytical column (5 µm particle size, 25 cm x 4.6 mm i.d.) fitted with a guard column. The mobile phase contained acetonitrile:ammonium acetate buffer pH 6.8, 0.05 M (75:25) and 10 mM tetraethylammonium perchlorate. The flow rate was 1.0 mL/min and compounds were detected at 266 nm. The retention time for the E2-Q+ was 9.5 min.

3. RESULTS AND DISCUSSION

Improved oral delivery of various pharmaceuticals may be possible through the use of cyclodextrins such as CMEβCD whose release characteristics are pH dependent. Uekama et al have for example found that CMEβCD can serve as a delayed-release type carrier for water soluble drugs with the nature of release depending on the pH environment [9]. In the current application, the ability of CMEβCD to increase oral bioavailability thorough protection of a water-insoluble drug in the stomach was considered. E2-CDS was readily included with CMEβCD generating system containing 25 mg E2-CDS/g powder. This is similar to the extent of incorporation obtained with HPβCD (~35 mg/g).

Administration of E2-CDS to rat p.o. using either a CMEβCD or HPβCD complex resulted in significant difference in compound uptake. As illustrated in Figure 1, E2-CDS in CMEβCD generated significantly higher brain levels of the corresponding CDS metabolite (E2-Q+) than did the HPβCD-containing vehicle.

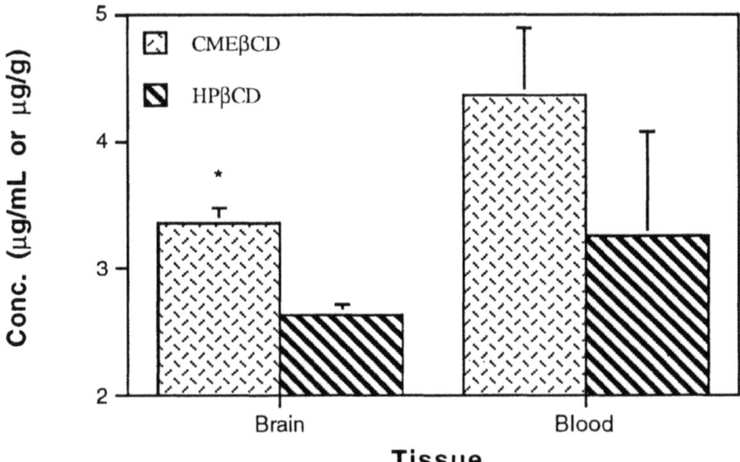

Figure 1. Brain and Blood Concentrations of E2-Q+ after Oral Doses of either E2-CDS complexed to HPβCD or CMEβCD. Tissues were taken at 4 h post-drug administration (* $p < 0.05$).

In addition, there was a tendency for blood levels to be increased as well. Interestingly, both cyclodextrins gave reasonably high blood and brain E2-Q+ level compared to i.v. dosing. For example, a 5 mg/Kg E2-CDS bolus resulted in 6 h brain levels of 3.66 μg/g [11]. These data are consistent with an oral bioavailability of 20-25% although direct comparison will require additional experimental information. In addition, previous work has suggested that oral bioavailability of E2-CDS is highly variable with interanimal responses ranging from 0 to 100% .

3. CONCLUSIONS

E2-CDS is a potentially useful brain-targeted estrogen. A broader range of applications for this nascent drug would clearly be possible if an oral dosage form existed. CMEβCD may be advantageous in this regard by generating a water insoluble complex in the stomach which dissolves only in the more basic environment of the small intestine. Data presented herein suggest that E2-CDS complexes of CMEβCD provide for improved oral bioavailability as compared to HPβCD systems and that the extent of oral bioavailability based on brain concentration of E2-Q+ is on the order of 20-25%

REFERENCES

[1] Grady, D., Rubin, S., Petitti, D., Fox, C., Black, D., Ettinger, B., Ernster, V., Cummings, S., Hormone therapy to prevent disease and prolong life in postmenopausal women, *Ann. Inter. Med.*, **117**, 1016-1037 (1992).

[2] Paganini-Hill, A., Henderson, V., Estrogen deficiency and risk of Alzheimer's disease in women, *Am. J. Epidemiol.*, **140**, 256-261 (1994).

[3] Brewster, M., Simpkins, J., Bodor, N., Brain-targeted delivery of estrogens, *Rev. Neurosci.*, **2**, 241-285 (1990).

[4] Brewster, M., Bartruff, S., Anderson, W., Druzgala, P., Bodor, N., Pop, E., Effect of molecular manipulation on the estrogenic activity of a brain-targeted estradiol chemical delivery system, *J. Med. Chem.*, **37**, 4237-4244 (1994).

[5] Brewster, M., Pop, E., Bodor, N., Chemical approaches to brain-targeting of biologically active compounds. In: *Drug Design for Neurosciences*, pp. 453-467. Kozikowski, A. (Ed.), Raven Press, New York, N.Y. (1993).

[6] Brewster, M., Estes, K., Loftsson, T., Perchalski, R., Derendorf, H., Mullersman, G., Bodor, N., Improved delivery through biological membranes 31: Solubilization and stabilization of an estradiol chemical delivery system by modified β-cyclodextrins, *J. Pharm. Sci.*, **77**, 981-985 (1988).

[7] Estes, K., Brewster, M., Bodor, N., Evaluation of an estradiol chemical delivery system (CDS) designed to provide enhanced and sustained hormone levels in the brain, *Adv. Drug Deliv. Rev.*, **14**, 167-175 (1994).

[8] Howes, J., Brewster, M., Harris, A., Griffith, W., Garty, N., Buccal and parenteral clinical evaluation of a brain-targeting estradiol chemical delivery system (E2-CDS), *Clin. Pharmacol. Therap.*, **57**, 172 (1995).

[9] Uekama, K., Horiuchi, Y., Irie, T., Hirayama, F., O-Carboxymethyl-O-ethylcyclomaltoheptaose as a delayed-release-type drug carrier: improvement of the oral bioavailbility of diltiazam in the dog, *Carbohydr. Res.*, **192**, 323-330 (1989).

[10] Pitha, J., Hoshino, Y., Effect of ethanol on formation of inclusion complexes of hydroxypropylcyclodextrins with testosterone and methyl orange, *Int. J. Pharm.*, **80**, 234-251 (1992).

[11] Brewster, M., Druzgala, P., Anderson, W., Huang, M., Bodor, N., Pop, E., Efficacy of a 3-substituted versus 17-substituted chemical delivery system for estradiol brain targeting, *J. Pharm. Sci.*, **84**, 38-43 (1995).

INTRAVENOUS AND BUCCAL 2-HYDROXYPROPYL-β-CYCLODEXTRIN FORMULATIONS OF E2-CDS - PHASE I CLINICAL TRIALS

M. E. BREWSTER[a,b,*], J. HOWES[a], W. GRIFFITH[c], N. GARTY[a], N. BODOR[b], W. R. ANDERSON[a,b] and E. POP[a,b].

[a]Pharmos, Corp., Alachua, FL, 32615 and Rehovot, Israel, [b]Center for Drug Discovery, University of Florida, Gainesville, FL 32610 and [c]Pharmakopius, Ltd., Goring-on-Thames, U.K.

ABSTRACT

Intravenous and buccal dosage forms were configured for E2-CDS, a brain-targeting estrogen delivery system, using 2-hydroxypropyl-β-cyclodextrin (HPβCD) and examined in a Phase I clinical protocol. Eighteen post-menopausal volunteers were administered a single oral dose of Progynova (estradiol valerate, 1.0 mg). Four days later the group was randomized to receive either i.v. E2-CDS (n=9, 0.6 to 2.5 mg in 20% HPβCD) or buccal i.v. E2-CDS (n=9, 2.0 to 16.0 mg in HPβCD administered as a solution). Subsequent to the rising dose study, nine of the volunteers were treated with estradiol (equimolar to 2.5 mg E2-CDS in HPβCD). All treatments were well tolerated. Both i.v. and buccal treatment lead to dose-dependent decreases in serum LH consistent with brain-targeted delivery. In the case of the buccal treatments, bioavailability was estimated at 20-25%.

1. INTRODUCTION

Estrogen depletion associated with the menopause produces a constellation of debilitating symptoms which range the gamut from vasomotor complaints which severely affect over one-third of all climacteric women to cognitive deficits [1,2]. While many of these complications can be alleviated with traditional hormone replacement therapy (HRT), fear of cancer and other metabolic disease prompt many women to avoid treatment. A recently completed survey suggests that over 30% of all prescriptions written for HRT are never filled due to these concerns. One approach to address the central components of postmenopausal symptoms without elevating peripheral concentrations is the chemical delivery system methodology. Such a system for estradiol (E2-CDS or estradiol 17β-(1-methyl-1,4-dihydronicotinate)) has been synthesisized and extensively tested [3-5]. Various preclinical evaluations have demonstrated the organ-targeting potential of the E2-CDS and therefore its potential usefulness as a therapeutic modality in certain subpopulations of menopausal women including those at risk to breast carcinomas. A very exciting potential use of the E2-CDS is in neurodegenerative diseases. Recent data have suggested that estrogens may improve the mental performance of elderly patients suspected of having Alzheimer's disease [6].

J. Szejtli and L. Szente (eds.), Proceedings of the Eighth International Symposium on Cyclodextrons, 507–510.

One difficulty in developing E2-CDS as a drug is its physicochemical profile. The drug is poorly water soluble and manifests limited aqueous stability. These untoward properties complicate the development of dosage forms intended for either i.v. or buccal administration. Cyclodextrins, and in particular, 2-hydroxypropyl-β-cyclodextrin, were found to improve E2-CDS solubility and stability [7]. These attributes as well as the safety profile of the excipient has allowed its application to formulations used in Phase I clinical trials intended to provide proof of concept for the E2-CDS [8,9]. We have extended these initial human studies in order to: (1) increase the number of volunteers exposed to E2-CDS, (2) collect further safety information on the E2-CDS in humans, (3) establish the feasibility for buccal dosing in humans and (4) demonstrate the central action of the E2-CDS relative to a both commercially available oral preparations and parenteral estradiol.

2. MATERIALS AND METHODS

2.1 Materials

Progynova (estradiol valerate, 1.0 mg tablet) was obtained from Schering Healthcare Ltd, Burgess Hill, West Sussex, UK. E2-CDS was complexed with HPβCD in a borate buffer (0.005 M sodium tetraborate) and lyophilized in 5 mL vials by University of Strathclyde, Glasgow and provided to two strengths: a 30.8 mg E2-CDS/vial dosage form (containing 1.17 g HPβCD) for buccal administration and a 3.1 mg E2-CDS/vial preparation (containing 0.6 g HPβCD) for i.v. dosing. The HPβCD was present at a concentration of 40% w/v in the buccal formulation and 20% in the i.v. preparation. Samples were diluted with water for injection (USP) and administered i.v. (at volumes between 0.5 and 4.0 mL) or buccally (at volumes between 0.14 and 1.14 mL).

2.2 Clinical Protocols

Studies were conducted by Pharmakopius, Goring-on-Thames, UK as a double-blind, randomized, parallel group study. The study designed included 18 woman who received an oral preparation of E2 valerate (Progynova®) on their first visit followed by clinical and endocrinological follow-up. The group was then randomized into two subgroups that were to receive E2-CDS either buccally or i.v. The doses in both arms of the study were administered in a rising fashion starting at 0.6 mg in the i.v. portion of the study and 2.0 mg in the buccal arm. The highest doses examined were 2.5 mg i.v. and 16 mg buccally. At the end of the experiment, the groups were recombined and nine of the eighteen woman were selected for an i.v. dose of E2 (solubilized in HPbCD equimolar to the 2.5 mg E2-CDS dose). For all treatments, clinical chemistry, hematology and hormone levels were determined and recorded over a prolonged time course.

3. RESULTS AND DISCUSSION

Both i.v. and buccal dosing of E2-CDS demonstrated a dose-related suppression of circulating LH levels (Figure 1). A comparison of buccal doses with biological effect suggested significant bioavailability on the order of 25%. A comparison of the area under the LH suppression curves and plasma E2 curves for E2-CDS and the commercially available oral preparation, Progynova, suggested that the E2-CDS was significantly more active (as measured by LH suppression) yet manifested less of a tendency to elevate serum E2. This was manifested after either buccal or i.v. administration. In a comparison of i.v. E2 versus i.v. E2-CDS, similar results to the initial trial were documented.

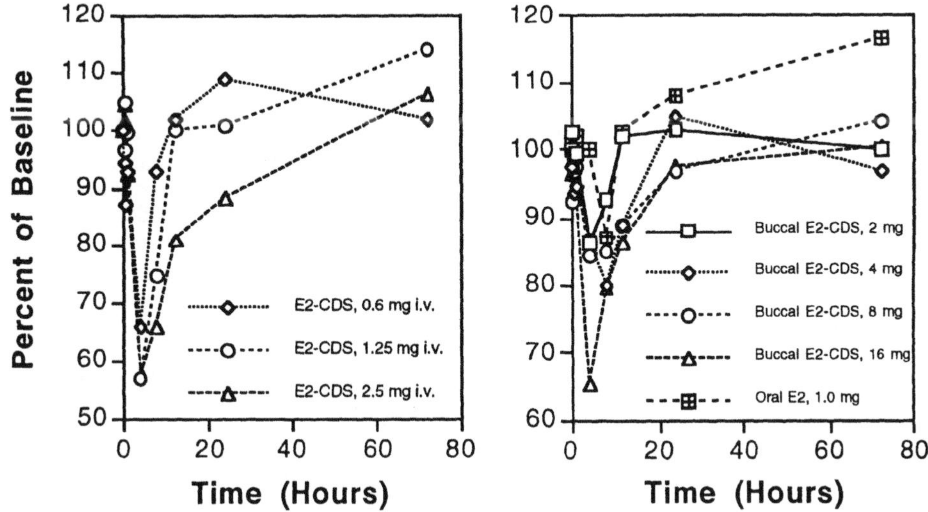

Figure 1. Effect of Various Doses of E2-CDS Administered either I.V. or Bucally on Serum LH levels in Post-Menopausal Female Volunteers

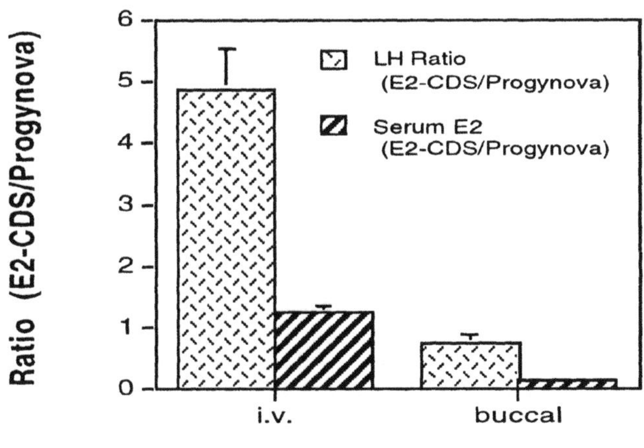

Figure 2. The Effect of I.V. or Buccal E2-CDS relative to Oral Progynova on LH Suppression or E2 Serum Elevation. Data are collected from all dosing groups and are corrected for baseline and dose.

Thus, while E2 and E2-CDS exerted similar effects on LH suppression, the E2-CDS elevated E2 levels by only a small fraction (4%) as compared to i.v. E2 administration. The results of this limited study indicated that the E2-CDS is acutely safe at levels far in excess of those contemplated for commercial use. In addition, the evaluation found that E2-CDS was bioavailable buccally and that, if anything, the doses administered overestimated the potential clinical need. A cursory pharmacokinetic evaluation of the data also suggested that peripheral estradiol levels can be easily manipulated through alterations in dosing amounts and frequency. This is related to the long half-life of E2-CDS in the CNS. Thus a dosing schedule can be devised that minimizes plasma E2 as a function of central E2 or that produces a particular peripheral level of E2.

3. CONCLUSIONS

E2-CDS has been demonstrated to be a useful, potent and long-lasting estradiol delivery system that targets the central nervous system. The drug substance can be prepared in high yield and purity. Data suggest that the best formulation for this potentially useful drug is a HPbCD complex which provided for both buccal/sublingual and intravenous administration. Clinical evaluations of the E2-CDS suggest a potent central effect with only marginal tendencies to elevate systemic estrogen levels. Importantly, peripheral levels can be easily manipulated using the delivery system as a function of dosing amount and frequency. Taken as a whole, the data suggest that the E2-CDS may be a useful and safe therapy for menopausal symptoms as well as estrogen-dependent cognitive deficits.

REFERENCES

[1] Grady, D., Rubin, S., Petitti, D., Fox, C., Black, D., Ettinger, B., Ernster, V., Cummings, S., Hormone therapy to prevent disease and prolong life in postmenopausal women, *Ann. Inter. Med.*, **117**, 1016-1037 (1992).

[2] Session, D., Kelly, A., Jewelewicz, R., Current concepts in estrogen replacement therapy in the menopause, *Fertil. Steril.*, **59**, 227-284 (1993).

[3] Brewster, M., Simpkins, J., Bodor, N., Brain-targeted delivery of estrogens, *Rev. Neurosci.*, **2**, 241-285 (1990).

[4] Brewster, M., Bartruff, S., Anderson, W., Druzgala, P., Bodor, N., Pop, E., Effect of molecular manipulation on the estrogenic activity of a brain-targeted estradiol chemical delivery system, *J. Med. Chem.*, **37**, 4237-4244 (1994).

[5] Brewster, M., Druzgala, P., Anderson, W., Huang, M., Bodor, N., Pop, E., Efficacy of a 3-substituted versus 17-substituted chemical delivery system for estradiol brain targeting, *J. Pharm. Sci.*, **84**, 38-43 (1995).

[6] Paganini-Hill, A., Henderson, V., Estrogen deficiency and risk of Alzheimer's disease in women, *Am. J. Epidemiol.*, **140**, 256-261 (1994).

[7] Brewster, M., Estes, K., Loftsson, T., Perchalski, R., Derendorf, H., Mullersman, G., Bodor, N., Improved delivery through biological membranes 31: Solubilization and stabilization of an estradiol chemical delivery system by modified β-cyclodextrins, *J. Pharm. Sci.*, **77**, 981-985 (1988).

[8] Howes, J., Bodor, N., Brewster, M., Estes, K., Eve, M., A pilot study with PR-63 in postmenopausal volunteers, *J. Clin. Pharmacol.*, **28**, 951 (1988).

[9] Estes, K., Brewster, M., Bodor, N., Evaluation of an estradiol chemical delivery system (CDS) designed to provide enhanced and sustained hormone levels in the brain, *Adv. Drug Deliv. Rev.*, **14**, 167-175 (1994).

EFFECT OF 2-HYDROXYPROPYL-β-CYCLODEXTRIN COMPLEXES OF THE NEUROSTEROIDS, ALFAXALONE, PREGNANOLONE AND PREGNENOLONE, ON VARIOUS CONVULSANT STIMULI IN THE MOUSE

M. E. BREWSTER[a,b,*], W. R. ANDERSON[a,b], T. LOFTSSON[c], N. BODOR[b] AND E. POP[a,b].

[a]Pharmos, Corp., Alachua, FL, 32615 and Rehovot, Israel, [b]Center for Drug Discovery, College of Pharmacy, University of Florida, Gainesville, FL 32610 and the [c]Department of Pharmacy, University of Iceland, Reykjavik, Iceland

ABSTRACT

Water-soluble complexes of the progesterone derivatives, alfaxalone, pregnanolone and pregnenolone were prepared and characterized. The complexes were tested in vivo in the mouse using two administration routes (i.v. and i.p.) and six convulsant stimuli. Intravenous administration of alfaxalone and pregnanolone inhibited convulsions associated with electrical and pentylenetetrazole treatment but not after picrotoxin, bicuculline or strychnine treatment. Pregnenolone was only effective in pentylenetetrazole threshold seizures and only at higher doses. On the other hand, i.p. treatment with the steroids revealed that both alfaxalone and pregnanolone were active against picrotoxin- and bicuculline-induced seizures while manifesting poor action against maximum electroconvulsive shock.

1. INTRODUCTION

Various synthetic and naturally occurring progesterone derivatives exert significant anesthetic, anxiolytic and anticonvulsant action even though these compounds are inactive as glucocorticoids, mineralocorticoids and sex steroids [1-4]. Many of these effects, especially the antiseizure manifestations, are thought to be related to the ability of these compounds to rapidly modulate neuronal excitability, most likely through a potentiation of GABA interaction with the $GABA_A$-benzodiazepine receptor complex. Based on these activities, several compounds including the progesterone A-ring reduction products, pregnanolone and pregnenolone as well as alfaxalone have been evaluated as potential antiepileptic agents.

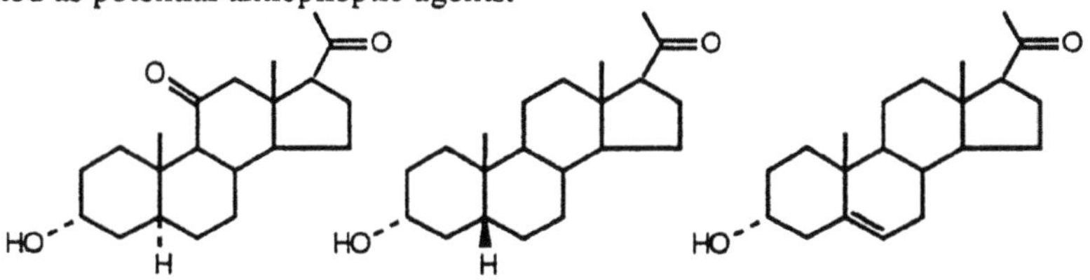

Alfaxalone (Alf) **Pregnanolone (Nan)** **Pregnenolone (Nen)**

J. Szejtli and L. Szente (eds.), Proceedings of the Eighth International Symposium on Cyclodextrons, 511–514.

Unfortunately, current formulations of these steroids are problematic. Alfaxalone is available as a Cremophor®-ethanol cocktail (for veterinary use) which is diluted with saline or other aqueous vehicles prior to i.v. administration. Cremophor is an electroneutral detergent that solubilizes poorly water-soluble agents through micellular formation. It is also associated with anaphylactoid reactions in sensitive individuals prompting the removal of the product intended for human use, Althesin®, from the market in the mid-1980's [4,6]. Pregnanolone has not been approved for human use but is available as an emulsion for investigational purposes [7].

We sought to generate safe and useful formulations for neurosteroids in general and for alfaxalone, pregnanolone and pregnenolone in particular using water-soluble cyclodextrins. 2-Hydroxpropyl-β-cyclodextrin was selected for feasibility investigations because of its proven ability to enhance the solubility of steroids, its low toxicological potential and because of the extensive human experience of this excipient.

2. MATERIALS AND METHODS

2.1 Materials

2-Hydroxypropyl-β-cyclodextrin (HPβCD) was obtained from Roquette (France) and was characterized by a degree of substitution of 4.2 by FAB-mass spectrometry. Alfaxalone was obtained from Gideon-Richter (Hungary) while pregnanolone and pregnenolone were procured from Sigma Chemical Co. (US). Solid complexes were prepared by adding an excess of the steroid of interest to a 43.5% aqueous solution of HPβCD. After equilibrating for 5 days at room temperature, the suspensions were filtered through 0.45 μm membranes and freeze-dried using a Labconco Model 18 lyophilizer. The solid complexes were then milled through a 60 mesh sieve (particle size ≤ 250 μm). The degree of drug incorporation was then determined by HPLC, as previously described [1], and was found to be: Pregnanolone•HPβCD (Nan•CD) = 101.3 mg/g (i.e., the complex was 10.1% pregnanolone by weight), pregnenolone•HPβCD (Nen•CD) = 56.0 mg/g (5.6 % w/w) and alfaxalone•HPβCD (Alf•CD) = 154.2 mg/g (15.4 % w/w).

2.2 Animal Studies

Anticonvulsant tests were completed using male ICR mice weighing 30-35 g (Harlan Sprague-Dawley, Indianapolis, Indiana, US). Animals received i.v. (via the lateral tail vein) or i.p. doses of Nan•CD, Nen•CD or Alf•CD followed by a convulsant stimulus. A stock solution of the appropriate steroid complex was prepared and diluted with a blank solution of HPβCD to provide the desired steroid dose at a constant level of cyclodextrin (22.5% w/v). For i.v. treatments, the volume of the injection was 3.0 mL/Kg (giving a HPβCD dose of 675 mg/Kg) and for i.p. administration, 10 mL/Kg (giving an HPβCD dose of 2.25 g/Kg). Steroid doses ranged from 0 to ~150 mg/Kg i.v. and 0 to ~300 mg/Kg i.p.

One of six convulsant stimuli were administered 15 min subsequent to neurosteroid dosing [8]. These included: (1) Maximum electroconvulsive shock (MES) -70 mA, 60 Hz, 0.2 sec. shock was delivered to the animal by means of pinna electrodes. (2) Maximum pentylenetetrazole seizures (MPS) - mice were given a 150 mg/Kg dose of pentylenetetrazole (Sigma Chemical Co., St. Louis, Missouri, US) i.p. in saline. (3) Pentylenetetrazole threshold seizure (PTS) - mice were administered 85 mg/Kg pentylenetetrazole s.c. in saline. (4) Bicuculline- (5) picrotoxin- and (6) strychnine-induced seizure (BIS), (PIS) and (SIS), respectively, - administration of a freshly prepared acidic solution of bicuculline at a dose of 2.70 mg/Kg s.c. in saline, picrotoxin at a dose of 3.15 mg/Kg, s.c. in saline or strychnine, 1.2 mg/Kg, s.c. in saline. Abolishment of tonic or clonic and other seizure manifestations were considered protection. Median effective doses (ED_{50}'s) of neurosteroids were determined by using the method of probits.

3. RESULTS AND DISCUSSION

3.1 Intravenous Studies

As demonstrated in Table I, both Alf•CD and Nan•CD protected mice against MES with estimated ED_{50} values of 13.8 and 23.1 mg/Kg, respectively, while the unsaturated derivative, Nen•CD was ineffective in this model. Similarly, mice were protected against MPS by both Alf•CD and Nan•CD while Nen•CD was not effective. By contrast, all three of the tested steroids inhibited PTS with the highest potency observed in the case of the Nan•CD (ED_{50} = 4.1 mg/Kg) followed by Alf•CD (ED_{50} = 8.4 mg/Kg) and Nen•CD (ED_{50} = 21 mg/Kg). At the i.v. doses employed in this study, none of the mice were protected by the three steroids against BIS, PIS or SIS.

Table I. Anticonvulsant Effects (ED_{50} in mg/Kg and μmol/Kg), Number of Dosing Groups (n) and the Square of the Correlation Coefficient (r^2) for Intravenous Alf•CD, Nan•CD and Nen•CD in Various Mouse Seizure Models.

Paradigm	Alf•CD mg/Kg, μmol/Kg (n, r^2)	Nan•CD mg/Kg, μmol/Kg (n, r^2)	Nen•CD mg/Kg, μmol/Kg (n, r^2)
MES	13.8, 41.6 (5, 0.969)	23.1, 72.4 (4, 0.952)	N.P.
MPS	17.0, 51.0 (6, 0.941)	14.2, 44.6 (6, 0.915)	N.P.
PTS	8.4, 25.3 (7, 0.832)	4.1, 12.9 (5, 0.897)	21.0, 66.4 (5, 0.940)
PIS, BIS, SIS	N.P.	N.P.	N.P.

N.P. - no protection

3.2 Intraperitoneal Studies

Intraperitoneal administration of the neurosteroids produced a different spectrum of action compared to i.v. treatment. As indicated in Table II, neither Nan•CD or Alf•CD exerted anticonvulsant action in MES when administered i.p. When Nan•CD was examined for its anti-MPS activity, it was found to inhibit seizure but only at doses 3-fold higher than those employed i.v. Thus, the ED_{50} increased from 14.2 mg/Kg to 45.7 mg/Kg. Unlike in the i.v. case, steroids administered i.p. did exert significant action against various chemical convulsants consistent with previously published reports [2]. Alf•CD and Nan•CD were equally effective in their inhibition of PIS with ED_{50}'s of 27.8 and 32.9 mg/Kg, respectively. In the case of the BIS paradigm, both steroid complexes tested were active with Alf•CD exerting an ED_{50} of 16.2 mg/Kg and Nan•CD and ED_{50} of 8.7 mg/Kg. Literature values for carbamazepine indicated that the neurosteroids tested were 50% more potent against PIS and 5-20-fold more potent against BIS [8].

Table II. Anticonvulsant Effects (ED_{50} in mg/Kg and μmol/Kg), Number of Dosing Groups (n) and the Square of the Correlation Coefficient (r^2) for Intraperitoneal Alf•CD and Nan•CD in Various Mouse Seizure Models.

Paradigm	Alf•CD mg/Kg, μmol/Kg (n, r^2)	Nan•CD mg/Kg, μmol/Kg (n, r^2)
MES	N.P.*	N.P.
MPS	N.D.	45.7, 143.3 (4, 0.999)
PIS	27.8, 83.7 (4, 0.958)	32.9, 103.2 (4, 0.912)
BIS	16.2, 44.8 (4, 0.974)	8.7, 27.4 (6, 0.870)

* 10% protection at 200 μmol/Kg, N.P. - no protection, N.D. - not determined

4. CONCLUSIONS

2-Hydroxypropyl-β-cyclodextrin proved to be an efficient and safe solubilizing agent for the neurosteroids examined allowing the production of prototype dosage forms to study the anticonvulsant action of these compounds uncomplicated by organic co-solvents or surfactants. The steroids, formulated in this way, exerted rapid and potent anticonvulsant activity suggesting that the bioavailability of the agents were in no way reduced by complexation. It is important to note that the steroids examined also exert anesthetic action resulting in reduced locomotor activity. For example, the ED_{50} of inhibition of MES, MPS and PTS by Alf•CD was found to be 41.6, 25.3 and 51.0 μmol/Kg i.v. In separated experiments, 25 and 50 μmol/Kg i.v. doses of alfaxalone (as Alf•CD) resulted in sedation of 100% of mice tested with average sleeping times of 12.5 and 6.2 min, respectively. These results are consistent with other data suggesting that there is poor separation between anesthetic and anticonvulsant action which may limit the usefulness of these compounds [3]. These neurosteroids do, however, manifest a different spectrum of activity as compared to carbamazepine which may be important in various drug-resistant epilepsies.

REFERENCES

[1] Brewster, M., Anderson, W., Loftsson, T., Huang, M., Bodor, N., Pop, E., Preparation, characterization and anesthetic properties of 2-hydroxypropyl-β-cyclodextrin complexes of pregnanolone and pregnenolone in rat and mouse, *J. Pharm. Sci.*, **84**, 1154-1159 (1995).

[2] Belelli, D., Bolger, M., Gee, K. Anticonvulsant profile of the progesterone metabolite 5a-pregnan-3a-ol-20-one, *Eur. J. Pharmacol.*, **166**, 325-329 (1989).

[3] Peterson, S., Anticonvulsant profile of an anesthetic steroid, *Neuropharmacol.*, **28**, 877-879 (1989).

[4] Luntz-Leybman, V., Frend, R., Collins, A. 5α-Pregnan-3α-ol-20-one blocks nicotine-induced seizures and enhances paired-pulse inhibition, *Eur. J. Pharmacol.*, **185**, 239-242 (1990).

[5] Estes, K., Brewster, M., Webb, A., Bodor, N. A non-surfactant formulation for alfaxalone based on amorphous cyclodextrin: activity studies in rats and dogs, *Int. J. Pharm.*, **65**, 101-107 (1991).

[6] Brewster, M., Esters, K., Bodor, N. Development of a non-surfactant formulation for alfaxalone through the use of chemically modified cyclodextrins, *J. Parent. Sci. Technol.*, **43**, 262-265 (1989).

[7] Brewster, M., Anderson, W., Estes, K., Bodor, N., Development of aqueous parenteral formulations for carbamazepine through the use of modified cyclodextrins, *J. Pharm. Sci.*, **80**, 380-383 (1991).

PHARMACOKINETIC STUDY OF ORALLY ADMINISTERED KETOCONAZOLE AND ITS MULTICOMPONENT COMPLEX ON RABBITS OF NORMAL AND LOW GASTRIC ACIDITY

A. GERLÓCZY[1], J. SZEMÁN[1], K. CSABAI[1], I. KOLBE[1], L. JICSINSZKY[1], D. ACERBI[2], P. VENTURA[2], E. REDENTI[2], J. SZEJTLI[1]
[1]Cyclolab R&D Lab.,Ltd
H-1525 Budapest P.O. Box 435, Hungary
[2]Chiesi Farmaceutici S.p.A.,
43100 Parma, Via Palermo 26/A, Italy

ABSTRACT

Multicomponent complex formation of ketoconazole (KC) with β-cyclodextrin (βCD) in the presence of tartaric or citric acids resulted in freely soluble complexes, more than 100 mg/ml dissolved KC concentration could be achieved at a final pH of 3.6. In vivo studies on rabbits have proven that multicomponent complex significantly improves the absorption and bioavailability of KC under both normal and achlorhydric conditions. Only multicomponent complex seems to be feasible for the effective treatment of patients suffering from achlorhydria.

1. INTRODUCTION

The bioavailability of orally administered KC, a potent member of systemically administered antifungal azole derivatives varies greatly from patient to patient according to the actual gastric pH, as KC can be solubilized only at extremely acidic pH.
Previous studies have shown that the aqueous solubility of drugs containing a piperidine group as a basic moiety - e.g. terfenadine - increased extremely in the presence of acids and cyclodextrins due to multicomponent complex formation (1). Simultaneously, also the solubility of βCD showed to be largely enhanced and was higher than the solubility of βCD in the presence of acid alone.
Similarly, multicomponent complex formation of KC - which contains a piperazine group - with βCD in the presence of tartaric or citric acids resulted in freely soluble complexes, more than 100 mg/ml dissolved KC concentration could be achieved at a final pH of 3.6.

J. Szejtli and L. Szente (eds.), Proceedings of the Eighth International Symposium on Cyclodextrons, 515–518.

It was supposed that the in vivo absorption of KC from the gastrointestinal tract is enhanced as well due to the increased solubility at neutral or slightly alkaline pH.

2. MATERIALS AND METHODS

2.1. Materials

KC was obtained from Chiesi Pharm. (Parma, Italy), β-cyclodextrin (βCD) was supplied by Wacker Chemie (Munich, Germany). Tartaric acid was purchased from Reanal (Budapest, Hungary).

2.2 Preparation of complexes

Multicomponent complex: 2.12 g KC (4 mmol), 0.61 g L(+)-tartaric acid (4 mmol) and 7.95 g βCD (water content: 13.66%, 6 mmol) were dissolved in 12 ml water and lyophilized to obtain solid product. KC content: 21.66%
Binary complex: 1.33 g KC and 7.76 g βCD were suspended in distilled water, stirred for 24 hours at room temperature, turraxed with Ultraturrax T25 for 3x2 minutes, and the white suspension was lyophilized. KC content:16.2%

2.3 Pharmacokinetic study

Pharmacokinetic studies were performed on New Zealand white male rabbits weighing 2.8 kg average. In the 1st experiment, rabbits having physiological gastric pH (1.7) were used. In the 2nd experiment, rabbits were treated with 1 mg/kg Omeprazol s.c., two times daily, altogether 12 times, in order to achieve an elevated gastric pH (6.5-8). Two hours following the last treatment, rabbits were treated orally with KC or its binary and multicomponent complexes. Test substances were administered in 0.5% methylcellulose solution per os. The dose related to KC was 15 mg/kg.

2.3. Determination of KC by HPLC method

Heparinized blood was taken from rabbits and KC was determined from plasma by a reverse phase HPLC method. A Hewlett-Packard 1050 equipment was used. KC was extracted from plasma samples by solid phase extraction using Bakerbond C18 (100 mg, No 7020-01) columns.
HPLC conditions: column: NUCLEOSIL 120-5 C18 (Macherey-Nagel), mobile phase: 700 ml methanol and 300ml triethylamine phosphate buffer (pH=7.6)
Flow rate: 1.0 ml/min., column temperature: 40 °C, detector wavelength: 231 nm

3. RESULTS AND DISCUSSION

At first, pharmacokinetics of KC following oral administration of KC and its multicomponent complex were compared on rabbits with normal, physiological gastric pH (Fig. 1, Table I.)

Table I. Pharmacokinetic parameters
(1st experiment, rabbits with normal gastric pH)

	KC	Multicomponent complex
C_{max} (μg/ml)	2.495	4.352
T_{max} (hour)	1.271	0.993
AUC_{0-5}* [(μg/ml)xhour]	9.073	12.771

* statistical significance: $p<0.01$

Both C_{max} and AUC_{0-5} values were higher in the multicomponent complex treated animals than those in KC treated ones (1.7 times and 1.4 times higher values, resp.). The T_{max} values were 1 and 1.3 hours, resp. Multicomponent complex increased the absorption of KC significantly compared to that of free KC substance.
In the second experiment, absorption of KC and its binary and multicomponent complex was compared on rabbits having a gastric pH of 6.5-8 (Fig. 1., Table II.). Absorption of orally administered KC was almost completely inhibited under high gastric pH conditions. On the contrary, KC was absorbed from both binary and multicomponent complexes. Plasma KC levels were 4-5 times higher in the multicomponent complex-treated group than those in the binary complex-treated animals in the first hour; the difference was even larger at later time points. The AUC_{0-8} value was more than 6 times higher in the multicomponent complex-treated rabbits. (AUC_{0-8} values : 0.053, 0.651 and 4.16 (μg/ml)xh for KC, binary and multicomponent complex, resp.).
For comparison, AUC values (all calculated for the 0-5 hour interval) of the rabbits in the two experiments are summarized in Table II.

Table II. AUC values (1st and 2nd experiment, 0-5 hour interval)

	1st experiment		2nd experiment		
	KC	Multicomponent complex	KC	Binary complex	Multicomponent complex
AUC_{0-5} [(μg/ml)xh]	9.073	12.771	0.053	0.651	3.858

It can be seen that under high gastric pH conditions only the multicomponent complex provides a KC absorption comparable to that in rabbits of normal gastric acidity.

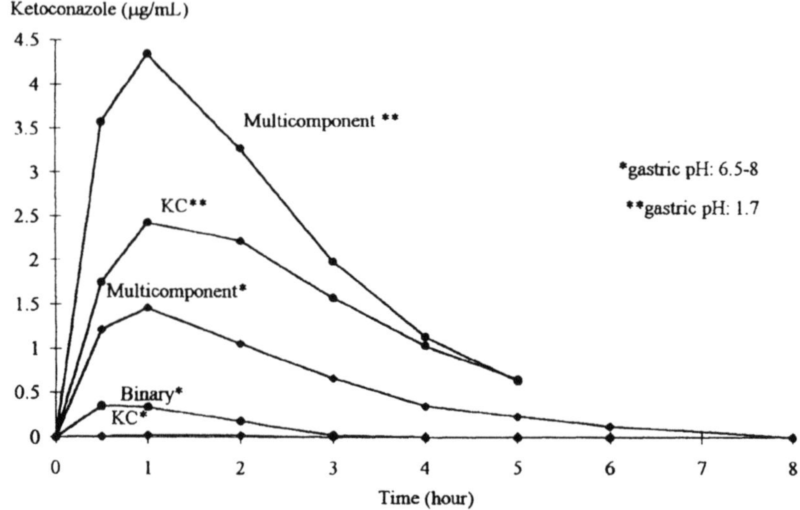

Fig. 1. Absorption of KC following oral administration to rabbits with physiological and elevated gastric pH

4. CONCLUSION

The bioavailability of orally administered KC in rabbits can be significantly enhanced by the new multicomponent type complex.
Multicomponent complex provides absorption of KC even at slightly alkaline pH conditions of the stomach and under these conditions is superior to the binary complex.

ACKNOWLEDGEMENT

Thank is due to Ms. K. Tóth, O. Nemes and Zs. Zachár for their contribution to this work.

REFERENCES

[1] P. Chiesi, P. Ventura, M. Pasini, E. Redenti, J. Szejtli, M. Vikmon: Highly soluble multicomponent complexes containing a base-type drug, an acid and CD. PCT Patent Appl. WO 94/16733.

Chapter 4.
INDUSTRIAL APPLICATION OF CYCLODEXTRINS

APPLICATION OF CYCLODEXTRINS IN BIOTECHNOLOGY

RAPHAEL BAR*
Dept. of Applied Microbiology,
The Hebrew University , Jerusalem, Israel

1. INTRODUCTION

Cyclodextrins (CDs) are increasingly being introduced in biotechnological applications. While Biotechnology is certainly a broad term, it implies here processes involving cultivation of cells (microbial, animal or plant cells) or formation of a product mediated by either cell culture or enzymes. All applications of CDs in biotechnology are firmly anchored onto their well known ability to interact and form complexes with organic substances and they have been recently reviewed extensively [1]. In some applications, these substances are exogenous and serve as substrates to the enzyme or cell, while in others, they are *in situ* generated as their product. Most often, these substrates or products are either water-insoluble, toxic or inhibitory to the biocatalyst (cell or enzyme) and their interaction with CDs results in an enhanced bioavailability or stimulatory effect on the enzymatic or metabolic activities. Most often, the net effect of CDs is therefore an intensified bioprocess. Yet, in every application in which CDs are contacted with enzymes or cells, interactions between the two are also possible by virtue of either their complexing ability and/or their surface activity. These interactions could result in altered metabolic and catalytic activities of the cell and enzyme, respectively. While these types of concurrent CD-biocatalyst and CD-substance interactions were reviewed in the former symposium, this presentation addresses the very applications of CDs in biotechnology. These applications include the following biotechnological domains:

- cultivation of cells in CD media
- biosynthesis by fermentation
- bioconversions in CD medium
- microbial degradations of organics in CD media

* Present address: Teva Pharmaceutical Industries Ltd., R&D Dept., P.O.Box 353, Kfar-Sava, Israel.

J. Szejtli and L. Szente (eds.), Proceedings of the Eighth International Symposium on Cyclodextrons, 521–526.

2. CULTIVATION OF CELLS IN CD MEDIA

Substituting serum for CD in cultivation of animal cells can be seen as the first biotechnology-oriented application of CDs and this was reported by Yamane et al [2]. Animal cells are usually grown in a medium supplemented with serum (up to 10%) as a source of sterol, fatty acids, hormones, vitamins and some other undefined growth factors. Cyclodextrins can substitute, when this is required, for the binding of cholesterol and fatty acids as well as for detoxifying actions of serum. To this purpose, Yamane *et al.*[2] demonstrated the growth of human lymphoblast cells in a serum-free medium supplemented with 100 mg/l α-CD-fatty acids (oleic and linoleic acids) complexes and 1 g/l α-CD to be comparable to that in a serum-supplemented identical medium. Subsequent studies aimed at producing antibodies in a CD-supplemented medium containing a reduced yet necessary amount of serum. Further progress was achieved by Minamoto et al[3]. who devised a serum-free but heat-sterilizable medium containing *inter alia* α-CD and cholesterol for the continuous cultivation of a human lymphoblastoid cell line. Interestingly, the choice of α-CD was made following the examination that ß-CD and Dimeb were less effective and even slightly toxic at higher concentrations.

The first attempt to include a CD in a nutrient medium for microbe cultivation involves the *in vitro* cultivation of *Mycobacterium leprae* [4]. Surface cultivation of yeast cells was successfully undertaken on a solid agar-hexadecane medium [5]. Subsequent studies employed either a natural or chemically modified CD for the cultivation of primarily fastidious bacteria such as *Mycobacteria* [6] and *Mycoplasma* [7] species on CD-complexed substrates. These studies are listed in Table 1. Of particular interest is the recent report on a method for cultivating the fastidious *Helicobacter* and *Campylobacter* species in a medium containing CDs [8].

3. BIOSYNTHESIS BY FERMENTATION

While in cultivation media CD interacts with an exogenous substance to be taken up by cells, here it interacts with product metabolite synthesized by cells such as toxins and antibiotics.

Imiaizumi et al. investigated the effect of natural and methylated CDs on the growth of *Bordetela pertussis* in defined media and on the production of *Pertussis* toxin [9]. The authors undertook a similar study on the production of filamentous hemagglutinin [10]. Among all CDs, DIMEB at 50 - 2000 mg/l was reported as the best stimulant for growth and for production of either protein. The manufacture of *pertussis* toxin using culture media containing hydroxyl-ß-CD and DIMEB was subsequently patented [11]. The activity of the extra-cellular adenylate cyclase of a *B. pertussis* culture (this enzyme seems to be esssential for bacterial virulence) was also reported to increase in the presence of 1 g/l DIMEB [12].

TABLE 1. Microbes cultivated in CD-containing media

Cell	*CD type*	*Purpose*
Activated sludge	β-CD	Degradation
Candida tropicalis	β-CD polymer	Degradation
Activated sludge	β-CD	Dègradation
Candida lypolitica, *C. tropicalis*	α-CD, β-CD	Yeast growth on solid medium
Mycobacterium leprae	DIMEB	Growth in synthetic liquid and solid media
M. leprae, *M. phlei*	DIMEB	Oxidation
Mycoplasma capricolum, *Acholeplasma laidlawii*	β-CD, HPBCD	Supply of lipids, serum substitute
M. paratuberculosis	DIMEB, natural CDs	Substitute for *mycobactin*
Pseudomonas putida	β-CD	Degradation

TABLE 2. Bioconversions in CD-containing media

Substrate	*Biocatalyst*
Hydrocortisone and other steriods	*Arthrobacter*, *Flavobacterium*, *Streptomyces*, *Curvularia*, *Ophiolobus* and *Mycobacterium* species
Olive oil, triolein	Lipase
Cholesterol, cholest-4-en-3-one, sitosterol	*Mycobacterium* species
Benzaldehyde, vanillin	*S. cerevisiae*
Benzaldehyde	*S. cerevisiae*
17β-Oestradiol	Phenol oxidase from plant cells
Androst-4-en-3,17-dione	*S. cerevisiae*
Deacetyllanatodise A	β-Glucosidase
Adenosine diphosphate	Adenylate kinase
Vitamin D	*N. autotrophica*
α,β-Unsaturated esters	*S. cerevisiae*
(25*R*)-5α-Spirostan-3β-ol	*N. restricta*
(25*R*)-5α-Spirostan-3β-ol-12-one	
p-Nitrophenol butyrate and caprilate	Lipase, esterase
Cholesterol	*R. erythropolis*
Cholesterol	*R. erythropolis* cells, free cholesterol oxidase
Ketopantolactone	*S. cerevisiae*
13-Ethylgon-4-en-3,17-dione (GD)	Free and immobilized cells of *Penicillium raistrickii*

The first CD- stimulated fermentation of an antibiotic was reported by Sawada et al.[13] who investigated the effect of natural and methylated CDs on the production of lankacidins by *Streptomyces* species. Among all CDs, ß-CD was the most effective in enhancing the production of lankacidin C without affecting the microbial growth nor being degraded. Additional CD-enhanced fermentations of antibiotics followed and these include prodigiosin by *Serratia marcescens* [14] and nonensin A and nigericin by several *Streptomyces* species [15].

4. BIOCONVERSIONS

A great deal of bioconversions mediated by either cells or enzymes were performed on a variety of organic substrates complexed by natural or chemically modified CDs and these are listed in Table 2.The first application of CDs in microbial conversions was performed on various steroids by Udvardy *et al* [16]. In the vast majority of bioconversions listed in Table 2, the CD-induced effect was that of either a larger loading of water-insoluble or toxic substrates or fast reaction kinetics. Among these, the CD-enhanced specific hydroxylation of vitamin D_3 by *Amycolata autotrophica* [17] has been commercialized in Japan.
Of a special interest is the recently reported CD-enhanced enantioselectivity of bioconversions. Indeed the enantiomeric excess of the yeast-catalyzed addition of an amine to an unsaturated ester [18] as well as that of the reduction of ketopantolactone by yeast [19] were shown to increase in the presence of β-CD.

5. MICRIOBIAL DEGRADATIONS

Toxicity and water insolubility are major obstacles to efficient biodegradations. A few studies employing CDs to alleviate or remove the obstacles were conducted and they are included in Table 1. The first attempt to emply CDs in microbial degradations is reported in the patent literature[20] and it involves an improved purification of wastewater supplemented with ß-CD from phenols and benzene. Another study was conducted by Olah et al.[21] in view of demonstrating the detoxifying power of ß-CD upon industrial wastewaters containing several pesticides or their decomposition products such as chlorophenols and nitrophenols. The idea of "damping" an expensive CD, which is *per se* certainly biodegradable and therefore contributes itself to the biochemical oxygen demand of the treated waters is not an appealing one. A more attractive and elegant approach would be to degrade pollutants at conditions which preserve CDs, thus enabling repeated uses or recycling of CDs. This approach has been adopted by Banky et al.[22] and by Schwartz [23].

Banky et al. [22] employed a ß-CD polymer to increase the efficiency of phenol degradation in wastewater by *Candida tropicalis*. To this purpose, the authors fixed the yeast cells with the poly-ß-CD in polymer beads in order to allow repeated uses. In a more recent report [24], ß-CD was incorporated in beads of immobilized microbial cells in order to enhance adsorption and degradation of various substances such as pesticides, polycyclic aromatic compounds and halogenated hydrocarbons.

Schwartz [23] employed a strain of *Pseudomonas putida* experimentally proven to be incapable of degrading CDs but capable of utilizing aromatic solvents as sole carbon and energy source. An aqueous suspension of the microbial cells is intoxicated by liquid toluene but degrades toluene vapors. The *Pseudomonas* strain was shown to degrade up to 10 g/l of the "solidified" ß-CD-complexed toluene, after which ß-CD re-dissolved and became available for re-use. When toluene was supplied as vapors, ß-CD added to the bacterial suspension enhanced its degradation as a result of a facilitated adsorption of the substrate vapors.

It appears then that if a microbial degradation system is to allow for use of either recoverable CDs or a catalytic amount of CD, it should be based on a pure mono culture of the degrading microorganism, previously tested for its inertness towards CDs. A future potential use of CDs could be in the combination of absorption and degradations of hydrophobic gaseous pollutants. While water-soluble gaseous contaminants are easily absorbed into water, little means are available for absorbing vapors of hydrophobic water-immiscible pollutants such as benzene, toluene and xylene (BTX). CDs can efficiently capture BTX from polluted stream [25] and the CD complexes could be fed into a bioreactor with a suitable degrading microorganism.

REFERENCES

[1]. Bar , R., *Applications of CDs in biotechnology*, in Comprehensive Supramolecular Chemistry: Cyclodextrins, Vol. 3, (Eds. Szejtli, J. and Osa, T.) in press.

[2]. Yamane, I., Kan, Minamoto,Y., Amatsuji Y., α-Cyclodextrin, a novel substitute for bovine albumin in serum-free culture of mammalian cells *Proc. Japan Acad.* , **57**(Ser B), 385-389 (1981).

[3]. Minamoto, Y., Ogawa, K., Abe, H., Iochi, Y., Mitsugi, K., Development of a serum-free and heat-sterilizable medium and continuous high-density cell culture, *Cytotechnol,*, **5**, 535-551 (1991).

[4]. Kato, L. Psychrophilic Mycobacteria in *M. leprae*-infected tissues. *Int. J. Lepr.* , **56** (4), 631-632 (1988).

[5]. Bar, R., A new cyclodextrin-agar medium for surface cultivation of microbes on lipophilic substrates,*Appl. Microb. Biotechnol.*, **32**, 470-472 (1990).

[6]. Ishaque, M., Water soluble palmitic-methylated cyclodextrin complex: a substrate oxidized by *Mycobacterium leprae*, *Int. J. Lepr.* **60**, 279-280 (1992).

[7]. Greenberg-Ofrath, N., Terespolosky, Y., Kahane, Bar, R., Cyclodextrins as carriers of cholesterol and fatty acids in cultivation of mycoplasmas, *Appl. Environ. Microbiol.* **59**, 547-551 (1993).

[8]. Figura, N., Bugnoli, M., Akivama, S., Nakao, Y., *Eur. Pat. Appl.* 540 897 (1993) (*Chem. Abstr.*, **119**, 7343).

[9]. Imaizumi, A., Suzui, Y., Ono, S., Sato, H., Sato, Y, Heptakis(2,6-O-dimethyl)ß-cyclodextrin, a novel growth stimulant for *Bortedella pertussis* phase I, *J. Clin. Microbiol* . **17**, 781-786 (1983).

[10]. Imaizumi, A., Suzuki, Y., Ginnaga, A., Sakuma, S., Sato Y, A new culturing method for production of filameantous hemagglutinin of *Bordetella pertussis.*, *J. Microbiol. Meth.* **2**, 339-347 (1984).

[11]. Okuma, K., *Jpn. Kokai Tokyo Koho* 03 056 422 (1991) (*Chem. Abstr.* **115**, 181 519)

[12]. Hozbor, D. F., Samo, A., Yantorno, O. M., Effect of mathyl-cyclodextrin on adenylate cyclase activity of *Bordetella pertussis, World J. Microbiol. Biotechnol.*, **7**, 309-315 (1991).

[13]. Sawada, H., Suzuki, T., Akiyama, S., Nakao, Y., Stimulatory effect of cyclodextrins on the production of lankacidin-group antibiotics by *Streptomyces* species, **26**, 522-526 (1987).

[14]. Bar, R., Rokem, J. S., Cyclodextrin-stimulated fermentation of prodigiosin by *Serratia marcescens, Biotechnol. Lett.*, **12**, 447-448 (1990).

[15]. Patzelt, H., Gaudet, V., Robinson, J. A., *J. Antibiot.*, **45**, 1806-1808 (1992).

[16]. Udvardy, N. E., Bartho, I., Hantos, G., Trinn, M., Vida, Z. S., Szejtli, J., Stadler-Szoke, A., Habon, I., Balas-Czurda, M., *Belg. Pat.* 894 501 (1983) (*Chem. Abstr.* , **99**, 4069).

[17]. Takeda, K., Asou, A., Matsuda, A., Kimura, K., Okamura, K., Okamoto, R., Sasaki, J., Adashi, T., Omura, S., *J. Ferment. Bioeng*, **78**, 380 (1994).

[18]. Rao, K. R., Nageswar, Y. V, D. , Sampath Kumar, H., M., *Tetrahedron Letters*, 32, 6611-6612 (1991).

[19]. Nakamura, K., Kondo, S., Kawai, Y., Ohno, A., *Tetrahedron: Assymetry*, 4, 1253-1254 (1993).

[20]. Farkas, P., Olah, J., Szetli, J., Szente, *Hung. Pat.* Appl. 2650/84 (1984) .

[21]. Olah, J., Cserhati, T., Szejtli, J., Beta-cyclodextrin enhanced biological detoxification of industrial wastewaters. *Water Res.*, **22**, 1345-1352 (1988).

[22]. Banky, B., Recseg, K., Novak., D., (1985) Application of water soluble beta-cyclodextrin in microbiological decomposition of phenol. *Magy. Kem. Lapja* , **40**, 189-192 (1985), (*Chem. Abstr.*, **103**, 165518).

[23]. Schwartz, A., Bar, R., Cyclodextrin-enhanced degradation of toluene and *p*-toluic acid by *P. putida* , Appl. Environ. Microbiol., **61**, 2727-2731 (1995).

[24]. Wildenauer, F. X., Franz, R., Menzel, R., Vetter, H., *Ger. Pat. 4 027 223 (1992), (Chem. Abstr.***117**, 55603).

[25]. Szejtli, J., Cyclodextrin Technology, Kluvwer Academic Publishers, Dordrecht, 1988, p. 393.

MODIFIED CDs-MEDIATED ENHANCEMENT OF MICROBIAL STEROL SIDECHAIN DEGRADATION

DONOVA M.V., DOVBNYA D.V., and KOSHCHEYENKO K.A.
Institute of Biochemistry & Physiology of Microorganisms, Rus.Acad.Sci., Pushchino, Moscow region, 142292, Russia

ABSTRACT

The effect of modified β-CDs (hydroxyethyl-, hydroxypropyl, carboxymethyl-, methyl-, glycidyl- substituted ones) on the selective sidechain degradation of natural sterols by *Mycobacterium sp.* strains forming androstenedione (AD), its 1,2-dehydrogenated (ADD), or 9α-hydroxylated (9α-OH-AD) derivatives was studied. A 2.5-6.5 increase of the androstenone yields with a 94-95% conversion of sitosterol (5-20 g/l) was recorded. The enhancement effect was found to depend on the mCD type, mCD/sterol molar ratio, and less on the sidechain structure of sterols. The ratio of the solubility enhancement and complex stability constants was found to determine the shifts in the ratio of major products at sitosterol to AD(D) conversion in the presence of HPCD.

1. INTRODUCTION

Selective microbial sidechain cleavage of natural sterols is a best way to produce primary products for majority of pharmaceutical steroids nowadays [1]. However, extremely poor sterol solubility in aqueous media (1.8-3.0 mg/l) [2] and the inhibition of bacteria growth, respiration and the sidechain degradation by some andostenones substantially limited the efficiency of the bioconversions [3]. Inclusion complexation of steroids in CDs seems to be the promising way to solve the problems. For microbial sterol sidechain degradation, the 1.7 - 3.0-fold increase of the specific activity was recorded in the presence of natural CDs (β-, γ-) at sterol concentration of 1 g/l [4]. However, the solubilizing activity of natural CDs (except γ-CD) towards sterols [5] and the androstenones is restricted by relatively low steroid concentrations (B-type of phase solubility diagrams) [6]. The latter can suppress the effective sterol bioconversions at higher sterol concentrations.

The present study was aimed at the investigation of the effect of chemically modified β-CDs on sterol sidechain degradations by *Mycobacterium sp.* strains forming useful steroidal precursors: androst-4-ene-3,17-dione (AD), androsta-1,4-diene-3,17-dione (ADD), or 9α-hydroxyandrost-4-ene-3,17-dione (9α-OH-AD).

J. Szejtli and L. Szente (eds.), Proceedings of the Eighth International Symposium on Cyclodextrons, 527–530.

2. MATERIALS AND METHODS

2.1. Materials

Sitosterol (91.4% of pure β-sitosterol) (Kaukas, Finland) was used as a substrate. Hydroxyethylated β-CD (HECD), hydroxypropylated β-CD (HPCD), and glycidylated β-CD (GECD) were obtained from the Moscow Research Institute of Organic Semiproducts and Dyes (Russia), randomly methylated β-CD (MeCD) was purchased from Wacker-Chemie GmbH (Germany), carboxymethylated β-CD (CMCD) was kindly gifted by Prof. A. Kestner (Tallinn, Estonia).

2.2. Methods

Sterol-transforming cultures - *Mycobacterium sp.* VKM Ac-1815D; *Mycobacterium sp.* VKM Ac-1816D and *Mycobacterium sp.* Ac-1817D were obtained from Russian Collection of Microorganisms (IBPhM RAS). Culture maintenance and precultivation was carried out as described earlier [7].

Biotransformations of sterols (5 g/l) were carried out at 220 rpm, 29°C, as described earlier [7]. The mCDs were added in molar ratio to substrate of (0.2-3.0) to 1. Sitosterol (20 g/l) to ADD conversion was carried out in a 10-l stirred bioreactor of ANKUM type (Russia) under the control of pH, temperature and oxygen supply using HPCD/sitosterol molar ratio of 1.2/1.

Solubilities of ADD, AD and HMPD in HPCD solutions were measured spectrophotometrically according to [6].

Steroids were extracted by ethyl acetate (5:1, v/v), and analysed by TLC, GC and HPLC methods as previously described [7]. Products identification was performed using MS with silization of some products.

Biomass was followed by measuring of protein content [7].

3. RESULTS AND DISCUSSION

3.1. Effect of mCDs on sterol transformations

The identification of the major products of sterol oxidation by *Mycobacterium sp.* strains in the presence of the mCDs showed the products were similar to those isolated during the transformations without mCDs. The mCDs (except GCD) mediated a 2.7-6.5 -fold increase of the goal products yields (AD, ADD, or 9α-OH-AD, depending on the strains used) at the increase of sitosterol conversion from 40 - 60% (in control) to 85-98%. The enhancement of sterol conversions was accompanied by remarkable shifts in the ratio of the major transformation products: the ADD/AD ratio shifted towards more ADD; 9α-OH-AD/9α-OH-testosterone (9α-OH-T) ratio - towards more 9α-OH-T; for the three process tested increase of corresponding C-22-steroids was observed (Table 1).

The results were in accordance with the solubilities of the major transformation products in mCDs solutions. The HPCD-mediated enhancement of solubility of ADD, AD and HMPD was calculated from phase solubility diagrams as (in mol/mol) 0.914, 0.572, and 0.419, respectively; the complex stability constants was determined as (in M^{-1}) 6390, 4880, and 48620, respectively.

TABLE 1. Effect of mCDs on sitosterol (5 g/l) conversion by mycobacterium strains.*

Strain	mCD	Transformation duration, h	Molar yield of product, %		
Mycobacterium sp. VKM Ac-1816D			ADD	AD	20-Hydroxymethyl-pregna-1,4-diene-3-one
	Control	96	16.4	1.5	0.9
	HPCD	96	83	3.4	13.4
	HECD	96	73	4.5	22.0
	MeCD	96	77	5.0	14.0
	CMCD	144	70	3.0	13.5
Mycobacterium sp. VKM Ac-1815D			AD	ADD	20-Hydroxymethyl-pregna-4-ene-3-one
	Control	120	11.5	0.5	0.8
	HPCD	100	73	10.3	15.0
	HECD	96	68	10.0	15.0
	MeCD	75	75	5.0	17.0
	CMCD	144	75	10.0	13.0
Mycobacterium sp. VKM Ac-1817D			9α-OH-AD	9α–OH-T	9α,20-Dihydroxy-methylpregna-4-ene-3-one
	Control	130	28	0.4	5.0
	HPCD	96	76	9.0	13.0
	HECD	96	76	9.0	16.0
	CMCD	120	65	n.d.	10.0

* Molar ratio of mCDs to sitosterol of 2/1 was used

3.2. Effect of mCD concentration

The HPCD-mediated enhancement effect on sitosterol to ADD conversion was found to increase with mCD/sterol molar ratio from 0.2/1 to 1/1, while no considerable changes in the process dynamics were observed at the increase of mCD concentration from 1/1 to 3/1.

3.2. Effect of sidechain structure of sterols

The enhancement effect of the mCDs was non-remarkably differed depending on the sidechain configuration of the substrate. For β-sitosterol, cholesterol, and stigmasterol conversion by *Mycobacterium sp.* VKM Ac-1816D in the presence of HECD enhancement factors of 6.4; 5.4, and 5.5, respectively, were obtained.

3.3. Sitosterol (20 g/l) to ADD transformation in a laboratory scale bioreactor

A molar yield of ADD of 64% (9.14 g/l) with a 94% sitosterol conversion for a 130 h transformation was recorded at sitosterol to ADD transformation in a bioreactor. The results demonstrate a high biotechnological potential of mCDs applications for C_{17}-ketosteroid productions.

4. CONCLUSION

The high enhancement effect of chemically modified CDs on sterol sidechain degradation processes forming useful steroidal precursors have been demonstrated in this study. The increase of sterol conversion with the high androstenones yields accompanied with the shifts in the ratio of the major products observed can be partially explained by the increase of the products solubilities followed by the decrease of the inhibitory product concentrations in mCDs media. However, the mechanism of the mCDs action on sterol sidechain cleavage is certainly much more complicated, and can include, in addition, the partial solubilizing of the substrate (and major intermediates), action of the mCDs as detergents and mass-transfer carriers, some physiological aspects of CD-biocatalyst interactions.

ACKNOWLEDGEMENTS

The authors thanks Dr. Arinbasarova A.Yu., for the valuable remarks, Dr. Kalinichenko A.N. from the Moscow Institute of Organic Semiproducts and Dyes, Nikolayeva V.M.; Mr. Baskunov B.P., Mrs. Scherbakova V.A. (IBPhM, Pushchino), for the assistance.

REFERENCES

[1] Szentirmai, A. Microbial physiology of sidechain degradation of sterols *J.Ind.Microb.*, **6**, 101-116 (1990)

[2] Haberland, E., Reynolds, I.A. Self-association of cholesterol in aqueous solutions. Proc.Nat.Acad.Sci.USA, **70**, 2313-2316 (1973)

[3] Dias, A.C.P., Cabral, J.M.S., Pinheiro, H.M. Sterol side-chain cleavage with immobilized *Mycobacterium* cells in water -immiscible organic solvents. *Enzyme Microb.Technol.*, **16**, 708-714 (1994)

[4] Hesselink, P.G.M., van Vliet, S., de Vries, H., Witholt, B. Optimization of steroid side chain cleavage by *Mycobacterium sp.* in the presence of cyclodextrins. *Enzyme Microbiol.Technol.*, **1**, 384-404 (1989)

[5] Jadoun, J., Bar, R. Microbial transformations in a cyclodextrin medium. Part.3. Cholesterol oxidation by *Rhodococcus erythropolis. Appl.Microbiol.Biotechnol.*, **40**, 230-240 (1993)

[6] Singer, Y., Shity, H., Bar, R. Microbial transformations in a cyclodextrin medium. Part.2. Reduction of androstenedione to testosterone by *Saccharomyces cerevisiae. Appl.Microbiol.Biotechnol.*, **35**, 731-737 (1991)

[7] Donova, M.V., Dovbnya D.V., Kalinichenko, A.N., Arinbasarova, A. Yu., Morozova, Z.V., Vagabova, L.M., Koshcheyenko K.A. Method of androstadienedione production. Russian Patent, N 2039824, (1995)

IN VITRO STUDY AND COMPUTATIONAL SIMULATION OF CYCLODEXTRIN INCLUSION COMPLEXES IN ENZYMATIC LIPID HYDROLYSIS

G. J. KOLOSSVÁRY(1), I. KOLOSSVÁRY(2) AND E. BÁNKY-ELŐD(3)
(1) Central Food Research Institute, Department of Enzymology, Herman O. u. 15., 1025 Budapest, Hungary
(2) Technical University of Budapest, Department of Chemical Information Technology, Szt. Gellért tér 4., 1521 Budapest, Hungary
(3) National Committee for Technological Development, Szervita tér 8., 1052 Budapest, Hungary

ABSTRACT

Based on the ability of β-cyclodextrin and its partially methylated derivative (dimethyl-β-cyclodextrin) to form inclusion complexes, the hydrolysis rate of triolein by pancreatic lipase in aqueous solution can be increased up to 3-7-fold. Two types of cyclodextrin inclusion complexes could be responsible for this effect, either the β-cyclodextrin - triolein or the β-cyclodextrin - oleic acid complex. Molecular dynamics simulations clearly suggest that β-cyclodextrin hosts only the liberated fatty acid thus accelerating lipolysis by decreasing product inhibition.

1. INTRODUCTION

Triglyceride hydrolysis as catalysed by lipases (EC 3.1.1.3) in an aqueous environment is a slow process. The hydrolysis of these apolar substrates is a classic example of heterogeneous biocatalysis. The water soluble enzyme must penetrate through an interfacial layer of lipid molecules to reach the substrate. Partial glycerides and soaps of free fatty acids liberated during lypolysis accumulate at the oil-water interface and additives which either affect the state of aggregation or the size of the surface area may have a marked effect on the reaction rate. In order to enlarge the size of the interface where the reaction takes place, and stabilize the microemulsion, different kinds of detergents have to be present in the reaction mixture, however this causes difficulties in deemulsification at the final product separation step. A different kind of additive is therefore desirable which can accelerate the lipolysis by another mechanism.

The present study is based on the special ability of cyclodextrins being capable to make inclusion complexes [1] either with the triglyceride itself, making the substrate more easily accessible for the enzyme, or with the liberated fatty acids, causing a decrease of product inhibition. β-cyclodextrin and its partially methylated derivative, heptakis-(2,6-di-O-methyl)-β-cyclodextrin, have been investigated in order to accelerate the lipolysis without the presence of surfactants.

J. Szejtli and L. Szente (eds.), Proceedings of the Eighth International Symposium on Cyclodextrons, 531–534.

To find out which mechanism yields the increased hydrolysis rate, we studied the three-dimensional structure of the BCD-triolein and the BCD-fatty acid complexes with extensive molecular dynamics simulations.

2. MATERIALS AND METHODS

2.1 Materials

Porcine pancreas lipase (triacylglycerol acylhydrolase, EC 3.1.1.3) was obtained from Serva (2.5 U/mg measured activity). Trioleoyl glycerol (99% purity) was from Loba Feinchemie, Austria. β-Cyclodextrin (cycloheptaamylose, BCD) and heptakis-(2,6-di-O-methyl)-β-cyclodextrin (DMCD) were the kind gift of Prof. Szejtli. The bovine serum albumin (BSA) and all the other chemicals were the products of Reanal, Hungary.

2.2 Determination of the degree of hydrolysis

The measurement of the triolein hydrolysis rate was carried out in an automatic titrimeter, type OP-506 (Radelkis, Hungary). The reaction mixture contained 1.0 ml triolein substrate, 1.0 ml of 100 mg/ml BSA, and 0.75-3.75 ml of 200 mg/ml DMCD. The emulsion was prepared in an Ultra-Turrax T25 instrument at 10,000 rpm for 3 min and than the reaction mixture was always completed to 15.0 ml with distilled water. The triolein hydrolysis was initiated with the addition of 1.0 ml of 5 mg/ml lipase solution, prepared in 0.005 M calcium chloride. The liberated fatty acids were titrated with 0.05 M NaOH and the corresponding NaOH consumption was monitored by pH-stat titration for two hours.

2.3 Computational details

In our computer simulations we used a variant of molecular dynamics (MD), termed stochastic dynamics (SD) [2] which is particularly suited for simulations in aqueous solution. The SD simulations were carried out using the Macromodel/Batchmin V5.0 computational chemistry program package [3] running on a Silicon Graphics Challenge computer.

The cyclodextrin inclusion complexes (see below) were built with Macromodel using the three-dimensional graphical user interface. The potential energy of the cyclodextrin complexes was calculated with the Macromodel variant (MM2*) of the authentic MM2 force field [4]. The aqueous solution was considered utilizing the so-called generalized Born surface area model (GBSA) [5].

The SD simulations were carried out at 300 K constant temperature. The time span of each of the simulations was 1 ns using 1 fs time steps. This gives a total of one million SD steps that required five to eight central processing unit (CPU) days to accomplish a single 1 ns simulation on the Silicon Graphics Challenge computer. Intermediate structures were saved in a file after every 1,000 SD steps to obtain a representative, sequential set of structures generated during the simulation.

3. RESULTS AND DISCUSSION

β-cyclodextrin and its partially methylated derivative (DMCD) were found to increase the hydrolysis rate of the triolein substrate, in aqueous solution in the presence of pancreas lipase enzyme.

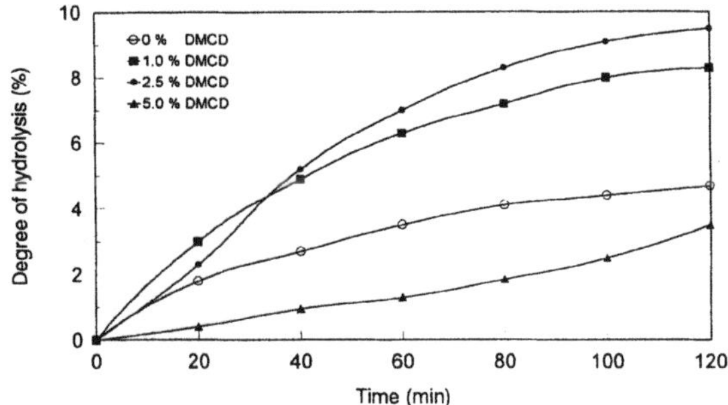

Fig.1- Effect of DMCD concentrations on the hydrolysis of triolein (37 °C, pH 9)

Figure 1 shows the hydrolysis rate of the triolein substrate in the presence of 1.0, 2.5 and 5.0 % DMCD respectively, at the optimum reaction parameters of the enzyme (37 °C, pH 9). The 1.0 and 2.5 % DMCD concentrations accelerated the lypolysis from the beginning, in comparison with the hydrolysis curve obtained without DMCD. During the 120 minute reaction the hydrolysis curve of the 5.0 % DMCD was running below the curve obtained without DMCD. However, we carried out an extended 6 hour reaction, where the 5.0 % DMCD curve crossed the 0 % DMCD curve at three hours and even after, the hydrolysis rate was still exponentially increasing.

Our computational study was aimed at finding a *qualitative* answer as to which of the catalytic mechanisms suggested previously is more likely responsible for the increased degree of lipolysis in aqueous solution found in the presence of different β-cyclodextrin derivatives [6]. The two potential catalytic mechanisms involve formation of an inclusion complex of the cyclodextrin ring with either the triolein making it more easily accessible for the enzyme in the aqueous phase, or with the liberated fatty acid preventing product inhibition. Dimethyl-β-cyclodextrin (DMCD) was shown to be more effective on lipolysis than the unsubstituted β-cyclodextrin, primarily due to its better solubility. Therefore, we concluded that for the present study the unsubstituted BCD was appropriate as host molecule to study the behavior of the two potential guest molecules, triolein and the corresponding liberated oleic acid.

The three-dimensional computer models clearly showed that BCD cannot envelop the whole triolein neither can host two fatty acid strings. The only way for BCD forming an inclusion complex with triolein is to accept either the middle string or one of the side strings. Figure 2 shows the energy-minimized structure of the BCD-triolein inclusion complex in aqueous solution used to initiate the dynamics simulation. At the first look, the cyclodextrin ring seems to host the coiled-up triolein perfectly, suggesting that the SD simulation would only reveal the fluttering of the triolein strings. However, the situation is different! Fluttering of the triolein strings does take place, but more importantly, the SD simulation clearly suggests that triolein gradually slips out of the grip of the cyclodextrin ring. Figure 3 captures the instant of slip-out after about 3.5 ns simulation time revealing the inherent instability of the BCD-triolein inclusion complex in aqueous solution.

On the other hand, our dynamics simulation suggests that the BCD-oleic acid inclusion complex shown in Figure 4, is stable in aqueous solution, at least through the duration of the 10 ns SD simulation. The BCD-oleic acid "movie" showed that the oleic acid was moving in and out of the cyclodextrin ring like the bolt in the nut, but neither extreme positions resulted in slip-out.

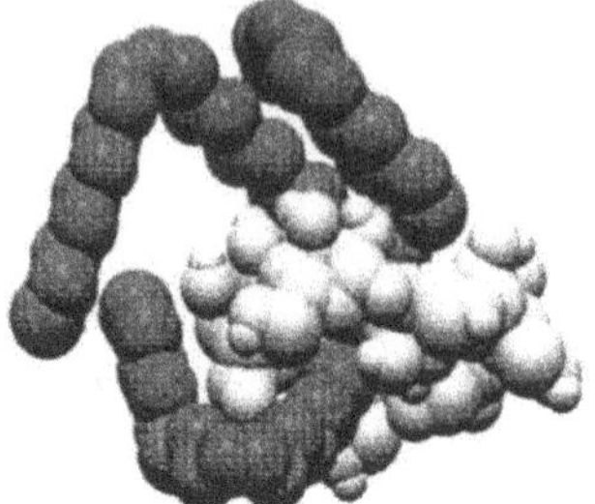

Fig. 2- Energy minimized BCD-triolein complex

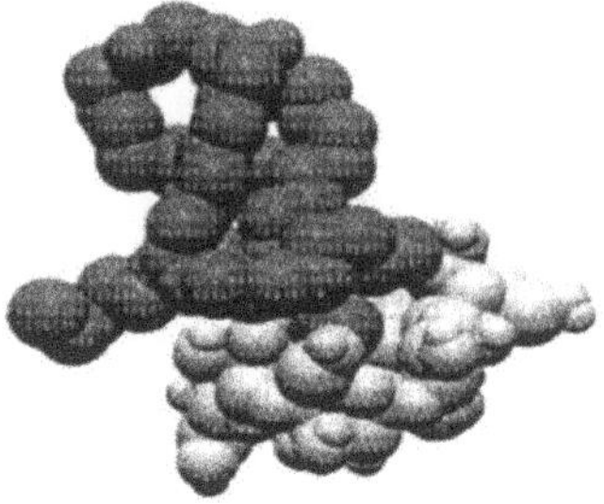

Fig. 3- BCD ring losing triolein

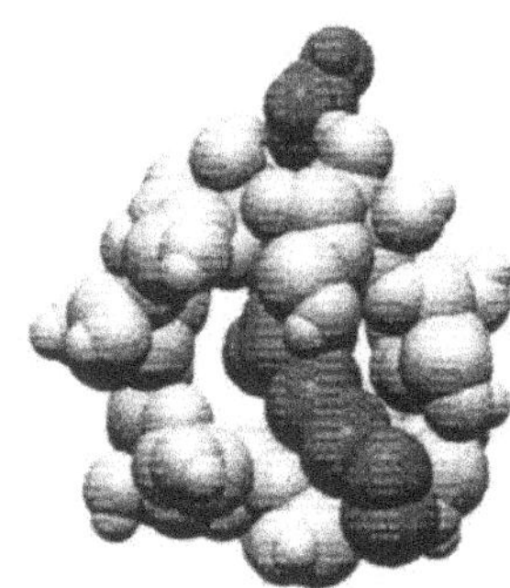

Fig. 4- Stable BCD-oleic acid inclusion complex

4. CONCLUSION

Our computer simulations clearly suggest that BCD cannot form a stable complex with triolein, but indeed, it can host the liberated oleic acid forming a stable inclusion complex in aqueous solution. Therefore we conclude that the increase of lipolysis found experimentally in the presence of BCD derivatives can be attributed to the removal of the liberated fatty acids from the reaction zone at the final stage of the enzymatic hydrolysis by the formation of a stable BCD-oleic acid inclusion complex.

REFERENCES

[1] Szejtli, J., Cyclodextrins and Their Inclusion Complexes, Akadémiai Kiadó, Budapest, 1982

[2] van Gunsteren, W.F., Berendsen, H.J.C., A Leap-frog Algorithm for Stochastic Dynamics, *Molec. Simul.*, **1**(3), 173-186 (1988)

[3] MacroModel V5.0: Mohamadi, F.;Richards, N.G.J.; Guida, W.C.; Liskamp, R.; Lipton, M.; Caufield, C.; Chang, G.; Hendrickson, T.; Still, W.C., MacroModel-An Integrated Software System for Modeling Organic and Bioorganic Molecules Using Molecular Mechanics, *J. Comput. Chem.*, **11**, 440-467 (1990)

[4] Allinger, N.L., MM2. A Hydrocarbon Force Field Utilizing V_1 and V_2 Torsional Terms, *J. Am. Chem. Soc.*, **99**, 8127-8134 (1977)

[5] Still, W.C., Tempczyk, A., Hawley, R.C., Hendrickson, T., Semianalytical Treatment of Solvation for Molecular Mechanics and Dynamics, *J. Am. Chem. Soc.*, **112**, 6127-6129 (1990)

[6] Kolossváry, G. J., Bánky-Elöd, E., Enhancement of Enzymatic Hydrolysis of Triolein in Aqueous Solution by Cyclodextrin Derivatives, *Biotechnol. Techn.*, **10**(2), 115-120 (1996)

DEGRADATION OF FATTY ACID-CYCLODEXTRIN COMPLEXES BY MYCOBACTERIA

M. ISHAQUE[1] AND V. STICHT-GROH [2]

[1] *Institut-Armand-Frappier, University of Quebec, Laval, P.Q., H7N 4Z3, Canada.*
[2] *Armauer-Hansen-Institut, Hermann-Schell Str. 7, Würzburg, Germany.*

ABSTRACT

Mycobacterium leprae, rich in fatty acids, has not been cultivated *in vitro*. Degradation of insoluble and water soluble fatty acid-cyclodextrin complexes by mycobacteria using manometric techniques was investigated. While water soluble fatty acid-cyclodextrin complexes were readily oxidized by *M. leprae*, *M. lepraemurium*, *BCG* and *M. phlei*, water insoluble fatty acids were not oxidized.The use of water soluble fatty acid-methylated cyclodextrin complexes is encouraged for metabolic studies as well as in culture media used for *in vitro* cultivation of non-cultivated *M. leprae*.

1. INTRODUCTION

Mycobacterium leprae bacilli are rich in lipids (1). In the host, large quantities of lipids are excreted by *M. leprae* into the environment. The most important membrane lipids are phospholipids which serve primarily as structural elements of the membranes. Sphingolipids, an important class of phospholipids, are present in most membranes of animal tissues. Three well known sphingolipids are sphingosine, sphingomyelin and cerebroside. The key to *in vitro* cultivation of thus far noncultivated *M. leprae* is to find the oxidizable substrate which can be incorporated into the culture medium to achieve its cultivation. Our earlier results (2) indicated that palmitate could serve as a source of carbon and energy for *in vitro* cultivation of *M. leprae*. Since sphingolipids occur naturally in membranes of animal tissues, it is likely that they are utilized by human leprosy bacilli for their growth and multiplication. However, palmitate and sphingolipids are insoluble in water and thus not readily available to the bacilli. A preparation of palmitate -heptakis-2,6-di-o-methyl-ß-cyclodextrin complex soluble in water was reported in 1992 (3). Also, water soluble preparations of sphingolipids in 40% methylated ß cyclodextrin have become available. Wheeler et al. (4) reported that *M. leprae* requires an exogenous sourece of fatty acids. A variety of fatty acids have been known to be metabolized by some mycobacteria (5). Water soluble copmlexes of fatty acids when incorporated in culture media, are dissociated and become easily biologically available

J. Szejtli and L. Szente (eds.), Proceedings of the Eighth International Symposium on Cyclodextrons, 535–540.

(6). In view of the importance of these fatty acids their metabolism by various mycobacteria was investigated.

2. MATERIALS AND METHODS

2.1. GROWTH OF MYCOBACTERIA AND PREPARATION OF CELL SUSPENSIONS

2.1.1. Growth of *M. leprae* and preparation of cell suspensions

Foot pads of nude mice previously infected with human leprosy bacilli were removed aseptically. The foot pads after surface decontamination with proviodine were cut into small pieces and minced with pestle and mortar in distilled water. The suspension was filtered through a nylon filter and purified *M. leprae* bacilli were obtained by differential centrifugation. Bacilli were suspended in 2% NaOH, diluted five-fold with distilled water and after 30 min of incubation, it was centrifuged at 8,000 x g for 10 min. The resulting pellet was washed twice with 0.05 M potassium phosphate buffer, pH 6.5 and resuspended in a small volume of the same buffer.

2.1.2. Growth of *in vivo* and *in vitro* grown *M. lepraemurium* and preparation of cell suspensions

Growth of *in vivo* and *in vitro* grown *M. lepraemurium* as well as preparation of cell suspensions was obtained as described earlier (7).

2.1.3. Growth of *M. phlei* and preparation of cell suspensions

M. Phlei were grown on Lowenstein-Jensen medium at 32°C for about 10 days. The growth was removed from the surface and after washing twice with 0.05 M potassium phosphate buffer, pH 6.5, an appropriate volume of cell suspension was prepared in the same buffer.

2.1.4. Growth of *BCG* and preparation of cell suspensions

Bacille Calmette and Guerin (*BCG*) was grown on Sauton medium for about 15 days . The cells were washed twice with 0.05 M potassium phosphate buffer, pH 6.5 and an appropriate volume of cell suspension was made in the same buffer.

2.2. SUBSTRATES

Sodium palmitate and sphingolipids, insoluble in water, were purchased from Sigma Chemical Co; U.S.A. Preparations of palmitate and sphingolipids such as sphingosine, sphingomyelin and cerebroside containing 10 mg of each in 40% methylated ß cyclodextrin were, respectively, provided by Dr. L. Kato, Cathrine Booth Hospital Centre , Montreal, Canada and Professors L. Szente and J. Szejtli, Budapest, P.O. Box 435, Hungary.

2.3. MEASUREMENT OF OXIDATION OF SOLUBLE AND INSOLUBLE FATTY ACIDS

Oxidation of insoluble and water soluble palmitate as well as of insoluble and water soluble sphingolipids was determined by using the standard manometric techniques (8). All possible controls such as suspension of live bacilli; boiled bacillary suspensions with or without substrates and cyclodextrin solution alone were used to ascertain that the fatty acids were indeed biologically oxidized.

2.4. EFFECT OF INHIBITORS ON THE OXIDATION OF SOLUBLE PALMITATE AND SPHINGOLIPIDS

To study the effect of specific inhibitors of the respiratory chain on the oxidation of soluble fatty acids, appropriate concentrations of each inhibitor were added in the cell suspension containing soluble fatty acids. After known period of incubation, O_2 uptake was measured.

3. RESULTS AND DISCUSSION

Preliminary experiments indicated that soluble palmitate was oxidized after a lag period of 2 hours by *M. leprae* and *M. lepraemurium.* Results reported in Table 1 were obtained after 2 hours of incubation of soluble and insoluble palmitate with bacillary suspensions of *M. leprae* and *M. lepraemurium.* Cell suspensions of all mycobacteria used exhibited considerable endogenous respiration. When soluble palmitate was the substrate, it yielded much higher O_2 uptake values than the endogenous activity. However, when insoluble palmitate was added in the cell suspensions, it was not oxidized by *M. leprae, M. lepraemurium* and *BCG* bacillary suspensions in a period of 6 hours (Table 1). Insoluble as well as soluble palmitate was actively oxidized by *M. phlei* cell suspensions.

TABLE 1. Oxidation of soluble and insoluble palmitate by mycobacteria

	µl O_2 uptake/ 6h					
	Soluble palmitate			Insoluble palmitate		
Mycobacteria	Endogenous	Total	Exogenous	Endogenous	Total	Exogenous
M. leprae	130	240	110	130	132	2
M. lepraemurium (In vivo)	127	300	173	127	130	3
M. lepraemurium (In vitro)	145	337	192	145	145	0
M. Phlei	240	660	420	240	645	405

Sphingolipids were preincubated with the cell suspensions of *M. leprae*, *M. lepraemurium* and *BCG* for 8 hours and then the O_2 uptake was determined. The results of a typical experiment on the oxidation of soluble and insoluble sphingolipids are shown in Table 2. When soluble sphingosine, sphingomyelin and cerebroside were the substrates, they were oxidized by bacillary suspensions of *M.leprae, M. lepraemurium* and *BCG.* However, the addition of insoluble (100 ug) sphingosine, sphingomyelin or cerebroside in cell suspensions of these mycobacteria showed no enhanced O_2 uptake over endogenous respiration for 4 hours. It may be emphasized that insoluble sphingolipids were not oxidized by these mycobacteria for a period of 24 hours. Both soluble and insoluble sphingolipids were readily oxidized by *M. phlei* cell suspensions.
Since ß Cyclodextrin was used to solubilize palmitate and sphingolipids, to ensure that if this compound had any effect on endogenous or substrate oxidation and also to confirm that the oxidation of fatty acids was not due to auto or chemical oxidation but was a biological phenomenon, a number of controls were used and O_2 uptake was measured.

Table 2. Oxdation of soluble and insoluble sphingolipids by mycobacteria

Sphingolipid	ųl O_2 uptake/ 4h							
	M. Phlei		*M. lepraemurium*		*BCG*		*M .leprae*	
	Endog.	Exog.	Endog.	Exog.	Endog.	Exog.	Endog.	Exog.
Sphingosine (Sol.)	52	153	48	38	85	55	35	45
Sphingosine (Insol.)	52	35	48	18	85	11	35	4
Sphingomyelin (Sol.)	52	105	48	50	43	49	35	40
Sphingomyelin (Insol.)	52	39	48	4	43	10	35	0
Cerebroside (Sol.)	52	126	48	65	43	51	35	50
Cerebroside (Insol.)	52	47	48	5	43	3	35	3

The results in Table 3 show that the cell suspensions of *M. phlei* alone or in the presence of cyclodextrin (CD) yielded the same amount of O_2 uptake. However, no detectable O_2 uptake was observed when CD or plamitate were used in the buffer. Cell suspensions containing CD when boiled for 5 min they were devoid of any activity. Cell suspensions of viable bacilli containing soluble palmitate exhibited 106 ųl of O_2 in one hour. No measurable O_2 uptake was observed when soluble palmitate was added to the heat killed cell suspensions. The same results were obtained when boiled *M. leprae* or other mycobacterial suspensions containing soluble palmitate or sphingolipids were used.

Table 3. Oxidation of cyclodextrin and soluble palmitate by *M. phlei* cell suspensions

System	ųl O_2 uptake/ h
Cell suspension 1 ml. + buffer (Endogenous)	40
Cell suspension 1 ml. + cyclodextrin 40 ug.	39
Buffer + cyclodextrin 40 ug.	0
Cell suspension 1 ml. , boiled + cyclodextrin 40 ug.	0
Buffer + soluble palmitate 20 ug.	106
Cell suspension 1 ml. , boiled + soluble palmitate 20 ug.	0

The effect of respiratory chain inhibitors on the oxidation of soluble palmitate and sphingosine is shown in Table 4. The results show that oxidations of soluble palmitate were quite sensitive to all the inhibitors of the respiratory chain. O_2 uptake due to soluble palmitate and sphingolipids was nearly completely (94-100 %) inhibited by the inhibitors of the respiratory chain. These results strongly suggest that the oxidations of both soluble palmitate and sphingosine carried out by viable *M. leprae* and *M. phlei* bacilli are mediated by the cytochrome systems.The presence of cytochrome in *M. leprae* (9), *M. lepraemurium* (10) and *M. phlei* (11) has previously been reported.

Table 4. Effect of inhibitors on the oxidation of palmitate and sphingosine by *M. phlei* and *M. leprae* cell suspensions

Inhibitor	Concn (mM)	% inhibition Palmitate *M. leprae*		Sphingosine *M. leprae*	
		M. leprae	*M. phlei*	*M. leprae*	*M. phlei*
None	-	0	0	0	0
Rotenone	0.05	95	97	100	98
Amytal	0.50	97	100	95	95
Atabrine	0.20	96	97	94	96
Antimycin A	0.10	95	98	97	100
Cyanide	0.20	100	100	95	97

Our results suggest that these fatty acids could be used as energy and carbon source. The use of water soluble fatty acid methylated cyclodextrin complexes have many advantages compared with insoluble compounds. Therefore, the use of these soluble fatty acids is encouraged for metabolic studies as well as in culture media used for *in vitro* cultivation trials of *M. leprae*. This mycobacterium is very slow growing. No substrate is known which can be used by *M. leprae in vivo* for their multipliacation.These fatty acids are heat stable; thus culture media containing water soluble palmitate or sphingolipids can be incubated for a long period of time and growth of *M. leprae* can be evaluated.

4. CONCLUSION

The results have shown that the methylated-ß-cyclodextrin used to solubilize fatty acids, is an innert substance and it had no inhibitory or stimulatory effect on the oxidation of fatty acids. Although, insoluble palmitate and sphingolipids were not oxidized in 6 hours, these fatty acids when solubilized in cyclodextrin were oxidized by *M. leprae*, *M. lepraemurium* and *BCG*. Based on these results, it is concluded that soluble fatty acid should be used to achieve *in vitro* cultivation of *M. leprae*.

ACKNOWLEDGEMENTS

This investigation was generously supported by the German Leprosy Relief Association, Würzburg, Germany and Le Secours aux Lepreux, Canada.

REFERENCES

[1] Ratledge, C., Lipids. Cell composition, fatty acid biosysnthesis. In: *The Biology of Mycobacteria; Volume 1, Physiology, Identification and Classification.* Ratledge, C. and Stanford, J., eds. London: Academic Press, 1982, pp.53-93.

[2] Ishaque, M. Direct evidence for the oxidation of palmitic acid by host-grown*Mycobacterium leprae*. *Res. Microbiol.* **140**, 83-93 (1989)

[3] Kato, L; Szejtli, J; Szente, L. Water soluble complexes of palmitic acid and palmitates for metabolic studies and cultivation of *Mycobacterium leprae.Int. J. Lepr.* **60**: 105-107 (1992)

[4] Wheeler, P.R., Bulmer, K., Ratledge, C. Enzymes for biosynthesis de *novo* and elongation of fatty acids in mycobacteria grown in host cells: is *Mycobacterium leprae* competent in fatty acid biosynthesis?. *J. Gen. Microbiol.* **36**: 211-217 (1990).

[5] McCarthy, C. Utilization of palmitic acid *by Mycobacterium avium. Infect. Immun.***4**: 199-204 (1971)

[6] Kato, L., Szejtli, J., Szente, L. Water soluble complexes of palmitic acid and palmitates for metabolic and cultivation trials of *Mycobacterium leprae.Int. J. Lepr.* **60**: 105-107 (1992)

[7] Ishaque, M. Palmitate oxidation by in *vivo* and *in vitro* grown *Mycobacterium* lepraemurium. Cytobios,**71**: 19-27 (1992)

[8] Umbreit, W.W., Burris, R.H., Stauffer, J.F. *Manometric Techniques*. Burgess Publ. Co. Minneapolis, Minn. (1964).

[9] Ishaque, M., Kato, L. The cytochrome system in *Mycobacterium lepraemurium. Can. J. Microbiol.* **20**: 943-947 (1974)

[10] Kato, L., Ishaque, M., Walsh, G.P. Cytochrome pigments in *Mycobacterium leprae* isolated from armadillos (Dasypus *novemcinctus* L.) *Microbios*,**12**: 41-50 (1975)

[11] Asano, A., Brodie, A. F. Oxidative phosphorylation in fractionated bacterial system. XIV. Respiratory chains of *Mycobacterium phlei. J.Biol. Chem.* **240**: 4280-4291 (1965).

CYCLODEXTRINS IN FABRIC CARE CONSUMER PRODUCTS

TOAN TRINH

The Procter & Gamble Company
Sharon Woods Technical Center
11520 Reed Hartman Highway
Cincinnati, OH 45241 USA

ABSTRACT

Cyclodextrins can be used to provide a long lasting freshness benefit on laundered fabrics. This benefit can be achieved by incorporating cyclodextrin/perfume complexes in granular detergents, in liquid fabric softeners, and, most effectively, in dryer-added fabric softeners. Such dryer-added fabric softener products are commercially available, and provide perfume benefits, such as in-wear long-lasting fabric freshness and in-use perfume blooming, that are recognized and appreciated by the consumer.

1. INTRODUCTION

In fabric care, cyclodextrins can be used either (1) to protect desirable materials, such as perfume or bactericides, in the product, then release them in use, or (2) to absorb undesirable materials, such as body malodor, environmental malodor such as cigarette odor, kitchen odor, and product malodor. This paper primarily discusses the odor-related improvement in fabric care.

Almost all laundry products contain some perfume, for two reasons: (1) the perfume provides some nice, fresh odor to fabrics after the clothes are washed and dried, and (2) since the products themselves often contain some ingredients that have an unpleasant odor, perfume is needed to mask that odor to make the products more appealing to the consumer. Consumer surveys always indicate that consumers want to obtain a long lasting fabric freshness benefit from home laundry. That freshness is usually provided by selecting the right perfume.

The key steps in the home laundry process are washing, rinsing and drying. There are laundry products that can be used in each step, e.g., detergents and bleaches are used in the washing step, liquid softeners in the rinsing step, and dryer-added softeners when there is a tumble drying step.

Detergents are the least effective means to deliver perfume to fabrics because the perfume can be lost at many stages. During storage, some perfume is lost due to evaporation after the box is open, and there may be degradation of some perfume

J. Szejtli and L. Szente (eds.), Proceedings of the Eighth International Symposium on Cyclodextrons, 541–546.

ingredients if the detergent also contains a bleach system. In the wash itself, a substantial amount of perfume is lost to the wash water, where it is solubilized by the surfactants. Some perfume is also degraded by any bleach that is present in the washing step. More perfume is lost to rinse water and in the drying step, especially tumble-drying, where there is a large volume of hot drying air [1].

2. PRIOR ART

2.1. Wash-Added Products

A 1981 German Patent Application by J. Kock [2] discloses a detergent composition which contains a cyclodextrin/perfume complex to improve perfume stability. This patent application addresses the problem of perfume loss in storage by evaporation and/or due to the presence of a bleach in the product. However, once this detergent is added to wash water, the perfume is released, and again the perfume will substantially be lost to the wash and the rinse water, and in the drying step.

A 1991 U.S. Patent issued to Nebashi et al. [3] discloses a detergent composition which contains a cyclodextrin/perfume complex in the form of rather large particle size granules. It discloses that the granules help to improve perfume stability in the product, and provide a better perfume odor in the wash water and on fabrics. However, a large part of the complexed perfume is released to the wash and rinse water, and it is expected that a substantial amount of the perfume is lost in the whole laundry process.

Another application of cyclodextrin in detergents is disclosed in a 1989 Japanese Laid-Open Patent Application by Yabe et al [4], which claims a detergent composition containing free cyclodextrins to remove the odor of enzymes which are present in said detergent composition.

2.2. Rinse-Added Products

Uncomplexed cyclodextrins are claimed for malodor control in a liquid softener composition. A 1988 Japanese Laid-Open Patent Application by Yoshimura et al [5] discloses liquid softener compositions containing different cyclodextrins to remove malodor from fabrics in-wear. It is not clear how the compositions are applied to fabrics. Apparently, the product is applied neat to the fabrics, and not in the rinse, because the patent application discloses an effective and identical odor control performance by four compositions which contain, respectively, α-, β-, γ-, and cationic β-cyclodextrins. If the products were dispensed to the rinse water, it would be expected that (1) the cyclodextrins would be rinsed away and would not be effectively deposited on fabrics to provide the desired odor control performance, and (2) even if some residual cyclodextrins were to be retained on fabrics after the rinse, different degrees of odor control effectiveness should be obtained, due to the vast difference in water solubility, charge effect, and, hence, deposition level of these different cyclodextrins.

The delivery of free perfume via a rinse-added liquid softener is more effective than that via a detergent, because the washing step where the perfume loss is the most severe, is avoided. However, the rinse water, which still has a sufficient amount of carry-over surfactants, also causes a significant loss of perfume.

Cyclodextrin/perfume complexes are normally not very useful in liquid softeners because the aqueous phase of the product and the presence of large amounts of organic materials in the liquid composition make the perfume complexes unstable. Furthermore, when the product is added to the rinse water, almost all complexed perfume is expected to be released, and thus there is no significant added benefit, as compared to a composition which contains just the free perfume.

3. DRYER-ADDED FABRIC SOFTENER SHEETS

Products that are added in the drying step provide the most effective means to deliver perfume to laundered fabrics. However, without protection, the perfume loss is still substantial, due to high temerature. Furthermore, most of the ingredients that are lost in the drying steps are the more volatile ingredients, which are believed to provide the "freshness" impression to fabrics.

3.1. Conventional Products

Dryer-added sheets are typically composed of a blend of fabric softener active and perfume impregnated on a sheet of non-woven fabric. The softener blend is a waxy solid at room temperature, but melts and is distributed to the laundered fabrics at the high temperature of the tumble dryer.

The perfume level in the product is normally chosen such that the initial fabric perfume odor, smelled immediately after the fabric is treated in the dryer, receives the best rating by the majority of consumers. However, the fabric perfume odor intensity decreases rapidly with time. Consumer research indicates that most consumers would prefer having a perfume odor that lasts longer.

One way to increase the perfume odor after extended storage times is to increase the perfume level in the softener blend to provide a higher odor intensity for a longer time. However, this would cause an initial odor which is too strong, which is a big negative to the consumers.

3.2. Products with Cyclodextrin/Perfume Complex

Another approach is to start with the current product in order to maintain the best initial odor, but also to incorporate a perfume reservoir to release extra perfume over time. It is found that cyclodextrin is a very effective means to deliver extra perfume in these products. The dryer-added fabric softener product that contains a cyclodextrin/perfume complex is disclosed in a series of U.S. Patents, including those issued in 1992 and 1993 to Gardlik, et al. [6]. Essentially, the cyclodextrin/perfume complex is incorporated into the softener blend, and is released in the tumble dryer to the fabrics, along with the softener active and free perfume.

At the beginning of the tumble drying cycle, the temperature of the fabrics is low, because the fabrics are still wet. At this stage, the cyclodextrin/perfume complex remains coated and protected by the fabric softener active, and is not significantly in contact with the wet fabrics. Therefore, the decomposition of the complex by the wet fabrics is minimized. Later in the drying cycle, when the fabrics become drier and their temperature is higher, the softener blend is transferred to the fabrics. Most of the molten softener active diffuses into the fabric fibers, and the cyclodextrin complex is then exposed. At this stage, although the complex is exposed, there is not enough moisture on the fabrics to destabilize the complex, and thus, the majority of the complex remains intact on fabrics, even at the end of the drying cycle. The complexed perfume releases slowly when the clothes are worn, and a much faster release occurs to give a perfume bloom when the fabrics are exposed to sources of water, such as in the case of bath and hand towels.

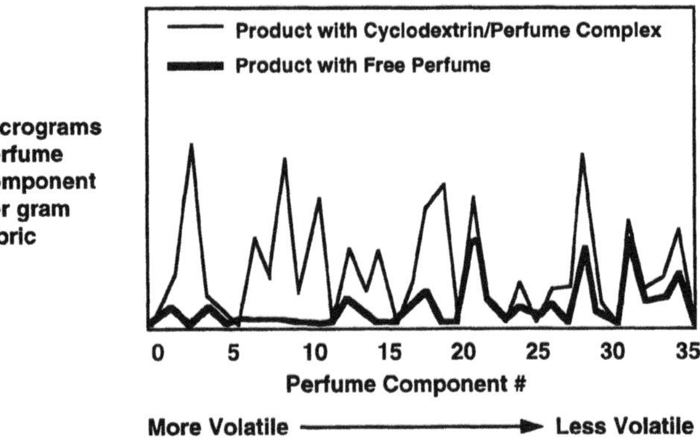

Fig. 1 Dryer Sheet Fabric Perfume Deposition via Cyclodextrin vs. Free Perfume.

Figure 1 shows gas chromatograms of perfume deposited on fabric from the dryer sheets. The heavy line is the perfume deposited on fabric from a sheet containing neat, free perfume. The thin line is the perfume deposited on fabric from a sheet containing the same amount of perfume, but complexed with β-cyclodextrin. It can be seen that most of the volatile ingredients are lost in the neat perfume product treatment, while they are retained almost quantitatively in the cyclodextrin complex product treatment.

Currently, product containing a cyclodextrin/perfume complex is commercially available in the United States and Canada, where automatic tumble dryers are the most numerous.

4. REAPPLICATIONS

The learning from the dryer sheet development is reapplied to detergent and liquid softener development. The key feature is a way to protect the cyclodextrin/perfume complex from the wash and/or rinse water. Two U.S. Patents issued in 1993 to Trinh, et al., [7], [8] disclose liquid softener and detergent compositions which contain protected, dryer-activated, cyclodextrin/perfume complex particles.

The protected cyclodextrin/perfume complex particles are matrix microcapsules composed of a cyclodextrin/perfume complex powder embedded in a highly hydrophobic, meltable wax matrix. The wax protects the cyclodextrin complex from decomposition by water in the products, and in the washing and the rinsing steps. In the tumble drying step, the wax melts to expose the cyclodextrin/perfume complex particles on the dried fabrics, similar to the case of the dryer sheets.

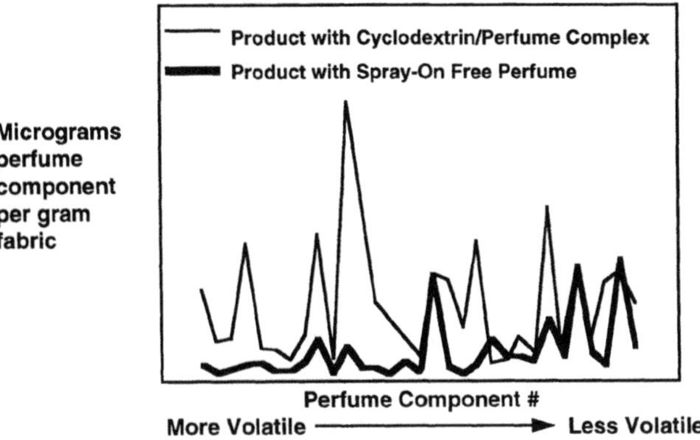

Fig. 2 Detergent Perfume Fabric Deposition via Cyclodextrin vs. Spray-On Free Perfume.

As an example, Figure 2 shows that fabrics washed with a detergent composition containing a protected cyclodextrin complex retain significantly more volatile perfume ingredients. It needs to be emphasized, however, that the protected cyclodextrin complexes only work well if the clothes are dried in a tumble dryer, so that the cyclodextrin complex can be exposed.

The protected cyclodextrin/perfume complex particles can be made by first blending the dried cyclodextrin/perfume complex powder with a suitable molten hydrophobic wax. This blend is then either cryogenically ground or spray-dried into small particles of suitable size.

Figure 3a is a scanning electron micrograph of cryogenically ground protected cyclodextrin complex particles. This particular batch has a particle size in the range of from about 150 microns to about 250 microns in diameter. Figure 3b is a scanning electron micrograph of protected cyclodextrin complex particles made by the spray-drying method. This particular batch has particles with sizes mostly in the range of from about 75 microns to about 150 microns in diameter.

It is found that the cryogenically ground protected cyclodextrin complex particles are more effective in delivering perfume to fabric than particles made by the spray-drying method. It is believed that the cryogenically ground particles are retained better on the

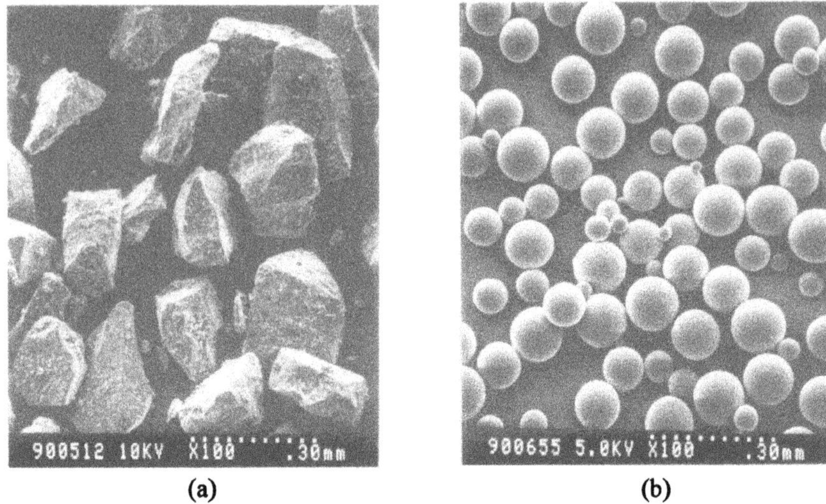

(a) (b)

Fig. 3 Protected Cyclodextrin Complex Particles:
(a) Cryogenically Ground, (b) Spray-Dried.

fabrics due to their irregular shape, as compared to the smooth and spherical particles made by the spray-drying method.

In summary, the commercial value of the cyclodextrin technology has been fully validated in dryer-added fabric softener products. One would expect that this is only the first of many such successes in large-volume fabric care consumer products.

REFERENCES

[1] In the United States, about 95% of home laundry loads are washed by machines, and about 85% of the loads are dried in tumble dryers.

[2] Kock, J., Storage-Stable, Powdered Detergent and Cleaning Agent Containing Perfume, German Patent Application No. 30 20 269 (1981).

[3] Nebashi, T., Yabe, S., Sai, F., Izumi, Y., Fujieda, T., Detergent Composition Containing Clathrate Granules of a Perfume-Clathrate Compound, U.S. Patent No. 4,992,198 (1991).

[4] Yabe, S., Nehashi, T., Sai, F, Powdered Detergent Composition, Japanese Laid-Open Patent Application No. 256,596 (1989).

[5] Yoshimura, M., Nakazawa, H., Kandori, H., Soft-Finishing Agent, Japanese Laid-Open Patent Application No. 165,498 (1988).

[6] Gardlik, J. M., Trinh, T., Banks, T. J., Benvegnu, F., Treatment of Fabric with Perfume/ Cyclodextrin Complexes, U.S. Patent Nos. 5,102,564 (1992) and 5,234,610 (1993).

[7] Trinh, T., Bacon, D. R., Benvegnu, F., Fabric Softener, Preferably Liquid, with Protected, Dryer-Activated, Cyclodextrin/Perfume Complex, U.S. Patent No. 5,234,611 (1993).

[8] Trinh, T., Bacon, D. R., Solid, Particulate Detergent Composition with Protected, Dryer-Activated, Water Sensitive Material, U.S. Patent No. 5,236,615 (1993).

CD DYE COMPLEXES AND THEIR USE IN DYEING PROCESSES

H.-J. Buschmann
German Textile Research Centre North-West e.V.
Frankenring 2, D-47798 Krefeld, Germany

ABSTRACT

The solubility of disperse dyes in aqueous solutions increases in the presence of cyclodextrins due to complex formation. The stability constants of the cyclodextrin-dye complexes can be calculated from the absorbance of the solutions. The extinction coefficients of the disperse dyes are measured in aqueous mixtures of organic solvents. The use of solid cyclodextrin complexes of disperse dyes in dyeing processes has some advantages compared with the conventional process using dyestuff dispersions.

1. INTRODUCTION

The complex formation of dyes with cyclodextrin has already been described in numerous articles [1-3]. However, water soluble the dye molecules have been used. Due to interactions with cyclodextrins the solubility of dyes nearly insoluble in water can be increased. This is known in principal from measurements with other organic substances, which are not soluble in water [3]. Up to now very little is known about the formation of cyclodextrin complexes with disperse

J. Szejtli and L. Szente (eds.), Proceedings of the Eighth International Symposium on Cyclodextrons, 547–552.

dyes and the use of these complexes in industrial processes [4].

2. EXPERIMENTAL

All dyes examined were industrial products. If necessary dispergators and other auxiliaries were removed from the dyes by extraction with organic solvents and recrystallization or sublimation. The cyclodextrins were used without further purification.

The solid disperse dyes were added to aqueous solutions of the cyclodextrins and these solutions were shaken for several days to ensure complete complex formation. After filtration and centrifugation of these solutions the absorbance was measured using a spectrophotometer (Cary 5E, Varian).

The measured absorbance A at a constant wavelength is given by the following equation:

$$E = (\epsilon_1[D]_{sat} + \epsilon_2[D\text{-}CD])/d \ ,$$

with the pathlength d of the optical cell, as the extinction coefficents of the dye ϵ_1 and the corresponding complex with a cyclodextrin ϵ_2, the concentration of saturated aqueous solution of the dye $[D]_{sat}$ and the concentration of the dye complex with a cyclodextrin [D-CD]. The extinction coefficents of the pure dyes were obtained from measurements of dye solutions in mixtures of water with dioxane, ethanol, methanol and acetone. Using the assumption that the extinction coefficents of the dye and the dye complexes with cyclodextrins are nearly equal the stability constants for the formation of the complexes of disperse dyes with cyclodextrins can be calculated.

Solid complexes of cyclodextrins with disperse dyes were prepared by dissolving the dye molecules in hot solutions of the cyclodextrins. The hot solutions were filtered and cooled. The precipitated complexes were isolated by filtration. These complexes were used for dyeing of polyester fibres without the use of further auxiliaries. For comparison polyester fibres were also dyed using a standard procedure and commercial disperse dyestuffs. The absorbance of the residual dye liquor and the Kubelka-Munk values (K/S) were measured spectrophotometrically.

3. RESULTS AND DISCUSSION

3.1 FORMATION OF CD-DYE COMPLEXES

The solubility of disperse dyes in aqueous solution increases due to the interactions with cyclodextrins. In Figure 1 the change of the absorbance of aqueous solutions containing different concentrations of α-, ß- and τ-cyclodextrins and the dye C.I. Disperse Orange 3 is shown.

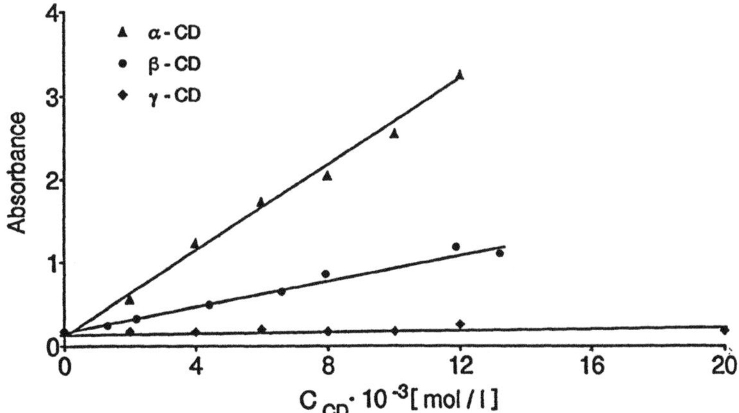

Fig.1 Absorbance of aqueous solutions containing different concentrations of α-, ß- and τ-cyclodextrins and C.I. Disperse Orange 3

Comparable results are obtained with all other disperse dyes examined. Under the assumption that only 1:1-complexes of cyclodextrins with disperse dyes are formed the stability constants can be calculated. These results are summarized in Table 1.

TABLE 1. Stability constants (log K, K in $dm^3 \cdot mol^{-1}$) for the complexation of disperse dyes by cyclodextrins in aqueous solution at 25 °C

Disperse dye	α-CD	ß-CD	γ-CD
Orange 3	3.24	2.70	0.93
Orange 11	1.54	2.23	1.36
Orange 29	<0.5	<0.5	<0.5
Yellow 42	2.19	2.27	2.59
Violet 1	1.65	2.54	1.79

Obviously the stability of the complexes formed depends on the chemical structure of the dye molecules and the cavity sizes of the cyclodextrins.

3.2 THE USE OF CD-DYE COMPLEXES IN DYEING PROCESSES

The simultaneous addition of cyclodextrins and disperse dyes in dyeing processes shows no effect, because only the monomolecular dissolved dye molecules can be adsorbed by the fibre and penetrate into it. The increase in solubility of the dye molecules due to complex formation is not sufficient. As a result the duration of the dyeing process has to be increased very much. This is not reasonable under economic considerations.

This situation changes completely if preformed dye-cyclodextrin complexes are used in dyeing processes. After dissolution the complex preferentially dissociates in aqueous solution. Lacking the possibility to form dye aggregates most of the monomolecular dissolved disperse dye molecules are adsorbed by the fibre quickly. Thus, the remaining concentration of the disperse dye in the aqueous phase after the dyeing process is rather small. This is shown in Figure 2 for a conventional dyeing process and for a process using the solid dye-cyclodextrin complexes. The concentration of the dye in solution influences the colour depth obtained on the dyed polyester fibres.

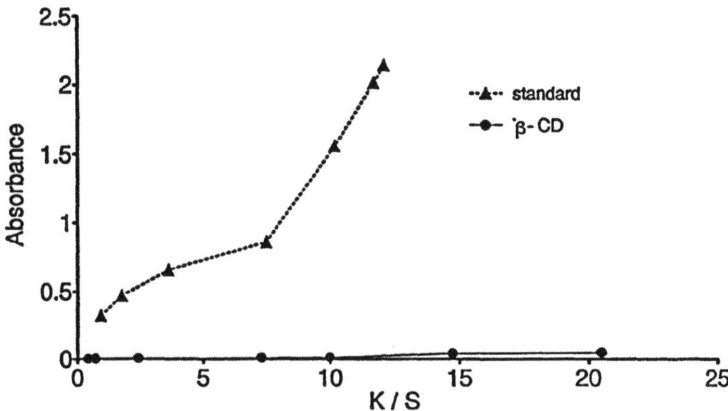

Fig.2 Absorbance of the residual dye liquor as function of K/S-values for dyeing with C.I. Disperse Orange 13 in a standard process and with the CD-dye complex

In a standard disperse dyeing process a high amount of these dyes remain unfixed in solution. Using the corresponding cyclodextrin complexes the concentration of the residual dyes is drastically reduced in the aqueous phase after the dyeing process.

4. CONCLUSIONS

The use of complexes of cyclodextrins with disperse dyes has some advantages compared with the standard procedure. More dye molecules are used for the dyeing of the fibre and as a result the concentration of the remaining dyes in the waste water is reduced. No further auxiliaries are required. The cyclodextrin molecules in the waste water do not cause any problems. The industrial use of cyclodextrin-dye complexes is favoured by ecological and economical reasons.

ACKNOWLEDGEMENTS

The supply with pure disperse dyes from BASF, Bayer AG and Ciba-Geigy and with cyclodextrins from Wacker Chemie is gratefully acknowledged. We are also grateful to the Forschungskuratorium Gesamttextil for their financial support for this research project (AIF-No. 8672). This support was granted from resources of the Federal Ministry of Economics via a supplementary contribution by the Association of Industrial Research Organizations (Arbeitsgemeinschaft Industrieller Forschungsvereinigungen, AIF).

REFERENCES

[1] Cramer,F., Über Einschlußverbindungen, I. Mitteil.: Additionsverbindungen der Cycloamylosen, Chem. Ber. **84**, 851-854 (1951).
[2] Cramer, F., Saenger, W. and Spatz, H.-Ch., Inclusion Compounds. XIX. The Formation of Inclusion Compounds of α-Cyclodextrin in Aqueous Solutions. Thermodynamics and Kinetics, J. Am. Chem. Soc. 89, 14-20 (1967).
[3] Szejtli, J., Cyclodextrin Technology, Kluwer Academic Publishers, Dordrecht,1988.
[4] Buschmann, H.-J., Knittel, D. and Schollmeyer, E., Möglichkeiten des Einsatzes von Cyclodextrin-Farbstoff-Komplexen in Färbeprozessen, Textilveredlung, in print.

BETA W7 MCT - NEW WAYS IN SURFACE MODIFICATION

H. Reuscher and R. Hirsenkorn
Wacker-Chemie GmbH, Hanns-Seidel-Platz 4,
D-81737 München, GERMANY

ABSTRACT

BETA W7 MCT is the first reactive cyclodextrin derivative manufactured on an industrial scale. It has a monochlorotriazinyl group as a reactive anchor well known from many reactive dyes. This derivative is able to form stable covalent bonds with nucleophilic groups and can be easily prepared in water in an effective one pot synthesis from cyanuric chloride and ß-cyclodextrin in a yield of approx. 90% based on the triazinyl group. The optimised degree of substitution of DS = 0.4 per anhydroglucose ensures a good complexation capacity even when this derivative is fixed to surfaces like textiles. This cyclodextrin derivative containing 2-3 reactive groups in the ring can be used as building block for new CD derivatives, as a crosslinking agent or as an excellent material for surface modification.

1. INTRODUCTION

Cyclodextrins and their derivatives are suitable for a wide variety of possible uses e.g. for stabilisation, masking or controlled release of hydrophobic substances [1,2]. With a cyclodextrin derivative bearing a reactive group it should be possible to incorporate these favourable properties of cyclodextrins permanently on a surface. Possible surfaces are e.g. cotton, paper or natural and synthetic polymers. For greatest stability the cyclodextrin should be fixed via a covalent bond. A reactive cyclodextrin would be also useful for the synthesis of new cyclodextrin derivatives. The selected reactive group in the new cyclodextrin derivative BETA W7 MCT (Fig. 1) is the monochloro-triazinyl group well known from many reactive dyes. This anchor is able to form the required stable covalent bond to nucleophilic groups.

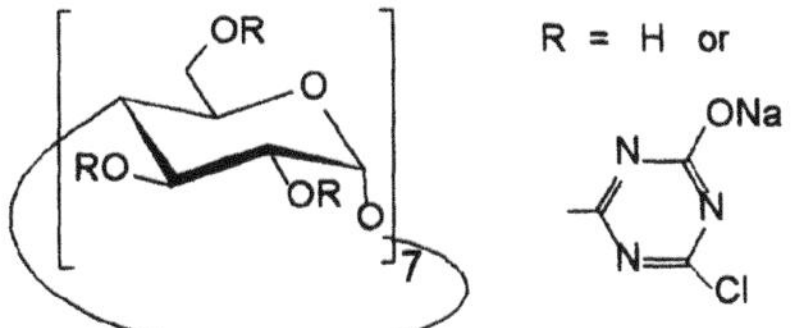

Fig. 1 Structure of BETA W7 MCT

J. Szejtli and L. Szente (eds.), Proceedings of the Eighth International Symposium on Cyclodextrons, 553–558.

2. MATERIALS AND METHODS

Beta-cyclodextrin is a product of Wacker-Chemie GmbH, Munich, Germany and cyanuric acid a product of SKW Trostberg AG, Trostberg, Germany. The synthesis of Beta W7 MCT was developed by Wacker-Chemie [3,4]. The degree of substitution of the reactive chlorine in the triazine group was calculated from ^{1}H-NMR after reaction of Beta W7 MCT with diethylamine in water [3,4]. The toxicological data have been carried out according to OECD guidelines. Measurements of the solubilisation capacity of MCT-derivatives with different degrees of substitution were carried out according to known methods from the literature.

3. RESULTS AND DISCUSSION

3.1. Synthesis

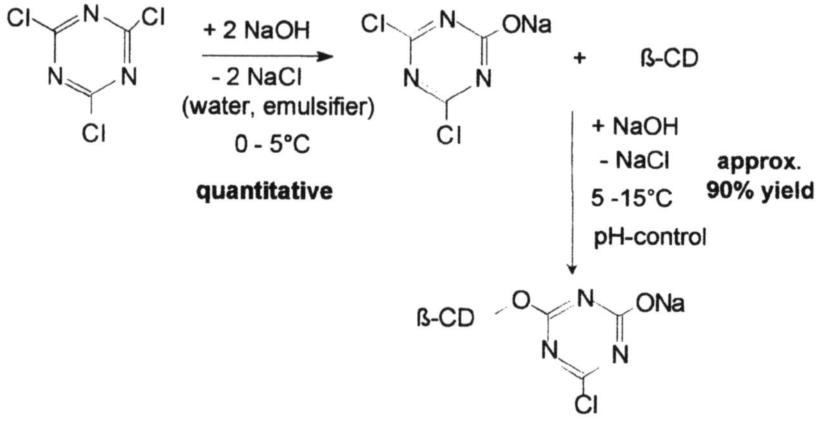

+ 2 NaOH
- 2 NaCl
(water, emulsifier)
0 - 5°C
quantitative
+ ß-CD
+ NaOH
- NaCl
5 -15°C
pH-control
approx. 90% yield
ß-CD

Fig. 2 Effective one pot synthesis of Beta W7 MCT

The first step in the effective one pot synthesis (Fig. 2) of BETA W 7 MCT is the preparation of a clear aqueous solution of dichlorotriazine sodium salt at low temperature starting from a dispersion of cyanuric chloride in water. Cyanuric chloride is an inexpensive chemical product available in very large quantities normally used for manufacturing e.g. agrochemical products or reactive dyes. It contains three chlorine atoms with different reactivity for nucleophiles. Under the chosen reaction conditions only one chlorine atom reacts with sodium hydroxide.

In the second reaction step of the synthesis the prepared dichlorotriazine compound reacts with ß-cyclodextrin in water at higher temperature under alkaline conditions and pH-control. When the degree of substitution of the reactive group decreases from DS = 1.0 to DS = 0.4 the achieved yield according to the triazine moiety increases from 78% to 88% due to a pre-complexation effect. After purification and spray drying BETA W7 MCT has an appearance of a stable white powder. Comparable Alpha- or Gamma-MCT derivatives are also available.

3.2. Properties

BETA W7 MCT is good soluble in water (> 30% (g/g)) and as a dry powder it is stable for a minimum of one year. In water hydrolysis occurs, the degree of hydrolysis and its reaction velocity mainly depends on temperature and pH-value. An aqueous solution of BETA W7 MCT e.g. is stable at pH = 8-9 and room temperature for at least 2 months. The hydrolysis becomes faster at higher temperature or at lower pH-value due to an autocatalytic effect of the released hydrogen chloride. The solubilisation capacity of BETA W7 MCT depends on the degree of substitution. With decreasing DS-value the solubilisation capacity increases. On the other hand a high fixation yield affords a high degree of substitution. Table 1 summarises these general properties:

Table 1 Beta W7 MCT - General properties

properties	benefit
reactive group	covalent bonding
Degree of substitution DS = 0.4	ensure high fixation yields ensure good complexing capacity
good water solubility (>30%)	easy handling

Table 2 reports a short overview of toxicological data known up to now. BETA W7 MCT has no irritating or sensitising effect according to OECD tests. Therefore no irritating or sensitising effect e.g. of MCT finished textiles is expected.

Table 2 Beta W7 MCT - Toxicological data

Test	System	Result
Acute Toxicity, oral (OECD No 401)	rat	LD50 > 2000 mg/kg
Primary skin irritation (OECD No 404)	rabbit	not irritating
Skin sensibilisation (OECD No 406)	guinea pig	not sensitising
Amestest (OECD No 471)	Salmonella typhimurium	not mutagenic

3.3. Possible applications of BETA W7 MCT

3.3.1. as building block

The formation of new cyclodextrin derivatives for special applications is readily achieved with this multifunctional agent. BETA W7 MCT reacts under slightly alkaline conditions with molecules bearing nucleophilic groups such as OH-, NH- or SH-groups. Fig. 3 shows as one example the reaction of BETA W7 MCT with amines. With this reaction it is e.g. possible to fix a hydrophobic anchor with a high affinity to hydrophobic surfaces covalently to the cyclodextrin. Potential applications could be the modification of synthetic textiles (e.g. polyester, polyamides, polyacrylics).

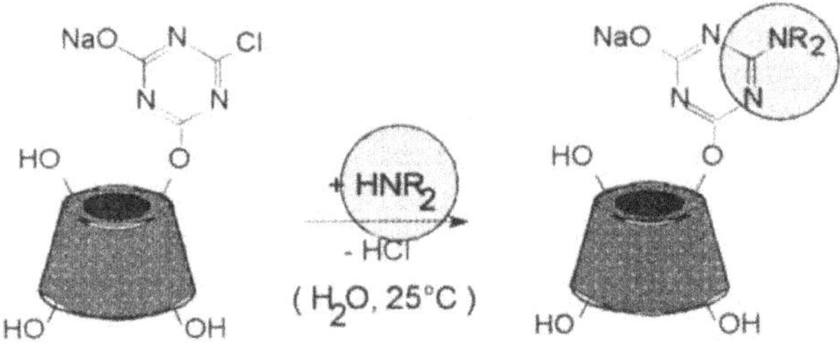

Fig. 3 Beta W7 MCT - Building block for new derivatives

3.3.2. as crosslinking agent

With a degree of substitution of DS = 0.4 BETA W7 MCT contains 2 -3 reactive triazine groups per cyclodextrin molecule. This is why this cyclodextrin derivative can also be used as a new formaldehyde free crosslinking agent with complexing capacity e.g. for the synthesis or modification of natural or synthetic polymers like starch, cellulose, polyallylamine or gelatine. With a higher degree of substitution BETA W7 MCT forms an insoluble cyclodextrin polymer by reacting with itself. Possible applications could be new chromatography materials or membranes for extraction purposes.

3.3.3. for surface modification

Surfaces like textiles or paper with nucleophilic groups can be modified with BETA W7 MCT using well established methods known from the textile industry. BETA W7 MCT has only a very low affinity to cotton, therefore it is impossible to finish cotton at high temperature in an exhausting process like some textile dyes. In this case the cyclodextrin derivative would react mainly with water and not with the textile. A possible way for textile finishing using BETA W7 MCT would be the so-called pad / roll process. Therefore an alkaline aqueous MCT solution is padded on the textile and squeezed. Fig. 4 shows this general procedure for textiles like cotton in a schematic representation. After drying the fixation of the reactive cyclodextrin takes place at elevated temperature. Final washing removes unreacted cyclodextrin.

In lab-trials using a drying oven a fixation time of 3-5 min at 150 °C was found to be suitable. Using contact heat this fixation time decreases to less than 1 min at 145°C or 10 - 20 sec. at 175°C. By this method a gravimetric fixation yield of approx. 80-90% could be reached.

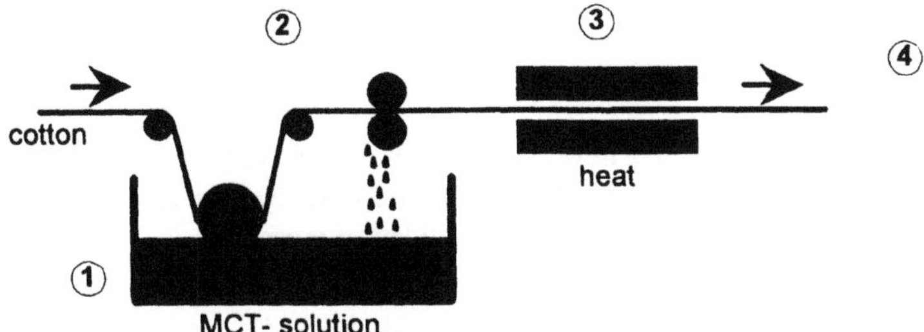

① Alkaline conditions (e.g. sodiumcarbonate)
② Padding and squeezing
③ Drying and fixation
④ Final washing

Fig. 4 Schematic representation for textile finishing with BETA W7 MCT

Measurements of the complexing capacity of MCT-finished cotton with hydrophobic substances like toluene indicate that the cavity of the bonded cyclodextrin is freely available and can be used for complexations. Increasing amounts of BETA W7 MCT on the textiles correspond to increasing amounts of complexed guest substances. The finishing is washfast and the components can be re-loaded after a washing step.

The cavity of the cyclodextrin can be utilised in a variety of ways. Active ingredients can be included in the cyclodextrin and released again in a controlled manner. A few possible applications of MCT-finished textiles are listed in table 3. For laundry odoring these textiles are able to complex fragrances. These components are released over a period of time after rewetting the textile. Simultaneously sweat absorption takes places when these textiles are worn. Another possible application of these textiles with new interesting properties would be an antimicrobial textile finishing for hospital hygiene.

Table 3 Possible applications of MCT-finished textiles

applications	examples
fragrance release	laundry odoring
odor absorption	sweat absorption
controlled release	antimicrobial finishing
stabilisation	active ingredients

4. CONCLUSION

With BETA W 7 MCT a new cyclodextrin derivative is commercially available in technical scale for a variety of possible uses (Fig. 5). The future will show the potential of this new derivative and its main application areas.

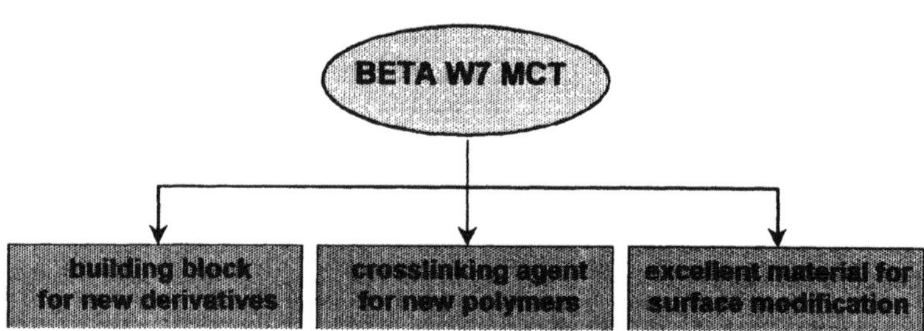

Fig. 5 Possible uses of BETA W7 MCT

REFERENCES

[1] Parrish, M.A., Cyclodextrins - a review, *Spec. Chem.* 7 366-380, 1987

[2] Saenger, W., *Angew. Chem.*, **92**, 343, 1980; *Angew. Chem. Int. Ed. Engl.*, **19**, 344, 1980

[3] Consortium für elektrochemische Industrie GmbH, Cyclodextrinderivate mit mindestens einem stickstoffhaltigen Heterocyclus, ihre Herstellung und Verwendung, Offenlegungsschrift DE 44 29 229 A1 (1996)

[4] Consortium für elektrochemische Industrie GmbH, Cyclodextrinderivate mit mindestens einem stickstoffhaltigen Heterocyclus, ihre Herstellung und Verwendung, Europäische Patentanmeldung EP 0 697 415 A1 (1996)

SURFACE MODIFICATION OF SYNTHETIC AND NATURAL FIBRES BY FIXATION OF CYCLODEXTRIN DERIVATIVES

U. DENTER, E. SCHOLLMEYER

German Textile Research Centre North-West e.V.
Frankenring 2, D-47798 Krefeld, Germany

ABSTRACT

It is demonstrated that cyclodextrin derivatives can be fixed permanently onto polymer surfaces by functional groups using conventional technologies of textile processing. As a result of the modification process some properties of the fibres are directly influenced. But also special effects may be obtained by complexing specific chemical substances, because the fixed cyclodextrin cavities do not loose their complexing power. Conceivable fields of application are medical or technical textiles and textiles for clothing.

1. INTRODUCTION

As a last step in textile processing a finish is done to get materials of specific quality [1]. A new concept for the modification of synthetic and natural fibres is based on the permanent fixation of supramolecular components, e.g. cyclodextrins, on the surfaces by functional groups using common technologies of textile processing [2,3]. The properties of the polymers may be directly influenced by the modification process. But also the fibres achieve new and particular properties by means of the inclusion of non-polar organic molecules into the fixed cyclodextrin cavities. Such effects are not obtainable by conventional finishing methods.

In the following conditions for the application of cyclodextrins and methods for the detection of the ligands on fibre materials will be discussed. Some examples for the textile properties attainable will be demonstrated.

J. Szejtli and L. Szente (eds.), Proceedings of the Eighth International Symposium on Cyclodextrons, 559–564.

2. MATERIALS AND METHODS

2.1. Materials

The cyclodextrin (CD) derivatives used are development products of *Wacker-Chemie GmbH* (abbreviations and average substitution degrees in brackets): monochloro-triazinyl-β-CD (MCT-β-CD, 0.4), dihydroxypropyl-ethylhexylglycidyl-β-CD (DHP-EHG-β-CD, 0.65/0.34), hydroxypropyl-butylglycidyl-β-CD (HP-BG-β-CD, 0.63/0.32), hydroxypropyl-hydroxyhexyl-β-CD (HP-HH-β-CD, 0.43/0.41), hydroxypropyl-phenyl-glycidyl-β-CD (HP-PG-β-CD, 0.90/0.35), o-cresylglycidyl-β-CD (o-CG-β-CD, 1.5), ethylhexylglycidyl-β-CD (EHG-β-CD, 1.4), hydroxypropyltrimethylammoniumchloride-β-CD (HPTMAC-β-CD, 0.5).

Fabrics of cotton, regenerated cellulose (CV), polyethylenterephthalate (PET), poly-amide-6.6 (PA-6.6) and polyacrylonitrile (PAN) are used as textile materials. All fabrics have been purified from accompanying substances.

2.2. Methods

Cotton samples are dipped into an alkaline solution (20 g/l sodium carbonate) of MCT-β-CD (20 g/l) at 25 °C and squeezed out under defined conditions. The fixation step is varied as follows: treatment with saturated steam (5 min at 100 °C), dry heat (5 min at 130 °C) or contact heat (3 min at 150 °C) or storage with exclusion of air (15 h at 25 °C or 4 h at 80 °C). Also cotton and CV fabric are printed with MCT-β-CD following a conventional reactive printing process (printing paste consisting of 40 g MCT-β-CD, 75 g urea, 20 g sodium hydrogen carbonate, 25 g alginate thickener and 840 g deionized water). The prints are dried (2 min at 100 °C) and fixed with saturated steam (8 min at 100 °C). All treated cellulose samples are rinsed with hot and cold deionized water.

Nonionic CD derivatives are applied to synthetic materials according to conventional exhaust dyeing methods for disperse dyes. The starting concentration is $5 \cdot 10^{-5}$ mol CD/g material and the liquor ratio is 1:50. The system is heated up with a constant rate (2 °C/min) to 130 °C (PET) or 100 °C (PA-6.6 and PAN). The total time of the treatment is 100 min. The cationic derivative HPTMAC-β-CD is applied to PAN fabric under the conditions of a basic dyeing process. The starting concentration is $5 \cdot 10^{-5}$ mol CD/g material and the liquor ratio is 1:50. The system is heated up with 2 °C/min to 80 °C and further with 0.7 °C to 95 °C. The total time of the treatment is 80 min. All samples are rinsed with hot and cold deionized water.

MCT-β-CD is directly detected on cellulose fibres by reflectance measurements in the UV region. PAN fabric which is treated with HPTMAC-β-CD is stained with the anionic indicator dye bromophenol blue ($1.5 \cdot 10^{-4}$ mol/l) in acidic solution (pH 3, acetic acid) at 25 °C, rinsed and investigated spectroscopically. Nonionic CDs are detected on synthetic fibres by the interaction with the anionic fluorescence dye ANS (ammonium salt of 8-anilinonaphthaline-1-sulfonic acid). Samples are treated with aqueous solutions of ANS (10^{-4} mol/l) for 15 min at 25 °C, squeezed out under defined conditions and dried. The fluorescence intensity of the fabrics is determined with a spectrofluorometer RF-5001PC (*Shimadzu*).

The modified fabrics are characterized as follows: a) 10 µl of water are put onto the polymer surface from a constant distance and the time for penetration is determined. b) C.I. Reactive Red 123 is applied to PET according to a normal reactive dyeing process. The fabric is dyed (starting dye concentration 50 mg/g) in presence of 50 g/l sodium sulfate at 40 °C. After 30 min 15 g/l sodium carbonate are added. The dyeing is continued for 80 min and the samples are rinsed with cold deionized water. c) The fabrics are treated with iodine vapour in a closed chamber during 2 h at 25 °C.

All absorption and reflectance measurements are done with a Cary 5E spectrophotometer (*Varian*).

3. RESULTS AND DISCUSSION

In **Table 1** feasible interactions between fibre polymers and chemicals of processing (e.g. dyes) are depicted. Considering those mechanisms CD derivatives have to meet some structural requirements for a successful fixation on polymer surfaces by conventional methods of dyeing. In the following several examples are discussed in detail.

TABLE 1. Interactions between fibre polymers and chemicals of processing (+ feasible interactions, - unfeasible or weak interactions; PES = polyester)

	Fibre material				
Fixation mode	cellulose	wool	PA	PES	PAN
ionic interactions	-	+	+	-	+
covalent bonds	+	+	+	-	-
van der Waals interactions	-	-	+	+	+

Cellulosic materials are treated with MCT-β-CD according to conventional reactive dyeing processes. The reaction mechanism is a nucleophilic substitution of the chlorine atoms at the CD triazine rings giving covalent bonds with cellulose. The successful application is proved by UV spectroscopic measurements. MCT-β-CD shows a strong absorption band between 220 and 260 nm. The fixation yield of the ligand depends on the fixation conditions. Best results are obtained by treating the impregnated fabrics at temperatures between 120 and 150 °C. High humidity should be avoided because of the competing hydrolysis of the reactive sites. Another possibility is to apply MCT-β-CD at cellulose surfaces by means of common reactive printing systems. The ligand cannot be removed from the fibres even by numerous washing processes.

Hydrophobic polymers (PET, PA-6.6, PAN) are impregnated with nonionic CD derivatives which contain substituents of partly hydrophobic character using conventional methods of disperse dyeing. The permanent fixation of the ligands is based on diffusion of ring substituents into the fibres forming hydrophobic interactions. The ligands are detected on the fibres by means of interactions with the anionic fluorescence dye ANS. Dye fluorescence depends on the surroundings and is increased by inclusion into the hydrophobic CD cavities. Concerning PET fabric best results are obtained by treatment with DHP-EHG-β-CD or HP-HH-β-CD according to a common exhaust dyeing process at 130 °C. The fixation yields on PAN and PA-6.6 fabrics are less compared to PET, but normally this is the same with disperse dyes. Another possibility is the reaction of such fibres which contain ionic end-groups with counter-ionic CDs corresponding to basic or acid dyeing processes. As an example the successful fixation of HPTMAC-β-CD on PAN is proved by the interaction of the ligand with the anionic indicator dye bromophenol blue.

Some properties of the polymers may be directly influenced by the CD modification process, e.g. hydrophily, physiological and electrostatic behaviour, reactivity and complexing power. In **Figure 1** the wettability of PET fabric which has been treated with different nonionic CD derivatives is demonstrated. The hydrophily of the material is influenced significantly by kind and number of the substituents combined with the applied CD rings. The penetration time of a drop of water into the fabric is decreased drastically after impregnation with water soluble or emulsifiable CDs compared to the untreated material. On the other hand the interaction with hydrophobic derivatives like EHG-β-CD increases the hydrophoby of the polymer.

Drop penetration time (sec)

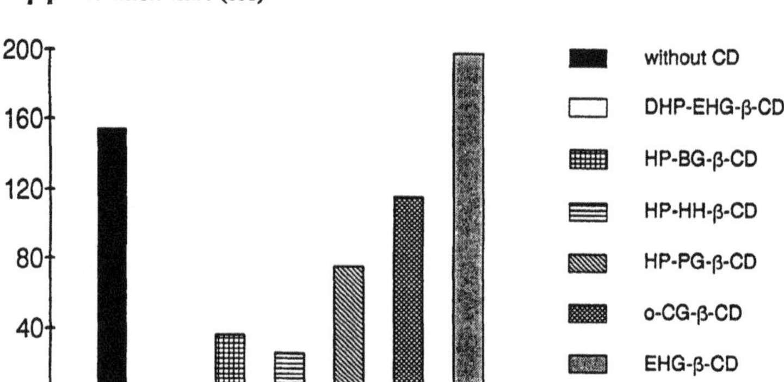

Fig. 1 Wettability of PET fabric

The application of CD derivatives causes a change of reactivity of the polymer surfaces. As an example this is demonstrated with reactive dyeings on PET fabric. **Figure 2** presents the absorption spectra of untreated and modified PET samples dyed with C.I. Reactive Red 123 according to a common reactive dyeing process. The untreated material is only slightly stained by the dye, but the dyeability of the modified fibres is significantly enhanced. This is caused by the increased number of reactive hydroxyl groups at the surface. The lower dyeability of PET treated with EHG-β-CD compared to PET impregnated with o-CG-β-CD is attributed to the less efficient wettability of the fibres (see **Figure 1**).

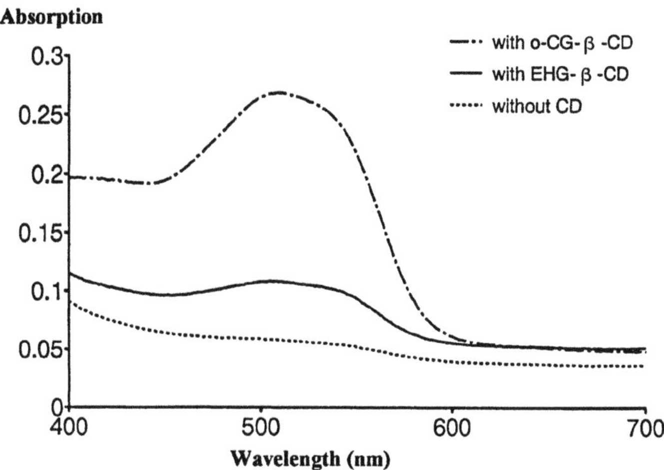

Fig. 2 Absorption spectra of PET fabric dyed with C.I. Reactive Red 123

To demonstrate the complexing power of the fixed CD cavities iodine vapour is used as a model system for building up inclusion complexes. The brown iodine colour is directly detected on the fabrics by spectroscopic measurements. It is observed that the samples which are treated with CDs absorb a higher amount of iodine compared to the untreated material. Also the host molecules are more slowly released from the modified fibres. So by the fixation at polymer surfaces the CD cavities do not loose their capability to complex non-polar organic molecules.

4. CONCLUSION

CD derivatives can be fixed permanently onto polymer surfaces by functional groups using simple methods of textile processing. As a result of the modification process some properties of the fibre materials are influenced. The fixed cavities do not loose their complexing power so that special effects may be obtained by forming inclusion complexes with specific chemical substances. On the other hand components of sweat or dirt can be disguised during the use of the treated fabrics. The fields of application for the new functional textiles are manifold, but the main purpose is to stabilize sensitive substances or to control the release and increase the biological availability of active agents (e.g. on medical textiles) by complex formation with fixed CDs (see **Table 2**).

TABLE 2 Some possible fields of application for CD modified textile materials

	Field of application		
Complexed substance	clothing	medical textiles	technical textiles
fungicide, bactericide		x	x
perfume	x		
pharmaceutical active agent		x	
UV absorbent	x		x
insect repellent agent	x		

ACKNOWLEDGEMENTS

We are grateful to the Forschungskuratorium Gesamttextil for their financial support concerning this research project (AIF-No. 9695). The support was granted from resources of the Federal Ministry of Economics via a supplementary contribution by the Association of Industrial Research Organizations (Arbeitsgemeinschaft Industrieller Forschungsvereinigungen, AIF). The supply with cyclodextrin derivatives by Wacker-Chemie is gratefully acknowledged.

REFERENCES

[1] Lewin, M., Sello, S.B., Handbook of Fiber Science and Technology: Vol. II, Chemical Processing of Fibers and Fabrics, Functional Finishes, Marcel Dekker, Inc., New York, Part A 1983, Part B 1984.

[2] Buschmann, H.-J., Knittel, D., Schollmeyer, E., German Patent DE 40 35 378 A 1, 1992.

[3] Knittel, D., Buschmann, H.-J., Schollmeyer, E., Neuartige Ausrüstungseffekte für Natur- und Chemiefasern, Maßgeschneiderte Eigenschaften, *Bekleidung + Textil*, **44**, No. 12, 34-40 (1992).

THE ACTION OF LIPOPHILIC UV ABSORBERS — SOLUBILIZED BY CYCLODEXTRIN — ON PHOTOFADING OF AQUEOUS SOLUTION OF AZOREACTIVE DYES

EDIT RÉMI[1], ÉVA FENYVESI[2], ISTVÁN RUSZNÁK[1], ANDRÁS VÍG[3]
[1]Technical University of Budapest, Department of Organic Chemical Technology, H-1111 Budapest, Műegyetem rkp. 3., Hungary
[2]CYCLOLAB Ltd., H-1525 Budapest, P.O.Box 435, Hungary
[3]Research Group of the Hungarian Academy of Sciences, TUB, H-1111 Budapest, Műegyetem rkp. 3., Hungary

ABSTRACT

Two representatives of water-insoluble UV absorbers (3-(4-methylbenzilidene)-camphor and 2-hydroxy-4-methoxybenzophenon), different in their chemical structure, could be solubilized by two pairs of studied CDs (BCD, GCD, RAMEB, AcGCD). The substituted CD entities were markedly more effective solubilizers than the unsubstituted ones. The rate of photofading of an aqueous azoreactive dye solution was significantly increased in the presence of the substituted CDs. Mixing UV absorbers, solubilized by the mentioned CDs, to the aqueous dye solution marked deceleration in photofading could be achieved.

1. INTRODUCTION

The lightstability of azoreactive dyes, rather popular in dyeing, is generally only medium or poor [1]. The light-fastness of dyed textiles principally can be improved by various UV absorbers. The practical realisation of this is hindered by the water-insolubility of UV absorbers used in polymer protection on one hand, while on the other, water-soluble UV absorbers cannot be fixed on fibres permanently.
CDs can improve the aqueous solubility of a rather great variety of lipophilic molecules [2].

J. Szejtli and L. Szente (eds.), Proceedings of the Eighth International Symposium on Cyclodextrons, 565–569.

The aim of the present work were as follows:
— solubilizing water-insoluble UV absorbers with CD-s
— application of the complex systems in improving lightstability of azoreactive dyes in aqueous solution.

2. MATERIALS AND METHODS

2.1 Materials

The used types CDs were: β-cyclodextrin (BCD), γ-cyclodextrin (GCD), acetyl-γ-cyclodextrin (AcGCD) and methyl-β-cyclodextrin (RAMEB). UV absorbers were: 3-(4-methylbenzylidene)-camphor (UV1) and 2-hydroxy-4-methoxybenzophenon (UV2). Hetero-bifunctional reactive azodye (RAD) of medium lightfastness was used in the experiments.

2.2. Measurement of the solubility of the UV absorbers in aqueous CD solutions

Excess amount (2 g/l) of UV absorbers was added to aqueous CD solutions of 1,25; 2,5; 5; 10; 20 g/l respectively. The stirring was continued at ambient temperature for 1 hour. The suspension was filtered through glass prefilter and the solution was tested spectrophotometrically after dilution with the 1:1 mixture of distilled water and ethanol.

2.3. The estimation of kinetics of photofading

UV absorbers were dissolved in the 20 g/l aqueous solution of AcGCD and RAMEB, respectively, and after filtration $5 \cdot 10^{-5}$ mole/l RAD was dissolved in the solution. A photofading device with high-pressure mercury vapour lamp was used, and the changes in optical density were measured spectrophotometrically.

3. RESULTS AND DISCUSSION

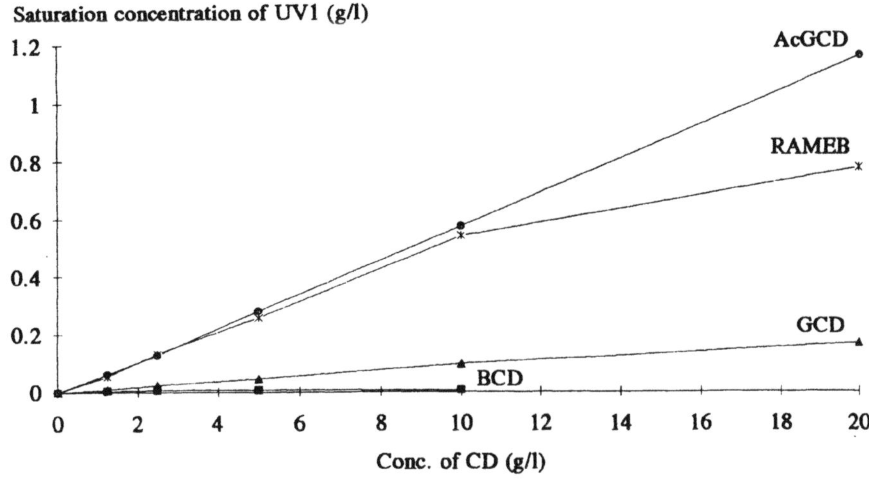

Fig. 1 Saturation concentration of UV1 in aqueous solution of different CD species

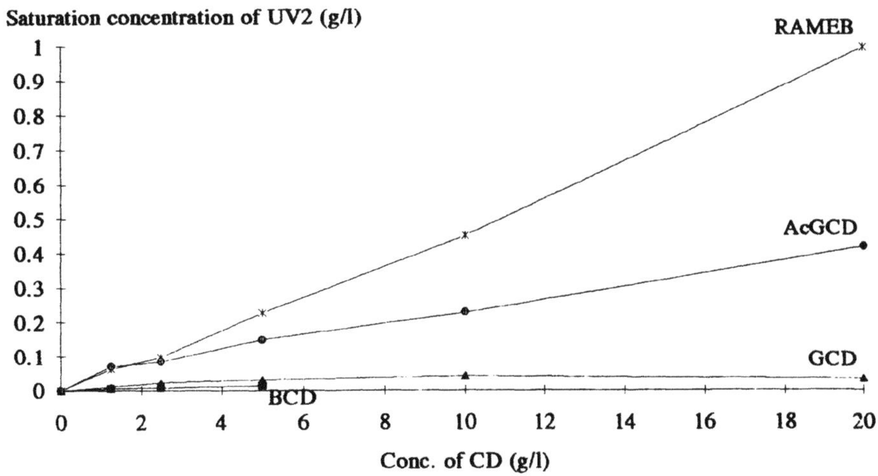

Fig. 2 Saturation concentration of UV2 in aqueous solution of different CD species

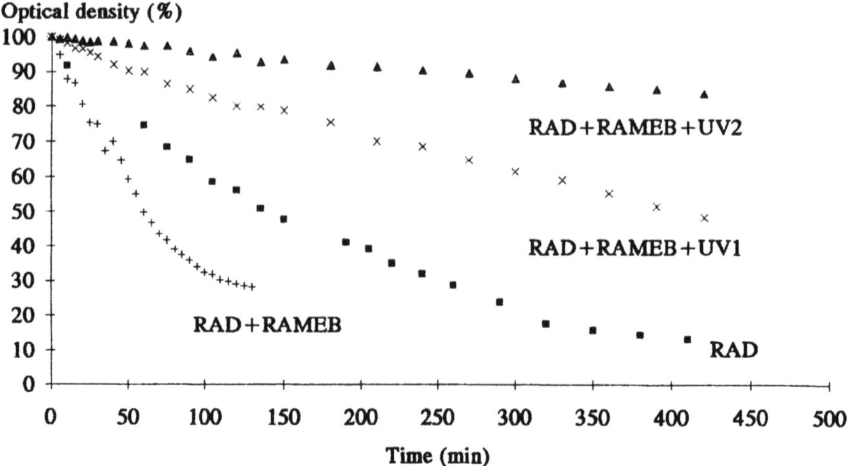

Fig. 3 Kinetics of photofading of RAD in aqueous solution (▪), influenced by the presence of RAMEB (+), RAMEB+UV1 (×) and RAMEB+UV2 (▴), respectively

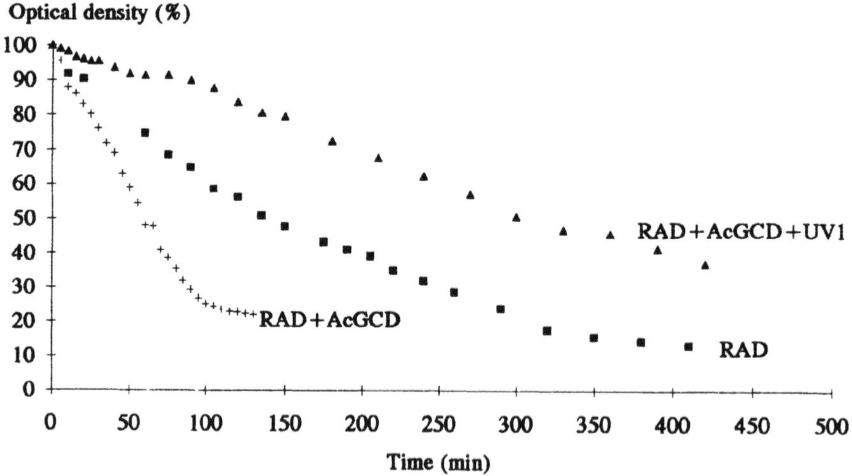

Fig. 4 Kinetics of photofading of RAD in aqueous solution (▪), influenced by the presence of AcGCD (+) and AcGCD+UV1 (▴), respectively

All the four studied CDs are effective solubilizing agents for the two investigated UV absorbers (Fig. 1, 2). Even with BCD 5, 1.5 fold increase in solubility of UV1 and UV2 was achieved, respectively. More UV absorbers can be solubilized by GCD than BCD. UV1 was the most efficiently solubilized by AcGCD, whereas UV2 by RAMEB (890, 110 fold, respectively). No correlation could be demonstrated between the solubility of two UV absorbers and the structural characteristic of studied CDs. Significant increase in solubility of both UV absorbers could be achieved by the substituted CDs than by the unsubstituted CDs.

Both substituted CDs (RAMEB, AcGCD) accelerated significantly the photofading of the dissolved dye when applied without (Fig. 3, 4). The combined systems including dissolved dye and UV absorbers solubilized by RAMEB or AcGCD generated significant deceleration in photofading of dye solution. The highest drop in rate of photofading (one tenth of the original one) could be observed by the application of UV2 solubilized by RAMEB. As AcGCD brought about significantly higher increase in the rate of photofading of the dye than RAMEB, it is obvious that deceleration in photofading caused by UV absorbers was more efficient in systems with RAMEB than those with AcGCD. (Remark: systems including RAD and UV2 solubilized by AcGCD could not be studied under exposure because unexpected precipitation occurred.)

4. CONCLUSION

The solubility of the UV absorbers can be significantly increased by CDs. The most effective solubilizers were AcGCD and RAMEB. Though the photofading of the azoreactive dye was accelerated by these CD derivativies, applying them as solubilizers of the UV absorbers a marked deceleration of photofading was observed.

REFERENCES

[1] Allen, N.S., Photofading mechanisms of dyes in solution and polymer media, *Rev. Prog. Coloration*, 17, 61-71 (1987)

[2] Szejtli, J., Cyclodextrin Technology, Kluwer Academic Publishers, Dodrecht, 1988

SORPTION OF TEXTILE DYES ON β–CYCLODEXTRIN-EPICHLORHYDRIN GELS

Y. SHAO[1], B. MARTEL[2], M. MORCELLET[2], M. WELTROWSKI[1], G. CRINI[3]

[1]Textile Technology Center 3000 Boullé J2S1H9 Saint Hyacinthe, Quebec, Canada ;
[2]Laboratoire de chimie macromoléculaire UA-CNRS351 Université des Sciences et Technologies de Lille 59650 Villeneuve d'Ascq, France
[3]Istituto scientifico di chimica e biochimica «G. Ronzoni » via G Colombo 81 20133 Milano, Italia

ABSTRACT

β–Cyclodextrin (CD), hydroxypropyl β–cyclodextrin (HPCD), poly(vinylalcohol) (PVOH) have been reacted with epichlorhydrin yielding different gels. The sorption capacity of the gels has been tested with acid, direct, mordant and reactive textile dyes as substrates, by the batch method. CD gels were more efficient than HPCD gels. No correlation has been observed between the performances of the CD gels and their respective crosslinking degree. The influence of pH is rather low whilst that of ionic strength is prominent on the sorption rate. The addition of anionic surfactants have either positive (CPA) or negative (SDS) effect on adsorption. The mechanism of sorption of dyes on gels could be physical adsorption in the polymer network and /or a host-guest inclusion complex formation

1. INTRODUCTION

Removal of colour from the effluent of the dyeing industry became ecological necessity ; now the laws impose more and more drastic norms for the content of the effluents in organic pollutants. Amongst the numerous techniques of depollution, sorption on natural sorbents [1] or derivatives [2] has been proposed. Many reports deal with the interaction of dyes with cyclodextrins [3] in solution. For that reason we propose to use cyclodextrin as a heterogeneous system to sorb dyes. Acid, direct, mordant and reactive dyes have been sorbed onto the synthesised gels ; the parameters that have been considered are the pH and the influence of some of the dyeing auxiliaries usually used in the industrial dyeing processes as sodium chloride or surfactants as CPA and SDS.

J. Szejtli and L. Szente (eds.), Proceedings of the Eighth International Symposium on Cyclodextrons, 571–574.

2. MATERIALS AND METHODS

Synthesis of the gels : We applied the technology developed by Komiyama et al.[4] with some modifications in order to increase the degree of crosslinking to the minimum required for mechanically stable gels. In a reactor vessel, 160 ml of a 50% wt aqueous sodium hydroxide solution containing 20 mg of $NaBH_4$ was heated to 50° . Meanwhile one hundred grams (88 mmol) of CD (gel CD1 to 3), HPCD (gel HPCD1), a mixture of PVOH and HPCD (gel HPCD2) were then dissolved. A desired amount of epichlorhydrin was added dropwise and the mixture was vigorously stirred (800-1000 rpm). After 2 hours the gel beads appeared. Then 300 ml of acetone were added with stirring and heating was continued for one hour. After cooling, the resin beads were poured into a 1l water beaker, filtered and washed with water and acetone, and dried in vacuum at 60°C for 20h. The ratio CD/epichlorhydrin is 1/10, 1/15, 1/20 for gel CD1, CD2 and CD3 respectively.
Adsorption measurements : 40 mg of gel (diameter between 0.125 and 0.25 mm) and 40 ml of dye solution (1.10^{-5} or 3.10^{-5} M) were put into a stopped erlenmeyer. The mixture was stirred for 24 hours and the absorbance of the supernatant was measured at the corresponding λmax with a UVIKON 930 spectrophotometer. The uptake percentage was calculated as following : (Ao- A)/Ao x 100 = % uptake.
Products : CD and HPCD were a gift from Roquette Freres ; HCl-NaOH-citric acid was used as the pH4 buffer solution and 0.1M borax was used as pH 9.2 buffer. The concentration of sodium chloride (Aldrich) was adjusted to 1M and that of CPA and SDS (Merk) to 2g/l.

3. RESULTS AND DISCUSSION

CD and HPCD gels have been tested for the sorption of Acid Blue 15 at neutral pH (=5.7) (figure 1). The two HPCD gels have a lower sorption capacity than the CD gels. The presence of poly(vinylalcohol) in gel HPCD 2 does not induce any noticeable difference with gel HPCD 1. No direct relation between the initial ratio β–CD/epichlorhydrin and the sorption capacity of the CD gels can be proposed. Gel CD 3 has been used in the following experiments.

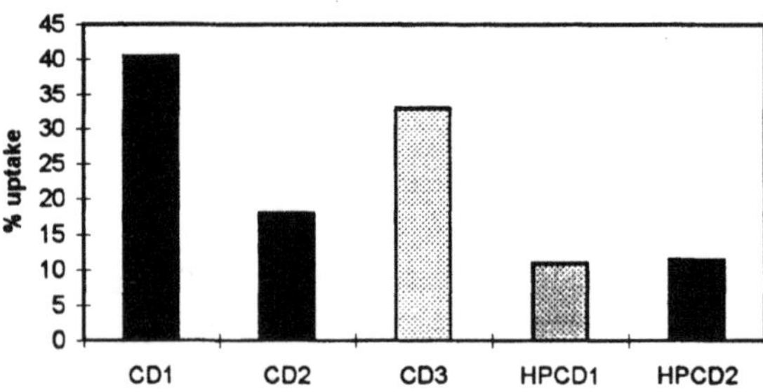

Fig. 1 : Sorption of Acid Blue 15 (10^{-5}M) on CD and HPCD gels at pH 5.7.

Figure 2 reports the results of the study of the influence of pH on the sorption capacity of Gel CD 3 . For most of the studied dyes in this experiment, the good results were obtained either in acidic or (and) in basic medium. A low sorption capacity was observed at the neutral pH. On the other hand, addition of sodium chloride to neutral solutions produces a strong increase of the performance of the gel. In most of experiments reported in figure 2, the capacity of the gel in the presence of NaCl medium reaches the optimum values measured in buffer solution. This observation has led us to conclude that the sorption capacity of the gel depends on ionic strength more than pH. Sodium chloride minimises electrical charge on the surface of the sorbent and increases the adsorption.

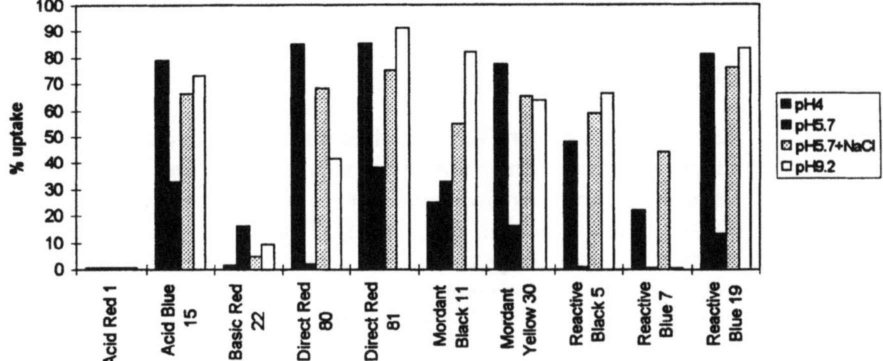

Fig. 2 : pH and salt effect on the sorption capacity of gel CD3 for various dyes (10^{-5}M)

Figure 3 reports experiments where dyes (3.10^{-5}M) have been mixed with auxiliaries in the range of concentration usually used in industrial conditions. NaCl, CPA and NaCl-CPA mixtures have a positive effect on the performances ; however, SDS decreases drastically the sorption. This can be explained by the insulation of the dye molecule inside the surfactant micelle (critical micellar concentration is reached in the conditions of the experiment). Furthermore, SDS is known to form inclusion compounds with β–cyclodextrin. Micellar dissolution and competition with SDS are the reasons of low the sorption of the dye on the gel. In our previous study [5], we found that some of the dyes (DR80, MY30, RB7, RB19) do not have specific interactions with β-CD in the solution (no modification of the visible spectra), but they still can be adsorbed by CD gels. This suggests that the sorption of dyes on gels is not only because of the host-guest inclusion complex formation, but also because of adsorption of dyes in the polymer network.

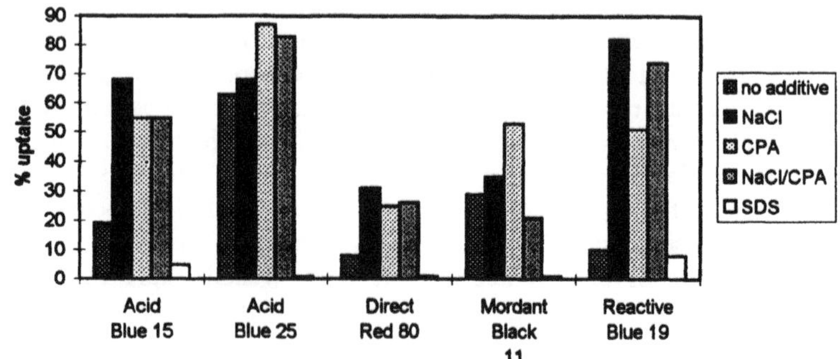

Fig.3 : Influence of sodium chloride and anionic surfactants on the sorption capacity for various dyes.

4. CONCLUSION

We observed that CD gels can sorb different dyes such as acid, direct, mordant or reactive dyes without specificity. The sorption capacity is reliable in the concentration range of the dyes in the experiments. We found that the dyeing auxiliaries (NaCl, CPA) can improve the sorption of the dye and that SDS has the opposite effect. The mechanism of sorption of dyes on gels could be physical adsorption in the polymer network and/or a host-guest inclusion complex formation. For reuse of the dye adsorbed gels, study on desorption of dyes from CD gels is on the way.

Special thanks to Mrs A. M. Cazé for her skilful assistance. This work was supported by : Agence pour la Défense de l'Environnement et la Maîtrise de l'Energie (ADEME), Ministère des Affaires Etrangères (France) , Ministère des Relations Internationales (Québec).

REFERENCES

[1] Ahmed M. N., Ram R.N., Removal of basic dye from waste water using silica as adsorbent, *Environ. Pollut.*, 77, 79-86 (1992)

[2] Hwang M. C., Chen K. M., Removal of color from effluent using Polyamide-epichlorhydrin-cellulose polymer. I :. Preparation and use in direct dye removal, *J. Appl. Polym. Sci.*, **48**, 299-311 (1993)

[3] Szejtli J., *Cyclodextrin and their inclusion complexes*, Budapest, 1982, Pp 162-175

[4] Komiyama M, Sugiura I.,Hirai H., Immobilised β-cyclodextrin catalyst for selective synthesis of 4-hydroxybenzoic acid, *Polym. J.*, 17, 1225-1227 (1985)

[5] Shao Y. et al., Interaction between β-cyclodextrin and water soluble dyes in solutions ; to appear.

APPLICATION OF γ-CYCLODEXTRIN FOR THE STABILIZATION AND/OR DISPERSION OF VEGETABLE OILS CONTAINING TRIGLYCERIDES OF POLYUNSATURATED ACIDS

M. Regiert, T.Wimmer and J.-P. Moldenhauer
Wacker-Chemie GmbH, Hanns-Seidel-Platz 4,
D-81737 München, GERMANY

ABSTRACT

To improve the storage stability of instable vegetable oils with a high content of polyunsaturated fatty acid triglycerides, these essential compounds can be complexed with native cyclodextrins. Only with γ-CD a nearly complete complexation of the oils was achieved as shown by complexation kinetics measurements. Storage trials of the insoluble CD-complexes followed by the determination of the peroxide value of the oils indicated that the best stabilization against autoxidation is obtained with γ-CD. An additional benefit of the complexation of triglycerides of polyunsaturated fatty acids with γ-cyclodextrin is the formation of stable dispersions of these oils in aqueous media.

1. INTRODUCTION

Liquid vegetable oils with a high content of polyunsaturated fatty acid triglycerides, like evening primrose, borage or blackcurrant oil are valuable substances, which are used for cosmetic applications, most of all for skin care to enhance elasticity and to decrease transepidermal loss of moisture [1]. In addition they are very important in the food sector for the supply with essential fatty acids.

Table 1. Typical fatty acid profile of the vegetable oils

fatty acid	double bonds	evening primrose oil	borage oil	blackcurrant oil
palmitic acid	0	6 - 10 %	9 - 13 %	6 %
stearic acid	0	1.5 - 3.5 %	3 - 5 %	1 %
oleic acid	1	6 - 12 %	15 - 17 %	10 - 12 %
linoleic acid	2	74.2%	40.4%	48 %
linolenic acid	3	8 - 12 %	19 - 25 %	30 %

J. Szejtli and L. Szente (eds.), Proceedings of the Eighth International Symposium on Cyclodextrons, 575–578.

The unfavorable property of these oils that prevents their broader application in cosmetics, pharmaceuticals and food is their low stability against atmospheric oxygen, especially in combination with the exposure to heat and daylight. The primary formation of toxic hydroperoxides by autoxidation induces further degradation to aldehydes, ketones and polymeric species, which account for a bad, rancid smell and for discolorization. The amount of peroxides is expressed by the peroxide value (POV). It is known from the literature that unsaturated fatty acids and the corresponding triglycerides can be complexed by α– and β–cyclodextrin [2,3,4]

2. MATERIALS AND METHODS

α-, β- and γ-cyclodextrin are products of Wacker-Chemie GmbH, Munich.
The vegetable oils used in this study were a gift of Görlich-Handels GmbH, Wasserburg or purchased in a pharmacie.
The complexes were prepared in suspension or in concentrated cyclodextrin solution or by kneading with the exclusion of light and oxygene. Kinetic measurements were carried out by determining the amount of the water soluble fraction of the reaction mixture. peroxide values (POV [meq/kg]) of the oils after storage of the complexes were measured by decomplexation in a methanol/hexane mixture at room temperature. and potentiometric thiosulfate titration after addition of iodid to the oils.

3. RESULTS AND DISCUSSION

3.1. Complexation

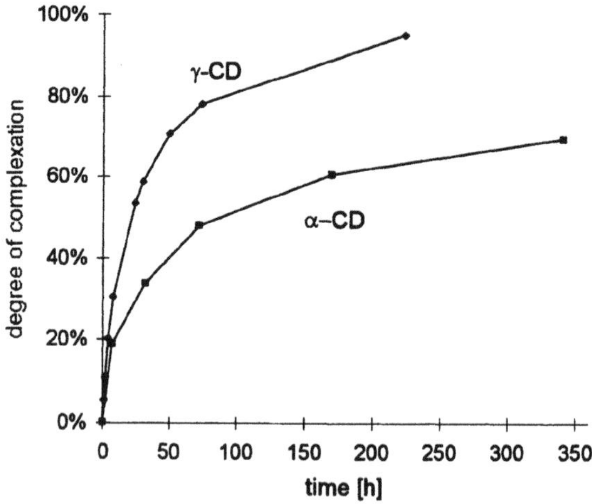

Fig.1 Kinetics of complexation in suspension at 25°C with excess of wheat germ oil

All three types of native CDs are forming insoluble complexes with vegetable oils.
The degree of complexation can be easily determined by measuring the water soluble fraction of the reaction mixture, because uncomplexed oil and cyclodextrin complex have an extremely low solubility in water. The kinetics of the complexation with α- and γ-CD were studied.
Surprisingly a complete complexation of the vegetable oils was only achieved with γ-CD, although it was recognized before [2] that for free fatty acids α-CD is more suitable than β-CD. Through the exclusion of light and oxygene during the complexation procedure the peroxid values of the oils were nearly unchanged. Optimal temperature for the complexation was found to be between 40 - 50 °C.
Stirring mixtures of γ-CD and vegetable oils with higher oil to CD ratio in water resulted in very stable o/w emulsions which can be used in cosmetics.

3.2. Stability tests

The rapid autocatalytic oxidation of uncomplexed evening primrose oil is clearly seen in Fig.2 . It could be demonstrated that high peroxide values of the oils are coincident with discolorization and rancid smell. It was also found that the oxidation was enhanced by exposure to daylight.

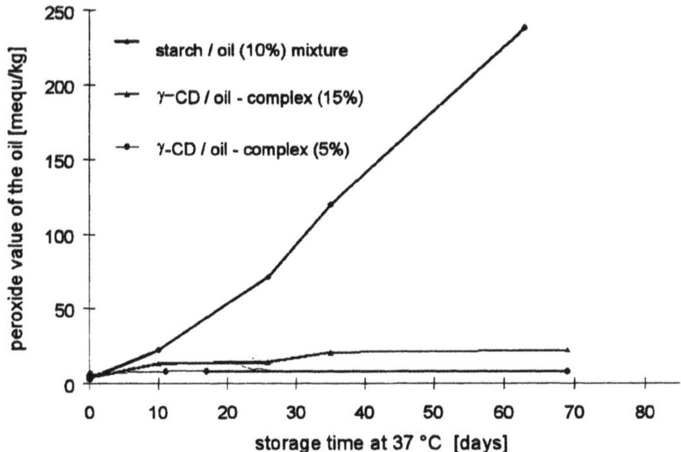

Fig. 2 Storage trials of evening primrose oil / γ-CD complexes in comparison to uncomplexed oil

A very good stability of the oil as its γ-CD complex was proved by a slight increase of the POV over a period of 2 month at elevated temperature. Complexes with higher content of γ-CD seem to be even more stable. Similar results were found with borage and wheat germ oil.

The stabilization of evening primrose oil by complexation with α-, β- and γ-CD was examined in a comparing storage trial at room temperature and daylight exposure over a period of 38 days. Under these conditions γ-CD proved to have the best stabilizing effect of the native cyclodextrins. After 15 days the POV of the α-CD-complexed oil was clearly higher than with γ-CD, while no significant stabilization was found with β-CD. In combination with the complexation kinetics this result indicates that the large cavity of γ-CD is most suitable for the inclusion of long chain triglycerides.

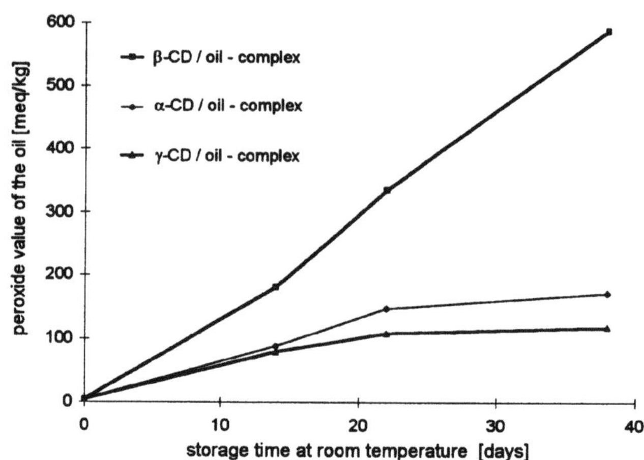

Fig. 3 Storage trials at daylight exposure of evening primrose oil complexes

4. CONCLUSION

As shown by complexation kinetics and stability tests γ-CD is the favorable CD to complex and to stabilize triglycerides of polyunsaturated fatty acids. With γ-CD also stable dispersions of these oils in aqueous systems were obtained.

REFERENCES

[1] Eggensperger H., Pflanzliche Wirkstoffe für Kosmetika, Melcher-Verlag, Heidelberg/München, 1995

[2] Szente L., Szejtli J., Szeman J., Fatty acid-cyclodextrin complexes: properties and applications, *J. Incl. Phen* **16**, 339-354 (1993)

[3] Hibino, T., Nakao, K., Okada, T., Sahashi, H., Stabilization of gamma-Linolenic acid by forming its inclusion compounds with cyclodextrins, Japanese Patent 87 84,041 , (1987)

[4] Murata, S., Hara, K., Manufacture of oil-cyclodextrin inclusion compounds for food and pharmaceutical preparations, Japanese Patent 87 263,143 , (1987)

OXIDATION STABILITY OF EICOSAPENTAENOIC AND DOCOSAHEXAENOIC ACID INCLUDED IN CYCLODEXTRINS

H. YOSHII[1], T. FURUTA, A. YASUNISHI, Y.-Y. LINKO[2] AND P. LINKO[2]

Department of Biotechnology, Tottori University, Tottori 680 Japan. [1]Department of Biochemical Engineering, Toyama National College of Technology, Toyama 939 Japan. [2]Department of Chemical Engineering, Helsinki University of Technology, Espoo, FIN-02150 Finland.

ABSTRACT

We have proposed a novel method for complexing of the ω-3 polyunsaturated fatty acids (PUFA), such as eicosapentaenoic and docosahexaenoic acids, with cyclodextrin as dry powders by a twin-screw kneader. The effect of various oxidation conditions on powdery PUFA were investigated. Further, the powdery PUFA developed was employed in the preparation of fish meal as a functional sea food paste.

1. INTRODUCTION

The ω-3 polyunsaturated fatty acids (PUFA) such as eicosapentaenoic and docosahexaenoic acids have important physiological functions. They are believed to have antithrombotic, cholesterol depressant, and antiallergenic properties, among others. These PUFAs are chemically quite reactive, requiring proper encapsulation in a powder form to protect against autoxidation[1]. We have developed a novel method for complexing of PUFAs with cyclodextrin as dry powders by a twin-screw kneader[2]. In the present study the effect of various oxidation conditions on powdery eicosapentaenoic ester and docosahexaenoic acids were investigated. It was also observed that carbohydrates could be used to further protect the PUFA cyclodextrin inclusion complexes. Further, the powdery docosahexaenoic acid developed was employed in the preparation of fish meal as a functional sea food paste.

2. MATERIALS AND METHODS

2.1 Materials

J. Szejtli and L. Szente (eds.), Proceedings of the Eighth International Symposium on Cyclodextrons, 579–582.

Reagent grade of α-, β-, and γ-cyclodextrin(CD) were from Ensuiko Sugar Refining Co. and Wacker Chemicals. Eicosapentaenoic acid ethyl ester (EPA: purity 82%) was gift from Nippon Suisan Co., and docosahexaenoic acid oil:triglyceride form (DHA: purity 45%) was from Maruha Co. Maltodextrin was gift from Nippon Starch Chemical Co. Other chemicals were of analytical grade.

2.2 Preparation of inclusion complex

The inclusion complex powders of EPA or DHA with cyclodextrin were prepared by a twin-screw kneader in a nitrogen atmosphere. Twenty to thirty grams of α-, β- or γ-CD were weighed and mixed with EPA or DHA, followed by adding distilled water to an initial moisture content of 30 or 50% on dry basis. Carbohydrate such as maltose, maltodextrin or pullulan was also added when the coating effect of carbohydrate were investigated. The mixture was kneaded in a twin screw kneader (KRC-S1, Kurimoto Steel Ltd.) at 60 to 50 °C for 30 min. The wet slurry was washed with diethyl ether and dried *in vacuo* at 20 to 40 °C for 15 hr.

2.3 Autoxidation of inclusion powders

About 0.1-3 g of the complex powder was placed in a glass bottle or plate, and stored at 50°C and a relative humidity of 75%. The effect of humidity was measured with a humidity-controlled gas flowing through the glass bottle by connecting plastic tubes. The powder was sampled at certain intervals for the determination of the residual EPA and the peroxide value (POV) of DHA. EPA was analyzed by gas chromatography (GC-14A, Shimadzu). POV value of DHA was quantified by iodometric titration method.

2.4 Preparation of fish meal paste ("Kamaboko") involved DHA powder

Surimi from walleye pollack was chopped with NaCl, water and DHA, by means of a speed cutter at 10°C for 5min. The salt-ground paste was stuffed into a plastic vessel with a cover, and heated at 90°C for 20 min in a water bath to prepare "heating gel". After cooling the gel with ice water, and stored in a refrigerator. POV of DHA in the gel was measured periodically by extracting DHA into chloroform.

3. RESULTS AND DISCUSSION

3.1 Autoxidation of powdery PUFA

The time courses of autoxidation of powdery PUFA are illustrated in Fig.1. As shown in Fig.1(a), the washed powdery EPA in α- or β-CD was quite stable to oxidation, and most EPA remained unoxidized. However, the amount of EPA included per unit mass of α- or β-CD was quite small, about 0.05 and 0.01, respectively. The resistances against autoxidation of the unwashed powders are also improved in comparison with liquid EPA ester, but are inferior to the washed powder. Powdery EPA in α-CD was rapidly

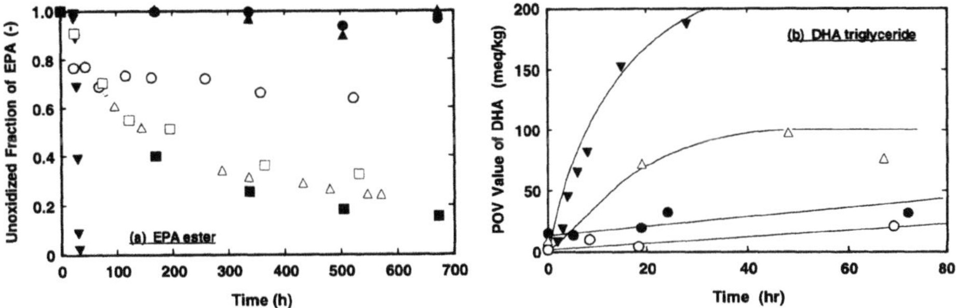

Fig. 1 Retardation of the autoxidation of powdery EPA ester and DHA triglyceride. (▼: Liquid EPA ester or DHA triglyceride. ○,●: in α-CD, △,▲: in β-CD, □,■: in γ-CD. Open symbol are unwashed powder. Closed symbols are washed powder.)

oxidized during an initial period, followed by a stable region against oxidation. This indicates that EPA truly included in α-CD may be quite resistant to oxidation. On the other hand, powdery EPA in β- or γ-CD were gradually oxidized as time increased. The oxidation time courses of DHA in α- and β-CD are shown in Fig.1(b), in which the oxidation degree was measured by the POV value. The powdery DHA in α-CD has about 8 times smaller POV value than liquid DHA. β-CD was also useful for elongation of DHA shelf-life. Though the unwashed DHA powder in α-CD has slightly higher initial POV value than the washed powder, oxidation did not progress abruptly.

3.2 Protecting autoxidation by adding carbohydrates

Since the mass fraction of PUFA truly included in CDs is small, much PUFA remained unincluded and adsorbed onto CD's molecules[2]. To protect these PUFA molecules

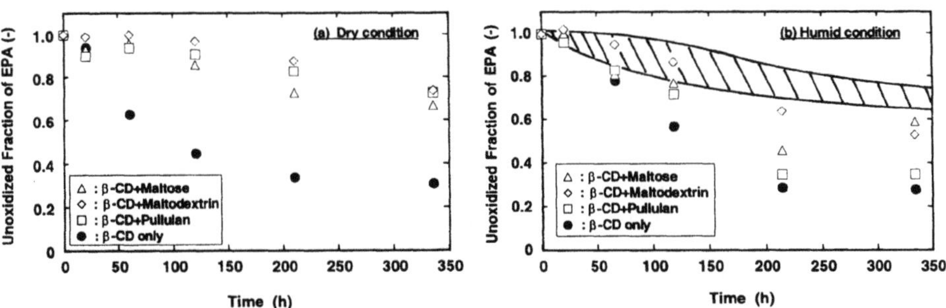

Fig. 2 Effectiveness of the coating with carbohydrate to the retardation of autoxidation of the complex powder between EPA ester and β-CD.

from oxidation, carbohydrates such as maltose, maltodextrin, and pullulan, were added for encapsulation. Figure 2(a) shows the effect of the carbohydrates on the resistance against autoxidation in β-CD. The hybrid powder had equivalent stability to the washed β-CD complex powder, though EPA content was much higher (0.08 g/g). The similar effects were observed for other CD and carbohydrate combination. The addition of a carbohydrate was especially helpful in the case of β-CD inclusion complexes. As shown in Fig.2(b), the oxidation was accelerated in humid condition. From Figs.2(a) and (b), it is found that carbohydrates were less effective under humid condition. Similar results were obtained for α- and γ-CD complex powder.

3.3 Oxidation of DHA powder involved in fish meal paste

Figure 3 shows the oxidation of DHA in the fish meal paste, which was stored at 10°C. Three different forms of DHA was used in the fish paste. In a liquid form DHA was oxidized rapidly, and the POV value exceeded 40 after 4 days of storage. POV value of

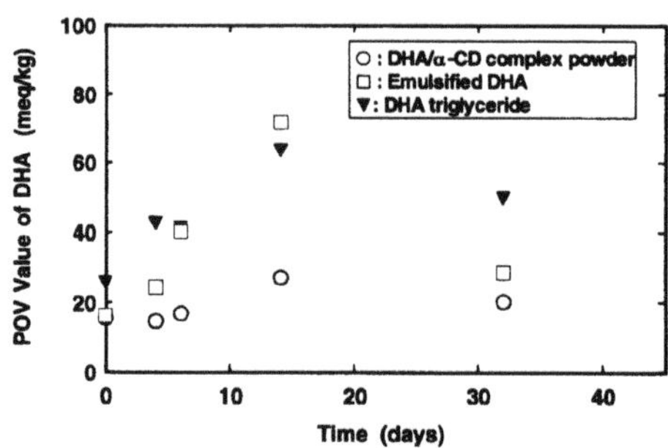

Fig.3 Shelf-life of DHA triglyceride of various forms in the fish meal paste.

the DHA emulsified with distilled water increased markedly during an initial period of oxidation. In the contrary, the powdery DHA in α-CD was very stable against oxidation, and the POV value did not increase above 30 during one month storage. This implied that the powdery DHA in α-CD was useful for a functional fish meal paste.

4. CONCLUSION

Both the powdery EPA ester and DHA triglyceride exhibited a marked resistance against autoxidation. It was also observed that carbohydrates could be used to further protect the PUFA cyclodextrin inclusion complexes from oxidation. The POV of powdery DHA in fish meal paste remained virtually unchanged for 20 days without any antioxidants.

REFERENCE

[1] Szente, L.., *Industrial Significance of Lipd/Cyclodextrin Interactions*. in Proc. 7th Intern. Cyclodexrin Sympo., 13-18, (1994)

[2] Yoshii, Y., Furuta, T., Kawasaki, K., Hirano, H., Morita, Y., Shiina, C., and Nakayama, S., Quantitative Analysis of α-Cyclodextrin Inclusion Complexes with Fatty Acid Methyl/Ethyl Ester by X-ray Diffractometry., *Oyo Toshitsu Kagaku*, **42**, 243-249 (1995)

A SHORT-CUT METHOD FOR ESTIMATING *l*-MENTHOL RETENTION INCLUDED IN CYCLODEXTRIN DURING DRYING A SINGLE DROP

T. FURUTA, H. YOSHII[1], T. FUJIMOTO, A. YASUNISHI, Y.-Y. LINKO[2] AND P. LINKO[2]

Department of Biotechnology, Tottori University, Tottori 680 Japan. [1]Department of Biochemical Engineering, Toyama National College of Technology, Toyama 939 Japan. [2]Department of Chemical Engineering, Helsinki University of Technology, Espoo, FIN-02150 Finland.

ABSTRACT

Good taste and flavor of the spray-dried powder products are the most important quality factors. In the present study, molecular encapsulation in cyclodextrin has been applied to prevent the loss of hydrophobic flavor compounds during drying. We have also proposed a new short-cut method for estimating flavor retention. We have applied this method to calculate the final retention of *l*-menthol encapsulated in a blend of β-cyclodextrin and maltodextrin during drying of a single droplet.

1. INTRODUCTION

The prevention of the loss of volatile ingredients is of utmost importance during spray drying of liquid food materials. Several methods have been developed for improving the flavor retention during spray drying. Previously, the retention of a trace amount of ethanol was examined experimentally, by using the drying process of a single droplet [1]. The selective diffusion theory[2] has been successfully applied to estimate the amount of ethanol retained during drying. In the present study, molecular encapsulation in cyclodextrin has been applied to prevent the loss of hydrophobic flavor compounds during drying. We have also proposed a new short-cut method to calculate the final retention of *l*-menthol encapsulated in a blend of β-cyclodextrin and maltodextrin during drying of a single droplet.

2. MATERIALS AND METHODS

2.1 Materials

J. Szejtli and L. Szente (eds.), Proceedings of the Eighth International Symposium on Cyclodextrons, 583–586.

Reagent grade of α-, β-, and γ-cyclodextrin(CD) were from Ensuiko Sugar Refining Co. and Wacker Chemicals, respectively. *l*-Menthol (menthol) was from Wako Chemical Co. Chloroform was from Kanto Chemical Co., and maltodextrin (MD) was from Nippon Starch Chemical Co. All chemicals were of reagent grade.

2.2 Drying Experiment of a Single Droplet

An aqueous maltodextrin solution, into which a trace amount of menthol (400 ppm) was dissolved, was used as a model liquid food. α-, β-, or γ-CD was added to the liquid to form an inclusion complex with menthol. The experimental equipment for drying a single droplet was the same as that used previously [1]. An up-stream hot air of uniform temperature, humidity and velocity was realized by the assembly of a packed bed humidifier, a computer-aided electric heater and a contraction nozzle. Precisely, 5μ*l* of the sample solution was measured with a micro-syringe, and a small droplet was formed at the tip of a glass filament and hung in the up-stream hot air. After drying for the designated time, the droplet was dissolved in a mini-vial containing 100μ*l* of distilled water, and menthol was extracted into the additional 50μ*l* of chloroform. The menthol content retained in the dehydrated drop was then determined by gas chromatography (GC-14A, Shimadzu). The retention of menthol was defined as the ratio of menthol concentration in the droplet before and after drying. Menthol in the solution was not completely included within cyclodextrins, but free and included states of menthol were present in an equilibrium. The equilibrium constant was measured by a head space method.

2.3 A short-cut method for estimating the final retention of menthol

A short-cut method based on the selective diffusion hypothesis was applied to estimate the final retention of menthol. For simplicity, the following assumptions were made.
(a) Free menthol in the droplet is in equilibrium with the complex one in cyclodextrin.
(b) Most menthol was lost during the surface evaporation period t_c.
If the apparent diffusion coefficient of menthol D_{app} and the radius of the droplet R are constant during the period, then the retention of menthol at time interval t_c is given by

$$\frac{C_M}{C_{M0}} = 1 - 6\sqrt{\frac{D_{app} \cdot t_c}{\pi R^2}} \tag{1}$$

The rate of dissociation of menthol out of the CD's cavity may be described as:

$$-\frac{dC_{\beta M}}{dt} = k\,(C_M^* - C_M) = k\left(\frac{C_{\beta M}}{K_{eq} \cdot C_\beta} - C_M\right) \tag{2}$$

where $C_M^* - C_M$ is the driving force of the release of menthol. The equation of change of the free menthol in the droplet is obtained as:

$$-\frac{dC_M}{dt} = 3\,(C_M + C_{\beta M0} - C_{\beta M})\sqrt{\frac{D_{app}}{\pi R^2 \cdot t}} \tag{3}$$

Since the interval of the surface evaporation period t_c was obtained experimentally by the measurement of the droplet temperature, we can get D_{app} value from Eq.(1). By substituting D_{app} into Eq.(3), Eqs.(2) and (3) could be solved simultaneously with the initial conditions $C_M = C_{M0}$ and $C_{\beta M} = C_{\beta M0}$ at t=0, which were obtained from the initial equilibrium data (head space analysis).

3. RESULTS AND DISCUSSION

Figure 1 shows the effectiveness of cyclodextrin in the retention of menthol, where five molar times of α-, β- or γ-CD was added in 10% aqueous MD solution. β-CD could markedly prevent the loss of menthol during drying. On the contrary, α- or γ-CD had not extra ability to retain menthol, and the final retention of menthol was nearly the same as that for solution without CD. This implies that the menthol molecules stably

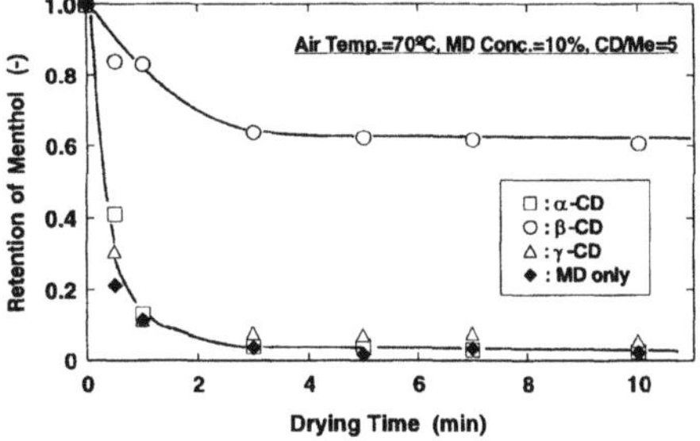

Fig. 1 Effectiveness of cyclodextrin to the retention of menthol during drying.

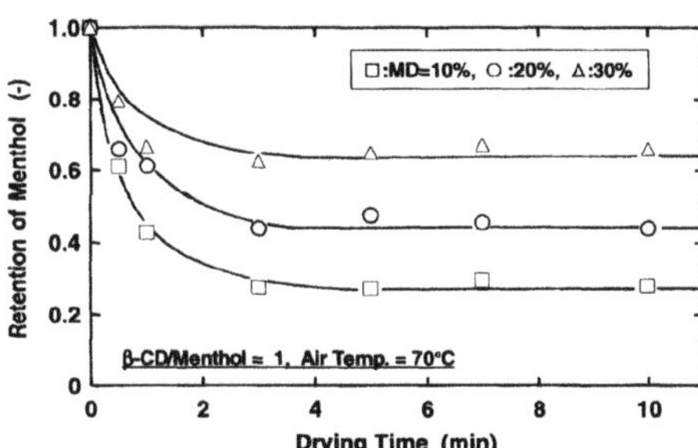

Fig. 2 Effect of MD content on the retention of menthol.

exist in the cavity of β-CD. The final retention of menthol was increased as the amount of β-CD added was increased. The effect of MD concentration on the menthol retention is illustrated in Fig.2 for the solution of a molar ratio of β-CD dissolved. It was found that the retention of menthol was also increased with an increase in the concentration of MD, when the content of β-CD was kept constant. These findings suggest that menthol in the droplet is present in two states; as an inclusion complex and in free state. These two states showed an equilibrium against each other. If the vapor pressure of menthol included in CD could be considered negligible, the equilibrium constant could be measured by a head space analysis. Figure 3 shows the relative concentration of menthol vapor in the head space as a function of the amount of each CD. The vapor pressure of menthol in the head space decreased markedly by the addition of β-CD. From these data, K_{eq} in Eq.(2) can be calculated. As the drying proceeds, free menthol in the droplet may move to the surface of the droplet by molecular diffusion, and evaporate into drying air. Then, the equilibrium between complexed and free menthol are break down, and the

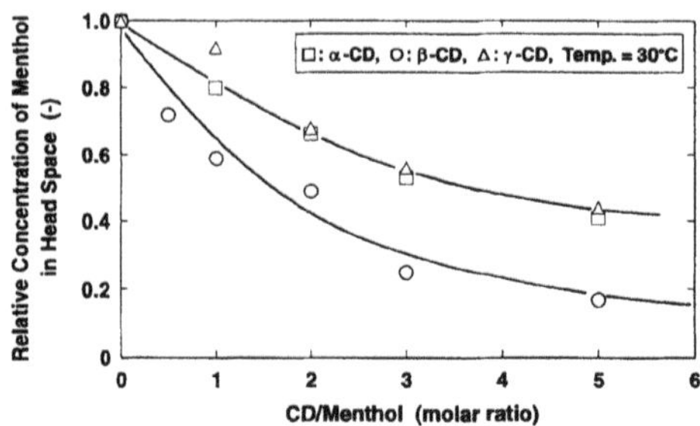

Fig. 3 Relative vapor pressure of menthol in the head space.

complexed menthol might be released from the cavity of CD. These postulations can explain the experimental results in Figs.1 and 2 qualitatively. The theoretical estimation of the final retention of menthol is shown by the solid lines in Figure 4, being compared with the experimental results at 70°C of the drying air temperature. The theoretical results show good agreement with the experimental results. Similar results were obtained at the drying air temperature of 50 and 90°C.

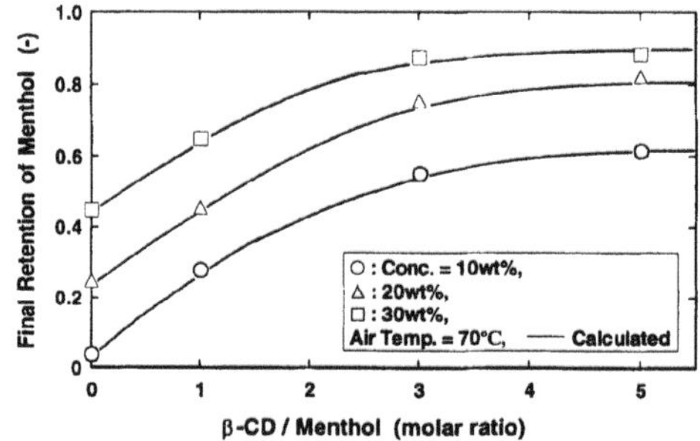

Fig.4 Comparison between the theoretical calculation and the experimentals of the final retention of menthol

REFERENCES

[1] Furuta, T., Tsujimoto, S., Okazaki, M. and Toei, R., *Effect of Drying on Retention of Ethanol in Maltodextrin Solution During Drying of a Single Droplet*, Drying Technology, **2**, pp.311-327 (1984)

[2] Kerkhof, P.J.A.M., A Quantitative study on the Effect of Process Variables on the Retention of Volatile Trace Component in Drying, Ph. D. Thesis, Technical University Eindhoven, The Netherlands, 1975

NOMENCLATURE

C_M: concentration of menthol, C_{M0}: initial value of C_M, $C_{\beta M}$: concentration of complexed menthol, $C_{\beta M0}$: initial vale of $C_{\beta M}$, C_M*: equilibrium value of C_M, Dapp: apparent diffusivity of menthol, Keq: equilibrium constant, k: release rate constant, R: radius of droplet, t: time, t_c: surface evaporation interval.

INTERACTION OF CYCLODEXTRINS AND THEIR POLYMERIC DERIVATIVES WITH SPEARMINT OIL: PREPARATION OF DURABLE FLAVORS FOR CHEWING-GUMS

F. Giordano[1], C. Pierin[1], M. Rillosi[2], G.P. Bettinetti[2], A. Gazzaniga[3], R. Cursano[3], E. Fossati[4]

[1]*Dipartimento Farmaceutico, Viale delle Scienze 78, 43100 Parma, Italy*

[2]*Dipartimento di Chimica Farmaceutica, Viale Taramelli 12, 27100 Pavia, Italy*

[3]*Istituto Chimico Farmaceutico, Viale Abruzzi 42, 00131 Milano, Italy*

[4]*Roquette Italia SpA, Via Serravalle 26, 15063 Cassano Spinola, Italy*

ABSTRACT

The interaction of spearmint oil with polymeric derivatives of cyclodextrins was investigated aiming to isolate solid phases suitable for the formulation of chewing gums. The physico-chemical characterization of materials was accomplished by means of thermal methods, uv spectroscopy, and by assessing the ability of releasing in vitro spearmint oil. Results show that preparations containing a swellable water insoluble beta-cyclodextrin polymer have satisfactory features of durable in vitro release of the included flavoring agent.

1. INTRODUCTION

The improvement of the organoleptic properties of chewing-gums, especially in terms of taste duration through an increase of the release time of aromas, is a stimulating task from a practical point of view (generally aromas are rather expensive materials) and also from a theoretical one, when considering the investigation of possible interactions between components of such formulations. The achievement of this goal is particularly desirable for medicated items (e.g. containing aspirin, antimicrobial agents, nicotin, etc.), showing problems of stability and acceptability.

The process of molecular encapsulation is commonly applied in food, chemical or pharmaceutical industries for the protection of the materials against external environmental factors. Increased resistance against oxidation or hydrolysis, reduction in volatility and remarkable stability upon long term storage at different ambient conditions as a result of complex or inclusion compound formation have been reported [1]. When dealing with flavoring agents the ideal candidate complexing agent should allow the

J. Szejtli and L. Szente (eds.), Proceedings of the Eighth International Symposium on Cyclodextrons, 587–590.

advantages of increased stability and volatility reduction of aromas, while showing moderate solubility in water and ability of remaining in the oral cavity for a sufficient time to mask bad taste of actives. Cyclodextrins have been demonstrated able of performing the first part of this objective by means of inclusion compound formation. Once in contact with aqueous media, however, their inclusion compounds show, as a rule, higher solubilities and dissolution rates than the guest alone, owing to particle size reduction, complexation, and increased apparent solubility, thus being washed off rapidly from the oral cavity by saliva. From this point of view some water insoluble beta-cyclodextrin polymeric derivatives with peculiar swelling ability, recently proposed as disintegrants for pharmaceutical tablets, could represent materials able to minimize the disadvantages of being rapidly removed, while keeping to some extent the complexing ability of the parent compound.

We report here the results of preliminary investigations on the interaction of betacyclodextrin and betacyclodextrin polymers with spearmint oil aiming to isolate solid phases suitable for the formulation of chewing-gums with the above mentioned features. The chemico-physical characterization of materials was accomplished by means of differential scanning calorimetry, termogravimetry, and u.v. spectroscopy. The ability of releasing in vitro spearmint oil from all solid products was also assessed. Results show that preparations containing a swellable water insoluble beta cyclodextrin polymer have satisfactory features of durable release of the included flavoring agent.

2. MATERIALS AND METHODS

2.1 Materials

Betacyclodextrin (ROQUETTE ITALIA SpA) LOTTO E0157
Swellable betacyclodextrin polymer CYL 135 (CYCLOLAB, UNGHERIA)
Disintegrant betacyclodextrin polymer CYL 132/2 (CYCLOLAB, UNGHERIA)
Spearmint oil 31762 COM.CS

2.2 Preparation of samples

The following Table reports codes, compositions and spearmint oil contents of the final products.

Table: Codes and compositions of formulations

CODES	COMPOSITIONS	SPEARMINT OIL (% by weight) *
KCWS1	BCD : WATER : SPEARMINT OIL	7.75
KCWS2	SWELLABLE BCD POLYMER : WATER : SPEARMINT OIL	5.74
KCWS3	DISINTEGRANT BCD POLYMER : WATER : SPEARMINT OIL	6.48
KCS	DISINTEGRANT BCD POLYMER : SPEARMINT OIL	15.63
KCSB	DISINTEGRANT BCD POLYMER : SPEARMINT OIL : BCD	15.54

* in the final product

KCWS1, KCWS2, KCWS3 samples were prepared by kneading in a china mortar with pestle the cyclodextrin under investigation, water, and spearmint oil in the 1:0.5:0.1 (g:mL:mL) ratio, until a homogeneous and creamy paste was obtained (about 20 min). The products were then equilibrated at room temperature and humidity under air stream. KCS samples were prepared in the same way using only the betacyclodextrin disintegrant polymer and spearmint oil (1:0.4, g:mL). In KCSB preparations the polymer was partially substituted by betacyclodextrin so that the composition of the ternary system (betacyclodextrin polymer:spearmint oil:betacyclodextrin) was 0.9:0.4:0.3, g:mL:g.

2.3 Analytical methods

Spearmint oil contents of products and dissolution samples were spectrophotometrically assessed at 241 nm ($A^{1\%\ 1\ cm}$ 428, water solution).
DSC and TGA analyses were performed with a Mettler TA 4000 apparatus equipped with DSC-25 and TG-50 cells. Samples were scanned in open Al pans (DSC) or open alumina containers (TGA).
TGA measurements were carried out on samples scanned from 25 to 250 °C at a heating rate of 5 K min^{-1}; isothermal TGA runs were also performed on samples (30 °C for 12 hours).
DSC was performed in the 25-250 °C temperature range, heating rate 10 K min^{-1}.
Karl Fischer titrations were performed with a Mettler Memo Titrator DL40GP.
Dissolution tests: USP XXII paddle method, 100 rpm, 25 °C , 1 liter of distilled water, manual sampling at fixed time intervals and spearmint oil assay as described above. Weighed samples of powdered kneaded products, corresponding to a constant amount of spearmint oil (12 mg), were then added.

3. RESULTS AND DISCUSSION

Isothermal TGA analysis performed on spearmint oil showed a weight loss of about 6% after 12 hours at 30 °C .
The satisfactory agreement between thermogravimetric measures and Karl Fischer titrations on the same samples evidences that the weight loss (in the 30-100 °C range) is attributable mainly to water, thus confirming the good stability of the flavour.
The interaction between cyclodextrins and spearmint oil is well confirmed by DSC patterns: while spearmint oil shows a characteristic peak around 165 °C, all the interaction compounds prepared by kneading have, in the same region, a flat profile. Figure 1 shows the dissolution profiles of all preparations. When betacyclodextrin was used (KCWS1 and KCSB formulations) the release of the aroma was almost complete within 5 minutes.
KCWS2 and KCWS3 products with water insoluble cyclodextrin polymers release spearmint oil much more slowly. In particular, a pseudo-zero order dissolution rate is evident when disintegrant betacyclodextrin polymer was used.

The release patterns of spearmint oil from samples prepared by two different procedures (KCWS and KCS), using the same disintegrant betacyclodextrin polymer together with a KCSB preparation, indicate that the use of water during the kneading process clearly allows the preparation of a product with satisfactory release properties.

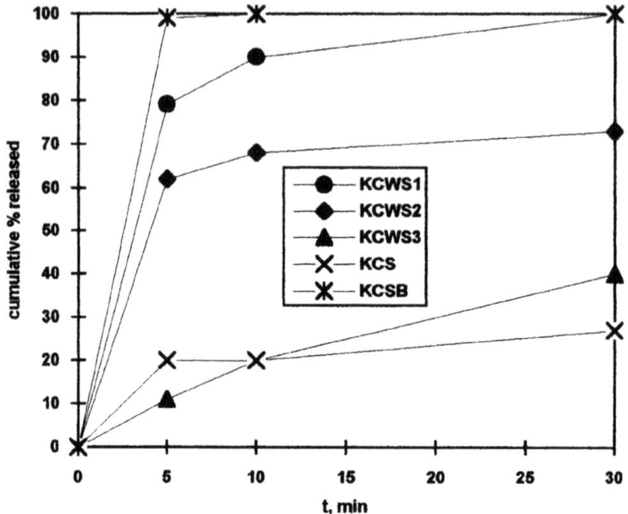

Fig. 1 Release profiles of spearmint oil from preparations

4. CONCLUSIONS

The preliminary results presented here show that swellable water insoluble polymers of cyclodextrins can form with spearmint oil inclusion compounds which slowly release the aroma, probably through a combined mechanism of inclusion within the cavity of the macrocycle and the formation of a swollen hydrogel network able to entrap hydrophobic molecules such as the components of spearmint oil.

ACKNOWLEDGMENTS

The financial support of MURST (Ministero dell'Università e della Ricerca Scientifica e Tecnologica) is gratefully acknowledged.

REFERENCES

[1] Szente L., Szejtli J., Stabilization of Flavors by Cyclodextrins, Chpt.16 in "Flavor Encapsulation", Ed. G.A. Reineccius, pp. 148-157. ACS Books, Am. Chem. Soc.,Washington, D.C, (1988).

EFFECT OF 2-HYDROXYPROPYL-CYCLODEXTRINS ON PERCUTANEOUS ABSORPTION OF *p*-HYDROXYBENZOIC ACID

Hajime Matsuda, Muneo Tanaka, Hidetoshi Arima &
Kaneto Uekama

Reseach Center, Shiseido Ltd, 1050 Nippa-cho, Kohoku-ku, Yokohama 223, Japan, School of Pharmacy, Tokyo University of Pharmacy and Life Science, 1432-1 Horinouchi, Hachioji, Tokyo 192-03, Japan, and Faculty of phamacetical Sciences, Kumamoto University, 5-1, Oe-honmachi, Kumamoto 862, Japan

ABSTRACT

A potential use of 2-hydroxypropyl-cyclodextrins (HP-CDs) to solubilize *p*-hydroxybenzoic acid methyl ester (methyl paraben) and to suppress its percutaneous absorption was examined, and compared with nonionic sufactant HCO-60 (Hydrogenated castor oil 60EO). HP-CDs significantly increased the solubility of methyl paraben in water, where the apparent 1:1 stability constant of the methyl paraben: HP-β-CD complex was determined to be 2150M^{-1}. The *in-vitro* cutaneous permeability of methyl paraben through an isolated skin of hairless mouse was suppressed by HP-CDs, thus promoting the bioconversion of methyl paraben to the less toxic metabolite, *p*-hydroxybenzoic acid (p-HBA), in the epidermis. These effects of HP-CDs were greater than those of HCO-60. HP-β-CD (2% w/v) reduced the *in-vivo* percutaneous absorption of methyl paraben by 66% 24h after the topical application of a solution containing [^{14}C] methyl paraben to hairless mouse skin. In addition, the percutaneous absorption of [^{14}C] HP-β-CD was confirmed to be extremely low. These results suggest that HP-β-CD is useful in liquid preparations of methyl paraben for topical application.

1.INTRODUCTION

p-Hydroxybenzoic acid methylester (methyl paraben) as an antiseptic is in a wide use in the field of drugs, cosmetics, foods and the like. The application of methyl paraben (MP) to aqueous preparations requires its solubilization with surfactant, because of its low water solubility.

From the viewpoint of safety, lower absorptivity of paraben and quick transfomation into its metabolite of better safety following its transcutaneous

J. Szejtli and L. Szente (eds.), Proceedings of the Eighth International Symposium on Cyclodextrons, 591–595.

application are desirable.
In the present study, we took notice of 2-hydroxypropyl cyclodextrins (2HP-CDs) as functional materials substitute for surfactants and examined the solubilizing power of MP and its transcutaneous absorptivity *in vivo* and *in vitro*, using 2HP-α-CD, 2HP-β-CD and 2HP-γ CD.

2.MATERIALS AND METHODS

Materials

Methyl paraben was purchased from Wako Pure Chemical Industries, Ltd(Osaka, Japan). [^{14}C] Methyl paraben (sp. act 2・4 MBq mg^{-1}) was obtained by esterification of [^{14}C] p-HBA (New England Nuclear Co.,MA, USA) using diazomethane. HP-β-CD was donated by Nihon Shokuhin Kako Ltd (Shizuoka, Japan); the average substitution degree of 2-hydroxypropyl groups per β-CD molecule was determined to be 4.8 by ^{1}H NMR (Pitha et al 1986). [^{14}C] HP-β-CD (sp. act. 257kBq mg^{-1}) was obtained from Pharmatec Inc. (FL,USA). All other chemicals and solvents were of analytical reagent grade.

Solubility measurements

A constant but excess amount of methyl paraben was added to an aqueous solution containing a given concentration of HP-β-CD. These were mixed by a magnetic stirrer at 20℃, and after equilibrium was attained (about 12h) the mixture was centrifuged at 3000g for 5min and the supernatant was filtered through a pipette with a cotton plug. After diluting the filtrate, methyl paraben was assayed by high-performance liquid chromatography (HPLC), under the following conditions using a JASCO 860-CO HPLC apparatus (Tokyo, Japan) : column, CAPCELL-PACK C_{18} SG 4.6×250mm (Shiseido, Tokyo, Japan) ; detection, UV at 256 nm; mobile phase, water: methanol=1:1 (v/v) with 0.1% phosphoric acid; injection, 10μ L; flow rate, 1.0mL min^{-1}.

In vitro cutaneous permeability

The dorsal epidermis of a female hairless mouse (8 weeks old) was attached to a perpendicular diffusion cell (Franz 1975) and the test was performed at 20℃. An aqueous solution of methyl paraben was used on the donor side and physiological saline was used on the receiver side. The concentrations of methyl paraben and HP-β-CD were 0.2% (w/v) and 2.0% (w/v), respectively. At appropriate intervals, the sample solution (0.8mL) was withdrawn from the receiver phase. The amounts of methyl paraben and p-HBA were determined by HPLC as described above.

In vivo percutaneous absorption

One hundred microlitres of 0.2% methyl paraben solution containing [^{14}C] methyl praben (0.0002% w/v, 480 kBq) in the absence and presence of HP-β CD (2.0%w/v) was applied onto the dorsal skin (2cm^2) of a female hairless mouse (8 weeks old) under sealed conditions using human adhesive test patches (Torii Yakuhin Co.,Tokyo, Japan). After the patches had been attached onto the skin for 24h, the radioactivity in the patches, in the atratum corneum collected by stripping, and in the epidermis and cutis of the skin ontained by peeling off the treated portion, was measured with a liquid scintillation counter (Beckman model LS-1701,Tokyo, Japan) according to the method of Iwata et al (1988). In addition, the amounts of methyl paraben excreted in the urine and faces and remaining in the body were obtained by driving of the CO_2, collecting and counting, after ashing. Furthermore, the radioactivity excreted by respiration was quantified as [^{14}C] CO_2. To measure the percutaneous absorption of HP-β-CD *in-vivo*, 100μL 2.0% [^{14}C]HP-β-CD (500 kBq) in the absence and presence of 0.2% methyl paraben was applied in a mannar similar to the above experiment. Likewise, the radioactivity excreted in the urine, faces and respiration and remaining in the body was quantified. Student's t-test was used for statistical evaluation of the date, and P values of <0.05 were considered to be statistically significant.

3. RESULTS AND DISCUSSION

3.1 solubilization

The solubility of methyl paraben linearly increased with increase in the amount of cyclodextrin added .
The slope of these straight lines was α, β and γ in the increasing order of solubility. Since the slopes for HP-γ-CD's were not less than 1, the formation of multiple complexes was expected.
We assumed 1:1 complex in the α and β series and 2:1 in the γ series, and calculated respective apparent stability constants.

3.2 *In-vitro* cutaneous permeability

The skin permeability to methyl paraben until 28th hour was investigated.
We utilized two controls in this experiment: one was an aqueous solution of methyl paraben alone, and the other was an HCO-60 solution as the representative of widely used surfactants.
The skin permeation of methyl paraben was remarkably depressed. Although the differences in solubility among α, β, and γ were not so large, the solubilities were found to be HP-α-CD, HP-β-CD, and HP-γ-CD in the increasing order of solubility. On the other hand, the methyl paraben control and HCO-60 series did not exhibit any solubility reduction but showed increasing tendecies.

3.3 Percutaneous absorption of methyl paraben and HP-β-CD

Plasters containing tracer amounts of ^{14}C-labeled methyl paraben solutions were applied onto the dorsal slin of hairless mice, and, after 24 hours, the radioactivity of each tested sites was measured.
The amount of remaining methyl paraben within the plaster, denotes that, in

contrast to the case of the absence of cyclodextrin, the remaining amount was reduced by 50%. We then investigated the absortption of labeled HP-β-CD, the host molecules, into the skin. As seen in Table 1, the intact skin did not practically absorb HP-β-CD, irrespective of whether with or without methyl paraben. However, in the case of application to tape-stripped skin from which the corneum had been removed, some degree of HP-β-CD migration was observed.This indicates that the corneum acted as a barrier against HP-β-CD permeation. From these findings, it is inferred that the skin permeation of cyclodextrin complexes is hindered, and that exclusively the free from of the paraben is involved in skin permeation.

Table 1 Percutaenous absorption of radioactivity corresponding to ^{14}C-HP-β-CD 24hr after topical application of plasters containing HP-β-CD solution onto dorsal skin of hairless mice[a]

system	skin condition	amount of HP-β-CD absorbed (% of dose)
with MP	intact	0.023±0.007
without MP	intact	0.018±0.003
without MP	stripped	23.987±7.364

[a] Each value represents the mean±S.D. of 5 hairless mice.

4. CONCLUSION

1) 2HP-CDs increased the solubility of methyl paraben (MP) by forming a soluble complex in water.
2) *In vitro* experimental results of skin permeability test revealed that 2HP-CDs remarkably suppressed MP skin permeability and accelerated intracutaneous transformation of MP into *p*- hydroxybenzoic acid.
3) 2HP-CDs suppressed transdermal absorption of MP, where the transdermal absorptivity from normal skin for 2HP-α-CDs itself was extremely low.
4) 2HP-β-CDs was quickly eliminated in urine following its subcutaneous administration.

The above results suggest that 2HP-β-CD is useful as an excellent functional material substisute for surfactants from the viewpoints of solubilizing power and safety.

REFERENCES

1. Brewster, M. E., Estes, K. S., Bodor, N. (1990)
An intravenous toxicity study of 2-hydroxypropyl-β-cyclodextrin, a useful drug solubilizer, in rats and monkeys. Int.J.Pharm. 59: 231-243
2. Higuchi, T., Connors, K. A. (1965)
Phase solubility techniques. Adv. Anal. Chem. Instr. 4: 117-212
3. Iwata, Y., Satoh, I., Akiu, S., Kobayashi, T., Fijiyama, Y. (1988)
Studies on the percutaneous absorption and the skin metabolism of arbutin. Proc. Jpn. Soc Invest. Dermatol. 12: 176-177
4. Pitha, J., Pitha. J. (1985)
Amorphous water-soluble derivatives of cyclodextrins, nontoxic dissolution enhancing excipients. J. Pharm. Sci. 74: 987-990
5. Pitha, J., Milecki, J., Fales, H., Pannell, L., Uekama, K. (1986)
Hydroxypropyl-β-cyclodextrin: preparation and characterization; effects on solubility of drugs. Int. J. Pharm. 29: 73-82
6. Tanaka, M., Nakamura, T., Ito, K., Taniguchi, K., Matsuda, H., Uekama, K. (1993)
Movement of skin permeation behavior of parabens by 2-hydroxypropyl-β-cyclodextrin. Nihon Koshouhin Kagakukai Zasshi 17: 185-190
7. Uekama, K., Ikeda, Y., Hirayama, F., Otagiri, M., Ikeda, M. (1980)
Inclusion complexation of p-hydroxybenzoic acid esters with α- and β-cyclodextrins; dissolution behaviors and antimicrobial activities. Yakugaku Zasshi 100: 994-1003
8. Yoshida, A., Arima, H., Uekama, K., Pitha, J., (1988)
Pharmaceutical evaluation of hydroxyalkyl esters of β-cyclodextrins. Int. J. Pharm. 46: 217-222
9. Tanaka, M., Iwata, Y., Kozuki, Y., Taniguchi, K., Matsuda, H., Arima, H., Tuchiya, S., (1995)
Effect of 2-Hydroxypropyl-β-cyclodextrin on percutaneous absorption of methyl paraben . J. Pharm. Phamacol., 47: 897-900

CAPTURE OF VOLATILE CHLORINATED HYDROCARBONS BY AQUEOUS SOLUTIONS OF BRANCHED CYCLODEXTRINS

I. UEMASU, S. KUSHIYAMA, and R. AIZAWA
National Institute for Resources and Environment
16-3 Onogawa, Tsukuba, Ibaraki 305, Japan

ABSTRACT

The use of aqueous solutions of branched cyclodextrins was examined in order to develop an effective method of capturing toxic volatile chlorinated hydrocarbons such as trichloroethylene and monochlorobenzene. From the experiments in which trichloroethylene diluted with nitrogen gas came into contact with aqueous solutions of branched cyclodextrin mixtures, it was found that absorption could be performed without the formation of inclusion complex solids, which should simplify the whole process of absorption and recovery.

1. INTRODUCTION

Volatile chlorinated hydrocarbons have been widely used as solvents, detergents, agents for chemical synthesis, etc. Since they are toxic and cause environmental pollution, the control of their emission from industrial sources is mandatory.

It is well known that cyclodextrins(CDs) have a strong affinity for halogenated hydrocarbons and easily form inclusion complexes, which are usually insoluble in water[1]. The use of aqueous CD solutions to recover the vapors of such organic compounds was reported more than ten years ago[2]. One of the difficulties consists in handling the solid inclusion complexes, which are usually very fine and sticky. In order to retard the production of insoluble complexes, some additives such as cetyl-trimethyl ammonium bromide were employed or the use of highly-soluble CD derivatives were examined.

We have been studying the application of branched CDs to the capture of volatile halogenated hydrocarbons[3]. Branched CDs, which are now commercially available, have high water solubilities and their inclusion complexes, too. In this study, we examined the solubilities of four volatile chlorinated hydrocarbons in aqueous solutions of branched CD mixtures and the absorption of trichloroethylene diluted with nitrogen gas into the CD solutions.

J. Szejtli and L. Szente (eds.), Proceedings of the Eighth International Symposium on Cyclodextrons, 597–600.

2. MATERIALS AND METHODS

2.1. Materials

Branched CD mixtures were obtained from Ensuiko Sugar Refining Co., Ltd.(Yokohama, Japan). Glucosyl-α-CD mixture(G-α-CD mixture) contains 45% monoglucosyl-α-CD(6-*O*-α-D-glucosyl-α-CD), 34% diglucosyl-α-CD and other. Maltosyl-β-CD mixture(M-β-CD mixture) contains 43% monomaltosyl-β-CD(6-*O*-α-maltosyl-β-CD), 44% dimaltosyl-β-CD, 9% trimaltosyl-β-CD and other. Carbon tetrachloride, trichloroethylene, tetrachloroethylene, monochlorobenzene, benzene, and diethyl ether were purchased from Wako Pure Chemical Industries, Ltd.(Tokyo, Japan). They were all guaranteed reagents and used without further purification. Distilled and deionized water was used.

2.2. Methods

MEASUREMENT OF SOLUBILITIES OF CHLORINATED HYDROCARBONS IN AQUEOUS SOLUTIONS OF BRANCHED CD MIXTURES A branched CD mixture was dissolved in water. 1.0 g of a chlorinated hydrocarbon was added to 5 mL of an aqueous solution of branched CD mixture. Then it was vigorously stirred in a thermostated bath. The reaction product was then placed in a glass tube and centrifuged at 3,000 rpm for 5 min. It separated into two transparent liquid layers, i.e., the chlorinated hydrocarbon phase and the aqueous phase containing CD and chlorinated hydrocarbon-CD inclusion complex. 3 mL of the aqueous solution was taken with a pipette and benzene was added as internal standard for analysis. It was transferred to a separatory funnel, with water and diethyl ether added, and shaken for a few minutes. The included chlorinated

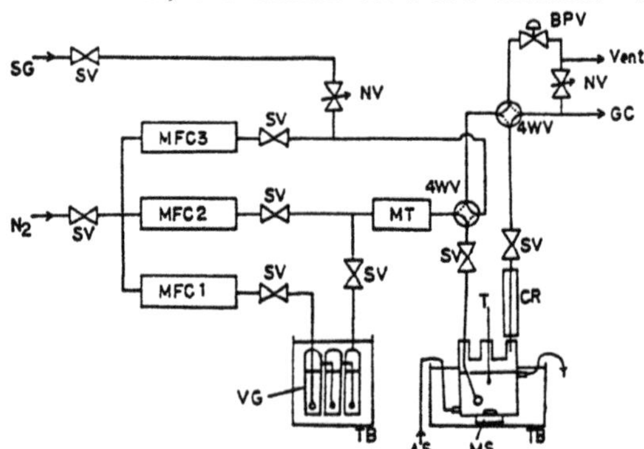

Fig. 1 Apparatus for Absorption Experiment
MFC: Mass flow controller, SV: Stop valve, NV: Needle valve, SG: Standard gas, 4WV: Four way valve, BPV: Back pressure valve, VG: Chlorinated hydrocarbon vaporizer, AS: Absorbing solution, MS: Magnetic stirrer, CR: Reflux condenser, T: Thermometer, TB: Thermostated bath, MT: Gas mixing tube, GC: Gas chromatograph

hydrocarbon and benzene was almost completely extracted into the ether layer. The ether solution was then subjected to gas chromatographic analysis.

ABSORPTION OF TRICHLOROETHYLENE GAS IN A CONTINUOUS FLOW SYSTEM The experimental apparatus used for gas absorption is illustrated in Figure 1. As shown in the figure, the concentration of trichloroethylene gas was changed by controlling the flow of N_2 via mass flow controllers 1 and 2. Then the diluted trichloroethylene gas was introduced into the aqueous solution of M-β-CD mixture. Fresh aqueous solution of the CD was continuously supplied. The concentration of trichloroethylene was measured every 3 min by GC. For comparison, a similar experiment was carried out using pure water in place of M-β-CD mixture solution.

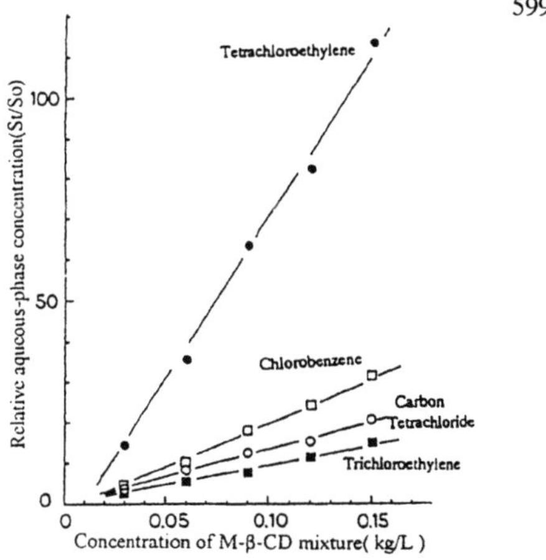

Fig. 2 Solubilities of Chlorinated Hydrocarbons in Aqueous Solutions of M-β-CD mixture (See text as for St and So.)

3. RESULTS AND DISCUSSION

3.1. Solubilities and partition coefficients

As can be seen in Figure 2, the solubilities of four chlorinated hydrocarbons increased almost linearly with increasing concentration of branched cyclodextrin mixture. The data were analyzed according to a linear partition model[4] expressed as follows;

$$S_t / S_o = 1 + K_{CW} X_{CD}$$

where St is the total aqueous-phase concentration of a chlorinated hydrocarbon, S_o is the aqueous solubility of the chlorinated hydrocarbon, K_{CW} is the partition coefficient of the chlorinated hydrocarbon between cyclodextrin and water, and X_{CD} is the concentration of

TABLE 1. Values of partition coefficient(log Kcw)

	G-α-CD mixture	Monoglucosyl -α-CD	M-β-CD mixture	Monomaltosyl-β-CD
Carbon tetrachloride	1.95 (0.999)	2.14 (0.997)	2.12 (0.998)	2.33 (0.998)
Trichloroethylene	1.72 (0.999)	1.84 (0.999)	1.98 (0.992)	2.06 (0.998)
Tetrachloroethylene	2.13 (0.984)	2.22 (0.968)	2.88 (0.995)	2.86 (0.999)
Chlorobenzene	2.07 (0.999)	2.10 (0.996)	2.35 (0.999)	2.28 (0.999)

Values of correlation coefficient for least-squares fitting are shown in parentheses.

cyclodextrin(kg/L). Table 1 shows that the Kcw values in the system using branched CD mixtures are not very different from those in the system using highly purified branched CDs. This indicates that branched CD mixtures have an inclusion characteristic almost equivalent to that of highly purified branched CDs. This is significant from a viewpoint of practical use since branched CD mixtures are much cheaper than highly purified branched CDs. Thus a branched CD mixture was used in the following absorption experiment.

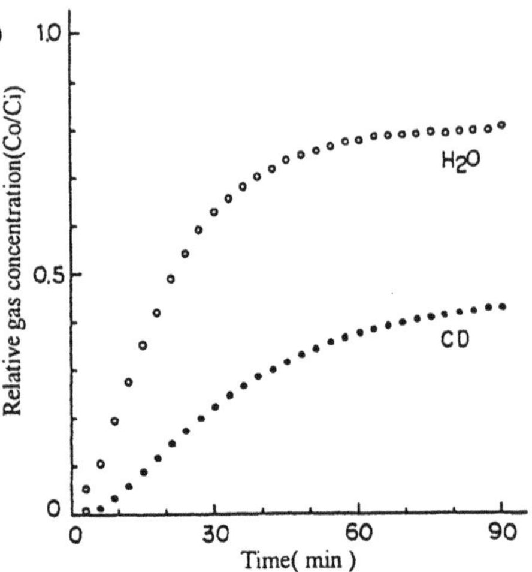

Fig. 3 Concentration Change of Trichloroethylene Gas after Contact with M-β-CD Mixture Solution or Pure Water (Ci and Co mean concentrations of trichloroethylene before and after contact with absorbent, respectively.)

3.2. Absorption in a continuous flow system

A preliminary result is shown in Figure 3. 3,000 ppm trichloroethylene gas diluted with N_2 was bubbled through 600 mL of 0.1 kg/L branched β-CD mixture solution at 25°C(GHSV: 10 h^{-1}) while 0.1 kg/L branched CD mixture solution was continuously provided(LHSV: 1 h^{-1}). Production of solid compounds was not observed in this experiment. From the comparison between the CD solution system and the H_2O system, it is clear that trichloroethylene gas was absorbed in the aqueous solution of branched CD mixture. It is expected that more effective absorption can be performed by optimizing the experimental conditions such as concentration of CD, flow rate of absorbing solution, the way of contacting gas with absorbing solution, etc. By the way, the branched CD mixture solution used for the absorption of trichloroethylene was heated to 70°C. Then the chlorinated hydrocarbon included by CD was released and the CD solution could be used to absorb trichloroethylene again.

REFERENCES

[1] French, D., Levine, M.L., Pazur, J.H., Norberg, E., Studies on the Schardinger dextrins. The preparation and solubility characteristics of alpha, beta and gamma dextrins, *J. Am. Chem. Soc.*, **71**, 353-356(1949)

[2] Szejtli, J., Cyclodextrin Technology, Kluwer Academic Publishers, Dordrecht, 1988

[3] Uemasu, I., Inclusion complexation of volatile chlorinated hydrocarbons in aqueous solutions of branched cyclodextrins, *J. Incl. Phenom.*, **17**, 177-185(1994)

[4] Wang, X., Brusseau, M.L., Solubilization of some low-polarity organic compounds by hydroxypropyl-β-cyclodextrin, *Environ. Sci. Technol.*, **27**, 2821-2825(1993)

REMOVAL OF PHENOLS FROM AQUEOUS SOLUTIONS IN THE PRESENCE OF HORSERADISH PEROXIDASE AND CYCLODEXTRIN DERIVATIVES

F. TROTTA, R.P. FERRARI, E. LAURENTI, G. MORAGLIO, A. TROSSI
Dipartimento di Chimica Inorganica, Chimica Fisica e Chimica dei Materiali, Università di Torino
Via P. Giuria 7, 10125 Torino, Italy

ABSTRACT

The hydroxylated aromatic compounds, present in the industrial waste waters, constitute a great environmental problem and many different processes, such as chemical or enzymatic oxidation, were reported in order to solve this problem. In this preliminary work we studied the influence of some cyclodextrin derivatives on the oxidation of phenols by horseradish peroxidase in presence of H_2O_2. We found that the cyclodextrin presence promote phenols removal from aqueous solution for a large set of compounds and we started to study the kinetic of the highly reactive 2,4,6-tribromophenol.

1. INTRODUCTION

Horseradish peroxidase (HRP), for its relatively low cost and its high reactivity towards many different organic and inorganic compounds, is an enzyme largely used in industrial and laboratory applications. In particular, the oxidation of some phenols and aromatic amines by the system HRP/H_2O_2 was reported in literature [1-4], and recently the relative production of insoluble polymers was used to remove phenols from industrial waste waters [5, 6].

The phenols low solubility in water and the presence of emulsions with organic solvents, which deplete enzyme activity, constitute the greatest problems in the enzymatic

J. Szejtli and L. Szente (eds.), Proceedings of the Eighth International Symposium on Cyclodextrons, 601–604.

treatment of phenol derivatives. Furthermore it is well known that cyclodextrin derivatives give inclusion compounds with substrates of suitable size and polarity, such as many hydroxylated aromatic compound, and this property is largely used in inverse transfer catalysis [7]. For these reasons we started to study the effect of the presence of β-cyclodextrin (β-CD) and of two cyclodextrin derivatives (methyl-β-cyclodextrin, Me-β-CD and hydroxypropyl-β-cyclodextrin, HP-β-CD) on the enzymatic oxidation of water solutions containing some different phenol derivatives. In this communication we report some qualitative preliminary results on the phenols degradation and the kinetic data on the oxidation of the highly reactive 2,4,6-tribromophenol (TBP).

2. MATERIALS AND METHODS

2.1 Materials

β-CD was a gift from Roquette Italia (Cassano Spinola, Italy); HP-β-CD and Me-β-CD were kindly supplied from Wacker Chemie (Germany). Horseradish peroxidase (E.C. 1.11.1.7) was aquired from Sigma Chemical Co. (USA) (Type VIA, RZ = 3.0, 1100 Sigma units/mg solid). Hydrogen peroxide (30%, ACS grade) was from Merck (Germany). All phenols, from Aldrich (USA), were on analytical grade and used without further purification. UV-visible spectra and kinetic measurements were recorded on a double beam UVIKON 930 spectrophotometer (Kontron Instruments, Italy).

2.2 Methods

The solutions of cyclodextrin derivatives were obtained by diluition of a stock solution of 2.5 wt % of CD in water. These solutions were saturated with the selected phenols and equilibrated overnight under magnetic stirring at room temperature; subsequently the excess of phenol was removed by filtration. UV-visible spectra and kinetics measurements were obtained introducing in a quartz cuvette 1 ml of CD solution, 1 ml of bidistilled water, 20 μl of HRP (1.19×10^{-5} M) and 80 μl of H_2O_2 (7.4×10^{-2} M). The reaction rate was followed by the increase of the absorbance of the reaction product at 359 nm.

3. RESULTS AND DISCUSSION

In Table 1 we report the results of our catalytic preliminary studies: the phenols were organized on the basis of their relative reactivity towards HRP/H_2O_2 system in the presence and in the absence of cyclodextrin derivatives.

TABLE 1.

phenols NON-reactive	phenols more reactive without CD	phenols more reactive with CD
2,6-di-*tert*-butyl-4-methylphenol	4-hydroxydiphenyl	2,4,6-tribromophenol
3-nitrophenol	2,4,6-tri-*tert*-butylphenol	2,4,6-trichlorophenol
dodecylgallate		3-chlorophenol
		1-naphtol
		2-naphtol

The first column accounts for the phenols unreactive towards HRP both in the presence or in the absence of the CD derivatives. These molecules seems to be not suitable as substrate for HRP under our experimental conditions.

The second group show an higher reactivity in the absence of CDs. We suppose that this behaviour could be ascribed to the protective action of the CD cavity which gives the substrate less available for the enzyme.

The third one (the most numerous) reports the phenols that have higher reaction rate in the presence of the CDs. In particular with 1- and 2-naphtol, insoluble polymeric materials were obtained, which can be easily removed from waste waters by filtration.

With the aim to study in detail the CD influence on these oxidations, we made a series of measures with TBP using different cyclodextrins and various experimental conditions.

The oxidative process of TBP in absence and in presence of β-CD, Me-β-CD or HP-β-CD was followed spectrophotometrically and in all the cases we obtained the same spectral pattern. In agreement with a reported oxidation mechanism [8] we suggest that TBP (**1**) was been oxidated to 2,6-dibromo-1,4-benzoquinone (**2**) as shown in the figure:

OH, Br, Br, Br — HRP / H_2O_2 → O, Br, Br, O

1 **2**

In Table 2 we report the reaction initial rates obtained for TBP in presence of different concentrations of the CDs. From these preliminary quantitative data we can argue that the oxidation initial rate seems depend on the CD concentration and that the catalytic efficiency of the three CDs is quite similar, but with a slight prevalence of the Me-β-CD promotion.

TABLE 2: TBP oxidation initial rate (M/min)

% of CD	β-CD	Me-β-CD	HP-β-CD
0.05	$3.4 \cdot 10^{-4}$	$5.2 \cdot 10^{-4}$	$3.8 \cdot 10^{-4}$
0.10	$4.3 \cdot 10^{-4}$	$6.3 \cdot 10^{-4}$	$4.4 \cdot 10^{-4}$
0.15	$4.7 \cdot 10^{-4}$	$6.4 \cdot 10^{-4}$	$4.8 \cdot 10^{-4}$
0.25	$7.4 \cdot 10^{-4}$	$7.7 \cdot 10^{-4}$	$6.0 \cdot 10^{-4}$

We could suppose that the cyclodextrin derivatives effect on the peroxidatic oxidation of TBP depends principally on the substrate solubility increase due to its favourite inclusion in the CD cavity.

In order to completely clarify this problem, further studies concerning the calculation of the inclusion constants and the investigation of the kinetic properties of a series of phenols are in progress.

4. REFERENCES

[1] Job, D., Dunford, H.B., Substituent effect on the oxidation of phenols and aromatic amines by horseradish peroxidase compound I, *Eur. J. Biochem.* **66**, 607-614 (1976)

[2] Berry, D.F., Boys, S.A., Oxidative coupling of phenols and anilines by peroxidase: structure-activity relationship, *Soil Sci. Am. Soc. J.* **48**, 565-569 (1984)

[3] Ferrari, R.P., Laurenti, E., Casella, L., Poli, S., Oxidation of catechols and catecholamines by horseradish peroxidase and lactoperoxidase: ESR spin stabilization approach combined with optical methods, *Spectrochim. Acta* **49A**, 1261-1267 (1993)

[4] Ferrari, R.P., Laurenti, E., Rossi, M., Studies on the horseradish peroxidase enzymatic activity and stability correlations by spectroscopic techniques, *Life Chem. Reports* **10**, 249-258 (1994)

[5] Huixian, Z., Taylor, K.E., Products of oxidative coupling of phenol by horseradish peroxidase, *Chemosphere* **28**, 1807-1817 (1994)

[6] Nakamoto, S., Machida, N., Phenol removal from aqueous solution by peroxidase-catalyzed reaction using additives, *Wat. Res.* **26**, 49-54 (1992)

[7] Trotta, F., Phtalic acid esters hydrolysis under inverse phase-transfer catalysis conditions *J. Mol. Catalysis.* **85**, L265-L267 (1993)

[8] Labat, G., Seris, J.-L., Meunier, B., Oxidative degradation of aromatic polluttants by chemical models of ligninase based on porphyrin complexes, *Angew. Chem. Int. Ed. Engl.* **29**, 1471-1473 (1990)

EXTRACTION OF PAHS AND PESTICIDES FROM CONTAMINATED SOILS WITH AQUEOUS CD SOLUTIONS

ÉVA FENYVESI, JULIANNA SZEMÁN, JÓZSEF SZEJTLI

CYCLOLAB CD R. & D. Laboratory, Ltd.

Dombóvári u. 5-7., Budapest, HUNGARY

ABSTRACT

A new method for the mobilization of organic pollutants from contaminated soils has been developed using aqueous CD solutions. The CD derivatives (methyl, hydroxypropyl, polymer) are able to desorb the lypophylic xenobiotics (pyrene, pentachlorophenol) sorbed on the particles of the soil and to solubilize these organics via inclusion complex formation and make them more accessible for the microorganisms. The CDs themselves do not pollute the environment being biodegradable.

1. INTRODUCTION

CDs have their potential in the reduction of environmental pollution, especially in waste water treatment [1]. Application of CDs in the decontamination of soils seems to be also promising [2].

Recently hydroxypropyl β–CD (HPBCD), was reported to enhance the transport of low-polarity organic compounds through soil [3]. It was found that HPBCD itself is not retarded by the soil, and it decreased the retardation of such compounds as anthracene, pyrene, and trichlorobiphenyl significantly.

The aim of the present work was to enhance the desorption of xenobiotics by aqueous CD solutions.

2. MATERIALS AND METHODS

2.1. Materials

Randomly methylated β–CD (RAMEB); *DS*=1.8 (Wacker Chemie GmbH, Munich), 2-hydroxypropyl β–CD (HPBCD); *DS*= 2.8 (Cyclolab, Budapest), β–CD polymer

J. Szejtli and L. Szente (eds.), Proceedings of the Eighth International Symposium on Cyclodextrons, 605–608.

(BCDPS) 55 % CD content, M_W = 4300 (CYL-310, Cyclolab, Budapest), γ-CD polymer (GCDPS) 57 % CD content, M_W = 3800 (CYL-309, Cyclolab, Budapest)

DS= degree of substitution per glucose unit, M_W= weight average molecular weight, measured by gel chromatography on ACA 54 Ultrogel column

2.2. Methods

Desorption experiments were performed on the following way: soil samples (garden soil or clay) were polluted with 0.5 mg/g anthracene, 50.5 μg/g pyrene or 28.4 μg/g perylene or 2.28 mg/g mixture of polycyclic aromatic hydrocarbons (the composition of the mixture is given in Table I) by incubating the sample with a solution containing the pollutant(s) dissolved in heptane then the samples were dried on air for 3 days. 1 g of each sample was stirred with 10 mL solvent (heptane or ethanol or aqueous CD solution) for 15 min at room temperature using magnetic stirrer, then it was left to sediment. In case of individual hydrocarbon pollutants the supernatant was measured spectrophotometrically after filtration and in case of mixtures with HPLC (apparatus: Hewlett-Packard 1050, column: Nucleosil 5 C18 PAH, Macherey Nagel, mobile phase: acetonitrile-methanol-water 55:25:20, UV detection at λ=250 and 300 nm). The extracted pollutant was related to the initial pollutant concentration of the soil.

Table I Composition of the PAH mixture

PAH	1-methyl naphthalene	fluorene	anthracene	fluoranthene	pyrene	perylene	coronene
Conc. (mg/g)	0.56	0.6	0.55	0.14	0.27	0.09	0.07

3. RESULTS AND DISCUSSION

The aqueous solubility of naphthalene and anthracene is increased with increasing concentration of various CD derivatives [4,5]. On the basis of the solubility enhancing effect RAMEB, HPBCD, BCDPS and GCDPS were selected for the soil-extraction measurements.

Soil samples were polluted separately with anthracene or pyrene or perylene and extracted with aqueous CD solutions of 20 % content of the CD derivative.

Table II Pollutants extracted from soil in percent of the pollutant content before extraction

	Anthracene	Pyrene	Perylene
heptane	74	53	79
20 % RAMEB	72	87	32
20 % BCDPS	84	61	25
20% GCDPS	52	95	17
20 % HPBCD	68	65	10

RAMEB was almost as effective in extracting anthracene from soil as heptane, and BCDPS exceeded even the performance of heptane. In case of anthracene the effectivity of RAMEB and GCDPS are nearly double compared to that of heptane. All the studied CD derivatives showed low affinity toward perylene.

Soil samples contaminated with a mixture of PAHs were extracted with 5 - 20 % RAMEB or 96 % ethanol. The extraction efficiency for each of the 7 components increased with increasing RAMEB concentration (Table III). The effect of 10 % RAMEB exceeded that of ethanol in case of naphthalene, fluorene and anthracene. Fluoranthene was extracted with 15 % RAMEB as effectively as with ethanol. For the other three components even 20 % RAMEB could not reach the effectivity of ethanol.

Table III Extraction of soil samples contaminated with a 7 component-PAH mixture

solvent	1-methyl naphthalene (μg/ml)	fluorene (μg/ml)	anthracene (μg/ml)	fluoranthene (μg/ml)	pyrene (μg/ml)	perylene (μg/ml)	coronene (μg/ml)
100 %*	56.5	59.9	54.8	14.3	26.3	8.4	7.1
5% RAMEB	7.1	29.1	18.4	1.9	2.4	<0.5	0.7
10% RAMEB	10.0	44.1	27.6	7.3	13.2	0.6	0.9
15% RAMEB	11.6	50.2	33.2	9.40	21.0	1.0	1.2
20% RAMEB	12.1	50.2	32.0	10.0	23.4	1.5	1.3
ethanol	7.93	36.4	24.5	9.1	25.1	5.9	6.5

* calculated composition of the extract if all the pollutants had been removed

A clay sample contaminated with 0.53 mg/g pentachlorophenol (PCP) was extracted with water, ethanol and aqueous solutions of CD derivatives. The results expressed as percent of the input contaminant are listed in Table IV.

Table IV Pentachlorophenol extracted from clay

solvent	water	20 % RAMEB	20 % HPBCD	20 % GCDPS	ethanol
PCP (%)	37.7	45.3	71.7	54.7	50.9

In this case hydroxypropyl β–CD (HPBCD) was found to be superior to the studied solvents.

4. CONCLUSIONS

Organic contaminants such as lower PAHs and pentachlorophenol can be extracted from soils by aqueous solutions of different CD derivatives. The most effective CD derivative should be found to the target pollutant or a mixture should be used for this purpose.

In the practice the aim would not be the extraction, but the mobilization only of the PAHs and the other xenobiotics. By enhancing the desorption of these organic pollutants probably their microbial decomposition will be facilitated.

ACKNOWLEDGMENTS

This study was supported in part by a grant of OTKA (T 014424) from the Hungarian Academy of Sciences. The technical assistance of Zs. Nagy and Zs. Zachár is greatly acknowledged.

REFERENCES

[1] Szejtli, J., Cyclodextrins in reduction of environmental pollution, *Minutes 6th Symposium CDs* (Ed. A. Hedges) Ed. de Sante, Paris, p.380 (1992)

[2] Szejtli, J.,. Fenyvesi, É., (Wacker Chemie), Extraction of organic pollutants from contaminated soil, EP 0 613 735 (1994)

[3] Brusseau, M. L., Wang, X., Hu, Q., Enhanced transport of low-polarity organic compounds through soil by cyclodextrin, *Environ. Sci. Technol.* **28**, 952 (1994)

[4] Fenyvesi, É., Szemán, J., Szejtli, J., Extraction of organic pollutants from contaminated soils with aqueous cyclodextrin solutions, 2nd Int. Symp. on Environmental Contamination in Centr. East Europe, Budapest, Symposium Proceedings, 1994, p.383

[5] Wang, X.., Brusseau, M. L., Solubilization of some low-polarity organic compounds by hydroxypropyl β-cyclodextrin, *Environ. Sci. Technol.* **27**, 2821 (1993)

POTENTIAL USE OF CYCLODEXTRINS IN SOIL BIOREMEDIATION

K. GRUIZ[a], É. FENYVESI[b], É. KRISTON[a], M. MOLNÁR[a] and B. HORVÁTH[a]

[a]*Technical University of Budapest, Dept. of Agric. Chem. Technology*
1111. Budapest, Gellért Sq. 4. HUNGARY
[b]*Cyclolab Ltd., 1117 Budapest Dombóvári út 4-6.*

ABSTRACT

Bioavailability and toxic effect of contaminants are the main limitations during soil bioremediation. Cyclodextrins may influence bioavailability of the contaminants during biodegradation and also toxicity of the pollutant on soil microbes and plants since their ability to form inclusion complex with organic compounds. The effect of cyclodextrins on bioremediation and on toxic effect of hydrocarbons was investigated by testing the activity of hydrocarbon degrading microflora and of plant growth. The effect of cyclodextrins could be demonstrated in both cases: biodegradation of hydrocarbons could be enhanced and toxic effect of hydrocarbons on plants and soil microbes could be decreased by adding cyclodextrins.

1. INTRODUCTION

Different *in situ* and *ex situ* physical, chemical and biological technologies can be used for remediation of soils contaminated by hydrocarbons. The role of biotechnologies in soil decontamination is to upgrade biodegradation by ensuring nutrients and oxygene for the indigenous or the inoculant's microflora and by reducing limiting factors [1,2].
In case of biodegradation of hydrocarbons **bioavailability** is one of the main limitations. Hydrocarbons are strongly bound to the surface or built into the organic structure of the soil particles. The strongly absorbed hydrophobic contaminant is hardly available for the microflora which lives in the hydrophilic biofilm. Some of the biotechnologies apply surfactants to improve contact between contaminant and microflora, or such microbes, which can produce surfactants by themselves. Contaminants, due to their toxic effect, can inhibit microbial activity and plant growth [3]. Even biodegrading microbes could be inhibited by too high concentrations of toxic organic contaminants, eg. oily wastes, polyaromatic hydrocarbons (PAH) or polychlorinated biphenyls [4].
The effect of cyclodextrin (CD) on soil bioremediation and soil toxicity was investigated.

2. MATERIALS AND METHODS

Soil was contaminated with Diesel oil, mineral oil, polycyclic aromatic hydrocarbons, or transformator oil containing polychlorinated biphenyls. The soils were originated from contaminated sites or were polluted artificially.

J. Szejtli and L. Szente (eds.), Proceedings of the Eighth International Symposium on Cyclodextrons, 609–612.

The intensity of biodegradation; the count of the living- and oil degrading microbes, the amount and the composition of the residual hydrocarbon contamination were investigated on the effect of cyclodextrin addition. RAMEB (DS: 1.8, Wacker Chemie) a randomly methylated β-cyclodextrin was applied in 0.1-5 % concentration range. The investigations were carried out in 100-500 g laboratory scale experiments, with the maximal duration of 3 months.
The content of extractables was measured by gravimetry after CCl_4 extraction, their composition was determined by gaschromatography and by liquid chromatography.
The toxic effect of soil contaminated with hydrocarbons was tested by measuring the growth and dehydrogenase enzyme activity of soil microbes and by seed germination and root growth inhibition of higher plants, like *Sinapis alba* (white mostard).

3. RESULTS AND DISCUSSION

The effect of cyclodextrin addition into the soil contaminated with hydrocarbons was investigated by studying the efficiency of biodegradation and the changes in toxicity.

3.1. Effect of cyclodextrin on oil biodegradation.

Loess soil contaminated artificially was used to investigate the effect of cyclodextrin on biodegradation of Diesel-oil and mineral oil. The results of long bioremediation experiments with and without cyclodextrin were compared.

TABLE 1. Changes in soil characteristics during bioremediation of hydrocarbons with and without cyclodextrins

	Cell number [cell/g soil]		Oil degrading cells [cell/g soil]		Extract [mg/kg]	
Sample	start	end	start	end	start	end
Diesel oil+inoculant	$2.7 \cdot 10^7$	$3.1 \cdot 10^7$	10^3–10^4	$>1.1 \cdot 10^5$	46220	7100
Diesel oil+CD+inoculant	$4.3 \cdot 10^7$	$10.0 \cdot 10^7$	10^3–10^4	$>1.1 \cdot 10^5$	46290	5900
Mineral oil+inoculant	$2.6 \cdot 10^7$	$1.4 \cdot 10^7$	10^3–10^4	$9.3 \cdot 10^4$	46280	19300
Mineral oil+CD+inoculant	$1.9 \cdot 10^7$	$2.9 \cdot 10^7$	10^3–10^4	$>1.1 \cdot 10^5$	47170	17400
Cyclodextrin	$6.9 \cdot 10^6$	$1.4 \cdot 10^7$	10^3–10^4	—	4000	400
Cyclodextrin+inoculant	$4.1 \cdot 10^6$	$6.5 \cdot 10^7$	10^3–10^4	$9.3 \cdot 10^4$	4000	500

The decrease of oil content is considerable during the 2.5 months period. CD resulted in further decrease in residual oil amount and increase in cell number compared with CD untreated. Cyclodextrin itself was degraded in the soil after 2.5 months.
The results of the gaschromatographic measurements give more detailed information about the composition of the extract and its oil content. Hydrocarbon under C17 were completely degraded. Table 2. showes the amount of hydrocarbons under carbon number C25. In case of Diesel oil the chromatograms show considerable biodegradation, but no significant difference between CD treated and untreated samples could be observed. In case of mineral oil, the treatment with cyclodextrin resulted in a highly increased biodegradation of hydrocarbons under C25. In spite of the increased biodegradation only slight decrease in the amount of the extractables was measured. It suggests that two contrary processes take place: an increase in the amount of extractables and a decrease in the amount of biodegradable hydrocarbons. The explanation for both changes may be a general increase in degradational activity of the soil microflora. This may due to CD, which effects not only the contaminant but also the organic compounds of the soil matrix.

TABLE 2. Gas chromatographic characterisation of the residual oil

Sample	Cromatographed amount [mg]	CH under C_{25} [mg]	CH under C_{25} [%]	Amount of the extract [mg/kg]
Diesel-oil+inoculant	20	2.4	12.0	7100
Diesel-oil+CD+inoculant	23	2.9	12.6	5900
mineral oil+inoculant	113	2.8	2.5	19300
mineral oil+CD+inoculant	469	1.1	0.2	17400

The chromatograms of residual mineral oils show the increased biodegradation by the decreased amount of residual oil, compared to an inner standard (IS).

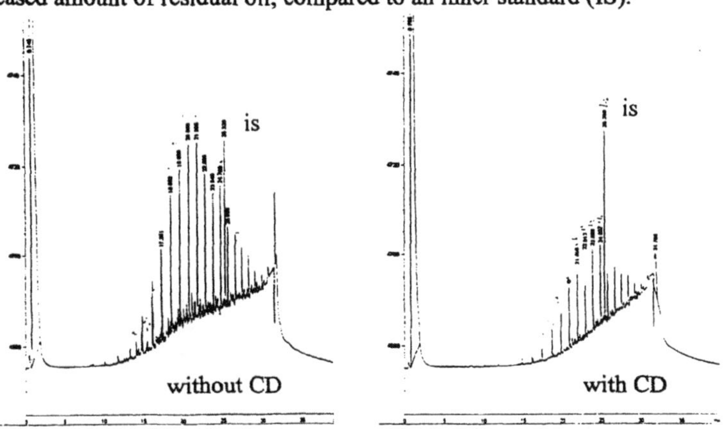

Fig. 1 Gas chromatograms of residual mineral oil, after CD enhanced biodegradation

3.2. Effect of cyclodextrin on biodegradation of PAHs in contaminated soil.
In case of soils derived from sites contaminated with PAHs the above mentioned phenomenon could be also indicated: changes in the amount of extractables does not correlate with the hydrocarbon content of the extract. The relationship depends on the soil type. The amount of the extracts could be either less or more after CD treatment, but the PAH content was always considerable lower. Also the cell number of oil degrading microbes in the soil increased to a higher value on the effect of CD.

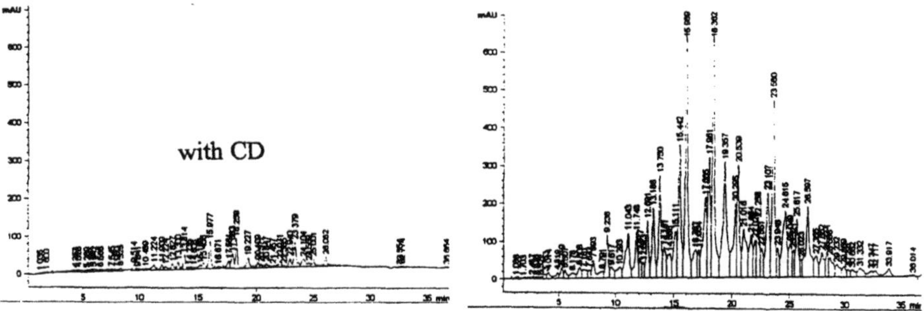

Fig. 2 HPLC of the residual polyaromatic hydrocarbons after biodegradation of soils treated and untreated with cyclodextrin

Two liquid chromatograms show the difference between the PAH content of the extracts of the untreated and the CD treated soils after 3 months biodegradation.
We suppose, that in the presence of cyclodextrin not only the contaminant, but also some components of the hardly available, structural organic matter are affected by the soil microflora, and degraded to smaller, more soluble and extractable molecules.

3.3. Effect of CD on soil toxicity: plant root growth in mineral oil and PAH contaminated soils.

In a soil contaminated with high, 30 000 ppm Diesel oil, root growth has been inhibited in a large scale: root size decreased from 50 mm to 5 mm. The addition of 0,1-5 % CD increased the root length from 5 mm to 20 mm, proportionally with CD concentration. Growth and enzyme activity of soil microbes showed the same tendency.
In case of lower contamination, 6500 ppm of transformator oil, root growth inhibition could be compensed by 0,5 % of CD. It suggests that a stochiometric relationship exists. It means that application of CD is realistic and effective in case of not too concentrated, but highly toxic contaminants.

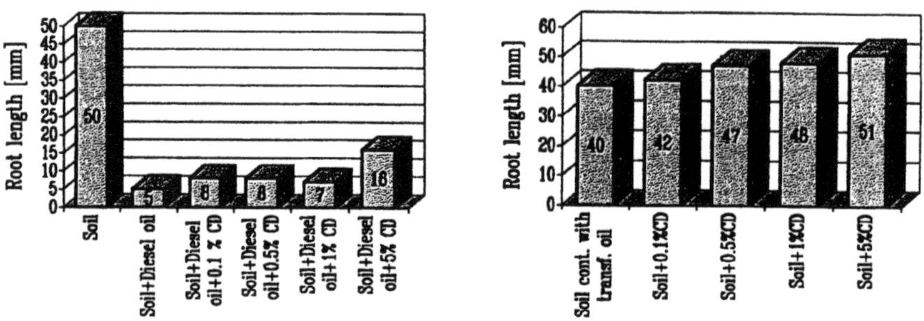

Fig. 3 Changes in the white mostard root growth inhibition in soils contaminated with Diesel oil and transformator-oil by addition of cyclodextrin

4. CONCLUSION

The application of CD in soil bioremediation resulted in an increase in the efficiency of biodegradation of hydrocarbons. Addition of cyclodextrin decreased soil toxicity, which positively influenced both degrading microflora and plants growing in contaminated soils. Both phenomena may be due to the encapsulating effect of the cyclodextrin.

REFERENCES

[1] Gruiz, K.: In-situ Bioremediation of Hydrocarbon in Soil: Pilot Tests and Field Experiments - In: Contaminated Soil '93, Eds.: F. Arendt, G.J. Annokkée, R. Bosman and W.J. van den Brink, pp. 1163-1164. Kluwer Academic Publ., The Netherlands, 1993

[2] Gruiz, K. and Kriston, É.: In-Situ Bioremediation of Hydrocarbon in Soil - *Journal of Soil Contamination* 4 (2) 163-173. 1995

[3] Horváth, B., Gruiz, K. and Sára, B.: Ecotoxicological testing of soil by four bacterial biotest - *Toxicol.Environ.Chem.* In press.

[4] Kriston, É. and Gruiz, K.: Pilot Tests for Bioremediation of Soil Contaminated with Different Types of Oil. A Comparative Study - In: Contaminated Soil '95 Eds.: W.J.van den Brink et al pp. 1271-72. Kluwer Academic Publ, 1995

PREPARATION AND CHARACTERIZATION OF PEEK-WC/β-CYCLODEXTRIN CARBONATE MEMBRANES

CECILIA PAGLIERO[2], MARIAPINA NATOLI[2],FRANCESCO TROTTA[3], AND ENRICO DRIOLI[1,2]

[1] *Institute of Research on Membranes and Modelling of Chemical Reactors*

[2] *Department of Chemical Engineering and Materials, University of Calabria, I-87036 Arcavacata di Rende (CS), Italy*

[3] *Department of Inorganic, Physical and Material Chemistry, University of Torino Via Pietro Giuria, 10125 Torino, Italy*

ABSTRACT

Membranes of PEEK/WC containing carbonate derivative β-cyclodextrin (β-CD) have been prepared and their catalytic behaviour for the p-nitrophenylacetate (PNPA) hydrolysis has been investigated. The effect of different average degree of β-CD carbonate substitution (DS), the amount of immobilized β-CD carbonate and operating stirring speeds have been examined under a constant substrate concentration and constant permeation rate. The performance and productivity of the β-CD membrane reactor indicate an improvement of CD stability.

[2]on leave from : Facultad de Quimica, Bioquimica y Farmacia, UNSL-CONICET, Chacabuco y Pedernera - 5700 San Luis - ARGENTINA

1. INTRODUCTION

In a previous study we have reported that O-octyloxycarbonyl β-cyclodextrin membrane improves the reaction rate in the PNPA hydrolysis reaction.[1].

In aqueous basic solutions cyclodextrins cleave phenyl acetates by acyl transfer from the ester to an ionized hydroxyl group of the cyclodextrin. The reaction takes place with an inclusion complex in which the phenyl group of the ester stays in the hydrophobic cavity of the CD [2].

Besides, in homogeneous system the catalytic action of CD derivatives has been found to decrease with increasing of the substitution degree [3]. Moreover, the β-CD acyclic carbonate seems to be not suitable for basic ester hydrolysis because they are not stable

J. Szejtli and L. Szente (eds.), Proceedings of the Eighth International Symposium on Cyclodextrons, 613–618.

at high pH, whereas the rate constant for the CD-catalyzed reaction is maximal at a pH of 12 - 13 [2].

However, we have found that the β-CD acyclic carbonate derivative is an efficient nucleophilic catalyst when incorporated in the PEEKWC membrane [1]. It seems to possess a high nucleophilicity towards the reagent p-nitrophenylacetate with exceptional lability of the intermediate on the reaction pathway leading to the formation of the product with a significant rate acceleration.

In the present study we have examined the influence of some variables on the catalytic performance of CD-membrane reactor such as amount of immobilized cyclodextrin, degree of carbonate substitution of cyclodextrin and different stirring speeds of reactor. We have prepared membranes based on PEEKWC and β-CD acyclic carbonate with DS 5 and 7. Three different amounts of CD derivative 2.5 wt %, 5 wt %, 7.5 wt % have been immobilized for each DS used. The CD derivative density and DS may help to understand how the cyclodextrinic catalytic properties depend on the change in the geometry of the active site and on the inclusion complex in the membrane structure. This study shows the performance of a novel design of catalytic membrane reaction in which the specific properties of a non conventional catalyst immobilized in a polymeric membrane can promote and extend new applications of these systems.

2. MATERIALS AND METHODS

2.1. Materials

PEEK-WC,poly(oxa-p-phenylene-3,3-phthalido-p-phenylenxoxa-p-phenylenenexoxi-p-phenyl ene) was supplied from the Chanchung Institute of Applied Chemistry, Academia Sinica.

β-cyclodextrin (β-CD) was supplied from Roquette Italia (Cassano Spinola - Italy). Its O-octyloxycarbonyl derivative was synthesized according to the literature and the average degree of substitution (DS 5 and 7) was determined via quantitative FT-IR analysis [4].

p-Nitrophenylacetate (PNPA) and p-nitrophenol (PNP) were purchased from commercial sources (Fluka Chemicals) and were used without purification.

2.2. Methods

The membranes were prepared following the traditional phase inversion process [5, 6] which permits the production of membranes with an asymmetric pore structure. The membrane forming system is composed of PEEKWC, O-octyloxycarbonyl β-CD derivative, DMF as solvent and water as non solvent. All components are only miscible in a concentration range between 15 wt% PEEKWC / 2.5 wt% O-octyloxycarbonyl β-CD derivative and 15 wt% PEEKWC/7.5 wt% O-octyloxycarbonyl β-CD derivative.

The solution was cast on a glass plate. The cast film (250 μm) was then immersed in a coagulation bath containing distilled water at 27 °C; the cast films were kept in water for 10 min and then transferred to fresh distilled water for 2 hours. All membranes were stored in water. The operational conditions for membrane preparation are listed in Table 1.

TABLE 1. Conditions of asymmetric PEEK-WC/β-cyclodextrin carbonate membranes (AM) preparation

Membrane	Polymer Solution % w/w	β-cyclodextrin carbonate % w/w		Solvent, DMF % w/w
		DS=7	DS=5	
AM-1	15	2.5	-	82.5
AM-2	15	5.0	-	80.0
AM-3	15	7.5	-	77.5
AM-4	15	-	2.5	82.5
AM-5	15	-	5.0	80.0

The hydraulic permeability coefficient (Lh) was calculated from measurement at 20°C of the mass flux obtained at various transmembrane pressure differences.

TABLE 2. Flux water and hydraulic permeability measurements of AM at P = 0.8 bar

Membrane	Flux water l/m^2h	Hydraulic Permeability l/m^2h bar
AM-1	154.6	193.2
AM-2	50.0	62.5
AM-3	31.8	38.6
AM-4	143.2	179.7
AM-5	45.5	56.8

The hydrolysis reaction was carried out in a phosphate buffer solution (pH = 8.4). In the typical experiment, 50 ml of phosphate buffer solution were placed in the stirred CD-membrane reactor and 0.25 ml of the standard solution of PNPA (0.02 M in acetonitrile) was added. During the reaction, samples (3 ml) of permeate and retentate were collected and concentrations of PNP was determined spectrophotometrically at 401nm with a Shimadzu UV-160A UV-VIS recording spectrophotometer, at room temperature.

3. RESULTS AND DISCUSSION

PNPA hydrolysis reaction is an ideal model reaction for its simple mechanism, widely investigated to obtain information on the catalytic and selectivity properties of CDs [7, 8, 9]; an improved reaction rate and productivity has been already observed when the reaction was carried out in the catalyzed β-CD carbonate membrane reactor in comparison to the alkaline batch reactor [1].
Figure 1 and table 3 show the reaction rate, productivity and conversion degree obtained for the five membrane reactors.

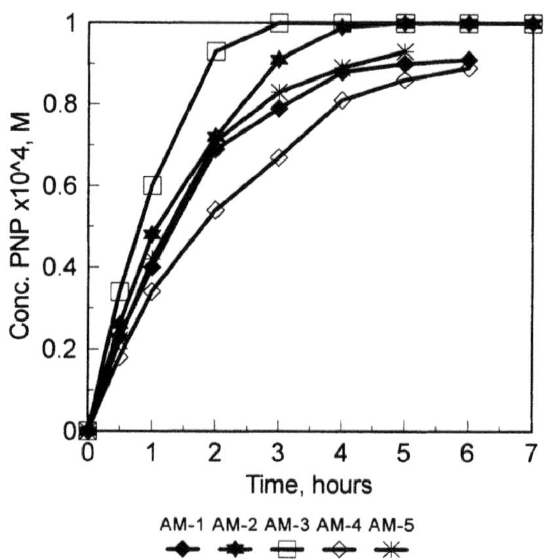

Fig. 1. PNPA hydrolysis *vs.* time carried out in a O-octyloxycarbonyl-β-CD stirred membrane reactor as a function of CD loaded membrane and DS.

We have used three concentration of CD derivative: 2.5 wt%, 5 wt% and 7.5 wt% for both CD with DS 5 and 7. In all cases the rate of reaction was measured from the quantity of PNP producted. The hydrolytic activity of the immobilized CD derivatives was linearly correlated with the increasing of CD density in the membrane, confirming a real catalytic activity of CD derivatives when entrapped in the membrane.
Results exibit that the initial reaction rate and productivity are higher for AM-1, AM-2 and AM-3 containing β-cyclodextrin-O-octyloxycarbonyl derivative with DS 7. The investigation of DS parameter has been carried out by correlating the CDs derivatives activity with their structural morphology when immobilized in the polymeric PEEKWC membrane. The higher activities of the membrane-loaded CD carbonate with DS 7 in

comparison to the membrane-loaded CD carbonate with DS 5, demonstrate that the catalytic activity is not affected by a moderate increasing in the substitution degree of the acyclic chains [9]. These results confirme the previous hypothesis that the acyclic carbonate chains are bonded preferentially to the primary hydroxyls side of the β-CD. The CD derivatives are thus able to form bonds with PEEK polymer, to include and to cleave the phenyl acetate [1].

TABLE 3. Initial reaction rate, productivity and conversion degree of membrane reactors

Membrane	Reaction rate mmol/ml·h	Productivity mmol/h after 3 hours	Conversion degree %
AM-1	0.45	4.62	77
AM-2	0.58	5.46	91
AM-3	0.74	6.00	100
AM-4	0.40	4.02	67
AM-5	0.50	4.98	83

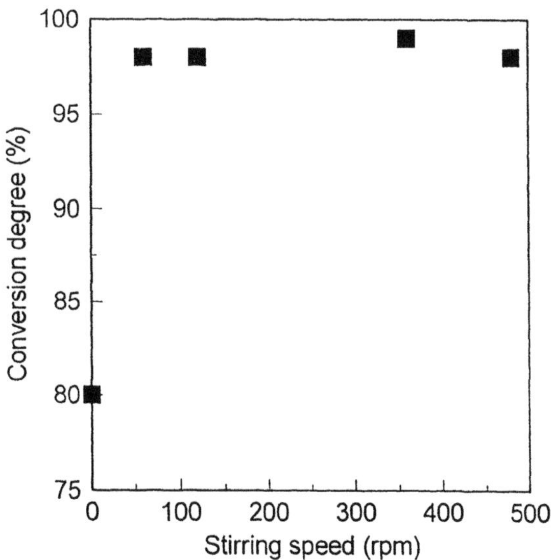

Fig. 2. Conversion degree of AM-3 reactor as a fuction of stirring speed.

The AM-3 had the highest catalytic activity; it was chosen to investigate the effects of stirring on the ester hydrolysis and on the reactor productivity. Conversion degree was measured, after five hours, at five stirring speed: 60, 120, 240, 360 and 480 rpm.
As shown in figure 2, the degree of conversion reveals that the CD-functionalized membrane does not exibit a significant changed catalytic activity as a function of the used stirring speed, although the conversion is always higher than the unstirred reactor system.

4. CONCLUSION

Immobilization in the polymeric PEEKWC membrane of CD derivatives with two different DS (5 and 7) and three different concentration (2.5 wt%, 5 wt% and 7.5 wt%) has been performed. The membranes showed a performance equal to 100% in terms of conversion degree when CDs with DS 7 and concentration of 7.5 wt% have been used. The quantification of initial reaction rates indicated that the CD-membrane activity depended on the CD loading and the maximum activity value has been obtained at a maximum CD loading.
The observed correlations between the reactor speed rate, cyclodextrin loading and productivity suggest that the optimum conditions can be obtained at a low economic impact.

REFERENCES

[1] Drioli, E., Natoli, M., Koter, I. and Trotta, F., An Experimental Study on a β-Cyclodextrin Carbonate Membrane Reactor in PNPA Hydrolysis, *Biothecnology & Bioengineering*, **46**, 415-420 (1995).

[2] Szejtli, J., Cyclodextrin inclusion complexes: Reaction catalysed by clyclodextrin, in Clyclodextrin Technology, Kluwer, Dordrecht (1988) Pp.112-117 and 366-371.

[3] Fujita, K., Shinoda, K. and Imoto, T., Selectivity variation in hydrolysis of phenyl acetates by simple modifications of β-cyclodextrin, *Tetrahed. Lett.*, **21**, 1541-1544 (1980).

[4] Trotta, F., Moraglio, G., Marzona, M. and Maritano, S., Aclyclic Carbonates of β-Cyclodextrin, *Gazz. Chim.*, **123**, 559-562 (1993).

[5] Kesting, R., Synthetic Polymeric Membranes 2nd ed., Wiley, New York (1985) P.7.

[6] Kimmerle, K. and Strathmann, H., Analysis of the Structure-Determining Process of Phase Inversion Membranes, *Desalination*, **79**, 283 (1990).

[7] Fujita, K., Akihiro, S., Taiji, I., Hydrolysis of phenyl acetates with capped β-cyclodextrins, *Bioorg. Chem.*, **4** (1980) 237-249.

[8] Kitaura, Y., Bender, M. L., Ester hydrolysis catalized by modified cyclodextrins. *Bioorg. Chem.*, **4**, 237-249 (1975).

[9] Van Etten, R. L., Sebastian, J.F., Clowes, G.A., Bender, M.L., Acceleration of phenyl ester cleavage by clycloamylose. A model of enzimatic specificity, *J. Am. Chem. Soc.*, **89**, 3242-3252 (1967).

[10] Bender, M.L., Komiyama, M., Cyclodextrin Chemistry, Springer-Verlag, Berlin (1978) Pp. 35.

β-CYCLODEXTRIN COMPLEX OF THE INSECTICIDE AZINPHOS-METHYL

M. Marzona*, R. Carpignano*, S. Girelli*, M. Dolci°

**Dipartimento di Chimica Generale e Organica Applicata - Università di Torino - Corso M. D'Azeglio 48 - 10125 Torino, Italy*
°Dipartimento di Valorizzazione e Protezione delle Risorse Agroforestali - Università di Torino - Via P. Giuria 15 - 10126 Torino, Italy

ABSTRACT

The complexation of the insecticide azinphos-methyl with β-cyclodextrin has been investigated in an attempt to develop better formulation and application methods.
An inclusion complex with a 3:2 β-cyclodextrin:azinphos-methyl molar ratio was obtained. The complex was characterized by solubility properties, elemental analysis, UV and proton NMR spectroscopy and inclusion constant. Enhancement of water solubility and suspensibility, disappearance of the unpleasant odour and reduction of dermal toxicity are the main results of the complexation. The insecticidal effectiveness of the complex, tested in the field against *Cydia pomonella* on apple trees and *Cydia molesta* on peach trees, was comparable with that of a commercial formulate of non included insecticide.

1. INTRODUCTION

Several complexations of pesticides with cyclodextrins are reported in the literature: enhancement of stability, absorption and persistency are some of the effects observed [1].
Azinphos-methyl, [S-3,4-dihydro-4-oxo-1,2,3-benzotriazin-3-ylmethyl O,O-dimethyl phosphorodithioate] (I) is a non-systemic insecticide and acaricide chiefly effective against biting and sucking insect pests.

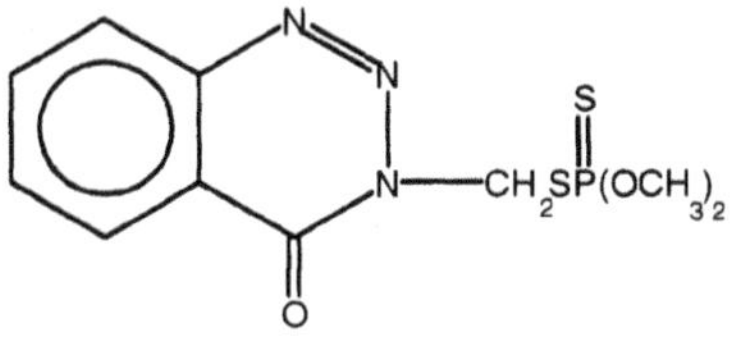

(I)

J. Szejtli and L. Szente (eds.), Proceedings of the Eighth International Symposium on Cyclodextrons, 619–622.

Formulated as a wettable powder or as an emulsifiable concentrate of strong unpleasant odour, it is widely used in our area (Piedmont , Italy) on several fruits and vegetables.
In an attempt to develop better formulations, the complexation of azinphos-methyl with β-cyclodextrin was investigated. The complex was characterized by its physico-chemical properties and tested in the field against *Cydia pomonella* on apples trees and against *Cydia molesta* on peach trees. Suspensibility, storage stability and toxicological properties were also determined.

2. MATERIALS AND METHODS

Azinphos-methyl E 103630 (99,9%) was purchased from dr. Ehrenstorfer, Augsburg (Germany). Complexation was carried out on technical azinphos-methyl (~95%) kindly supplied by Bayer.
β-cyclodextrin (>99%) (Kleptose) was purchased from Roquette S.p.A. All the other reagents were from Merck.
Spectrophotometric determinations were performed on an ATI UNICAM UV2 spectrophotometer and NMR spectra on JEOL EX400.

3. RESULTS AND DISCUSSION

3.1 Determination of stability constant

The apparent stability constant of the complex was determined in aqueous solution according to the solubility method [2] on samples shaken for 7 days at 25 °C. The concentration of the dissolved insecticide was determined, after filtration, by UV spectrophotometry. From the AL type solubility isotherm, K_d = 1758 mol^{-1} L was calculated.

3.2 Preparation and characterization of the complex

The complex was prepared by suspension method using a 2:1 β-cyclodextrin (200 g /L H_2O):azinphos-methyl molar ratio. After stirring for 3 days at ambient temperature the precipitate was filtered off, washed with diethyl ether to remove the uncomplexed insecticide and dried by lyophilization.
The complex, m.p. > 230 °C, odourless, was characterized by physical properties, NMR and UV spectra and elemental analysis.
Water solubility of azinphos-methyl at room temperature rised from 28 mg/L to 105 mg/L, with a 3.8 fold increase, which is consistent with complex formation.
UV spectra in deionized water of azinphos-methyl (λ_{max} 226 nm) and its cyclodextrin complex were essentially identical. Also the proton NMR spectra of the compound and of its complex in DMSO-d_6, compared with the proton NMR spectrum of β-

cyclodextrin, did not show significant changes in the chemical shifts of the host and the guest molecules.
Elemental analysis of the complex showed a 3:2 β-cyclodextrin:azinphos-methyl ratio, in accordance with the spectrophotometric determination of the azinphos-methyl content (15.9%).

3.3 **Storage stability**

The complex was stored solid in a dry and fresh place for 7 months: melting point, water solubility and insecticide content, determined spectroscopically, remained the same.

3.4 **Water suspensibility**

Suspensibility tests were performed according to the CIPAC MT 15.1 method. The complex, previously lyophilized, reduced to a particle size of 2-3 μm, showed a suspensibility of 84% in deionized water; in the same conditions the suspensibility of the commercial formulate Gusathion wettable powder was 96%.

3.5 **Toxicity studies**

A preliminary study on the acute oral toxicity of the complex on rats (male and female) led to determine a LD_{50} value between 35 and 70 mg/kg, corresponding to 5.6 and 11.2 mg/kg of azinphos-methyl respectively. LD_{50} reported in the literature [3] is about 10 mg/kg.
It may be concluded that oral toxicity does not change significantly in the complexation.
Acute dermal toxicity of the complex, in comparison with technical azinphos-methyl, was evaluated on Sprague Dawley rats (male and female), in compliance with OECD-GPL in the testing of chemicals. No systemic toxic effect was observed for the complex even at the dose of 4000 mg/kg, showing therefore LD_{50} higher than 4000 mg/kg. The comparative technical azinphos-methyl caused the typical sign of the acute poisoning induced by the organophosphorous compounds, without mortality, at the dose of 1784 mg/kg. The results indicate that the complexation with β-cyclodextrin reduces the dermal toxicity of azinphos-methyl.

3.6 **Insecticidal activity**

The activity of the complex was tested in the field relative to that of a commercial formulate, Gusathion 25%, water dispersible powder.
The results obained against *C. pomonella* on apple trees and against *C. molesta* on peach trees are reported in Tables 1 and 2, respectively. Symbol A refers to treatments carried out by following fixed schedules, symbol B to treatments carried out by following indications given by capture traps. Percentages followed by the same small letter are not significantly different.

TABLE 1. Comparison between the activity of β-cyclodextrin:azinphos-methyl complex and a commercial formulate (Gusathion) against *Cydia pomonella* on apple trees

Assessment date	Percentage of damaged fruits				
	Blank	Gusathion A	Complex A	Gusathion B	Complex B
26.6.95	4.25 a	0.25 b	0.25 b	0 b	0.25 b
16.8.95	32.5 a	0.5 b	2 b	2.5 b	5 b
22.9.95	36.7 a	2.2 b	5.8 bc	7.3 bc	10.4 b

TABLE 2. Comparison between the activity of β-cyclodextrin:azinphos-methyl complex and a commercial formulate (Gusathion) against *Cydia molesta* on peach trees

Assessment date	Percentage of damaged fruits		
	Blank	Gusathion	Complex
12.9.94	30 a	7.5 b	2 b

The results show that in both cases the activity of the complex was comparable with that of the commercial formulate.

4. CONCLUSIONS

The β-cyclodextrin complex of the insecticide azinphos-methyl with a molecular ratio 3:2 has been prepared and characterized.
Complexation resulted in enhancement of water solubility and suspensibility, high storage stability, disappearance of the unpleasant odour and reduction of dermal toxicity. The complex, tested in the field against *C. pomonella* on apples trees and *C. molesta* on peach trees showed an activity comparable with that of a commercial formulate of non included insecticide.

ACKNOWLEDGEMENTS

The research has been carried out on behalf of Regione Piemonte, Assessorato alla tutela dell'Ambiente.

REFERENCES

[1] Szejtli J., Cyclodextrin Technology, Kluwer Academic Publishers, Dordrecht, 1988
[2] Higuchi T., Connors K.A., Phase solubility techniques, *Adv. Anal. Chem. Instr.*, **4,** 68-72 (1965)
[3] Worthing R.C., Hance R.J. ed., The Pesticide Manual, 9th ed., The British Crop Protection Council, Farnham, 1991

PREPARATION AND CHARACTERIZATION OF 2,4-D COMPLEXES WITH α-CYCLODEXTRIN

J.I. PEREZ-MARTINEZ, J.M. GINES, M.J. ARIAS, J.R. MOYANO, E. MORILLO*, A. RUIZ-CONDE#, P.J. SANCHEZ-SOTO#, C. NOVAK@

Departamento de Farmacia y Tecnología Farmacéutica. Facultad de Farmacia. Universidad de Sevilla. 41012 - Sevilla, Spain

**Instituto de Recursos Naturales y Agrobiología, C.S.I.C., Apdo. 1052, 41080 - Sevilla, Spain*

#Instituto de Ciencias de Materiales. Centro Mixto C.S.I.C. - Universidad de Sevilla, Apdo. 1115. 41080 - Sevilla, Spain.

@Technical University of Budapest. Institute of General and Analytical Chemistry. St. Gellert ter 4. Budapest, H-1521, Hungary.

ABSTRACT

The interaction of the herbicide 2,4-D with α-CD investigated by phase solubility, XRD and DSC indicate that 2,4-D forms an inclusion complex with α-CD by coprecipitation and spray-drying methods. These complexes exist in solution with 1:1 and 1:2 stoichiometry and in the solid phase with 1:2 stoichiometry. The calculated values of stability constant were $K_{1:1}$ and $K_{1:2}$ were: $K_{1:1} = 94.5\ 10^{-3}$ mM^{-1} and $K_{1:2} = 3.48\ 10^{-3}$ mM^{-2}.

1. INTRODUCTION

In recents years, the CD have received considerable attention in various applied fields, such as biotechnology, organic chemistry, drugs, foods and pesticides. In particular, inclusion complexes using α-CD are gaining very wide use in preparative and applied chemistry, besides agriculture and pharmaceutical industry.The principal objective of the present work is to investigate the possibility of improving the aqueous solubility and dissolution properties of a systemic herbicide [1], the 2,4-dichlorophenoxyacetic acid (2,4-D) *via* complexation or inclusion with α-CD. The formation of such inclusion compounds or complexes is confirmed by a variety of methods and techniques, such as solubility determination, differential scanning calorimetry (DSC) and X-ray diffraction (XRD).

J. Szejtli and L. Szente (eds.), Proceedings of the Eighth International Symposium on Cyclodextrons, 623–626.

2. MATERIALS AND METHODS

2.1. Materials

2,4-D (purity 99 %) was supplied by Sigma (St. Louis, Missouri, USA) and α-CD (99 %) by Roquette (Lestrem, France).

2.2. Methods

Preparation of solid inclusions: The preparation of solid complexes of 2,4-D with α-CD was performed by three methods: coprecipitation, kneading and spray-drying.

Phase solubility determinations: The solubility determinations were performed according to the method reported by Higuchi and Connors [2].

X-Ray Diffractometry (XRD): X-ray powder diffraction diagrams of different samples were obtained using a X-ray diffractometer Siemens, model Kristalloflex D-500.

Differential Scanning Calorimetry (DSC): DSC was carried out with a Mettler apparatus with FP85 furnace, FP80 HT temperature control unit and FP89 HT software under static air atmosphere, at a heating rate of 10 °C / min.

3. RESULTS AND DISCUSSION

3.1. Inclusion complexation in solution:

According to Higuchi and Connors [2], the phase solubility diagram, can be classified as B_s type. An insoluble complex was formed.

The diagram shows a plateau region before the descending part of the curve. On the basis of the length of the plateau, it is possible to estimate the stoichiometry of the complex precipitating from the solutions. Therefore, the 2,4-D content of the complex formed in the plateau region, $[S]_C$, is equal to the total 2,4-D added to system $[S]_T$, minus 2,4-D in solution at the start of the plateau region $[S]_S$. From, the data of Fig. 2, it follows

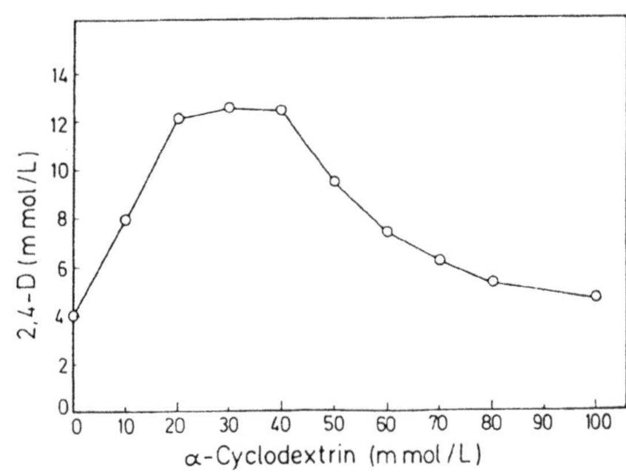

Fig. 1.- Phase solubility diagram of 2,4-D-α-CD.

$$[S]_C = [S]_T - [S]_S = 10.51 \text{ mM}$$

The α-CD content of the complex in the same region, $[L]_C$, is equal to that entering into the complex during this interval, $[L]_C$ = 40 mM - 20 mM = 20 mM.

The stoichiometric ratio of complex 2,4-D-α-CD is then

$$[S]_C / [L]_C = 10.51 \text{ mM}/20\text{mM} = 0.5$$

thus indicating a 1:2 stoichiometric ratio. Consequently, the precipitated complex has the formula S_1L_2, indicating that two α-CD molecules are available for the inclusion of one 2,4-D molecule, because of the small cavity size.

On the other hand, the solubility of α-CD complex estimated from the initial straight line portion (8 mM) is significantly different from that at high α-CD concentrations (4mM). This results can be interpreted in accordance with others authors [3] on the basis of other complex formation with 1:1 stoichiometry. Certainly, if the complex responsible for the initial rise in the solubility diagram is the same complex finally precipitated, the increase in 2,4-D concentration must be equal to the final 2,4-D concentration. This condition is not observed for the 2,4-D - α-CD system because these concentrations are 8mM and 4mM respectively (see Fig. 1). These results suggest that in the system must be involved the formation of two or more distinct complexes, one or two of these being responsible for the initial rise in solubility, while another is precipitated at the later stages, as shown the phase solubility diagram .

A more adequate description of the initial rise in solubility diagram (B_S type) may be achieved assuming the formation of the two complexes SL and SL_2 characterized by $K_{1:1}$ and $K_{1:2}$ stability constants, according to the following equilibrium:

$$S + L \overset{K_{1:1}}{\rightleftarrows} SL + L \overset{K_{1:2}}{\rightleftarrows} SL_2$$

where [S], [L], [SL], and $[SL_2]$ are the solubility of the 2,4-D, α-CD, 1:1 complex and 1:2 complex respectively, and $K_{1:1} = [SL] / [S] [L]$ and $K_{1:2} = [SL_2] / [SL] [L]$.

The material balance equations are $S_T = [S] + [SL] + [SL_2]$ and $L_T = [L] + [SL] + [SL_2]$ where S_T and L_T are the total concentrations of substrate and ligand, respectively.

$[S] = S_0$, is the solubility equilibrium of 2,4-D in absence of CD

Combinating the above equations and assuming that the extent of complexation is fairly small, it may be permissible to set $[L] \approx [L_T]$, and obtain the following equation, according to [4]

$$(S_T - S_0) / L_T = K_{1:1} S_0 + K_{1:1} K_{1:2} S_0 L_T$$

A plot of $(S_T - S_0) / L_T$ against L_T should be linear. From the slope and intercept, may calculated $K_{1:1} = 94.5\ 10^{-3}\ \text{mM}^{-1}$ and $K_{1:2} = 3.48\ 10^{-3}\ \text{mM}^{-2}$

3.2. Inclusion complexation in the solid state:

3.2.1. D.S.C.

Supporting evidence for complex formation was obtained from thermal analysis studies (Fig. 2). The thermograms of kneaded and the physical mixture of 2,4-D with α-CD show an endothermic peak at about 140 °C, corresponding to the melting point of the 2,4-D. In contrast, this peak was not observed in the case of coprecipitated and spray-dried complex. These results evidence that the coprecipitated and spray-dried samples are true inclusion complexes.

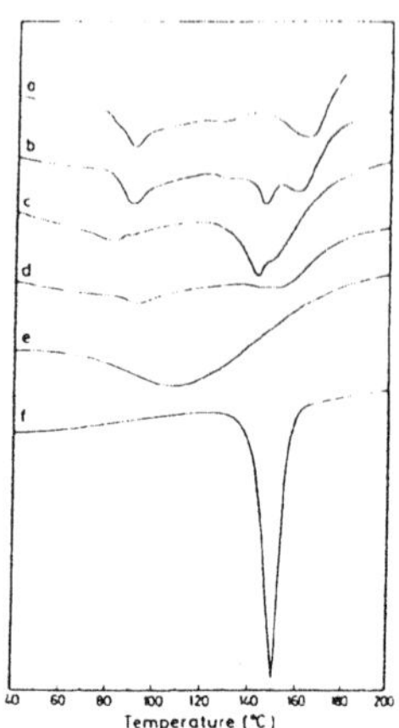

Fig. 2.- DSC of a) α-CD; b) physical mixture; c) kneaded; d) coprecipitated; e) spray-dried and f) 2,4-D.

3.2.1. XRD

The diffraction pattern (Fig. 3) of the physical mixture and the kneaded system are simply the superposition of each isolated component. In contrast, the coprecipitated and spray-dried samples exhibits a different diffraction pattern with very few peaks. This indicates that these systems are less crystalline than the isolated compounds. The above results support that 2,4-D and α-CD form a true inclusion complex in solid state, suggesting that a new solid phase is formed in the coprecipitate and spray-dried products.

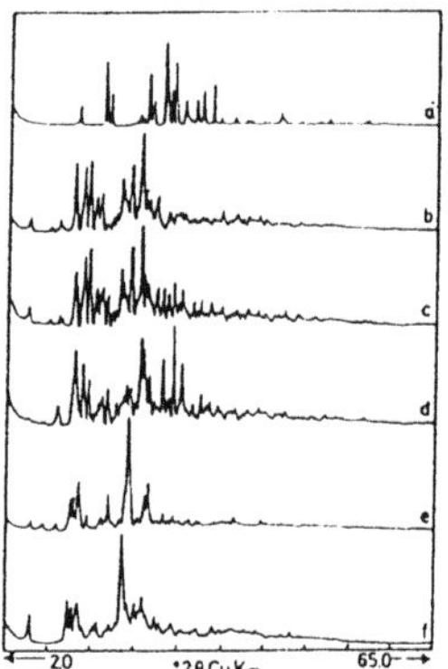

Fig. 3.- XRD of a) 2,4-D, b) α-CD, c) physical mixture, d) kneaded, e) coprecipitated *and f) spray-dried.*

REFERENCES

1. Hermosín M.C., Ulibarri M.A., Mansour M. and Cornejo J., *Fresenius Environ. Bull.*, 1993, **1**, 472.
2. Higuchi T. and Connors K.A., *Adv. Anal. Chem. Instr.*, 1965, **4**, 117.
3. Kikuchi M. , Uemura Y., Hirayama F., Otagiri M. and Uekama K. *J. Inclus. Phenom.*, 1984, **2**, 623.
4. Duchêne D., *Cyclodextrins and their industrial uses*, Ed. de santé. Paris, 1987.

MOLECULAR CALCULATIONS AND THERMODYNAMICAL CONSIDERATIONS ON THE SOLUBILITY ENHANCEMENT OF TRIFLUMIZOLE BY CYCLODEXTRIN COMPLEXATION

G. KÖHLER, H. VIERNSTEIN, P. WOLSCHANN

Institut für Pharmazeutische Technologie und Institut für Theoretische Chemie und Strahlenchemie der Universität Wien

A-1090 Wien, Althanstraße 14, Österreich

ABSTRACT

Complexation of pesticides (fungicides, herbicides, insecticides, etc.) with cyclodextrins results in advantageous modifications of their properties like e.g. enhanced solubility and therefore a better activity combined with a reduced application concentration. The activity of the fungicide Triflumizole encapsulated in β-cyclodextrin was investigated and thermodynamical parameters of the association process were determined.

1. INTRODUCTION

The solubilizing effect of various cyclodextrins is of particular interest in the pharmaceutical industry and chemical technology. Solubility enhancement of pesticides and improved activity is caused by complexation of mostly hydrophobic compounds with convenient host molecules. Such inclusion complexes are studied widely [1-4]. The stability of the inclusion complexes is determined by the molecular shape of the guest molecule and the complementary molecular surface of the cavity additional to attractive van der Waals and electrostatic forces. As a quantitative measure of the complex stability the association constant K can be taken together

J. Szejtli and L. Szente (eds.), Proceedings of the Eighth International Symposium on Cyclodextrons, 627–631.

with the corresponding thermodynamical parameters. The solvent is of main importance for the building and the stability of the inclusion complexes.

Triflumizole (TF), (Picture 1), is a fungicidal substance with systemical effect against plant mycosis like powder mildew, rust and scab (black spot-desease) on cereals, arboricultures and vinicultures.

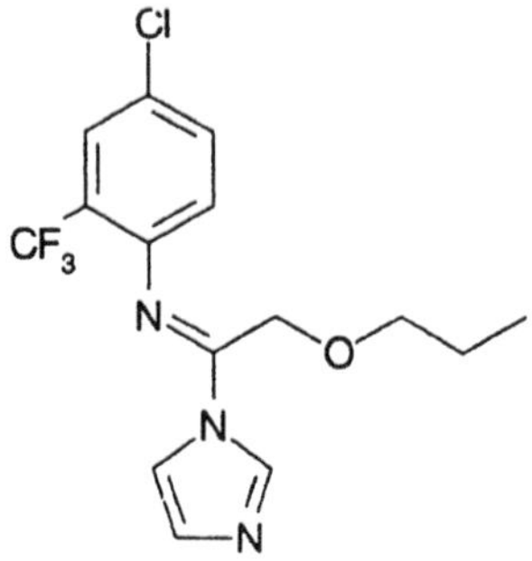

Cl

CF_3

N

O

N

N

Picture 1: Triflumizole

Due to the poor wettability and extremely low solubility in water Triflumizole is to be applicated on plants as suspension. To improve the solubility as well as the activity of the conazole fungicide solid Triflumizole/β-cyclodextrin complexes were prepared by precipitation in a molar ratio of 1:2, leading to complete encapsulation of the active ingredient (AI).

2. MATERIALS AND METHODS

Triflumizole, (E)-4-chloro-α,α,α-trifluoro-N-(1-imidazol-1-yl-2-propoxy-ethylidene)-*o*-toluidine (JUPAC), (E)-1-[1-[[4-chloro-2-(tri-fluoro-methyl)phenyl]-imino]-2-propoxyethyl]-1H-imidazole (C.A.), CAS Nr. *99 387-89-0,* is a systemic fungizide and has a curative and protective action against Gymnosporangium and Venturia spp. in pome fruits, against powdery Erysiphaceae in fruits and vegetables, against Fusarium, Fulvia and Monilinia spp. as well as Helmintosporium, Tilletia and Ustilago spp. in cereals [3]. The compound was obtained from Nippon Soda Co. Ltd. (Japan) with a purity of >99%. It forms colourless crystals, m.p. 63,5º, v.p. 0.0014 mPa (25ºC); solubilities (20ºC): in water: 12,5mg/l; in trichloromethane: 2.22g/l; hexane 17.6g/l. β-cyclodextrin was provided by Roquette Fréres (Lestrem, France) as Kleptose® with a humidity of 14%(w/w).

The stability constant of the association complex between Triflumizole and β-cyclodextrin was estimated by different methods: By an iterative procedure, where calculated and experimentally determined extinctions at various concentrations are fitted together [5], and by another method based on the monitoring of changes in solubility of Triflumizole by addition of various concentrations of β-cyclodextrin. These measurements were carried out by suspending excess amounts of Triflumizole in solutions of β-cyclodextrin in the concentration range of 4.10^{-4} to 6.10^{-3}M. After stirring the samples at constant room temperature for some hours, the concentration of dissolved Triflumizole was determined again by electron absorption spectroscopy.

3. RESULTS AND DISCUSSION

The saturation concentration of the compound in water was estimated as $4,2.10^{-5}$ mol/l. The electronic absorption spectra of Triflumizole and the association complex are different, aqueous solutions of Triflumizole show absorptions of 234,5nm (ε=25800) and 293,8 (ε=4050), which shift to 362,2nm (ε=25400) and 3990nm (as shoulder) for the inclusion complex. The respective association constant determined by different methods was estimated to 470±20 M^{-1} in aqueous solution. The overall reaction enthalpy and the overall reaction entropy were calculated from the temperature dependence of the equilibrium constant (ΔH = -22,4 kJ/mol; ΔS = -24 J/K.mol). Small amounts of organic cosolvents like ethanol or dioxane diminish the association constant drastically parallel to an enhanced solubility of pure Triflumizole. The driving force of the inclusion of Triflumizole into the hydrophobic interior of β-cyclodextrin is therefore the hydrophobicity of the substrate. Complexation decreases the total hydrophobic surface compared to the isolated molecules and leads to a rather stable associate.

The structures of the host-guest complex were estimated by molecular mechanics calculations, using the MM3 force field [5-7]. A broad variety of conformations can be observed for the cyclodextrin rim, due to the single bonds between the glucose subunits, which allow some deformation motions of the cone. The flexibility of the cyclodextrin ring structures as well as of the association complexes were analyzed by a stochastical search [4]. The association process itself was investigated by modified simulated annealing procedures [4, 8].

The fungicide activity of the Triflumizole/β-cyclodextrin complex was investigated on apples on small plot trials with 4 replications. The treatment was done with a 20 l handsprayer 7 times in the growing period using 0,5 to 0,9 l water per tree. After the growing period the apples were harvested and representative samples were taken to determine the attack of scab. The biological activity of β-cyclodextrin containing preparations against leaf scab on appletrees was also tested in comparison to conventional commercial products. The results of complex activity tests against fruit scab on apples and against leaf scab on appletrees are shown in tables 1 and 2.

	1. Replication %attached (total)	2. Replication %attached (total)	3. Replication %attached (total)	4. Replication %attached (total)	Average %attached (total)
Triflumizole/β-CD-complex (10mg% AI)	1,36 (661)	1,18 (680)	1,48 (677)	1,27 (551)	1,32 (643)
Condor ® (12,5mg% AI)	3,77 (531)	1,71 (702)	2,99 (703)	0 (449)	2,12 (596)
untreated	15,5 (625)	-	-	-	15,5 (625)

Table 1: Activity against fruit scab on apples *(Arlet, Gold Delicius, Jonagold, Empire)*

	1. Replication % leaf surface attached	2. Replication % leaf surface attached	3. Replication % leaf surface attached	Average % leaf surface attached
Triflumizole/β-CD-complex (10mg% AI)	1,15	0,17	0,27	0,53
Tank mix (35mg% AI)	0,57	0,26	0,83	0,55
untreated	17,93	18,10	17,75	17,93

Table 2: Activity against leaf scab on appletrees *(Gold Delicius)* Tank mix = 0,0125% Triflumizole + 0,0225% Dodine (standard mix)

The in vivo investigations demonstrate that clathrates have the same activity against leaf scab at a level of 0,01% AI as a standard tank mix of Triflumizole and Dodine with a combined AI rate of 0,035%.

By applicating Triflumizole/β-cyclodextrin complexes against fruit scab on different sorts of apples again complexes showed a higher fungicide activity compared with a standard formulation.

The data from these studies suggest that a treatment with fungicides as cyclodextrin complexes can save drug substance; as fungicide/cyclodextrin association complexes show higher biological activity than the pure drug, ecological as well as economical benefits can be achieved by using low pesticide doses in plant-protecting formulations.

4. CONCLUSION

The fungicide Triflumizole forms inclusion complexes with β-cyclodextrin caused by the hydrophobicity of the compound and the interior of the host molecule. The equilibrium constant K of the association reaction is in such a order of magnitude, that the solubility of Triflumizole in water is considerably enhanced and that the availbility of the free compound is given. These properties make the association complex useful for a commercial application as fungicide, which has been proven by extensive activity tests.

ACKNOWLEDGEMENTS

The authors want to thank Fa. Kwizda for financial support.

REFERENCES

[1] Szente, L., Szejtli, J., Cyclodextrins in Pesticides, in *Comprehensive Supramolecular Chemistry*, Vol.3, Pergamon 1996

[2] Viernstein, H., Reiter-Mondik, S., Wolschann, P., Solubility enhancement of Triflumizole by host-guest interaction with β-cyclodextrin, *Chemical Monthly* **125**, 681-689 (1994)

[3] Reiter-Mondik, S., Preparation and Characterization of Fungicide-Cyclodextrin systems; Thesis, University of Vienna, Inst. of Pharm. Technology (1995)

[4] Klein, Ch., Köhler, G., Mayer, B., Mraz, K., Reiter-Mondik, S., Viernstein, H., Wolschann, P., Solubility and Molecular Modeling of Triflumizole-β-cyclodextrin Inclusion Complexes, *J. Incl. Phenomena* **22**, 15-32 (1995)

[5] Burkert, U., Allinger, N.L., *Molecular Mechanics*, ACS Monograph 177, Washington DC (1982)

[6] Allinger, N.L., *QCPE Buuletin*, (1989)

[7] Allinger, N.L., Yuh H.Y., Lii, Jenn-Huei, Molecular Mechanics. the MM3 Force Field for Hadrocarbons. *J. Amer.Chem.Soc.* **111**, 8551-8575

[8] Marconi, G., Monti, S., Mayer, B., Köhler, G., Circular Dichroism of Methylated Phenols Included in β-Cyclodextrin. an Experimental and Theoretical Study, *J.Phys.Chem.*, **99**, 3943-3950 (1995)

Chapter 5.

ANALYTICAL SEPARATIONS BY CYCLODEXTRINS

USE OF CYCLODEXTRINS OR THEIR DERIVATIVES IN CAPILLARY ELECTROPHORESIS AND HIGH PERFORMANCE LIQUID CHROMATOGRAPHY

Salvatore Fanali

Istituto di Cromatografia del C.N.R., Area della Ricerca di Roma, P.O.Box 10 - 00016 Monterotondo Scalo, Roma, Italy.

Abstract
The resolution of enantiomers represents an interesting topic of research, especially in the pharmaceutical field. Their separation can be performed using several analytical techniques and among them high performance liquid chromatography (HPLC) and capillary electrophoresis (CE) are the most popular at this moment. Cyclodextrins or their derivatives firstly successful employed in gas chromatography (GC) and liquid chromatography (LC) have been widely studied in CE and resulted to be good resolving agents for a wide number of analytes, mainly of pharmaceutical interest. In this paper the basic principles of enantiomers separation by CE, the parameters affecting the enantioselectivity and resolution are discussed. A comparison between CE and HPLC for optical isomers separation is done.

1 Introduction

Capillary electrophoresis (CE) is a recent analytical technique with high efficiency and high resolution power that allow separations in very short time. This is obtained using a relative high electric field (200-1000 V/cm) applied to electrolytes (BGE) contained in narrow fused silica tubes.

In the electrophoretic process analytes are injected as a narrow zone and the component of the mixture are moving toward the detector under the influence of the electric field; the separation is achieved only if analytes possess different velocity (different electrophoretic mobility).

The resolution of two compounds (1 and 2) depends from efficiency N and selectivity $\frac{\Delta \mu}{\mu_m}$ where $\Delta\mu = \mu_2 - \mu_1$, $\mu_m = \frac{\mu_2 + \mu_1}{2}$, μ the electrophoretic mobility.

$$R = 0.25 \sqrt{N} \frac{\Delta \mu}{\mu_m} \qquad (1)$$

From equation 1 it is clearly shown that the selectivity plays a stronger influence than efficiency on the resolution. Thus, in CE, several studies have been focused on the improvement of such parameters in order to optimize the separation of compounds with similar physico-chemical properties.

J. Szejtli and L. Szente (eds.), Proceedings of the Eighth International Symposium on Cyclodextrons, 635–640.

2. MATERIALS AND METHODS

Permethylated β-cyclodextrin bonded to dimethylpolysiloxane via a 6-mono-octamethylene spacer (CHIRASIL-DEX) (cf. Fig. 1) has been prepared as described previously in detail [2]. The chromatographic investigations were performed with commercially available instrumentation.

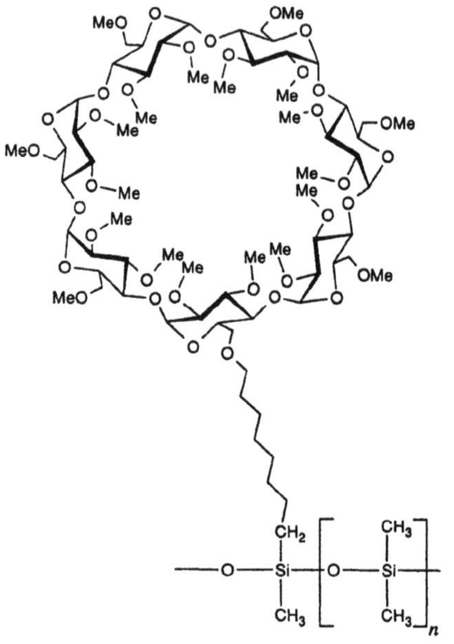

Fig. 1. Structure of CHIRASIL-DEX

3. RESULTS AND DISCUSSION

3.1. Unified enantioselective capillary chromatography

The GOLAY equation of the chromatographic process in a coated capillary predicts that the optimum efficiency (H_{min}) of a capillary column is independent of the nature of the mobile phase while the optimum velocity (v_{opt}) depends solely on the diffusion coefficient (GC > SFC > OTLC ~ CEC) (cf. Fig. 2). Thus, toward the goal of *unified enantioselective capillary chromatography* we demonstrate that, *e.g.*, hexobarbital can be separated into enantiomers employing a single 1m x 0.05 mm (i.d.) capillary column coated with immobilized CHIRASIL-DEX (film thickness 250 nm) [2] by four independent methods, *i.e.* by GC, SFC, OTLC and CEC (Fig. 3) [3].

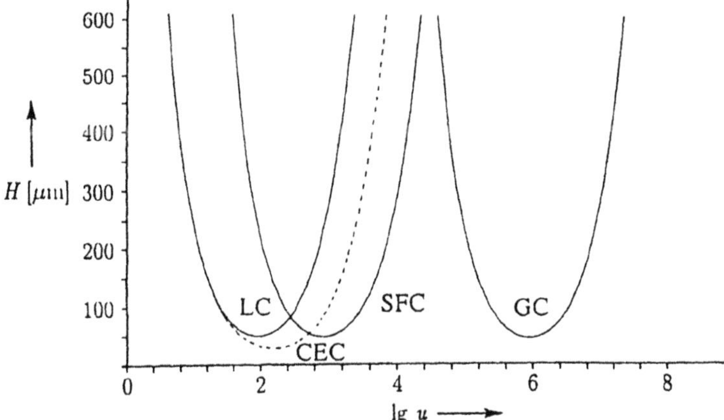

Fig. 2. Height equivalent of a theoretical plate H versus *log* of mobile phase velocity v (neglecting C') for an 0.05 mm (i.d.) capillary ($k = 5$, $D_m(\mu m^2 sec^{-1}) = 10^7(GC), 10^4(SFC), 10^3(OTLC, CEC)$, $f(k)_{parabol} = (1 + 6k + 11k^2)/(96(1 + k)^2)$ [note that in CEC, H_{min} and v_{opt} are different due to the nearly flat peak profile (dotted curve), $f(k)_{flat} = k^2/(16(1 + k)^2)$].

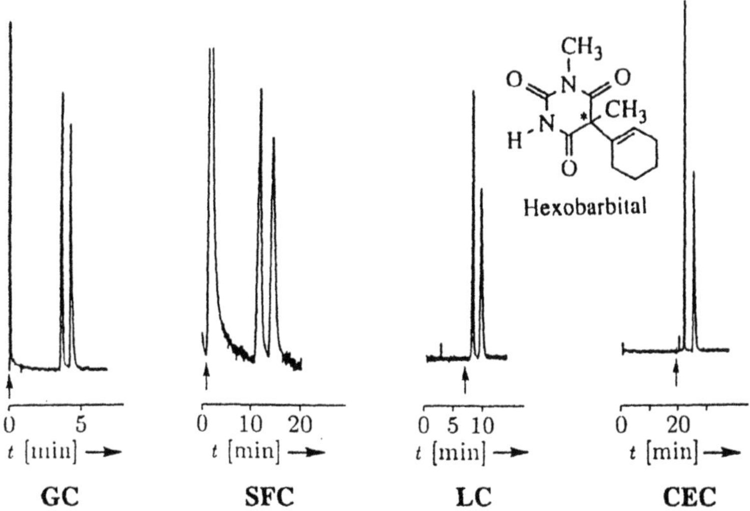

Fig. 3. Enantiomer separation of hexobarbital on a 1 m x 0.05 mm fused silica column coated with CHIRASIL-DEX (film thickness 250 nm) by GC, SFC, OTLC and CEC [3].
The arrow in the chromatograms indicates the dead time t_M.

For hexobarbital, inspection of Fig. 3 shows that CEC is superior to GC, SFC and OTLC in respect to all parameters important in enantiomer separation, *i.e.*, the chiral separation factor α, peak resolution R_S and efficiency N. Enantiomer separation by capillary electrochromatography (CEC) [4] on a chiral stationary phase merits further comments. Thus, $\alpha = 3$ (Fig. 3, right) is the highest value ever observed for CHIRASIL-DEX. The retention factor k is very low, *e.g.*, k = 0.1, for the first eluted peak resulting in a further improvement of both H_{min} (efficiency) and v_{opt} (speed of analysis).

3.2. Miniaturization

Supported by theoretical considerations it is anticipated that OTLC and CEC employing chiral stationary phases will strongly benefit from smaller column diameters. While the merit of miniaturization in the GC mode is already evident in Fig. 3 (left), further reduction of column length and film thickness of CHIRASIL-DEX allows enantiomer separations in less than one minute for many chiral analytes. The decrease of sample capacity (injection) is outweighed by an increase in the signal to noise ratio (detection).

3.3. Effect of cyclodextrin dilution

In CHIRASIL-DEX, the cyclodextrin selector resides in an apolar environment, thus decreasing analysis times for very polar compounds. Contrary to undiluted CD stationary phases, the chiral separation factor α_{dil} (R/S denote enantiomers) is concentration-dependent in CHIRASIL-DEX coated columns according (1) [5]

$$\alpha_{dil} = \frac{K_R\ m + 1}{K_S\ m + 1} = \frac{R'_R + 1}{R'_S + 1} \qquad \text{and } R' = \frac{r}{r_o} - 1$$

with K = association constant, m = molality of CD in polysiloxane, R' = retention-increase or chemical capacity factor, r = rel. retention on CD column, r_o = rel. retention on pure polysiloxane.

Thus, the α vs. CD concentration curves level off at higher weight percentages. The optimum is already reached at low CD concentrations when the chemical association is strong (*i.e.*, large K). The relationship between α and the weight of the CD in the polysiloxane matrix implies that any improvement of enantioselectivity would only be gained in case of weakly associated analytes. CHIRASIL-DEX used in this work typically contains 24 % (w/w) permethylated CD (m = 0.22). Thus, statistically one out of 60 silicon atoms in the polymer chain carries a CD moiety.

3.4. Multidimensional techniques

The prerequisite of CHIRASIL-DEX in unified enantioselective chromatography its immobilizability on the fused silica surface. This feature is also important when utilizing highly sensitive detection devices (GC-MS(SIM), ECD). Here, the multidimensional GC analysis of chiral atropisomeric polychlorinated biphenyls (PCBs) is demonstrated as an example [6]. In Fig. 4 'heartcutting' of the congeners PCB 132 and 153 from a commercial Clophen A 60 mixture and subsequent separation on CHIRASIL-DEX is shown. This method allows the screening of chiral PCBs in biological matrices for expected enantiomeric bias. Significantly, PCB 132 in mother liquor was found to deviate from the expected 1:1 enantiomeric ratio [7].

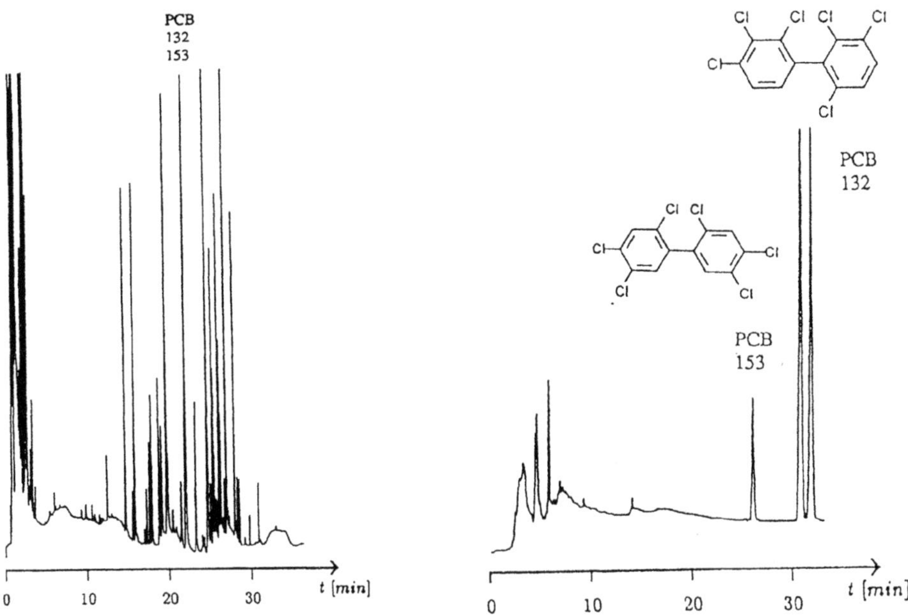

Fig. 4. MDGC of a PCB mixture and 'heart-cutting' of PCB 132 (stable enantiomers) and PCB 153. Left: *achiral column* (30 m x 0.22 mm fused silica coated with DB5). Right: *chiral column* (10 m x 0.25 mm fused silica coated with immobilized CHIRASIL-DEX (film thickness 250 nm).

3.5. High performance liquid chromatography

CHIRASIL-DEX coated on silica can also be employed in the normal HPLC [3] and micro-packed HPLC mode. The coating of the surface by a polysiloxane-anchored cyclodextrin moiety represents a new development involving the effective blocking of polar sites at the support.

4. CONCLUSION

In conclusion, immobilized CHIRASIL-DEX can be employed as universal stationary phase in chromatography. The chiral polymer shows extended lifetimes and is configurationally and thermally stable at least up to 250°C. Incidentally, it may also be used as a chiral polymer coating for sensor and transducer devices.

ACKNOWLEDGEMENTS

The author thanks Deutsche Forschungsgemeinschaft and Fonds der chemischen Industrie for financial support. The author is indebted to his co-workers named in the references for their invaluable contribution to inclusion chromatography involving cyclodextrin derivatives.

REFERENCES

[1] Schurig, V., Nowotny, H.-P.
Gas chromatographic separation of enantiomers on cyclodextrin derivatives
Angew. Chem., Int. Ed. Engl., **29**, 939 (1990)
and references quoted therein

[2] Jung, M., Schurig, V.
Extending the scope of enantiomer separation by capillary supercritical fluid chromatography on immobilized polysiloxane-anchored permethyl-cyclodextrin
J. High Resolut. Chromatogr., **16**, 215 (1993)

[3] Schurig, V., Jung, M., Mayer, S., Fluck, M., Negura, S., Jakubetz, H.
Unified enantioselective capillary chromatography on a Chirasil-Dex stationary phase – advantages of column miniaturization
J. Chromatogr., A, **694**, 119 (1995)

[4] Mayer, S., Schurig, V.
Enantiomer separation by electrochromatography in open tubular columns coated with Chirasil-Dex
J. Liquid Chromatogr., **16**, 915 (1993)

[5] Jung, M., Schmalzing, D., Schurig, V.
Theoretical approach to the gas chromatographic separation of enantiomers on dissolved cyclodextrin derivatives
J. Chromatogr., **552**, 43 (1991)

[6] Schurig, V., Glausch, A.
Enantiomer separation of atropisomeric polychlorinated biphenyls (PCBs) by gas chromatography on Chirasil-Dex
Naturwissenschaften, **80**, 468 (1993)

[7] Glausch, A., Hahn, J., Schurig, V.
Enantioselective determination of chiral 2,2',3,3',4,6'-hexachlorobiphenyl (PCB 132) in human milk samples by multidimensional gas chromatography electron capture detection and by mass spectrometry
Chemosphere, **30**, 2079 (1995)

UNIFIED ENANTIOSELECTIVE CHROMATOGRAPHY INVOLVING A PERMETHYLATED β-CYCLODEXTRIN BONDED TO A DIMETHYLPOLYSILOXANE (CHIRASIL-DEX)

V. SCHURIG

Institute of Organic Chemistry of the University

Auf der Morgenstelle 18, D-72076 Tübingen, Germany

ABSTRACT

In CHIRASIL-DEX, permethylated β-cyclodextrin is bonded to a dimethylpolysiloxane via a 6-mono-octamethylene spacer. The polymer, when coated and immobilized onto the interiour surface of fused silica capillaries, permits the separation of enantiomers by gas chromatography, supercritical fluid chromatography, open-tubular liquid chromatography and electrochromatography.

Toward an unified approach, a single column (1 m x 0.05 mm) can be employed for the enantiomer separation of hexobarbital on CHIRASIL-DEX by all four methods. CHIRASIL-DEX can also be coated onto silica and used for enantiomer separation by conventional and capillary HPLC.

1. INTRODUCTION

Cyclodextrin (CD) derivatives are useful chiral stationary phases for enantiomer separation by various methods of chromatography [1]. Permethylated β-cyclodextrin bonded to dimethylpolysiloxane via a 6-mono-octamethylene spacer (CHIRASIL-DEX) (cf. Fig. 1) can be coated and immobilized on the interiour surface of narrow bore fused silica capillaries.

The use of these columns for the enantiomer separation by gas chromatography (GC), supercritical fluid chromatography (SFC), open-tubular liquid chromatography (OTLC) and capillary electrochromatography (CEC) will be outlined.

J. Szejtli and L. Szente (eds.), Proceedings of the Eighth International Symposium on Cyclodextrons, 641–647.

Among the different approaches done we would like to remark some of them that can serve here as examples.

1) The separation of oligonucleotides is a widely studied topic and has been performed adding to the BGE sieving media (polyacrylamide, agarose etc.), here the separation is based on the mass difference because due to the similar mass/charge ratio, oligonucleotides possess similar electrophoretic mobility.

2) Uncharged compounds cannot be separated in free zone electrophoresis (CZE) but the addition of surfactant forming micelles (e.g., sodium dodecyl sulphate, SDS) can be helpful for the analysis of such analytes. Here the compounds are moved to the detector by the electroosmotic flow and selective interactions with the charged micelles (SDS) cause retardation and then separation.

3) The separation of metal cations such as lanthanides is not possible by CE when using buffers such phosphate, acetate etc. because their similar mobility. We have shown that the use of α-hydroxyisobutyric acid (α-HIBA) to the BGE caused a selective retardation of the analytes allowing their complete resolution.

4) Finally we would like to focus the attention to the successful separation of optical isomers where a wide number of chiral selectors have been employed in order to selectively modify the electrophoretic mobility of the analytes. Among them chiral surfactants, proteins , cyclodextrins and their derivatives are the most useful in CE [1-3].

The aim of this paper is to discuss the use of cyclodextrins or their derivatives in capillary electrophoresis for optical isomers separation, comparison with high performance liquid chromatography will be also done.

2 Cyclodextrins and their derivatives

Cyclodextrins (CDs) have been widely used in capillary electrophoresis either added to the background electrolyte or bound to the capillary wall or included to a gel for the separation of different classes of compounds including positional or optical isomers mainly in pharmaceutical analyses. Most of the work has been performed in aqueous buffers where the main separation mechanism is represented by inclusion complexation in which the analyte penetrate the CD's cavity (inclusion-complexation) and secondary bonds between analytes and hydroxyl or hydroxyl-modified groups on the CD rim are formed

Among the native CDs used in analytical chemistry (α-, β- and γ-CD), β-CD seems to be the best complexing agent but, due to its low solubility in aqueous solvents its applicability in CE is reduced. Its solubility can be improved modifying the BGE with ,e.g., methanol, ethanol, urea etc.

However in most cases β-CD derivatives have been studied allowing the separation of a wide number of enantiomers. CD derivatives resulted to be very effective for chiral separations by CE because the modification of hydroxyl groups causes several changes on the structure and properties of the native CD. Among the changes we can remark: i) the increase of solubility, ii) possibility to form different stereoselective bonds with analytes, iii) introduction of charged/chargeable groups on the CD structure. In order to illustrate the advantages for using modified CDs let we consider as an example dimethylated-β-CD (Me-β-CD) and sulfobutylated-β-CD (SBE-β-CD). In both cases the solubility is very high compared to that of the parent one (e.g., 57 % and 1.85 % for

Me-β-CD and β-CD, respectively). Furthermore Me-β-CD is more hydrophobic than β-CD (possesses a deeper cavity). The SBE-β-CD, with sulfobutyl groups on its structure, is negatively charged in a wide range of pH and this property makes this chiral selector very useful in CE for the separation of either charged and uncharged analytes. Besides the separation of uncharged compounds using uncharged CDs has been performed modifying the BGE with surfactants like sodium dodecyl sulphate (SDS), the use of charged CDs extends the applicability of CE for the separation of such class of compounds. Table 1 shows the most used CDs in capillary electrophoresis.

The list of the separated enantiomers compounds by CE using CDs or their derivatives is wide and includes aminoacids and their derivatives, antidepressants, antipsychotics, hypnotics, barbiturates, anaesthetics, adrenergic drugs, cardiotonics, dopamine agonists, bronchodilators, β-blockers, antihypertensive, antiarrhythmics, anticholinergics, antiinflammatory etc. [1-4].

Table 1. The most used cyclodextrins in CE

uncharged	charged
α-cyclodextrin (α-CD)	methylamino-β-CD (6^{A}-MeNH-β-CD)
β-cyclodextrin (β-CD)	4-sulfobutyl-ether-β-CD (SBE-β-CD)
γ-cyclodextrin (γ-CD)	mono-(6-β-aminoethylamino-6-deoxy)-β-CD
heptakis-2,6-di-O-methyl-β-CD (di-OMe-β-CD)	carboxymethyl-β-CD (CM-β-CD)
heptakis-2,3,6-tri-O-methyl-β-CD (tri-OMe-β-CD)	carboxymethyl-β-CD polymer
hydroxyethyl- or hydroxypropyl-β-CD (HE- or HP-β-CD)	phosphate-β-CD
soluble β-CD polymer	

3. Selection of operational mode

Figure 1 shows the electrophoretic scheme separation of two enantiomers L and D by CE using a) uncharged CD and b) negatively charged CD.

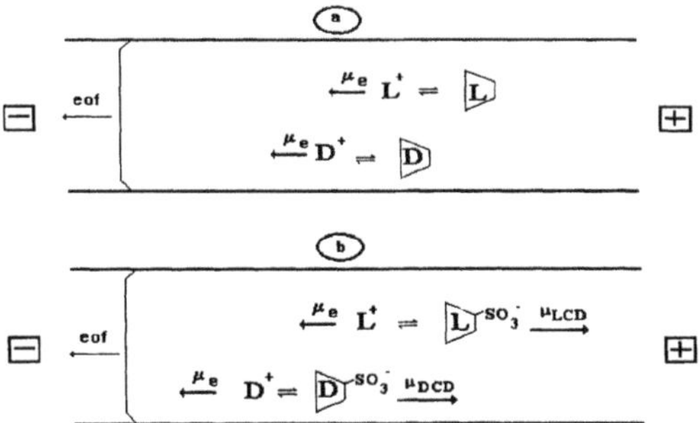

Fig. 1. Electrophoretic scheme of enantiomers separation using CDs

As depicted in figure 1 the two enantiomers, positively charged, possess their own electrophoretic mobility and during the electrophoretic run they are reversibly

interacting with the CD forming complexes that are modifying the effective mobility allowing their separation. The best operational conditions are usually obtained when the difference of mobility of free enantiomer and complexed one are maximized and this is obtained when the charged CD is used. The presence of a strong electroosmotic flow is usually negative for either resolution and selectivity; good enantiomers resolution has been achieved employing coated capillaries that can offer reduction/suppression of electroosmotic flow and reduction of analytes adsorption.
In some instance, e.g., when charged CDs are used peak dispersion can be obtained either due to strong interaction with analytes or because the etherogeneous chiral selector. The problem can be solved reducing the concentration of the CD and/or using a counter-flow mode (only part of the capillary is filled with the chiral selector).
With the reported examples we would like to point out that the choice of the operational mode in chiral separations by CE is of paramount importance in order to achieve the best results.

4. Resolution and selectivity

Inclusion complexation and stereoselectivity are influenced by several experimental parameters, e.g., CD type and concentration, applied voltage, capillary temperature and length, ionic strength, organic solvent, electroosmotic flow, polymeric additives etc. Some effect will be discussed in this section with the aim to select the appropriate experimental conditions when enantiomers have to be separated using cyclodextrins.

CD type and concentration

The first requirement for inclusion complexation is the fitting of the analyte into the CD cavity and thus its selection has to be done considering the shape of the analyte and the dimension of CD. As an example let we consider the chemical structure of terbutaline and nicergoline. Terbutaline contains only one aromatic ring and one asymmetric carbon at α position while nicergoline is formed by more than one aromatic rings and several chiral centres. The electrophoretic analysis at pH 2.5 of the two compounds using native CD revealed that α-CD was not able to separate both racemic mixtures in their enantiomers because its cavity is too small in order to guest the two compounds. β-CD allowed the resolution of terbutaline while nicergoline was partly resolved and γ-CD was not able to guest terbutaline (the molecule is too small) but was very effective for nicergoline. Previously we showed that for the enantiomeric separation of terbutaline either β-CD and dimethylated-β-CD can be successfully used in CE but the latter is forming strongest inclusion complexes and is more stereoselective than the former. The complexation of terbutaline increased by increasing the concentration of both β-CD and its derivative, while the resolution showed a maximum at 15 mM and 5 mM for β-CD and dimethylated-β-CD, respectively.

Concerning the CD type it is important to consider the degree of substitution (DS) of the cyclodextrin used, this effect was demonstrated for the chiral separation of some hydroxyacids using different modified β-CD with methyl amino or hydroxypropyl groups where the presence of more than one substituent can strongly affect the chiral recognition. The CD type can also influence the migration order of the two enantiomers, this effect was observed for the separation of dansyl-norvaline enantiomers using trimethylated-β-CD and β-CD.

The optimum concentration of CD for enantiomeric separation has been also discussed theoretically by Wren and Rowe and found that is dependent from the formation constants of the complexes formed during the electrophoresis (equation 1)

$$\left[C_{CD}\right] = \frac{1}{\left[K_{(+)}K_{(-)}\right]^{1/2}} \qquad (1)$$

where C_{CD} is the optimum concentration of the CD used, K the formation constants of the two enantiomers (+) and (-).

Analytes shape

The shape of the analytes is another important parameter to be considered for the inclusion-complexation with cyclodextrin. We studied this effect analyzing tryptophan derivatives (methyl, hydroxyl) where the position of the substituting group on the indole ring strongly influenced the stability of the complex formed and thus the chiral recognition.

The shape of the enantiomeric pairs can be modified, e.g., derivatizing with hydrophobic achiral compounds that can be included into the CD cavity. Amino acids as dansyl-, naphthalene-2,3-dicarboxaldheide- (NDA), naphthylamide derivatives and monosaccharides, derivatized as naphtalene sulphonate, were resolved in their enantiomers using cyclodextrins.

The composition of the background electrolyte

Other parameters to be controlled in CE for chiral separations are the BGE's ionic strength, concentration, pH, as well as the presence of polymeric additives and the content of the organic solvent.

BGE with low *ionic strength* can generally cause a reduction of either migration time and resolution and peak tailing. Electromigration dispersion is responsible for this negative effect and can be avoided using a BGE at a concentration 10^2 times higher than that of the sample, furthermore the peak shape can be controlled selecting a co-ion with similar electrophoretic mobility of the analyte.

The addition of an *organic solvent* to the BGE could produce a negative effect on the binding constant of the inclusion-complex with cyclodextrins, generally will be reduced due to the competition between the organic additive and the analyte. Anyway its use can improve the enantiomers resolution. We showed that propranolol and flurbiprofen enantiomers were base-line resolved adding methanol to the BGE containing β-CD and 2,3,6-tri-O-methyl-β-CD, respectively.

The *pH* of the BGE can strongly influence both resolution and selectivity because this parameter modify both the electroosmotic flow and the charge of analytes.

Strong electroosmotic flow should generally be avoided because the time spent by the analyte for inclusion-complexation equilibrium is reduced. A coated capillary but shorten than uncoated one could be used in order to either improve the resolution and to perform the analysis in a reasonable time.

For a wide number of applications in the separation of enantiomers of basic compounds acidic pH have been used. At this pH the electroosmotic flow (eof) is reduced increasing the resolution. Similar effect can be obtained using additives such as hydroxyethylcellulose (HEC), hydroxypropylmethylcellulose (HPMC), polyvinyl alcohol (PVA) etc. The addition of polymers will reduce the eof, increase the viscosity and reduces adsorption of analytes.

Capillary temperature

An efficient temperature control is recommended in order to avoid loss in resolution when cyclodextrins are used as chiral selectors for enantiomers separations.

The increase of temperature is usually causing a decrease of the diffusion coefficient, buffer viscosity and thus a decrease of the migration time. Furthermore the change of the capillary temperature can strongly influence the stability of the inclusion-complex formed between analyte and cyclodextrin (generally an increase of temperature is causing a decrease of the binding constant). For the separation of sympathomimetic drugs we found that the optimum temperature was 15-25 °C [1-4].

5. CD in HPLC

The use of insoluble CD polymers in liquid chromatography (LC) has been documented since 1965 for the separations of enantiomers as well as nitrophenol isomers. However such resins had some drawbacks when employed in HPLC, e.g., high pressure and organic solvents could not be used. Thus silica stationary phases-bounded CD have been developed. The first generation resulted to be unstable due to the presence of amino silica derivative and thus Armstrong in 1985 bound the CD to the silica gel via a 6-10 atom spacer in absence of nitrogen linkages. Since this time CD columns were commercially available. This stationary phases have been successfully used for the separation of structural and geometrical isomers as well as enantiomers such as amines, amino acids or their derivatives, barbiturates etc.. In the second generation of CD columns modified CDs were employed in the synthesis extending the applicability to enantiomers separations. Another interesting approach has been shown dynamically coating a silica column with permethylated-β-CD that could be used in either reversed-phase and normal-phase. Finally due to the properties of CDs (stability, no light absorption etc.), in several cases this compounds have been added to the mobile-phase and the separation of enantiomers, diastereoisomers as well as structural isomers has been shown. The addition of CDs to the mobile-phase presents several advantages over the use of CD-columns, e.g., i) selectivity and resolution can be increased, ii) different CDs are commercially available, iii) detector response could be improved. Of course it is also necessary to evaluate the costs because volumes of mobile-phase used in HPLC are not low and thus this system could result more expensive than that one employing CD columns. In HPLC, several parameters can be modified in order to improve selectivity and resolution, e.g., the CD type, the composition of the mobile-phase (pH, organic solvent, type and concentration of buffer), temperature [5, 6].

6. Conclusions

The different separation modes, namely zone electrophoresis (CZE), micellar electrokinetic chromatography (MEKC), gel electrophoresis (CGE) and isotachophoresis (ITP) combined with the high resolution power and high efficiency makes CE complementary and even competitive with other separation methods.

Among the other advantages that CE can offer in comparison with other techniques such as HPLC is noteworthy to remark

i) relatively small volumes of either samples and buffers (nL and μL, respectively) are requested

ii) expensive chiral columns can be avoided because the chiral selector can be easily added to the BGE

iii) the separation can be highly reproducible changing the buffer with the chiral selector after each run.

However for preparative purposes HPLC must be preferred.

References

[1] Nishi, H. and Terabe, S., Optical resolution drugs by capillary electrophoretic techniques, J. Chromatogr. A, 694, 245-276 (1995)

[2] Fanali, S., *Use of cyclodextrins in capillary electrophoresis*, in Capillary electrophoresis technology, (Ed. Guzman, N.A.) New York, Marcel Dekker Inc., 1993 Pp. 731-752

[3] Fanali, S., Cristalli, M., Vespalec, R. and Bocek, P., *Chiral separations in capillary electrophoresis*, in Advances in Electrophoresis (Eds. Chrambach, A., Dunn, M.J. and Radola, B.J.), Weinheim, VCH Verlagsgesellschaft mbH, 1994, Vol 7 Pp. 1-86.

[4] Fanali, S., *An introduction to chiral analysis by capillary electrophoresis*, Bio-Rad Laboratories Bulletin 1973 US/EG RevA, 1995, Pp. 1-50

[5] Ward, T.J. and Armstrong, D.W., *Cyclodextrin-stationary phases*, in Chromatographic chiral separations, (Eds. Zief, M. and Crane, L.J.) New York-Basel, Marcel Dekker Inc., 1988 Pp. 131-163

[6] Li, S. and Purdy, W.C., Cyclodextrins and their applications in analytical chemistry, Chem. Rev., 92, 1457-1470 (1992).

PERSPECTIVES OF CHIRAL CAPILLARY ELECTROPHORESIS USING PHOSPHATED CYCLODEXTRINS AS ADDITIVES

Z. JUVANCZ[1], K. E. MARKIDES[2], L. JICSINSZKY[3]
[1]*EGIS Pharmaceuticals. Ltd., Dept. of Pharmacokinetics, H-1106 Kereszturi út 30-38 Budapest, Hungary.*
[2]*Uppsala University, Dept. of Anal. Chem., S-751 21 Uppsala, Sweden.*
[3] *Cyclolab Ltd., H-1525, Budapest, Hungary.*

ABSTRACT
The acidic, phosphate substituted cyclodextrins (CD)s are excellent chiral selective agents for several neutral and positively charged molecules in CE. They are multimodal chiral selectors, and useful in a broad pH (2-9) range. Organic modifiers can drastically influence their selectivity in various ways. A good chiral separation is a result of compromise among various analysis parameters.

1. INTRODUCTION
Enantiomers can show rather different biological activity, therefore their stereoselective synthesis and analysis have become essential for pharmaceutical and food industry [1]. The capillary electrophoresis (CE) is a rapidly growing effective separation method which can determine enantiomer ratio in broad range (down to 0.1%) even in biological matrices [2]. The CDs are the far most popular chiral selectors in CE, mostly as buffer additives [3]. The chiral separation can be expected, if the migration speed of enantiomers in complexed and free states are different, and the stability constants of complexes of R and S isomers are not same. The selectivity is improved using oppositely charged analytes and chiral selectors, because they have opposite increments in their electrophoretic mobility [4]. The phosphated CDs are also excellent for such purposes [5, 6]. In this report, phosphated CDs (α, β, γ) were used as additives for separations of basic drugs in CE. The effect of type and concentration of CDs, pH, organic modifier were studied. Some guidelines and precautions are given for the use of these effective chiral selectors.

2. Materials and Methods
α-,β-, γ- phosphated CDs were donated by Cyclolab (Budapest, Hungary), other chemicals were bought from Fluka (Busch, Switzerland). The test compounds were products of Hassle (Möndal, Sweden). A 37 cm x 50 μm i.d. FSOT column was placed into a Beckman P/ACE 2100 capillary electrophoresis instrument (Fullerton, USA) with 214 UV detection. 100 mMol phosphate background buffers (BGE) were adjusted

J. Szejtli and L. Szente (eds.), Proceedings of the Eighth International Symposium on Cyclodextrons, 649–652.

to appropriate pH values. The numbers show migration times (min) in the electropherograms.

3. Results and Discussion

The short path length of the on-column UV detection allows detection at 214 nm even applying 40 mMol phosphated CD content of electrolyte in spite of these CDs have a cut off above 240 nm.

All of the tested enantiomers were separated with more than one type of phosphated CDs. The migration order of enantiomers (e.g. Metoprolol) can be reversed using different phosphated CDs.

The phosphated CDs are well soluble in water up to 150 mMol. Much lower concentrations of these selectors are enough for good separations in most cases. Even 3 mMol and 7 mMol γ-CD phosphate produce separation of rac. Tocainide with R_S:1.5, and 4.7 values, respectively [6]. Rac. Desmethyl Tocainide shows also a good separation using of a low concentrations of these selectors (Fig. 1).

The chiral recognition features of phosphated CDs were observed in a broad range of pH (2-9). The ionized phosphated CDs can separate neutral and ionized enantiomers. For example, the ionized phosphated α-CD can separate well the rac Metoprolol at pH: 6.8 in its ionised state (Fig. 2A) and at pH: 8.3 in its neutral state too (Fig. 2B).

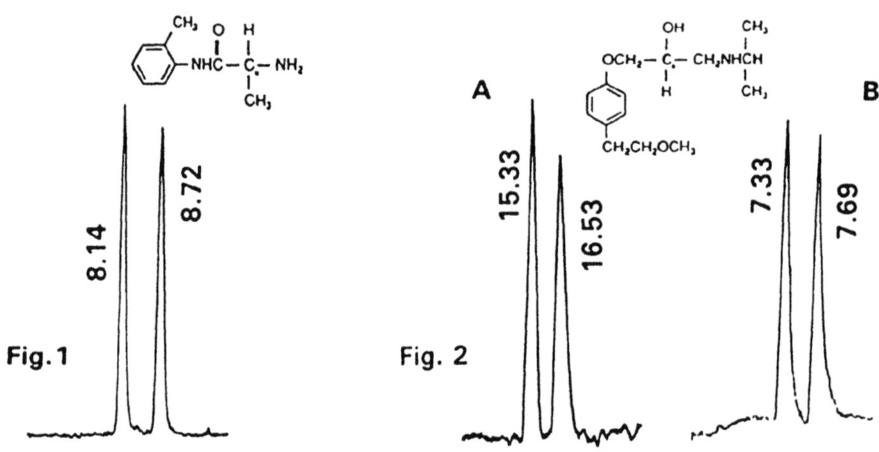

Fig. 1, 7 mMol γ-CD phosphate is enough for the complete separation of rac. Desmethyl Tocainide. Conditions: column: 37 cm x 0.05 mm i. d. FSOT, BGE: 100 mMol phosphate, 7 mMol γ-CD phosphate, 25 kV, UV: 214 nm , pH: 6.8.

Fig. 2, The rac. Metoprolol is separated with ionised as well as neutral α-CD phosphate chiral additive. Conditions: column: 37 cm x 0.05 mm i. d. FSOT, BGE: 100 mMol phosphate, 20 kV, UV: 214 nm. **A**; pH: 6.8, 10 mMol α-CD phosphate, **B**; pH: 8.3, 15 mMol α-CD phosphate.

The neutral state of phosphated CDs can also separate ionized enantiomers. It is interesting to note, the selectivity of these CD can belong to their ionised state. For example, the resolution value of rac. Disopyrmide is 3.5 at pH: 2 (neutral 5 mMol γ-CD phosphate), but the resolution value of same enantiomers is only 0.7 at pH: 3 (mostly ionised 5 mMol γ-CD phosphate). The ionized γ-CD phosphate, however, reserves a decreased selective toward to rac. Disopyramide (R_S:1.05, pH: 4.9, 15 mMol γ-CD phosphate). On the other hand, the selectivity of γ-CD phosphate does not depend on its charged state in the case of rac. Tocainide [6]. These example show that, the phosphated CDs have multimodal chiral recognition characters.
The pH value of electrolyte is important not only from point of view of ionised states of components. Some enantiomers can be well separated at low pH (e.g. Disopyramide, Tocainide), but in others cases (e.g. Metoprolol) an elevated pH is recommended.
An unwanted sticking effect of α-CD phosphate was observed analysing strongly interacting enantiomers at low pH values. The rac. Tocainide and rac. Metoprolol were coinjected and analysed at pH: 4.9 applying 15 mMol at pH: 4.9 (Fig. 3A). Under these conditions, the separation of rac. Tocainide is appropriate (R_S: 1.4), but the peaks of rac Metoprolol does not appear in the electropherogram under 30 min analysis time. Increasing the pH value to 8.3 (other conditions are same) results a good separation of rac. Metoprolol (R_S: 2.0) but the separation of rac Tocainide is only moderate (Fig. 3B). The negative peak corresponds to the electroosmosis flow (EOF) in Fig. 3B. The lack of peaks of rac. Metoprolol is consequence of sticking effect of α-CD phosphate (Fig. 3A.). The α-CD phosphate sticks strongly to the wall of column stopping the migration of strongly complexed Metoprolol at pH 4.9[5,6]. At higher pH values, the sticking effect becomes weaker, because the increased silanol content of column wall repels the α-CD phosphate. The interactions between the α-CD phosphate and Metoprolol are also decreased at higher pH values. These two effects result in a good separation of rac Metoprolol (Fig. 3B). The sticking effect influences the separation of weakly complexed enantiomers only in limited extent. The weakly complexed enantiomers can be separated at low pH value too (Fig. 3A). .
The sticking effect, however, can be eliminated using a high pH values, a low concentration of selectors or adding an organic modifier even in the cases of the strongly complexed enantiomers. On the other hand, it is important an intensive washing step (e.g. 5 min water flush) between each analysis to eliminate the stacked layer from the column wall.
An organic modifier can influence the resolution of enantiomers not only in way. Without methanol modifier no peak of rac. Metoprolol was observed (Fig. 3A), but 5 % methanol results in a R_S: 1.5 value and 10% methanol in a R_S = 1.2 value. (pH: 4.9, 15 mMol α-CD phosphate). On the other hand, 0, 5, 10 percentages of methanol modifiers result in 1.4, 1,0, 0.8 resolution values respectively for rac. Tocainide (pH: 4.9, 15 mMol α-CD). An organic modifier competes with the analyte for the cavity of CD. This competition shifts the stability equilibrium CD- analyte complexes, influencing the resolution values of enantiomers.

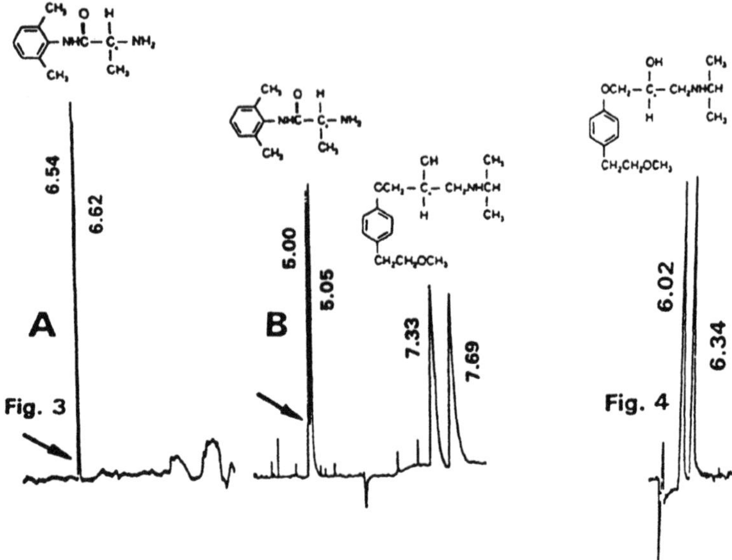

Fig. 3 The coinjection of rac. Tocainide and rac Metoprolol produce a good separation for rac. Tocaidine, but the peaks of rac Metoprolol can not be recognised at all at pH: 4.9 (**A**). The peaks of rac. Tocainide partly, the peak of rac. Metoprolol are totally separated at pH 8.3 (**B**). The arrows show the valley point of peaks of rac. Tocainide. Conditions: column: 37 cm x 0.05 mm i. d. FSOT, BGE: 100 mMol phosphate, 15 mMol α-CD phosphate UV: 214 nm, **A**; pH: 4.9, 25 kV, **B**; pH: 8.3, 20 kV.

Fig. 4 A good separation of rac Metoprolol is result of compromises among the analysis parameters. Conditions: column: 37 cm x 0.05 mm i. d. FSOT, BGE: 100 mMol phosphate, 15 mMol α-CD phosphate + 5 % MeOH, pH: 8.3, 20 kV, UV: 214 nm.

4. Conclusion

The phosphated CDs are effective chiral selective agents with broad application ranges in CE. They have multimodal chiral recognition characters in a broad, 2-9 pH range. They can separate neutral and ionized enantiomers in their neutral and ionized state. Their unwanted sticking feature can be eliminated with appropriate running parameters. A good separation is the result of optimization of competing factors (Fig. 4).

Acknowledgements

The authors thank for the financial support of Hassle and OTKA No. T14326 grant.

References

[1] E.J. Ariens, Trends Pharmacol Sci. 7, 200 (1986).
[2] R. Weinberger, Practical Capillary Electrophoresis (Academic Press, Boston, 1993).
[3] H. Nishi, S. Terabe, J. Chromatogr. **A 694**, 245 (1995).
[4] T. Schmidt, H. Engelhardt, Chromarographia 37, 475 (1993).
[5] Z. Juvancz, Cyclodextrin News 9, 181 (1995).
[6] Z. Juvancz, K.E. Markides, L. Jicsinszky (submitted)

ENANTIOMERIC SEPARATION OF DICHLORPROP BY CAPILLARY ELECTROPHORESIS USING β-CYCLODEXTRIN DERIVATIVES AS CHIRAL RESOLVING AGENTS

F. TROTTA[1], O. ZERBINATI[2], C. GIOVANNOLI[2], C. BAGGIANI[2], G. GIRAUDI[2], G. MORAGLIO[1]

[1]*Dipartimento di Chimica Inorganica, Fisica e dei Materiali, Università di Torino, via P.Giuria 7, 10125 Torino, Italy*

[2]*Dipartimento di Chimica Analitica, Università di Torino, via P.Giuria 5, 10125 Torino, Italy*

ABSTRACT

In this work chemical derivatives of β-cyclodextrins (β-CD) have been considered and the inclusion constants with the racemic herbicide dichlorprop, only the (+) isomer of which being herbicidally active, have been determined. The experimental results showed that β-CD functional substituent affect both the inclusion constant values and the resolving power.

1. INTRODUCTION

Cyclodextrins have been extensively used as chiral selectors for several separation techniques, including capyllary electrophoresis (CE). The enantioselective complexation depends on the dimensions of the cavity and also by the nature of cyclodextrin's substituents. Among the possible enantiomeric molecules of interest, dichlorprop, (±)-2-(2,4-dichlorophenoxy)propionic acid has been considered, since it is used as herbicide. Commercial dichlorprop is a racemate only the (+) isomers is herbicidally active, therefore the resolution the two optical isomers can be of interest in agricultural and environmental applications. This work takes into account different cyclodextrin derivatives, among which two new carbonate derivative of β-cyclodextrin have been considered, in order to find the best resolving agent for the racemate considered.

2. MATERIALS AND METHODS

α-cyclodextrin was purchased from Fluka, methyl-β-cyclodextrin and hydroxypropyl-β-cyclodextrin were kindly supplied by Wacker Chemie. C6-capped-β-cyclodextrin and

J. Szejtli and L. Szente (eds.), Proceedings of the Eighth International Symposium on Cyclodextrons, 653–656.

ethyl carbonate-β-cyclodextrin were synthesised according to the procedures described in the literature [1]. Racemic dichlorprop and its pure D(+) enantiomer (dichlorprop-p) were purchased from Alltech. All other chemicals were purchased from Merck.
CE was performed using a BioFocus 2000 CE system (BIO-RAD) equipped with a polyacrilamide-coated 24 cm, 25 μm i.d. capillary thermostatted at 20 °C
All separations were carried out by operating in reversed-polarity mode (cathode on the injection side), with 7 kV migration voltage and 206 nm detection. The running electrolyte was a 100 mM acetic acid/sodium acetate, pH 5.0 buffer, to which the different chiral additives tested were added. $1 \cdot 10^{-5}$ M dichlorprop water solutions were prepared immediately before injection. Cyclodextrin concentrations ranging from 0 to 25 mM were added to the running buffer and the resulting migration times of dichlorprop enantiomers were used for calculating the inclusion constants by fitting them to the following equation:

$$\mu_A = \frac{\mu_f + \mu_c K_{A-CD}[CD]}{1 + K_{A-CD}[CD]}$$

where μ_A is the apparent electrophoretic mobility of the enantiomer, μ_f is the mobility of the free enantiomer, μ_c is the apparent electrophoretic mobility of the complexed enantiomer, K_{A-CD} is the formation constant of the complex and [CD] is the cyclodextrin concentration [2]. Sigmaplot 2.0 was used for executing mathematical calculations.

3 RESULTS AND DISCUSSION

Figure 1 shows the enantiomeric resolution of dichlorprop obtained with 8 mM ethylcarbonate-β-CD. The effect of the concentration of this cyclodextrin on migration times is plotted in Figure 2. The inclusion constants calculated for the six cyclodextrins experimented are listed in Table I.

Table I. Inclusion constants

chiral resolving agent	D(+) inclusion constant	L(-) inclusion constant
α-cyclodextrin	402	347
β-cyclodextrin	257	257
C6-capped-β-CD	96	98
Ethylcarbonate-β-CD	77	82
Methyl-β-CD	81	41
Hydroxypropyl-β-CD	79	87

Baseline resolution of dichlorprop enantiomers was observed with the newly synthesised ethylcarbonate-β-CD, with native α-cyclodextrin and with the commercially available methyl-β-CD and hydroxypropyl-β-CD, while only partial resolution was given by the C6-capped-β-CD. It can be observed that the inclusion constant values of α-cyclodextrin are approximately one order of magnitude larger than those of all the derivatized β-

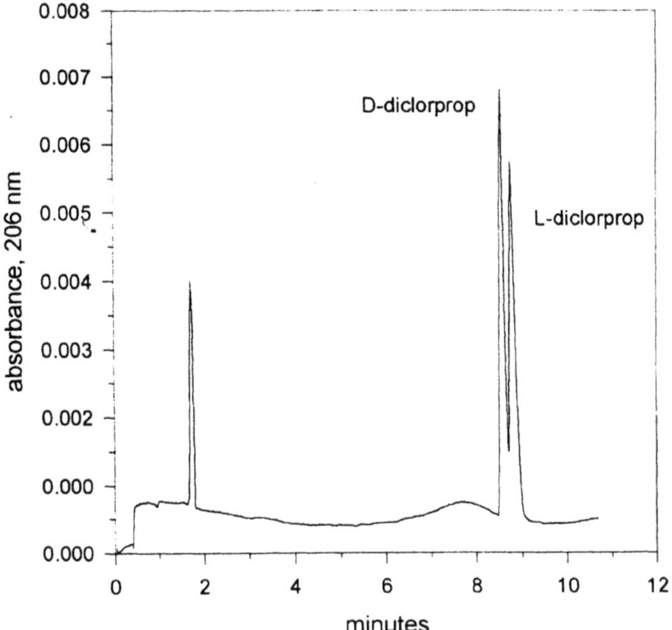

Fig. 1. Electropherogram of dichlorprop racemate with 8 mM ethylcarbonate β-CD

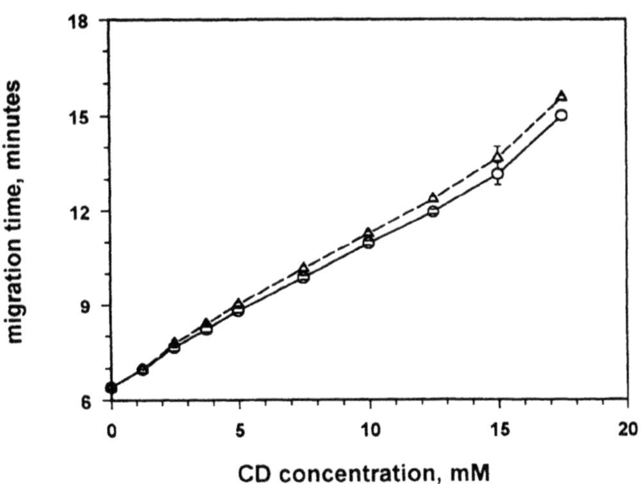

Fig. 2. Dependence of migration times on ethylcarbonate-b-CD concentration. Circles: L-enantiomer
Triangles: D-enantiomer

cyclodextrins, which can still produce enantiomer separation despite their lower complexation ability, due to the relatively large difference in the inclusion constants of the two enantiomers. Complexing ability does not necessarily indicate chiral resolving properties: in fact native β-cyclodextrin, although is a stronger complexing agent than its derivatives, does not separate dichlorprop enantiomers.
C6-capped-β-CD did not give satisfactory performances; in fact, the relative difference in the inclusion constants of the two enantiomers is quite small and, in addition, its solubility in the running buffer (12.5 mM) is smaller than those of the remaining CD's.
The lower complexation constant could be explained with steric factors involving the C6 bridge on the smaller edge of the β-CD, which could not permit arrangement of the molecules optimal for evidencing interaction of the chiral centres.
The order of elution of the L- and D-enantiomers is not the same among the cyclodextrins studied; in particular, the L-dichlorprop is eluted first with α-cyclodextrin, while it is eluted last with ethylcarbonate-β-CD. This fact can be attributed to differences in the energies of interaction of the optical antipodes with the host molecule, but the interpretation of the experimental evidence should require quite sophisticate molecular mechanics models which are not yet optimised for this purpose [3].

4 CONCLUSION

A new ethylcarbonate derivative of β-CD was found to resolve the two optical antipode as effectively as native α-CD or methyl- β-CD or hydroxypropyl- β-CD. Although the complexation constants of the β-CD derivatives were found to be smaller than the one with the native β-CD, chiral resolution was still possible due to the relatively large difference between the complexation constants of the two optical antipodes. Both native β-CD and a C6-capped-β-CD gave unsatisfactory resolution of the racemate, thus evidencing the specific character of the interaction between analyte and resolving agent needed to achieve optical resolution.

REFERENCES

[1] Trotta, F., Moraglio, G., Marzona M., Maritano, S., *Gazz. Chim. It.*, 123, 559 (1993)
[2] Palmasdottir, S., Edholm, L.-E., *J. Chromatogr. A*, 666 (1994) 337-50
[3] Black, D. R., Parker, C. G., Zimmerman, S. S., Lee, M. L., *1st Electronic Computational Chemistry Conference*, November 7-18, 1994

USE OF ELECTROSTATIC INTERACTION FOR CHIRAL RECOGNITION. ENANTIOSELECTIVE COMPLEXATION OF ANIONIC BINAPHTHYLS WITH PROTONATED AMINO-β-CYCLODEXTRIN

Koji Kano,* Takashi Kitae, and Hiroshi Takashima
Department of Molecular Science and Technology, Faculty of Engineering, Doshisha University, Tanabe, Kyoto 610-03, Japan

ABSTRACT

The studies using NMR and capillary zone electrophoresis clearly reveal that protonated 6^{A}-amino-6^{A}-deoxy-β-cyclodextrin recognizes the chirality of anionic binaphthyl derivatives such as 1,1'-binaphthyl-2,2'-diyl phosphate and 1,1'-binaphthyl-2,2'-dicarboxylate through electrostatic interaction between the host and the guest and simultaneous inclusion.

1. INTRODUCTION

Chiral recognition is one of the recent topics in cyclodextrin (CDx) chemistry [1]. Three-point attachment model has been accepted as a mechanism which explains chirality-recognizing host-guest complexation [2]. Hydrogen-bonding interaction has widely been used to achieve chiral discrimination through this mechanism [3]. In general, however, hydrogen bond is hardly formed in water because of strong hydration to hydrogen-bonding sites of both host and guest molecules. Electrostatic force is expected to act as a point interaction in place of hydrogen bonding. The present study deals with the chiral recognition of anionic, chiral guests by a cationic CDx. We used protonated 6^{A}-amino-6^{A}-deoxy-β-CDx (6-Amino-β-CDx) as the cationic host and 1,1'-binaphthyl-2,2'-diyl phosphate (BNP) and 1,1'-binaphthyl-2,2'-dicarboxylate (BNC) as the anionic guests having the axial chirality. To the best of our knowledge, there are three examples with the chiral recognition by aminated CDxs. Nardi et al. [4] reported that the enantiomers of mandelic acid and its related compounds are separated by capillary zone electrophoresis (CZE) using 6^{A}-*N*-methylamino-6^{A}-deoxy-β-CDx and 6^{A},6^{D}-di-*N*-methylamino-6^{A},6^{D}-dideoxy-β-CDx as the chiral separators. Brown et al. [5] found a weak ability of protonated 6-Amino-β-CDx to discriminate between the enantiomers of 2-phenylpronionate. Kano et al. [6] demonstrated that the conformational enantiomerism of bilirubin induced by heptakis(6-amino-6-deoxy)-β-CDx is

J. Szejtli and L. Szente (eds.), Proceedings of the Eighth International Symposium on Cyclodextrons, 657–662.

achieved by a hydrogen-bonding interaction between the NH_3^+ groups of the cyclodextrin and the CO_2^- groups of bilirubin.

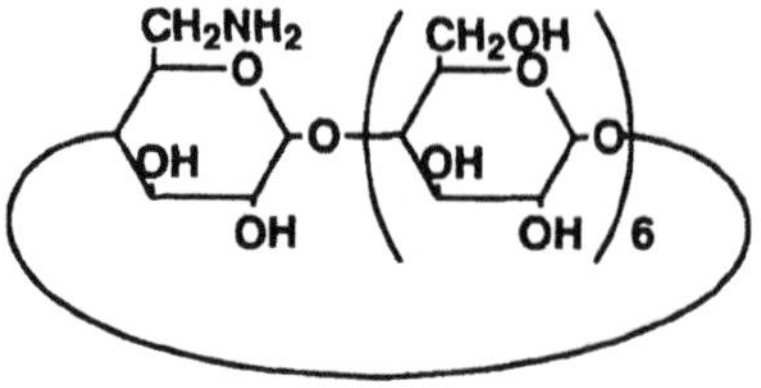

6-Amino-β-CDx 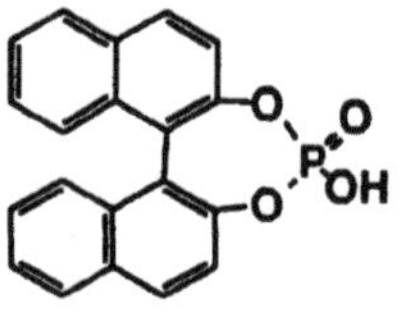 BNP

BNC 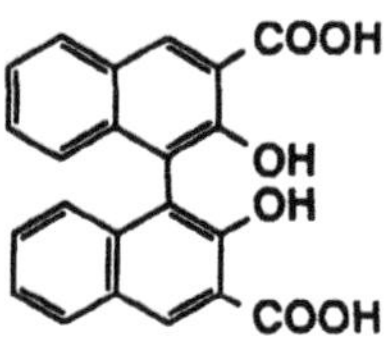 HBNC 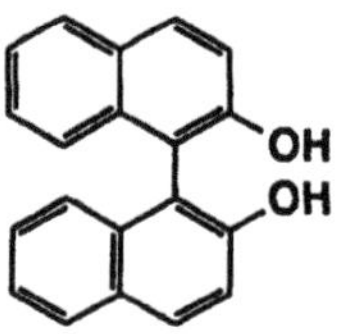 BN

The present study reveals that protonated 6-Amino-β-CDx is a good host for recognizing the axial chilarities of the binaphthyl derivatives. The cooperative effect of the electrostatic and van der Waals and/or hydrophobic interactions seems to be very important to form stable inclusion complexes of 6-Amino-β-CDx.

2. EXPERIMENTAL

6-Amino-β-CDx was prepared according to the procedures described in the literature [7]. β-CDxs (Nacalai) commercially obtained was washed with THF using a Soxhlet extractor to remove an antioxidant. Optically active and racemic BNP and 1,1'-bi-2-naphthol (BN, Aldrich) were commercially obtained. The preparation of (±)-, (*S*)- and (*R*)-2,2'-dihydroxy-1,1'-binaphthyl-3,3'-dicarboxylic acids (HBNC) were described before [8]. 1,1'-Binaphthyl-2,2'-dicarboxylic acid (BNC) was the same as previous one [9].

The CZE experiments were carried out by using a Jasco capillary electropherograph system CE-800 having a 50 μm (I.D.) x 300 mm fused silica capillary cartridge (non-coated). The capillary was filled with a buffer solution with CDx and the sample in the same buffer solution was introduced into the capillary by applying the potential for 10 s. The electropherogram was taken at the same potential using a Jasco 875-CE UV-Vis detector. The NMR spectra were taken with a JEOL JNM A-400 (400 MHz) in D_2O using sodium 3-(trimethylsilyl)propionate-d_4 (Aldrich) as an external standard.

The molecular mechanics-molecular dynamics (MM-MD) calculations were carried out as described previously [9].

3. RESULTS AND DISCUSSION

3.1. CZE of Binaphthyl Derivatives Using 6-Amino-β-CDx as Separator

CZE is very convenient tool to know the system where chiral host enantioselectively complexes with chiral guest [9, 10]. The electropherogram of a mixture of (±)-BNC and (±)-BNP is shown in Fig. 1. The migration time (t) of BNP is shorter than that of BNC, indicating that the binding constant (K) for the BNP complex is larger than that for the BNC complex (*vide infra*). At the same time, it can be predicted that protonated 6-Amino-β-CDx prefers the anionic forms of (*S*)-BNP and (*R*)-BNC as the guests. The results of CZE of other binaphthyl derivatives are summarized in Table 1. The enantiomers of BNC can be separated by β-CDx in the pH 3.08 to 5.50 range. Of course, no separation occurs for undissociated BNC at lower pH range. At pH above 5.50, dissociated BNC is not included in the β-CDx cavity leading to unification of the peaks of the BNC enantiomers. In the case of 6-Amino-β-CDx, however, the separation

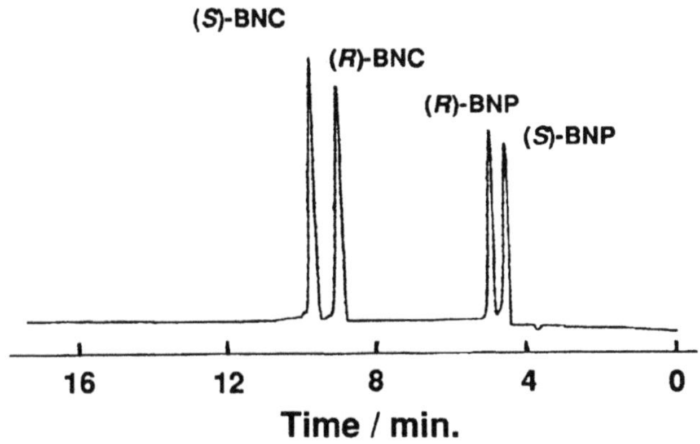

Fig. 1 Electropherogram of a mixture of (±)-BNP and BNC (4×10^{-4} M).
Separation solution: 6-Amino-β-CDx (0.01 M)
Buffer: 0.033 M phosphate buffer at pH 7.21
Applied voltage: 13.1 kV (30.1 μA)

can be achieved at wider pH range. Even at higher pH range, the BNC anion can be bound to protonated 6-Amino-β-CDx through electrostatic interaction. Therefore, the enantiomers of BNC are separated each other by CZE up to pH 7.70. At pH above 7.70, the amounts of protonated 6-Amino-β-CDx become small resulting in the disappearance of electrostatic interaction between the host and the guest. Although it was very difficult to separate the optical isomers of HBNC by CZE using β-CDx, 6-Amino-β-CDx makes it possible. Since BN has no charge, no separation occurs in CZE

TABLE 1. Chiral separation of the binaphthyl derivatives (4×10^{-4} M) by CZE using 6-Amino-β-CDx (0.01 M) as a chiral separator

Sample	CDx	pH	Voltage/kV	Current/μA	t_S/min	t_R/min	α
BNP	β-CDx	5.40	12	13	12.2	13.0	1.12
	6-Amino-β-CDx	5.54	12	26	6.80	8.20	1.18
BNC	β-CDx	4.15	21	10	7.40	6.20	1.20
	6-Amino-β-CDx	6.33	21	10	7.40	6.20	1.20
HBNC	6-Amino-β-CDx	8.35	8	6	12.3	13.0	1.05
BN	6-Amino-β-CDx	2.40	12	21	8.20	10.6	1.29

using β-CDx. The enantiomeric separation of BN, however, is realized by CZE using 6-Amino-β-CDx at the low pH range, where the electroosmotic flow is negligible.

3.2 NMR Study on Chiral Recognition of BNP by 6-Amino-β-CDx

Figure 2 shows the 6-Amino-β-CDx-induced shifts in the ^{1}H NMR signals due to (*R*)- and (*S*)-BNPs in D_2O at pD 6.0. Although it is very difficult to deduce the structure of

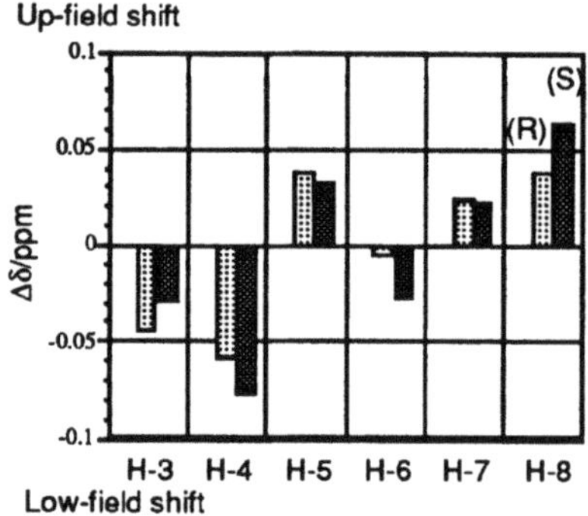

Fig. 2 6-Amino-β-CDx-induced shifts in the ^{1}H NMR signals of BNP in D_2O ([BNP] = 0.001 M, [6-Amino-β-CDx] = 0.01 M)

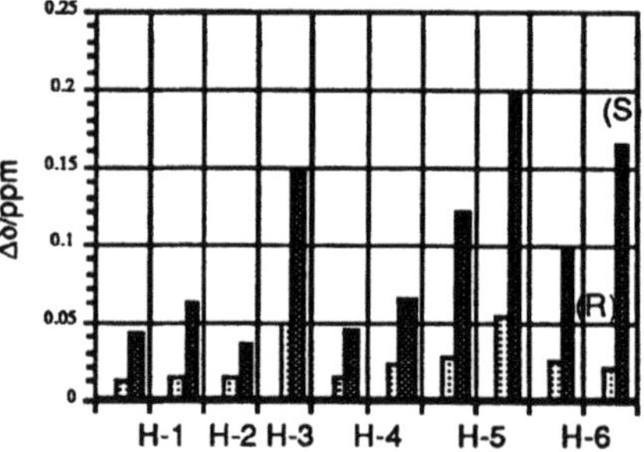

Fig. 3 BNP-induced shifts in the ^{1}H NMR signals of 6-Amino-β-CDx in D_2O ([BNP] = 0.001 M, [6-Amino-β-CDx] = 0.01 M)

the complex from the results exhibited in Fig. 2, it can be assumed that the structure of the (*S*)-enantiomer is similar to that of the (*R*)-enantiomer. The BNP-induced shifts of the ^{1}H NMR signals due to 6-Amino-β-CDx is shown in Fig. 3. The signals of the protons at the 3-, 5-, and 6-positions remarkably shift to higher magnetic fields due to the ring-current effect by the aromatic ring of the guest, suggesting that BNP is bound to both sides of the cyclodextrin. Namely, it seems that the BNP-6-Amino-β-CDx complex is not so simple as to be shown by a sole structure.

The binding constants for complexation of the enantiomers of BNP and BNC with 6-Amino-β-CDx were determined from the ^{1}H NMR-titration curves. The results are

TABLE 2. Binding constants for complexation of 6-Amino-β-CDx with BNP and BNC in D_2O at pD 6

Sample	K/M^{-1}	R/S
(*S*)-BNP	3474	3.5
(*R*)-BNP	998	
(*S*)-BNC	214	0.52
(*R*)-BNC	415	

revealed in Table 2. The order of the *K* values is in good agreement with that predicted from the results of CZE (*vide supra*). Although dissociated BNC cannot be bound to β-CDx, it forms the complex with protonated 6-Amino-β-CDx. This is one of the advantages of the use of electrostatic interaction in the host-guest chemistry in water.

3.3 Structures of Complexes

The plausible structures of the complexes of (*S*)- and (*R*)-BNPs and 6-Amino-β-CDx are obtained from molecular dynamics and molecular mechanics calculations. The information on the charges was obtained from the MOPAC calculations and the effects of water molecules were considered in calculations. The results are shown in Fig. 4.

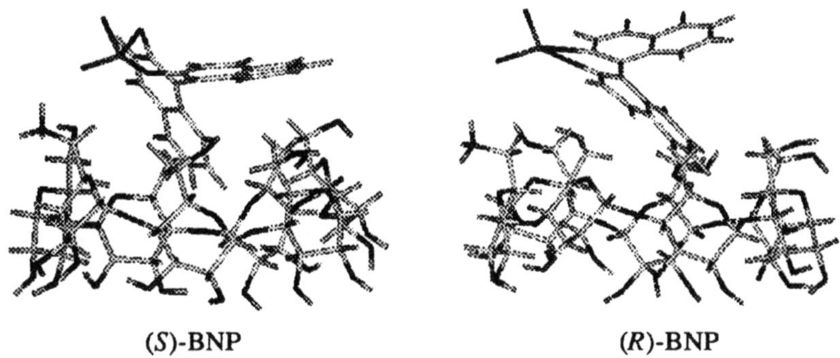

Fig. 4 Calculated structures of the complexes of the (*S*)- and (*R*)-BNP anions and protonated 6-Amino-β-CDx

In both complexes, the electrostatic attraction is observed between the NH_3^+ group of 6-Amino-β-CDx and the phosphate anion group of BNP. In the case of (*S*)-BNP, a naphthalene moiety of BNP penetrates somewhat deeply into the cyclodextrin cavity.

Meanwhile, more shallow binding is observed for the (*R*)-enantiomer. The ROESY spectra indicated the correlation between the protons at the 3-, 5-, 7-, and 8-positions of (*S*)-BNP and the protons at the 5- and 6-positions of 6-Amino-β-CDx. In contrast with this, only the protons at the 7- and 8-positions of (*R*)-BNP correlate with the protons at the 5- and 6-positions of the host. Such difference may be interpreted in terms of that (*S*)-BNP penetrates more deeply into the CDx cavity compared with (*R*)-BNP.

Of course, the chiral recognition cannot be achieved only by an electrostatic interaction between host and guest. In the present study, the electrostatic binding occurs only at one position. Therefore, the simultaneous inclusion of the guest into the host cavity is essential for the chiral recognition.

ACKNOWLEDGMENT

This work was supported by a Grant-in-Aid for Science Research from the Ministry of Education, Science and Culture, Japan.

References

[1] Kano, K. *Selectivities in cyclodextrin chemistry*, in Bioorganic Chemistry Frontiers, (Ed. Dugas, H., Schmidtchen, F. P.) Springer-Verlag, Berlin, 1993, Vol. 3, Pp. 1-23

[2] Daffe, V., Fastrez, J., Cyclodextrin-catalysed hydrolysis of oxazol-5(4H)-ones. Enantioselectivity of the acid-base and ring-opening reactions, *J. Chem. Soc., Perkin Trans. 2*, 789-796 (1983); Boehm, R. E., Martire, D. E.; Armstrong, D. W., Theoretical considerations concerning the separation of enantiomeric solute by liquid chromatography, *Anal. Chem.*, **60**, 522-528 (1988)

[3] Rebek, Jr., J. *Recent progress in molecular recognition*, in Topics in Current Chemistry, (Ed. Weber, E.) Springer-Verlag, Berlin, 1988, Vol. 149, Pp. 189-210.

[4] Nardi, A., Eliseev, A., Bocek, P., Fanali, S., Use of charged and neutral cyclodextrins in capillary zone electrophoresis: enantiomeric resolution of some 2-hydroxy acids, *J. Chromatogr.*, **638**, 247-253 (1993)

[5] Brown, S. E., Coates, J. H., Duckworth, P. A., Lincoln, S. F., Easton, C. J., May, B. L., Substituent effects and chiral discrimination in the complexation of benzoic, 4-methylbenzoic and (RS)-2-phenylpropanoic acids and their conjugate base by β-cyclodextrin and 6^A-amino-6^A-deoxy-β-cyclodextrin in aqueous solution: Potentiometric titration and ^1H nuclear magnetic resonance spectroscopic study, *J. Chem. Soc., Faraday Trans.*, **89**, 1035-1040 (1993)

[6] Kano, K., Arimoto, A., Ishimura, Conformational enantiomerism of bilirubin and pamoic acid inducd by protonated aminocyclodextrins, T., *J. Chem. Soc., Perkin Trans. 2*, 1661-1666 (1995)

[7] Melton, L. D., Slessor, K. N., Synthesis of monosubstituted cyclohexaamyloses, *Carbohydr. Res.*, **18**, 29-37 (1971)

[8] Kano, K., Tamiya, Y., Otsuki, C., Shimomura, T., Ohno, T., Hayashida, O., Murakami, Y., Chiral recognition by cyclic oligosaccharides. Enantioselective complexation of binaphthyl derivatives with cyclodextrins, *Supramol. Chem.*, **2**, 137-143 (1993)

[9] Kano, K., Minami, K., Horiguchi, K. Ishimura, T., Kodera, M. Ability of non-cyclic oligosaccharides to form molecular complexes and its use for chiral separation by capillary zone electrophoresis, *J. Chromatogr. A*, **694**, 307-313 (1995)

[10] Li, S. F. Capillary Electrophoresis, Principles, Practice and Applications, Elsevier, Amsterdam, 1992

INVESTIGATIONS INTO THE GC SEPARATION OF ENANTIOMERS ON 3-TRIFLUOROACETYL-2,6-DIPENTYL-γ-CYCLODEXTRIN

Separation of the components of cyclodextrin derivatives

I.D. SMITH* & C.F. SIMPSON[1]
SmithKline Beecham Pharmaceuticals, Old Powder Mills, Leigh, Kent TN11 9AN, UK
[1]Dept of Chemistry, Birkbeck College, 29 Gordon Square, London WC1H 0PP, UK

ABSTRACT

The separation of enantiomers by gas chromatography has been investigated on a fused silica capillary column coated with octakis(3-*O*-trifluoroacetyl-2,6-di-*O*-*n*-pentyl)-γ-cyclodextrin. The objective of this work is to propose possible mechanisms for the stereoselectivity of this stationary phase by rationalising the observed behaviour of relatively simple structurally-related compounds (alcohols and some of their fluoroacyl derivatives), characterising the cyclodextrin derivative and carrying out suitable molecular modeling experiments. Recent work on developing methods to characterise the stationary phase will be presented.

1. INTRODUCTION

The objective of this work was to develop HPLC methods on 300Å pore-size 5 μm silica reversed-phase columns to separate the components of cyclodextrin derivatives and allow fractions to be collected and characterised. A specific goal is to use this approach to characterise the 6-trifluoroacetyl-2,3-dipentyl-γ-cyclodextrin used as the GC stationary phase for the separation of alcohols and some of their derivatives[1,2]. Suitable separations of components of cyclodextrin derivatives were achieved and preliminary results are presented.

J. Szejtli and L. Szente (eds.), Proceedings of the Eighth International Symposium on Cyclodextrons, 663–666.

2. MATERIALS AND METHODS

The HPLC system used consisted of a Waters 616 pump, 600S system controller and 717 autosampler with a Sedex 55 evaporative light-scattering detector. All experiments were performed at ambient temperature, with a mobile phase flow rate of 1 mL min^{-1} and the detector nebuliser at 80°C.

The HPLC columns used were Spherisorb 5CX C18, 15 cm x 4.6 mm(Phase Separations Ltd) and Vydac C18, 25 cm x 4.6mm (Hichrom Ltd).

Mobile phases were of acetonitrile, methanol and tetrahydrofuran (HPLC grade, Romil Chemicals) mixed with HPLC grade water (18.2MΩ resistivity). Pentyl-γ-cyclodextrin was prepared by a procedure similar to that described by Armstrong[3]. The other analytes were purchased from the Aldrich Chemical Company and Technicol Ltd.

3. RESULTS AND DISCUSSION

3.1 Hydroxyethyl-β-cyclodextrin

Hydroxyethyl-β-cyclodextrin components were separated with both acetonitrile and methanol gradients. The choice of solvent has an effect on selectivity (Figure 1).

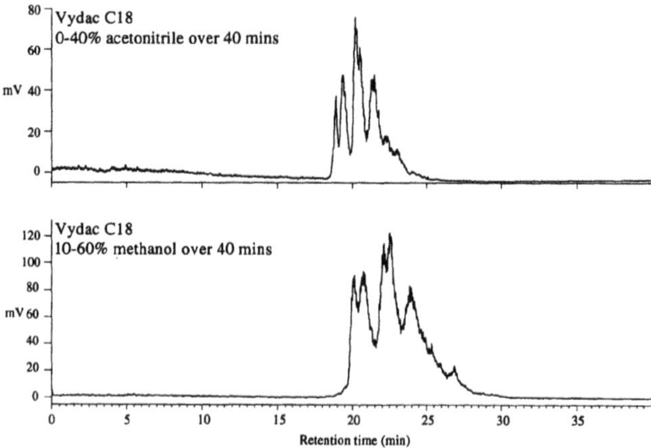

Figure 1 : Comparison of acetonitrile and methanol gradients on hydroxyethyl-β-cyclodextrin.

3.2 Methyl-β-cyclodextrin

There are a large number of components within the profiles obtained and material from different suppliers showed some differences in the profiles (Figure 2).

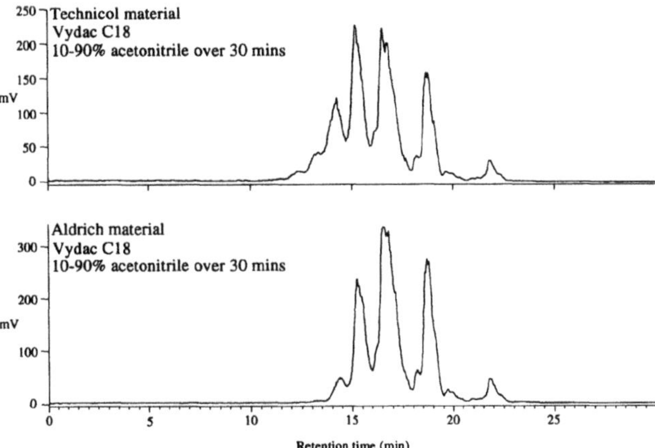

Figure 2 : Comparison of methyl-β-cyclodextrin from Technicol Ltd and Aldrich Chemicals

3.3 Triacetyl-β-cyclodextrin

This material was separated into three components on both columns by acetonitrile and tetrahydrofuran gradients, the selectivity being affected by the solvents (Figure 3).

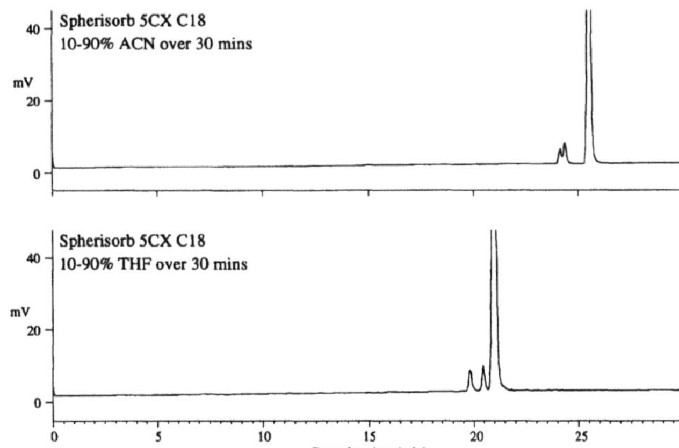

Figure 3 : Comparison of acetonitrile and tetrahydrofuran gradients on triacetyl-β-cyclodextrin.

3.4 Pentyl-γ-cyclodextrin

This material was separated into a number of multi-component peaks. The two columns showed different retention properties but similar selectivities (Figure 4).

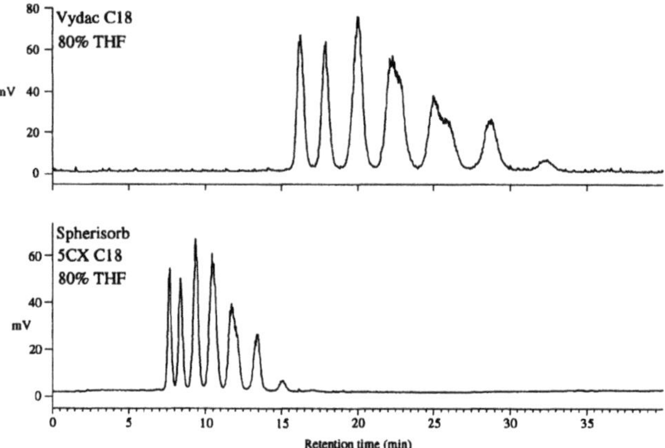

Figure 4 - Comparison of pentyl-γ-cyclodextrin on Spherisorb and Vydac columns.

4. CONCLUSION

Preliminary work has shown reversed-phase HPLC has been shown to be a useful technique for the separation of the components of cyclodextrin derivatives. Retention and selectivity are affected by the choice of both the HPLC phase and the mobile phase organic modifier. This work is continuing and will be fully reported in due course.

5. REFERENCES

[1] Smith, I.D., Simpson, C.F., Investigation into the GC separation of enantiomers on a trifluoroacetylated cyclodextrin. I. Effect of analyte structure on stereoselectivity for alcohols, *J High Resol Chromatogr*, **15**, 800-806 (1992).

[2] Smith, I.D., Simpson, C.F., Investigation into the GC separation of enantiomers on a trifluoroacetylated cyclodextrin. II. Effect of derivatisation on stereoselectivity towards alcohols, *J High Resol Chromatogr*, **16**, 530-535 (1993).

[3] Armstrong, D.W., Li, W., Pitha, J., Reversing enantioselectivity in capillary gas chromatography with polar and non-polar cyclodextrin derivative phases, *Anal Chem*, **62**, 214-217, (1990).

HPLC OF STRUCTURAL ISOMERS USING CYCLODEXTRIN-POLY(ALKYLAMINE) SYSTEMS COATED ON SILICA BEADS

G. CRINI[1], B. MARTEL[1], G. TORRI[2], M. MORCELLET[1]
[1]*Laboratory of Macromolecular Chemistry, URA CNRS 351, University of Lille, 59655 Villeneuve d'Ascq Cedex France*
[2]*Istituto Scientifico di Chimica e Biochimica « G. Ronzoni », Via G. Colombo 81, 20133 Milano Italy*

ABSTRACT

Various stationary phases for HPLC were prepared by coating a βCyclodextrin (CD) grafted polymer on silica beads. Three differents polyamines were used as the polymer support namely poly(ethyleneimine) (PEI), poly(allylamine) (PAA) and poly(vinylamine) (PVA). CD was grafted on the amino groups through a tosyl derivative. Two different methods were used for the preparation of the materials. They lead to difference in the mobility of CD in the coating layer as revealed by CP MAS ^{13}C NMR. The two classes of supports also differ by their efficiency in the separation of nitrophenol isomers. The concentration of CD in the interface layer and the nature of the polymer are also important factors.

1. INTRODUCTION

A variety of synthetic approaches have been proposed for the preparation of βCD-containing stationary phases in HPLC. One of these is the coating of a modified polymer on the surface of silica beads [1]. Using a polyamine as supporting polymer we prepared two classes of supports : in the first case (method I) the successive steps were : (a) grafting of CD on the polymer in homogeneous phase (b) coating on silica and (c) cross linking with epichlorhydrin or butanediol diglycidyl ether (BUDGE). In the second case (method II) the steps were : (a) coating of the polyamine (b) cross linking (c) grafting of CD in heterogeneous phase. These two methods differ in many ways : the conformations of the polyamine and of the CD-grafted polyamine are very different in the coating solvent and this should influence the way in which they interact with the silanol groups of silica (number of hydrogen bonds, fractions of tails, loops and trains, amount of polymer and CD per unit of surface area). Moreover, during cross linking, only amino groups are concerned (method II) or both amino groups and OH's of CD (method I). Lastly, when

J. Szejtli and L. Szente (eds.), Proceedings of the Eighth International Symposium on Cyclodextrons, 667–670.

method II is used, less amino groups are available because of coating and cross linking and the reaction occurs in heterogeneous conditions.
These parameters should influence the mobility and accessibility of CD at the interface between the solid and the solution and thus the efficiency of the supports in HPLC separations. This paper presents some data on their solid state NMR analysis and some results in the separation of nitrophenol isomers.

2. MATERIALS AND METHODS

2.1 Materials

Lichrospher Si-100 (10 mm, 100 Å) was used as silica beads. Poly(ethyleneimine) PEI, partially branched (molecular weight 50-60,000), and poly(allylamine) (PAA) (Molecular weight 60,000) were purchased from Aldrich. Poly(vinylamine) (PVA, M=36,600) was prepared in our laboratory [2]. βCD was supplied by Roquette Frères, France.

2.2 Methods

The preparation of the stationary phases was described elsewhere [3-5]. In all cases, the monotosyl derivative of CD was used as the intermediate for coupling with the amino groups of the polymer. Coating was done either in water or methanol. The calculation of the amount of CD on the supports (expressed in µmole/ g of support) was done after determination of the reducing sugars with tetrazolium blue [6].
Viscosimetric measurements were carried out with a Schott apparatus. HPLC measurements were done with a Waters and a Merck systems. NMR data were obtained with a CXP-300 Bruker spectrometer.

3. RESULTS AND DISCUSSION

It was first checked that the grafting of CD onto the polyamine changes the conformation of the polymer in solution.

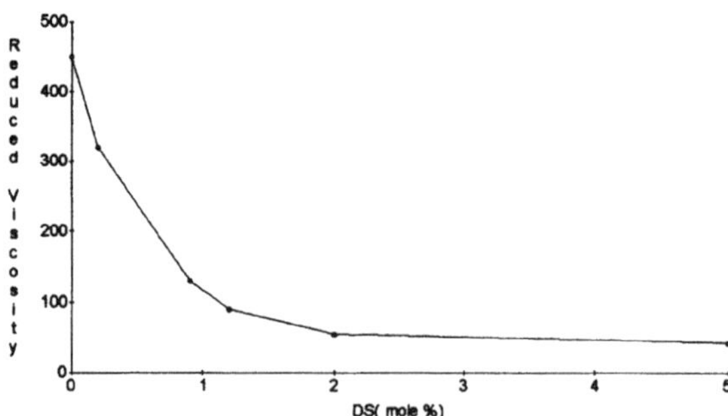

Figure 1. Viscosity change when grafting CD onto PVA.

Fig. 1 shows that for PVA, the viscosity in water decreases strongly even for rather low amounts of CD. The grafted polymer collapses probably by the formation of hydrogen bonds between the amino groups and the hydroxyl groups of CD and/or between CD's. It is clear that this should induce differences in the coating steps between method I and II. For example, when coating is done in water, the amount of adsorbed polymer decreases from about 100 mg/g for pure PVA to about 70 for a substitution degree of 5% (leading to 53 μmole/ g of support). Similar results were obtained with methanol as coating solvent. Taking into account that the average molecular weight of the polymer residue strongly increases by substitution with CD, this means that the number of molecules interacting with the surface has decreased. Differences are also observed concerning the desorption of the polymer which is more important for method I for example during the intermediate washing steps. Thus cross linking is an important step.

CP MAS ^{13}C NMR measurements were also carried out. Figure 2a and 2b show the spectra of supports prepared from PVA by methods I and II respectively and containing nearly the same amount of CD (~60 μmole/g). Similar results were obtained with PEI and PAA. It seems clear that the mobility of the CD cavity is very different. Relaxation times measurements are in progress on these samples.

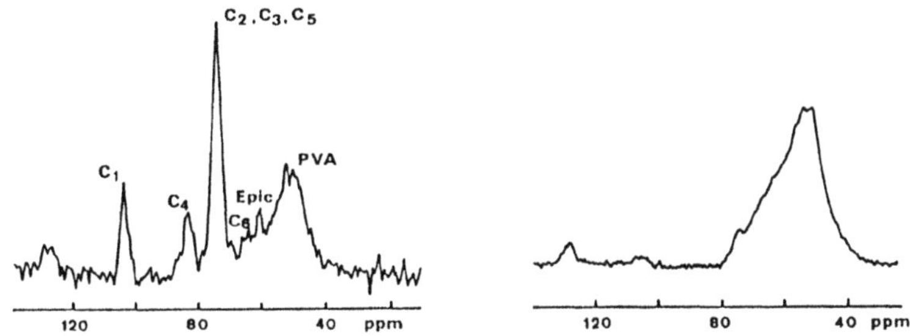

Figure 2 CP MAS ^{13}C NMR of PVA/CD prepared by method I (left); method II (right)

In HPLC, the height H of the theoretical plate depends of at least three parameters concentration of CD; nature of the polymer; method of preparation of the support.

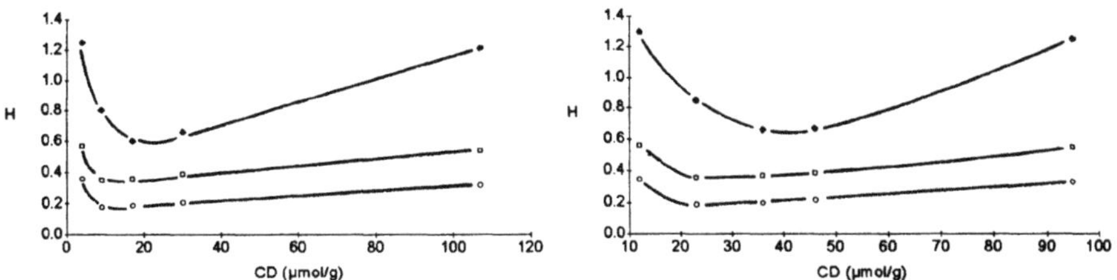

Figure 3 Effect of the concentration of CD on the height of the theoretical plate for meta (○), ortho (□) and para (◆) isomers of nitrophenol. Samples prepared by method II : PEI (left); PVA (right)

Fig 3 shows that, when increasing the CD concentration, H first decreases because more and more CD cavities are active in the separation process, then increases again after a minimum. This could mean that steric and hydrogen bond interactions between CD's prevents a good access of the substrates when the polymer layer is enriched in CD. The position of this minimum is lower for PEI and could be related to its branched, less compact structure. Comparison of Figs.3 and 4 also points out the influence of the method of preparation. H is surprisingly much lower (the phase is more efficient) when using method I.

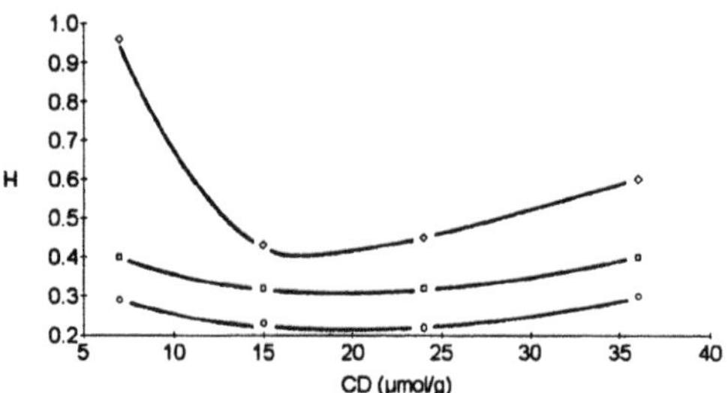

Figure 4 Effect of the concentration of CD on the height of the theoretical plate for meta (○), ortho (□) and para (◆) isomers of nitrophenol. Stationary phases prepared by method I with PVA.

4. CONCLUSION

The results above clearly show that the method of preparation of CD stationary phases is of extreme importance. Additional experiments are needed for a better understanding of the accessibility of the CD cavity at the polymer/solution interface, including relaxation measurements by NMR and EPR spectroscopies. The determination of the fractions of loops, tails and trains in the different polymer layers would probably be interesting. HPLC measurements with other sets of geometric isomers could also be useful.

ACKNOWLEDGMENTS

Thanks to Anne Marie Cazé and Yahya Lekchiri for assistance during this work.

REFERENCES

[1] Sébille B., Thuaud N., Piquion J., Behar N., Determination of association of βCD and βCD bearing polymers with drugs by CLHP, *J. Chromatogr.*, **409**, 61-69 (1987)

[2] Hart R., Synthèse de la poly(vinylamine), Hydrolyse du poly-N-carbamate de t-butyle, *Makromol. Chem.*, **32**, 51-56 (1959)

[3] Crini G., Lekchiri Y., Morcellet M., Separation of structural isomers using cyclodextrin-polymers coated on silica beads, *Chromatographia*, **40**, 296-302 (1995)

[4] Crini G., Torri G., Lekchiri Y., Martel B., Janus L., Morcellet M., High performance liquid chromatography of structural isomers using a cyclodextrin-poly(allylamine) coated silica column, *Chromatographia*, **41**, 424-430 (1995)

[5] Crini G., Torri G., Morcellet M., High performance liquid chromatography of structural isomers using cyclodextrin-poly(vinylamine) coated silica columns, *J. Chromatogr. Sci.*, in press

[6] Jue C; K;, Lipke P; N., Determination of reducing sugars in the nanomole range with tetrazolium blue, *J. Biochem. Biophys. Methods*, **11**, 109-115 (1985)

MODIFICATION OF THE FACE SELECTIVITY IN ASYMMETRIC INDUCTION BY CYCLODEXTRINS THROUGH THE FORMATION OF THREE-COMPONENT INCLUSION COMPOUNDS

E. RENARD,[1] A. DERATANI,[2] F. DJEDAINI-PILARD,[3] and B. PERLY[3]
[1] *Laboratoire de Physico-Chimie des Biopolymères, UMR 27, CNRS - Université Paris Val de Marne, 2 rue H. Dunant, 94320 Thiais France*
[2] *Laboratoire des Matériaux et Procédés Membranaires, UMR 9987, CNRS - ENSCM - Université Montpellier II, 2 place E. Bataillon, 34095 Montpellier cedex 5 France*
[3] *Laboratoire de RMN Haute Résolution, DRECAM-SCM, CEA, CE Saclay, 91191 Gif-sur Yvette France*

ABSTRACT

Stoichiometric amounts of triethylamine (TEA) were found to enhance the chiral induction by β-cyclodextrin (β-CD) in the reduction of acetophenone (ACPH) by aqueous $NaBH_4$. The enantioselectivity obtained depends upon the molar ratio β-CD:ACPH:TEA. Evidence for the formation of a three-component inclusion compound was obtained from detailled ^1H and ^2H NMR studies. The restriction of the molecular motion of the prochiral center probably accounts for the strong enhancement of the chiral induction observed.

1. INTRODUCTION

Cyclodextrins are known to induce regio- and enantioselectivity in organic reactions carried out in water [1]. Molecular size matching of the substrate with the host and geometry of binding in inclusion compounds are responsible for the selectivity observed. Regioselectivity can be achieved in excellent yields since a part of the guest molecule is shielded by the walls of the CD cavity in such a way that the reagent attack is selectively directed towards the more accessible position. On the contrary, it appears more difficult to create an asymmetric micro-environment through the complexation by CDs so that an enantiofacial discrimination occurs. In fact, few exceptions of complete or nearly complete chiral induction have reported in the literature [2-5].
To explore whether a decrease of the degree of freedom of the included molecule can improve the face selectivity, the reduction of the β-CD:ACPH complexe was investigated in the presence of chemically inert species as potential co-guest. The idea was to form three-component inclusion compounds in order to impose more steric constraints in the micro-environment of the prochiral site. We found that asymmetric inductions can be dramatically affected by a adding third components (tertiary amines,

J. Szejtli and L. Szente (eds.), Proceedings of the Eighth International Symposium on Cyclodextrons, 671–674.

amides or alcohols) in stoichiometric ratios with respect to the substrate [6]. The aim of this work is to demonstrate the formation of a ternary association when the reduction of ACPH included in β-CD was perfomed in the presence of TEA and to understand how the co-guest TEA can affect the enantioface selectivity.

2. MATERIALS AND METHODS

2.1. Reduction by $NaBH_4$

The three-component system β-CD:ACPH:TEA with 1:1:1 molar ratio was prepared by adding an equimolar amount (3mmol) of ACPH (0.35ml) and TEA (0.42ml) to a suspension of 3.4 g (3 mmol) of dry β-CD in 40 ml of 0.2M Na_2CO_3. After stirring overnight, the resulting slurry was reacted with $NaBH_4$ under vigourous stirring for further 24 h at room temperature. After neutralization (HCl), the mixture was extracted with diethyl oxide. The organic layer was washed with water and dried over $MgSO_4$. After evaporation of the solvent, 1-phenyl ethanol was purified by preparative TLC (CH_2Cl_2). Enantiomeric excesses (e.e.) were determined from the integral ratios of signals obtained by GLC analysis on chiral capillary column (CYDEX-B, Scientific Glass Engineering). Runs were triplicated. The reduction of other molar ratios were conducted according the same procedure using the desired amount of each compound.

2.2 NMR analysis

^1H NMR experiments were performed at 500MHz using a Brüker AMX 500 spectrometer. In all cases, the samples were prepared in D_2O and measurements carried out at 298K. Chemical shifts are given relative to external TMS (0ppm) and calibration was performed using the signal of the residual protons of the solvent as a secondary reference. ROESY experiments were obtained using the program provided by the Brüker library with a 300msec spin-lock time. ^2H NMR spectra were obtained in ^2H depleted water (Euriso-Top) using a Brüker MLS 300 spectrometer operating at 46 MHz for ^2H.

2.3 Molecular modelling

Molecular models were obtained using the Insight II software programme running on a Silicon Graphics Iris Indigo station.

3. RESULTS AND DISCUSSION

3.1 Chiral induction

Reductions of ACPH by $NaBH_4$ were carried out in carbonate buffer on the mixture of the three compounds β-CD, ACPH, and TEA in the desired molar ratio. It can be seen from Table 1 that a stoichiometric molar fraction of TEA dramatically affects the enantioselective formation of 1-phenyl ethanol induced by β-CD. The increase of e.e.

values was accompagnied by an inversion of the face selectivity. Both effects obviously reveal the contribution of TEA in the chiral induction mechanism.

TABLE 1. Enantioselectivities obtained for $NaBH_4$ reduction of β-CD:ACPH complexes in the presence of different amounts of TEA

Molar ratio β-CD:ACPH:TEA	1:1:0	1:1:1	2:1:0	2:1:1	2:1:2	2:1:3	2:1:10
Chemical yield (%)	60	55	61	50	54	48	49
E.e. (%)	6	15	5	43	48	39	6
Configuration	S(-)	R(+)	S(-)	R(+)	R(+)	R(+)	R(+)

In the absence of TEA, only low e.e. values of S(-) 1-phenyl ethanol were obtained. It can be noticed that better results were observed in the presence of two molar equivalents of β-CD. This is probably due to the lowering of the free ketone concentration by shifting the complexation equilibrium. The maximum e.e. value was reached for the molar ratio 2:1:2. Beyond this ratio, the e.e. decreased likely because TEA in excess oompetes with ACPH for inclusion.

3.2 Evidences for the formation of a three-component complex

^1H NMR spectra show the formation of an inclusion compound in the binary ACPH:β-CD (Fig. 1a) and in the ternary system ACPH:β-CD:TEA (Fig. 1b). Upfield shifts of CD protons were observed upon addition of ACPH, the most important being those of protons H-3 and H-5 located in the CD cavity. The presence of TEA did not induced significant variations of these signals. On the other hand, the signal of the ACPH methyl group was changed into a very complex multiplet. The most probable explanation is to consider that a strong immobilization of the methyl group at the time scale of the NMR analysis. ^2H NMR technique was used to determine unambigously the molecular dynamics. In the binary system, ACPH retains a large mobility in the cavity (Fig.2a, sharp signal). In the presence of TEA, the linewidth increased to a value very similar to that obtained for β-CD (Fig. 2b). This implies that the included ACPH molecule has no or very little internal mobility in the cavity when TEA is present.

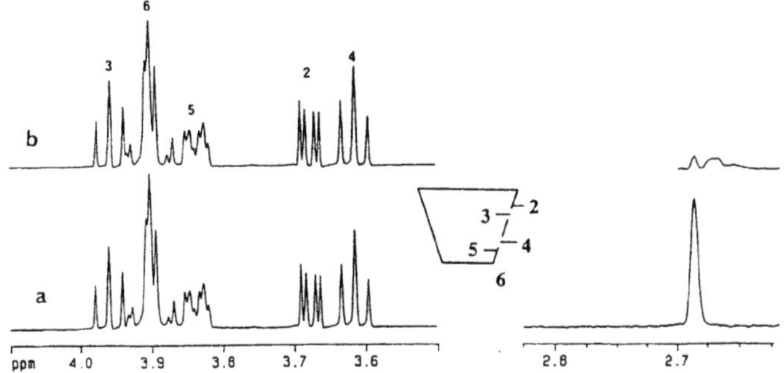

Fig. 1. 500 MHz ^1H NMR spectra (D_2O, 298 K) (a) β-CD 4 mM, ACPH 5 mM; (b) + TEA 12.5mM

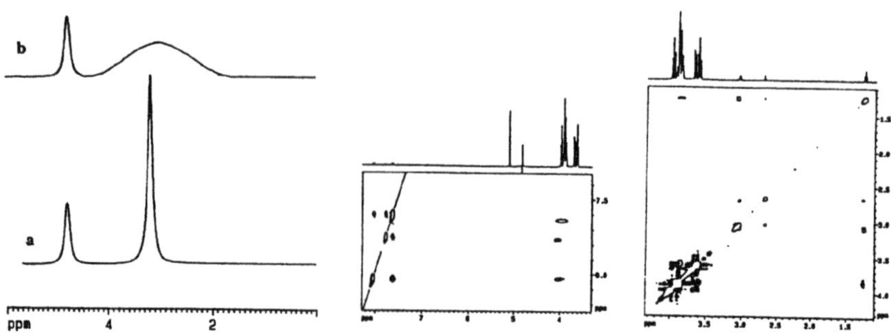

Fig. 2. 46 MHz ^{2}H NMR spectra Fig. 3. ROESY (a) aromatic protons; (b) non-aromatic protons

To get a deeper insight in the possible structure of the ternary complex, a two-dimensional ^{1}H NMR experiment was performed (Fig. 3). All non-diagonal peaks are indicative of spatial proximities between protons. Dipolar interactions can be observed between the ortho protons of ACPH and H-5, H-6 protons of β-CD. This result indicates that inclusion proceeds by the primary hydroxyl side. Besides, it can be seen interactions between the methyl group of TEA and H-6 protons of β-CD and, at the same time, the methyl group of ACPH. From these data, we propose a possible structure for the ternary association ACPH:β-CD:TEA where TEA partially binds to β-CD, the ethyl group included strongly limiting the mobility of ACPH within the cavity.

4. CONCLUSION

Chemically inert species were demonstrated to affect the chiral induction mediated by CDs. The locking effect of TEA can be related to a simple model of allosteric effect in enzyme catalysis. It can be supposed that dramatic improvements of selectivity could be achieved with a co-guest of the appropriate shape and structure. The powerful combination of NMR experiments and molecular modelling provides an useful way for the screening of the appropriate co-guest molecule.

REFERENCES

[1] Takahashi, K., Hattori, K., Asymmetric reactions with cyclodextrins, *J. Incl. Phenom.*, **17**, 1-24 (1993)

[2] Tanaka, T.,Sakuraba, H., Nakanishi, H., Asymmetric halogenation and hydrohalogenation of olefins in crystalline cyclodextrin complexes, *J. Chem. Soc., Chem. Commun*, 947-8 (1983)

[3] Sakuraba, H., Ushiki, M., Highly enantioselective addition of S(-)lithiomethyl 1-naphthyl sulfoxide to ketones, *Tetrahedron Lett.*, **31**, 5349-52 (1990)

[4] Sakuraba, H., Inomata, N., Tanaka, Y., Asymmetric reduction of ketones with crystalline cyclodextrin complexes of amine-boranes, *J. Org. Chem.*,**54**, 3482-84 (1989)

[5] Kawagiri, Y., Motohashi, N., Strong asymmetric induction without covalent bond connection, *J. Chem. Soc., Chem. Commun..*, 1336-37 (1989)

[6] Deratani, A., Maraldo, T., Renard, E., Can the presence of chemically inert species modify the face selectivity in asymmetric induction by cyclodextrins, *J. Incl. Phenom.*, in press

DETERMINATION OF DRUG-CYCLODEXTRIN BINDING CONSTANTS BY CAPILLARY ZONE ELECTROPHORESIS

JÁNOS GYIMESI(1), ÉVA SZÖKŐ(2), KÁLMÁN MAGYAR(2)
AND LAJOS BARCZA(1)
(1) Department of Inorganic and Analytical Chemistry, L. Eötvös University, Pázmány P. sétány 2; Budapest, H-1117 Hungary
(2) Department of Pharmacodynamics, Semmelweis University of Medicine, Nagyvárad tér 4; Budapest, H-1089 Hungary

ABSTRACT

Binding constants of the optical isomers of Deprenyl® (selegiline) and its potential metabolites with (2,6-di-O-methyl)-β-cyclodextrin were determined using electrophoretic mobility data gained from separations performed by capillary electrophoresis, and absorbancies obtained from spectrophotometric experiments. To calculate equilibrium constants 1:1 complex formation have been assumed. The comparison of the equilibrium constants calculated from different methods shows similar values in their order of magnitude. Their difference may probably be explained by the different media of the measurements. The effect of the structure of compounds on chiral discrimination were also elucidated.

1. INTRODUCTION

Cyclodextrins are widely used in capillary electrophoresis (CE) in order to separate drug molecules, especially chiral isomers. The mechanism of the separation is mainly based on the different binding constants of the isomers with cyclodextrins.

To determine such binding constants have various interests:
-the optimum concentration of the chiral selector for enantiomer resolution in CE depends on the magnitude of the complex formation constants and the different mobilities of the free and complexed analytes [1,2].
-knowing these binding constants, experimental conditions to achieve chiral separation can well be predicted.

In this work a simple equilibrium model was used to calculate the formation constants. Enantiomer pairs of Deprenyl® and its metabolites were separated in running buffers containing various concentrations of the chiral selector. The corrected mobilities were used for the estimation of the binding constants.

The formation constants of some compounds were also determined by spectrophotometric method in order to compare data derived from CE and spectrophotometry, as well as from the literature.

J. Szejtli and L. Szente (eds.), Proceedings of the Eighth International Symposium on Cyclodextrons, 675–678.

2. MATERIALS AND METHODS

2.1 Materials

The enantiomers of the following standard compounds were used in the sample mixtures: amphetamine (A), propargyl-amphetamine (PA), methamphetamine (MA), p-fluoro-methamphetamine (pFMA), deprenyl (D), p-fluoro-deprenyl (pFD), which compounds were kindly provided by Chinoin Pharmaceutical and Chemical Works (Budapest, Hungary). Pseudoephedrine (PSE), norephedrine (NE) and ephedrine (E) obtained from Sigma (Budapest, Hungary). All other materials were of analytical grade.

2.2 CE apparatus and separation conditions

A CRYSTAL 300 (ATI, UNICAM, Cambridge, UK) capillary electrophoresis system equipped with a variable-wavelength UV absorbance detector set at 190 nm was used. The separations were performed in a 70 cm x 75 μm ID uncoated fused silica capillary, the length to the detection window was 55 cm. Samples were introduced by electrokinetic injection (3KV, 12s). The applied field strength was 300 V/cm, the separation temperature 21 °C. Axxiom 727 software was used for data collection.

20 mM Tris-phosphate pH 2.7 containing 0.5% hydroxypropylmethyl-cellulose (HPMC), and various concentration of heptakis (2,6-di-O-methyl)-β-cyclodextrin (DIMEB) and methanol was used as running buffer [3].

2.3 Spectrophotometry

Conditions: pH=2.7 was adjusted with formic acid, ionic strength =20 mM adjusted with NaCl. Temperature:25±1°C.
The absorbances was measured with Spectromom 195D at λ=520 nm.
The binding constants were determined based on a competing reaction [4] between cyclodextrin and two ligands (methyl orange and drug molecules studied). Upon addition DIMEB to a red-coloured acidic solution of methyl orange, the absorbance decreases on account of the formation of a yellow deprotonated methyl orange-cyclodextrin complex. In the presence of another potential ligand molecule, only a fraction of the cyclodextrin cavities can interact with the methyl orange, resulting in a weaker discoloration effect.

2.4 Viscosity

Viscosity was measured using a Ubbelode suspended level capillary viscometer held in a water bath at 25.0°C.

3. RESULTS AND DISCUSSION

The electrophoretic mobilities of the enantiomer pairs were measured in the 0-48 mM concentration range of the chiral selector, (2,6-di-O-methyl)-β-cyclodextrin.

Application of DIMEB in the buffer decreased the electrophoretic mobility of each test compound. This indicated the inclusion complex formation of separands with the chiral additive, as complexes formed have lower electrophoretic mobilities, than the free analyte. Besides the complex formation, the increasing viscosity of the running buffer containing higher concentration of DIMEB also influences the mobility, and this effect has to be taken into correction.
The sructure of the investigated drugs and the calculated binding constants on the basis of CE and spectrophotometry are shown in Table1.
Comparing the structure of the test compounds and the calculated binding constants the following were found:
-drugs having hydroxyl group at the α position showed weaker interaction;
-the fluoro substitution of the aromatic ring decreased the values of binding constants;
-the binding constants of compounds substituted at the nitrogen were higher in all cases.
The chiral discrimination was found to be the best for compounds substituted by OH group at α position. These compounds have two chiral centres.

The binding constants obtained by CE and spectrophotometry are the same order of magnitude and show chiral discrimination. The observed differences in the values of binding constants obtained from different methods are probably due to the different buffer composition of the measurements. Although the pH and ionic strength were kept constants, the buffer chemicals and additives were unlike. On the bases of these results, the binding constants obtained from CE data are close to the thermodynamic equilibrium constants in a given media. This means that chiral capillary electrophoresis separations can be used to determine binding constants of enantiomers with cyclodextrin derivatives. CE separations provide several advantages compared to other methods used to determine binding constants. These comprise the simple and fast measurements, which can be performed even in fairly complex media.
The constants derived from other measurements can be used as an approximation to optimize the conditions of CE separation.

ACKNOWLEDGEMENTS

The experiments were performed using capillary electrophoresis instrument provided by the National Scientific Fund (OTKA) Grant No A 178. This study was also supported by National Scientific Fund (OTKA) grant No F017568 and T017749.

REFERENCES

[1] Wren, S.A.C. and Rowe, R.C.:Theoretical aspects of chiral separation in capillary electrophoresis, *J. Chromatography*, **603**, 235 (1992)

[2] Penn, S.G., Bergström, E.T., Goodall, D. M. and J. S. Loran:Capillary electrophoresis with chiral selestors, *Anal.Chem.*,**66**, 2866 (1994)

[3] Szökő,É., Magyar,K.:J.:Chiral separation of deprenyl and its major metabolites using cyclodextrin-modified capillary zone electrophoresis, *Chromatography. A*, **709, 157** (1995)

[4] Buvári, Á., Szejtli, J. and Barcza, I.:Complexes of short-chain alcohols with β-cyclodextrin, *J Inclusion Phen.*, **1**, 151(1983)

[5] L. Barcza, L., Szejtli, J. and Buvári, Á.:Spectrophotometric determination of stability constants of cyclodextrin complexes, *Ann. Univ. Sci. Budapest*, **XVI**, 11(1980)

[6] Rekharsky, M. V., Goldberg, R.N., Schwarz, F.P., Tewari, Y.B., Ross, P.D., Yamashoji, Y., Inoue, Y.: Thermodynamic and Nuclear magnetic resonance study of the interactions of α- and β-cyclodextrin with model substances, *J. Am. Chem. Soc.*, **117**, 8830 (1995)

Table 1. The chemical structure of the analytes and the calculated bindindg constants

Name	Structure	Binding constants from CE data K_1	K_2	Binding constants. from photometry K_1	K_2	Literature[a] K_1	K_2
Deprenyl	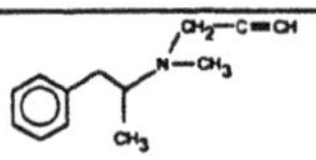 CH_2—C≡CH N—CH_3 CH_3	42.3±4.3	153.3±6.9	156.8	269.5		
Propargyl-amphetamine	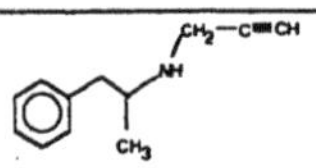 CH_2—C≡CH NH CH_3	K_1=K_2= 151.5±6.6					
p-Fluoro-deprenyl	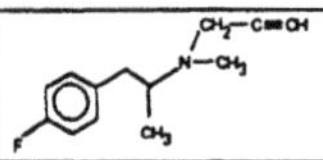 CH_2—C≡CH N—CH_3 CH_3 F	123.4±4.5	133.2±5.6				
Metham-phetanine	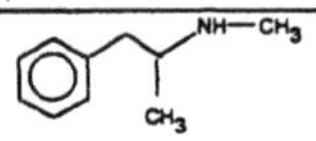 NH—CH_3 CH_3	107.0±3.9	111.9±3.2				
p-Fluoro-metham-phetanine	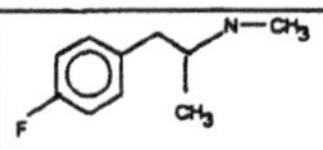 N—CH_3 CH_3 F	98.9±4.1	100.2±3.9				
Amphetamine	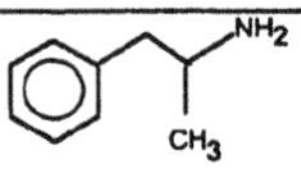 NH_2 CH_3	89.1±1.0	94.5±3.3				
Ephedrine	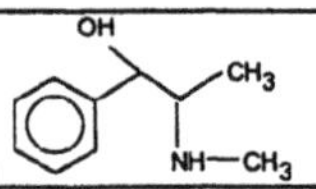 OH CH_3 NH—CH_3	57.6±2.5	63.4±3.1	98.4	130	79.2	71.3
Ψ-Ephedrine	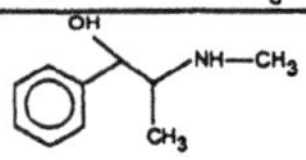 OH NH—CH_3 CH_3	53.8±4.0	77.5±4.5	60.4	114.3	68.9	96.7
Norephedrine	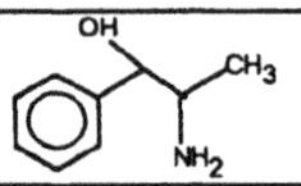 OH CH_3 NH_2	43.8±4.0	52.6±4.1				

Notes: K1 and K2 are the binding constants of (-) and (+) enantiomers, respectively. K = [EC]/ [E] [C] where E is one of the enantiomer C is the cyclodetxtrin and EC is the complex equbilibrium concentration.

a) from ref 6.;The binding constants with β- cyclodextrin were determined by calorimetric titration in phosphate buffer at pH 6.9 .

SUBJECT INDEX

GPSR Compliance
The European Union's (EU) General Product Safety Regulation (GPSR) is a set of rules that requires consumer products to be safe and our obligations to ensure this.

If you have any concerns about our products, you can contact us on

ProductSafety@springernature.com

In case Publisher is established outside the EU, the EU authorized representative is:

Springer Nature Customer Service Center GmbH
Europaplatz 3
69115 Heidelberg, Germany

www.ingramcontent.com/pod-product-compliance
Ingram Content Group UK Ltd.
Pitfield, Milton Keynes, MK11 3LW, UK
UKHW021333070726
13610UKWH00011B/47